Jane Su 510-896-6058
su-j@abgenix.com
Room 2403

AF682684

PACIFIC SYMPOSIUM ON BIOCOMPUTING 2002

PACIFIC SYMPOSIUM ON
BIOCOMPUTING 2002

Kauai, Hawaii
3–7 January 2002

Edited by

Russ B. Altman
Stanford University, USA

A. Keith Dunker
Washington State University, USA

Lawrence Hunter
University of Colorado Health Sciences Center, USA

Kevin Lauderdale
Stanford University, USA

Teri E. Klein
Stanford University, USA

World Scientific
New Jersey • London • Singapore • Hong Kong

Published by

World Scientific Publishing Co. Pte. Ltd.

P O Box 128, Farrer Road, Singapore 912805

USA office: Suite 1B, 1060 Main Street, River Edge, NJ 07661

UK office: 57 Shelton Street, Covent Garden, London WC2H 9HE

British Library Cataloguing-in-Publication Data
A catalogue record for this book is available from the British Library.

BIOCOMPUTING
Proceedings of the 2002 Pacific Symposium

ISBN 981-02-4777-X

Printed in Singapore by World Scientific Printers

PACIFIC SYMPOSIUM ON BIOCOMPUTING 2002

The seventh Pacific Symposium on Biocomputing (PSB) marks the first PSB held following the tragic events of September 11, 2001 in New York, Pennsylvania and Washington DC. These events have affected the world at large and cannot go unnoticed by the computational biology community. The organizers would like to add their condolences to those who suffered. In spite of technical and personal difficulties that individuals incurred, we are happy able to put forth these proceedings.

PSB is sponsored by the International Society for Computational Biology (http://www.iscb.org/). Meeting participants benefit once again from travel grants from the U.S. Department of Energy, the National Library of Medicine/National Institutes of Health, Applied Biosystems and Boston College. We gratefully acknowledge the hardware contributions from Compaq.

We thank Professor David Botstein in advance for his plenary address on *Extracting Biologically Interesting Information from Microarrays* and Professor Rebecca Eisenberg for her plenary address on *Bioinformatics, Bioinformation and Biomolecules: the Role and Limitations of Patents*. Kevin Lauderdale has gone beyond the call of duty and once again expertly created the printed and online proceedings. Al Conde has ensured that the hardware and network systems are functional. We would especially like to acknowledge the contributions of the session organizers who solicited papers and reviews, and ensured that the quality of the meeting remains high. The session organizers (and their associated sessions) are:

Inna Dubchak, Lior Pachter and Liping Wei (Genome-wide Analysis and Comparative Genomics)

Peter Karp, Pedro Romero and Eric Neumann (Genome, Pathway and Interaction Bioinformatics)

Willi von der Lieth (Expanding Proteomics to Glycobiology)

Lynette Hirschman, Jong C. Park, Junichi Tsujii, Cathy Wu and Limsoon Wong (Literature Data Mining for Biology)

Isaac Kohane, Clay Stephens, Julie Schneider and Francisco De La Vega (Human Genomic Variation: Disease, Drug Response, and Clinical Phenotypes)

Scott Stanley and Benjamin Salisbury (Phylogenetic Genomics and Genomic Phylogenetics)

Peter Clote, Gavin Naylor, and Ziheng Yang (Proteins: Structure, Function and Evolution)

The PSB organizers and session leaders relied on the assistance of those who capably reviewed the submitted manuscripts. A partial list of reviewers is provided elsewhere in this volume. We thank those who have been left off this list inadvertently or who wish to remain anonymous.

Aloha!

Pacific Symposium on Biocomputing Co-Chairs *October 1, 2001*

Russ B. Altman
Stanford University

A. Keith Dunker
Washington State University

Lawrence Hunter
University of Colorado Health Sciences Center

Teri E. Klein
Stanford University

Thanks to reviewers . . .

Finally, we wish to thank the scores of paper reviewers. PSB requires that every paper in this volume be reviewed by at least three independent referees. Since there is a large volume of submitted papers, that requires a great deal of work from many people, and we are grateful to all of you listed below, and to any whose names we may have accidentally omitted.

Aram Adourian
Laura Almasy
Orly Alter
Chris Amos
Mike Bada
Pierre Baldi
Serafim Batzoglou
Jadwiga Bienkowka
Eckart Bindewald
Erich Bornberg-Bauer
Phil Bradley
Richard Broughton
Michael Brudno
Andrea Califano
Matt Callow
Roland Carel
Vincent J. Carey
Simon Cawley
Hue Sun Chan
Joseph Chang
Andrew Clark
Julio Collado-Vides
Josep Comeron
Olivier Couronne
Derek Dimcheff
Chris Ding
Roland Dunbrack
Jeremy Edwards
Jodi Vanden Eng
Niklas Eriksen
George Estabrook
Andras Fiser
Jennifer Gleason
Richard Goldstein
Susumu Goto
Douglas Greer
Igor Grigoriev
Mark Grote
Ivo Gut
Alexander J. Hartemink
Lynette Hirschman
Steve Holbrook
David Paul Holden
John Holmes
Roderick V. Jensen
Ruhong Jiang
Kenneth Karol
Peter Karp
Ju Han Kim
Jessica Kissinger
Alex Lancaster
Jobst Landgrebe
Rick Lathrop
Hans-Peter Lenhof
Jin-Long Li
Weizhong Li
Pat Lincoln
Jan Liphardt
Irene Liu
Xiaole Liu
Gaby Loots
Joanne Luciano
Andrew Martin
Kate McKusick
William Newell
Magnus Nordborg
Gary Nunn
Matej Oresic
Christos Ouzounis
Ivan Ovcharenko
Jong Park
Peter Park
Hugh Pasika
Len Pennacchio
Yitzhak Pilpel
Tom Plasterer
Darrent Platt
David Pollock
John Quackenbush
Mark Rabin
Marco Ramoni
Aviv Regev
Michael Reich
Markus Ringnér
Pedro Romero
Vincent Schachter
Steffen Schulze-Kremer
Jody Schwartz
Thomas Seidl
Imran Shah
Ron Shamir
Roded Sharan
Victor Solovyev
Terence Speed
Paul Spellman
Scott Stanley
Robert Stuart
Jane Su
Xiaoping Su
Zoltan Szallasi
Amos Tanay
Debra Tanguay
Glenn Tesler
Denis Thieffry
Glenys Thomson
Jeff Thorne
Martin Tompa
Jun'ichi Tsuji
Jacques van Helden
Mike Walker
Teresa Webster
Simon Whelan
Kelly Ewen White
Glenn Williams
Limsoon Wong
Cathy Wu
Yu Xia
Dong Xu
Ying Xu
Chen-Hsiang Yeang
John Yin
Ping Zhan
Ge Zhang
Yingdong Zhao

CONTENTS

GENOME, PATHWAY AND INTERACTION BIOINFORMATICS

Session Introductions and Peer Reviewed Papers

HUMAN GENOME VARIATION: DISEASE, DRUG RESPONSE, AND CLINICAL PHENOTYPES

FRANCISCO M. DE LA VEGA
*Applied Biosystems, 850 Lincoln Centre Drive,
Foster City, CA 94404, USA*

ISAAC S. KOHANE
*Children's Hospital Informatics Program & Harvard Medical School,
300 Longwood Avenue, Boston, MA 02115, USA*

JULIE A. SCHNEIDER and J. CLAIBORNE STEPHENS
*Genaissance Pharmaceuticals, Inc.,
Five Science Park, New Haven, CT 06511, USA*

With the completion of a rough draft of the human genome sequence in sight, researchers are shifting to leverage this new information in the elucidation of the genetic basis of disease susceptibility and drug response. Massive genotyping and gene expression profiling studies are being planned and carried out by both academic/public institutions and industry. Researchers from different disciplines are all interested in the mining of the data coming from those studies; human geneticists, population geneticists, molecular biologists, computational biologists and even clinical practitioners. These communities have different immediate goals, but at the end of the day what is sought is analogous: the connection between variation in a group of genes or in their expression and observed phenotypes. There is an imminent need to link information across the huge data sets these groups are producing independently. However, there are tremendous challenges in the integration of polymorphism and gene expression databases and their clinical phenotypic annotation

This is the third session devoted to the computational challenges of human genome variation studies held at the Pacific Symposium on Biocomputing[1,2]. The focus of the session has been the presentation and discussion of new research that promises to facilitate the elucidation of the connections between genotypes and phenotypes using the data generated by high-throughput technologies. Nine accepted manuscripts comprise this year's original work presented at the conference.

A major incentive for collecting genetic variation data is to use this information to identify genomic regions that influence disease susceptibility or drug response. In this volume, Zhang *et al.* outline a new approach to identify clinically relevant genes that produce quantitative phenotypes. Although similar methods have been developed to measure the strength of association between haplotypes and binary (case-control) data, Zhang *et al.*'s method is particularly valuable because many important clinical phenotypes display quantitative inheritance. On the other hand, the manuscript of Moore and Hahn introduce a novel computational approach using cellular automata (CA) and parallel genetic algorithms to identify combinations of SNPs associated with clinical outcomes. They use a simulated dataset of a discordant sib-pair study design to demonstrate that the CA approach has good power to identify high-order nonlinear interactions with few false-positives. Given the current uncertainties on the genetic architecture underlying complex disease[5], it is critical to develop new approaches, such as the CA advanced by the authors, that can test for association in the presence of allelic heterogeneity[6] and epistatic interactions between loci.

Large quantities of DNA sequence variation data is needed to better understand the contribution of genetics to human disease, drug response, and clinical phenotypes. In order to insure the quality of these data, fully automated genotyping processes are required: from assay design, assay validation, assay interpretation, quality control, to data management and release. One of the major challenges involved in developing a streamlined, high-throughput genotyping is creating appropriate software to support the system. In their conference paper, Heil *et al.* describe the components of a successful, ultra high-throughput genotyping process developed at Celera Genomics. Their approach could be an excellent starting point for those involved in developing similar infrastructures elsewhere.

How to properly store and combine complex biological data is an extremely important subject in the post-genome era. Among the challenges to develop an efficient data or knowledge base are the diversity of semantics, potential uses, and data sources. Ontologies have been successfully applied in the past to develop knowledge base systems to store complex data, such as the Gene Ontology for gene annotations[3], and RiboWeb[4] for capturing experimental results in scientific literature. The contributions of Rubin *et al.* and Oliver *et al.* to this conference present a successful application of ontologies on genotype-phenotype data in relation to clinical drug response. The approach used in "PharmGKB" presented by the authors address many of the complex problems arising when retrieving data from diverse genomics and clinical databases, and when updating links to external database domains. Their methodology may be very helpful for making the diverse genomics data better suited for scientific analysis.

Molecular profiling is a tool that is gaining acceptance to classify tissue samples and other clinical outcomes based on gene and potentially protein expression profiles. Its accuracy depends on the appropriate analysis of the resulting datasets, and typically involves multivariate statistics and other machine learning techniques. The paper of Ben-Hur *et al.* describes an algorithm to investigate the stability of the solutions of clustering algorithms. The authors apply their method to the hierarchical clustering of microarray and synthetic data. On the other hand, Ghosh applies a regression analysis to data that has been first transformed by Singular Value Decomposition (SDV), for uncovering possible relations between microarray expression data of tumor samples and tumor diagnosis. The problem is a novel application for SVD, which has been recently applied to microarray data in a different but complementary approach. The paper of Potter and Draghici addresses a clinically important problem: classification of HIV protease's resistance to IC90 drug solely from protein sequences. Their contribution shows that improved accuracy can be achieved by combining SOFM classifiers.

As high-throughput genotyping and expression-measurement methodologies are applied to large populations, the opportunity now exists to use existing clinical phenotypic annotations (i.e., the extended medical record) in the analysis of the relationship between genotype/haplotype variation and phenotype. Typically, however, the forward link is sought, leading from genetic variation data to the inference of clinical phenotypes. The paper of Malin and Sweeney in this volume offers instead a reverse approach, allowing the inference of genetic variability data based on clinical phenotypes. In this unusual approach, clinical/hospital/claims data is brought together with phenotype/genotype through the use machine learning techniques to predict the underlying genotype.

Acknowledgments

We would like to acknowledge the generous help of the anonymous reviewers that supported the selection process for this session, as well as the panelists that joined us to discuss the challenges in this field.

References

1. F. M. De La Vega, and M. Kreitman. "Human genome variation" In: *Pacific Symposium on Biocomputing 2000*, R.B. Altman *et al.* (Eds.). World Scientific Press, Singapore (2000).
2. F.M. De La Vega, M. Kreitman, and I. S. Kohane. "Human genome variation: Linking genotypes to clinical phenotypes" In: *Pacific Symposium on Biocomputing 2001*, R.B. Altman *et al.* (Eds.). World Scientific Press, Singapore (2001).
3. The Gene Ontology Consortium. "Creating the gene ontology resource: design and implementation" *Genome Res.* **11(8)**, 1425-1433 (2001).
4. R.O. Chen, R. Feliciano, R.B. Altman. "RIBOWEB: linking structural computations to a knowledge base of published experimental data" In *Proc Int Conf Intell Syst Mol Biol* **5**, 84-87 (1997).
5. A.F. Wright and N.D. Hastie. "Complex genetic diseases: controversy over the Croesus code" Genome Biology **2(8)**, comment 2007.1–2007.8 (2001).
6. J.K. Pritchard. "Are Rare Variants Responsible for Susceptibility to Complex Diseases?" *Am. J. Hum. Genet.* **69**,124–137 (2001).

A stability based method for discovering structure in clustered data

Asa Ben-Hur*, Andre Elisseeff† and Isabelle Guyon*
BioWulf Technologies LLC
*2030 Addison st. Suite 102 †305 Broadway (9th Floor)
Berkeley, CA 94704 New-York, NY 10007

Abstract

We present a method for visually and quantitatively assessing the presence of structure in clustered data. The method exploits measurements of the stability of clustering solutions obtained by perturbing the data set. Stability is characterized by the distribution of pairwise similarities between clusterings obtained from sub samples of the data. High pairwise similarities indicate a stable clustering pattern. The method can be used with any clustering algorithm; it provides a means of rationally defining an optimum number of clusters, and can also detect the lack of structure in data. We show results on artificial and microarray data using a hierarchical clustering algorithm.

1 Introduction

Clustering is widely used in exploratory analysis of biological data. With the advent of new biological assays such as DNA microarrays that allow the simultaneous recording of tens of thousands of variables, it has become more important than ever to have powerful tools for data visualization and analysis. Clustering, and particularly hierarchical clustering, play an important role in this process. [1,2,3]

Clustering provides a way of validating the quality of the data by verifying that groups form according to the prior knowledge one has about sample categories. It also provides means of discovering new natural groupings. [4] Yet there is no generally agreed upon definition of what is a "natural grouping." In this paper we propose a method of detecting the presence of clusters in data that can serve as the basis of such a definition. It can be combined with any clustering algorithm, but proves to be particularly useful in conjunction with hierarchical clustering algorithms.

The method we propose in this paper is based on the stability of clustering with respect to perturbations such as sub-sampling or the addition of noise. Stability can be considered an important property of a clustering solution, since data, and gene expression data in particular, is noisy. Thus we suggest stability as a means for defining meaningful partitions. The idea of using stability to evaluate clustering solutions is not new. In the context of hierarchical clustering, some authors have considered the stability of the whole hierarchy. [5] However, our experience indicates that in most real world cases the complete dendrogram is rarely stable. The stability of partitions has also been addressed. [6,7,8] In this model, a figure of merit is assigned to a partition of the data according to average similarity of the partition to a set of partitions obtained by clustering a perturbed dataset. The optimal number of clusters (or other parameter employed by the algorithm) is then determined by the maximum value of the average similarity. But we observed in several practical instances that considering the average, rather than the complete distribution was insufficient. The distribution can be used both as a tool to visually probe the structure in the data, and to provide a criterion for choosing an optimal partition of the data: plotting the distribution for various numbers of clusters reveals a transition between a distribution of similarities that is concentrated near 1 (most solutions highly similar) to a wider distribution. In the examples we studied, the value of the number of clusters at which this transition occurs agrees with the intuitive choice of the number of clusters. We have developed a heuristic for comparing partitions across different levels of the dendrogram that make this transition more pronounced. The method is useful not only in choosing the number of clusters, but also as a general tool for making choices regarding other components of the clustering algorithm. We have applied it in choosing the type of normalization and the number of leading principal components. [9]

Many methods for selecting an optimum number of clusters can be found in the literature. In this paper we report results that show that our method performs well when compared with some of the more successful methods reported in recent surveys.[10,11] This may be explained by the fact that our method does not make assumptions about the distribution of the data or about cluster shape as most other methods;[11,10] only our method and the gap statistic can detect the absence of structure. Our method has advantages over information-theoretic criteria based on compression efficiency considerations and over related Bayesian criteria[12] in that they are model free, and work with any clustering algorithm. Some clustering algorithms have been claimed to generate only meaningful partitions, so do not require our method for this purpose.[4,13] We also mention the method of Yeung *et al.*[14] for assessing the relative merit of different clustering solutions. They tested their method on microarray data; however, they do not give a way of selecting an optimal number of clusters, so no direct comparison can be made.

The paper is organized as follows: in Section 2 we introduce the dot product between partitions and express several similarity measures in terms of this dot product. In Section 3 we present our practical algorithm. Section 4 is devoted to experimental results of using the algorithm. This is followed by a discussion and conclusions.

2 Clustering similarity measures

In this section we present several similarity measures between partitions found in the literature,[15,7] and express them with the help of a dot product. We begin by reviewing our notation. Let $X = \{\mathbf{x}_1, \ldots, \mathbf{x}_n\}$, and $\mathbf{x}_i \in \mathbb{R}^d$ be the dataset to be clustered.

A *labeling* $\mathcal{L}$ is a partition of X into k subsets $S_1, \ldots, S_k$. We use the following representation of a labeling by a matrix C with components:

$$C_{ij} = \begin{cases} 1 & \text{if } \mathbf{x}_i \text{ and } \mathbf{x}_j \text{ belong to the same cluster and } i \neq j\,, \\ 0 & \text{otherwise}\,. \end{cases} \tag{1}$$

Let labelings $\mathcal{L}_1$ and $\mathcal{L}_2$ have matrix representations $C^{(1)}$ and $C^{(2)}$, respectively. We define the dot product

$$\langle \mathcal{L}_1, \mathcal{L}_2 \rangle = \left\langle C^{(1)}, C^{(2)} \right\rangle = \sum_{i,j} C^{(1)}_{ij} C^{(2)}_{ij}\,. \tag{2}$$

This dot product computes the number of pairs of points clustered together, and can also be interpreted as the number of common edges in graphs represented by $C^{(1)}$ and $C^{(2)}$, and we note that it can be computed in $O(k_1 k_2 n)$.

As a dot product, $\langle \mathcal{L}_1, \mathcal{L}_2 \rangle$ satisfies the Cauchy-Schwartz inequality: $\langle \mathcal{L}_1, \mathcal{L}_2 \rangle \leq \sqrt{\langle \mathcal{L}_1, \mathcal{L}_1 \rangle \langle \mathcal{L}_2, \mathcal{L}_2 \rangle}$, and thus can be normalized into a correlation or cosine similarity measure:

$$cor(\mathcal{L}_1, \mathcal{L}_2) = \frac{\langle \mathcal{L}_1, \mathcal{L}_2 \rangle}{\sqrt{\langle \mathcal{L}_1, \mathcal{L}_1 \rangle \langle \mathcal{L}_2, \mathcal{L}_2 \rangle}}\,. \tag{3}$$

This similarity measure was introduced by Fowlkes and Mallows. [7] Next, we show that two commonly used similarity measures can be expressed in terms of the dot product defined above. Given two matrices $C^{(1)}, C^{(2)}$ with 0-1 entries, let N_{ij} for $i, j \in \{0, 1\}$ be the number of entries on which $C^{(1)}$ and $C^{(2)}$ have values i and j, respectively. The *matching coefficient* [15] is defined as the fraction of entries on which the two matrices agree:

$$M(\mathcal{L}_1, \mathcal{L}_2) = \frac{N_{00} + N_{11}}{N_{00} + N_{01} + N_{10} + N_{11}}\,. \tag{4}$$

The *Jaccard coefficient* is a similar ratio when "negative" matches are ignored:

$$J(\mathcal{L}_1, \mathcal{L}_2) = \frac{N_{11}}{N_{01} + N_{10} + N_{11}}\,. \tag{5}$$

The matching coefficient often varies over a smaller range than the Jaccard coefficient since the N_{00} term is usually a dominant factor. These similarity measures can be expressed in terms of the labeling dot product and the associated norm:

$$\begin{aligned} J(\mathcal{L}_1, \mathcal{L}_2) &= \frac{\langle C^{(1)}, C^{(2)} \rangle}{\langle C^{(1)}, C^{(1)} \rangle + \langle C^{(2)}, C^{(2)} \rangle - \langle C^{(1)}, C^{(2)} \rangle} \\ M(\mathcal{L}_1, \mathcal{L}_2) &= 1 - \frac{1}{n^2} \| C^{(1)} - C^{(2)} \|^2 \end{aligned}$$

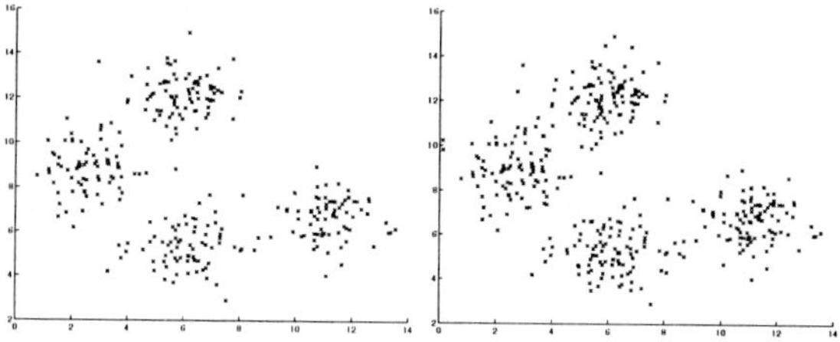

Figure 1: Two 250 point sub-samples of a 400 point Gaussian mixture.

This is a result of the observation that $N_{11} = \langle C^{(1)}, C^{(2)} \rangle$, $N_{01} = \langle 1_n - C^{(1)}, C^{(2)} \rangle$, $N_{10} = \langle C^{(1)}, 1_n - C^{(2)} \rangle$, $N_{00} = \langle 1_n - C^{(1)}, 1_n - C^{(2)} \rangle$, where 1_n is an $n \times n$ matrix with entries equal to 1. The above expression for the Jaccard coefficient shows that it is close to the correlation similarity measure, as we have observed in practice.

3 The model explorer algorithm

When one looks at two sub-samples of a cloud of data points, with a sampling ratio f (fraction of points sampled) not much smaller than 1 (say $f > 0.5$), one usually observes the same general structure (Figure 1). Thus it is reasonable to postulate that a partition into k clusters has captured the "inherent" structure in a dataset if partitions into k clusters obtained from running the clustering algorithm with different sub-samples are similar, i.e. close in structure according to one of the similarity measures introduced in the previous section. "Inherent" structure is thus structure that is stable with respect to sub-sampling. We cast this reasoning into the problem of finding the optimal number of clusters for a given dataset and clustering algorithm: look for the largest k such that partitions into k clusters are stable. Note that rather than choosing just the number of clusters, one can extend the scope of the search for a set of variables where structure is most apparent, i.e. stable. This is performed elsewhere. [9]

We consider a generic clustering algorithm that receives as input a dataset (or similarity/dissimilarity matrix) and a parameter k that controls either directly or indirectly the number of clusters that the algorithm produces. This input convention is applicable to hierarchical clustering algorithms: given k, cut the tree so that k clusters are produced. We want to characterize the stability for each value of k. This is accomplished by clustering sub-samples of the data, and then computing the similarity between pairs of sub-samples according to similarity between the labels of the points common to both sub-samples. The result is a distribution of similarities for each k. The algorithm is presented in Figure 2.

The distribution of the similarities is then compared for different values of k

Input: X {a dataset}, $k_{\max}$ {maximum number of clusters}, num_subsamples {number of subsamples}
Output: $S(i,k)$ {list of similarities for each k and each pair of sub-samples }
Require: A clustering algorithm: cluster(X,k); a similarity measure between labels: $s(L_1,L_2)$

$f = 0.8$
for $k = 2$ to $k_{\max}$ **do**
  for $i = 1$ to num_subsamples **do**
    sub_1 =subsamp(X,f){a sub-sample with a fraction f of the data}
    sub_2 =subsamp(X,f)
    L_1 =cluster(sub_1,k)
    L_2 =cluster(sub_2,k)
    Intersect= $sub_1 \cap sub_2$
    $S(i,k) = s(L_1(Intersect), L_2(Intersect))$ {Compute the similarity on the points common to both subsamples}
  end for
end for

Figure 2: The Model explorer algorithm.

(Figure 3). In our numerical experiments (Section 4) we found that, indeed, when the structure in the data is captured by a partition into k clusters, many sub-samples have similar clustering, and the distribution of similarities is concentrated close to 1.

Remark 3.1 For the trivial case $k = 1$, all clusterings are the same, so there is no need for any computation in this case. In addition, the value of f should not be too low; otherwise not all clusters are represented in a sub-sample. In our experiments the shape of the distribution of similarities did not depend very much on the specific value of f.

4 Experiments

In this section we describe experiments on artificial and real data. We chose to use data where the number of clusters is apparent, so that one can be convinced of the performance of the algorithm. In all the experiments we show the distribution of the correlation score; equivalent results were obtained using other scores as well. The sampling ratio, f, was 0.8 and the number of pairs of solutions compared for each k was 100. As a clustering algorithm we use the average-link hierarchical clustering algorithm.[15] The advantage of using a hierarchical clustering method is that the same

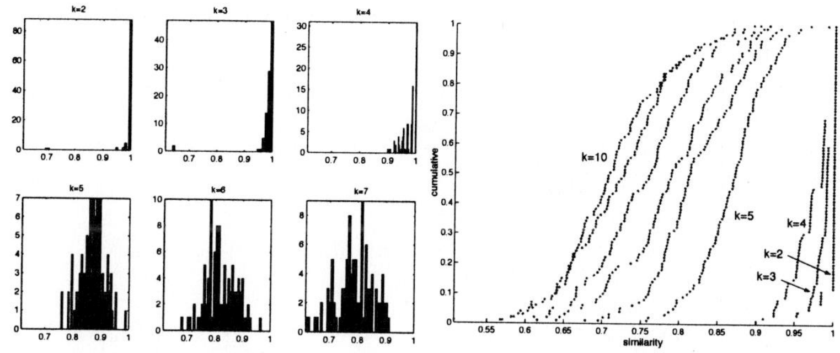

Figure 3: Left: histogram of the correlation similarity measure; right: overlay of the cumulative distributions for increasing values of k.

set of trees can be used for all values of k, by looking at different levels of the tree each time. To tackle the problem of outliers, we cut the tree such that there are k clusters, each of them not a singleton (thus the total number of clusters can be higher than k). This is extended to consider partitions that contain k clusters, each of them larger than some threshold. This helps enhance the stability in the case of a good value of k, and de-stabilizes clustering solutions for higher k, making the transition from highly similar solutions to a wide distribution of similarities more pronounced.

We begin with the data depicted in Figure 1, which is a mixture of four Gaussians. The histogram of the score for varying values of k is plotted in figure 3. We make several observations regarding the histogram. At $k = 2$ it is concentrated at 1, since almost all the runs discriminated between the two upper and two lower clusters. At $k = 3$ most runs separate the two lower clusters, and at $k = 4$ most runs found the "correct" clustering which is reflected in the distribution of scores still concentrated near 1. For $k > 4$ there is no longer one preferred solution, as is seen by the wide spectrum of similarities. We remark that if the clusters were well separated, or the clusters arranged more symmetrically, there would not have been a preferred way of clustering into 2 or 3 clusters as is the case here; in that case the similarity for $k = 2, 3$ would have been low, and increased for $k = 4$. In such cases one often observes a bimodal distribution of similarities.

The next dataset we considered was the yeast DNA microarray data of Eisen *et al.*[1] We used the MYGD functional annotation to choose the 5 functional classes that were most learnable by SVMs,[16] and that were noted by Eisen *et al.* to cluster well.[1] We looked at the genes that belong uniquely to these 5 functional classes. This gave a dataset with 208 genes and 79 features (experiments) in the following classes: (1)

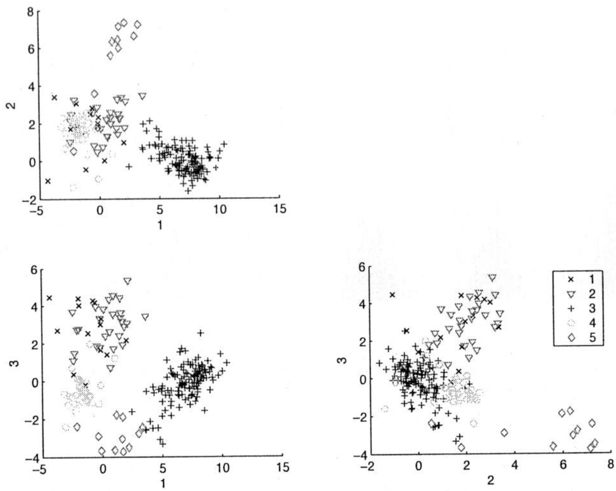

Figure 4: First three principal components of the yeast microarray data. The legend identifies the symbols that represent each functional class. Class number corresponds to the numbers given in the listing of the classes in the text.

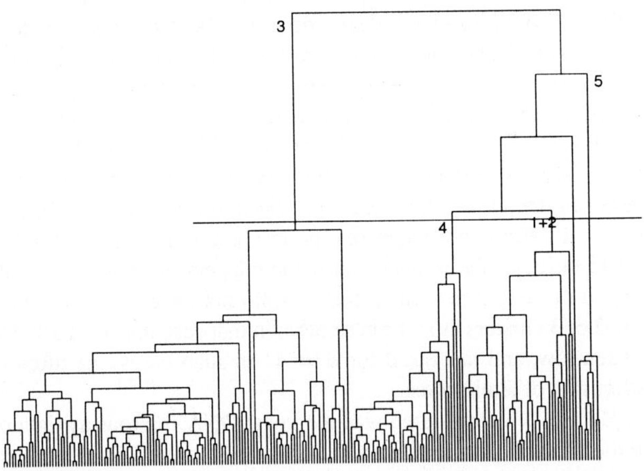

Figure 5: Dendrogram for yeast microarray data. Numbers indicate the functional class represented by each cluster. The horizontal line represents the lowest level at which partitions are still highly stable.

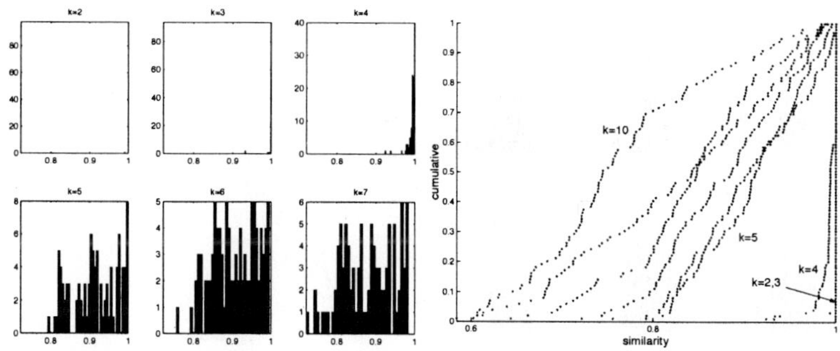

Figure 6: Left: histogram of the correlation similarity measure for the yeast gene expression data for increasing values of k. For $k = 2$ all similarities were equal to 1. Right: overlay of the cumulative distribution functions.

Tricarboxylic acid cycle or Krebs cycle (14 genes), (2) Respiration chain complexes (27 genes), (3) Cytoplasmaticribosomal proteins (121 genes), (4) proteasomes (35 genes), and (5) Histones (11 genes). The first three principal components were then extracted (see Figure 4). One can clearly see four clusters in the data; these agree well with the MYGD classes, with classes 1 and 2 strongly overlapping. The distribution and histogram of scores is given in Figure 6. We observe the same behavior as in the Gaussian mixture data. Between $k = 4$ and $k = 5$ there is a transition from a distribution that has a large component near 1, to a wide distribution that is similar to the distribution on random data (see below). Since there was a singleton cluster, the total number of clusters is 5. The clusters agree well with the protein classes that were assigned to the genes in the MYGD database, with the exception that clusters 1 and 2 were clustered together. The way the dendrogram was cut to produce the partition is illustrated in Figure 5. Looking at the dendrogram one might think that further splitting of cluster 3 is justified. However, the length of the edge in the dendrogram can be misleading: in the case of average linkage the length of the edge is proportional to the average distance between clusters, and since cluster 3 is large, a long edge does not necessarily imply a well separated sub-cluster. At high levels of the hierarchy long edges generally result simply because the clusters become larger, even if the data contains no structure. When clusters 1 and 2 are assigned the same label, the similarity between the clustering and the known classification is 0.97. We note that principal component analysis (PCA) not only allows visualization of the data, it enhances cluster structure reflected in the stability and also improves the agreement of the clustering with the MYGD classes. [9]

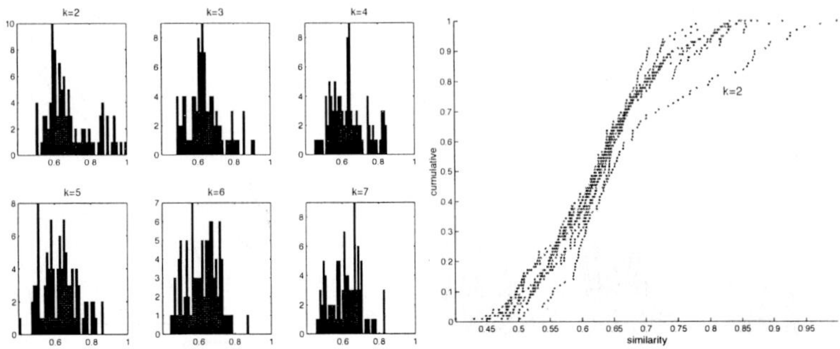

Figure 7: Left: histogram of the correlation score for 208 points uniformly distributed on the unit cube; right: overlay of the cumulative distributions of the correlation score.

A run on data uniformly distributed on the unit cube is shown in Figure 7. The distributions are quite similar to each other, with no change that can be interpreted as a transition from a stable clustering to an unstable one.

These examples indicate a simple way for identifying k; choose the value where there is a transition from a score distribution that is concentrated near 1 to a wider distribution. This can be quantified, e.g. by a jump in the area under the cumulative distribution function or by a jump in $P(s_k > \eta)$, where s_k is the random variable that denotes the similarity between partitions into k clusters, and η is a constant. A value of $\eta = 0.9$ would work on the set of examples considered here.

The results of our method are compared in Table 1 with a number of other methods for choosing k. We used most of the methods tested by Tibshirani *et al.* against their gap statistic method. [11] They are among the methods tested by Milligan and Cooper. [10] Jain's method uses the quotient between the in-cluster average distance and out-of-cluster average distance, averaged over all the clusters. The optimal number of clusters is chosen as the k that minimizes this quantity. The method of Calinski and Harabsz is similar, but uses a different normalization, and the squared distances. The silhouette statistic is based on comparing the average distance of the point to members of other clusters with the average distance of a point to members of its own cluster. A point is "well clustered" if it is closer on average to the members of its own cluster than to points of other clusters. The silhouette statistic is the average of the point silhouettes, and k is chosen to maximize it. The KL (Krzanowski and Lai), Hartigan, and gap statistic methods use criteria that are based on the k-dependence of a function of the within-cluster sum-squared distances. Almost all the methods were successful on the Gaussian mixture data; this is to be expected since some of the

Table 1: Number of clusters obtained by various methods for choosing the number of clusters. Subsamp denotes our method.

problem	Jain	Silhouette	KL	CH	Hartigan	gap	**subsamp**	true
4Gaus	6	4	9	4	4	4	4	4
Microarray	4	5	2	2	3	6	5	5
Random	7	9	5	2	9	1	1	1

methods are specifically constructed for such data. The microarray data proved more difficult. We note that our method gave the same results when all the variables rather than just the first three principal components were clustered, whereas the gap statistic did not give a result when all the variables were clustered. The gap statistic is based on a comparison of the within cluster sum-squared distance of the given clustering with an average obtained over random data. Perhaps the comparison with random data does not scale well to very high dimensionality. All the methods we tested, other than the gap statistic and our own method cannot detect a lack of structure: they produce a meaningful result only if it is known beforehand that the number of clusters is greater than 1. When these methods are run on data with no structure they still provide (erroneously) a result. Running these methods on sub-samples of the data can provide the information required to rule out the hypothesis of no structure: intuitively, for data with clear clusters the result is likely to remain the same, while for data with no structure the criterion is likely to be unstable, and fluctuate across sub-samples.

5 Discussion

In the set of experiments we ran, only the gap statistic method performed as well as our method. Since the gap statistic is based on a sum-squared distances criterion, it is biased toward compact clusters; our method has no such bias. Further work should include a more systematic experimental analysis to differentiate the two methods. Both methods are the most computationally expensive, requiring running the clustering algorithm a number of times. Our method can be used not only to choose the number of clusters, but also as a comparative tool that can help in choosing other aspects of the clustering such as normalization. [9] Our algorithm is most efficient with hierarchical clustering, since once a dendrogram is computed, varying the number of clusters is achieved at little additional computational expense.

The datasets analysed in this paper were chosen for illustrative purposes for having a distinct structure. One might argue that many real world datasets do not have such an obvious number of clusters. Indeed, partitions obtained on a large set of variables (e.g. thousands of genes from DNA microarrays) are usually unstable. We see that as a symptom that prior knowledge is needed to select meaningful subsets of variables (genes) that can yield stable clusters.

Our method is related to the bootstrap and jackknife methods in its use of sampling to estimate a statistic.[17] However, in our case, sampling is used as perturbation that *generates* the statistic. (Alternatively, one can add noise to the data instead of sub-sampling.) We also note that generating pairs of clusterings can be performed in various ways: comparing pairs of clustered subsamples as done here; comparing clustered subsamples to a reference clustering; or dividing the data into two, clustering both parts, and obtaining a second clustering of each part by assigning its points to clusters according to the nearest cluster center of the other part. Stability of a classifier is a notion that was applied in supervised learning as well.[18] It was shown that stability can be used to bound the prediction error of a classifier: the more stable the classifier, the more likely it is to perform well. In future work we plan to extend this theoretical framework to the case of unsupervised learning. Finally, the notion of stability can be applied in other types of data analysis problems whose objective is to detect structure in data, e.g. extraction of gene networks, or ranking of genes according to their predictive power.

6 Conclusion

Determining the optimum number of clusters in data is an ill posed problems for which many solutions have been proposed. None of them is widely used, and the level of their performance is data dependent.[10] In this paper we propose to use the distribution of pairwise similarity between clusterings of sub-samples of a dataset as a measure of the stability of a partition. The number of clusters at which a transition from stable to unstable clustering solutions occurs can be used to choose an optimal number of clusters. In all the experiments we ran, the results coincide with the intuitive choice. Whereas most model selection methods give a result without attaching a level of confidence to it, the sharpness of the transition from stable to unstable solutions can give information on how well defined the structure in the data is, and unlike most other methods it can provide information on the lack of structure. In another study we have found it useful as a comparative tool that can help in choosing various aspects of the clustering such as the number of principal components to cluster, and which normalization to use. Thus we view our method as a general exploratory tool, and not just as a way of selecting an optimal number of clusters.

1. M. Eisen et al, "Genetics cluster analysis and display of genome-wide expression patterns" *Proc. Natl. Acad. Sci. USA* 95, 14863–14868 (1998).
2. J. Quackenbush, "Computational anaysis of microarray data" *Nature Reviews Genetics* 2(6), 418–427 (2001).
3. R. Shamir and R. Sharan, "Algorithmic approaches to clustering gene expression data" In T. Jiang, T. Smith, Y. Xu, and M.Q. Zhang, editors, *Current Topics*

in Computational Biology (MIT Press, 2001).
4. G. Getz, E. Levine, and E. Domany, "Coupled two-way clustering analysis of gene microarray data" *Proc. Natl. Acad. Sci USA* 94, 12079–12084 (2000).
5. S.P. Smith and R. Dubes, "Stability of a hierarchical clustering" *Pattern Recognition* 12, 177–187 (1980).
6. E. Levine and E. Domany, "Resampling method for unsupervised estimation of cluster validity" *Neural Computation*, to appear.
7. E.B. Fowlkes and C.L. Mallows, "A method for comparing two hierarchical clusterings" *Journal of the American Statistical Association* 78(383), 553–584 (1983).
8. M. Bittner et. al, "Molecular classification of cutaneous malignant melanoma by gene expression profiling" *Nature* 406(3), (2000).
9. A. Ben-Hur and I. Guyon, "Detecting stable clusters using principal component analysis" In *Methods in Molecular Biology* (Humana Press, to be published).
10. G.W. Milligan and M.C. Cooper, "An examination of procedures for determining the number of clusters in a data set" *Psychometrika* 50, 159–179 (1985).
11. R. Tibshirani, G. Walther, and T. Hastie, "Estimating the number of clusters in a dataset via the gap statistic" *J. Royal. Statis. Soc. B*, to appear.
12. C. Fraley and A.E. Raftery, "How many clusters? which clustering method? answers via model-based cluster analysis" *Computer Journal* 41 548–588 (1998).
13. A. Ben-Hur, D. Horn, H.T. Siegelmann, and V. Vapnik, "A support vector method for hierarchical clustering" In *Advances in Neural Information Processing Systems 13* 367–373 (MIT Press, 2000).
14. K.Y. Yeung, D.R. Haynor, and W.L. Ruzzo, "Validating clustering for gene expression data" *Bioinformatics* 17(4), 309–318 (2001).
15. A.K. Jain and R.C. Dubes, *Algorithms for clustering data* (Prentice Hall, Englewood Cliffs, NJ, 1988).
16. M.P.S. Brown, W.N. Grundy, D. Lin, N. Cristianini, C. Sugnet, M. Ares, and D. Haussler, "Knowledge-based analysis of microarray gene expression data by using support vector machines" *Proc. Natl. Acad. Sci. USA* 97(1), 262–267 (2000).
17. B. Efron, *The Jackknife, the Bootstrap and Other Resampling Plans* (SIAM, Philadelphia, 1982).
18. O. Bousquet and A. Elisseeff, "Algorithmic stability and generalization performance" In *Advances in Neural Information Processing Systems 13* (MIT press, 2000).

SINGULAR VALUE DECOMPOSITION REGRESSION MODELS FOR CLASSIFICATION OF TUMORS FROM MICROARRAY EXPERIMENTS

DEBASHIS GHOSH

Department of Biostatistics, University of Michigan
1420 Washington Heights, Ann Arbor, MI 48109-2029
ghoshd@umich.edu

An important problem in the analysis of microarray data is correlating the high-dimensional measurements with clinical phenotypes. In this paper, we develop predictive models for associating gene expression data from microarray experiments with such outcomes. They are based on the singular value decomposition. We propose new algorithms for performing gene selection and gene clustering based on these predictive models. The estimation procedure using the regression models occurs in two stages. First, the gene expression measurements are transformed using the singular value decomposition. The regression parameters in the model linking the principal components with the clinical responses are then estimated using maximum likelihood. We demonstrate the application of the methodology to data from a breast cancer study.

1 Introduction

DNA biochips have the potential of significantly impacting the study of human disease. By simultaneously gauging the expression of thousands of genes in clinical specimens, a wealth of data points is generated coalescing to form a molecular fingerprint of a disease process. Such experiments have been performed on acute leukemias, lymphomas, breast cancers and cutaneous melanomas.[1,2,3] Obtaining large-scale gene expression profiles of tumors should theoretically allow for the identification of subsets of genes that function as prognostic disease markers or biologic predictors of therapeutic response.

Most primary analyses have utilized hierarchical clustering techniques.[4] However, in many instances, there is external clinical information (such as survival time or tumor type) available. Typically, the investigators use these variables in secondary analyses. For many molecular profiling studies, the scientific goal appears to be finding candidate genes that successfully discriminate between disease classes based on the clinical phenotype. These genes can then be screened for further follow-up studies using immunohistochemical techniques such as tissue microarrays.[5]

Some preliminary work has been put forward correlating gene expression data with clinical outcomes.[6,7] However, these approaches have been univariate and ignore correlations between genes. A problem with joint modeling of gene effects on clinical outcomes is that the number of genes is typically much larger than the number of samples profiled. In statistical terminology, the dimension space of the predictors is much larger than that of the independent samples. Consequently, it is not possible to calculate regression parameter estimates using traditional statistical procedures.

In this paper, we develop a regression framework based on the singular value decomposition for correlating gene expression data with clinical phenotypes. We explore the use of these models for three goals: prediction, gene selection and clustering. We propose novel algorithms for accomplishing the latter two tasks. While the framework presented here can be generalized, we are motivated by the specific problem of modeling the association between gene expression data with type of tumor. Singular value decomposition has been applied to other areas of microarray data analysis. [8,9,10] In the statistical literature, singular value decomposition analysis is known as principal components analysis; we will use the two terms interchangeably throughout the article. Regression modeling using SVD has been done with great success in other areas of application, such as chemometrics.[11] A complication in the current setting that does not arise in other applications is that the clinical outcome may not be continuous; our proposal here involves using categorical regression models[12] for associating the gene expression measurements with tumor type. We demonstrate the procedure using data from a recently published breast cancer study.[13] Because of space limitations, we refer the interested reader to the following URL for more details regarding this project and the analysis of the breast cancer data:

http://www.sph.umich.edu/~ghoshd/SVD/.

2 Methods

Before describing the regression model for correlating gene expression profiles with tumor phenotype, we introduce some notation. Let $\mathbf{X}_i$ denote the p-dimensional column vector of gene expression measurements for the ith subject, $i = 1, \ldots, n$. Note that p will typically be much larger than n. For $i = 1, \ldots, n$, we define Y_i to be the tumor type for the ith individual; this will take values $0, 1, \ldots, J-1$, where J is the number of tumor types. The class $Y = 0$ will be known as the reference category or reference tumor type. We will assume that the $\mathbf{X}_i$ are standardized across chips to have mean zero and variance one for each gene.

2.1 Regression model and estimation

We formulate the effects of gene expression on tumor type using the following multinomial logistic regression model:

$$\log \frac{P(Y_i = r)}{P(Y_i = 0)} = \beta_{r0}^T \mathbf{X}_i, \tag{1}$$

where $P(A)$ is the probability of the event A, $\mathbf{a}^T$ is the transpose of the vector or a matrix $\mathbf{a}$, and β_{r0} is a p-dimensional vector of unknown regression coefficients, $r = 1, \ldots, J-1$. The model is quite general in that separate gene effects are specified for each of the J(J-1)/2 tumor comparisons. More structure can be imposed by placing constraints on β_{r0} $(r = 1, \ldots, J-1)$. For example, we could set $\beta_{r0} = \beta_0$ for all r. This corresponds to a one-unit change in expression level for any gene having the same effect for discriminating any two tumor classes.

In a typical microarray experiment, it is not possible to estimate the parameters in (1) using standard statistical methods because p is much larger than n. We propose using the singular value decomposition to reduce the dimension of β_{r0}. If we let $\mathbf{X}$ denote the $p \times n$ matrix $[\mathbf{X}_1 \cdots \mathbf{X}_n]$, then the singular value decomposition leads to the following decomposition of $\mathbf{X}$:

$$\mathbf{X} = \mathbf{UDV}, \tag{2}$$

where $\mathbf{U}$ is $p \times n$ matrix, and $\mathbf{D}$ and $\mathbf{V}$ are $n \times n$ matrices. The columns of $\mathbf{U}$ are orthonormal, i.e. $\mathbf{U}^T\mathbf{U} = \mathbf{I}_n$, the $n \times n$ identity matrix. The diagonal matrix $\mathbf{D}$ contains the ordered eigenvalues of $\mathbf{X}$ on the diagonal elements so that $\mathbf{D} = \text{diag}(d_1, \ldots, d_n)$, where $d_1 \geq d_2 \geq d_3 \geq \cdots \geq d_n \geq 0$. We will assume without loss of generality that $d_i > 0$ for $i = 1, \ldots, n$. Finally, $\mathbf{V}$ is the $n \times n$ singular value decomposition factor matrix and has both orthonormal rows and columns. The algorithms used to compute the singular value decomposition are typically iterative and quite computationally efficient.[14]

The effect of the singular value decomposition is to project high-dimensional multivariate data into a lower dimensional subspace. By plugging (2) into (1), we obtain the following model:

$$\log \frac{P(Y_i = r)}{P(Y_i = 0)} = \gamma_{r0}^T \mathbf{W}_i, \tag{3}$$

where γ_{r0} $(r = 0, \ldots, J-1)$ is a $n \times 1$ vector of regression coefficients and $\mathbf{W}_i$ $(i = 1, \ldots, n)$ is the ith column of the $n \times n$ matrix $\mathbf{W} \equiv \mathbf{DV}$. It can be shown that β_{r0} in (1) and γ_{r0} in (3) are linked by the following relationship: $\gamma_{r0} = \mathbf{U}^T\beta_{r0}$.

By transforming the regression model from (3) into (1), we have reduced the dimension of the space for the predictor variables from p to n. This makes the problem computationally tractable, i.e. model (3) can be fit using traditional statistical estimation procedures. We use the method of maximum likelihood to estimate γ_{r0} $(r = 0, \ldots, J-1)$.

2.2 Gene selection and clustering based on SVD regression

Ultimately, we are interested in determining which genes have the greatest ability in discriminating between disease classes defined by the clinical phenotype. This corresponds to ranking the components of β_{r0} $(r = 0, 1, \ldots, J-1)$. It would be desirable if we could backtransform the estimators of γ_{r0} in order to derive estimators of β_{r0} in (1) $(r = 0, 1, \ldots, J-1)$. However, this is not possible because the mapping from β_{r0} to γ_{r0} (defined by $\mathbf{U}$) is a many to one mapping, so the inverse mapping is not well-defined.

Our proposal is to rank the p genes using the vector of gene scores $\mathbf{s}_r = \mathbf{U}\widehat{\gamma}_r$ $(r = 0, 1, \ldots, J-1)$. This gives a measure of the p genes to discriminate between the rth category relative to the reference category. If one were to adopt a Bayesian framework for model (1), one can show that with a suitable choice of prior on the regression parameters, $\mathbf{s}_r$ is asymptotically equivalent to the posterior mode of β_{r0}.[15] However, our interest is in ranking the values of $\mathbf{s}_r$, not in performing formal inference. An advantage of this proposed gene selection scheme relative to previous approaches is that potential correlation between the genes is taken into account.

The variance-covariance matrix of the $\mathbf{s}_r$ $(r = 0, \ldots, J-1)$ can be standardized to yield a correlation matrix, which can then be used as an input in a hierarchical clustering algorithm. The clustering algorithm attempts to find relationships between these discriminating genes and is based on the assumption that mutual coexpression potentially implies a common regulatory mechanism or that the genes might be involved in the same pathway. This clustering procedure utilizes the clinical phenotype information in a sensible fashion. Previous clustering methods have failed to take this external information into account.[4]

2.3 Filtering genes

Typically in microarray experiments, the number of potential predictor genes will be on the order of thousands. In studies involving gene expression, it seems biologically plausible that only a fraction of the set of genes on the chip have real biological activity. Consequently, certain authors have suggested that reducing the initial number of variables under consideration leads to improved predictive performance.[15,16] With the breast cancer data, we study the use of an initial preprocessing in order to filter out a subset of the original set of genes. We fit an analysis of variance (ANOVA) model of gene expression measurement versus tumor class individually for each gene. For each ANOVA model, we calculate an overall F-statistic; this yields a set of p F-statistics. We then take the M genes with the largest F-statistics as the potential predictor variables in the model. The effect of this variable selection is to eliminate genes whose power in discriminating between tumor types is not significantly above the experimental variability in the gene expression measurements. An empirical study of the effect of M on the predictive performance on the singular value decomposition regression modeling is given in the application to the breast cancer data.

It has been noted in the literature that variable selection is an inherently unstable procedure.[17] This instability will be even more apparent here because of the relatively small values of n. In order to stabilize the performance of the variable selection described in the previous paragraph, we also examined the use of bagging methods.[18] This method involves creating B perturbed versions of the original dataset by resampling from the set of independent samples B times. For each dataset, we rank the genes by the values of the F-statistic. We then compute the average rank of each gene over the B datasets and take those with the M highest averages. We break ties using random jittering.

2.4 Choosing number of principal components

A major issue in the application of singular value decomposition regression modeling to high-dimensional data is determining how many principal components to use in model (3). There are many ways of performing this variable selection.[11] We have employed leave-one-out cross-validation. In this procedure, one sample is removed from the dataset at a time. For a fixed number of principal components, say k, the regression model is fit to the remaining data. Based on the estimated model, the model is used to predict the tumor type of the withheld sample. An error measure is then calculated based on Hamming distance. We repeat this training procedure leaving out each of the other samples from the dataset one at a time; this yields an estimate of the classification error rate. This is done for every possible value of k; the value of k that yields the smallest classification error rate is then chosen. Leave-one-out cross-validation is a popular method in situations with small samples where no test data are available. We note that this is a data-driven rule for selecting the number of principal components to use in the modeling.

3 Application

In this section, we apply the proposed methodology to data from a study of BRCA1- and BRCA2-positive tumors.[13] In this study, 23 biopsy specimens of primary breast tumors were collected. Seven had BRCA1 germ-line mutations, and eight had BRCA2 germ-line mutations. In addition, another eight samples were collected that had neither BRCA1 nor BRCA2 germ-line mutations; these were treated as sporadic cases of breast cancer. The goal of the study was to determine if there were differences in global gene expression profiles that could be used to discriminate the three classes of cancer (BRCA1, BRCA2 and sporadic).

While we will not go into the details of the analysis performed by Hedenfalk et al., we do wish to make two points. First, univariate statistical methods were used in order to determine the ability of genes to discriminate between the tumor types. Second, the analysis of the data was divided into two subgroup analyses. The first subgroup comparison was between BRCA1-positive and sporadic tumors; the second involved comparing BRCA2-positive and sporadic tumors. While this analysis approach seems reasonable in terms of the scientific goals of the study, it is potentially statistically more efficient to incorporate the correlations between the three tumor classes as well as the genes in order to incorporate correlations between genes. In the discussion that follows, we take the sporadic tumor class to be the reference category.

We first focus on the performance of the principal components regression modeling in terms of the classification error rate, defined using Hamming distance. In particular, we look at the effect of varying M. The results are summarized in Figure 1. Based on Figure 1, the optimal number of principal components varies on M; however, it does not appear to be possible to derive a general rule. For example, for $M = 25$, we have one misclassification using the singular value decomposition procedure with 11 principal components in the model. Comparable optimal misclassification rates can be obtained using $M = 1500$ and $M = 3226$. Using cross-validation, the choice of the number of principal components will depend on the particular dataset. We also examined the effect of the bagging variable selection procedure described in the paper (data not shown). The bagging variable selection tends to improve the predictive performance of the singular value regression models; we refer the interested reader to our website for these results.

We now illustrate the ranking and clustering procedures based upon the SVD regression modelling. For the purposes of discussion, we take $M = 100$. Based on Figure 1, the number of principal components for $M = 100$ that minimizes the classification error rate is $k = 2$. We subsequently fit model (3)

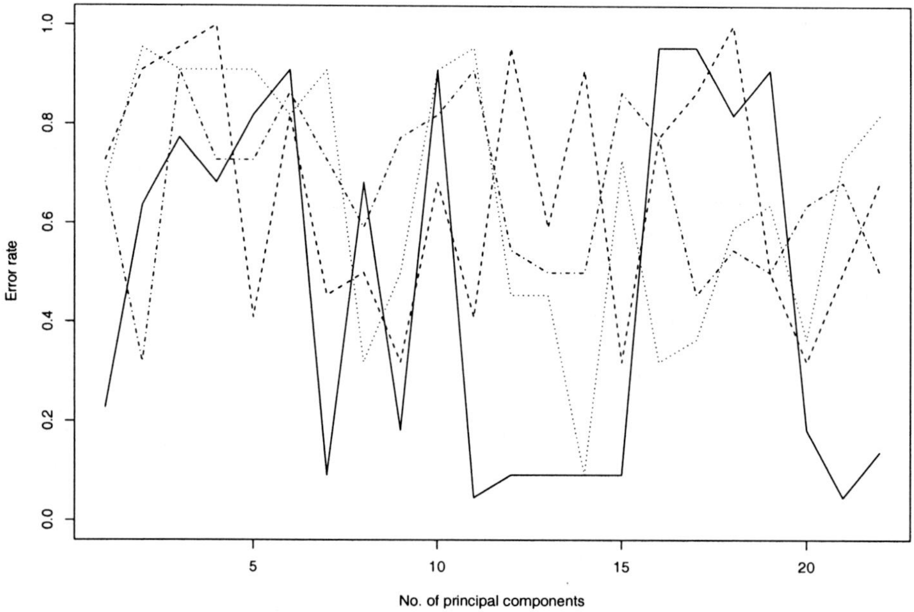

Figure 1: Plot of estimated classification error rates (based on Hamming distance) versus number of principal components. Solid line: $M = 25$; dashed line: $M = 100$; dotted line: $M = 1500$; dashed/dotted line: $M = 3226$.

with two principal components and estimate the regression parameters using maximum likelihood estimation. Based on fitting the model and the back-transformation described in Section 2.2, we can rank the genes in terms of their ability to discriminate between these three classes of tumors. A ranking of the top 20 genes from the subset of $M = 100$ and their corresponding gene scores for discriminating BRCA1-positive tumors from sporadic tumors is given in Table 1. A similar table of the top genes for discriminating BRCA2-positive tumors from sporadic tumors can be found at the website. Many of the genes on this list overlap with the discriminatory genes found by Hedenfalk et al., but there are also genes that do not make their list.

Finally, we wish to examine potential relationships between the genes in Table 1. One way to do this would be to simply cluster the genes using average linkage hierarchical clustering.[4] We do not present the resulting dendrogram here; it can be found at our website. However, if we now use the estimated variance matrix of the gene scores from the SVD regression model based on two principal components as the basis of the hierarchical clustering, this yields the dendrogram in Figure 2. In particular, we find that there are two distinct groupings with the second dendrogram, but this increase in separation comes at the price of losing the finer substructure between the genes . The reason for this because the estimates of the gene scores are highly correlated. Consequently, most of the off-diagonal entries of the distance matrix used in the hierarchical clustering algorithm are close to one. However, the initial separation between the genes is greater using this method compared to that from performing hierarchical clustering on the gene expression data where the tumor class is not taken into account (data not shown).

4 Discussion

In this article, we have developed a singular value decomposition regression modelling approach for correlating gene expression profiles with tumor class in microarray settings. This methodology is important for determining the diagnostic and predictive ability of microarray technology in clinical settings. While we have focused mainly on a categorical response (tumor type), the ideas in this article can be applied to other types of clinical phenotypes, such as censored failure times, using different regression models in lieu of (1). Singular value decomposition regression models have a rich tradition in other fields of application, but the presence of noncontinuous clinical phenotypes introduce new issues in statistical modelling.

We utilized SVD regression modeling for three purposes. First, predictive models were constructed in the situation where the dimension of predictors is

Table 1: List of ranked genes and gene scores for discriminating BRCA1-positive tumors from sporadic breast cancer tumors.

Clone	Gene	Score
823775	guanine nucleotide binding protein (G protein), alpha inhibiting activity polypeptide 3	0.204
364840	ESTs, Moderately similar to mouse Dhm1 protein [M.musculus]	0.194
44180	alpha-2-macroglobulin	0.175
32231	KIAA0246 protein	0.172
81518	apelin; peptide ligand for APJ receptor	0.171
417124	APEX nuclease (multifunctional DNA repair enzyme)	0.167
839594	ribosomal protein L38	0.155
239958	DKFZP586G1822 protein	0.154
234150	myotubularin related protein 4	0.151
73531	nitrogen fixation cluster-like	0.150
204897	phospholipase C, gamma 2 (phosphatidylinositol-specific)	0.148
725860	transcription factor AP-2 gamma (activating enhancer-binding protein 2 gamma)	0.146
246524	CHK1 (checkpoint, S.pombe) homolog	0.144
429135	suppression of tumorigenicity 13 (colon carcinoma) (Hsp70-interacting protein)	0.143
307843	ESTs	0.142
22230	collagen, type V, alpha 1	0.131
50413	armadillo repeat gene deletes in velocardiofacial syndrome	0.130
81331	fatty acid binding protein 5 (psoriasis-associated)	0.129
341130	retinoblastoma-like 2 (p130)	0.128
810551	low density lipoprotein-related protein 1 (alpha-2-macroglobulin receptor)	0.127

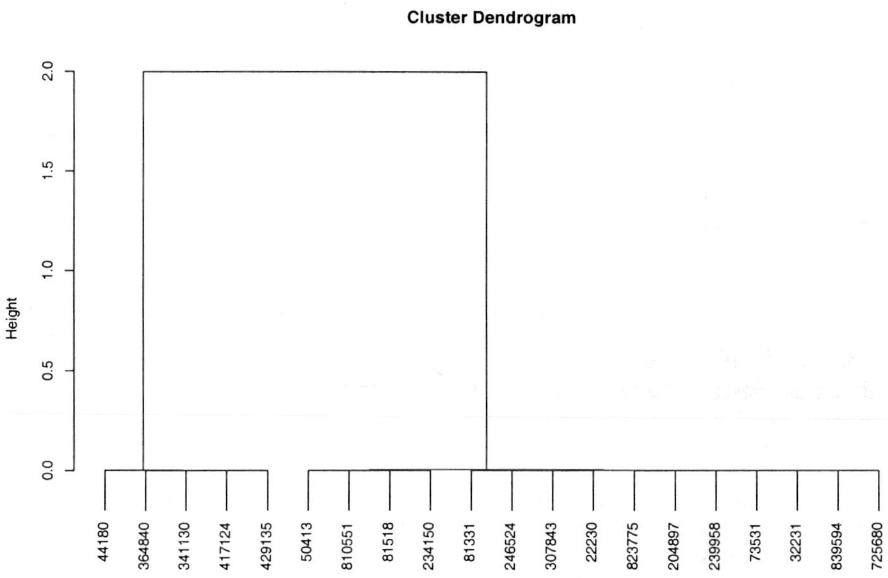

Figure 2: Hierarchical clustering dendrogram of genes from Table 1 based on gene scores. Average linkage clustering used.

much larger than that of the independent samples. Second, it provided the basis for ranking genes in terms of their discriminative abilities. Finally, the parameter estimates from the principal components regression method were used to cluster genes. Based on the analysis of the breast cancer data, we found that the SVD regression approach is successful for prediction and variable selection. However, it is problematic for clustering in terms of finding finer structural relationships among genes.

As was mentioned in the Introduction, singular value decomposition regression models have been applied in other disciplines; one unique challenge here is that the outcome measure is not continuous. A major advantage of this method is that it can accommodate the scenario where the number of predictors is larger than the number of independent samples. However, other predictive modelling methods exist in this setting, such as partial least squares and ridge regression.[11] It would be very useful to compare these methods in terms of their predictive modelling capabilities and is a current area of focus of our research. However, it should be noted that it does not appear to be straightforward to develop gene selection and clustering schemes based on partial least squares.

Because gene expression data are highly multivariate, they are inherently complex. This research has also demonstrated that multiple levels of data analysis are needed in order to perform classification of tumors using microarray data.

References

1. Golub T. R. *et al.* (1999). Molecular classification of cancer: class discovery and class prediction by gene expression monitoring. *Science* **286**, 531–537.
2. A. A. Alizadeh *et al.* (2000). Distinct types of diffuse large B-cell lymphoma identified by gene expression profiling. *Nature* **403**: 503–511.
3. M. Bittner *et al* (2000). Molecular classification of cutaneous malignant melanoma by gene expression profiling. *Nature* **406**: 536–540.
4. M. B. Eisen, P. T. Spellman, P. O. Brown, and D. Botstein. (1998). Cluster analysis and display of genome-wide expression patterns. *Proc Nat Acad Sciences* **95**, 14863–14868.
5. J. Kononen *et al.* (1998). Tissue microarrays for high-throughput molecular profiling of tumor specimens. *Nat Med* **4**, 844–847.
6. P. J. Park, M. Pagano and M. Bonetti. (2000). A nonparametric scoring algorithm for identifying informative genes from microarray data. In *Proc Pac Symp Biocomputing.*

7. V. G. Tusher, R. Tibshirani and G. Chu. (2001) Significance analysis of microarrays applied to ionizing response. *Proc Nat Acad Sciences* **98**, 5116–5121.
8. S. Raychaudhuri, J. M. Stuart and R. Altman. (2000). Principal components analysis to summarize microarray experiments: application to sporulation time series. In *Proc Pac Symp Biocomputing.*
9. N. S. Holter *et al.* (2000). Fundamental patterns underlying gene expression profiles: simplicity from complexity. *Proc Nat Acad Sciences* **97**, 8409–8414.
10. O. Alter, P. O. Brown and D. Botstein (2000). Singular value decomposition for genome-wide expression data processing and modeling. *Proc Nat Acad Sciences* **97**, 10101–10106.
11. I. E. Frank and J. H. Friedman (1993). A statistical view of some chemometric regression tools (with discussion). *Technometrics* **35**, 109–135.
12. A. Agresti. *Categorical Data Analysis.* (1990). New York: John Wiley and Sons.
13. I. Hedenfalk *et al.* (2001) Gene expression profiles in hereditary breast cancer *NEJM***344**, 539–548.
14. G. H. Golub and C. F. van Loan. *Matrix Computations.* (1996). Baltimore: John Hopkins University Press.
15. West, M. (2001). Bayesian regression analysis in the "large p, small n" paradigm. Technical Report, Institute of Statistics and Decision Sciences, Duke University.
16. Dudoit, S., Fridlyand, J. and Speed, T. P. (2001). Comparison of discrimination methods for the classification of tumors using gene expression data. Technical Report, Department of Statistics, University of California at Berkeley.
17. L. Breiman (1996). Heuristics of instability and stabilization in model selection. *Ann Stat* **24**, 2350–2383.
18. L. Breiman (1996). Bagging predictors. *Mach Learn* **24**, 123–140.

AN AUTOMATED COMPUTER SYSTEM TO SUPPORT ULTRA HIGH THROUGHPUT SNP GENOTYPING

HEIL J, GLANOWSKI S, SCOTT J,
WINN-DEEN E, MCMULLEN I, WU L, GIRE C, SPRAGUE A
Celera Genomics, 45 W. Gude Drive,
Rockville, MD 20850, USA
Jeremy.Heil@celera.com, Stephen.Glanowski@celera.com

Celera Genomics has constructed an automated computer system to support ultra high-throughput SNP genotyping that satisfies the increasing demand that disease association studies are placing on current genotyping facilities. This system consists of the seamless integration of target SNP selection, automated oligo design, in silico assay quality validation, laboratory management of samples, reagents and plates, automated allele calling, optional manual review of autocalls, regular status reports, and linkage disequilibrium analysis. Celera has proven the system by generating over 2.5 million genotypes from more than 10,000 SNPs, and is approaching the target capacity of over 10,000 genotypes per machine per hour using limited human intervention with state of the art laboratory hardware.

1. INTRODUCTION

Since the completion of the human genome sequence by Celera Genomics[1] and The Human Genome Project[2], efforts have turned to analyzing and comparing the results. One outcome of such analysis, a set of over 3 million putative Single Nucleotide Polymorphisms (SNPs), will be used for disease association, eventually replacing the current RFLP and STRP linkage analysis screening sets.[3]

Unfortunately, SNP facilities must generate many times the number of genotypes that STR facilities are currently producing as a consequence of the lower informativeness of SNPs compared to STRs. Moreover, until recent efforts to validate large numbers of evenly spaced markers become fruitful, SNP laboratories must validate markers *ad hoc* before genotyping with clinical samples.[3,4,5]

Celera has created a complete software solution that minimizes manual intervention at each step of the genotyping process. This system may serve as a template for future designs and aid others in realizing the benefits of automation.

2. LABORATORY HARDWARE

The process of designing universally usable software is too often derailed by the need to customize programs for particular laboratories and specific instruments. While Celera has developed many system components independent of the hardware to simplify its duplication at other facilities, some components are highly specialized. It is therefore worthwhile to mention the laboratory equipment that our system was intended to support.

The 5' nuclease allelic discrimination method employed by Applied Biosystem's TaqMan® platform was chosen to allow for the least human labor while in the laboratory[6]. Unlike other methods that may require hybridization to chips[7] or separate allele reactions[8], TaqMan® PCR preparation merely entails adding a pre-made Master Mix containing buffer, deoxyribonucleotides, and DNA polymerase to the sample template and SNP specific oligonucleotides. TaqMan® has already established a successful presence in large scale SNP genotyping facilities[9,10].

TaqMan® chemistry employs two allele specific probes for each SNP in addition to the common PCR primers. Each probe contains a 5' fluorescent dye, commonly VIC or FAM, to detect the presence of the specific allele, and a 3' quencher to absorb fluorescence when the allele is not present. The result is much like any microarray or molecular beacon technology, one of the dyes will fluoresce for homozygous alleles and both dyes will fluoresce for heterozygotes (Figure 1).

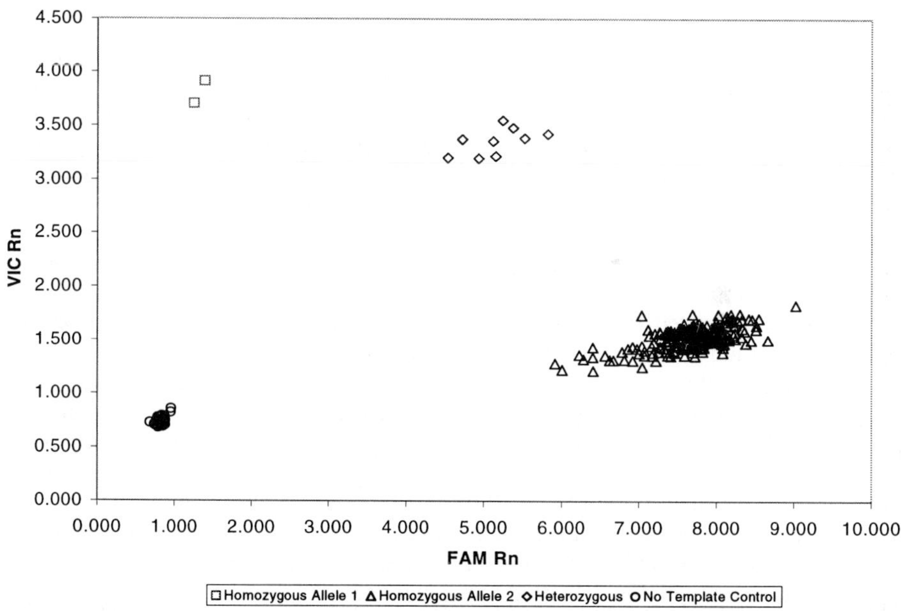

Figure 1. The typical analyzed TaqMan® plate showing four genotype clusters for a particular SNP. Each data point represents one sample and is plotted by the intensity measured from each of two flurescent dyes. Clusters of points are classified as being homozygous for either allele, heterozygous, or no amplification.

The ABI Prism® 7700 and ABI Prism® 7900HT Sequence Detection Systems are used for endpoint analysis of 96 and 384 well plates, respectively, to record the fluorescence of each well's PCR product. The latter is bundled with an 84 plate robot for long term hands-free automation.

Roughly 42 dual plate GeneAmp® PCR System 9700 thermal cyclers are needed to keep one 7900HT supplied with an adequate number of PCR plates for continuous operation. Each piece of hardware, plate, oligo, person, and sample is barcoded in order to record the time, technician, and component of every transaction. Any genotype's history can be accessed to evaluate the performance of particular components in the system.

3. DATA MANAGEMENT

All information gathered throughout the genotyping process is stored in a central ORACLE database. The repository is purposely divided into project management and laboratory schemas.

The project schema offers the ability to manage the abstract entities such as SNP, sample donor, or genotype. This includes 'creating' the project by indicating the intended customer and loading the desired SNP information. Users control what SNP is ordered, scanned, considered validated, possibly discarded or re-designed, and delivered to the customer. In addition, numerous reports can be generated regarding the current progress of a SNP, failure rate of samples, or allele frequencies per population.

The project schema is specifically designed for fast data analysis. It allows for efficient phenotype relations to both donors and SNPs, and has the ability to store haplotypes constructed from specific SNP alleles after analysis. The schema may also track literature references for individual SNPs and donors.

In contrast, the laboratory partition consists of tracking every detail of the process taken by the actual physical components mirrored in the project partition. Samples are received, barcoded, and placed into plates and freezers. Oligos are received, diluted, assigned into sets, and also placed into freezers. Plates are arrayed with particular samples and oligos for specific projects. Each well is scanned and possibly re-scanned many times in order to assure a high level of accuracy. However, only the 'final' genotype is copied to the project partition where it may eventually be delivered to the customer.

The advantage of having common but separated partitions is that the laboratory space provides a tracking environment in which experiments can be re-arrayed, re-run, and reviewed multiple times. The project partition remains uncluttered with details as analysis requires a compact schema designed for speed and clarity. This integration of LIMS and data analysis provides for segregated storage to satisfy each schema's different requirements, while keeping the data in one repository for the ability to track an individual genotype's entire history.

The versatility of the database schema has further proven itself by also supporting a large scale resequencing laboratory by adding relatively few tables. This combines SNP discovery, validation, and genotyping into one central repository.

4. PRE-PROJECT SETUP

4.1 SNP Selection

Genotyping services are generally contracted with either a predetermined set of SNPs or particular locus of interest.

If a set of markers is provided, their flanking context is mapped to the Celera Human Genome and any discrepancies or adjacent SNPs masked out. Resequencing has shown that unexpected clustering of dye intensities (Figure 2) are caused by unknown SNPs residing within the probe or primer sequences, making accurate genotype calling difficult. It is advantageous to gather all possible information about the sequence surrounding the SNP before attempting to design a SNP assay.

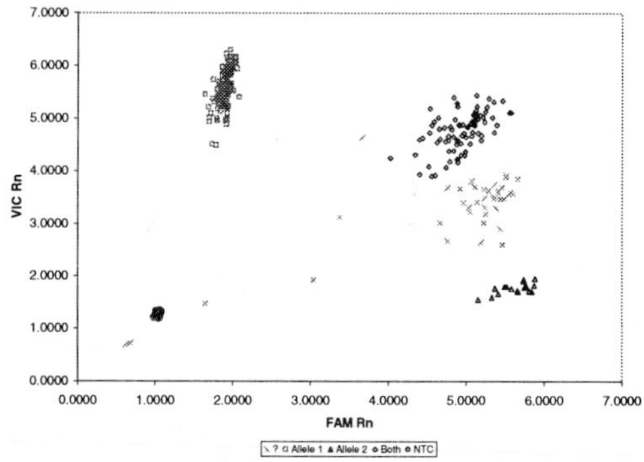

Figure 2. Undesired genotype clustering attributed to other unknown SNPs in the probes sequence.

In the case where a locus is targeted, PERL scripts have been written to select SNPs based on desired distance and coverage constraints using Celera's comprehensive RefSNP database[1,11]. RefSNP not only includes SNPs extracted from Celera's genomic assemblies, but also externally generated SNPs from TSC, NCBI, HGMD, and literature articles that are then mapped to Celera's Human Genome.

4.2 Assay Design

The result of either selection method is a 'clean' flanking sequence for each target SNP that is fed into a modified version of Applied Biosystem's Primer Express™ program[12]. This program has been ported to JAVA and adapted to find customized multi-allelic MGB TaqMan® probes with no user interaction, allowing for thousands of assays to be designed within a matter of minutes. In addition to a faster run time, each assay is designed consistently using predetermined parameters eliminating the human bias that can occur in manual designs.

4.3 Assay Quality Control

Pseudo-SNPs are a common problem that arises from misassemblies, paralogs, or repeat elements and can needlessly squander genotyping resources[9]. Similar sequences from different regions in the genome may erroneously align due to matching at all but a few bases. These differing bases are then incorrectly assumed to be SNPs. If a pseudo-SNP is genotyped, every sample will appear to be heterozygous since each sample contains both of the pseudo-alleles (Figure 3). Celera has already identified several internal, external and literature cited SNPs that are actually pseudo-SNPs.

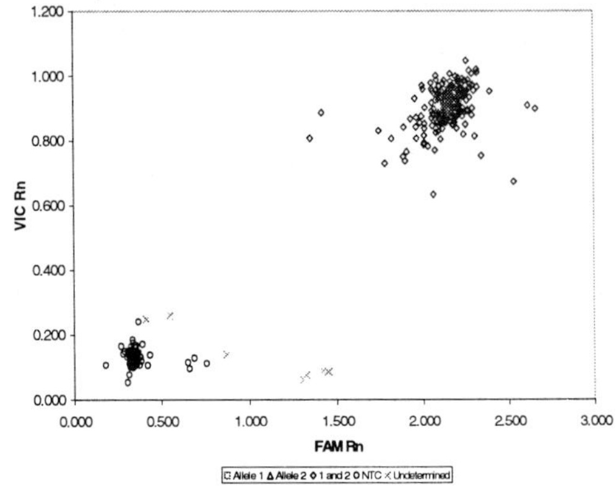

Figure 3. Pseudo-SNP resulting in all samples appearing heterozygous.

Each oligo of an assay is BLASTed against the Celera Human Genome and results parsed with a PERL script using the GNU BioPerl module[13]. Classifying the BLAST hits according to an allowable number of base mismatches and the number of alternative locations matched in the genome provides an estimation of primer specificity. Primers with a low specificity are automatically discarded or sent for assay redesign.

Pairs of forward and reverse primers are also tested to see if they may form alternative PCR products other than the one desired. These are alternative annealing locations of the forward and reverse primers close enough together and in proper orientation to form a viable PCR product.

Although primer sets with one low-specificity primer forming single products have succeeded in laboratory tests, there is often some loss of signal strength. However, assays forming multiple products of similar length will consistently produce less than desirable results and are systematically discarded or redesigned.

4.4 Assay Order and SNP Tracking

The finalized project, SNP, and sample information is loaded into the database. Using a JAVA GUI to access the database (Figure 4) users may generate a formatted oligo purchase order. Details, current status, and past history for any project, SNP, or plate are available to review or update from this screen.

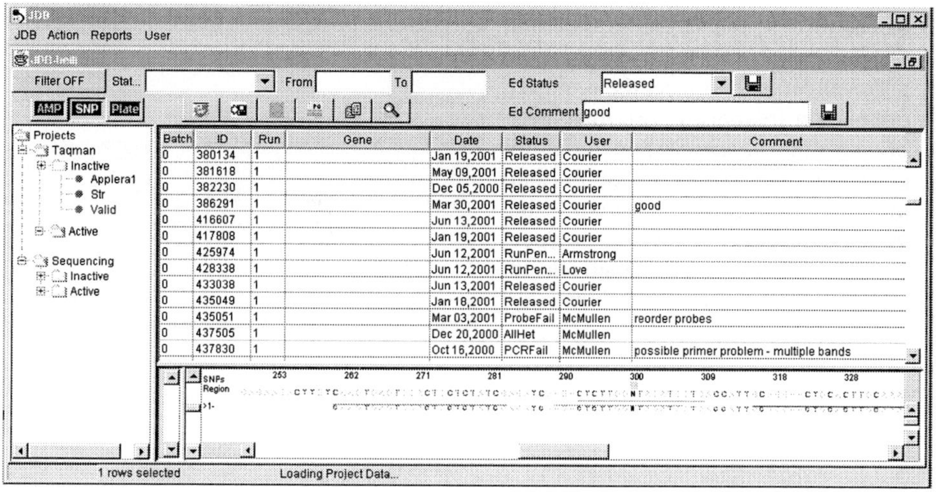

Figure 4 Managing each SNP and plate status.

5. LABORATORY INFORMATION MANAGEMENT SYSTEM (LIMS)

5.1 Check In

Individual oligos arriving from vendors are immediately barcoded and scanned into the database using a Microsoft VBA GUI (Figure 5). The two primers and two probes of each SNP are grouped and stored together in what is referred to as an oligo set. Oligo sets may also be ordered in one tube, or in 96 well microtiter plates. Scanning received oligos into the database tracks inventory and allows for a nightly report to be generated that notifies lab managers of sets ready to be run the following day.

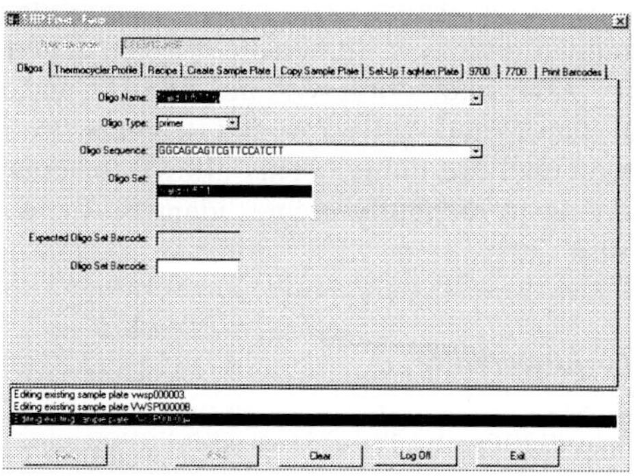

Figure 5 Checking oligos into the database.

Samples are arrayed in 96 or 384 well plates and a map of the plate entered into the database (Figure 6). Putative SNPs must first be validated using a combination of diversity panels from the Coriell Cell

Repository[14] and several internal Celera samples. Only SNPs that pass validation and meet the required population frequencies are then used on the clinical samples in order to conserve clinical DNA. SNPs that are determined to be nonpolymorphic are used to reduce the number of false-positives in Celera's SNP calling algorithms.

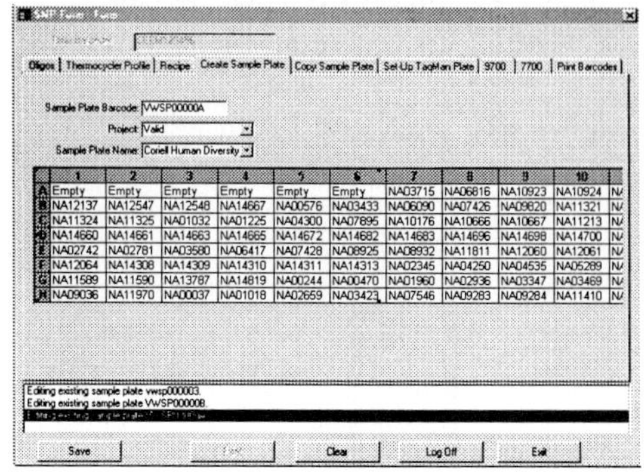

Figure 6 Editing the array of samples on a plate

5.2 Preparation and Scanning

When a SNP is ready to be run, TaqMan® Universal PCR Master Mix is prepared using the oligo set and added to a daughter sample plate prepared by the Protedyne robot. The plate is thermal cycled using a GeneAmp® 9700. Each step is logged in the LIMS, allowing software to automatically trigger and create the SDS binary file required by the 7900HT. Therefore, the laboratory staff need only to place the plate into one of the 7900HT's stackers and select the pre-created file in the ABI robot program, rather than manually create the SDS file using the SDS software.

Our software recognizes the scanned data file from the 7900HT and automatically passes it to the SDS Multicomponent Analysis software. A multicomponent file is created containing the dye intensities of each well and is subsequently passed to a customized autocaller program. As discussed in more detail below, the program identifies the genotype clusters and assigns appropriate calls to the wells. The putative genotypes are loaded into the database for either manual review or immediate release, depending on the confidence of the autocaller.

User interaction with the 7900HT and SDS package is limited via a combination of automated software and triggers to detect and predict what the laboratory personnel are doing. This allows for continuous scanning by the 7900HT without having to manually create, identify, locate, analyze, call genotypes, or export data files.

5.3 Autocalling

Automating the process of annotation is the challenge of any genotyping facility, as using laboratory staff to manually review each genotype is inefficient and costly[9,15,16]. Celera overcame this problem by developing a novel and highly accurate allele caller with the ability to flag plates that do not meet predetermined signal thresholds or adhere to common genetic principles.

K-means clustering has recently been applied to the classification of SNP genotypes[8]. Although Celera initially experimented with similar methods, examining the results of several plates showed K-nary clustering methods were not adaptable enough to accommodate the variety of possible outcomes. Output from a TaqMan® assay is typically four clusters falling into separate quadrants of a rectangle (Figure 1). However, autocalling must also be able to correctly identify situations such as two clusters (Figure 7), three clusters (Figure 8), four scattered clusters (Figure 9), and five clusters (Figure 2). In any non-ideal circumstance, K-nary clustering would consistently force the data into four clusters.

Other machine learning methods, such as neural networks and decision trees, were also tested on a set of 80 plates, but failed due to the classification of points based on previous observations. The vastly variable difference in dye intensities between any two plates confused the training process. For example, the heterozygote cluster in Figure 8 is centered at (5.0, 2.5) while in Figure 9 it is at (3.0, 1.7). Normalizing the coordinates does not alleviate this problem.

A novel alternative algorithm was developed that uses the relative position of samples in polar coordinates to make calls. By considering the inherent genetics qualities of the data, filters allow questionable SNPs to be flagged for manual review. Celera is currently seeking a patent covering this method.

This program was tested on a collection of 1,007 plates, consisting of a

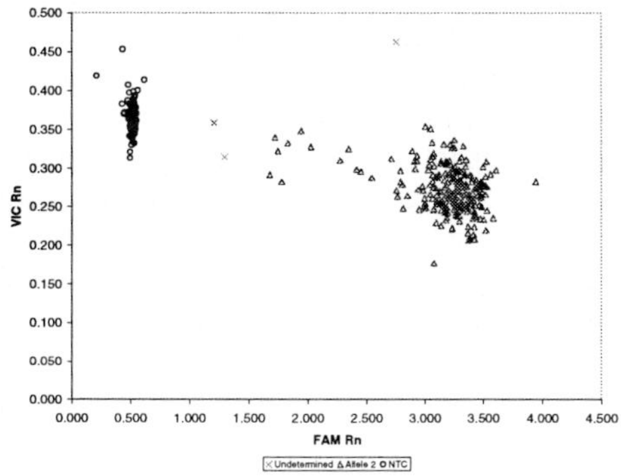

Figure 7. All homozygous genotypes producing two clusters.

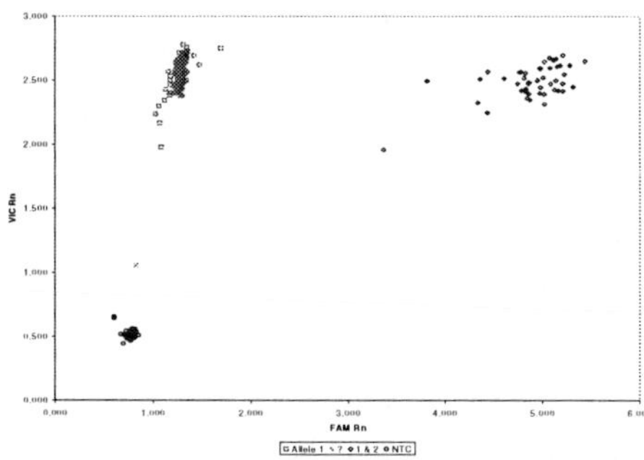

Figure 8. A SNP with no rare allele homozygotes produces three clusters.

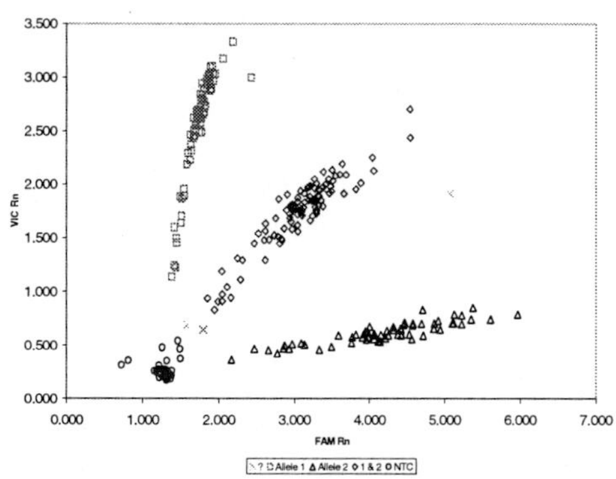

Figure 9. Clusters not well defined. The most difficult to autocall accurately.

mix of 96 and 384 wells, from differing SNP assays. The accuracy of the algorithm, assuming human annotation as truth, was 88.9%. However, the method was able to flag 237 low quality files to be reviewed manually. Once these are removed from the assessment, the algorithm scored 99.3% on the remaining 770 files. This translates into 68,467 correctly autocalled genotypes out of 68,920.

The program was further tested with 472 384-well plates generated during the breaking in of a newly installed 7900HT. Although laboratory protocols were also being tested at the same time, the algorithm averaged 68.4% accuracy on all plates, and flagged 294 plates for human review. The accuracy for the non-flagged plates was 98.6%.

The ability to recognize a low confidence plate greatly improves accuracy to 99%, despite the slightly less than perfect accuracy of the automatic calls when considering all genotypes. Even on the plates that were flagged for human review, only a handful of samples per plate require changes as human reviewers are supplied with automated calls to start from, and make corrections instead of tediously manual calling every well of a plate from scratch.

Although Celera's current policy is to review all autocalled plates manually, the near future will see confident calls finalized automatically. This will reduce the time and effort required to manually analyze, call, and export results from the SDS system even more

5.4 Manual Review

All calls, even on plates flagged with low confidence, are automatically loaded into the LIMS database schema. Celera has developed software to function similarly to the ABI Prism® SDS software package (Figure 10) allowing users to save changes directly in the database.

Using customized software allows for several internal quality control checks to be performed on the data while the user is making calls. These include tests for allele frequency, known blind controls, and possible plate mislabeling. Warnings are also generated for samples whose stored genotype differs from the current call.

Once a call has been finalized for a particular sample and passed quality checks, the corresponding genotype is created in the project partition where it can efficiently be accessed for analysis

6. ANALYSIS

Genotypes are periodically delivered to the customer formatted in tab delimited files, table dumps, or XML. After a contracted waiting period, the genotypes are also released to Celera's RefSNP database to increase its intrinsic value for other customers.

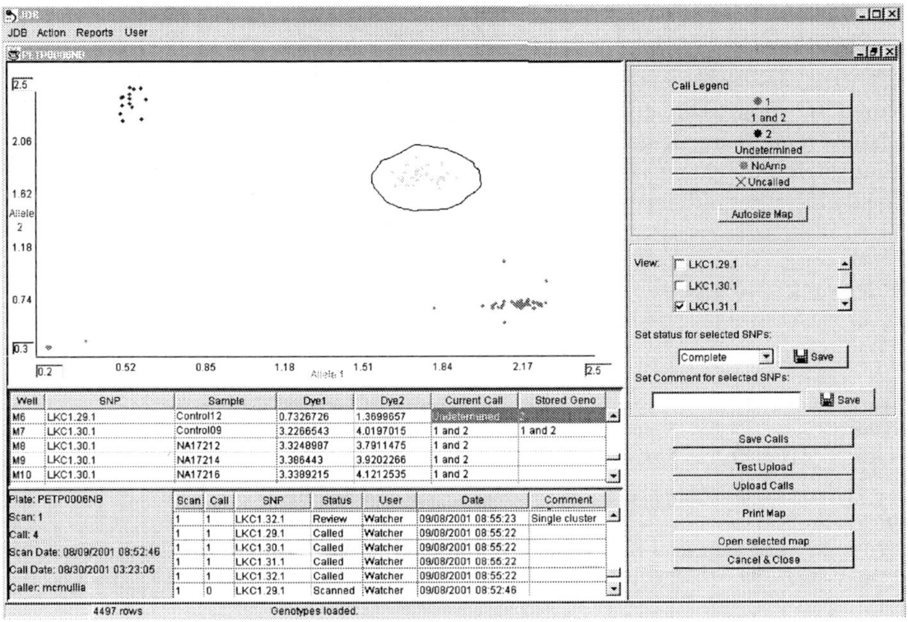

Figure 10. Interface used to manually edit autocalls. Each SNP on a plate is called separately, and any controls are compared to their stored genotypes. Calls are quality tested and then saved as genotypes.

To enable analysis of the high volumes of genotyping data being generated, PERL scripts using the GNU DBI module are utilized to extract genotype calls for a given set of SNP sites.[17] Where phenotype data is available for samples, it is integrated with the genotype calls. A comma-delimited format is generated for a PERL script to use χ^2 analysis and Maximum Likelihood in identifying regions of Linkage Disequilibrium (Figure 11). Vector files are also generated for import into Rulequest Research's C5.0 data mining tool,[18] WEKA's machine learning software,[19] and customized haplotype inference programs[20].

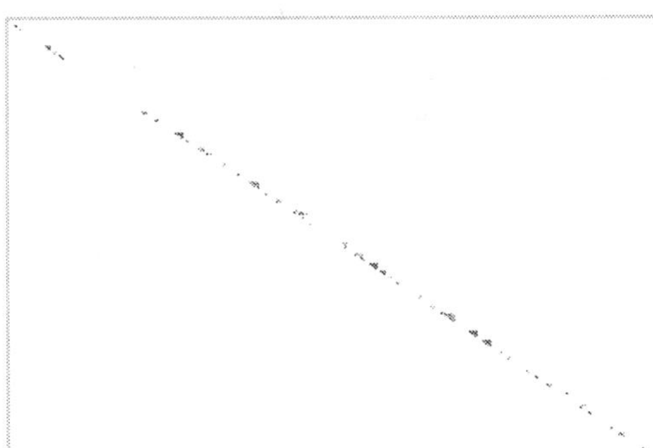

Figure 11. LD map generated from database SNPs.

7. CONCLUSION

Estimates show that automation during the Human Genome Project increased output by more than 40-fold.[21] While the world of genotyping may not have attained quite that magnitude of efficiency as of yet, Celera has developed a facility using high-throughput software that significantly reduces human intervention and greatly increases efficiency.

8. REFERENCES

1. Venter et al., "The Sequence of the Human Genome", Science, Volume 291, Number 5507, pp 1304-1351, (2001)
2. International Human Genome Sequencing Consortium , "Initial Sequencing and Analysis of the Human Genome", Nature, Volume 409, pp 860-921, (2001)
3. E Lai, "Application of SNP Technoloies in Medicine", Genome Research, Volume 11, pp 927-929, (2001)
4. E Massod, "As Consortium Plans Free SNP Map of Human Genome", Nature, Volume 398, pp545-546, (1999)
5. J Weber, J Che, D David, J Heil, J Opolka, C Volkmann, M Doktycz , K Beattie, "Identification and Analysis of Human Short Insertion/Deletion Polymorphisms", Am. Society of Hum. Gen.. Oct (1998)
6. K Livak, J Marmaro, J Todd,, "Towards Fully Automated Genome-Wide Polymorphism Screening" , Nature Genetics, Volume 9, pp 341-342, (1995)
7. D Wang et al., "Large Scale Identification, Mapping, and Genotyping of Single Nucleotide Polymorphisms in the Human Genome", Science, Volume 280, pp 1077-1082, (1998)
8. C Mein et al,, "Evaluation of Single Nucleotide Polymorphism Typing with Invader on PCR Amplicons and its Automation", Genome Research, Volume 10, pp330-343, (2000)
9. R Koustubh et al., "High Throughput Genotyping with Single Nucleotide Polymorphisms", Genome Research, Volume 11, pp 1262-1268, (2001)
10. J Hampe et al, "An Integrated System for High Throughput Taqman Based SNP Genotyping", Bioinformatics, Volume 17, pp 654-655, (2001)
11. http://www.celera.com
12. http://www.appliedbiosystems.com
13. http://www.bioperl.org
14. http://locus.umdnj.edu/
15. M Perlin, G Lancia, S Ng, "Toward Fully Automated Genotyping: Genotyping Microsatellite Markers By Deconvolution", Am. J. Hum. Gen., Volume 57, pp 1199-1210, (1995)
16. C Zhoa, J Heil, W Dickenson, L Ott, J Weber et al., "A Computer System for Large Scale STRP Genotyping", Am. J. of Hum. Gentics., Volume 61, p A302, (1997)
17. http://www.perl.org
18. J Quinlan, "The Morgan Kaufmann Series in Machine Learning", Morgan Kaufmann Publishers, ISBN 1-55860-238-0, October (1992)
19. I Witten, E Frank, "Data Mining: Practical Machine Learning Tools and Techniques with Java Implementations", Morgan Kaufmann Publishers, ISBN 1-55860-552-5, (2000)
20. A Clark, "Inference of haplotypes from PCR-amplified samples of diploid populations.", Mol Biol Evol., Mar 7, pp111-122, (1990)
21. D Meldrum , "Automation For Genomics, Part One", Genome Research, Volume 10, pp 1081-1092, (2000)

INFERRING GENOTYPE FROM CLINICAL PHENOTYPE THROUGH A KNOWLEDGE BASED ALGORITHM

B.A. MALIN, L.A. SWEENEY, Ph.D.
School of Computer Science, Carnegie Mellon University
Pittsburgh, PA 15213, USA

Genomic information is becoming increasingly useful for studying the origins of disease. Recent studies have focused on discovering new genetic loci and the influence of these loci upon disease. However, it is equally desirable to go in the opposite direction – that is, to infer genotype from the clinical phenotype for increased efficiency of treatment. This paper proposes a methodology for such inference. Our method constructs a simple knowledge-based model without the need of a domain expert and is useful in situations that have very little data and/or no training data. The model relates a disease's symptoms to particular clinical states of the disease. Clinical information is processed using the model, where appropriate weighting of the symptoms is learned from observed diagnoses to subsequently identify the state of the disease presented in hospital visits. This approach applies to any simple genetic disorder that has defined clinical phenotypes. We demonstrate the use of our methods by inferring age of onset and DNA mutations for Huntington's disease patients.

1 Background

1.1 Genotype-Phenotype Relationships

Over the past decade, growing interest has surfaced in recognizing relationships between the genotype and clinical phenotype of an individual. It is believed that more efficient treatment of many diseases can be achieved through tailoring drug administration to specific genotypes.[1] With this in mind, one must consider that the etiology of many diseases resides in a combination of genetic predispositions, environmental variables, and random chance. Genetic influence varies among diseases, ranging from a weak influence on Alzheimer's disease to a deterministic effect on sickle cell anemia.

This paper addresses the relationship in single gene mutation diseases known as simple Mendelian traits. The traits are an interesting study from the standpoint that their DNA mutation is considered the direct cause of the disease and permit a wide range of genotype-phenotype relationships. For example, the autosomal recessive disease cystic fibrosis is caused by mutations in the cystic fibrosis transmembrane conductance regulator gene, of which over 750 mutations have been documented. While cystic fibrosis' clinical expression is variable, the phenotypes have been demonstrated to relate to particular mutations of the gene.[2]

Though the relationship between the genotype and some aspect of the clinical phenotype is known, the relationship is often obscured in standardized medical information. This paper describes a generally applicable method for discovering the particulars of this relationship from standardized medical information as they relate to individual patients. We demonstrate our general method to determine clinical phenotype of the autosomal dominant disorder known as Huntington's disease in a group of patients. Huntington's disease is caused by a CAG trinucleotide repeat expansion of the HD gene, a feature relatively independent of the observed clinical features.[3] Rather, the size of the repeat harbors an inverse exponential relationship to the age of onset of the disease, a feature of the clinical phenotype not recorded in general medical information.[4] Our methods therefore, are utilized to infer non-recorded features of the clinical phenotype (such as age of onset) from standardized hospital information to reveal characteristics of the genotype.

1.2 Knowledgebase and Statistical Learning Approaches

Nowadays knowledge-based systems, which gained tremendous popularity in the 1980's, and data mining techniques, which gained tremendous popularity in the 1990's, are often viewed as rival approaches. The era of expert systems began with exciting systems like DENDRAL, which inferred molecular structure from information provided by a mass spectrometer[5], and MYCIN, which diagnosed blood infections[6]. But the era of expert systems ended with disillusionment as the costs of constructing real-world knowledge-based systems far exceeded their perceived benefits. Excitement shifted to neural networks and statistical data mining techniques, in part, because they provided results using standardized learning models with little or no explicit representation of domain knowledge required. But as the problem space becomes more complex, the advantages of a knowledge-based approach become apparent. Examples include games such as chess, checkers, and backgammon where efforts to learn an evaluation method using knowledge of the game are much more successful then methods void of domain knowledge[7].

In this paper, we tackle a problem in which a neural network, for example, could be used if training data were available; but in this environment, we assume no training data are available. Our approach constructs an initial knowledge-based model with minimal effort by relating diagnoses to a disease's symptoms and relating those symptoms, in turn, to stages of the disease. The represented knowledge is not from a domain "expert" but from simple extractions drawn from common literature and so, these initial mappings may be inconsistent. We therefore calibrate the model by generalizing, specializing and partitioning the initial mappings as needed. Finally, we apply the model to infer genotype for patients.

2 Methods

Materials needed for this approach are standardized hospital data, general facts about the clinical presentation of the disease, and for a sample of individuals, known features of the clinical phenotype. Following are descriptions of these materials, which are necessary for use with our approach.

INDIV1	AGE1	DOB1	SEX1	ZIP1	ADMIT1	{DIAGNOSES}
INDIV2	AGE2	DOB2	SEX2	ZIP2	ADMIT2	{DIAGNOSES}
INDIV2	AGE2	DOB2	SEX2	ZIP2	ADMIT2	{DIAGNOSES}
INDIV2	AGE2	DOB2	SEX2	ZIP2	ADMIT2	{DIAGNOSES}
INDIV3	AGE3	DOB3	SEX3	ZIP3	ADMIT3	{DIAGNOSES}

Figure 1. Longitudinal medical profiles from clinical information databases. Multiple visit profile is shaded.

2.1. Inference Algorithm Definition

First, we consider the collections of inpatient hospital visits. The National Association of Health Data Organizations reported that 44 of 50 states have legislative mandates to gather hospital-level data on each patient visit. As shown in Figure 1, patient demographics, hospital identity, diagnosis codes, and procedure codes are among the attributes stored with each hospital visit. Previous research has demonstrated that publicly available discharge data permits the formation of genetic population subsets.[8]

INPUT	Patient profiles of clinical data (basic hospital visit information)
ASSUMES	Disease is known to be temporal or constrained to an exclusionary status of clinical phenotype for the duration of a profile
Step I	Manually map diagnoses to symptoms to clinical phenotype states and diagnoses to clinical phenotype states
do	
Step II	Automatically adjust symptoms to cover clinical phenotype states
Step III	Learn accuracy of diagnosis codes using the defined mappings
Step IV	Automatically classify each patient visit into a clinical phenotype state based on the mappings
Step V	Align the predicted clinical phenotype states for each set of patient visits to optimally respect temporal or disease stage constraints
until	predictions converge
OUTPUT	Specific inferences and/or constraints regarding genotype of patient

Figure 2. Genotype inference algorithm.

Second, we consider the current corpora of knowledge about particular diseases. General information about symptoms and clinical presentation of the disease are needed. Such information can be found in articles on MEDLINE, on authoritative web sites devoted to the disease, and in general medical texts. From the study of this information, the symptoms of the disease in question can be defined. Symptoms are defined narrowly enough, so that diagnoses map to a single symptom. With the previous information available, we present an algorithm for knowledge representation and inference in Figure 2. Steps of the algorithm are discussed in the following subsections.

2.2 *Manual Mapping of Diagnoses and Symptoms (Step I)*

The simplest model is the direct mapping of each diagnosis to its corresponding disease states. Such a model permits the use of a neural network for parameter estimation for a defined model as depicted in Figure 3a. However, such learning tools require ample training data. Yet, this study uses a small amount of data and no training data, and as such, we attempt to use known knowledge about a disease to tune the model by adding an internal layer of nodes. Each node in the additional layer represents a group of diagnoses, as shown in Figure 3b. The question becomes, "What criteria is best for grouping the raw diagnoses?" Three possibilities are explored.

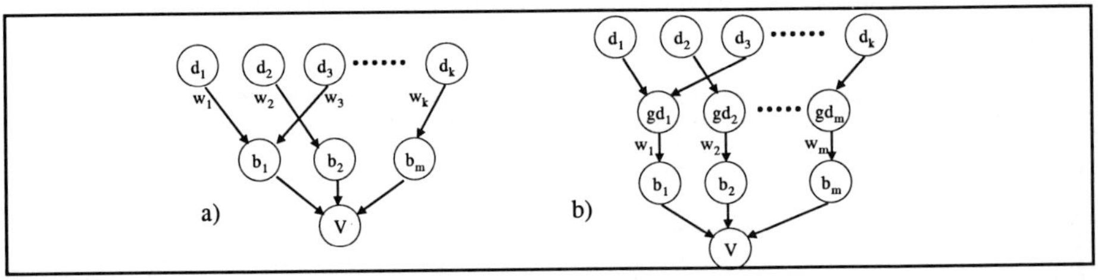

Figure 3. Clinical phenotype inference models. **a)** Diagnoses (d_i) are directly mapped to vectors describing clinical phenotype states (b_i). **b)** Diagnoses are grouped (gd_i). Basis for grouping may be generalization of code, semantic text description, defined symptoms, or some other binning feature. *V* is the final combined result.

One possibility is to generalize on diagnosis codes. In the case of ICD-9 diagnosis codes, codes having the same leftmost digits are semantically related. The fewer the number of leftmost digits found in common between codes, the greater the number of codes to which those digits refer. By "generalizing" diagnosis codes, codes that share the same left most digits are grouped together. A second possibility uses the textual descriptions that accompany the ICD-9 diagnosis codes. String matching words or phrases found in the text descriptions provide the basis for grouping diagnosis codes together. Finally, a third possibility identifies symptoms based on published articles and books about the disease in question. Each symptom then has an associated set of diagnoses not dependent on the coding structure or common words in the description. In all three possibilities, diagnosis codes are grouped into sets, which we generally refer to as "symptoms." Tools for performing each of these ways of grouping diagnosis codes into symptoms were constructed.

Each clinical phenotype may be thought of as a different state of the disease. Of the states, only one may be presenting at any particular point in time. Thus, the disease can be characterized as a vector of 0s and 1s with k positions, where k is the number of distinct phenotypes associated with the disease. From published descriptions of the disease, groups of diagnoses ("symptoms"), which are reported as being related to states of the disease, are designated a 1 in each vector position that corresponds to those states and a 0 in all other vector positions. Each symptom therefore has a non-zero "state vector" associated with it that identifies the states in which the symptom is expected to appear. For example, the vector [1,1,0,0] represents a disease having 4 states and the symptom related to this vector has diagnoses that only appear in the first two states. Each b_i in Figure 3 represents such a vector.

At this point, we have an initial knowledge-based model that relates each diagnosis code to states of the disease through a symptom as shown in Figure 3b. We now look at an alternative mapping directly from diagnosis codes to disease states as shown in Figure 3a. Using literature about the disease, each diagnosis code appearing in the hospital visits is directly mapped to corresponding disease states without the use of symptoms. These are manual mappings that are not calibrated by actual observations or mapped by a real domain expert. They provide a second guess at the relationship between diagnosis and disease states.

2.3 Mapping Adjustment (Step II)

The question becomes, "Do the mappings consistently define the clinical phenotype?" For example, a hospital visit with diagnoses that when binned into their respective symptoms (using the model in Figure 3b) could provide the vectors {[0,1,1,0] , [0,1,1,0] , [0,0,1,0]}, while those same diagnoses mapped directly to disease states (using the model in Figure 3a) could yield only the fourth state. Using the vectors resulting from the symptoms, the fourth clinical phenotype would never be considered. Thus, the scenario exists where a symptom may under-represent the states provided in the diagnosis-to-state mappings. Similarly, the situation could occur where a symptom over-represents the diagnosis-to-state mappings. If a symptom vector presents [0,1,1,0], but the corresponding diagnosis-to-state mappings appear only in the second clinical phenotype, then the symptom falsely assumes the third phenotype.

We address these scenarios, with a method to update the symptoms, such that our model becomes consistent in its mappings. Our approach involves finding maximally specific mappings in the space defined by the initial brute-force mappings. This approach builds on prior work in the area of concept learning using general-to-specific ordering[9]. Let Sy_i be the state vector for the ith symptom. Let D_i be the state vector for the set of diagnoses mapped to Sy_i, but whose state is determined from the diagnosis to disease state mappings. In each position j of the vector D_i, the number is 1 if any diagnosis, within D_i, maps to clinical state j, and 0 otherwise. For each Sy_i and D_i, there are four possible scenarios each with a specific action as defined in Table 1.

Table 1. Symptom refinement based on diagnoses vector comparison

Scenario	Example	Action
$D_i = Sy_i$	$D_i = [0,1,0,0]$, $Sy_i = [0,1,0,0]$	None
$D_i < Sy_i$	$D_i = [0,1,0,0]$, $Sy_i = [0,1,1,1]$	$Sy_i = D_i$
$D_i > Sy_i$	$D_i = [0,1,1,1]$, $Sy_i = [0,1,0,0]$	
Not($D_i = Sy_i$) And Not($D_i < Sy_i$) And Not($Sy_i < D_i$)	$D_i = [1,1,0,0]$, $Sy_i = [0,1,1,0]$	Partition symptoms

The first scenario, D_i=Sy_i, is trivial, the symptom does not change. When there are more states found for a symptom than the diagnoses provide, D_i>Sy_i, or the symptom covers too many states, D_i<Sy_i, we set Sy_i equal to D_i. The final scenario, when the two vectors are unequal we partition Sy_i into several symptoms and redefine the state mappings. The partitioning rule is explained with an accompanied example. Let D_i= [1,1,0,0] and Sy_i = [0,1,1,0]. Diagnoses within D_i that are contained by Sy_i (≤) remain mapped to Sy_i. So, diagnoses that provide [0,1,0,0], [0,0,1,0], or [0,1,1,0], remain mapped to Sy_i. Next, create new symptoms with the vectors defined to be equal to the largest range of states spanning the remaining diagnoses. Thus, if there existed diagnoses with vectors [1,0,0,0] and [1,1,0,0] a new symptom would be created with the vector [1,1,0,0]. Yet, if the remaining diagnoses all had the vector [1,0,0,0], the new symptom would have a vector of [1,0,0,0]. Partitioning modifies the structure of the model. We now have mappings of diagnoses to symptoms and symptoms to states that are generally specific to the mapping of diagnoses to states.

2.4 Learning Diagnosis Weights (Step III)

For any particular hospital visit, we are interested in what state of the disease a patient is presenting. To accomplish this, we must determine how accurate a diagnosis code is for predicting any particular state. Initially we cannot make such a determination because we have no training data. Hospital visits are not initially classified as representing a particular phenotype of the disease. So initially, we assume all mappings are equally accurate. On subsequent iterations however, hospital visits are classified as representing certain disease states thereby revealing some diagnosis codes as being more accurate then others. The accuracies of diagnosis code mappings, once hospital visits are classified, are determined as follows.

From the classified hospital visits, the frequency of each diagnosis is calculated for each state. The frequency vectors are compared with the state vectors from symptoms to determine the accuracy of each diagnosis code in the prediction of disease state. A vector containing the frequency of diagnosis codes appearing in the hospital visits and a symptom's state vector are incomparable. For comparison, the frequency vector is thresholded to transform it into vector of 0s and 1s. The threshold was calculated as the average number of non-zero counts per stage:

$$\text{Threshold} = \frac{\sum_{i\in|S|} x_i}{\sum_{i\in|S|} \theta(x_i)}, \qquad \theta(x) = \begin{cases} 0, & \text{if } x = 0 \\ 1, & \text{if } x > 0 \end{cases}$$

where x represents the frequency counts for state i, S is the number of states, and $\theta(x)$ is the indicator function. Each position of the frequency vector is changed to a bit, where a 1 is assignment to frequencies greater than the threshold and 0 otherwise. For example, the frequency vector [1,9,1,1] would provide a threshold of 3 ((1+1+9+1) / (1+1+1+1)), and the frequency vector after thresholding would be [0,1,0,0]. All frequency data is now on the order of a string of bits. The corresponding vectors are termed "sample vectors."

From the sample and symptom vectors, we can determine an accuracy score for each diagnosis code. Because each vector position has a binary choice, we consider the accuracy score as the following. True positives and true negatives add a score of +1, while false positives add –1. The score is normalized by the total number of states k. Formally, accuracy is defined as:

$$\text{Accuracy} = \frac{\sum_{i\in|S|} \phi(t_i - f_i)}{|S|}, \quad \phi(x) = \begin{cases} -1, & \text{if } x = -1 \\ 0, & \text{if } x = 1 \\ +1, & \text{if } x = 0 \end{cases}$$

where t and f representative of the positions in the state and thresholded frequency vectors, respectively, and $\phi(x)$ a bit function. By the above definition, the accuracy of any particular feature must reside in the range [(k-1)/k , 1]. The lower bound is dependent on the nature of assigning a state pattern for a symptom. there must be a minimum of one position in the vector that is defined as 1.

2.5 Patient Visits Mapped to Disease States (Step IV)

The question becomes, "How do we use these accuracy values to relate distinct hospital visits to disease states?" Consider the feed forward schematic depicted in

Figure 4. The useful information from a hospital visit, diagnosis codes (though procedure and discharge status codes could also be used), denoted as *d,* are mapped to their respective symptoms, denoted as *s*. Rather than feed the raw frequency vector associated with a symptom, the accuracy value of each constituent diagnosis code is used. This is represented by w_i, which recognizes the degree to which the hospital visit belongs to the disease state based on the appearance of the diagnosis. At each symptom, the maximum accuracy $max\ w_i$, a scalar value, is used to scale the symptom's state vector. The scaling is simple multiplication of a vector by a scalar, which converts the state vector into a vector of 0s and the max weight. For example, let s_i have a state vector [0,1,1,0] and the corresponding set of diagnoses weights that fed forward to s_i be [0.3, 0.4, 0.3, 0.7]. Because the max weight is equal to 0.7, the weighted state vector is [0, 0.7, 0.7, 0].

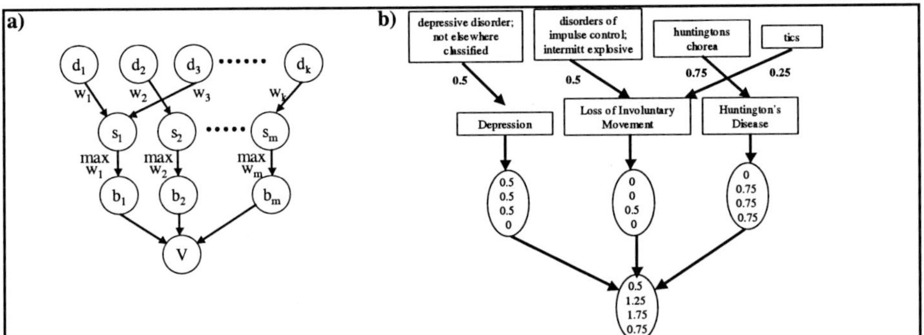

Figure 4. **a)** Weighted inference model. d_i= diagnosis, s_i= symptom, b_i=weighted clinical phenotype vector, V = final weighted clinical phenotype vector for a hospital visit. **b)** An example relating a hospital visit to disease states.

The weighted vectors, noted as b_i in Figure 4a, for each symptom are then combined via vector addition. Since the vectors are of the same dimension, with each position corresponding to the same state of the disease, the vector addition results in a vector the same size as the number of disease states. The final vector *V*, is a weighted score of the certainty in each state that a particular hospital visit exhibits. This score is determined for each hospital visit independently.

2.6 Temporal Constraints for Optimizing Disease State Alignments (Step V)

Now that each hospital visit has been converted into a weighted vector of disease states, we must determine how to relate visits belonging to a single patient with time dependent aspects. There exist some diseases that have a defined progression pattern, or one that may be inferred. One example of such progression inference has been demonstrated with partially observable Markov decision processes for studying heart disease.[11] Other diseases, such as Huntington's disease, are currently untreatable and therefore have a direct progression towards death. This section of the algorithm is not necessary for clinical information inference of diseases with an unknown progression status, such as cystic fibrosis. However, because progression of disease is an issue that helps define the current state of some diseases, time constraints must be taken into account.

Consider the profile depicted in Figure 5a. Each row corresponds to a state *s* of the disease. We are attempting to determine the optimal alignment of states for this profile. If we assume that the max value in each *v* corresponds to the actual state, then the predicted progression would be as depicted in Figure 5b. However, if the disease could only have a forward progression without remission, we would have to consider the maximum sum of single vector positions under this constraint. The result is depicted in Figure 4c. Yet, to prevent impossible stage alignments for longitudinal medical profiles, we must utilize knowledge about the disease.

a)	v_1	T_1	v_2	t_2	v_3	t_3	v_4	t_4	v_5
s_0	5.8		1.3		0.8		0.5		2.7
s_1	2.5	1.5	3.5	1	5.1	1	1.3	3	1.2
s_2	0.9		6.2		1.3		1.8		3.4
s_3	0		5.0		1.8		0.8		7.5

	Alignment				
b)	0	2	1	2	3
c)	0	1	1	2	3
d)	2	2	2	2	3

Figure 5. a) Sample patient profile with the time elapsed shown between each vertical state vector. The time elapsed is in years. **b-d)** Various statement alignments as the result of varying degrees of time constraints. In **b)** max cell values are assumed to be stage of disease; in **c)** linear forward progression constraint is enforced, and in **d)** time dependency for each stage is considered.

Many different state sequences could have given rise to the observed sets of diagnoses. For example, the state sequences could have been generated from the underlying states (1,1,1,1,1), (2,2,2,2,2), (2,2,3,3,3), or (1,1,2,2,2), though there would be many more possible alignments of the sequence. The difference between these alignments is that they would present diagnoses at each visit with different probabilities. We are interested in the path through the sequence alignment that provides the highest probability by considering the most probable state sequence given the set of compressed tuple vectors *V*, the set of time constraints *T*, and a state path through the profile.

The diagnosis-state mappings are updated from the hospital visit classifications and then steps II through IV of the algorithm repeat until no further refinement in the classification of hospital visits is realized. Metrics were defined to determine convergence.

3 Results on Huntington's Disease

Materials included hospital discharge data from the State of Illinois, for the years 1990 through 1997. There were approximately 1.3 million hospital discharges per year. Collection information has compliance with greater than 99% of discharges occurring in hospitals in the state.[8] As a sample set, a Huntington's disease registry from Rush Presbyterian Hospital of Chicago was used. The registry consisted of demographic information and the age of onset of the disease for each listing.

Longitudinal medical profiles were constructed as described in previous reports.[8] Profile construction was performed with an estimate of 100% uniqueness and identifiability of individuals based on {*ZIP*, *date of birth*, *gender*}. The resulting profiles were crossed with the registry. The resulting join, yielded a sample of 22 individuals, with a total number of 69 hospital visits.

The literature review provided a list of symptoms related to four stages. Clinically, there are three stages of the disease known to exist; an early stage, middle stage, and late stage, as well as the asymptomatic period of the disorder. It is worth noting there are two types of Huntington's disease that have different progression rates. One type is a juvenile onset that presents before the age of 20, while the other is normal adult presentation above the age of 20. Furthermore, the disease has an untreatable forward progression toward death. There is no remission, thus you could not backtrack from the middle to the early stage of the disease.

Step I of the approach involves mapping symptoms to phenotype states. Based on literature review, a list of 36 symptoms was constructed with each symptom mapping to any of 4 possible stages of the disease. Diagnosis codes and discharge status codes were identified and then mapped to the generalized diagnoses based on the methods described in section 2.2. For Step II, partitioning the symptom model resulted in a total of 45 symptoms.

Table 2. Comparison of Models for Generalizing Diagnoses

Model	Number of Nodes	Number of States Encountered
Diagnoses (direct)	156	477
Generalized Codes	60	791
Text Semantics	8	975
Symptoms	45	752

As shown in Table 2, the diagnosis model, which maps each diagnosis directly to its disease state, resulted in 477 states being encountered. Of the generalized models, we found the symptom model to be more specific than the models resulting from generalizing codes or using textual descriptions. Henceforth, we continue with the symptom model.

Step III of the approach involves learning corresponding weights for the diagnosis and discharge status codes. Step IV of the approach involves automatically classifying each patient visit into a known disease state. Finally, with each hospital visit independently classified as exhibiting a state of the disease, we had to align visits pertaining to the same patient in order to respect the temporal constraints specific to Huntington's Disease. The method used is described in the next paragraphs.

For Huntington's disease, the expected length of time is approximately 5 years per stage for the adult type, and 3 years for the juvenile type.[10] Furthermore, there is a linear progression of the disease. Once the patient reaches stage (or state) 2, remission to stage 1 is not possible. For the adult type, time up to the age of onset is defined as stage 0, from onset up to five years afterwards defines stage 1, five to ten years defines stage 2, greater than ten years defines stage 3.

Based on knowledge about the particular disease, we can construct a set of rules governing the time-dependency of the disease that overlaps the stages in time to account for transitions in the disease as well as in age reporting. The noted ranges are: 1 to 6 years for stage 1, 4 to 11 years for stage 2 and 9 to 15 years for stage 3.

Based on these constraints, we predicted the age of onset for each patient in the following manner: list the times for each visit as an inequality in the time ranges noted. This provides a set of inequalities, one inequality for each visit. Expand the list by reporting the time lapses between visits. Solve the resulting set of inequalities to get time bounds and then modify hospital visit classifications accordingly. Final results, converging after three iterations of the algorithm, appear in Figure 6.

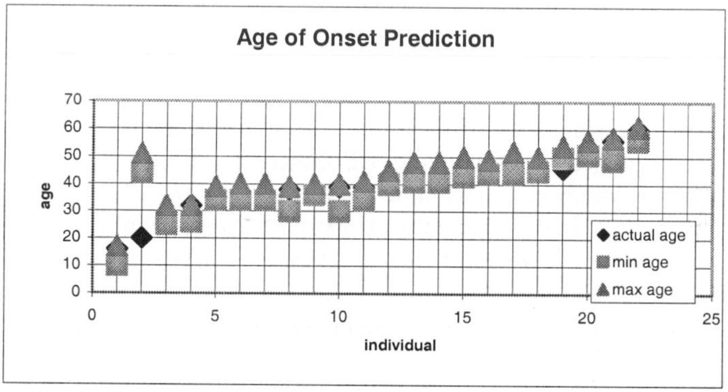

Figure 6. Results from Huntington's data. Age of onset is accurately predicted in 20 of the 22 cases.

The relationship between the age of onset and the trinucleotide repeat size has been shown to have an inverse relationship. The relationship has an r^2 regression value of 0.73 as noted in previous works.[12] The equation used to determine the relationship between age of onset and CAG repeat size is ln(age of onset) = 5.4053 - 0.0377*(trinucleotide repeat size). Out of the Rush Presbyterian dataset, there were 3 subjects that we knew the repeat size for. Predictions of the repeat size yielded matches.

4 Discussion

Our approach is biased by the initial disease and symptom mappings, but such bias is analogous to having incorrect knowledge in a knowledge base or incorrect classifications assigned in training data. The iterative nature of our approach further refines its models to better fit the data but cannot always overcome initial bias.

The methodology described above is general enough to be compatible with many single gene disorders. For each of these diseases, the number of states will be dependent on the number of clinical types.

References

1. F.M. De La Vega, M. Kreitman, and I.S. Kohane. "Human genome variation: linking genotypes to clinical phenotypes" In: *Pacific Symposium on Biocomputing 2001*, R.B. Altman et.al (Eds.) (World Scientific, Singapore, 2001).
2. M.R. Knowles, K.J. Friedman, and L.M. Silverman, "Genetics, Diagnosis, and Clinical Phenotype" Ed. J.R. Yankaskas and M.R. Knowles. (Lippincott-Raven, Philadelphia, 1999).
3. R.R. Brinkman, et al., "The likelihood of being affected with Huntington disease by a particular age, for a specific CAG size." *Am. J. Hum. Genet.* **60**. 1202-1210 (1997).
4. A.R. La Spada. "Trinucleotide repeat instabilitiy: genetic features and molecular mechanisms." *Brain Pathol.* **7**. 943-963 (1997).
5. B.G. Buchanan, et al. "Heuristic DENDRAL: a program for generating explanatory hypotheses in organic chemistry." In *Machine Intelligence 4,* B. Meltzer et. al (Eds.) (Edinburgh University Press, Edinburgh, 1969).
6. E.H. Shortliffe. *Computer-Based Medical Consultations: MYCIN*. Elsevier/North-Holland, Amsterdam, London, New York (1976).
7. G. Tesauro. "Practical issues in temporal difference learning" *Machine Learning*, 8 (3-4):257-277 (1992).
8. B.A. Malin, and L.A. Sweeney. "Determining the Identifiability of DNA Database Entries" *Proc AMIA Symp*. 537-541 (2000).
9. T.M. Mitchell, Machine Learning (McGraw-Hill, New York, 1997).
10. O. Quarrell, Huntington's Disease: The Facts. (Oxford University Press, New York, 1999).
11. M. Hauskrecht, and H. Fraser. "Modeling Treatment of Ischemic Heart Disease with Partially Observable Markov Decision Processes" *Proc AMIA Symp*. 538-532 (1998).
12. S.E. Andrew, et. al. "The relationship between trinucleotide (cag) repeat length and clinical features of Huntington's disease" *Nature*. **4**. 398-403 (1993).

A CELLULAR AUTOMATA APPROACH TO DETECTING INTERACTIONS AMONG SINGLE-NUCLEOTIDE POLYMORPHISMS IN COMPLEX MULTIFACTORIAL DISEASES

JASON H. MOORE, Ph.D., LANCE W. HAHN, Ph.D.
Program in Human Genetics, Department of Molecular Physiology and Biophysics, 519 Light Hall, Vanderbilt University Medical School, Nashville, TN 37232-0700, USA
Moore@phg.mc.Vanderbilt.edu

The identification and characterization of susceptibility genes for common complex multifactorial human diseases remains a statistical and computational challenge. Parametric statistical methods such as logistic regression are limited in their ability to identify genes whose effects are dependent solely or partially on interactions with other genes and environmental exposures. We introduce cellular automata (CA) as a novel computational approach for identifying combinations of single-nucleotide polymorphisms (SNPs) associated with clinical endpoints. This alternative approach is nonparametric (i.e. no hypothesis about the value of a statistical parameter is made), is model-free (i.e. assumes no particular inheritance model), and is directly applicable to case-control and discordant sib-pair study designs. We demonstrate using simulated data that the approach has good power for identifying high-order nonlinear interactions (i.e. epistasis) among four SNPs in the absence of independent main effects.

1 Introduction

The idea that epistasis or gene-gene interaction plays an important role in human biology is not new. In fact, Wright[1] emphasized that the relationship between genes and biological endpoints is dependent on dynamic interactive networks of genes and environmental factors. This idea holds true today. Gibson[2] stresses that gene-gene and gene-environment interactions must be ubiquitous given the complexities of intermolecular interactions that are necessary to regulate gene expression and the hierarchical complexity of metabolic networks. Indeed, there is increasing statistical and epidemiological evidence that epistasis is very common[3]. For example, in a study of 200 sporadic breast cancer subjects, Ritchie et al.[4] demonstrated a statistically significant interaction among four polymorphisms in three estrogen metabolism genes in the absence of any independent main effects. Further, Nelson et al.[5] found that epistatic effects of lipid genes on lipid traits was very common.

Despite the importance of epistasis in human biology there are few statistical methods that are capable of identifying interactions among more than two polymorphisms in relatively small sample sizes. For example, logistic regression is a commonly used method for modeling the relationship between discrete predictors such as genotypes and discrete clinical outcomes[6]. However, logistic regression, like most parametric statistical methods, is limited in its ability to deal with high dimensional data. That is, when high-order interactions are modeled, there are many contingency table cells that have no observations. This can lead to very large coefficient estimates and standard errors[6]. One solution to this problem is to collect very large numbers of samples to allow robust estimation of interaction effects. However, the magnitudes of the sample sizes that are often required are prohibitively expensive. An alternative solution is to develop new statistical and computational methods that have improved power to identify multilocus effects in relatively small sample sizes.

Several groups have addressed the need for new methods by developing data reduction approaches[4,5]. These approaches reduce the dimensionality of the data by pooling multilocus genotypes into a smaller number of groups. For example, the multifactor dimensionality reduction (MDR) method of Ritchie et al.[4] pools multilocus genotypes into high risk and low risk groups effectively reducing the dimensionality of the genotype predictors from n dimensions to one dimension. The new one-dimensional multilocus genotype variable is evaluated for its ability to classify and predict disease status using cross-validation and permutation testing. This has been shown to be an effective strategy for identifying gene-gene interactions, however, there is a loss of information in the data reduction step. An alternative strategy is to use pattern recognition that has the advantage of considering the full dimensionality of the data. In this paper, we describe a new pattern recognition approach for identifying gene-gene interactions that takes advantage of the emergent computation features of cellular automata and intelligent search features of parallel genetic algorithms. Using simulated single-nucleotide polymorphisms (SNPs) in a discordant sib-pair study design, we demonstrate that the CA approach has good power for identifying high-order nonlinear interactions with few false-positives.

2 Overview of Cellular Automata

Cellular automata (CA) are discrete dynamic systems that consist of an array of cells, each with a finite number of states. The state of each cell changes in time according to the local neighborhood of cells and their states as defined by a rule table. The simplest CA consist of one-dimensional arrays, although some, such as the Game of Life[7], are implemented in two or more dimensions. Figure 1A illustrates an example of a simple one-dimensional CA iterated through several time steps along with the simple rule table that governs its behavior. In this section, we review how CA can be used to perform computations and then how we exploit this feature to perform multilocus pattern recognition.

2.1 Emergent Computation in Cellular Automata

An intriguing feature of CA is their ability to perform emergent computation[8,9]. That is, the local interactions among spatially distributed cells over time can lead to an output array that contains global information for the problem at hand. For example, Mitchell et al.[9] have used CA to perform density estimation. In that application, a one-dimensional, two-state (1 or 0) CA is given an initial array of states. The goal is to identify a CA rule set such that the CA converges to an array of all 1's if the density of 1's is greater than 0.5 and to an array of all 0's if the density of 1's is less than 0.5. They found that the CA is able to perform this computation through a series of spatially and temporally extended local computations. This emergent computation feature of CA forms the basis of our proposed method for identifying patterns of genotype variations associated with common complex multifactorial diseases.

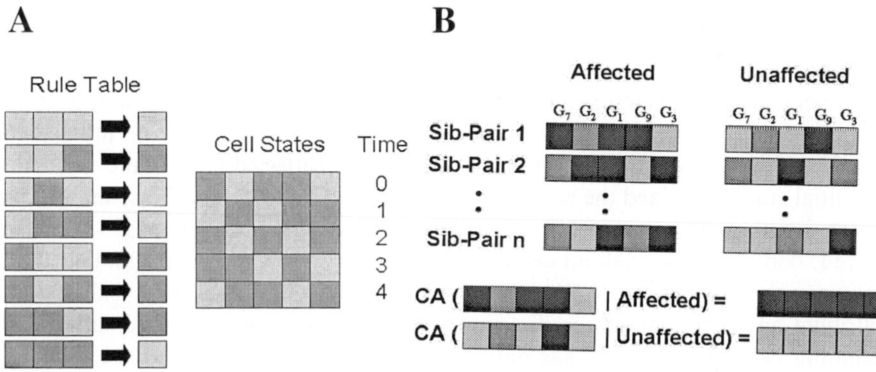

Figure 1. (A) The rule table, cells, and cell states for a simple one-dimensional CA iterated through four time steps. Note that the state of a given cell at time t is determined by the states of that cell and the states of adjacent cells at time t-1 as defined by the rule table. (B) General approach to using CA to perform emergent computation on combinations of genetic variations from affected and unaffected siblings. The goal is to identify a CA model that is able to produce one type of output (e.g. all black cells states) if the genetic variation input is from an affected sibling and another type of output (e.g. all gray cell states) if the input is from an unaffected sibling.

2.2 Description of our Cellular Automata Approach to Pattern Recognition

We begin with a description of the discordant sib-pair study design, a commonly employed design for identifying common complex disease susceptibility genes using single-nucleotide polymorphisms in human populations. With this approach, sib-pairs in which one sibling is affected with the disease and the other is unaffected are ascertained and genetic variations measured in each. The benefits of this study design are twofold. First, many common complex diseases have a late age of onset thus limiting the ability to collect parental samples which are useful for some types of statistical tests. Second, using unaffected sibs as controls instead of unrelated subjects prevents false-positive results due to population stratification (e.g. by chance you select unaffected controls of a different ethnic/genetic background). This is because, by definition, unaffected sibs come from the same ethnic or genetic background as the affected sib. Traditional statistical methods such as the sibship transmission/disequilibrium test (Sib-TDT)[10] compare observed differences in the frequencies of alleles (i.e. a single genetic variation from one of the two chromosomes) or genotypes (i.e. combination of two alleles) among affected and unaffected sib-pairs with the expected difference of zero under the null hypothesis that the particular genetic variation is not associated with the disease. TDT-type statistics have reasonable power for identifying genetic variations that have moderate to large effects on disease risk. This is evident from studies of linkage disequilibrium among single-nucleotide polymorphisms in and near the *APOE* gene in Alzheimer disease[25]. However, TDT statistics have low power for identifying genetic variations whose effects on disease risk are fully or partially through interaction with other genetic variations. This is because, in its original form, the TDT is a univariate statistic that considers only one genetic variation at a time.

We have developed a CA approach to identifying patterns of genotype variations associated with disease using the discordant sib-pair, or case-control, study design. The general approach is to identify a set of CA operating features that is able to take an array of genotypes as input and produce an output array that can be utilized to classify and predict whether siblings are affected or unaffected (see Figure 1B). In this initial study, we fixed the number of cells in the CA to five. Thus, the CA is presented with a maximum of five unique and/or redundant genotypes. We also allowed 'don't care' or wildcard cells to be included. This permits less than five genotypes to be evaluated. Wildcard cells all have the same state and do not contribute any information for discriminating affected from unaffected sib pairs. Assuming each genetic locus has only three possible genotypes, we used a binary encoding with '01' for the first genotype, '10' for the second genotype, '11' for the third genotype, and '00' for the wildcard. Thus, each array presented to the CA consisted of 10 bits with two bits encoding the state of each of the five cells. We used a simple nearest-neighbor rule table that is implemented by looking at the state of the cell in question and the adjacent cell states as is illustrated in Figure 1A. With three cells forming a rule and four different states per cell, there are 4^3 or 64 possible rule inputs with four possible output states for each. An important feature of CA is the number of time steps or iterations. This will govern the amount of spatial and temporal information processing that can performed. In this study, we allowed a maximum of 128 iterations for each CA. This maximum number of iterations was selected to allow enough time for parallel processing of all the information in an input array without an excessive number of iterations that might complicate interpretation. Thus, there are three essential components to the CA model. First, the correct combination of genetic variations must be selected for initiating the CA cell states. Second, the appropriate rule table that specifies the information processing must be selected. Finally, the number of time steps or iterations for the CA must be selected. We used the genetic algorithm machine learning methodology to optimize selection of these three model components (described below).

How is the output array of the CA used to perform classification and prediction? We first count the number of 1s in the binary encoded output array of the CA run on each set of genotypes for each affected and each unaffected sib in the sample. A classifier is formed by using a frequency histogram of the number of 1s among affected sibs and unaffected sibs. Each histogram bin is labeled affected or unaffected depending on whether the number of 1s represented by that bin were more frequently observed among affected or unaffected sibs. For example, consider the case where 100 discordant sib pairs were evaluated. Suppose the number of CA output arrays that contained three 1s was 20 for affected sibs and 10 for unaffected sibs. This bin would be labeled affected and thus the 10 unaffected sibs would be misclassified. This would contribute 0.05 to the overall misclassification rate. This is performed for each bin and a total classification error is estimated by summing together the individual rates for each bin.

3 Cellular Automata Optimization using Parallel Genetic Algorithms

3.1 Overview of Parallel Genetic Algorithms

Genetic algorithms (GAs), neural networks, case-based learning, rule induction, and analytic learning are some of the more popular paradigms in machine learning[11]. Genetic algorithms perform a beam or parallel search of the solution space that is analogous to the problem solving abilities of biological populations undergoing evolution by natural selection[12, 13]. With this procedure, a randomly generated 'population' of solutions to a particular problem are generated and then evaluated for their 'fitness' or ability to solve the problem. The highest fit individuals or models in the population are selected and then undergo exchanges of random model pieces, a process that is also referred to as recombination. Recombination generates variability among the solutions and is the key to the success of the beam search, just as it is a key part of evolution by natural selection. Following recombination, the models are reevaluated and the cycle of selection, recombination, and evaluation continues until an optimal solution is identified.

As with any machine learning methodology[11], GAs are not immune to stalling on local optima[14]. To address this issue, distributed or parallel approaches to GAs have been implemented[15]. Here, the GA population is divided into sub-populations or demes. At regular iterative intervals, the best solution obtained by each sub-population is migrated to all other sub-populations. This prevents individual sub-populations from converging on a locally optimum peak because new highly fit individuals are periodically arriving to increase the population diversity. In biology, it is believed that evolution progresses faster in semi-isolated demes than in a single population of equal size[16]. Indeed, there is some evidence that parallel GAs actually converge to a solution much faster than serial or single-population GAs[15]. This superlinear speedup may be due to additional selection pressure from choosing migrants based on fitness[15].

Genetic algorithms have been applied to microarray data analysis[17, 18] and are ideally suited for selecting polymorphisms and optimizing CA[9].

3.2 Solution Representation and Fitness Determination

The first step in implementing a GA is to represent the solution or model to be optimized as a one-dimensional binary array or chromosome. For the CA, we needed to encode five genetic variations and/or wildcards, the CA rule table, and the number of CA iterations (see Figure 2). Each of the genetic loci and the number of CA iterations were represented using a total of six 32-bit integers with a modulo operation used to constrain the integer to the desired range (0-19 for the genetic loci (described in Section 4) and the wildcard and 0-127 for the number of iterations.) As previously described, each CA cell has four possible states and each rule depends on the state of three cells. Encoding this set of 64 rules, with each rule producing one of four two-bit states as output, requires four 32-bit integers. In total, the GA manipulated 10 32-bit integers for a total chromosome length of 320 bits (see Figure 3). Fitness of a particular CA model is defined as the ability of that model to classify siblings as affected or unaffected. Thus, the goal of the GA is to identify a CA model that minimizes the misclassification error. Implementation using cross validation is described below.

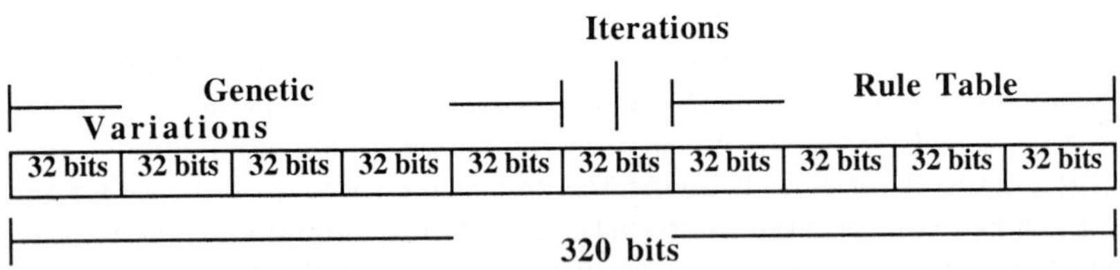

Figure 2. Encoding of the genetic algorithm chromosome. The first five 32-bit segments encode genetic variations and/or wild cards while the sixth 32-bit segment encodes the number of iterations. The last four 32-bit segments encode the CA rule table.

3.3 *Parallel Genetic Algorithm Parameters*

We implemented a parallel genetic algorithm with two sub-populations or demes undergoing periodic migration. Each GA was run for a total of 200 iterations or generations with migration of the best solutions to each of the other sub-populations every 25 iterations. Sub-population or deme sizes of 10, 50, 100, 200, 500, and 1000 were used for the analysis of all 50 simulated datasets (see Section 4). We used a standard recombination frequency of 0.6 and a standard mutation frequency 0.02[13].

3.4 *Hardware and Software*

The parallel GA used was a modification of the Parallel Virtual Machine (PVM) version of the Genetic ALgorithm Optimized for Portability and Parallelism System (GALLOPS) package for UNIX[19]. This package was implemented in parallel using message passing on a 110-processor Beowulf-style parallel computer cluster running the Linux operating system. Two processors were used for each separate run. Information about obtaining the CA and GA software can be found at http://phg.mc.vanderbilt.edu/software.

3.5 *Implementation*

The goal of the GA is to minimize the classification error of the CA model. However, from a genetics perspective, the goal is to identify the correct functional genetic variations. That is, from a pool of many candidate genes, the goal is to find those that play a role in influencing risk of disease. We used a 10-fold cross-validation strategy[24] to identify optimal combinations of genetic variations. Cross-validation has been a successful strategy for evaluating multilocus models in other studies of common complex diseases[4]. Here, we ran the GA on each 9/10 of the data and retained the CA models that minimized the misclassification rate. Across the 10 retained models, we selected the combination of polymorphisms that was observed most frequently. The reasoning is that the functional set of genetic variations should be identified consistently across different subsets of the data[4]. To determine statistical significance of the observed cross-validation consistency, we permuted the data 1,000 times to determine the empirical distribution of cross-validation consistency were the null hypothesis true. We rejected the null hypothesis of no association when the upper-tail Monte Carlo p-value was ≤ 0.05. We also estimated the general prediction error of the best set of retained models using each of the independent 1/10 of the data.

4 Data Simulation

The goal of the simulation study was to generate a series of datasets in which the probability of a sibling being affected is dependent on epistasis or interaction among a set of genetic variations. We first simulated 100 sibling pairs each with 20 unlinked (i.e. on different chromosomes) genetic variations using the Genometric Analysis Simulation Program (GASP)[20]. Each genetic variation had two alleles with frequencies 0.6 and 0.4. Four of the 20 genetic variations served as the functional genetic loci. The remaining 16 genetic variations serve as potential false-positives. Thus, the goal of the CA was to identify the correct four functional genetic variations from the total of 20 candidates.

The epistatic interaction among the four functional genetic variations was accomplished using a Boolean network (Figure 3). Alleles (e.g. A_1 and A_2) were encoded as either 1 for an A allele or 0 for an a allele. The allele combinations at each genetic locus (A-D) that contribute to disease risk were as follows; $A_1 = 1$ and $A_2 = 1$, $B_1 = 1$ and $B_2 = 0$, $C_1 = 0$ and $C_2 = 0$, $D_1 = 0$ and $D_2 = 0$. The logic functions AND, NOR, INHIBITION, and XOR[26] were selected such that the Boolean network would produce a one or affected status if a particular subject had one and only one of the four specified allele combinations. The XOR function has been described as an epistasis model[21] and is often used to evaluate pattern recognition approaches because of its inherent nonlinearity. With this model, the probability of being affected is one if the sibling has one and only one of the allele combinations or genotypes listed above. Inheriting more than one of the allele combinations listed above is protective against disease. The independent main effects of each genetic variation are minimal under this epistasis model.

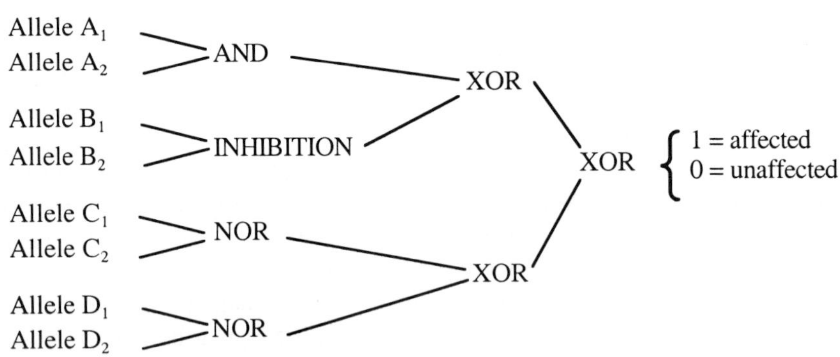

Figure 3. Boolean network used to simulate the epistasis effects on disease risk. The alleles of each single-nucleotide polymorphism combine to form genotypes using the AND, NOR, and INHIBITION functions. The genotypes are then combined using XOR functions that output affected status. Each sibling is at increased risk of being affected if they inherit one, and only one, of the four genotypes. Inheriting more than one of the particular genotypes is considered protective. With this epistasis model, the independent main effects of each genotype are minimized.

5 Data Analysis

The goal of the statistical analysis was to determine whether the cross-validation consistency (see Section 3.5) for a particular combination of genetic variations is expected by chance if the null hypothesis were true. A particular genetic variation was considered statistically significant if it was identified in six or more of the 10-fold cross-validation datasets. As described, this decision rule was determined empirically by permuting the data 1,000 times and selecting the number of 10-fold cross-validation datasets for which loci would be identified less than 5% of the time. This corresponds to a statistical significance level of 0.05. The power of the CA approach is reported as the number of simulated datasets out of 50 in which the correct four functional genetic variations were identified in six or more of the cross-validation datasets as described above.

6 Results

We first analyzed the single-locus effects of each of the 20 simulated SNPs in each dataset of 100 discordant sib-pairs using the traditional Sib-TDT statistic[10]. The power to detect the independent effects of each of the four functional loci was less than 50% while the power to detect each of the false-positive loci was close to the expected false-positive rate of 5%. Thus, a traditional approach would have missed each of the four functional loci more than half the time with occasional false-positives. This indicates that none of the four SNPs have a large independent main effect on disease risk. This is consistent with the simulation strategy used (see Section 4).

Table 1. Power (datasets out of 50) of CA to identify each locus.

Sub-Population Size	Locus A	Locus B	Locus C	Locus D	False Positives
10	4	0	16	1	27
50	50	20	50	34	0
100	50	26	50	33	0
200	50	32	50	37	0
500	50	30	50	40	0
1000	50	34	50	40	0

Table 1 summarizes the number of datasets (out of 50 total) in which each of the functional loci (A-D) were identified using the multi-locus CA approach. For GA deme sizes of 50 or greater, the CA yielded 100% power for identifying loci A and C. That is, these two genetic loci were always found. When the deme sizes were 1000, the power to detect locus B was 68% while the power to detect locus D was 80%. The power to detect the four functional loci was greatly improved over that of the Sib-TDT by considering combinations of SNPs using the CA approach. It should be noted that a GA deme size of 10 yielded low power and many false-positives. However, when a GA deme size of 50 or greater was used, no false-positives were identified. These results suggest that the CA pattern recognition approach is useful for identifying combinations of genetic variations that influence disease risk primarily through gene-gene interactions.

7 Discussion

We have introduced a cellular automata (CA) approach to identifying patterns of variations in multiple SNPs associated with risk of common complex multifactorial diseases. The development of this CA approach was motivated by the limitations of the generalized linear model for detecting and characterizing gene-gene[3] and gene-environment[22] interactions.

The CA approach shares many of the same advantages of the multifactor dimensionality reduction (MDR) approach[4]. For example, the CA approach is nonparametric. As Ritchie et al.[4] describe, this is an important distinction from traditional parametric statistical methods that rely on the generalized linear model. For example, with logistic regression, as each additional main effect is included in the model, the number of possible interaction terms grows exponentially. The number of orthogonal regression terms needed to describe the interactions among a subset, k, of n biallelic loci is n choose k multiplied by 2 raised to the power of k[23]. Thus, for 20 polymorphisms we would need 40 parameters to model the main effects (assuming two dummy variables per biallelic locus), 1,560 parameters to model the two-way interactions, 79,040 parameters to model the three-way interactions, 1,462,240 parameters to model the four-way interactions, etc. Thus, fitting a full model with all interaction terms and then using backward elimination to derive a parsimonious model would not be possible. The CA approach avoids the pitfalls associated with using parametric statistics such as logistic regression for modeling high-order interactions.

An additional advantage of the CA approach is that it assumes no particular genetic model (i.e. model-free). That is, no mode of inheritance needs to be specified. This is important for diseases such as cardiovascular disease and depression in which the mode of inheritance is unknown and likely very complex. In its current form, this approach can be directly applied to case-control and discordant sib-pair study designs. Extension to other family-based control study designs such as triads should also be possible.

As with MDR[4], an advantage of the CA approach is that false-positive results due to multiple testing are minimized. Indeed, we detected no false-positives in the present study when a GA deme size of 50 or greater was used. This is primarily due to the cross-validation strategy used to select optimal models. Data reduction and pattern recognition approaches are good at identifying complex relationships in data, even when those relationships are due to chance or false-positive variations. However, the real test of any approach is its ability to make predictions in independent data[24]. Cross-validation divides the data into 10 equal parts allowing 9/10 of the data to be used to develop a model and the independent 1/10 of the data used to evaluate the predictive ability of the model[24]. Optimal models are selected solely on their ability to make predictions in independent data. Once a final predictive model has been selected, only then is the null hypothesis of no association tested via permutation testing. It is this combined cross-validation and permutation testing approach that minimizes false-positives due to multiple looks at the data[4].

There are several advantages of the CA approach over data reduction approaches such as MDR. First, there is no loss of information. The CA considers the full dimensionality of the data whereas methods such as MDR seek to reduce the dimensionality of the data and in doing so lose information. Additionally, the CA does not have the same limitation as MDR for making predictions during cross-validation in high-dimensional data[4]. This is because the rule set that is generated is general enough to accept any combination of genotypes to make a prediction of disease risk.

Despite these important advantages, there are also several disadvantages to the CA approach. Most importantly, CA models are very difficult to interpret. Understanding the relationship among the multilocus SNP genotypes requires interpreting the spatial and temporal information processing that is occurring in the CA to produce a predictive output. Although Mitchell et al.[9] have made progress in this area, there is clearly a lot of work remaining. An additional disadvantage over traditional methods is that the CA approach is very computationally intensive. Selection of SNPs and optimization of the CA parameters requires a machine learning strategy such as the parallel GA. Effective implementation of GAs requires at least several workstations that are part of Beowulf-style parallel computer system. However, it should be noted that these systems are very inexpensive to set up since they use commodity-priced off-the-shelf components and freely available software.

In conclusion, we have introduced a new approach to identifying patterns of SNP variations in complex multifactorial diseases. This approach takes advantage of the emergent computation features of one-dimensional CA and the intelligent search features of parallel GAs. We anticipate that the results of this study will open the door for investigations of using CA to identify combinations of SNPs that interact in a non-additive or nonlinear manner to influence risk of common complex multifactorial diseases.

References

1. S. Wright, *Proc. 6th Int. Conf. Genet.* **1**, 356 (1932)
2. G. Gibson, *Theor. Popul. Biol.* **49**, 58 (1996)
3. A.R. Templeton in *Epistasis and Evolutionary Process*, Eds. M. Wade et al (Oxford University Press, New York, 2000)
4. M.D. Ritchie et al, *Am. J. Hum. Genet.* **69**, 138 (2001).
5. M.R. Nelson et al, *Genome Res.* **11**, 458 (2001).
6. D.W. Hosmer and S. Lemeshow, *Applied Logistic Regression* (John Wiley & Sons Inc., New York, 2000)
7. M. Gardner, *Sci. Amer.* **223**, 120 (1970)
8. M. Sipper, *Evolution of Parallel Cellular Machines* (Springer, New York, 1997)
9. M. Mitchell et al, *Physica D*, **75**, 361 (1994)
10. R.S. Spielman and W.J. Ewens, *Am. J. Hum. Genet.* **62**, 450 (1998)
11. P. Langley, *Elements of Machine Learning* (Morgan Kaufmann Publishers, Inc., San Francisco, 1996)
12. J.H. Holland, *Adaptation in Natural and Artificial Systems* (University of Michigan Press, Ann Arbor, 1975)
13. D.E. Goldberg, *Genetic Algorithms in Search, Optimization, and Machine Learning* (Addison-Wesley, Reading, 1989)
14. W. Banzhaf et al, *Genetic Programming: An Introduction* (Morgan Kaufmann Publishers, San Francisco, 1998)
15. E. Cantu-Paz, *Efficient and Accurate Parallel Genetic Algorithms* (Kluwer Academic Publishers, Boston, 2000)
16. S. Wright S, *Genetics* **28**, 114 (1943)
17. J.H. Moore and J.S. Parker, *Lec. Notes Artificial Intel.* **2167**, 372 (2001)
18. J.H. Moore and J.S. Parker in *Methods of Microarray Data Analysis* (Kluwer Academic Publishers, Boston, in press)
19. http://garage.cps.msu.edu
20. A.F. Wilson et al, *Am. J. Hum. Genet.* **59**, A193 (1996)
21. W. Li and J. Reich, *Hum. Hered.* **50**, 334 (2000)
22. C.D. Schlichting and M. Pigliucci, *Phenotypic Evolution: A Reaction Norm Perspective* (Sinauer Associates, Inc., Sunderland, 1998)
23. M.J. Wade in *Epistasis and Evolutionary Process*, Eds. M. Wade, B. Brodie III, J. Wolf (Oxford University Press, New York, 2000)
24. B.D. Ripley, *Pattern Recognition and Neural Networks* (Cambridge University Press, Cambridge, 1996)
25. E.R. Martin et al, *Am. J. Hum. Genet.* **67**, 383 (2000)
26. D. Kaplan and L. Glass, *Understanding Nonlinear Dynamics* (Springer-Verlag, New York, 1995)

ONTOLOGY DEVELOPMENT FOR A PHARMACOGENETICS KNOWLEDGE BASE

DIANE E. OLIVER, DANIEL L. RUBIN, JOSHUA M. STUART, MICHEAL HEWETT, TERI E. KLEIN, AND RUSS B. ALTMAN

Stanford Medical Informatics, Stanford University School of Medicine, 251 Campus Drive, MSOB X-215, Stanford, CA 94305-5479

oliver@smi.stanford.edu, rubin@smi.stanford.edu, stuart@smi.stanford.edu, hewett@smi.stanford.edu, klein@smi.stanford.edu, altman@smi.stanford.edu

Research directed toward discovering how genetic factors influence a patient's response to drugs requires coordination of data produced from laboratory experiments, computational methods, and clinical studies. A public repository of pharmacogenetic data should accelerate progress in the field of pharmacogenetics by organizing and disseminating public datasets. We are developing a pharmacogenetics knowledge base (PharmGKB) to support the storage and retrieval of both experimental data and conceptual knowledge. PharmGKB is an Internet-based resource that integrates complex biological, pharmacological, and clinical data in such a way that researchers can submit their data and users can retrieve information to investigate genotype–phenotype correlations. Successful management of the names, meaning, and organization of concepts used within the system is crucial. We have selected a frame-based knowledge-representation system for development of an ontology of concepts and relationships that represent the domain and that permit storage of experimental data. Preliminary experience shows that the ontology we have developed for gene-sequence data allows us to accept, store, and query data submissions.

1 Introduction

In the quest to understand the impact of genetic factors on drug response, researchers must integrate data produced from laboratory experiments, computational methods, and clinical studies. Different individuals have different responses to the same medications. With the draft sequence of the human genome and increased understanding of metabolic enzymes, drug transporters, and drug receptors, there is great promise for improving our understanding of how genetic variation affects variation in drug efficacy and toxicity. The sheer volume of biological data, the uncertainties associated with those data, and the complexity of the relationships among concepts present challenges for structuring and managing pharmacogenetic data and knowledge.

We are participating in the Pharmacogenetic Research Network and Knowledge Base consortium, a group of investigators funded by the National Institutes of Health (NIH) to study pharmacogenetics. The goal of NIH in funding this consortium is two-fold: (1) to create a network of multidisciplinary, collaborative research groups that study pharmacologically significant genetic variation, and (2) to build a knowledge base that is available to the research community and that can stimulate hypothesis-driven research[1]. Investigators in the consortium conduct studies to identify genetic polymorphisms, to assess functional variation of variant proteins, and to relate clinical drug responses to genetic variation[2]. The pharmacogenetics knowledge base (PharmGKB) stores the data, and is publicly accessible over the Internet (http://www.pharmgkb.org).

A principal challenge for PharmGKB is to integrate complex biological data, pharmacological data, and clinical data in such a way that researchers can contribute results. There are also ethical issues associated with studying populations from different ethnic backgrounds, maintaining confidentiality of data, and addressing issues of intellectual property[3]. However, the focus of this paper is limited to the problem of modeling biological, pharmacological, and clinical data to support the goal of linking genotype to phenotype.

Developing such a resource is challenging because (1) the data are complex, (2) the data come from diverse sources and must be integrated into a single system, (3) terms that identify clinical and biological concepts must be used consistently throughout the software modules and by different users, (4) knowledge is constantly changing, (5) a mixture of experimental data and knowledge about the domain must coexist in the knowledge base, and (6) data and knowledge within the system must be consistent with data stored in external public databases.

These problems are exacerbated by a lack of standard terminologies for clinical and biological concepts and by a lack of standard representations for the experimental data being submitted. However, successful management of names, meaning, and organization of concepts is crucial, and hence, an ontology of formally specified concepts and relationships is central to the design of our system.

2 What is an ontology?

Due to ambiguity associated with the term *ontology*, we first address the question of what an ontology is, and clarify our use of the term for PharmGKB. The word *ontology* was originally used by philosophers to describe a branch of metaphysics concerned with the nature and relations of being[4]. The artificial-intelligence community later adopted the term, but has debated its meaning. Guarino[5] states that in the most prevalent use of the term, an ontology refers to an engineering artifact, constituted by a specific vocabulary used to describe a certain reality, plus a set of explicit assumptions regarding the intended meaning of the vocabulary words. He says that in the simplest case, an ontology describes a hierarchy of concepts related by subsumption relationships, whereas in more sophisticated cases, suitable axioms are added to express other relationships between concepts and to constrain interpretation of those concepts. We adopt this definition, and recognize that it leaves open a number of questions regarding how an ontology should be designed, implemented, and used.

Two types of ontologies that are relevant to PharmGKB are (1) controlled vocabularies and (2) domain ontologies. A controlled vocabulary is a collection of terms organized in a hierarchy intended to serve as a standard nomenclature[6,7,8]. The purpose of a controlled vocabulary is to provide a common set of terms that users of a single system can share (e.g., MeSH in Medline[9]), or that users can share across multiple systems (e.g., Gene Ontology[7]). A controlled vocabulary usually contains concepts, but no instances. A domain ontology, described by Musen[10], is a set of classes and associated slots that describe a particular domain. The purpose of this type of ontology is to serve as a knowledge-base schema, analogous to a database schema. However, unlike a database schema, the domain ontology may also contain classes that are not intended to have instances, but that represent concepts organized in a hierarchy to serve as a controlled vocabulary. When a knowledge-base developer adds instances to classes in a domain ontology, the result is a knowledge base. The domain ontology itself does not contain instances.

For PharmGKB, we need a controlled vocabulary to specify shared names, synonyms, and meanings of concepts, and we need a knowledge-base schema to specify how experimental data are represented and stored. We also need other domain-relevant classes and instances that support various PharmGKB applications. Thus, we need to represent experimental data and domain knowledge, and all of the classes modeled for these purposes contribute to the domain ontology.

3 Methods

We discuss here the two primary tasks that are important in content development for the PharmGKB ontology: (1) modeling experimental data, and (2) modeling domain knowledge. The distinction between data and knowledge is fuzzy, but for our purposes, the data that result from experimental studies are *data*, and the controlled-vocabulary concepts that are used as values for the experimental data, that provide classification or synonym information, or that provide supportive relationship information for applications are *knowledge*. In general, we are taking a bottom-up approach for developing the ontology for the experimental data, and a top-down approach for developing the knowledge.

We have chosen to use a frame-based knowledge-representation system to store both experimental data and domain knowledge. We are using Protégé to build the PharmGKB knowledge base. Protégé[11] is a frame-based knowledge-representation system that offers classes, slots, facets, instances, and slot values as the building blocks for representing knowledge. *Classes* are data structures that may or may not have instances; they have *slots* (sometimes called attributes or roles) that establish relationships between classes. Classes are organized in a hierarchy, and each class has at least one parent (except the root, which has no parent). In PharmGKB, we make the restriction that non-root classes have only one parent. Slots have *slot values* that may or may not be inherited. Slots also have *facets* that specify cardinality and data-type constraints on the slot value (e.g., string, integer, enumerated symbols, or instance of another class).

3.1 Modeling Experimental Data

The research goals and data requirements of the first five research groups submitting data to PharmGKB provide an initial framework for modeling experimental data. Each research group is studying a set of genes, and each gene of interest codes for a protein that is thought to have an effect on a phenotypic response to a drug. Each protein (e.g., enzyme, transporter, or receptor) affects one or more drugs studied by the research group. To characterize genotypes, investigators look for polymorphisms in the genes of interest in human DNA samples. To link genotype to phenotype, the researchers select phenotypic observations that they can measure at the molecular, cellular, or clinical level, and that they can correlate with genotype.

Table 1 shows the five groups that are providing initial data, and summarizes their research interests[12,13,14]. The groups focus their research in different ways—for example, one group may focus on particular enzymes or transporters, and another may focus on a particular drug class or disease. However, all groups collect data to link genotype to phenotype. In our bottom-up approach to ontology development, we are modeling the PharmGKB ontology to fit the data collected by these research centers. We will expand the model as needed to accommodate additional kinds of experimental data provided by other research centers.

3.2 Modeling Domain Knowledge

Analysis of the areas of interest and data of the five groups suggests broad categories that are appropriate in modeling domain knowledge in pharmacogenetics. Table 2 shows several high-level categories that are useful for organizing controlled-vocabulary concepts in PharmGKB, and gives examples of entities from the researchers' areas of interest that fall into these general categories[12,13,14]. Additional modeling is required to refine the entities into a multi-level classification hierarchy.

Naming conventions are required for biological entities such as genes, alleles, and proteins, for pharmacological entities such as drugs and metabolites, and for clinical entities such as diseases, symptoms, laboratory tests, and test results. Fortunately, standards do exist in certain areas. The Enzyme Commission has established an enzyme nomenclature[15], and the Human Gene Nomenclature Committee has established rules for gene names and maintains the HUGO gene nomenclature[16]. However, a gene sequence may have multiple accession numbers in Genbank, a sequence can be identified by either a GenBank accession number or a LocusLink ID, and although there may be names for certain alleles of a particular

Table 1. Research groups that are providing data initially to PharmGKB and their areas of research. Each research group is identified by the principal investigator of the project and by the primary institutional affiliation of the group.

Research Group	Research Areas
Richard Weinshilboum (Mayo Clinic)	Enzymes involved in phase II metabolism of drugs, especially in methylation and sulfate conjugation reactions, and the impact of genetic variation in genes that encode these enzymes (e.g., TPMT, HNMT, COMT)
Kathleen Giacomini (University of California at San Francisco	Transporter genes, including genes that encode the serotonin transporter (SERT) and vesicular monoamine transporter (VMAT2), and the impact of variation in these genes on efficacy and toxicity of antidepressants
Mark Ratain (University of Chicago)	Pharmacogenetics of anticancer agents, with an emphasis on topoisomerase inhibitor drugs, such as irinotecan and etoposide
David Flockhart (University of Indiana)	Metabolism of tamoxifen by cytochrome P450 enzymes, and effects of genetic variation on pharmacokinetics, clinical efficacy, and adverse effects of tamoxifen
Scott Weiss (Harvard University)	Genetic factors that influence patient response to three classes of drugs used in asthma: (1) inhaled beta agonists, (2) inhaled steroids, and (3) leukotriene modifiers

gene, there may not be a name for every sequence variant of a gene that occurs in the population. In clinical medicine, standards are even less clear. Nevertheless, standards are emerging, and we are evaluating controlled vocabularies maintained by others to determine their suitability for PharmGKB. When necessary, we will develop our own approach, but will do this only when no standards exist.

The two parts of the ontology—the knowledge-base schema for experimental data and the domain conceptual knowledge—are integrated in PharmGKB to support user queries. The classes and instances that form the knowledge may be

Table 2. Examples of categories and entities in these categories, based on data supplied by the five groups.

Category	Entities
Genes	TPMT, HNMT, COMT, SLC6A4, SLC18A2, CYP3A4, CYP2C9, CYP2D6, UGT1A1, IL2, IL4, IL13, IL2RG
Proteins	
Enzymes	cytochrome P450 isoenzymes, methyltransferases, sulfotransferases, UDP-glucuronosyltransferases
Transporters	serotonin transporter (SERT), vesicular monoamine transporter (VMAT2)
Receptors	interleukin receptors, cholinergic receptor
Drugs	selective estrogen receptor modulators (SERMs), inhaled beta agonists, inhaled corticosteroids, leukotriene antagonists, topoisomerase inhibitors, selective serotonin receptor antagonists (SSRIs)
Diseases	asthma, depression, breast cancer, leukemia, colon cancer
Functional Studies	
In vitro	enzyme kinetic studies, measurement of levels of immunoreactive protein
In vivo	pharmacokinetic studies, clinical studies of drug efficacy and adverse effects

used in a variety of ways. For example, the domain conceptual knowledge offers the following features:

1. *Controlled-vocabulary terms.* Researchers who enter experimental data as PharmGKB instances must use names of entities in slot values that are the same as the names used by others who enter or query data. Thus, the domain knowledge provides a shared nomenclature, enforced by our data-acquisition methods.

2. *Alternative names.* Entities may have synonyms or near synonyms, and maintenance of alternative names in the system assists a users in searching for a concept.

3. *Accession numbers.* Accession numbers that are unique identifiers for entities in external databases are stored in PharmGKB to facilitate communicate with those databases. Examples of relevant external databases that have their own coding schemes of identifiers are Genbank, LocusLink, Refseq, PubMed, and Online Mendelian Inheritance in Man.

4. *Classification hierarchy.* The classification hierarchy allows users to browse for terms of interest by navigating up or down the hierarchy. It also permits the user to formulate a query in terms of a single high-level concept, and to apply that query automatically to multiple lower-level concepts that are subsumed by the concept in the original query.

5. *Nonhierarchical relationships between concepts.* Slots can be considered nonhierarchical links between classes. Such links can provide additional knowledge that supports browsing or querying. In PharmGKB, examples of useful associations include gene–enzyme links (a particular gene encodes a particular enzyme), gene–transporter links (a particular gene encodes a particular transporter), drug–metabolite links (a particular drug has a particular set of metabolites), and drug–enzyme links (a particular drug is metabolized by a particular enzyme).

4 Scenario of Use

To demonstrate use of the domain-knowledge hierarchy in conjunction with queries for experimental data, we present a scenario in which a researcher enters experimental data and a user later retrieves the data. Suppose the researcher has completed a pharmacokinetic study on the drug irinotecan and has collected genotype data from individuals in the study. Pharmacokinetic data collected include blood levels of the drug and its metabolites, as well as summary parameters that describe the rise and fall of drug and metabolite levels in the blood. The researcher selects the drug and metabolites by navigating through a display of the controlled-vocabulary hierarchy. Alternatively, he enters the drug and metabolite names as text, and the system confirms whether or not the drug and metabolites are known to PharmGKB. In addition, the system verifies that the metabolites are indeed metabolites associated with irinotecan.

Later, a user querying PharmGKB might search for information on studies of topoisomerase inhibitors. Since irinotecan is categorized in the knowledge hierarchy as a topoisomerase inhibitor, the system returns data from an irinotecan pharmacokinetic study. Another user might search for data on the drug Camptosar, which is the trade name for irinotecan. Since the knowledge hierarchy stores information about trade names that correspond to generic names, the system again returns information about the irinotecan pharmacokinetic study.

Figure 1 shows a portion of the ontology that reflects the knowledge-base schema for experimental data. Figure 2 shows a portion of the ontology that reflects domain conceptual knowledge. In the pharmacokinetic study, the drug is administered to an individual. The individual is identified by a PharmGKB subject identifier. Information about the event in which the subject is given a dose of the drug is stored in an instance of the class *DrugDosingEvent*. The value of the slot *Drug* in this instance would be *Irinotecan*. The entity *Irinotecan* is part of the

```
Class: DrugDosingEvent
   Slot: DisplayName
   Slot: Drug
   Slot: PharmGKBSubjectID
   Slot: RouteOfAdministration
   Slot: DoseAmount
   Slot: DoseAmountUnits
   Slot: ProtocolTimePoint
```

Figure 1. Information about a drug-dosing event. The *DrugDosingEvent* class is part of the knowledge-base schema for experimental data.

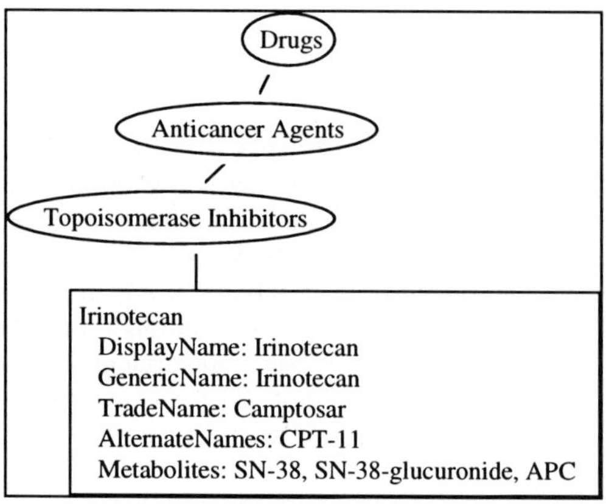

Figure 2. Domain knowledge about the drug irinotecan. In this fragment of the ontology, *Irinotecan* is an instance of the class *Topoisomerase Inhibitors*, and that class exists in a hierarchy of *Drugs*.

controlled vocabulary of drugs stored in the domain knowledge. As shown in Figure 2, *Irinotecan* is classified as one of the *Topoisomerase Inhibitors*, and its trade name, *Camptosar*, is stored in a slot. There is also a slot for metabolites of irinotecan, which enables the system to validate the fact that the metabolites entered as text by the user are indeed metabolites of irinotecan. Thus, in this example, the domain knowledge provides (1) values for experimental data, (2) constraints for data validation, (3) categorical classification support for queries, and (4) synonym support for queries.

5 Preliminary Experience

The initial task that we addressed in our ontology development was the creation of classes that would support submission of gene-sequence data. Network investigators submit data that describe the experiments performed, as well as the associated results. Data that describe experiments include information about the investigators who performed the study, genes studied, primers and methods used, and gene regions analyzed. Results include polymorphisms discovered, the frequency of those polymorphisms in particular populations, and summary single nucleotide polymorphism (SNP) data from pooled samples of DNA. The ontology contains all classes and slots necessary to support the automatic submission of SNPs to the NIH-supported dbSNP resource. When a user submits SNP data to PharmGKB and the SNP is not yet present in dbSNP, PharmGKB automatically sends a new SNP submission to dbSNP.

Prior to release of our sequence-submission software, we conducted a test with collaborators from the Mayo Clinic. The Mayo group produced a sample set of submissions to describe experimental data collected in their studies of the HNMT gene. They incorporated their data into the XML format required for direct XML submissions. The required format is specified by the PharmGKB XML schema (http://www.pharmgkb.org/xml-schemas.html), and that schema corresponds directly to the PharmGKB ontology.

In the development of the ontology for gene-sequence submissions, we encountered a number of subtleties of definition that had to be clarified before ontology developers and data submitters reached a shared understanding. We describe here several of the most important constructs in the resulting ontology.

A *reference sequence* is a specified sequence of bases against which variations are compared. A reference sequence is associated with a gene and may (but need not) correspond exactly to a sequence already deposited in GenBank. However, a reference sequence is not required to contain the entire gene structure associated with a gene. Reference sequences can be different molecule types (DNA, RNA, or protein). The only restriction is that they consist of a contiguous series of monomers. Gaps in the sequence or fragments pasted together are not allowed.

A *sequence coordinate system* is required to ensure agreement about how to identify a particular position in a reference sequence. The sequence coordinate system specifies which base is labeled +1, and indicates whether the base that precedes position +1 is numbered 0 or -1.

A *region of interest* identifies a segment of a reference sequence that is of interest to the investigators. It may be a subsequence of the reference sequence, or it may be the entire reference sequence.

```
<ForwardPcrPrimer>
<DisplayName>Exon 5 Forward Primer</DisplayName>
<FirstAnnealingPositionInPrimer>19</FirstAnnealingPositionInPrimer>
<FirstAnnealingPositionInRegion>6</FirstAnnealingPositionInRegion>
<LastAnnealingPositionInPrimer>41</LastAnnealingPositionInPrimer>
<LastAnnealingPositionInRegion>28</LastAnnealingPositionInRegion>
<Sequence>TGTAAAACGACGGCCAGTAGGAGTATCTAGCCCAAGCAATA</Sequence>
</ForwardPcrPrimer>
```


Figure 3. Sample input data from Mayo test submission in XML format. This sample input shows the submission of a forward PCR primer used, and is stored as experimental data in PharmGKB.

A *simple nucleotide difference* (SND) defines a position in a region of interest of a reference sequence where bases differ from the bases in the corresponding location of a tested sequence. A SND is *simple* because the bases in the variant segment must be contiguous, rather than located in different parts of the genome. A SND differs from a single nucleotide polymorphism (SNP) in that there is no frequency restriction in the definition of a SND. In contrast, when scientists perform SNP detection assays, it is common practice to filter out SNPs that have allele frequencies that are less than some threshold percentage (e.g., 10 percent). Also, in the spirit of dbSNP, a SND can refer to an insertion, a deletion, or a variable number of repeats, as well as to a single nucleotide difference. The convention in PharmGKB for specifying where the difference is located is to identify the position in the reference sequence that precedes the variant site (the position upstream in the 5' direction). This approach provides a consistent method for describing variant positions across all polymorphism types.

Figure 3 shows a representative sample of data from the Mayo test submission. It shows the submission of a forward PCR primer used in an experiment. Annealing positions are based on a numbering scheme that was previously specified in a sequence coordinate system for the reference sequence.

6 Concluding Remarks and Future Work

Given the potential impact of pharmacogenetic research and the vast quantities of data that are likely to result from efforts to link genotype to phenotype, the NIH has begun a program that encourages collaboration among investigators and that mandates public sharing of data. The value of PharmGKB as a resource for sharing pharmacogenetic experimental data and knowledge lies not only in its commitment to public dissemination of data, but also in its demonstration of the use of knowledge representation techniques to organize pharmacogenetic knowledge and data. There is currently no standard data model for pharmacogenetic knowledge, and without standards for names and meanings of terms, it is difficult to share information in computer-based systems. Thus, the ontology effort is essential to the success of this project, and may contribute to ontology development done by others who work in this area.

Our ontology development process is a process of iterative development and communication between bioinformatics professionals and other collaborators, including molecular biologists, chemists, clinical pharmacologists, and clinicians. Our bottom-up approach to modeling experimental data allows us to take a staged-delivery approach in software development. We can provide software that is usable to a few groups initially, and then extend it in a controlled fashion. However, our top-down approach to knowledge modeling also encourages us to consider the broader picture in the early stages.

Our ontology is comprised of the data model for experimental data, and the domain conceptual knowledge that provides controlled-vocabulary information and other knowledge that supports queries. These two parts are integrated in PharmGKB, but it is useful to distinguish them because the former is essential for communication with our collaborators who submit data, and the latter is essential for management of shared concepts in the system. Together, these two parts form the ontology that may be reusable in other settings in the field of pharmacogenetics.

Future work on the PharmGKB ontology includes (1) expansion of content to broaden the scope, (2) enhancement of constraint representation in the ontology to support automated or semi-automated data validation, (3) extension of change logging features to facilitate change management, (4) development of merging techniques to support the process of merging the production version of the knowledge base with the development version when a new version is released, and (5) enhancement of methods that help users to query PharmGKB in an intuitive manner to obtain genotype–phenotype associations.

Acknowledgements

PharmGKB is financially supported by grants from the National Institute of General Medical Sciences (NIGMS), the Human Genome Research Institute (NHGRI) and the National Library of Medicine (NLM) within the National Institutes of Health (NIH). This work is supported by the NIH/NIGMS Pharmacogenetics Research Network and Database grant U01GM61374, and by Stanford University's Children's Health Initiative. JMS is supported by National Library of Medicine grant LM07033.

References

1. Long RM, Giacomini KM. Announcement. June 1, 2001 http://www.nigms.nih.gov/pharmacogenetics/editors.html
2. RFA GM-00-003, April 7, 2000 http://grants.nih.gov/grants/guide/rfa-files/RFA-GM-00-003.html
3. MA Rothstein, PG Epps "Ethical and legal implications of pharmacogenomics" *Nature Review Genetics*, **2**, 228-231 (2001)
4. *Webster's New Collegiate Dictionary*, 9th edition, Ontology, p. 825 (Springfield, MA: Merriam-Webster, 1991)
5. N Guarino, "Formal ontology and information systems" *Proceedings of FOIS '98*, Trento, Italy, June 6-8, 1998. (Amsterdam, IOS Press, 1998) pp. 3-15
6. C Price, M O'Neil, TE Bentley, PJB Brown, "Exploring the ontology of surgical procedures in the Read Thesaurus" *Methods of Information in Medicine* **37**, 420-5 (1998)
7. The Gene Ontology Consortium, "Gene ontology: tool for the unification of biology" *Nature Genetics* **25**, 25-9 (2000)
8. D Fensel, "Ontologies and electronic commerce" *IEEE Intelligent Systems* January/February, 8 (2001)
9. "Medical Subject Headings" http://www.nlm.nih.gov/mesh/meshhome.html
10. MA Musen, "Domain ontologies in software engineering: Use of Protégé with the EON architecture" *Methods of Information in Medicine* 37(4–5), 540-50 (1998)
11. "Welcome to the Protégé project" http://www.smi.stanford.edu/projects/protege/
12. "PharmGKB Investigators" http://www.pharmgkb.org/investigators.html
13. "Query PharmGKB" http://www.pharmgkb.org/PharmGKB/query
14. "Pharmacogenetics Research Network and Knowledge Base First Annual Scientific Meeting" April 25, 2001 http://pub.nigms.nih.gov/pharmacogenetics
15. "Enzyme nomenclature" http://www.chem.qmw.ac.uk/iubmb/enzyme
16. "HUGO Gene Nomenclature Committee" http://www.gene.ucl.ac.uk/nomenclature

A SOFM Approach to Predicting HIV Drug Resistance

R. Brian Potter[a], Sorin Draghici
Department of Computer Science, Wayne State University, Detroit, MI 48202

The self-organizing feature map (SOFM or SOM) neural network approach has been applied to a number of life sciences problems. In this paper, we apply SOFMs in predicting the resistance of the HIV virus to Saquinavir, an approved protease inhibitor. We show that a SOFM predicts resistance to Saquinavir with reasonable success based solely on the amino acid sequence of the HIV protease mutation. The best single network provided 69% coverage and 68% accuracy. We then combine a number of networks into various majority voting schemes. All of the combinations showed improved performance over the best single network, with an average of 85% coverage and 78% accuracy. Future research objectives are suggested based on these results.

1 Introduction

1.1 Overview

The *human immunodeficiency virus (HIV-1)*, the causative agent of *acquired immune deficiency syndrome (AIDS)*, has been the subject of extensive research in recent years. A good, although somewhat dated introduction to AIDS research is provided by Watson, et. al.[1]

HIV-1 infection has been approached via many treatment pathways. One of the first was the use of *Azidothymidine (AZT)* to inhibit the synthesis of the HIV provirus in vivo. Unfortunately, the HIV virus was able to mutate in order to resist AZT, eventually overcoming its therapeutic benefits. Two other popular methods of treating the HIV virus are by attacking the reverse transcriptase responsible for synthesizing the DNA provirus from the retroviral RNA, and by inhibiting the HIV protease responsible for splicing the primary polyproteins produced by the HIV virus into the active proteins necessary for its replication. Both of these approaches also eventually fail due to mutation of the viral genome, leading to protease inhibitor resistant viral strains. Most current therapies involve combinations of drugs aimed at inhibition of both the reverse transcriptase and the protease.

Artificial neural network (ANN) based self-organizing maps were developed by Kohonen.[2] SOFM algorithms belong to the unsupervised learning, competitive network class of ANNs. An input vector is introduced to the network, after which a winning neuron is determined and the weight vectors of all neurons within a specified neighborhood of the winning neuron are updated.[3] In this

[a]Please send correspondence to this author at potterrb@ieee.org.

way, SOFMs are useful for clustering related patterns together. When patterns in the training set are labelled, clusters containing these labelled patterns can then be used to identify unknown patterns.

This laboratory has previously applied SOFM clustering to the HIV drug resistance problem.[4] Resistance to the protease inhibitor Indinavir was studied first by applying supervised learning techniques to protein structural data for various HIV protease mutants to predict Indinavir IC90 values. Only limited success was obtained, primarily due to an insufficient number of mutations with corresponding Indinavir IC90 values available from the literature with which to train the classifier. An SOFM was used to segment the same data into clusters of Indinavir-resistant mutants and non-resistant mutants based on structural features. We were able to divide all reported HIV mutants into several categories based on their 3-dimensional molecular structures and the pattern of contacts between the mutant protease and Indinavir. Our classifier shows reasonable prediction performance, being able to predict the drug resistance of previously unseen mutants with an accuracy of between 60% and 70%. We believe that this performance can be greatly improved once more data becomes available. The results support the hypothesis that structural features of the HIV protease can be used in antiviral drug treatment selection and drug design.

The goal of this research is to build a SOFM to predict the resistance of known mutations of HIV protease to Saquinavir, a protease inhibitor related to Indinavir that is also approved for use in the treatment of HIV infection. No attempt is made to understand the mechanism or reasons why certain mutation are or are not resistant to Saquinavir, only to predict such resistance based solely on the amino acid sequence of HIV protease mutants, a small number of which have reported Saquinavir IC90 values. Our hope is that this early work will ultimately enable clinicians to prescribe HIV treatments based on drug resistance predictions.

1.2 Related Work

Self-organizing maps have been used successfully in a wide variety of life science applications. Kaartinen et.al. have successfully used a SOFM to discriminate between human blood plasma lipoprotein lipids (LDL and HDL cholesterol, triglycerides) and furthermore to cluster plasma samples into different lipoprotein lipid risk profiles.[5] Makipaa et. al. have applied SOFMs to the clustering and subsequent classification of blood glucose data from insulin-dependent diabetic patients.[6] Santos-Andre and Roque da Silva combined a SOFM with a multi-layer perceptron to provide radiologists with a "second opinion" in the diagnosis of breast cancer.[7] Christodoulou and Pattichis have developed medical diagnostic systems for the assessment of electromyographic (EMG) signals necessary for the diagnosis and monitoring of patients with neuromuscular disorders, and carotid plaques based on ultrasound images of patients with pulmonary disease. The systems were comprised of multiple SOFM classifiers whose results were combined using majority voting and SOFM-derived confidence measures.[8,9] Finally, Golub et. al. were able to distinguish between acute myeloid leukemia (AML) and acute lymphoblastic leukemia (ALL) from SOFM clustering of gene expression monitoring data.[10]

2 Experimental Detail

2.1 Data Preparation

Only thirty-two patterns (HIV Protease mutants) were found in the literature with reported IC90 drug resistance values.[11] These patterns were supplemented with 910 reported HIV protease mutants obtained from the Los Alamos National Laboratory HIV Sequence Database (http://hiv-web.lanl.gov/), along with the wild type HIV protease sequence.

Netprep, a command line Java program, was written to convert the amino acid sequence of a protein or peptide segment (a string of alpha characters) into normalized numeric patterns suitable for input to a neural network. The input to Netprep is a file containing one peptide sequence per line, with each residue separated by a comma. The first pattern in the file is the wild type. For each residue, all of the patterns are compared to the wild type. Patterns that match the wild type at that residue are assigned a value of zero. Residues that differ from the wild type are ordered by frequency of occurrence. They are then assigned a value between 0 and 1 based on dividing (0,1] into n equal increments, where n is the number of different mutations from the wild type for that residue. For instance, if the wild type is V, and there are four mutations across all of the input patterns, say N, L, I, and A, N may be assigned a value of 1, L a value of .75, I a value of .5, and A a value of .25. Once these numeric assignments are made, each pattern is normalized and written to an output file.

The researcher may optionally specify at runtime a percentage of the patterns to withhold from training. All the patterns are processed as described above, after which the appropriate number of patterns to be withheld are randomly selected and output to a separate holdout file. The remaining patterns are used as input to the neural network. For the research described in this paper, ten percent of the 911 unclassified patterns were withheld. The 32 patterns with resistance values were all used, as described in the next section.

We were interested not in predicting the resistance of a particular mutant, but rather in classifying a mutant as having high, medium, or low resistance to saquinavir. We defined low resistance for a mutant as having less than a four-fold resistance to saquinavir as compared to the resistance of the wild type. High resistance was defined as greater than ten-fold resistance to saquinavir as compared to the resistance of the wild type. Having defined these cutoffs, twelve of the 32 patterns with IC90 values were classified as having low resistance, three with medium resistance (between 4- and 10-fold resistance), and the remaining patterns classified as exhibiting high resistance. The actual range of resistance values was from 0.33-fold to 269.33-fold (see Table 1[b]).

2.2 Training

A leave-one-out cross-validation strategy was used due to the scarcity of classified patterns. Thirty-one of the 32 patterns with resistance values were added to 800+ patterns remaining after holdout on the data set obtained from Los Alamos. The patterns with resistance values allowed us to identify clusters of mutants as high, medium or low resistance to saquinavir. Clusters with conflicting assignments were classified as 'mixed', and those with no assignment were classified as 'none'.

In all, 36 networks were trained a total of 32 times (one for each leave-one-out pattern to be tested), for a total of 1152 runs. See Table 3 for a complete listing of the networks. To summarize, networks with output matrices of 12x12, 10x10, 8x8, 6x6, 5x5, 4x4, and 3x3 were trained using initial learning rates of 0.9-0.5 and initial neighborhoods corresponding to the dimensionality of their output matrix (e.g., an initial neighborhood of 12 for the 12x12 matrix). All networks except one trained using 10 iterations. The 10x10 matrix was also trained using 50 iterations, an initial learning rate of 0.7, and an initial neighborhood of 10. The results of this test were then compared to the same conditions and 10 iterations to see if increasing the number of iterations would improve the performance of the network.

3 Results and Discussion

3.1 Single Network Performance

Once each network was trained, the lone test pattern was run through the network. If the pattern was assigned to a 'mixed' cluster or to one with no

[b]All mutations were obtained from Winters, et. al.[11], except as noted.

Mutation	IC_{90}(uM)	Fold resistance
Wild Type	0.03	1
L10I K14R N37D M46I F53L A71V – G73S V77I L90M	8.08	269.33
L10I E35D M36I R41K I62V L63P – A71V G73S I84V L90M I93L	6.00	200
L10I I15V M36I G48V I54V I62V V82A	1.18	39.33
L10I I15V M36I G48V I54V I62V	0.92	30.67
K14R I15V N37D F53L A71V G73S L90M	0.58	19.33
K14E M36V G48V L63P A71V T74S V82A	0.58	19.33
I15V R41K L63P A71T G73S L90M	0.37	12.33
G48V L63P T74A	0.80	26.67
K20I M36I L63P A71T G73S L90M	0.42	14
L10I E35D R41K I62V L63P A71V – G73S I84V L90M I93L	0.34	12.67
K14R R41K L63P V77I L90M I93L	0.21	7
L10I K20M L63P A71T V77I L90M I93L	0.20	6.67
N37D R57K D60E L63P A71V G73S – L90M I93L	0.20	6.67
I15V D30N E35D M36I R41K L63P	0.03	1
L63P T74S L90M	0.09	3
L63P L90M	0.08	2.67
K14R R41K L63P V77I I93L	0.07	2.33
L10V I62V G73S L90M	0.07	2.33
L63P T74A V77I	0.07	2.33
L63P L90M	0.06	2
N37D L63P A71V G73S L90M I93L	0.06	2
L10I L63P A71T V77I I93L	0.06	2
I15V E35D R41K L63P	0.06	2
K14R/K L63P I93I/L	0.06	2
K14E L63P A71V	0.06	2
I15V	0.04	1.33
L63P	0.05	1.67
L10I L63T A71T	0.02	0.67
L63P A71V L90M	0.02	0.67
L63A	0.01	0.33
G48V I54V L90M[12]	1.50	50
G48V I84V L90M[12]	0.90	30

Table 1: Resistance values of HIV Protease mutants to Saquinavir. The fold resistance was calculated as a ratio between the IC90 value of the mutant and the IC90 value of the wild type.

x	L	M	H
L		FP	FP
M	FN		FP
H	FN	FN	

Table 2: Truth table for determining false positives and false negatives. Actual classifications are on the left, classifications predicted by the SOFM are across the top.

label, then the pattern was not classified. Otherwise, a predicted resistance classification would be assigned based on the label of the cluster in which the pattern was placed. We defined a false positive (FP) as a mutation that was classified as being more resistant than it actually was based on its n-fold resistance value. For instance a false positive condition exists if the mutant's IC90 value as reported causes the mutant to be defined as low resistance (i.e., the IC90 of the mutant is less than four-fold more resistant to saquinavir than the wild type) and the network assigns to that mutant a label of medium or high resistance. Conversely, if a mutant is reported as more resistant than the label assigned by the network, a false negative (FN) condition exists. Table 2 summarizes this logic as a truth table.

For each network, the 32 test patterns are identified as correctly classified, FP, FN, or not classified (if they are assigned to a 'mixed' or unlabelled cluster). Then the coverage and accuracy of the network is calculated. Coverage is defined the ratio of test patterns that were classified (i.e., assigned to a labelled cluster) to total test patterns. Accuracy is defined as the ratio of patterns that were *correctly* classified to the total number patterns classified. For our purposes, both are expressed as percentages. A third number that has been calculated for each network is what we call the network's score:

$$\text{Score} = \text{Coverage}*\text{Accuracy}*100$$

The score allows us to compare networks based on a single number. Obviously, there are other ways one may calculate a score that weights the contribution of coverage and accuracy differently. For our purposes, we will treat them as equal contributions to the overall score of the network, although we will also discriminate by coverage before attempting to find the network with the best accuracy.

Our results are summarized in Table 3. The network with the best overall performance and also the best coverage was the 8x8 output matrix with an initial learning rate of 0.6. The most accurate network was the 8x8 output matrix with an initial learning rate of 0.5. This network produced 100% accuracy, but provided only 31% coverage. Note that there are other networks

Output	Learn Rate	Nbrhood	Iterations	Coverage	Accuracy	Score
12x12	0.9	12	10	50%	75%	38
12x12	0.8	12	10	41%	62%	25
12x12	0.7	12	10	47%	60%	28
12x12	0.6	12	10	28%	56%	16
12x12	0.5	12	10	38%	42%	16
10x10	0.9	10	10	53%	76%	40
10x10	0.8	10	10	31%	60%	19
10x10	0.7	10	10	41%	62%	25
10x10	0.7	10	50	28%	44%	12
10x10	0.6	10	10	53%	71%	38
10x10	0.5	10	10	44%	50%	22
8x8	0.9	8	10	53%	65%	34
8x8	0.8	8	10	41%	62%	25
8x8	0.7	8	10	38%	58%	22
8x8	0.6	8	10	69%	68%	47
8x8	0.5	8	10	31%	100%	31
6x6	0.9	6	10	31%	80%	25
6x6	0.8	6	10	31%	80%	25
6x6	0.7	6	10	41%	85%	35
6x6	0.6	6	10	41%	62%	25
6x6	0.5	6	10	41%	85%	35
5x5	0.9	5	10	25%	88%	22
5x5	0.8	5	10	22%	86%	19
5x5	0.7	5	10	9%	33%	3
5x5	0.6	5	10	19%	100%	19
5x5	0.5	5	10	25%	75%	19
4x4	0.9	4	10	9%	100%	9
4x4	0.8	4	10	9%	100%	9
4x4	0.7	4	10	6%	100%	6
4x4	0.6	4	10	6%	100%	6
4x4	0.5	4	10	13%	75%	10
3x3	0.9	3	10	0%	N/A%	0
3x3	0.8	3	10	0%	N/A%	0
3x3	0.7	3	10	0%	N/A%	0
3x3	0.6	3	10	0%	N/A%	0
3x3	0.5	3	10	3%	100%	3

Table 3: Summary of Results. Values listed for learning rate and neighborhood are initial values. Score = Coverage*Accuracy*100.

Output Matrix	Coverage	Accuracy	Score
12x12	41%	59%	25
10x10	44%	64%	29
8x8	46%	71%	32
6x6	37%	78%	29
5x5	20%	76%	16
4x4	9%	95%	8
3x3	1%	100%	1

Table 4: Average performance of networks by size of output matrix.

which produced 100% accuracy, but all of these networks exhibited very poor coverage (less than 10%) and were rejected from serious consideration.

Overall, it was observed (see Table 4) that the networks with 8x8 output matrices performed best (average score of 32) and also provided the best coverage (average of 46%). Networks with 12x12, 10x10, 6x6 and 5x5 output matrices also performed reasonably well. The networks with smaller output matrices had very high accuracy, but their coverage was quite poor (again, less than 10%). It was also observed that increasing the number of iterations during training did not improve network performance, but actually degraded performance for the test case (10x10 output matrix, 0.7 initial learning rate, 50 iterations).

3.2 Majority Voting Schemes

The performance of the best network allowed for better-than-random accuracy (68%) and acceptable coverage of 69%. The most accurate network had 100% success for those patterns that it was able to classify, but provided only marginal coverage at 31%. Certainly for such a critical application as predicting HIV drug resistance, we would want better performance.

One possibility is to make use of multiple networks at once using a majority voting scheme. In majority voting, the results of presenting a pattern to a number of networks is tallied, and the majority classification is taken as correct. In situations where one or more networks fail to classify the pattern (e.g., the pattern is assigned to a 'mixed' or unlabelled cluster), only the outputs of the networks that successfully classify the pattern are used. In the case of a tie (there were none for the schemes that we explored), the lowest drug resistance classification was selected. That is, we considered the risk of trying a drug treatment that did not work to be lower than the risk of missing a potentially effective drug treatment.

Voting Scheme	Coverage	Accuracy	Score
Majority of 6 Most Accurate	84%	85%	71
Majority of Best + 3 Most Accurate	88%	79%	70
Majority of 4 Best Score	84%	70%	59
Best Single Network[c]	69%	68%	47
Most Accurate Single Network	31%	100%	31

Table 5: Comparison of scores for various majority voting schemes.

Three schemes were tested and compared to the best single network and the most accurate single network. The first scheme was a combination of the six most accurate networks: 8x8-0.5, 6x6-0.7, 6x6-0.5, 5x5-0.9, 5x5-0.8, and 5x5-0.6 (the number after the dash is the initial learning rate). The second scheme combined the best single network with the three most accurate networks: 6x6-0.7, 6x6-0.5, and 8x8-0.5. Again, those networks with 100% accuracy but very low coverage (the networks with 4x4 and 3x3 output matrices) were ignored. Our final scheme combine the results of the four networks with the best overall scores: 8x8-0.6, 10x10-0.9, 10x10-0.6, and 12x12-0.9.

Perrone claims that the performance of a combiner (e.g., a majority voting scheme) is never worse than the average of the individual classifiers, but not necessarily better than the best classifier.[13] In our case, all of the majority voting schemes outperformed the single best network (see Table 5). The average coverage across the three voting schemes was 85%, the average accuracy of the three was 78%, and the average score was 67. This represents a significant improvement over the single best network (69%, 68%, and 47, respectively).

4 Conclusions and Further Work

This research explored the possibility of using self-organizing feature maps to predict drug resistance in HIV-1 infected patients based only on the peptide sequence of the HIV protease mutant strain. This differs from previous work which attempted to predict drug resistance based on structural features of the HIV protease.[4] This paper shows that the single best classifier found produces acceptable results (69% coverage and 68% accuracy), but to produce a predictive system suitable for clinical use, multiple networks configured in a majority voting scheme may be necessary. The best scheme was the six most

[c] Best single network was 8x8 output matrix, 0.6 initial learning rate, initial neighborhood of 8, 10 iterations; most accurate single network was 8x8 output matrix, 0.5 initial learning rate, initial neighborhood of 8, 10 iterations

accurate networks, with coverage of 84%, accuracy of 85%, and a score of 71. All majority voting schemes outperformed the single best network.

There are many opportunities for further research on using SOFMs for predicting drug resistance. In the case of HIV drug resistance, there are additional drugs (e.g., Indinavir and Nelfinavir) and drug combinations that may be explored. The difficulty with this work and work with other HIV treatments is the lack of publicly available clinical data (IC90 values). Christodoulou and Pattichis have also incorporated the use of confidence measures for weighting individual network results in majority voting schemes[8], which may be applied to the HIV drug resistance problem. Finally, SOFMs may be applied to the treatment of other retroviral diseases such as human T-cell leukemia virus (HTLV-1) and hairy cell leukemia (HTLV-2), as well as DNA viruses such as Hepatitis-B and Herpes.

References

1. James D. Watson, Michael Golman, Jan Witkowski, and Mark Zoller. *Recombinant DNA, 2nd Ed.*, pages 485–509. Scientific American Books, New York, 1992.
2. T. Kohonen. *Self-Organization Maps.* Springer-Verlag, Berlin Heidelberg, 1995.
3. Martin T. Hagan, Howard B. Demuth, and Mark Beale. *Neural Network Design*, pages 14.10–14.16. PWS Publishing Company, Boston, 1996.
4. Sorin Draghici, Lonnie Cumberland, and Ladislau C. Kovari. Correlation of hiv protease structure with indinavir resistance: a data mining and neural network approach. In *Proceedings of SPIE 2000*, volume 4057-40, Orlando, Florida, 2000.
5. Jouni Kaartinen, Yrjo Hiltunen, P.T. Kovanen, and Mika Ala-Korpela. Application of self-organizing maps for the detection and classification of human blood plasma lipoprotein lipid profiles on the basis of 1h nmr spectroscopy data. *NMR in Biomedicine*, 11:168–176, 1998.
6. Mikko Makipaa, Pekka Heinonen, and Erkki Oja. Using the som in supporting diabetes therapy. Helsinki University of Technology, Finland, June 4-6, 1997.
7. A.C.R. Santos-Andre, T.C.S.; da Silva. A neural network made of a kohonen's som coupled to a mlp trained via backpropagation for the diagnosis of malignant breast cancer from digital mammograms. In *IJCNN '99*, volume 5, pages 3647–3650, 1999.
8. C. I. Christodoulou and C. S. Pattichis. Medical diagnostic systems using ensembles of neural sofm classifiers. In *Proceedings of ICECS '99*,

volume 1, pages 121–124, 1999.

9. C. I. Christodoulou and C. S. Pattichis. Unsupervised pattern recognition for the classification of emg signals. *Biomedical Engineering, IEEE Transactions on*, 46(2):169–178, Feb 1999.
10. T.R. Golub, D.K. Slonim, P. Tamayo, C. Huard, M. Gaasenbeek, J.P. Mesirov, et al. Molecular classification of cancer: Class discovery and class prediction by gene expression monitoring. *Science*, 286:531–537, 1999.
11. Mark A. Winters, Jonathan M. Schapiro, Jody Lawrence, and Thomas C. Merigan. Human immunodeficiency virus type 1 protease genotypes and in vitro protease inhibitor susceptibilities of isolates from individuals who were switched to other protease inhibitors after long-term saquinavir treatment. *Journal of Virology*, 72(6):5303–5306, 1998.
12. Raymond F. Schinazi, Brendan A. Larder, and John W. Mellors. Mutations in retroviral genes associated with drug resistance: 1999-2000 update. *International Antiviral News*, 7(4):46–69, 1999.
13. M. P. Perrone. Averaging/modular techniques for neural networks. In M. A. Arbib, editor, *The Handbook of Brain Theory and Neural Networks*, pages 126–129, Cambridge, Massachusetts, 1999. MIT Press.

AUTOMATING DATA ACQUISITION INTO ONTOLOGIES FROM PHARMACOGENETICS RELATIONAL DATA SOURCES USING DECLARATIVE OBJECT DEFINITIONS AND XML

DANIEL L. RUBIN, MICHEAL HEWETT, DIANE E. OLIVER, TERI E. KLEIN, AND RUSS B. ALTMAN

Stanford Medical Informatics, MSOB X-215,
Stanford, CA 94305-5479 USA
E-mail: rubin@smi.stanford.edu, altman@smi.stanford.edu

Ontologies are useful for organizing large numbers of concepts having complex relationships, such as the breadth of genetic and clinical knowledge in pharmacogenomics. But because ontologies change and knowledge evolves, it is time consuming to maintain stable mappings to external data sources that are in relational format. We propose a method for interfacing ontology models with data acquisition from external relational data sources. This method uses a declarative interface between the ontology and the data source, and this interface is modeled in the ontology and implemented using XML schema. Data is imported from the relational source into the ontology using XML, and data integrity is checked by validating the XML submission with an XML schema. We have implemented this approach in PharmGKB (http://www.pharmgkb.org/), a pharmacogenetics knowledge base. Our goals were to (1) import genetic sequence data, collected in relational format, into the pharmacogenetics ontology, and (2) automate the process of updating the links between the ontology and data acquisition when the ontology changes. We tested our approach by linking PharmGKB with data acquisition from a relational model of genetic sequence information. The ontology subsequently evolved, and we were able to rapidly update our interface with the external data and continue acquiring the data. Similar approaches may be helpful for integrating other heterogeneous information sources in order make the diversity of pharmacogenetics data amenable to computational analysis.

1 Introduction

1.1 Pharmacogenetics and the need to connect diverse data

Connecting genotype and phenotype data is the quest of pharmacogenetics*—a discipline that seeks to understand how inherited genetic differences among people influence their response to drugs. Discovering important relationships between genes and drugs could lead to personalized medicine, where drug therapy is customized according to the genetic constitution of the patient. Thus, there is great interest in rapidly acquiring genotype and phenotype data in many individuals, and clinical trials in the future will routinely collect genotype as well as phenotype information.[1]

Modern experimental methods such as high-throughput DNA sequencing techniques and gene-expression microarrays are contributing detailed genetic and phenotypic information at a rapid rate.[2,3] These abundant and diverse data are a rich source for developing a comprehensive picture of relationships among genes and

*We will consider the term "pharmacogenomics" to be equivalent to "pharmacogenetics."

drugs, but they also create new and complex problems for data integration and interpretation. The plethora of diverse databases having genomic,[4-7] cellular,[8] and phenotype information[9] exacerbates this complexity. Even within a given class of database, such as those containing genetic sequence data, the organization, terminologies, and data models differ.[6,7,10] It is difficult to integrate heterogeneous databases, and standards are not easily adopted.[3]

In response to the need for an integrated resource for pharmacogenetics research, the National Institutes of Health funded the Pharmacogenetics Research Network and Knowledge Base initiative, including the pharmacogenetics knowledge base (PharmGKB).[11] The goal of the PharmGKB project is to develop a knowledge base that can become a national resource containing high quality publicly-accessible pharmacogenetics data that connects genotype, molecular/cellular phenotype, and clinical phenotype information. The challenge for PharmGKB is to integrate a wide scope of genetic and phenotypic information.

1.2 Integrating data in ontologies

To integrate diverse genetic, cellular phenotypic, and clinical information, it is necessary to develop a data model that specifies the pertinent concepts, the semantics of these concepts, and the relationships among them. Because biological understandings evolve, and new types of information continue to emerge after a database design is established, the data model changes. However, when the data model changes, the links to outside sources of data must be updated, which can be a time-consuming process.

Ontologies are models that describe concepts and the relationships among them, combining an abstraction hierarchy of concepts with a semantic network of relationships. Ontologies are flexible and highly expressive, and have been useful for building knowledge bases in biology,[12-15] as well as in the PharmGKB project.[16]

A disadvantage of ontologies is that network and hierarchical data models are very different from flat tabular relational models, and ontologies are not easily integrated with relational data sources; yet the latter are predominant in most biology databases[4,7,17] and experimental laboratories today. This is not a problem when the ontologies are relatively stable, do not change once data acquisition begins, and are manually curated to ensure integrity of the data.[14,15] But while developing the ontology for PharmGKB, it became clear that it will continue to change as our understanding of the concepts and relationships in pharmacogenetics data evolves. Furthermore, many biomedical scientists think about their data in terms of tables (a relational view), not in terms of ontologies. Our challenge, therefore, is to develop a robust interface between relational data acquisition and the PharmGKB ontology. We also sought a method that would automate updating this interface when the ontology changes.

1.3 XML and data exchange

Extensible Markup Language (XML[18]) is useful as a data representation scheme[19-21] and for exchanging data between resources and databases.[22-24] XML provides a general framework for exchanging data between resources because it is extensible, readable by humans, unambiguously parsed by computers, and can be formally defined using a document type definition (DTD) or XML schema. XML schema[25] is a more powerful language for defining XML formats. XML schema is superior to a DTD for expressing constraints because XML schemas specify not only the structure but also the data type of each element and attribute. XML schemas are written in XML, and thus are self-describing and easier to understand than a DTD. XML schemas are also extensible, permitting authors to develop customized constraints.

Data integration requires access to a variety of data sources through a single mediated schema. A major difficulty with integrating data from outside sources is the laborious manual construction of semantic mappings between the source schemas and the mediated schema. It is also necessary to validate the incoming data against the legal ranges for each field in the importing database. If we were to develop an XML schema to serve as the mediating schema, this would address the problem of validating the structure and content of incoming data. But we would still need to have a way of defining the content in the XML schema. Ideally, the XML schema should be defined from information in the PharmGKB ontology. We have developed a method for using an ontology to define a mediating XML schema.

2 Method

2.1 Overview of our method

Our method consists of several components that are shown schematically in Figure 1. The first component is the PharmGKB ontology, which contains the concepts (classes) that describe the domain of pharmacogenetics, and it also models the relationships among the classes (Figure 2, left side). Data are stored in the ontology by creating instances of these classes and storing the data in the appropriate slots (named attributes that store data) associated with the instances. To specify a relationship between instances, we connect them by assigning one instance to the slot value in the other instance. For example, a PCR assay submission has relationships to two instances: a forward PCR primer and a reverse PCR primer (Figure 2, right side). This relationship allows us to specify the particular primers used in a PCR assay.

The second component of our system is the XML schema (Figure 1), which is derived from the ontology and used as an interface between data acquisition and the ontology. The ontology contains a declarative representation of data constraints that are used to define validation constraints on incoming data, and to create the XML schema. This component includes an XML parser that validates incoming XML

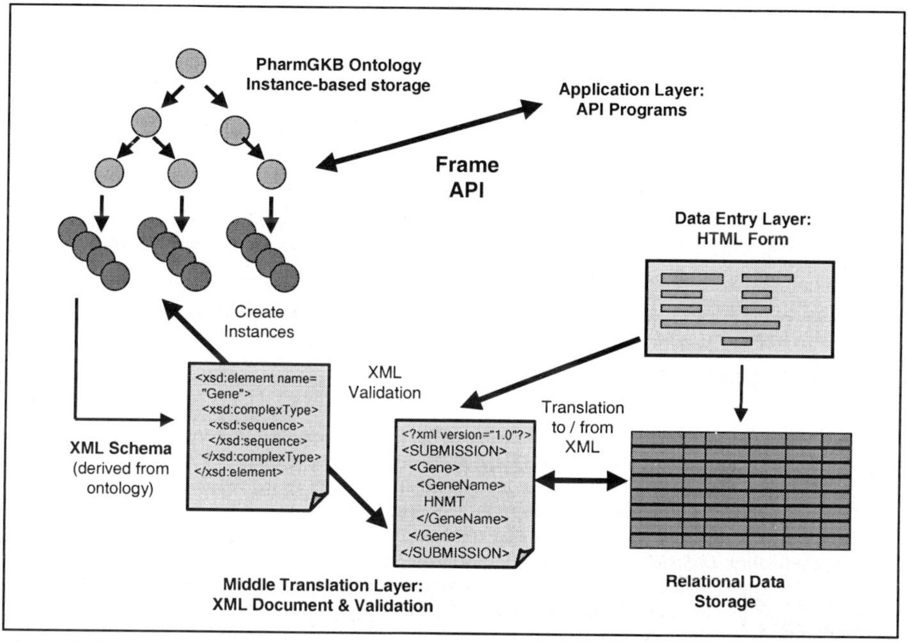

Figure 1. Model for data acquisition in PharmGKB. The PharmGKB ontology (above left) is a network of interrelated classes (upper circles) and instances (lower circles), which store data in slots (not shown). Data to be integrated from an external sources (either web forms or relational schemas) is transmitted in an XML document whose syntax is specified by an XML schema (the latter is derived from the ontology). The data in the XML document is stored in instances in PharmGKB that are created when the document is processed by the XML parsing module.

documents against the schema, creates new instances in the ontology and assigns their slot values from data in the parsed XML document, and creates the necessary links among the instances.

The third component in our method is an XML translator that converts external incoming relational data from an HTML web form into XML. It is also possible to submit data directly from a relational data source if the data are put into an XML document that is valid against the XML schema.

2.2 *Ontology model of genetic information and data validity constraints*

We initially developed and refined the PharmGKB ontology of genetic sequence data through a process of iterative refinement, where we evaluated the data currently available in genetic sequence databases as well as sample data from two study centers in the PharmGKB network, built a preliminary model, and subsequently re-evaluated and revised the ontology. The ontology was developed using the Protégé

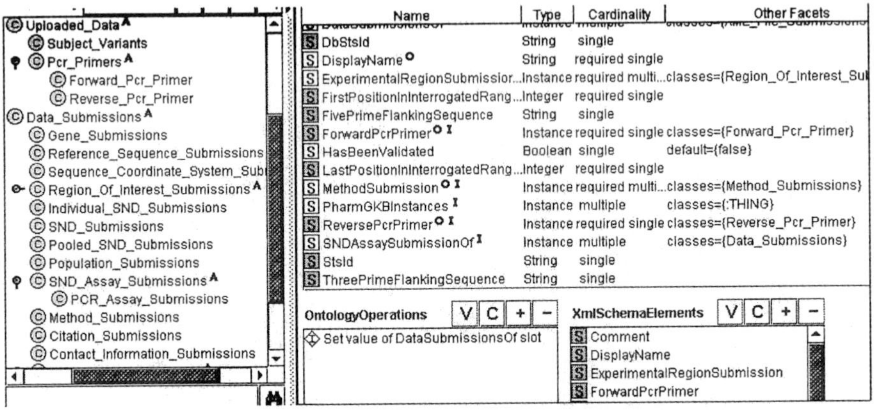

Figure 2. View of part of the PharmGKB model in the Protégé graphical user interface demonstrating both the ontology and constraints that specify the XML schema. The left panel displays the hierarchy of classes making up the ontology. Each class has slots that store data in the ontology. The slots for the "*PCR_Assay_Submissions*" class are shown (right top panel). Constraints on the values for data submitted are stored with the slots that store that data (*MethodSubmission* is an instance, it is required, and is multiple cardinality; *DbStsId* is a string, single-cardinality, and not required; these constraints are stored with each slot as seen in the right top panel). Some of the slots have values that are instances from other classes; for example, the slot "*ForwardPCRPrimer*" has a range of "*Forward_PCR_Primer*", the latter being another class in the ontology (seen in the top of the left panel). Some of the slots in the ontology are used for administrative purposes; those that are used for data acquisition from outside sources are listed in the lower right panel, "XmlSchemaElements" in the order required in the XML document.

suite of tools.[26] Protégé has a graphical user interface for editing ontologies. It is designed for rapidly evolving knowledge bases, which made managing changes to the ontology easier for us. The tool set also made the ontology readily available to application programs that use the ontology.

The ontology includes slots that contain data submitted to PharmGKB ("XML schema slots") and slots that are used for internal purposes in the knowledge base ("administrative slots"). For example, in the *PCR_Assay_Submissions* class (Figure 2), the *StsId* slot contains an STS identifier; the *HasBeenValidated* slot is used internally by PharmGKB to ascertain whether existing instances of *PCR_Assay_Submission*s have passed higher-level data validations.

After the ontology was built, we added these declarative constraints to the ontology (they define the XML schema used to validate data submitted to PharmGKB):

- A list of XML schema slots and the order they are to appear in XML documents
- The required data type for each XML schema slot (integer, string, instance, etc.)
- The cardinality (single or multiple) of each XML schema slot
- A flag indicating if a value is required or optional for each XML schema slot.

Figure 2 (right panel) shows how these constraints are represented in the ontology. Constraints such as data type, cardinality, and whether the data are optional or required are stored with the slot that will contain the corresponding data. Our method uses the following convention for naming XML elements: class and slot names are the same in the XML schema. The names of slots and classes are globally unique in PharmGKB, which prevents naming conflicts. Thus, the ontology in Figure 2 can be interpreted as a declarative representation of an XML schema.

2.3 *Creating the XML schema*

In order to generate an XML schema from the ontology, there must be a convention for naming and organizing the XML elements and attributes. To preserve the desired close connection between the ontology class/slot structure and the XML schema, we organized the XML schema into a set of nested elements having no attributes. The name of the outermost XML element is always the name of a class in the ontology, and each of a series of sub-elements is given the same name as the corresponding slot in that class. The data being submitted is contained within these sub-elements (Figure 3A).

Once the ontology is built and the constraints on data values are declared, the XML schema is sufficiently determined, and it can be compiled directly from the ontology (Figure 3). There is actually more than one way to write equivalent XML

```
A<xsd:element name="PCR_Assay_Submissions">
  <xsd:complexType>
    <xsd:sequence>
     <xsd:element name="Comment" type="xsd:string" minOccurs="0" maxOccurs="1"/>
     <xsd:element name="DisplayName" type="NonblankString" minOccurs="1" maxOccurs="1"/>
      <xsd:element name="ExperimentalRegionSubmission" type="NonblankString"
      minOccurs="1" maxOccurs="1"/>
     <xsd:element ref="ForwardPcrPrimer" minOccurs="1" maxOccurs="1"/>
     <xsd:element ref="ReversePcrPrimer" minOccurs="1" maxOccurs="1"/>
     <xsd:element name="MethodSubmission" type="NonblankString" minOccurs="1" maxOccurs="1"/>
     <xsd:element name="FirstPositionInInterrogatedRange" type="NonblankInteger" minOccurs="1" maxOccurs="1"/>
    <xsd:element name="StsId" type="xsd:string" minOccurs="0" maxOccurs="1"/>
     </xsd:sequence>
  </xsd:complexType>
</xsd:element>
```

```
B<xsd:simpleType name="NonblankString">
  <xsd:restriction base="xsd:string">
    <xsd:minLength value="1"/>
  </xsd:restriction>
</xsd:simpleType>
```

```
C<xsd:simpleType name="NonblankInteger">
    <xsd:restriction base="xsd:integer">
      <xsd:minInclusive value="0"/>
    </xsd:restriction>
  </xsd:simpleType>
```

Figure 3. **A**: An excerpt of the XML schema defining the format and constraints for submitting PCR assay data (not all the element definitions are shown). Note that the name of the outermost element matches the name of a class in the ontology (Figure 1, left panel), while the names of the sub-elements match the names of the XML schema slots in the ontology (Figure 1, right panel). For each of these sub-elements, the data type, cardinality, and required/optional status matches that specified in the ontology (Figure 1, right panel). **B, C**: XML schema defining custom data types: a string that must not be blank (B) and an integer value that is required (C).

schemas, so we cannot say the XML schema is completely determined. In Figure 3B, for example, specification of the required value constraints could have been placed within the XML schema elements that use them, without needing a separate declaration. The alternative ways of writing the XML schema convey the same constraints, so we assert that to the extent that it encodes constraints on content and data validations, the XML schema is sufficiently determined.

For this study, we generated the XML schema by copying the content and data constraints from the ontology directly into the XML schema; we are developing a program to generate the XML schema automatically from the ontology. The current XML schema for PharmGKB is available at http://www.pharmgkb.org/xml-schemas.html.

2.4 *Data acquisition*

Data acquired from external relational data sources must be put into an XML document that uses the syntax specified by XML schema. Generally, this is a direct mapping from columns in a relational table to the appropriate elements in the XML schema. Because the organization of the XML schema parallels the structure of the ontology, creating an XML document involves collecting the data pertaining to each class in the ontology for which data is to be submitted. For example, to submit data for a PCR assay, a single *PCR_Assay_Submissions* element and its sub-elements are

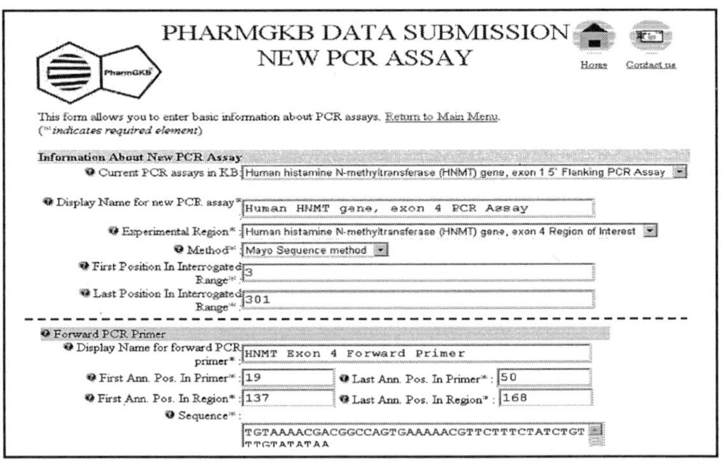

Figure 4. Portions of the HTML web form used for submitting PCR assay data to PharmGKB. Data in the form of strings and numbers are directly entered on the form. Values representing instances in PharmGKB are selected from pull-down menus that list all the relevant objects currently in PharmGKB. If a new object needs to be created, there is either a separate web form, or there are additional fields on the same web form for that purpose. Top half of figure shows the top of the form. Lower half of figure shows fields for entering information for new forward and reverse (not shown) PCR primers.

created (Figure 3A), and all the necessary data can be provided in a flat list that is similar to relational structures. Note that some submissions refer to preexisting instances in PharmGKB (e.g., a *PCR_Assay_Submissions* instance refers to forward and reverse primers). All instances have a slot named "*DisplayName*" which is used as the handle to the instance. If the value of an XML element is an instance in PharmGKB, then the *DisplayName* of the instance is provided. If that instance needs to be created at the time of the submission, the data for that instance is provided either by nesting additional elements in the XML file (as is the case for forward and reverse primers in Figure 3A), or as an additional XML element preceding the one that refers to it. In general, relational data to be input into PharmGKB can be directly mapped to a set of XML elements.

We created a set of HTML data entry forms to simplify the task of entering data into PharmGKB. The types of forms follow the types of classes in the ontology (Figure 1): There are separate forms for submitting genes, sequences, PCR assays, etc. In cases where a submission will create more than one instance in the ontology, all the required fields are supplied on the form. For example, for PCR assay submissions, there are fields for the specifics of the assay (Figure 4, top of figure) as well as for the forward and reverse primers (Figure 4, bottom of figure). The parsing module creates the necessary instances and links them (the instance for the assay is linked to instances for the forward and reverse primers).

2.5 *Ontology evolution and propagating changes*

The challenge of using an interface is updating it when the ontology changes. Our method automates the process of updating the XML schema interface. In our ontology design, there are two kinds of slots (see section 2.2): XML schema slots and administrative slots. If the change in ontology structure affects only administrative slots, then there will be no change in the XML schema or data acquisition. If the change affects XML schema slots, then a new XML schema must be created. Because the XML schema is directly determined from the ontology, changes to XML schema slots in the ontology can be directly transferred to the XML schema. At the time a new version of the ontology is created, a new version of the XML schema can be immediately produced, so new data can be submitted to PharmGKB using the new version of the XML schema. All XML schemas have a required element that stores a version number so that all incoming XML documents can be identified with respect to version of the schema.

3 Evaluation

We have tested our approach by implementing it in a production system. PharmGKB accepts data from multiple study centers. They submit data either through web forms (Figure 4) or by direct submission of XML files. The study

Nucleotide	Location	Sequence Change	WT/WT	WT/Variant	Variant/Variant	Frequency of Variant Allele
-463	5' FR	T→C	36	42	12	0.367
-430	5' FR	G→A	86	4	0	0.022
-376	5' FR	T→C	89	1	0	0.006
314*	Exon 4	C→T	74	15	1	0.094

Comment	Display Name	Position Preceding Variation	Subject Identifier	Subject Variants Of	Variant
	HNMT snp 4 Individual 126745291	271	126745291	HNMT snp 4 Individual	C/T
	HNMT snp 4 Individual 126745304	271	126745304	HNMT snp 4 Individual	T/T

Figure 5. Display of a summary of the polymorphism data in PharmGKB (right) after importing the data (left). Although this display appears similar to the format of the raw data, the data is actually stored as a set of linked instances in the PharmGKB ontology. This is a partial display of the imported data.

centers provided copies of their raw data; this confirmed that they organize and store their data in a tabular format (Figure 5). We tested the ability of our system to acquire their data by requesting one of the study centers to submit the same data in an XML file. Because the data model in the XML schema is similar to a flat file structure and the XML element names describe the data they contain, tabular relational data was directly translated into XML. In an initial draft of their XML submission, some of the required data values were missing—this was discovered when the XML document was validated against the XML schema.

After the omissions were corrected, the file was successfully imported into PharmGKB. We subsequently submitted a query to PharmGKB to view some of the polymorphism data for exon 4 of HNMT (Figure 5). This confirmed that the data had been successfully imported. While PharmGKB reports are tabular, the data is stored in the ontology as a set of interlinked instances; the links are automatically created while parsing the XML document.

Our ontology evolved after we began collecting data; occasionally, a new field was added, or the constraints on a data value type changed. When this happened, we generated a new XML schema after modifying the ontology and published it on the PharmGKB web site. To date, this approach to automating the updating of the XML schema interface has been successful and appears to be scaling well.

4 Discussion

Pharmacogenetics spans a broad range of information which must be synthesized in order to find possible connections between genotype and drug response. Ontologies are useful for modelling complex domains such as pharmacogenetics, and their benefits in bioinformatics have been previously described.[14,15] Most of the existing biology data resources are databases rather than knowledge bases: they describe miscellaneous objects according to the database schema, but no representation of the general concepts and their relationships is given.[27] Because of the large number of diverse concepts and relationships among them in pharmacogenetics, the PharmGKB data model is based on an ontology.[16]

Our work addresses the problem of creating a robust interface between an ontology and data acquisition for that ontology, such that when the ontology changes, the process of updating the interface can be automated. Our approach involves (1) using an XML schema to define the mapping from data acquisition to the ontology, (2) encoding the constraints that define the XML schema directly into the ontology, and (3) designing the XML schema to have related data are grouped together so that users submitting data can map their relational data directly into an XML document.

The approach taken to data integration in databases has been to either create a data warehouse[28] or create mappings between the sources.[29] Static mappings applied to ontologies would be difficult to maintain as the ontology changes. In our method, we establish a "common data model," specified in XML schema, shared by the ontology and an external relational data source (the study centers). Common data models have been used previously with relational databases.[21] We chose XML because it is self-describing, flexible, it can closely reflect ontology models, and it can facilitate semantic interoperability.[30]

Data acquisition for an ontology is usually done by a user who creates instances and fills in their slot values directly.[14,15] Collecting data as instances makes sense if one has an intimate understanding of the ontology and the user's model of the data is instance-based. But scientists who collect experimental data usually think in terms of tables, not in terms of instances in an ontology. When submitting data on PCR assays, the primers are part of the information about the assay; in the PharmGKB ontology, the primers are separate data objects. It is simpler for the user to submit data about primers and PCR assays together, rather than submitting primer information before sending the other data about the PCR assay.

Our solution is to provide an XML schema interface to the ontology that maps directly to the experimental data being collected. Our XML schema nests elements from classes having related information beneath the main submission class. For example, for PCR assays, the elements related to primer submissions are nested beneath those for PCR assay submissions. In this way, the user has a submission interface to PharmGKB that looks relational (Figure 4) while preserving the information required to store the data in a rich hierarchical ontology. We are not aware of a similar approach taken for integrating ontologies with external information.

The benefit of our method is that we can automate the process of updating our interface to data acquisition when the ontology changes—we simply update the XML schema. Because the XML schema is defined from metadata in the ontology, changes to the ontology can be immediately ported into a new version of the XML schema. We are also developing software to make this happen automatically. The user submitting data will still have to update the mappings from their data to the new XML schema, but the XML schema interface resembles a tabular representation that is closer to their own than the ontology.

Our evaluation to date is preliminary. We have shown our approach is feasible and has been successful with real data from one of the study centers. We plan to perform a more complete evaluation of our methodology, a task that will be possible as more study centers begin submitting data to PharmGKB.

A limitation of our method is that changes to XML schemas are generally not backward compatible with older XML documents that were created according to a previous version of the XML schema. This means that older XML documents that have previously been processed cannot be re-processed under the new schema. In addition, our method requires all users to keep current with the latest version of the XML schema. These limitations are typical of any system that declares a standard interface between two different components. However, the benefits of having a standard interface generally outweigh these limitations. In particular, the benefit of being able to integrate outside information in PharmGKB is vital to the project. Furthermore, a new version of the XML schema is automatically defined as the ontology changes, so the effort of maintaining a current interface is much less than the work that would be involved in manually establishing new mappings between the data and the ontology as the ontology changes.

In conclusion, we have developed a method for integrating an ontology of pharmacogenetics with data input from external sources. Our method allows us to preserve a relational view of the data in creating our interface, and it uses XML that keeps the data in a clear, human-readable format. Our approach appears promising with respect to being able to preserve the link between the ontology and external sources even as the ontology evolves and changes. We will use this method for integrating PharmGKB with other resources, and our methods could be applicable to data integration for ontologies in other domains.

Acknowledgments

The PharmGKB is financially supported by grants from the National Institute of General Medical Sciences (NIGMS), Human Genome Research Institute (NHGRI) and National Library of Medicine (NLM) within the National Institutes of Health (NIH) and the Pharmacogenetics Research Network and Stanford University's Children's Health Initiative. This work is supported by the NIH/NIGMS Pharmacogenetics Research Network and Database (U01GM61374). The authors also wish to thank Katrina Easton for her assistance in implementing this work.

References

1.The SNP Consortium Ltd., (2000). Available at http://snp.cshl.org/news/user_survey.pdf.
2.P.O. Brown & D. Botstein, *Nature Genetics* **21**, 33-7 (1999).
3.N. Williams, *Science* **275**, 301-2 (1997).
4.S.T. Sherry et al., *Nucleic Acids Research* **29**, 308-11 (2001).

5.M.P. Skupski et al., *Nucleic Acids Research* **27**, 35-8 (1999).
6.S.I. Letovsky et al., *Nucleic Acids Research* **26**, 94-9 (1998).
7.D.A. Benson et al., *Nucleic Acids Research* **28**, 15-8 (2000).
8.D. Jacobson & A. Anagnostopoulos, *Trends in Genetics* **12**, 117-118 (1996).
9.A. Hamosh et al., *Human Mutation* **15**, 57-61 (2000).
10.C. Harger et al., *Nucleic Acids Research* **28**, 31-32 (2000).
11.National Institute of General Medical Sciences, National Institutes of Health, (2001). Available at http://www.nigms.nih.gov/funding/pharmacogenetics.html.
12.R. Stevens, C.A. Goble & S. Bechhofer, *Briefings in Bioinformatics* **1**, 398-414 (2000).
13.P.G. Baker et al., *Bioinformatics* **15**, 510-20 (1999).
14.P.D. Karp et al., *Nucleic Acids Research* **26**, 50-3 (1998).
15.R.B. Altman et al., *IEEE Intelligent Systems & Their Applications* **14**, 68-76 (1999).
16.T.E. Klein et al., *The Pharmacogenomics Journal* (2001 -- in press).
17.S. Schulze-Kremer, "Integrating and Exploiting Large-Scale, Heterogeneous and Autonomous Databases with an Ontology for Molecular Biology" in *Molecular Bioinformatics, Sequence Analysis - The Human Genome Project* (ed. H. Lim) 43-56 (Shaker Verlag, Aachen, 1997).
18.T. Bray, J. Paoli & C.M. Sperberg-McQueen, (1998). Available at http://www.w3c.org/TR/1998/REC-xml-19980210.html.
19.S. Staab et al., AIFB, University of Karlsruhe, Technical Report 401 (2000).
20.S. Bowers & L. Delcambre, ECDL 2000 Workshop on the Semantic Web, 2000). Available at http://www.ics.forth.gr/proj/isst/SemWeb/proceedings/session1-1/html_version/.
21.D. Gardner et al., *Journal of the American Medical Informatics Association* **8**, 17-33 (2001).
22.F. Achard, G. Vaysseix & E. Barillot, *Bioinformatics* **17**, 115-125 (2001).
23.G.C. Xie et al., *Bioinformatics* **16**, 288-289 (2000).
24.P. Tarczy-Hornoch et al., *Journal of the American Medical Informatics Association* **7**, 267-76 (2000).
25.XML Schema Working Group, (2001). Available at www.w3.org/TR/xmlschema-1/; www.w3.org/TR/xmlschema-2/.
26.M.A. Musen et al., Proceedings of the Conference on Intelligent Information Processing (IIP 2000) of the International Federation for Information Processing World Computer Congress (WCC 2000), Beijing (2000).
27.C.D. Hafner & N. Fridman, *Proceedings of the International Conference on Intelligent Systems for Molecular Biology; ISMB* **4**, 78-87 (1996).
28.O. Ritter et al., *Computers & Biomedical Research* **27**, 97-115 (1994).
29.T. Etzold, A. Ulyanov & P. Argos, *Methods in Enzymology* **266**, 114-28 (1996).
30.S. Decker et al., *IEEE Internet Computing* **4**, 63-74 (2000).

ON A FAMILY-BASED HAPLOTYPE PATTERN MINING METHOD FOR LINKAGE DISEQUILIBRIUM MAPPING

SHUANGLIN ZHANG
Department of Epidemiology and Public Health, Yale University School of Medicine, New Haven, CT 06520, USA
Department of Mathematical Science, Michigan Technological University, Houghton, MI 49931, USA
Email: shuzhang@mtu.edu

KUI ZHANG
Department of Mathematics, University of Southern California, Los Angeles, CA 90089, USA
Email: kuizhang@hto.usc.edu

JINMING LI AND HONGYU ZHAO
Department of Epidemiology and Public Health, Yale University School of Medicine, New Haven, CT 06520, USA
Email: {jinming.li, hongyu.zhao}@yale.edu

Linkage disequilibrium mapping is an important tool in disease gene mapping. Recently, Toivonen et al. [1] introduced a haplotype mining (HPM) method that is applicable to data consisting of unrelated high-risk and normal haplotypes. The HPM method orders haplotypes by their strength of association with trait values, and uses all haplotypes exceeding a given threshold of strength of association to predict the gene location. In this study, we extend the HPM method to pedigree data by measuring the strength of association between a haplotype and quantitative traits of interest using the Quantitative Pedigree Disequilibrium Test proposed by Zhang et al. [2]. This family-based HPM (F-HPM) method can incorporate haplotype information across a set of markers and allow both missing marker data and ambiguous haplotype information. We use a simulation procedure to evaluate the statistical significance of the patterns identified from the F-HPM method. When the F-HPM method is applied to analyze the sequence data from the seven candidate genes in the simulated data sets in the 12th Genetic Analysis Workshop, the association between genes and traits can be detected with high power, and the estimated locations of the trait loci are close to the true sites.

Key words: Linkage disequilibrium mapping, data mining, quantitative trait, extended pedigree

1 Introduction

Linkage disequilibrium mapping (LDM) is a powerful method for the identification of disease genes. With the completion of the Human Genome Project, many genetic markers can be identified and genotyped within a very short physical distance, and LDM methods that use a set of markers simultaneously through the consideration of haplotypes across a set of markers may be more powerful than the methods that examine each individual marker separately. Various statistical methods have been proposed to locate disease mutation site based on LD around a disease susceptibility (DS) gene [3, 4, 5, 6, 7, 8]. The power of these methods, as well as their ability to identify the correct position of the DS gene, has been shown to be better than the traditional method based on the LD of two markers. However, most of these methods have been developed under explicit assumptions on the mode of inheritance of the disease and the population history of the studied population, and the effects of violations of these assumptions on the analysis of real data are not well understood. Recently, Toivonen et al. [1] proposed haplotype pattern mining (HPM), a technique that uses data mining methods in LD-based gene mapping. The HPM method aims to identify recurrent haplotype patterns and the haplotype patterns are sorted by the strength of their association to the disease. This method, applicable to data consisting of independent high-risk and normal haplotypes, works with a non-parametric statistical model without any genetic model assumption and allows for missing and erroneous markers within the haplotypes. Toivonen et al. [1] showed that the localization power of the method is high, even when the association is weak. However, there are three limitations for the method described by Toivonen and colleagues. First, related individuals cannot be analyzed in the same analysis because their method is only applicable to case-control data. Secondly, the method is only applicable to binary trait. Thirdly, their approach is purely descriptive and the statistical significance of the observed patterns cannot be assessed.

In this article, we introduce a Family-based Haplotype Pattern Mining (F-HPM) method that extends the HPM method. To allow simultaneous use of related individuals with quantitative trait from an extended pedigree, we employ the Quantitative Pedigree Disequilibrium Test (QPDT) statistic [2] to measure the strength of association between a haplotype and a quantitative trait. We then use a simulation method to assess the statistical significance for the observed patterns. When we apply the F-HPM method to analyze the sequence data of the seven candidate genes from the simulated data sets in the 12th Genetic Analysis Workshop (GAW12), the estimated locations of the trait loci are very close to the true sites and the genes having association with certain traits can be detected with high power.

2 Methods

The idea behind the F-HPM method, as well as the HPM method, is that haplotype patterns close to the DS locus are likely to have stronger association than haplotypes further away. Based on pedigree data that includes genotypes at a set of markers and the quantitative traits with possible missing values of the individuals, there are four steps in the F-HPM method: (1) reconstruct each individual's haplotypes across a set of markers and define the haplotype patterns; (2) for each haplotype pattern P, calculate the QPDT statistic [2] to detect if there is a *strong* association between P and a quantitative trait; (3) calculate the proportion of strongly associated haplotypes around a candidate locus L; and (4) use a simulation procedure to estimate the statistical significance for the observed association. We describe these four steps in detail in the following discussion.

2.1 Haplotype inference and haplotype pattern

Even with large pedigrees, we may not be able to infer the haplotypes of the individuals unambiguously, especially for the case that the haplotypes are across a large number of markers and there are missing genotype data for some individuals in the pedigree. For uncertainties in haplotype inferences, one method would be to estimate the probabilities for all compatible haplotypes. However, such probabilities depend on many parameters related to the population structure under study, as well as the parameters related to the disease model that we usually have little knowledge about. In our haplotype inference, we use the program HAPLORE (unpublished results; http://bioinformatics.med.yale.edu) to reconstruct each individual's haplotypes that include possible ambiguous data at certain markers. The algorithms implemented in HAPLORE are similar to those discussed by Wijsman [9]. Suppose that chromosome region we examine consists k markers, we denote a haplotype of an individual by a vector $H = (b_1,\dots ,b_k)$, where b_i is either an allele at marker i if the haplotypes can be reconstructed unambiguously at this marker, or is a symbol "*" if the haplotypes cannot be reconstructed unambiguously.

We example the association by looking for haplotype pattern that consists of a set of nearby markers, not necessarily consecutive ones. A haplotype pattern P around marker L is defined as a vector $P=(p_{L-l},\dots, p_{L-1}, p_L, p_{L+1}\dots,p_{L+r})$, where each p_i is either an allele of the ith marker or the "don't care" (missing symbol) "*" , however, the candidate marker L cannot have a missing symbol. A haplotype pattern P occurs in a given haplotype $H = (b_1,\dots ,b_k)$ if $1\leq L-l< L+r\leq k$ and $p_i =b_i$

or $p_i = *$ for all i, $L\text{-}l \leq i \leq L+r$. We use three parameters, the candidate marker L, the number of markers included in haplotype pattern N and the maximum number of the markers with missing data or "don't care" M to control the haplotype pattern. For example, for a given haplotype vector (2, 3, 5, 7, *, 6, 5, 8, 9, 1), all the haplotype patterns with parameters L=6, N=3 and M=2 that occur in this haplotype are (7, *, 6), (*, *, 6), (*, 6, 5), (*, 6, *), (6, 5 ,8), (6, *, 8), (6, 5, *) and (6, *, *).

2.2 Quantitative pedigree disequilibrium test (QPDT)

The QPDT is a TDT type test that allows quantitative traits and arbitrary pedigree structures [2]. The QPDT uses the following three types of nuclear families in an extended pedigree:

(I) Families with both parents available and at least one parent being heterozygous at the marker being studied.

(II) Families with one available parent and one or more offspring where all the offspring have the same genotypes.

(III) Families with at most one available parent and multiple offspring where at least two siblings have different genotypes.

When a haplotype pattern P is studied, we treat P as one allele, denoted by A, and the other haplotype patterns as another allele, denoted by B. Let X_i denote the number of A alleles carried by the ith child and $\bar{X}$ denote the mean number of A alleles among all the offspring in this nuclear family. For the first type of nuclear families, define X_{im} = 1 (or -1) if the mother is heterozygous and transmits allele A (or B) to the ith child, and X_{im} = 0 if the mother is homozygous. We similarly define X_{if} for the father. For the second type of nuclear families, we only consider offspring-parent pairs with genotypes (BA, BB) or (AA, AB), and offspring-parent pair with genotypes (BB, BA) or (BA, AA). The first genotype in the bracket is the offspring's genotype and second genotype in the bracket is the available parent's genotype. We define $X_{(1)}$ = 1 if the genotypes for the offspring-parent pair are (AB, BB) or (AA, AB), $X_{(1)}$ = -1 if the genotypes for the parent-offspring pair are (BB, AB) or (AB, AA), and $X_{(1)}$ = 0 for other genotypes of the offspring-parent pair. For the details on the analysis of the second type of nuclear families, see Sun et al. [10, 11]. Define random variables U_1, U_2, and U_3 as the covariance between the trait values and the genotypes for the first, second, and third types of nuclear families:

$$U_1 = \sum_{i=1}^{t} (Y_i - \bar{Y})(X_{im} + X_{if}),$$

$$U_2 = \sum_{i=1}^{t} (Y_i - \bar{Y}) X_{(1)}$$

and

$$U_3 = \sum_{i=1}^{t} (Y_i - \bar{Y})(X_i - \bar{X}),$$

where t is the number of offspring in a nuclear family, and Y_i is the trait value of the ith child for a quantitative trait of interest. Under the null hypothesis of no linkage or no linkage disequilibrium, $E(U_1) = E(U_3) = 0$. However, under null hypothesis of no linkage or no linkage disequilibrium, $E(U_2)$ is equal to 0 under one of the following two conditions:

A1. Males and females with the same genotype at the marker locus have the same mating preference.
A2. Father and mother in each nuclear family are equally likely to be missing given that one parent is missing.

Even if both of the above two assumptions are violated, we can modify U_2 such that $E(U_2) = 0$ under the null hypothesis [11]. In what follows, we assume $E(U_2) = 0$.

For an extended pedigree, let n_1, n_2, and n_3 denote the number of the first, second, and third types of nuclear families, respectively. Define

$$D = \frac{1}{n_1 + n_2 + n_3} \left(\sum_{j_1=1}^{n_1} U_{j_1,1} + \sum_{j_2=1}^{n_2} U_{j_2,2} + \sum_{j_3=1}^{n_3} U_{j_3,3} \right),$$

where $U_{j_1,1}$, $U_{j_2,2}$, and $U_{j_3,3}$ are the covariances between the trait values and the genotypes for the j_kth nuclear family of the kth type. For n independent extended pedigrees, let D_l denote the random variable D defined for the lth extended pedigree, then under the null hypothesis of no linkage or no linkage disequilibrium, $E(D_l) = 0\ (l = 1, 2, ..., n)$, $\mathrm{Var}(\sum_{l=1}^{n} D_l) \approx \sum_{l=1}^{n} D_l^2$ and the test statistic $T = \frac{\sum_{l=1}^{n} D_l}{\sqrt{\sum_{l=1}^{n} D_l^2}}$

is asymptotically normally distributed with mean 0 and variance 1. This test statistic is the QPDT introduced by Zhang et al. [2].

2.3 Measure of degree of association at a locus L

For a marker location L, we measure the degree of association between the haplotypes nearing L and the trait of interest as follows. Let N denote the number of markers we include in a haplotype pattern (including locus L), and M denote the maximum number of missing markers allowed in a haplotype pattern. Define Ω to be the set of haplotype patterns with respect to parameters N, M, and marker location L. We say that haplotype pattern P is "strongly associated" with the trait if $|T| \geq x$, where T is the QPDT statistic and x is an association threshold. In our analysis, we set $x = 1.96$ so that a strong association is approximately equivalent to setting statistical significance level at 5% for each haplotype pattern P. Intuitively, haplotype patterns near the trait locus are likely to have stronger association than haplotype patterns further away from the trait locus. Therefore, the trait locus is likely to be located at the site, where there is a high proportion of strongly associated haplotype patterns [1].

For a given marker L, we compute the frequency of strongly associated haplotype patterns around this marker as

$$f(L) = \frac{\text{The number of strongly associated haplotype patterns in } \Omega}{\text{The number of haplotype patterns in } \Omega}. \qquad (1)$$

For each marker L, we use $f(L)$ as a measure of the degree of evidence for association. If we assume that a trait locus exists in the region being examined, we can predict the location of the trait locus to be close to the markers with higher $f(L)$ values. In our analysis, we estimate the trait locus at the marker that gives the largest value of $f(L)$.

2.4 Statistical significance assessment of the observed measure of association

To test the null hypothesis that the region being examined is not associated with the trait of interest, we use $T_{\max}=\max_L f(L)$ as the test statistic for the null hypothesis. We adopt the simulation procedure proposed by Monks and Kaplan [12] to evaluate the statistical significance of the test statistic, and note that simply permuting trait values among the individuals is not a valid procedure. We describe the procedure in the following. For the first type of the nuclear families, under the null hypothesis of no association, the probability that a heterozygous parent transmits marker allele A and B with equal probabilities. Thus, if the mother is heterozygous, then, X_{im} is equally likely to be 1 and −1. If there is only one child, then our simulation procedure randomly assign X_{im} as being equal to 1 or −1 with equal probability. Complications arise when more than one child in the family is available. These complications are a result of linkage between the marker and the trait locus. In the presence of linkage, children with shared marker alleles will have similar quantitative traits, even in the absence of association. This can be taken into account by simultaneous randomization of X_{im} (and, similarly, of X_{if}), for heterozygous parents across the sibship. Let $U_{1m} = \sum_{i=1}^{t}(Y_i - \bar{Y})X_{im}$ and $U_{1f} = \sum_{i=1}^{t}(Y_i - \bar{Y})X_{if}$.

This procedure is equivalent to randomizing the sign of U_{1m} and the sign of U_{1f} with equal probability and then calculate the value $U_1 = U_{1f} + U_{1m}$. Similar procedures are used to simulate genotypes under the null hypothesis for the other two types of families that is equivalent to randomization of the sign of U_2 and the sign of U_3 with equal probability. For each simulated data set, we randomly give the sign of U_{1m}, U_{1f}, U_2, and U_3, and recalculate test statistic T and then $f(L)$ and $T_{\max}$. Note that $U_1 = U_{1m} + U_{1f}$. We can then derive the empirical distribution of $T_{\max}$ based on the calculated test statistics through a set of simulated data sets.

3 Results

3.1 Data Sets

We evaluate the performance of the proposed F-HPM method using the sequence data from the seven candidate genes ($G_1,\ldots,G_7$) in the simulated data sets in GAW12 for the isolated population scenario. Two of the seven candidate genes affect one or two of the five quantitative traits ($Q_1,\ldots,Q_7$). Table 1 summarizes the relationships between the genes and the traits and the sites of the functional alleles. There are multiple functional alleles within G_2, with changes in either regulatory elements or in the first or second base-pair of a codon, leading to amino acid substitutions. The simulation data set in GAW12 contains 50 replications for the isolated population. For each replication, the data consists of 23 pedigrees with 1497 individuals in total.

Table 1. The relationships between the candidate genes and the quantitative traits.

Gene	Length (kb)	Influence on the quantitative trait(s)	Location(s) of the functional allele(s)
G_1	20	None	None
G_2	13	Q_5	Multiple sites
G_3	16	None	None
G_4	20	None	None
G_5	17	None	None
G_6	17	Q_1 and Q_2	5782
G_7	20	None	None

3.2 Results on the associations between candidate genes and traits

We apply the F-HPM method to analyze associations between the seven candidate genes and the five quantitative traits. All of the 50 replications of the simulated data sets in the isolated population are used to investigate the false-positive rates and the power of the F-HPM method. Only polymorphic markers whose major allele frequency is less than 95% are used. In the F-HPM method, we set the association threshold at $x = 1.96$, set the maximum number of markers in a haplotype pattern (including marker L) to be 7 ($N = 7$), and allow up to 6 markers with missing information (denoted by "*") in a haplotype pattern ($M = 6$). For example, the haplotype may contain locus L and 6 markers on either side of L. We vary L from the first polymorphism to the last one in the entire gene, and calculate $f(L)$ for every marker and $T_{\max}$ for the gene. For each gene and each replication, we simulate 200 samples to evaluate the statistical significance of the observed test statistic. The power comparisons between the F-HPM method and two other methods, the QPDT [2] and QST [12] (a score test of linkage for quantitative traits using haplotypes in extended pedigrees and using hierarchical clustering method to group the haplotypes into two groups), based on these 50 replications are summarized in Table 2. Because Q_3 and Q_4 have no associations with any of the genes, they are not shown in Table 2.

According to Table 1, G_1, G_3, G_4, G_5, and G_7 have no association with all the quantitative traits, G_2 is associated with Q_5 only and G_6 is associated with Q_1 and Q_2. It can be seen from Table 2 that the false positive rate of our method is within the 95% confidence interval of the nominal level, i.e. 5%. For the power comparison, the QPDT testing one marker at a time has the lowest power in all the cases. For testing the association and linkage between G_1 with single mutation and the traits Q_1 and Q_2, the two haplotype methods F-HPM and QST have similar power. However, for detecting G_2 with multiple functional mutations, the F-HPM is more powerful.

Table 2. The power comparisons of the three tests: F-HPM and QPDT (after Bonferroni correction) for the associations and QST for linkage between the seven candidate genes and the five quantitative traits at statistical significance level 5%. There are three true gene-trait associations and the power for these three pairs is denoted in bold face font.

Gene	Q_1			Q_2			Q_5		
	F-HPM	QPDT	QST	F-HPM	QPDT	QST	F-HPM	QPDT	QST
1	0.04	0	0.08	0.10	0	0.14	0.06	0	0.14
2	0.10	0	0.02	0.08	0	0.04	**0.68**	**0.16**	**0.54**
3	0.00	0	0.02	0.02	0	0	0.10	0.02	0.02
4	0.08	0	0	0.02	0	0.10	0.02	0	0
5	0.04	0	0	0.06	0.02	0.02	0.08	0	0.08
6	**0.92**	**0.64**	**0.98**	**0.82**	**0.26**	**0.80**	0.04	0	0.04
7	0.08	0	0.02	0.02	0	0.06	0.08	0	0.02

For every candidate gene, as we vary the marker location along the gene, we obtain a curve of $f(L)$ for each replication. In Figure 1(a), we present the $f(L)$ curves for the association test between G_6 and Q_1 using the first five replicated samples. Although there are variations, the highest peak is near the true site of the functional allele. In Figure 1 (b), we present the $f(L)$ curves for the association test between G_2 and Q_5. Because there are multiple functional polymorphisms in G_2, the signal is not as strong as that in Figure 1 (a).

We estimate the trait locus at the marker L with the highest $f(L)$. The histograms for the estimated locations for Q_1 in G_6 and Q_5 in G_2 for those replications in which the trait value has significant associations with the gene are given in Figure 2. For G_6, the estimated locations of the trait locus for Q_1 are at site 6805 for 32 replications out of 46 significant samples. This estimate is ~ 1kb from the true location site 5782. For G_6, the estimated locations of the trait locus for Q_2 have a similar pattern. For G_2, the estimated locations of the trait locus for Q_5 are in sub-regions around sites 715, 4977, and 12411. The three sites are all within the regulatory regions where the true functional alleles are. Therefore, even there are multiple functional alleles in this gene, the F-HPM method is able to identify the locations of these functional alleles.

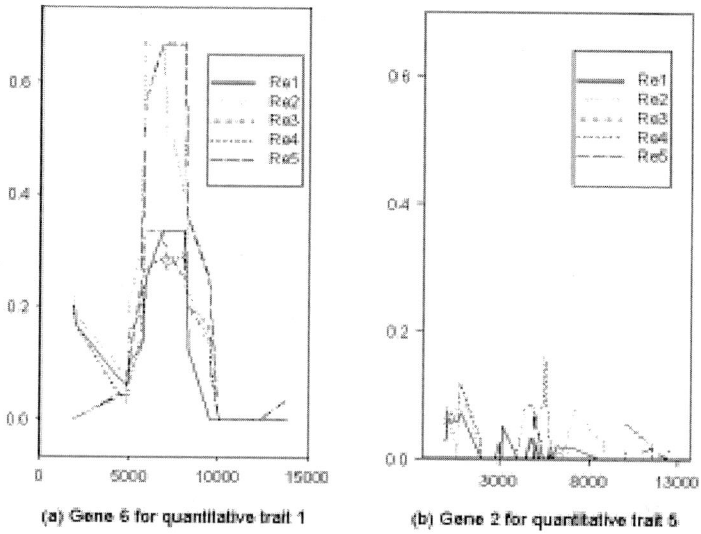

Figure 1. Frequency of strongly associated (with Q_1 for G_6 and with Q_5 for G_2) haplotype patterns versus polymorphic site location.

4 Discussion

We have proposed the F-HPM method to allow simultaneous use of related individuals with quantitative trait from an extended pedigree. This method works with a non-parametric statistical model without any genetic model assumption and allows for missing and erroneous markers within the haplotypes. It tests the association between a set of markers and the quantitative traits and predicts the location of the DS gene at the same time. When we apply the F-HPM method to analyze the sequence data of the seven candidate genes from the simulated data sets in GAW12, the estimated locations of the trait loci are very close to the true sites and the genes having association with certain traits can be detected with higher power comparison with the QPDT [2], the single marker method. For detecting genes with multiple functional mutation, the F-HPM method is more powerful than the QST, another haplotype method.

In the application of the F-HPM method, we need to specify the number of markers included in a haplotype pattern, the number of missing data markers allowed, and the association threshold. The optimal choices of these parameter values need further study, although the method seems to be quite robust with respect to the parameter values for the data analyzed here. From the applications of the F-HPM method to the simulated data sets, we feel that this approach represents a promising method to map complex disease genes.

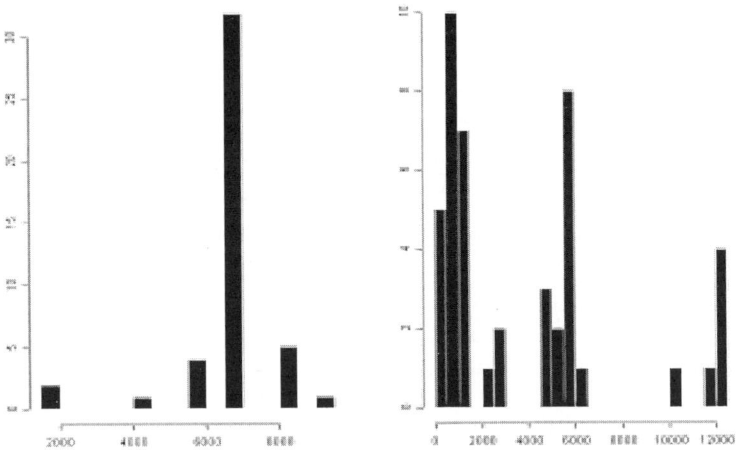

(a) Gene 6 for Trait 1 (b) Gene 2 for Trait 5

Figure 2. Histograms of estimated locations for Q_1 versus G_6 and Q_5 versus G_2.

5 Acknowledgements

Supported in part by grants GM59507 and HD36834 from NIH. We thank Dr. MacCluer for providing us the simulated data from GAW12. GAW is supported by NIH grant GM31575 from NIGMS.

References

1. Toivonen H. T. T., Onkamo P., Vasko K., Ollikainen V., Sevon P., Mannila H., Herr M. and Kere J., Data mining applied to linkage disequilibrium. *Am. J. Hum. Genet.* **67** (2000) pp 133–145.

2. Zhang S., Zhang K., Li J., Sun F. and Zhao H., Test of linkage and association for quantitative traits in general pedigree: The quantitative pedigree disequilibrium test. *Genet. Epdemiol.* (2001) in press. Edited by Wijsman E. M., Almasy L., Amos C. I., Borecki I., Falk C. T., King T. M., Martinez M. M., Meyers D., Neuman R., Olson J. M., Rich S., Spence M. A., Thomas D. C., Vieland V. J., Witte J. S. and MacCluer J. W. Analysis of complex genetic traits: Applications to asthma and simulated data.
3. Terwilliger J. D., A powerful likelihood method for the analysis of linkage disequilibrium between trait loci and one ore more polymorphic marker loci. *Am. J. Hum. Genet.* **56** (1995) pp 777–787.
4. Devlin B., Risch N. and Roeder K., Disequilibrium mapping: composite likelihood for pairwise disequilibrium. *Genomics* **36** (1996) pp 1–16.
5. Lazzeroni L. C., Linkage disequilibrium and gene mapping: an empirical least-squares approach. *Am. J. Hum. Genet.* **62** (1998) pp 159–170.
6. McPeek M. S. and Strahs A., Assessment of linkage disequilibrium by the decay of haplotype sharing, with application to fine-scale genetic mapping. *Am. J. Hum. Genet.* **65** (1999) pp 858–875.
7. Service S. K., Temple Lang D. W., Freimer N. B. and Sandkuijl L. A., Linkage-disequilibrium mapping of disease genes by reconstruction of ancestral haplotypes in founder populations. *Am. J. Hum. Genet.* **64** (1999) pp 1728–1738.
8. Zhang S. and Zhao H., Linkage Disequilibrium Mapping in populations of variable size using the decay of haplotype sharing and a stepwise-mutation model. *Genet. Epidemiol.* **19** (2000) pp S99–S105.
9. Wijsman E. M., A deductive method of haplotype analysis in pedigrees. *Am. J. Hum. Genet.* **41** (1987) pp 356–373.
10. Sun F. Z., Flanders W. D., Yang Q. and Khoury M. J., Transmission disequilibrium test (TDT) when only one parent is available: the 1-TDT. *Am. J. Epidemiol.* **150** (1999) pp 97–104.
11. Sun F. Z., Flanders W. D., Yang Q. and Zhao H., Transmission/disequilibrium test for quantitative traits. *Ann. Hum. Genet.* **64** (2000) pp 555–565.
12. Monks S. A. and Kaplan N. L., Removing the sampling restrictions from family-based test of association for a quantitative-trait locus. *Am. J. Hum. Genet.* **66** (2000) pp 576–592.
13. Li J, Wang D, Dong J, Jiang R, Zhang K, zhang S. Zhao H, Sun F., The Power of Transmission Disequilibrium Tests for Quantitative. *Genet. Epdemiol.* (2001) in press. Edited by Wijsman E. M., Almasy L., Amos C. I., Borecki I., Falk C. T., King T. M., Martinez M. M., Meyers D., Neuman R., Olson J. M., Rich S., Spence M. A., Thomas D. C., Vieland V. J., Witte J. S. and MacCluer J. W. Analysis of complex genetic traits: Applications to asthma and simulated data.

GENOME-WIDE ANALYSIS AND COMPARATIVE GENOMICS

INNA DUBCHAK
Lawrence Berkeley National Laboratory,
MS 84-171, Berkeley, CA 94720
ildubchak@lbl.gov

LIOR PACHTER
Department of Mathematics
UC Berkeley
Berkeley, CA 94720
lpachter@math.berkeley.edu

LIPING WEI
Nexus Genomics, Inc.
390 O'Connor St.
Menlo Park, CA 94025
wei@nexusgenomics.com

One of the key developments in biology during the past decade has been the completion of the sequencing of a number of key organisms, ranging from organisms with short genomes, such as various bacteria, to important model organisms such as *Drosophila Melanogaster*, and of course the human genome. As of February 2001, the complete genome sequences have been available for 4 eukaryotes, 9 archaea, 32 bacteria, over 600 viruses, and over 200 organelles (NCBI, 2001). The "draft" human sequence publicly available now spans over 90% of the genome, and finishing of the genome should be completed shortly. The large amount of available genomic sequence presents unprecedented opportunities for biological discovery and at the same time new challenges for the computational sciences. In particular, it has become apparent that comparative analyses of the genomes will play a central role in the development of computational methods for biological discovery. The principle of comparative analysis has long been a leitmotif in experimental biology, but the related computational challenge of large-scale comparative analysis leads to many interesting algorithmic challenges that remain at the forefront of computational biology research.

The papers in this track represent a cross-section of the ongoing efforts to bridge the gap between comparative computational genomics and biology, and have been selected both for their algorithmic content, and the relevance of their application.

We have also included papers on whole genome analyses, that describe exciting new discoveries that have been made possible only recently with the availability of so much genomic sequence. The breadth of applications that have been addressed is testimony to the explosion of activity in the field of comparative genomics (more generally computational genomics), and the success that it is already heralding in biology. It is clear that there is a lot of low hanging fruit.

The computational perspective of biology has traditionally been that one ought to be able to sequence a genome, then annotate the genes and regulatory elements thereby obtaining the proteins and some understanding of their regulation. Subsequently, solution of the protein folding problem, it was hoped, would elucidate the structure and hence function of the proteins, and then cures for diseases would follow. QED.

Of course every step along the way has proven to be nontrivial, in a highly nontrivial way! Even the sequencing of genomes has proven to be difficult, and the paper by Mulyukov and Pevzner addresses one of the core problems in the new field of assembly, namely the resolution of repeats. Rather than simply suggesting different heuristics and hacks for dealing with repeats in the context of an algorithm, they introduce a beautiful twist to the problem by providing an algorithmic solution for developing experimental assays for resolution of repeats. Hopefully computational biologists will take note that computational methods can interact with experimental biology in a very direct way.

Given that one has genomic sequence at hand, and as we have mentioned there is already plenty of it, the next step is to annotate the sequence. The paper by Kel-Margoulis et al. looks at the important (and difficult) problem of detecting regulatory elements in sequences. They have developed a method of looking for clusters of regulatory elements in genes that are functionally related. Also related to regulatory site detection is the paper by Sze, Gelfand and Pevzner on finding weak motifs in DNA sequences. The annotation of splice sites, while easier than that of regulatory elements, remains an unsolved problem, and the intriguing paper by Patterson, Yasuhara and Ruzzo discusses the possibility of a relationship between pre-mRNA structure and splice sites that might help in splice site detection. Biologists, lacking much hard experimental evidence, have debated the connection between structure of RNA and splicing for some time, so a fresh computational analysis is welcome and perhaps even overdue. Finally, the paper by Holmes and Rubin on the detection of RNA genes in sequences is a beautiful and natural generalization of single organism stochastic context free grammar approaches to two organisms.

Taking a more global view, and looking at protein sequences, the paper by Cline et al. compares protein families in humans, worms, flies and yeast. Also looking at biology from a whole genome perspective is the paper by Imoto, Goto and Miyano which addresses the problem of constructing genetic networks. Such whole genome studies aimed at mapping out the functional and regulatory relationships between proteins are an exciting development that has been heretofore impossible because of the lack of sequence. There are three papers dealing with evolutionary biology: the paper by Goldberg, McCouch and Kleinberg addresses the important problem of generating comparative genomic maps, and the technical paper by Wu and Gu on computing reversal distance is a nice example of some of the algorithms which can lead to estimates of evolutionary distance from such maps. The immediate application of computing reversal distance is the accurate construction of phylogenetic trees (although of course there are other measures of distance which may be even more useful), and the paper by Nakhleh et al. discusses an effective approach to rapidly and accurately constructing phylogenetic trees.

Alignment algorithms lie at the heart of comparative genomics and we have included two important papers that directly address technical issues that are critical in obtaining alignments. The paper by Chiaromonte, Yap and Miller is focused on the problem of obtaining accurate alignments. In particular, they look at the question of how to score nucleotide substitutions in alignments, and how to generate scoring matrices. The work has immediate application to the alignment methods underlying PIPMaker, which is a widely used alignment tool. The paper by Yamaguchi and Maruyama looks at how to generate alignments quickly, which is just as important as generating them accurately. In this interesting paper, they suggest the application of specialized hardware and discuss the associated algorithms.

Finally, the paper by Volkmuth and Alexandrov is an exciting addition in that it proposes a novel way for utilizing comparative genomic information for learning about folding. We believe that such creative ways of finding predictive power in comparative information, coupled with visualization tools such as the new one described by Gherbi and Herisson, hold the promis e of exciting and important biological discoveries based on genomic analysis.

The session co chairs are grateful to the reviewers for their help in choosing the best contributions from a large number of excellent submissions.

SCORING PAIRWISE GENOMIC SEQUENCE ALIGNMENTS

F. CHIAROMONTE
Department of Statistics, Penn State,
University Park, PA 16802
chiaro@stat.psu.edu

V.B. YAP
Department of Statistics, UC Berkeley,
Berkeley, CA 94720
yapvb@stat.berkeley.edu

W. MILLER
Department of Computer Science and Engineering, Penn State,
University Park, PA 16802
webb@bio.cse.psu.edu

The parameters by which alignments are scored can strongly affect sensitivity and specificity of alignment procedures. While appropriate parameter choices are well understood for protein alignments, much less is known for genomic DNA sequences. We describe a straightforward approach to scoring nucleotide substitutions in genomic sequence alignments, especially human-mouse comparisons. Scores are obtained from relative frequencies of aligned nucleotides observed in alignments of non-coding, non-repetitive genomic regions, and can be theoretically motivated through substitution models. Additional accuracy can be attained by down-weighting alignments characterized by low compositional complexity. We also describe an evaluation protocol that is relevant when alignments are intended to identify all and only the orthologous positions. One particular scoring matrix, called *HOXD70*, has proven to be generally effective for human-mouse comparisons, and has been used by the PipMaker server since July, 2000. We discuss but leave open the problem of effectively scoring regions of strongly biased nucleotide composition, such as low G+C content.

1 Introduction

Most sequence alignment programs employ an explicit scheme for assigning a score to every possible alignment. This provides the criterion to prefer one alignment over another. Alignment scores typically involve a score for each possible aligned pair of symbols, together with a penalty for each gap in the alignment. For protein alignments, the scores for all possible aligned pairs constitute a 20-by-20 *substitution matrix.* Amino acid substitution scores are well understood in theory [2,3], and the scores most used in practice are the PAM matrices of Dayhoff [7,13] and the newer BLOSUM series [16]. The landmark studies by Dayhoff and colleagues introduced "log-odds" scores, and connected the choice of a substitution matrix with the evolutionary distance separating two sequences.

Fewer papers have dealt with scoring schemes for alignments of DNA sequences [27,6]. A sophisticated scheme based on extensive analysis of evolutionary substitution patterns in human and rodent sequences was developed by Arian Smit, and used in the initial version of the PipMaker network server [25]. This scheme utilized non species-symmetric scores (a human `A` with a mouse `C` is not scored the same as a human `C` with a mouse `A`) to account for accelerated substitution rates in rodents [20]. Moreover, the scheme provides distinct scores for each of three ranges of G+C content (the percentage of letters that are either `G` or `C`) to account for dependence of patterns of nucleotide substitution on the latter [11]. We describe a simple log-odds technique for DNA substitution scores, reminiscent of the BLOSUM approach. Gap penalties are ignored.

A major issue in developing alignment software for genomic DNA sequences is experimental evaluation [22]. It is frequently difficult to tell which of two methods performs better in practice, in part because of the scarcity of data for which a "correct answer" is known, and in part because of disagreement on what a "correct answer" means. One may try to find protein-coding regions, regions with biologically relevant functions, or simply regions that can be reliably aligned. Perhaps the most attractive goal would be to align functional regions. However, there are very few large regions (indeed, probably none) of mammalian genomic sequence where all functional segments are known, which makes it difficult or impossible to reliably measure a program's success at attaining this ideal.

A more accessible goal is to align all detectably orthologous positions (nucleotide pairs derived from the same position in the most recent common ancestral species by substitution events). Functional regions may then be identified by other programs searching the resulting alignment for segments with special properties, such as particularly high levels of conservation [30]. The *blastz* alignment program used by PipMaker [25] takes this approach. We give a protocol for evaluating alignment software of this sort.

2 Substitution Scores

Following a common approach in protein alignment, we determine nucleotide substitution scores by identifying a set of trusted aligned symbol pairs and using log-odd-ratios [8]. To find a "training set" of nucleotide pairs, i.e., the columns of trusted alignments, we align human and mouse sequences on a pre-selected human region, using a very simple alignment program and scoring scheme. We start by deleting from the human region all interspersed repeats and low-complexity regions, as determined by the RepeatMasker program (Smit and Green, unpublished) with default settings, and all annotated protein-coding segments. The reduced human sequence is then aligned with the full mouse sequence using a variant of the *blast* program [1,24]. Alignments are scored by match = 1, mismatch = −1, and we retain only the matched segments that occur in the same order and orientation in both species. The program computes only gap-free alignments, commonly called *high-scoring segment pairs.*

Using the resulting training set, we apply the algorithm of Figure 1. Gap-free alignments in which nucleotide identity exceeds $max\_pct = 70\%$ (say) are discarded, so as to exclude strongly conserved portions from our analysis. (This is the step most reminiscent of the BLOSUM approach.) The hope is to accurately model moderately conserved regions, with the belief that strongly conserved ones can be found with any approach.

```
global int n1(1..4), n2(1..4), m(1..4, 1..4)     (initially all zeros)

for each gap-free local alignment do
    if the percent identity ≤ max_pct then
        for each column, x-over-y, of the alignment do
            observe(x,y)
npairs ← n1(A) + n1(C) + n1(G) + n1(T)
for x ∈ {A, C, G, T} do
    q1(x) ← n1(x)/npairs
    q2(x) ← n2(x)/npairs
    for y ∈ {A, C, G, T} do
        p(x, y) ← m(x, y)/npairs
for x ∈ {A, C, G, T} do
    for y ∈ {A, C, G, T} do
        s(x, y) ← log( p(x,y) / (q1(x) × q2(y)) )  (scale so largest entry is 100)

procedure observe(x,y)
    infer(x, y)
    infer(compl(x), compl(y))                (for strand symmetry)
    infer(y, x)                              (for species symmetry)
    infer(compl(y), compl(x))                (for strand and species symmetry)

procedure infer(x,y)
    m(x, y) ← m(x, y) + 1
    n1(x) ← n1(x) + 1
    n2(y) ← n2(y) + 1
```

Figure 1: Algorithm to determine a matrix $s(x, y)$ of nucleotide substitution scores. The complement of nucleotide x is denoted $compl(x)$.

The score of the alignment column x-over-y is the log of an "odds ratio"

$$s(x,y) = \log\left(\frac{p(x,y)}{q_1(x)q_2(y)}\right) \tag{1}$$

where $p(x,y)$ is the frequency of x-over-y in the training set, expressed as a fraction of the observed aligned pairs, and $q_1(x)$ and $q_2(y)$ denote the background frequencies of nucleotides x and y as the upper and lower components (resp.) of those same pairs. Frequencies actually include also aligned pairs "inferred" from the observed ones. For each x-over-y, we infer $compl(x)$-over-$compl(y)$, where *compl* denotes nucleotide complement (so $compl(A) = T$). This makes the scores strand symmetric, i.e., invariant under reverse complementation of the two sequences. Moreover, for each x-over-y we infer y-over-x. This makes the scores species symmetric ($s(x,y) = s(y,x)$) so that the same matrix can be used for human-mouse and mouse-human alignment, but the algorithm in Figure 1 can be used to compute two asymmetric matrices deleting the statements enforcing symmetry from the *observe* procedure. In applications, we find that species symmetric matrices work about as well as asymmetric ones (see Section 4). To permit use of integer arithmetic, we normalize the scores $s(x,y)$ so that the largest is 100, then round to the nearest integer.

Here we give substitution matrices calculated on three different human-mouse training sets. The regions were chosen to approximately span the range of G+C content seen in the human genome. In all three cases, *max_pct* was set to 70%.

	CFTR[9] matrix 37.4% G+C				HOXD matrix 47.5% G+C				hum16pter[10] matrix 53.7% G+C			
	A	C	G	T	A	C	G	T	A	C	G	T
A	67	−96	−20	−117	91	−114	−31	−123	100	−123	−28	−109
C	−96	100	−79	−20	−114	100	−125	−31	−123	91	−140	−28
G	−20	−79	100	−96	−31	−125	100	−114	−28	−140	91	−123
T	−117	−20	−96	67	−123	−31	−114	91	−109	−28	−123	100

Any of these matrices can then be used in the traditional way: to evaluate the relative likelihood that a gap-free alignment correctly matches related sequences, as opposed to unrelated ones, we read its column scores off the matrix, and add them together.

One possible refinement of this approach would be to compute a custom matrix each time sequences are aligned; that is, derive the training set from the region undergoing alignment itself. This could be easily and efficiently implemented within many existing alignment programs. For instance, the *blastz* program used by PipMaker operates in three phases, similar to those of the gapped *blast* program[4]: (1) find short exact matches, (2) determine ungapped extensions of the exact matches, (3) for sufficiently high-scoring ungapped matches, compute alignments allowing for gaps. Step (2) is relatively inexpensive, typically taking about 10% of the execution time. An initial set of ungapped alignments can be computed with generic substitution scores and used as described above to determine a locus-specific scoring matrix. Then phase (3), perhaps preceded by an iteration of phase (2), can utilize the customized scores. One might even consider several iterations of phase (2) and re-computation of substitution scores.

Another possible refinement of our approach would be to segment long genomic regions undergoing alignment into relatively homogeneous subregions (e.g., with respect to G+C content or, more directly, with respect to patterns of substitution frequencies) and use different substitution matrices in each subregion. Lastly, our approach could be generalized to the estimation of 16-by-16 matrices accounting for dependence of nucleotide substitution on adjacent nucleotides (e.g. `CG` tending to become `TG` or `CA`, and other similar effects[17]). Precedent for this can be found also in protein sequence alignment, with 400-by-400 substitution matrices[14].

2.1 Modeling substitution

An alternative method of deriving substitution scores from a training region is to view its gap-free alignments as independent realizations of a *reversible time-continuous Markov chain*. This models the substitution process linking the segments of each gap-free alignment through a common ancestral segment. The process is characterized by a 4 by 4 rate matrix calibrated to produce on average 1% substitutions per unit time, and the segments of each gap-free alignment are separated by an alignment-specific divergence time, which roughly corresponds to the percent identity. Thus, a unique process is viewed as generating alignments with different degrees of identity through different divergence times.

Numerical maximization of the likelihood function of this model provides estimates of the rate matrix, say Q, and of the divergence times, say t_ℓ, $\ell = 1, 2, \ldots$. Q can be used to estimate frequencies and background frequencies for a generic divergence time t as:

$$
\begin{aligned}
&p_t(x,y) = \pi(x)\exp\{Q(x,y)t\} \qquad (2)\\
&q_{1,t}(x) = \sum_y p_t(x,y) = \pi(x) \qquad q_{2,t}(y) = \textstyle\sum_x p_t(x,y) = \pi(y)
\end{aligned}
$$

where $\exp\{Q(x,y)t\}$ estimates the chance of y substituting x over t time units, and $\pi(x)$, x = `A`, `C`, `G`, `T` the chance of x in the stationary distribution of the process. Using these quantities in equation (1), we can compute a t-dependent scoring matrix $s_t(x,y)$. Although we could produce alignment-specific substitution matrices setting $t = t_\ell$, we produce a single matrix from the training region as follows. The frequencies $p(x,y)$ from the algorithm in Figure 1, considered as a whole, define a rate matrix $\bar{Q}$ and an "overall" divergence time $\bar{t}$. Setting $t = \bar{t}$ in equation (2) we obtain frequencies $p_{\bar{t}}(x,y)$, $q_{1,\bar{t}}(x)$, $q_{2,\bar{t}}(y)$, and scores $s_{\bar{t}}(x,y)$. Thus, when comparing these scores with the $s(x,y)$ obtained directly from $p(x,y)$, $q_1(x)$, $q_2(y)$, we are actually comparing two rate matrices, Q and $\bar{Q}$, using the same divergence time $\bar{t}$.

If we restrict attention to gap-free alignments with percent identity $\leq 70\%$, numerical likelihood maximization of the reversible Markov chain model on the CFTR, HOXD and hum16pter training regions gives scoring matrices practically indistinguishable from the ones generated by the algorithm in Figure 1, and lends a strong theoretical motivation to this simple procedure.

2.2 Score adjustment for low-complexity regions

Whatever the selected training regions and estimation procedures, the log-odds score contribution $s(x,y)$ of a pair x-over-y observed in the regions undergoing alignment is high if the pair occurs more often than by chance *in the training data.* But the compositional complexity of the regions under alignment may differ substantially from that of the training data. In particular, low compositional complexity in the regions under alignment may increase chance occurrence of pairs that are relatively rare in the training data, and hence misleadingly inflate the scores of some gap-free alignments.

A simple approach to adjust for such an effect is to multiply the score of each gap-free alignment ℓ by the relative *entropy* characterizing its top segment:

$$H(\ell) = \frac{-\sum_x q_{1,\ell}(x)\log q_{1,\ell}(x)}{\log 4}$$

where $q_{1,\ell}(x)$, $x =$A, C, G, T, are the background frequencies in the top segment.

As $H(\ell)$ ranges in $[0,1]$, this adjustment achieves the desired effect of downweighting misleadingly high scoring alignments (the fact that it works counterintuitively for low scoring ones – e.g., an alignment with negative score will have a lower adjusted score the higher its relative entropy – is inconsequential).

Since all the substitution matrices we are considering have positive entries only along the main diagonal, a high scoring alignment ℓ will have an abundance of matches: the alignment frequencies $p_\ell(x,y)$ will be largely concentrated on x-over-x pairs, and very small on mismatches; $p_\ell(x,y)$, $x \neq y$.

Consequently

$$-\sum_{x} q_{1,\ell}(x)\log q_{1,\ell}(x) \approx \sum_{x,y} p_{\ell}(x,y)\log\left(\frac{p_{\ell}(x,y)}{q_{1,\ell}(x)q_{2,\ell}(y)}\right)$$

so that adjusting the score of a high scoring alignment by $H(\ell)$ is approximately the same as adjusting it by its relative *expected quantity of information.* The expected quantity of information benchmarks pair occurrences in ℓ against the background frequencies of ℓ itself, so it can be interpreted as the "score of an alignment against itself": alignments that score high against the training set will be down-weighted if they score poorly against themselves. Although it is possible, in principle, to use directly the expected quantity of information instead of the entropy, the latter has the advantage of involving computations on only one of the sequences undergoing alignment.

3 Evaluation Procedure

We now describe a simple protocol for evaluating alignment software and substitution scores with respect to the goal of aligning orthologous positions.

Orthologous human and mouse genomic regions believed to be free of large-scale rearrangements such as gene duplications (small inversions shouldn't matter) differ due to nucleotide substitutions, small-scale insertions/deletions, and insertion of interspersed repeats. When the regions are aligned, the true matches appear along a diagonal path in the dot-plot, with spurious matches off the diagonal. To a first approximation, paired nucleotides on that path can be considered correct, and paired nucleotides off the path incorrect, treating overlapping alignment with some care. (The path can be determined by any of several methods [31,32].)

We again used the primitive *blast* program [24] for gap-free alignments. With each of a variety of scoring schemes, we determined the gap-free alignments that scored above various thresholds (denoted K below), then divided the aligned nucleotide pairs into correct (for the maximal chain of properly ordered matches) and incorrect (all other matches).

For instance, aligning the human CD4 region and its known mouse ortholog using the HOXD matrix (which might more properly be called HOXD70, to emphasize dependence on *max_pct* = 70%), we obtain the two dot-plots in Figure 2. The left panel shows alignments scoring at least 2000, and the right one those scoring above 3000. Note that increasing the threshold substantially reduces the number of spurious matches (off-diagonal), but at the cost of slightly reducing the number of putatively correct matches (on the diagonal path), a typical sensitivity-specificity tradeoff.

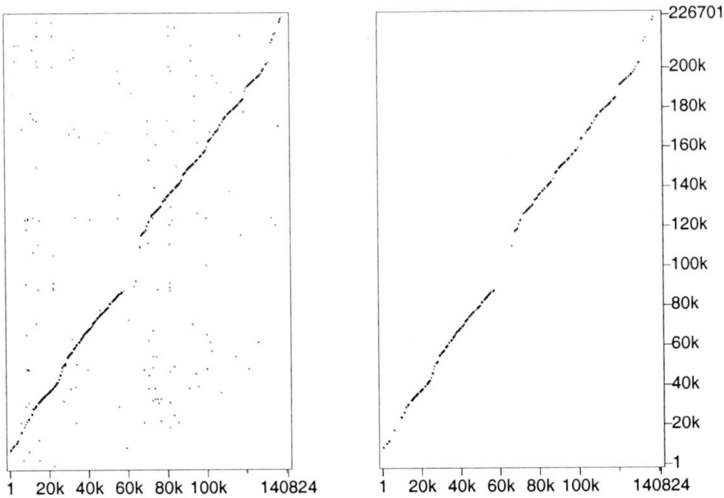

Figure 2: Dot-plots for alignments of the human and mouse CD4 loci using the HOXD matrix, for $K = 2000$ (left) and $K = 3000$ (right).

Using CD4 as our test region, the following table reports exact counts of correct and incorrect nucleotide pairs for four different scoring schemes, at six different threshold K levels - the last column refers to the HOXD matrix with subsequent entropy adjustment of alignment scores.

	unit (±1)			hum16pter			HOXD			HOXD+entropy	
K	right	wrong	K	right	wrong	K	right	wrong	K	right	wrong
20	47751	7942	2000	52919	12230	2000	53021	10007	1800	54054	5863
22	45862	1690	2200	50544	4526	2200	50468	3084	2000	51646	1799
24	43378	495	2400	48370	2017	2400	48246	941	2200	49480	540
26	41614	378	2600	46403	870	2600	46416	697	2400	46997	277
28	40227	252	2800	44326	204	2800	44794	272	2600	45625	65
30	38970	34	3000	42675	65	3000	43126	101	2800	43890	65

The six K levels considered for each scheme are different because of the different maximal value of a one-column score. This is 1 for the unit matrix, 100 for the hum16pter and HOXD matrices, and about 90 after correcting HOXD scores for entropy. Thus, the rows of the table are comparable, with threshold levels corresponding to maximal scoring contiguous matches of length 20, 22, 24, etc. The same information is summarized in Figure 3, plotting correct versus incorrect counts for each scoring scheme, at the various threshold levels.

The unit matrix lies well below all others, at all threshold levels. Among

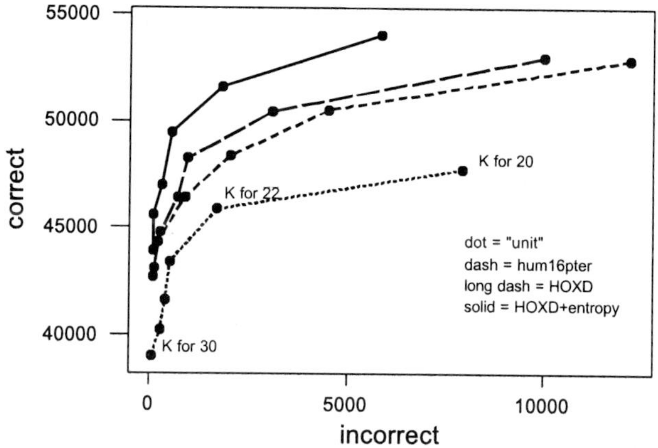

Figure 3: Correct vs. incorrect matches for human-mouse alignments of the CD4 region. Each curve corresponds to a scoring scheme, and comprises six threshold levels.

the non-unit schemes, the HOXD matrix with entropy adjustment is uniformly better, and more markedly so at low threshold levels. The HOXD and hum16pter matrices have comparable performances for high thresholds, but at low thresholds HOXD reduces the number of "false positives" with respect to the hum16pter. Thus, HOXD performs better despite originating from a region with G+C content less like that of CD4.

In addition to inspecting graphs like those described above, we summarized the comparison of two scoring schemes at a genomic locus with a single number, as follows. Consider a first scheme, say the HOXD matrix. We focus on "corner thresholds", i.e. thresholds that, if decreased by 1, produce a strictly larger number of incorrect matches. These are the relevant values because any non-corner K would be automatically discarded in favor of $K-1$, producing the same number of incorrectly aligned nucleotides and a larger or equal number of correctly aligned ones. For instance, at the CD4 locus, the HOXD matrix with $K = 2433$ gives 48796 correct versus 991 incorrect matches, while passing to $K = 2432$ gives 48796 versus 1047. Thus $K = 2433$ is a corner threshold for the HOXD matrix when aligning our CD4 sequences.

A second scoring scheme can then be compared to HOXD on its corner thresholds; exploring thresholds at which a second scoring matrix applied to CD4 produces about 1000 incorrect matches, we found 47001 correct versus

1106 incorrect at threshold 2350, and 46876 versus 981 at threshold 2351. Thus, at this corner, the HOXD matrix identifies about 1800-1900 more correct nucleotides for the same cost. We have software that performs this inspection at each corner threshold, and reports won-lost-tie counts. The won/lost ratio provides a single quantity to summarize the relative performance of two scoring schemes at a given genomic locus.

4 Experimental Results

We compared the HOXD matrix to eight other matrices, and to HOXD with the correction for entropy. The table below provides won/loss ratios on HOXD corner thresholds for nine genomic regions, each named for a gene that it contains – we always deleted interspersed and simple repeats from the human sequence (RepeatMasker with default settings). HOXD was superior on each comparison with a ratio larger than 1.

The table's second column gives the region's G+C content. Columns 3 and 4 refer to match/mismatch matrices; the unit matrix, and a match = 19, mismatch = −16 matrix suggested on theoretical grounds by Stephen Altschul as being the most appropriate match/mismatch choice for human-rodent comparisons. Columns 5 and 6 refer to the hum16pter and CFTR matrices. Then come three matrices proposed by Arian Smit for human-mouse comparisons in genomic regions of approximately 37%, 43% and 50% G+C content, respectively. The next-to-last column refers to an asymmetric version of the HOXD matrix, computed removing species and strand symmetrization from the algorithm in Fig. 1. Last comes the HOXD matrix with entropy adjustment.

Region	%G+C	± 1	$\frac{+19}{-16}$	16pt	CFTR	S37	S43	S50	asym	entro
MYO15 [21]	55.1	∞	∞	1.75	2.06	0.11	0.0	0.0	0.122	0.26
CD4 [5]	51.1	∞	83.0	15.8	∞	∞	2.42	11.0	∞	0.0
MECP2 [28]	48.6	∞	∞	13.3	∞	∞	4.2	∞	0.23	0.30
CECR [12]	47.3	30.0	9.3	19.7	14.5	11.4	6.6	5.9	3.3	0.051
SCL [15]	46.4	2.0	1.2	5.2	∞	7.3	∞	∞	0.79	0.85
BTK [23]	43.2	∞	0.87	13.0	1.54	5.5	13.0	13.0	6.0	0.4
Mnd2 [18]	41.4	11.0	0.50	12.0	∞	∞	12.0	12.0	12.0	0.083
FHIT [26]	38.4	∞	58.0	2.58	∞	117.0	18.7	117.0	0.24	0.035
SNCA [29]	36.3	∞	∞	1.3	∞	∞	0.28	0.53	1.0	0.0

These results allow us to draw some conclusions, and identify some open questions. Match/mismatch scores, which ignore the higher probability of transitions (conversion between `A` and `G`, or between `C` and `T`)) with respect to transversions (any other nucleotide substitution), can be substantially improved upon: The HOXD matrix does distinctly better than the unit matrix on all test regions, and better than the $+19/-16$ matrix on most. Another clear point is that our simple approach to down-weighting low-complexity regions improves performance: HOXD with entropy correction does distinctly better than HOXD itself on all regions.

Certain ambiguities remain. The asymmetric version of HOXD does better than HOXD on some regions, but worse on others, and the performance does not appear to be related to G+C content. HOXD does better than hum16pter on all regions, including those with high G+C, and better than CFTR on all regions, including those with low G+C. Similarly, HOXD does better than S37 on low G+C regions, better than S43 on medium G+C regions, and better than S50 on all high G+C regions except the most extreme, MYO15. On this region, though, *all* S matrices do better than HOXD. The lowest G+C region SNCA provides another seemingly paradoxical situation, since S43 and S50 do better than HOXD while S37 does worse.

Figure 3 shows how, on the CD4 test region, HOXD reduces the number of "false positives" with respect to hum16pter. While computing the won/loss ratio for the HOX–CFTR comparison on the FHIT test region, we observed that "false positives" identified by CFTR tended to have nucleotide composition different from that of "true positives" and of FHIT as a whole. For instance, at one threshold we obtained A:31.7%, C:18.2%, G: 19.1%, T: 31.0% for correct, and A: 39.7%, C:10.8%, G:34.0%, T:15.4% for incorrect alignments.

Excluding its own asymmetric and entropy corrected versions, the HOXD matrix wins 56 out of 63 comparisons. As reported by Lander *et al.* [19] (p.883), the HOX clusters are characterized by the lowest density of interspersed repeats in the human genome, making correct local alignments relatively easy to produce, even in segments with nucleotide identity below 70%. Moreover, the alignment-specific divergence times estimated with our reversible Markov chain model do not present a sizeable correlation with alignment-specific G+C content within HOXD (the correlation coefficient is 0.013, compared to 0.242 in hum16pter and −0.593 in CFTR). These factors and others may explain why local alignments from the HOXD region provide particularly effective training data for computing a single log-odds score matrix that performs well in a variety of contexts.

However, several aspects of our analysis strongly suggest that further improvements in scoring genomic DNA sequence alignments will likely be generated by exploiting G+C content and other local compositional properties.

Acknowledgments

Stephen Altschul suggested the +19/ − 16 match/mismatch matrix and David Haussler made several helpful recommendations. Our work was supported by grant HG02238 from the National Human Genome Research Institute.

References

1. S. Altschul *et al.*, *J. Mol. Biol.* **215**, 403 (1990)
2. S. Altschul, *J. Mol. Biol.* **219**, 555 (1991)
3. S. Altschul, *J. Mol. Evol.* **36**, 290 (1993)
4. S. Altschul *et al.*, *Nucleic Acids Res.* **25**, 3389 (1997)
5. M. Ansari-Lari *et al.*, *Genome Research* **8**, 29 (1998)
6. W. Bains, *DNA Sequence-J.DNA Sequencing and Mapping* **3**, 267 (1993)
7. M. Dayhoff *et al.*, in *Atlas of Protein Sequence and Structure*, M. Dayhoff ed., p. 345, 1978.
8. R. Durbin *et al.*, *Biological Sequence Analysis: Probabilistic Models of Proteins and Nucleic Acids*, Cambridge University Press, 1998.
9. R. Ellsworth *et al.*, *Proc. Natl. Acad. Sci. USA.* **97**, 1172 (2000)
10. J. Flint *et al.*, *Human Molecular Genetics* **10**, 371 (2001)
11. M. Francino, H. Ochman, *Nature* **400**, 30 (1999)
12. T. Footz *et al.*, *Genome Research* **11**, 1053 (2001)
13. D. George *et al.*, *Methods in Enzymology* **183**, 333 (1990)
14. G. Gonnet *et al.*, *Biochem. Biophys. Res. Comm.* **199**, 489 (1994)
15. B. Gottgens *et al.*, *Genome Research* **11**, 87 (2001)
16. S. Henikoff, J. Henikoff, *Proc. Natl. Acad. Sci. USA* **89**, 10915 (1992)
17. S. Hess *et al.*, *J. Mol. Biol.* **236**, 1022 (1994)
18. W. Jang *et al.*, *Genome Research* **9**, 53 (1999)
19. E. Lander *et al.*, *Nature* **409**, 806 (2001)
20. W. Li *et al.*, *Mol Phylogenet. Evol.* **5**, 182 (1996)
21. Y. Liang *et al.*, *Genomics* **61**, 243 (1999)
22. W. Miller, *Bioinformatics* **17**, 391 (2001)
23. J. Oeltjen *et al.*, *Genome Research* **7**, 315 (1997)
24. S. Schwartz *et al.*, *Nucleic Acid Research* **19**, 4663 (1991)
25. S. Schwartz *et al.*, *Genome Research* **10**, 577 (2000)
26. T. Shiraishi *et al.*, *Proc. Natl. Acad. Sci. USA.* **98**, 5722 (2001)
27. D. States *et al.*, *METHODS: A companion to Methods in Enzymology* **3**, 66 (1991)
28. K. Reichwald *et al.*, *Mammalian Genome* **11**, 182 (2000)
29. J. Touchman *et al.*, *Genome Research* **11**, 78 (2001)
30. N. Stojanovic *et al.*, *Nucleic Acids Res.* **27**, 3899 (1999)
31. W. Wilbur and D. Lipman, *Proc. Natl. Acad. Sci. USA* **80**, 726 (1983)
32. Z. Zhang *et al.*, *J. Comput. Biol.* **1**, 217 (1994)

STRUCTURE-BASED COMPARISON OF FOUR EUKARYOTIC GENOMES

MELISSA CLINE, GUOYING LIU, ANN E. LORAINE, RONALD SHIGETA, JILL CHENG, GANGWU MEI, DAVID KULP, and MICHAEL A. SIANI-ROSE
Affymetrix, Inc. 6550 Vallejo Street,
Emeryville, CA 94608 USA
E-mail: melissa_cline@affymetrix.com

The field of comparative genomics allows us to elucidate the molecular mechanisms necessary for the machinery of an organism by contrasting its genome against those of other organisms. In this paper, we contrast the genome of *homo sapiens* against *C. Elegans, Drosophila melanogaster, and S. cerevisiae* to gain insights on what structural domains are present in each organism. Previous work has assessed this using sequence-based homology recognition systems such as Pfam [1] and Interpro [2]. Here, we pursue a structure-based assessment, analyzing genomes according to domains in the SCOP structural domain dictionary. Compared to other eukaryotic genomes, we observe additional domains in the human genome relating to signal transduction, immune response, transport, and certain enzymes. Compared to the metazoan genomes, the yeast genome shows an absence of domains relating to immune response, cell-cell interactions, and cell signaling.

1 Introduction

To date, there are hundreds of completed genomes for prokaryotes, but only five for eukaryotes: *homo sapiens, Caenorhabditis elegans, Drosophila melanogaster, Saccharomyces cerevisiae,* and *Arabidopsis thaliana.* Eukaryotes exhibit more genomic complexity. Their genomes are larger, their proteins contain a wider variety of domains, and the domains appear in a greater number of combinations [3]. Yet compared to other eukaryotes, the human genome exhibits even more complexity. The human proteome contains a greater number of domains and domain combinations than other eukaryotic proteomes [2]. Compared to other sequenced eukaryotes, the human genome yields more immune response proteins, epithelial proteins, and olfactory [2], plus more proteins related to neural development and function, signaling, homeostasis, and apoptosis [1].

While some amount has been written about the eukaryotic genomes from a functional standpoint[1, 2], little has been published in terms of comparative structural genomics. Historically, structural analysis has been used largely as an intermediate step towards the goal of functional analysis, although this has its limitations[4, 5]. A structure-based analysis is worthwhile in its own right. First, while structural similarity is a clearly defined concept, functional similarity is more ambiguous. The many definitions of functional similarity include common domains, common EC numbers, similar keywords in Medline abstracts, and similar binding sites; whether or not two proteins are considered functionally similar depends on which definition of functional similarity is used. Second, structural classification schemes capture different information than functional classification schemes [6]. A comparison of SCOP[7] and Pfam[8] domains found that 70% of the Pfam families have corresponding entries in SCOP, while 57% of the SCOP families have corresponding entries in Pfam[9]. Third, structural classification schemes can be organized hierarchically, yielding a convenient way to summarize the content of a genome. Fourth, SCOP has been used in previous studies to examine genomes [5, 10-13], and the total number of folds in SCOP have been carefully examined [14, 15]. Thus, functional and structural genomic analyses are good complements.

To perform a structural genomics analysis, we applied the GRAPA method [16]. GRAPA features a library of HMMs based around the SCOP hierarchy, with each HMM optimized for family recognition, a contrast to other methods optimized for superfamily recognition [17, Karplus, 1998 #16]. In GRAPA, a protein is assigned to a SCOP family by comparing the distance scores of each candidate for each of the HMMs within a SCOP superfamily. We used this library to analyze four eukaryotic genomes: *homo sapiens, C. elegans, drosophila melanogaster,* and *S. Cerevisiae.* To put our results into perspective with prior work, we applied the same genomes to the Pfam library. Using these results, we contrasted the human genome to other eukaryotic genomes in both a functional and structural perspective.

2 Methods

2.1 Genome Gene Sets

A set of protein sequences covering the Golden Path of the human genome (October 7, 2000 freeze http://genome.ucsc.edu/) was generated by the Genie [18] programs suite [Kulp, D. & Wheeler, R., published in http://genome.ucsc.edu/], with the repeat regions masked out. The data set consists of three sets of amino acid sequences: (1) proteins from Genbank whose associated mRNA sequences could be mapped to genomic sequence (2) proteins predicted by *AltGenie*, an enhanced Genie program which predicts alternatively splice transcripts using merged mRNA/EST-to-genomic alignments and (3) proteins predicted by *StatGenie*, a purely *ab initio* gene-finder. These sequences are non-redundant; none of the included genes overlap the same genomic region. In cases where there were many genes overlapping the same region, the one with the longest CDS (translation) was kept [Williams, A., unpublished data]. This set, known as annot10, contains 59,378 protein sequences.

The non-redundant complete proteome set of SWISS-PROT plus TrEMBL entries for *Drosophila melanogaster* (13844 entries), *Caenorhabditis elegans* (18870 entries), and *Saccharomyces cerevisiae* (6186 entries) were obtained on June 15, 2001 from the EBI proteome analysis site (http://www.ebi.ac.uk/proteome/).

2.2 Model Building and Search Method

We searched for domains in the target genome according to GRAPA, a battery of Hidden Markov Models (HMMs) generated for each member of the SCOP hierarchy. GRAPA characterizes each SCOP protein domain found in ASTRAL[19], a service that allows the user to select proteins from SCOP according to various criteria. We selected all non-redundant proteins from SCOP version 1.53, yielding 4369 entries. For each entry, a hidden Markov model (HMM) was built using the Target99 protocol [20] with the Sequence Alignment and Modeling system (SAM 3.0) system[21]. Multiple species were included to capture characteristics of both mammalian and non-mammalian proteins.

Each gene was scored against each SCOP family, yielding an e-value. In practice, the e-values generated by a model are dependent in part on that model, with shorter models yielding higher e-values. This is consistent with the definition of the e-value: the number of equivalent or better scores that might arise by chance from non-homologous sequences, given a database of the same size. If a model is shorter, its best hits will be shorter; a shorter hit is more likely to occur by random chance than a longer one.

Because the e-value interpretation is dependent on the HMM, there is no single E-value that can be applied to all models. Instead, for each HMM, the DISTSIEVE program examines a model's set of e-values and determines a reasonable e-value cutoff by curve analysis. In general, the hits to an HMM will include a small number of hits with low e-values, corresponding to the training sequences and their homologs, followed by a large increase in the number of hits as the threshold rises to include the false positives. DISTSIEVE examines the hits to each model and identifies some e-value threshold beyond which the number of hits increases rapidly. The hits selected by DISTSIEVE are then assigned to whatever model in each superfamily they score against best. The performance of this method is comparable to PSI-BLAST, with a family recognition specificity of approximately 95% and sensitivity of approximately 75%.

For comparison with previous work, we searched Pfam with the same sequences. We applied the genomes to Pfam 6.2, recording all hits, which exceeded each model's, gathering threshold. The gathering threshold is defined by the Pfam authors as the scoring threshold above which they would admit a new sequence to the Pfam alignment. There is one gathering threshold per model, and it is set manually.

To establish a correspondence between the SCOP and Pfam domain definitions, we followed the method applied in previous work [9] and scored the SCOP domain sequences against all Pfam models. When the score of a SCOP sequence exceeded the gathering threshold for a Pfam model, we noted the hit as a potential correspondence. We then examined the potential correspondences by hand for the domains emphasized in this paper.

3 Results and Discussion

We have studied the types of structural domains found within four genomes. Table 1 lists the number of different SCOP domain families found within each genome. As expected, we see in Table 1 that the human genome is the most complex, with 442 families of structural domains represented. In rough terms, this approximates the number of different functions performed by the genes in the organism's genome.

Table 1. Number of different domain families found in each organism.

Organism	Number of Domain Families
Human	442
Fly	389
Worm	378
Yeast	245

Table 2 lists the twenty most frequent SCOP domain families in the human genome, and lists the number of occurrences of each domain within each organism. The corresponding Pfam domain and rank in the human genome is provided for comparison; in general, these numbers are equivalent to those previously published [22, Lander, 2001 #18, Venter, 2001 #32]. Mostly, the top-ranked domains are similar for SCOP and Pfam. One contrast concerns Leucine Rich Repeats, LRRs. These short sequence motifs appear in many different types of proteins and many different SCOP domains, including Internalin B, B LLR domain, U2A'-like, and Rna1p. Pfam, in contrast, includes a separate LLR model. Another contrast concerns proteases. Eukaryotic and prokaryotic proteases, which are among the more common SCOP domains, have no direct equivalent in Pfam. The closest Pfam model, the trypsin domain, appears to be more specific and would likely not rank in the top twenty. Immunoglobulin domains appear as a top hit under Pfam (rank = 4), while SCOP finds many such hits via the V set domains (antibody variable domain-like) (rank=19). Closer examination revealed that the SCOP model is more specific.

Table 2. Number of human proteins (H), C. elegans (W), Drosophila melanogaster (F), and S. cerevisiae (Y) sorted by most frequently occurring SCOP v 1.53 families in humans. The rank of the corresponding Pfam model (if any) is shown for comparison.

	SCOP Family Id	H	F	W	Y	SCOP family	Pfam rank equivalent
1	4.130.1.1	349	220	422	113	Serine/threonine kinases	pkinase (3)
2	7.37.1.1	296	185	56	27	Classic zinc finger, C2H2	zf-C2H2 (1)
3	4.130.1.2	246	34	399	114	Tyrosine kinase	pkinase (3)
4	3.32.1.8	184	128	147	72	G proteins	ras (20)
5	1.111.2.1	135	84	92	18	Ankyrin repeat (SH3-domain superfamily)	ank (5)
6	2.64.3.1	115	21	11	15	Trp-Asp repeat (WD-repeat)	WD40 (8)
7	3.9.2.1	99	119	35	8	Internalin B LRR domain	LRR (9)

8	3.9.2.3	97	108	46	9	U2A'-like Leucine Rich Repeat Fold, RNA recog.	LRR (9)
9	3.32.1.13	90	58	69	52	Extended AAA-ATPase domain (DNA helicases bacterial/yeast)	helicase_C (26) and DEAD (27)
10	3.9.1.2	86	103	15	12	Rna1p (in Leucine Rich Fold)	LRR (9)
11	2.44.1.2	82	208	12	0	Eukaryotic proteases	-
12	2.44.1.1	78	200	3	1	Prokaryotic proteases	-
13	1.4.1.1	74	96	78	7	Homeodomain	homeobox (14)
14	4.37.1.1	73	74	98	3	BTB/POZ domain (zinc finger)	BTB (24)
15	1.23.1.1	65	7	51	9	Nucleosome core histones	histone (29)
16	3.9.1.1	64	57	11	10	Ribonuclease inhibitor (LRR fold)	LRR (9)
17	4.82.1.1	63	39	65	0	SH2 domain	SH2 (25)
18	3.9.2.2	63	104	59	8	Rab geranylgeranyltransfer-ase a-subunit, N-terminal (C2 domain-like Fold)	C2 (22)
19	2.1.1.1	61	70	26	0	V set domains (Ab variable domain-like) Ig superfamily	Ig (4)
20	3.32.1.9	61	41	36	11	Motor proteins (nucleoside triphosphate hydrolase family)	myosin head motor domain (52)

The Zinc finger C3HC4 type RING domain (rank = 13) in the Pfam top twenty, while the corresponding SCOP RING model has a rank of ninety. Further examination would be required to determine whether these Pfam hits appeared in the SCOP C2H2 (rank = 2) or BTB/POZ zinc finger (rank = 14) domains. In the case for the Pfam EF-HAND domain (rank = 15), SCOP has broken the EF-hand superfamily into seven families, which would yield fewer hits per family and therefore make them less likely to appear in the top twenty models.

One implication of the data in Table 2 is the prevalence of signaling proteins in the human genome: kinases, proteases, G proteins, and so forth. This reflects the importance and variety of signaling mechanisms within higher-order organisms such as humans. Much of this emphasis on signaling proteins is also evidenced in the worm genome; much less is present in the genomes of fly and yeast.

The top twenty Pfam entries (data not shown) with no corresponding SCOP entry, correspond to transmembrane proteins (7tm_1) and other entries for which there are no solved 3D structures.

Table 3. Top structural domain families common to Metazoan genomes and absent from yeast.

	SCOP family designation	H	F	W	Y	SCOP family
1	2.44.1.2	82	208	12	0	Eukaryotic proteases
2	4.82.1.1	63	39	65	0	SH2 domain
3	2.1.1.1	61	70	26	0	V set domains (antibody variable domain-like)
4	2.1.2.1	57	40	25	0	Fibronectin type III
5	2.1.1.4	50	117	56	0	I set domains
6	2.1.6.1	48	15	15	0	Cadherin
7	4.154.1.1	41	34	224	0	C-type lectin domain
8	7.3.10.1	39	13	13	0	EGF-type module
9	1.116.1.1	36	18	185	0	Nuclear receptor ligand-binding domain
10	3.57.1.1	34	3	30	0	Integrin A (or I) domain
11	7.39.1.2	34	22	246	0	Nuclear receptor
12	2.2.5.1	32	18	11	0	p53-like transcription factors
13	2.34.1.2	31	51	41	0	Interleukin 16
14	7.17.1.2	29	7	4	0	Transforming growth factor (TGF)-beta
15	4.37.1.2	29	12	46	0	Tetramerization domain of Potassium Channels

Table 3 lists the structural domain found most frequently in human, fly, and worm, and not found in yeast. Not surprisingly, eukaryotic proteases figure prominently. This family of proteins is exclusive to eukaryotes and includes trypsin, chymotrypsin, neuropsin, collagenase, thrombin, carboxypeptidase, elastase, enteropeptidase, heparin binding protein, beta-tryptase, chathepsin G, coagulation factors VIIa and Xa, kallikrein A, tonin, Nerve Growth Factor, Factor D, plasminogen activator, activated protein c, myeloblastin, and plasminogen.

Many of the structural domains in Table 3 are involved in immune responses. These include proteases, SH2 domains, V-set domains, I-set domains, and nuclear receptors. Both V-set and I-set domains belong to the Immunoglobulin V set domain (antibody variable domain-like) superfamily. Fibronectin Type III family contains many immune system receptors. C-type lectin domains are found in Natural Killer cell receptors and other immune system cell-recognition cell-surface receptors. While these motifs would not have the same immunological function in Fly and Worm, as in humans, the structural elements are clearly present in these genomes.

Proteins specifically associated with multicellular organisms include: cadherin (cell adhesion protein), and integrin-A (or I) domains are involved in cell-cell interaction.

Other structural domains in Table 3 are involved in cell signaling. Cell signaling involves both extracellular signaling proteins, cell surface receptors, and signaling pathways which transmit the signal within the cell. EGF-type modules are found in epidermal growth factor and many other growth factor and hormone extracellular signaling proteins. Transforming growth factor-beta (TGF-beta) and the cytokine Interleukin 16 are intercellular signaling proteins. Cell surface receptors include I-set domains found in Natural killer cell receptors. Intracellular signaling cascades are often regulated by SH2 (Src homology 2) domains; they interact with high affinity to phosphotyrosine-containing target peptides in a sequence specific and phosphorylation-dependent manner. Furthermore, intracellular signaling proteins that affect transcription include the nuclear receptors. In addition P53 transcription factors are involved in tumor suppression, a feature one would more likely expect in multicellular organisms.

Of particular interest are the high number of I set domains found in Fly (117 verses 50 in human). Also, the high number of C-type lectin in Worm (224 verses 41 in human) and nuclear receptors (246 verses 34 in human) suggests evolutionary branching.

Table 4. Top ten SCOP superfamilies unique to human with relevant Gene Ontology function and process annotations. The column labeled H contains the number of human genes in the indicated SCOP superfamily.

SCOP super-family Id	H	SCOP superfamily title	GO Molecular Function	GO Molecular Process
11.27.1	40	4-helical cytokines	Signal transducer; ligand; growth factor	Cell growth & main-tenance; stress response
24.18.1	30	MHC antigen-recognition domain	Signal transducer;transmem-brane receptor	cell growth & main-tenance; stress response
32.11.1	26	Lipase/lipooxygen ase domain	Enzyme; hydrolase acting ester bonds	cell growth & main-tenance; protein metabolism and modification
44.9.1	16	Interleukin 8-like chemokines	Enzyme; transferase for phosphorus groups	cell growth & main-tenance; response to abiotic stimulus
54.72.1	12	Bactericidal permeability-increasing protein	Defense/immunity protein; antimicrobial response protein	cell growth & main-tenance; response to external stimulus
67.43.1	11	B-box zinc-binding domain	Enzyme; transferase for phosphorus groups	cell growth & main-tenance; developmental processes; metabolism
74.16.1	10	Cystatin/monellin	Enzyme inhibitor; proteinase inhibitor	cell growth & main-tenance; response to external stimulus
87.24.1	10	TNF receptor-like	Signal transducer; transmembrane receptor	cell growth & main-tenance; response to external stimulus

97.31.1	10	GLA-domain	Enzyme; hydrolase	cell communication; signal transduction
107.23.1	9	TB module/8-cys domain	Structural protein; ligand binding or carrier	cell growth & maintenance; response to external stimulus

To see what structures are unique to higher-order organisms such as humans, Table 4 lists the top ten superfamily domains that are unique to the human genome. Most of these domains are involved in immune responses. Mapping of the SCOP families unique to humans back to LocusLink[23] and then into the Gene Ontology[24] (GO) graphs, allows one to rapidly clarify the role these proteins play. In Table 4, one can readily see the relationship between the SCOP structure superfamily name and the Gene Ontology Molecular Function and Process categories. The proteins unique to humans are involved in signal transduction (both ligand and receptor), enzymes (including various growth factors and cytokines, oxidoreductases, transferases, and hydrolases), miscellaneous defense and immunity proteins, transporter proteins, structural proteins, and ligand binding or carrier proteins.

The superfamily with the most hits by far is SCOP 1.27.1, 4-helix bundle cytokines, with 40 genes; all appear to be unique to human. This structural class, including long-chain cytokines, short-chain cytokines, and Inteferons/interleukin-10, is responsible in humans for mediating immune response across different organs and tissues and is responsible for much our highly evolved immune system. The second largest set, SCOP 4.18.1 superfamily, consists of the MHC antigen recognition domain, which are involved in training immune system cells to recognize foreign proteins. The fourth largest set, SCOP 4.9.1 superfamily consists of the Interleukin-8-like chemokines, another set of signaling proteins involved in lymphocyte trafficking. Interestingly, the incorporation of GO into this process, lets us identify more genes as potential cytokines by the highly nested functional notation of the GO graph. In addition to the explicitly cytokine superfamilies (1.27.1 and 4.9.1), other domains with implicit cytokine activity can be found: (1) SCOP superfamily 7.25.1, under the heparin-binding domain from vascular endothelial growth factor; (2) SCOP superfamily 4.36.3 alpha/beta-hammerhead pyrimidine nucleoside phosphorylase C-terminal domain; (3) SCOP superfamily 3.21.1 pyrimidine nucleoside phosphorylase central domain; and (4) SCOP superfamily 1.48.2 pyrimidine nucleoside phosphorylase N-terminal domain (methionine synthase domain-like). Interestingly these last three distinct domains, all part of pyrimidine nucleoside phosphorylase, are unique to human, indicating a whole human protein consisting of three separate unique structural domains. In human beings, thymidine phosphorylase (TP) performs metabolic functions in degradation of various drug compounds as well as being overexpressed in many tumor types and linked to angiogenesis. While the exact role that TP plays is not biochemically characterized, it seems that humans have adapted TP to a signaling role. The enzyme's absence in lower metazoans implies that TP may have been the result of a lateral gene transfer from a bacterium.

Table 5. Genes with disproportionate numbers between genomes. These figures represent genes that are annotated with greater than or equal to 20% sequence Identity to the SCOP seed sequence (after being annotated by GRAPA HMM scoring. This is intended to ensure that all hits are real.

SCOP Id	Human	Fly	Worm	Yeast	SCOP family
7.39.1.2	34	22	210	0	Nuclear receptors
2.44.1.2	71	163	3	0	Eukaryotic proteases
2.44.1.1	60	156	2	0	Prokaryotic proteases
1.23.1.1	63	5	49	7	Nucleosome core histones
1.23.1.2	16	6	30	6	Archaeal histone
1.4.5.12	9	1	8	0	Histone H1/H5
4.37.1.2	26	10	23	0	Tetramerization domain of potassium channels
4.145.1.3	6	13	32	4	Protein serine/threonine phosphatase
4.154.1.1	21	4	22	0	C-type lectin domain
1.111.6.1	3	2	2	1	Protein prenylyltransferase
3.9.2.2	28	39	13	1	Rab geranylgeranyltransferase alpha-subunit, N-terminal domain

By examining SCOP families disproportionately represented among the three metazoans (see Table 5), we can identify some points where other families of proteins might have fulfilled some of the signaling functions observed only in mammals. Examples of this include cytokines and immune signaling pathways. The large number of nuclear receptors (7.39.1.2) in worm suggests that the worm might rely on hormonal small molecules for development rather than intracellular signaling proteins. This disproportionate number of nuclear receptors has been observed before[2]. Another example concerns the expansion of histone components (SCOP families 1.23.1.1, 1.4.5.12, and 1.23.1.3) *in C. elegans*. This suggests that perhaps gene regulation in *C. elegans* might be more reliant in histone packing and modification for cell signaling. The appearance of K-channel associated domains (4.37.1.2) and signaling/transport protein phosphatases (4.145.1.3) also show diversity in signaling/transport pathways in *C. elegans*

The worm also has a plethora of C-type lectins (4.154.1.1), which are responsible for mediating intracellular contact information through surface carbohydrates.

Expansions of protease families (2.44.1.2) and families involved in protein prenylation (1.111.6.1 and 3.9.2.2) in *Drosophila* suggest an emphasis on regulated, post-translational protein modifications in fly. Protein prenyl transferases attach hydrophobic prenyl groups to nuclear lamins as well as signaling molecules including ras superfamily members and the gamma subunits of trimeric G-proteins, all of which require this modification to attach to membranes and interact with effector molecules. Regulated prenylation has been demonstrated for farnesyl transferase in human [25] but has not been demonstrated for Rab prenyl transferase, which prenylates Rab proteins, a subcategory within the Ras superfamily and which help regulate discrete steps in the secretory pathway. Rab prenyl transferase is present in yeast, fly, worm, and human and requires a regulatory subunit (Rab Escort Protein) to bind and present the protein substrate to the α/β catalytic subunits of the enzyme [26, 27]. Interestingly, both fly and human possess a large number of proteins (fly: 104, human: 63) recognized by the single SCOP-HMM model trained on a structural motif present in the N-terminal region of the Rab prenyl transferase alpha subunit (3.9.2.2), a section of the protein proposed to interact with REP [28]. Expansion of the protease families in fly has already been discussed [2], but over-representation of proteins in fly and human exhibiting structural homology to protein prenyl transferases motifs has not been reported until now.

4 Conclusions

The SCOP structural domain hierarchy allows us to characterize diverse genomes in a complementary manner to previous work involving Pfam and Interpro. The two methods provide a consistent "view" of the most highly represented genes in the human, fly, worm, and yeast genomes.

Structure-based genome comparison, as provided by SCOP domain families, allows us to track the appearance or elimination of structural domains. Coupled with a system such as Gene Ontology, this method allows us to find related genes related via common structure that diverge in function or in expression pathway. We find that metazoans diverge from the yeast genome by a set of structure-based domains consistent with previous observations. These novel domains correspond to the well-characterized eukaryotic proteases, and processes and function associated with immune response-related, cell-cell interaction, and cell signaling pathways. Interestingly, some families of proteins which vary widely between genomes (e.g., nuclear receptors) appear much more within one genome than within another. These allow us to track evolutionary biases due to different uses of the same basic structural features.

To date, few of the genomes sequenced are eukaryotic. However, as shown here and in previous work [2, 3, 12], eukaryotic genomes exhibit a different composition of domains than prokaryotic genomes. Thus, if genomic databases are largely prokaryotic, researchers should bear in mind that such databases will have a decreased representation of certain classes of proteins important in eukaryotes and absent in prokaryotes, such as those involved in intercellular signaling, immune response, and cell-cell interactions.

5 References

1. Venter, J.C., et al., *The sequence of the human genome.* Science, 2001. **291**(5507): p. 1304-51.
2. Lander, E.S., et al., *Initial sequencing and analysis of the human genome.* Nature, 2001. **409**(6822): p. 860-921.
3. Apic, G., J. Gough, and S.A. Teichmann, *Domain combinations in archaeal, eubacterial and eukaryotic proteomes.* J Mol Biol, 2001. **310**(2): p. 311-25.
4. Devos, D. and A. Valencia, *Practical limits of function prediction.* Proteins, 2000. **41**(1): p. 98-107.
5. Wilson, C.A., J. Kreychman, and M. Gerstein, *Assessing annotation transfer for genomics: quantifying the relations between protein sequence, structure and function through traditional and probabilistic scores.* J Mol Biol, 2000. **297**(1): p. 233-49.
6. Gerstein, M. and R. Jansen, *The current excitement in bioinformatics-analysis of whole-genome expression data: how does it relate to protein structure and function?* Curr Opin Struct Biol, 2000. **10**(5): p. 574-84.
7. Murzin, A.G., et al., *SCOP: a structural classification of proteins database for the investigation of sequences and structures.* J Mol Biol, 1995. **247**(4): p. 536-40.
8. Bateman, A., et al., *The Pfam protein families database.* Nucleic Acids Res, 2000. **28**(1): p. 263-6.
9. Elofsson, A. and E.L. Sonnhammer, *A comparison of sequence and structure protein domain families as a basis for structural genomics.* Bioinformatics, 1999. **15**(6): p. 480-500.
10. Teichmann, S.A., C. Chothia, and M. Gerstein, *Advances in structural genomics.* Curr Opin Struct Biol, 1999. **9**(3): p. 390-9.
11. Gerstein, M. and H. Hegyi, *Comparing genomes in terms of protein structure: surveys of a finite parts list.* FEMS Microbiol Rev, 1998. **22**(4): p. 277-304.
12. Wolf, Y.I., et al., *Distribution of protein folds in the three superkingdoms of life.* Genome Res, 1999. **9**(1): p. 17-26.
13. Wolf, Y.I., N.V. Grishin, and E.V. Koonin, *Estimating the number of protein folds and families from complete genome data.* J Mol Biol, 2000. **299**(4): p. 897-905.
14. Govindarajan, S., R. Recabarren, and R.A. Goldstein, *Estimating the total number of protein folds.* Proteins, 1999. **35**(4): p. 408-14.
15. Zhang, C. and C. DeLisi, *Estimating the number of protein folds.* J Mol Biol, 1998. **284**(5): p. 1301-5.
16. Shigeta, R., et al., *Generalized Rapid Automated Protein Analysis (GRAPA): annotating the human genome based on SCOP domain-derived hidden Markov models.* submitted, 2001.

17. Gough, J., et al. *Optimal Hidden Markov Models for All Sequences of Known Structure.* in *Currents in Computational Molecular Biology 2000.* 2000.
18. Reese, M.G., et al., *Genie--gene finding in Drosophila melanogaster.* Genome Res, 2000. **10**(4): p. 529-38.
19. Brenner, S.E., P. Koehl, and M. Levitt, *The ASTRAL compendium for protein structure and sequence analysis.* Nucleic Acids Res, 2000. **28**(1): p. 254-6.
20. Karplus, K., C. Barrett, and R. Hughey, *Hidden Markov models for detecting remote protein homologies.* Bioinformatics, 1998. **14**(10): p. 846-56.
21. Hughey, R. and A. Krogh, *Hidden Markov models for sequence analysis: extension and analysis of the basic method.* Comput Appl Biosci, 1996. **12**(2): p. 95-107.
22. Rubin, G.M., et al., *Comparative genomics of the eukaryotes.* Science, 2000. **287**(5461): p. 2204-15.
23. Pruitt, K.D. and D.R. Maglott, *RefSeq and LocusLink: NCBI gene-centered resources.* Nucleic Acids Res, 2001. **29**(1): p. 137-40.
24. Ashburner, M., et al., *Gene ontology: tool for the unification of biology. The Gene Ontology Consortium.* Nat Genet, 2000. **25**(1): p. 25-9.
25. Goalstone, M.L., et al., *Insulin signals to prenyltransferases via the Shc branch of intracellular signaling.* J Biol Chem, 2001. **276**(16): p. 12805-12.
26. Armstrong, S.A., et al., *cDNA cloning and expression of the alpha and beta subunits of rat Rab geranylgeranyl transferase.* J Biol Chem, 1993. **268**(16): p. 12221-9.
27. Andres, D.A., et al., *cDNA cloning of component A of Rab geranylgeranyl transferase and demonstration of its role as a Rab escort protein.* Cell, 1993. **73**(6): p. 1091-9.
28. Zhang, H., M.C. Seabra, and J. Deisenhofer, *Crystal structure of Rab geranylgeranyltransferase at 2.0 A resolution.* Structure Fold Des, 2000. **8**(3): p. 241-51.

CONSTRUCTING COMPARATIVE GENOME MAPS WITH UNRESOLVED MARKER ORDER

DEBRA GOLDBERG

Center for Applied Mathematics, Cornell University, Ithaca NY 14853
E-mail: debra@cam.cornell.edu

SUSAN MCCOUCH

Department of Plant Breeding, Cornell University, Ithaca NY 14853

JON KLEINBERG

Department of Computer Science, Cornell University, Ithaca NY 14853

Comparative genome maps are a powerful tool for interpreting the genomes of related organisms. The species maps which are the input to the process of constructing comparative maps are often themselves constructed from incomplete or inconsistent data, resulting in markers (or genes) whose order is not fully resolved. This incomplete marker order information is often handled by placing markers whose relative order cannot be reliably inferred together in a bin which is mapped to a common location. Previous automated and manual methods have handled such markers in an *ad hoc* or arbitrary way. We present efficient algorithms for comparative map construction that provide a principled method for handling unresolved marker order. The algorithms are based on a technique for efficiently computing a marker order that optimizes a natural parsimony criterion; in this way, they also yield a working hypothesis about the original incomplete data set.

1 Introduction

Comparative mapping is based on the observation that the order of homologous genes along the chromosomes of related species is often conserved. *Colinearity* (conservation of gene order), and to a lesser extent *synteny* (neighborhoods containing a number of homologous gene pairs) in chromosomal regions of different species suggests that these chromosomal segments are likely to be *homeologous* (derived from a common ancestral linkage group). Comparative maps identify colinearity or synteny between genomes of different species, and allow us to exploit the research accumulated for each of the species under consideration to gain new insights into issues including gene characterization, phylogenetic relationships, and principles of chromosome evolution.

Work in comparative mapping dates back as far as studies of Sturtevant and Weinstein in the 1920's[11,12], and it has grown into a very large area of research. We refer the reader to O'Brien et al[7] for a general review of comparative studies in mammals, Paterson et al for plants[8], and Römling and

Tümmler[9] for bacterial genomes. Indeed, comparative genomics has proven so useful in understanding human genetics that it has been termed "the key to understanding the human genome project"[2].

Despite the considerable amount of research in this area, there has been relatively little algorithmic work aimed at formalizing what is meant by a *comparative map* in a mathematical sense, or at providing a precise means for computing such a map from input data. In the absence of such a framework, the maps produced by different labs have been constructed on an individual and largely *ad hoc* basis, making it difficult to reason about these different maps from a common set of principles. Motivated by this state of affairs, Nadeau and Sankoff challenged the community in 1998 "to devise objective methods that reduce the arbitrary nature of comparative map construction"[6].

In recent work[4], we have proposed a formal model of comparative mapping as a *chromosome labeling problem*, in which the goal is to divide chromosomes into contiguous segments for which there is significant evidence of common ancestral linkage groups. We provide background on this model in Section 2.1, where we show a natural labeling criterion for chromosomal segments, based on a trade-off between parsimony and consistency, under which the optimal labeling can be computed efficiently. This approach is distinct from sequence-alignment methods, which work on a much more localized scale, and also distinct from algorithms for inferring chromosomal rearrangement scenarios, which essentially start with a structure like a comparative map, and propose hypotheses about evolutionary history. Our framework is perhaps most similar to work of Sankoff, Ferretti, and Nadeau[10] which seeks to find non-contiguous segments that "cover" all homologous genes in a dataset.

1.1 The present work: Unresolved marker order

In this paper, we propose an algorithmic approach for addressing the ubiquitous problem of *unresolved marker order* in comparative map construction. Genetic maps are constructed from linkage analysis of finite mapping populations, and frequently there are markers which cosegregate in all individuals, so their relative order cannot be determined. Analogously, in physical maps, two markers which are contained in exactly the same set of clones also cannot be ordered precisely. In addition, the raw data may be ambiguous or inconsistent, due to experimental or statistical error, leading to markers whose relative order cannot be determined with a sufficient degree of confidence. Existing computational techniques for comparative map construction, however, have generally relied on the tacit assumption that there is a completely specified linear order on markers; and this assumption is present in our previous work. Since this assumption rarely holds in practice, the resolution of marker order has essentially been dealt with in an *ad hoc* or arbitrary way.

Here we develop a principled method for handling unresolved marker order within our model of comparative mapping. We work with a standard representation for markers whose order cannot be determined: the markers are partitioned into bins, or *megaloci*; markers within the same megalocus are considered to have an unknown relative order, but there is a total order on the megaloci themselves. Thus, the resulting dataset looks linearly ordered, except that in place of individual markers we have a sequence of megaloci.

We provide an efficient algorithm that simultaneously constructs a comparative map and an ordering of the markers in each megalocus. These two tasks are inter-related, in the sense that the megalocus orders are computed so as to optimize a natural parsimony criterion for the map. Our main technical result is to show that these optimal orders can be computed in polynomial time, and, indeed, by an algorithm that performs well in practice. Indeed, we will see that the algorithm can actually run faster in practice on an input with megaloci than on a totally ordered set of markers of the same size; this is essentially because the megaloci serve as a "compressed" representation which the algorithm can manipulate at a high level. We supplement our algorithms with a set of results showing comparative map construction based on this method for mouse-human data. We note that while our optimal orders thus provide a canonical hypothesis about marker order, which can serve as a basis for further lab work, we do not claim that they represent the "correct" or "true" order — essentially, we simply do not have enough information in these settings to identify such a correct order.

A number of studies use representations not based directly on megaloci, and our approach can be adapted to handle several of these as well. We briefly discuss one such extension in Section 3.2.

2 Algorithms

We cast comparative mapping as a *labeling* problem, as in our previous work[4]. We begin with two genomes, the *base* and the *target*, and we wish to *label* segments of the target using names of linkage groups from the base. In this section, we describe our underlying algorithms in detail. First we give some background, including notation and a review of the previously-published algorithms which form the foundation for this work. Then we develop a linear megalocus algorithm, which is extended to a stack megalocus algorithm in the final subsection. In Section 3, we discuss an implementation of this algorithm, and show some results from a comparative analysis of the human and mouse genomes, with human as the base and mouse as the target.

2.1 *Chromosome Labeling: An Approach to Comparative Mapping*

We fix a chromosome in the target genome, and let $M = \langle 1, 2, \ldots, n \rangle$ denote the sequence of comparatively mapped markers (genes) in order on this chromosome. We divide the base genome into linkage groups (usually chromosome arms) which will serve as labels for the target genome. Thus, we have a label set $L = \{c_1, c_2, \ldots, c_k\}$, where k is the number of linkage groups in the base genome. A comparative map is viewed as an assignment of labels to the markers of the target genome, i.e. a function $f : M \to L$.

We assume each marker i has been comparatively mapped to a single linkage group ℓ_i in the base genome (i.e. each marker i has a single homolog in the base genome, and it is located on ℓ_i). We say marker i has *type* ℓ_i. Markers that have not been comparatively mapped in the base genome are not informative for our purposes. We define a simple distance function $\delta(\cdot, \cdot)$ on pairs of labels as follows: $\delta(a, b) = 0$ if $a = b$; and $\delta(a, b) = 1$ if $a \neq b$. We extend this definition if one of the parameters is a set A, so that $\delta(A, b) = 0$ if $b \in A$; and $\delta(A, b) = 1$ otherwise. In the context of a labeling we say a marker i *matches* its label if $\delta(\ell_i, f(i)) = 0$; we call it a *mismatched marker* otherwise.

In previous work, we cast comparative mapping as the problem of computing an *optimal* labeling of the marker set[4]. In our basic *linear model*, the optimization criterion was based on balancing a *mismatch* penalty m and a *segment boundary* or *segment opening* penalty s. We refer the reader to Figure 1 for details. Only the ratio s/m affects the resulting optimal labelings, so this ratio is essentially the only tunable parameter in the algorithm; intuitively, s/m gives a minimum number of matching markers required to consider opening a new segment. We have found that in practice, the results produced by the algorithm are generally stable over a fairly wide range of parameter values.

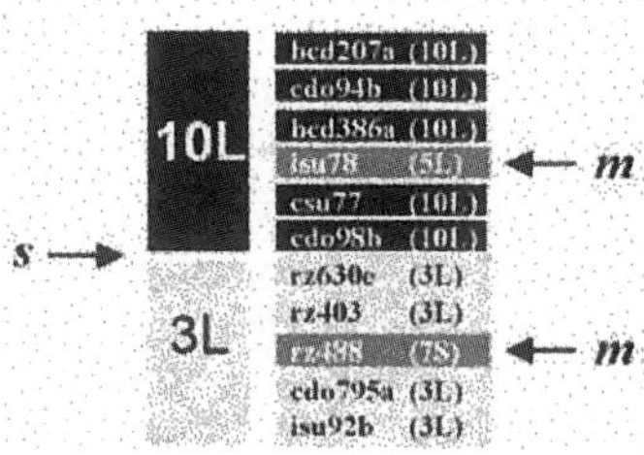

A potential labeling is scored by assessing the penalty m for each mismatched marker, and the penalty s for each segment boundary. Formally, the objective function $Q(f)$ is:

$$s\,(|\{i : f(i) \neq f(i+1)\}|) + m\,(|\{i : f(i) \neq \ell_i\}|)\,.$$

The objective function is minimized in an optimal labeling.

Figure 1. Diagram showing scoring scheme for linear model.

To compute an optimal labeling in this model, we use a dynamic programming formulation in which $S[i, a]$ denotes the optimal cost for labeling the suffix of M beginning at position i subject to the condition that $f(i) = a$. We initialize $S[n, a] = m \cdot \delta(\ell_i, a)$, and compute the optimal solution using a recurrence relation as follows:

$$S[i,a] \quad = \quad m \cdot \delta(\ell_i, a) \quad + \quad \min_{b \in L} \left(S[i{+}1, b] \quad + \quad s \cdot \delta(b, a)\,\right). \qquad (1)$$

We extended the linear model to a stack model[4] which allows labels to be remembered as though pushed and popped from a stack. In the stack model, a label can still change at a segment boundary by being replaced with another label (with associated penalty s) as in the linear model, but a label can also change by having another label pushed on top of it, which can later be popped off to recall the earlier label. This is demonstrated in Figure 2a. Pushing a label also incurs a penalty of s, but popping is nearly free, incurring only a small penalty ϵ. This corrects the linear algorithm's problem with long-range dependencies, which impedes the labeling of "*aba*" label patterns generated by insertions and other important chromosome rearrangement events.

An optimal labeling in the stack model can also be computed using a dynamic programming algorithm, in which $S[i, j, a]$ denotes the optimal cost of a labeling f of the subsequence of M which starts at position i and ends at position j, subject to the condition that $f(i) = a$. We initialize $S[i, i, a] = m \cdot \delta(\ell_i, a)$, and make use of the following recurrence:

$$S[i,j,a] \quad = \quad \min \begin{pmatrix} m \cdot \delta(\ell_i, a) \; + \; \min_{b \in L} \; (S[i{+}1, j, b] \; + \; s \cdot \delta(b, a)\,), \\ \min_{i<k<j} \; (S[i, k, a] \; + \; S[k{+}1, j, a] \; + \; \epsilon\,) \end{pmatrix} \qquad (2)$$

See Figure 2b for a graphical view of how push/pop is accomplished.

We note that the balance between minimizing mismatches and minimizing stack operations reflects the notion of parsimony discussed in the introduction. While we do not go into the details here, our objective function can be viewed as arising from a maximum *a posteriori* approach with a prior probability term favoring labelings that involve a small number of stack operations. We also note that the process of pushing and popping on a stack is suggestive of the biological process of insertion; this idea is discussed in our earlier work[4].

a. b.

Figure 2. Graphical representation of: a. the stack model, and b. push/pop as implemented in recurrence of stack algorithm.

2.2 Linear Megalocus Model

We now extend our labeling framework to the case in which the input contains *megaloci*, the types of unordered sets of markers discussed in the introduction. The goal is to produce an order for the markers in each megalocus so that the resulting totally ordered set of markers has a labeling with as low a score as possible (under the linear or stack models respectively). The main difficulty here is that the ordering problems in the different megaloci can interact in complex ways, since we must produce a labeling for the full ordered set of markers. Despite this, we show that an order yielding an optimal labeling can be computed efficiently for both the linear and stack models.

We begin with the linear model, since the algorithm for the stack model will build on this. For each megalocus, we consider the set of markers belonging to the megalocus as a *supernode* Z. Within Z, there is an optimal ordering that clusters markers of the same type contiguously. Thus we will search only for solutions of this form: we seek an ordering over these *clusters*, rather than over the markers themselves. Figure 3a depicts an example of a supernode with four clusters, each consisting of markers of the same type.

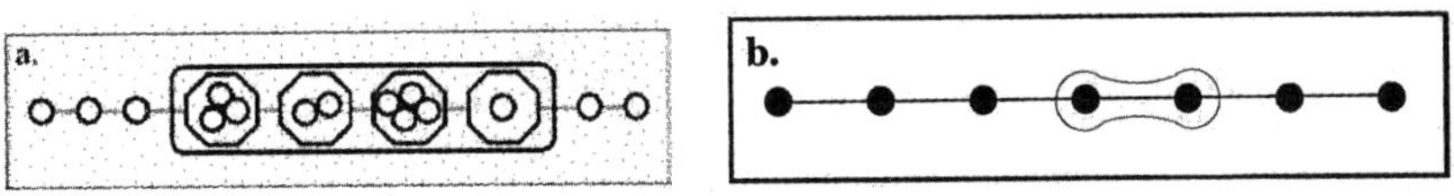

Figure 3. Diagram showing: a. markers (circles) arranged into clusters (octagons) within a supernode (rectangle), and b. High-level view of the map as viewed by the algorithm.

The key idea in the algorithm is that the clusters selected for the beginning and end of the supernode order are the ones that determine how the labeling of the chromosome before and after the supernode will interact with the labeling of the markers in the supernode. Once these two extreme clusters are selected, the remaining clusters can be ordered essentially arbitrarily. In keeping with this idea, we create a representation of the chromosome in which the supernodes and markers outside supernodes are totally ordered, and each supernode is represented by *two* consecutive positions in the order; the first of these positions will be assigned a label for the beginning of the supernode, and the second will be assigned a label for the end of the supernode. (See Figure 3b.) This pair of labels will be enough for us to determine an optimal ordering within the supernode by a post-processing step. Specifically, clusters within the supernode corresponding to the first supernode label (if any) will be placed at the beginning, clusters corresponding to the second label will be placed at the end, and the remainder will be ordered arbitrarily. If the two supernode labels are the same, then the markers matching this label can all be placed at the end, and all other clusters will be considered mismatches.

The full details of the algorithm and its correctness proof are somewhat complex, and due to the space limitations we can only sketch them here. The reader is referred to the Ph.D. thesis of the first author[3] for these details. Let $n' \leq n$ denote the number of positions in the modified map after each supernode has been replaced by a pair of positions. Let S denote the set of indices of supernode start positions, and E denote the set of indices of supernode end positions. If a is a label and $j \in S$ (so $j+1 \in E$), we define $n_j(a)$ to be the number of markers of type a in the supernode associated with start position j, and n_j to be the total number of markers in this supernode. We define $\ell_j = \ell_{j+1}$ to be the set of labels containing a homolog of a marker in the supernode associated with position j; i.e. $\ell_j = \{a \in L | n_j(a) \geq 1\}$.

The optimal labeling is constructed from a dynamic programming recurrence that follows the recurrence used in the basic linear model. As before, $S[i,a]$ denotes the optimal cost for labeling the suffix beginning at position i subject to $f(i) = a$. For markers outside supernodes, and for supernode end positions, this is built from $S[i+1,\cdot]$ as before. To deal with supernode start positions we include a cost for labeling the markers "hidden" inside the supernode by our representation. This cost can be determined from the labels for the supernode start and end positions, by augmenting the recurrence with a *hidden marker penalty* $p(i,\cdot,\cdot)$ defined for supernode start positions $i \in S$.

Thus, with $S[n',a] = m \cdot \delta(\ell_i, a)$, the recurrence is as follows:

$$S[i,a] \;=\; m \cdot \delta(\ell_i,a) \;+\; \min_{b \in L} \left(S[i+1,b] \;+\; s \cdot \delta(b,a) \;+\; p(i,a,b)\,\right). \quad (3)$$

It remains to define the hidden marker penalty $p(i,a,b)$. To prevent the implicit placement of a marker at both the start and end of a supernode, the case in which the two supernode indices receive the same label must be considered separately from the case in which they receive different labels, resulting in a two-case structure for $p(i,a,b)$:

$$p(i,a,b)) = \begin{cases} \sum\limits_{c \neq a,b} (\min(s, m \cdot n_i(c)) \;+\; m \cdot \Delta_i(a,b) & : \text{ for } i \in S,\ a \neq b \\ \sum\limits_{c \neq a} (m \cdot n_i(c)) \;+\; m \cdot \Delta_i(a,a) & : \text{ for } i \in S,\ a = b \\ 0 & : \ i \notin S. \end{cases} \quad (4)$$

Markers which match either of the pair of supernode labels (a or b) will not impart any mismatch penalty, and segment opening penalties associated with these labels are handled explicitly by the recurrence, so we need only consider the "hidden" markers which don't match either of the labels a or b.

For $c \neq a, b$, the first terms in the definition of $p(i,a,b)$ give the total cost attributed to markers of type c due to mismatched markers *or* a segment boundary penalty. The definition for the case $a = b$ does not hinder homeologies from being labeled within a supernode; rather it requires there be distinct labels at the two ends of the supernode whenever markers in the supernode should be labeled with at least two labels.

The function Δ_i used in computing $p(i,a,b)$ adjusts for the effect of assigning mismatch penalties (m) both in the recurrence (for the first and last positions in the supernode) and in the function p. It is defined as follows. For $i \in S$ we define $\mu_i(a,b) = (\delta(\ell_i,a) + \delta(\ell_i,b))$, which is the number of mismatch penalties assessed by the recurrence for a supernode labeled with a and b at its ends. Note that $\mu_i(a,b) \in \{0,1,2\}$. The notation $min_{\mu_i(a,b)}$ indicates to sum the $\mu_i(a,b)$ smallest values.

$$\Delta_i(a,b) = \begin{cases} 0 & : \mu_i(a,b) = 0 \\ -2 & : \mu_i(a,b) > 0, a = b \\ 0 & : \ell_i \in \{\{a\},\{b\}\}, a \neq b \\ -2 & : (*) \\ \min_{\substack{\mu_i(a,b) \\ c \neq a,b \\ c \in \ell_i}} \begin{Bmatrix} 0 : m \cdot n_i(c) \geq s \\ -1 : m \cdot n_i(c) < s \end{Bmatrix} & : \text{otherwise} \end{cases} \tag{5}$$

Condition $(*)$ is invoked when the first three conditions do not apply, the entire supernode has length $< s$ and consists exclusively of markers of a single type c, and c is not among the labels at the ends of the supernode.

By computing $p(i,a,b)$ for all i, a, and b prior to the recurrence loop and appropriate ordering of operations, this algorithm has running time $O(k^2 n)$, which is the same computational complexity as the original linear model. Since we view the label set as having fixed constant size, this is a running time linear in the number of markers.

2.3 Stack Megalocus Model

We now extend the stack model to also allow rearrangement of markers within megaloci. Since the order of clusters internal to a supernode (i.e. those that don't match the label at either end) is not explicitly determined by the recurrence, we do not allow pushing or popping with internal markers of a supernode; this maintains the stack structure through the megaloci. Given this, the algorithm for the stack megalocus model is based on the function $p(i,a,b)$ defined above for the linear megalocus model, together with a dynamic programming recurrence in which $S[i,j,a]$ has the same meaning as in the basic stack model. We initialize $S[i,i,a] = m \cdot \delta(\ell_i,a)$, and use the following recurrence:

$$S[i,j,a] = \min \left\{ \begin{array}{l} m \cdot \delta(\ell_i, a) \; + \; \min\limits_{b \in L} \left(S[i+1,j,b] + s \cdot \delta(b,a) + p(i,a,b) \right) \\ \min\limits_{\substack{i<k<j \\ k \notin S}} \left(S[i,k,a] + S[k+1,j,a] \right) + \epsilon \end{array} \right\} \tag{6}$$

The correctness of the algorithm is established by arguing that an appropriate hidden marker penalty computation $p(\cdot,\cdot,\cdot)$ for a given supernode is included exactly once in a subproblem if and only if the subproblem includes both the supernode start and end positions (and thus includes all the markers of the supernode). The algorithm has running time $O(kn^3)$, which is the same computational complexity as the original stack model. In practice, this algorithm is actually *faster* than the basic stack algorithm for a totally ordered marker set of the same size; for the running time is more precisely $O(k(n')^3)$, and the reduction in the number of elements in the modified map more than makes up for the additional processing for each element.

3 Results and Discussion

3.1 Computational results

The stack megalocus algorithm was implemented in Java and executed on a Sun Ultra-Sparc 10 running Solaris. The implementation was verified using synthetic data. We tested the stack megalocus algorithm with mouse-human data taken from the Mouse Genome Database[1]. The resulting comparative maps compared favorably with the mouse-human maps published by the Human Genome Project[5], despite the fact that they were produced from different input data. Chromosomes with up to 260 markers ran in about 10 seconds. The total processing time for all 19 mouse autosomes is about two minutes. Results were displayed using an OpenDX visualization program, as explained in Figure 4. Due to memory limitations, one mouse chromosome could only be processed after the label set L was manually reduced to only those chromosomes possible in an optimal labeling. We are exploring many space-efficiency options, but are not too concerned since mouse-human is the densest comparative data set, and computing power will improve as more data accumulates.

To provide a sense for the types of analysis one obtains from our stack megalocus algorithm, we have extracted the labeling of three small chromosomal regions in the mouse genome from a full genome analysis. Results from the original stack algorithm and the stack megalocus algorithm are shown and contrasted in each case. In these cases, as in many portions of the genome, the

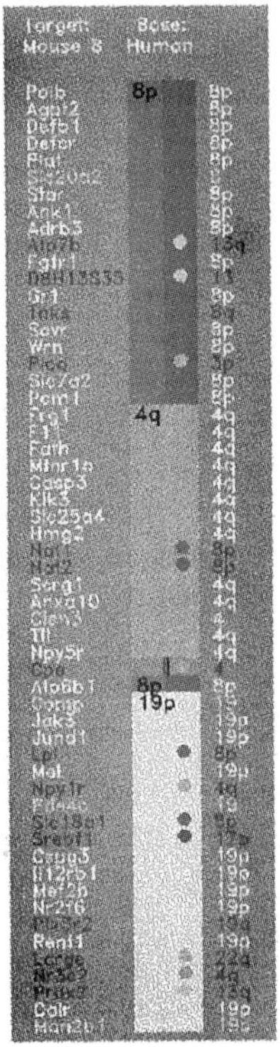

In the accompanying figure, the column of marker names on the left are the mouse chromosome 8 markers which have known homologs (actually orthologs) in human. The shaded rectangles to the right of these marker names show the labeling assigned by one of our comparative mapping methods, colored by chromosome. The actual visualization is in full color for optimally distinguishing among labels; the label name itself is displayed at the top of each rectangle. A translucent band is overlaid over the left half of these rectangles to indicate the arm. The rightmost column shows the linkage group of the homolog. Some of the homologs are mapped to a centromeric region, and others are mapped only to a chromosome (the arm is unknown); depending on the precise location of these, they may match the linkage group of either chromosome arm. Marker names and homolog locations of markers which match their assigned label are shown in white, mismatches are shown as black, and those which *may* be matches are shown in gray. A circle color-coded by the chromosome of the homolog is overlaid on the labeling rectangles, providing another way to visualize most mismatches (mismatches involving the two distinct arms of one chromosome are not apparent this way). Certain of the segments (colored rectangles) are connected with an intervening black bar, the result of a post-processing heuristic that indicates portions of the chromosome not considered clearly homeologous to any human linkage group.

Figure 4. Results of the original stack model for a portion of mouse chromosome 8.

rearrangement of markers in a megalocus allows a significantly more parsimonious labeling and provides a hypothesized canonical order for these markers.

Figure 5a shows a region of mouse chromosome 5 where rearrangement of markers in the same megalocus has allowed for a map with fewer mismatches. Figure 5b shows a region of mouse chrosome 8 where parsimonious rearrangement of co-located markers has resulted in a map where mismatched markers can be placed between labeled segments, which is preferable. Figure 5c shows a region of mouse chromosome 13 where rearrangement has also enabled the formation of an additional labeled segment. In this case the proposed segment is a small segment that pushed to a larger segment further down the chromosome, which could be suggestive of a chromosomal rearrangement such as an insertion or inversion. Again, this is a hypothesis that can be tested by further lab investigations (for example, by placing additional markers in this region or sequencing an area around this region in both genomes).

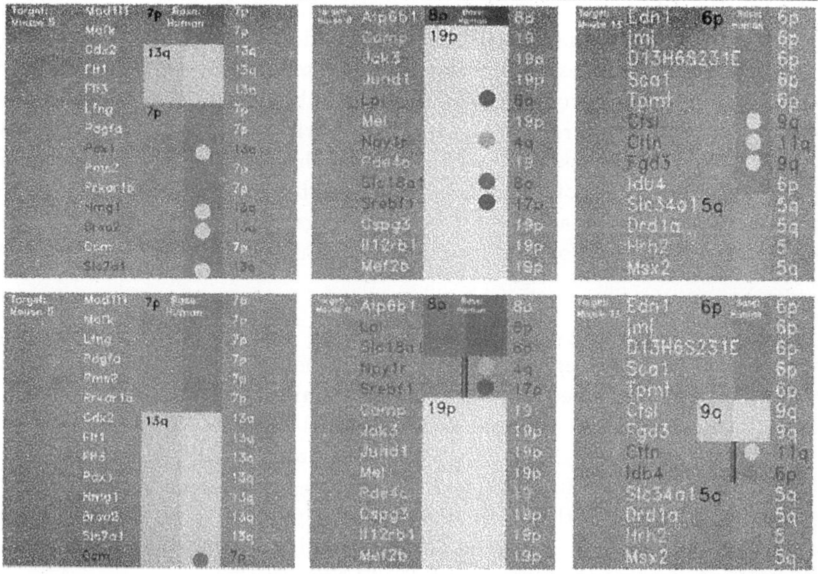

a. Mouse chromosome 5   b. Mouse chromosome 8   c. Mouse chromosome 13

Figure 5. **Detail of 3 mouse chromosomes in mouse-human comparisons. The upper figures shows results from the stack algorithm, and the lower figures show results from the stack megalocus algorithm.**

3.2 Discussion

This paper seeks to lay a principled foundation for comparative mapping studies in the presence of uncertain marker order. We use marker order and not distance between markers, and have not incorporated species-specific information, so that our algorithms work for a wide variety of species, for genetic and physical as well as high- and low-resolution species maps. We impose no assumptions about evolutionary mechanisms. Results from these algorithms can form the basis of hypotheses to guide further lab studies.

Some maps, such as most versions of the human map, do not use the megalocus representation; instead each marker is assigned an interval where it is likely to be located. The intervals of two markers overlap if and only if their relative order cannot be resolved, and the relative order of markers may be neither completely known nor completely unknown. For these, there is a simple modification to the stack algorithm that assigns a boundary point at the beginning and end of each interval, and assigns a fractional weight to the subinterval between consecutive boundary points for each marker whose interval spans it, such that the weight attributed to each marker sums to one.

We are in the process of putting together a web site where our suite of comparative mapping programs will be made publicly available. DeCAL (Detecting Common Ancestral Linkage segments) will provide access to both the stack algorithm and the stack megalocus algorithm. We are also investigating algorithmic extensions of this work; in particular, it is an interesting open question to extend the approach for pairwise genome comparison discussed here to one that allows for the simultaneous comparison of multiple genomes.

3.3 Acknowledgments

We thank Chris Pelkie and Vic Goldberg for their help with this work. The first author was supported in part by NSF Training Grant DEB-9602229 and the Packard Foundation Fellowship of the third author. The second author was supported in part by USDA National Research Initiative grant 94-37310-0661 and Cooperative State Research Education and Extension Service NYC 149-401. The third author was supported in part by a David and Lucile Packard Foundation Fellowship, an ONR Young Investigator Award, and NSF Faculty Early Career Development Award CCR-9701399.

References

1. JA Blake *et al. Nucleic Acids Research* 29(1):91-94, 2001.
2. MS Clark. *Bioessays* 21(2): 121-130, 1999.
3. D Goldberg. Algorithms for the construction of comparative genome maps. Ph.D. thesis, Cornell University, 2001.
4. D Goldberg, S McCouch, J Kleinberg. in *Comparative Genomics*, D Sankoff, JH Nadeau, eds., *Series in Computational Biology* Vol 1, Kluwer Academic Press, 2000.
5. Mouse and Human Genetic Similarities. Human Genome Project, Image Gallery. http://www.ornl.gov/hgmis/graphics/slides/98-075r2jpg.html
6. J Nadeau and D Sankoff. *Trends in Genetics*, 14(12):495-501, 1998.
7. SJ O'Brien *et al. Science* 286:458-481, 1999.
8. AH Paterson *et al. Plant Cell* 12(9):1523-1539 2000.
9. U Römling and B Tümmler. *J Biotechnology* 33:155-164, 1994.
10. D Sankoff, V Ferretti, JH Nadeau. *J Comp Bio* 4(4):559-565, 1997.
11. AH Sturtevant. *Proc. Nat. Acad. Sci.* 7:235-237, 1921.
12. A Weinstein. *Proc. Nat. Acad. Sci.* 6:625-639, 1920.

REPRESENTATION AND PROCESSING OF COMPLEX DNA SPATIAL ARCHITECTURE AND ITS ANNOTATED GENOMIC CONTENT

RACHID GHERBI AND JOAN HERISSON

Gesture and Image group
LIMSI-CNRS, Université Paris-Sud,BP 133, F-91403 ORSAY CEDEX, FRANCE*
http://www.limsi.fr
E-mail: gherbi@limsi.fr, herisson@limsi.fr
Phone: +33 1 69 85 81 64 or 66
Fax: +33 1 69 85 80 88

This paper presents a new general approach for the spatial representation and visualization of DNA molecule and its annotated information. This approach is based on a biological 3D model that predicts the complex spatial trajectory of huge naked DNA. With such modeling, a global vision of the sequence is possible, which is different and complementary to other representations as textual, linguistics or syntactic ones. The DNA is well known as a three-dimensional structure. Whereas, the spatial information plays a great part during its evolution and its interaction with the other biological elements This work will motivate investigations in order to launch new bioinformatics studies for the analysis of the spatial architecture of the genome. Besides, in order to obtain a friendly interactive visualization, a powerful graphic modeling is proposed including DNA complex trajectory management and its annotated-based content structuring. The paper describes spatial architecture modeling, with consideration of both biological and computational constraints. This work is implemented through a powerful graphic software tool, named ADN-Viewer. Several examples of visualization are shown for various organisms and biological elements.

1 Introduction and state of the art

People use very often the standard textual format (or a near representation), considering DNA as a succession of letters (A, C, G or T) that represent the nucleotides of the molecule. With such plat representation, it is very hard to perceive any global pertinent information of DNA sequence. Even if t is possible to structure the sequence as a hypertext [3.], the user has finally only a local textual point of view at each level of the document.

The DNA is well known as a three-dimensional structure [10.]. Whereas, the spatial information plays a great part during its evolution and its interaction with the other biological elements [4. ,5. ,8. ,11. ,12. ,13. ,14. ,15. ,16. ,17. ,18. ,19.].

In this context, it appears essential to design software tools focused on the representation, visualization and interactive exploration of the three-dimensional information of DNA. Besides, it is necessary to add functionalities in order to perform quantitative measures and to systemize the processing of various types of spatial information (curvature, compactness, geometric distances, etc.). Based on biological 3D conformation models of naked DNA, this paper presents our work aiming at building virtual three-dimensional information from DNA sequences. This

construction will allow biologists to visually scan and characterize the spatial architecture of DNA of any size. This involves computational consideration, in particular for computer graphics algorithms (scene management algorithms, user interaction facilities, reliable and pertinent visualization and representation, etc.).
Some present and past Biological studies are concerned with the 3D structure of DNA. However, these studies are very specific to particular sequence elements and limited to small size of DNA sequences. The most used model is Bolshoy and Trifonov one [4.]. From computer science point of view, a few tools can visualize DNA sequences. They operate using a prediction algorithm, but they can visualize sequences that do not exceed on thousand of nucleotides (700 for *DNAtools*© [7.]). Many of these tools are developed by biologists teams and are exclusively dedicated to some particular biological problems. As far we know, there is no application that allows representing the spatial conformation of complete chromosomes.

This paper firstly explains the mechanism of the trajectory prediction. Then, a description about graphical representation and visualization of DNA sequences is presented. Some of the user interface functionalities are listed and finally many examples of visualization for different organisms are shown.

2 Biological interest of the DNA spatial information

The analysis of the nucleic sequences is a problem of a great complexity because of the superposition of various "signals" in the DNA. It is necessary to carry out genomic analysis with various and complementary approaches. The interest to model the structural aspects of the nucleic acids was charged for a long time and significant progress were made during these last years. This subject, far from becoming exhausted, does not stop growing rich starting from the experimental knowledge acquired on the DNA [11. ,4. ,8. ,13.].

One cannot define in an exact and complete way the existence of a functional promoter. Thus, even for simple models as *S. cerevisiae*, it is difficult to affirm by data processing analysis that an ORF (Open Reading Frame) of small size is actually expressed. Several assumptions explaining the putative role of the curve in the regulation regions exist now. The curved DNA can form large loops around the RNA polymerase, and thus increases the affinity of the DNA complex. It was shown that even a very low curve could increase enormously the affinity of the connection proteins-DNA, which led Suzuki and Yagi [15.] to put forth the assumption that the local curve can be used to precisely adjust the forces of interaction between the promoters and the regulation factors. It was also supposed that the curve of the DNA gathered the components of the transcriptional complex, spread out along the molecule of DNA. The curve and/or the structure in super-helix of the DNA results from a torsion stress which affects the energy of fusion of the DNA and the unfolding of the double helix, thus making it possible to assist (even to replace) corresponding initiating proteins.

3 Prediction of the 3D trajectory of DNA

DNA is a double strands (*Watson* and *Crick* [10.]) structure representing a double helix. The strands are anti-parallel and complementary (A is associated with T and C with G and *vice versa*. This association is called *base pair* (*bp*).

To build the 3D trajectory of DNA, two kinds of input data are necessary: DNA sequence in textual format and a 3D conformation model [1.]. Several 3D models exist. The most used are Bolshoy's [4.] and Cacchione's [9.] ones. The algorithm used for the 3D prediction of the trajectory of DNA is described in a previous work [1.].

4 Graphical representation and user interaction

4.1 Genomic and genic representations

When the sequence length is greater than few hundreds nucleotides or the observer is far from the DNA, he does not need to see all its details, but only some global information (shape, compactness, its curvature, etc.). In addition, for huge sequences (cf. figures pages 7, 8, 9 and 10), it is essential to reduce to the utmost of one's ability the representation. That is why, in genomic representation, we represent a nucleotides by a simple successive linked points and only the two strands are displayed (cf. Figure 1).

Figure 1. Sequence of 11 bp: genomic representation on the left, genic one on the right.

If the user wants to visualize or access to a little part of the sequence, he needs to see more details than in genomic representation. The genic representation provides supplementary information on nucleotides themselves. Indeed, each nucleotide is represented by a colored sphere (cf. Figure 1). This representation is suitable to visualize DNA sequences as genes, transposons, promoters regions, etc.

4.2 User interaction

ADN-Viewer integrates several powerful and friendly interaction capabilities. The user can manipulate and animate (geometric rotations and translations) the molecule using the mouse. Besides, the keyboard allows the user to control many parameters within the application. Some of interaction functionalities are for example fragment extraction (cf. Figure 2), the choice of any of existing conformation models, etc.

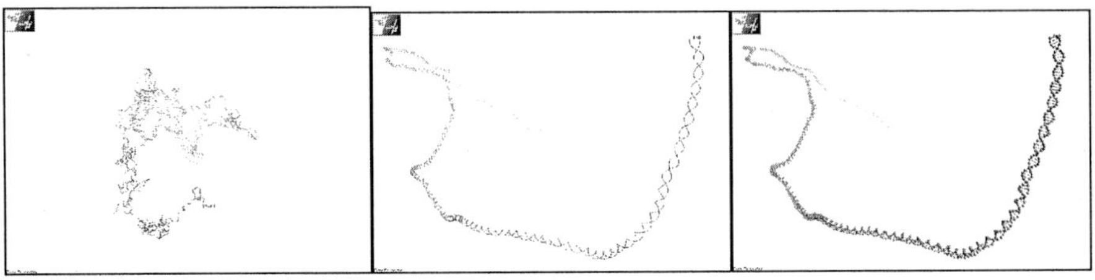

Figure 2. Sequence of 300 Kbp and an extracted part in genomic & genic representations.

5 Visualization and scene management

Several modes of visualization are provided within *ADN-Viewer.* The user can visualize a sequence according to different points of view. This functionality is very useful because, in general way, two 3D objects can have the same spatial properties in one view whereas they are different in the other ones. *ADN-Viewer* offers four different points of view: a front view using perspective projection; a front, left side and under views using orthogonal projection.

One can also visualize the molecule bulk in the three-dimensional space by displaying its bounding box (cf. Figure 3). This functionality is also used to compute quantitative compactness values (span, volume, density, etc.).

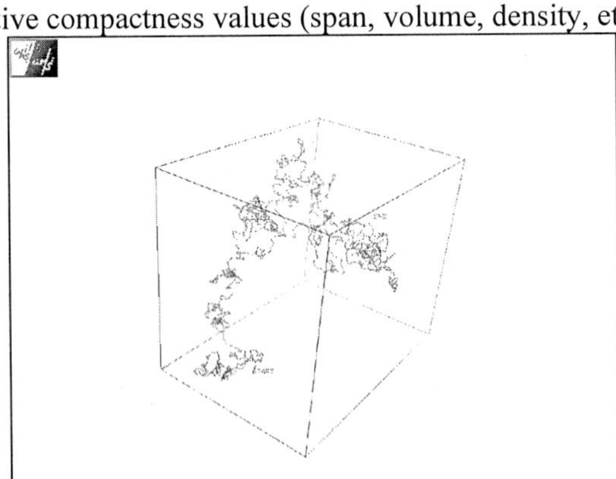

Figure 3. The bounding box gives bulk information on the sequence.

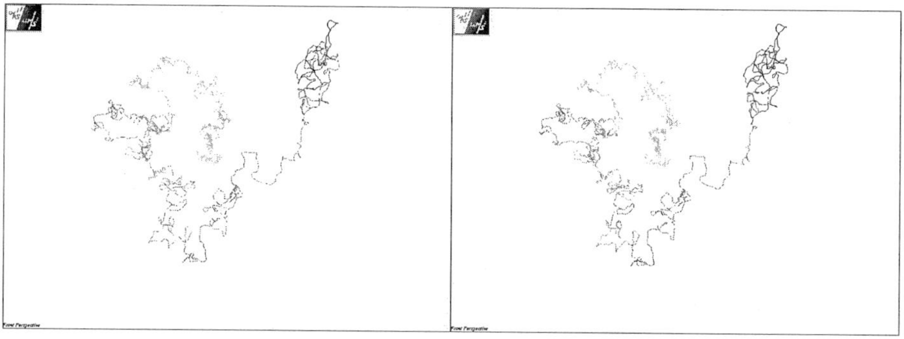

Figure 4. The *Yeast chrIII* original sequence and its sampled one by a factor value 20.

If the 16th chromosome of *Yeast* is considered as an example, it counts about 948061 bp. This implies to display about two millions of nucleotides (2 nucleotides per plate) at each movement of the molecule performed by the user. As we search a real time animation and interaction with the molecule, it is unthinkable to visualize such amount of points on classic workstation. It is thus necessary to filter as possible as the graphical information flow. Nevertheless, this process must not modify the trajectory of DNA. For the two kinds of representations (cf. section 4), sampling algorithms are applied on the sequence.

In the case of genomic representation, only the global shape of the trajectory is taken into account. The goal of the sampling algorithm is to reduce the number of displayed. The difficulty is to find the best sampling value that does not modify the trajectory but reduces the number of points allowing ADN-Viewer to function in interactive time. Presently, if the user changes the point of view of the DNA we must adapt the sampling according to the distance that separates the observer from the molecule. The Figure 4 illustrates an example of considerable sampling that does not modify the three-dimensional structure of the DNA.

For the genic representation, we use the same previous algorithm but applied on the number of displayed spheres and on their level of detail. This is also done according to the observer-molecule distance.

6 Modeling and visualization of annotated DNA sequences

The visualization of DNA sequence is very fruitful but not enough and it is necessary to carry out quantitative studies on the DNA molecule and its contents (density of genes, compactness cartography, curvature maps, spatial distribution, DNA-proteins interaction...). The trajectory of the DNA can inform us about its contents (genes, introns, exons, transposons...). Within the framework of this paper, we are interested in particular on genes in order to investigate their spatial relationship. The proteins that come to be fixed on a gene (upstream of it) for the transcription phase need to reach it directly. They must thus pass through the "swell of wool" formed by the DNA molecule. It thus appeared to us interesting to study the space of manœuvre available to these proteins to reach their target. For that, we are undertaking a comparative study of the distances that separate genes in order to reveal which areas are rich in genes. We can also study the correlation between parametric distances and spatial ones along the sequence. This work is in progress and some first results are shown by Figure 6 for genes content within chromosomes.

The visualization of DNA content needs interfacing ADN-Viewer with annotated accessible genomic databases or databanks. Nevertheless, there is no standard format used to describe a genomic sequence. In order to temporary bypass this problem, ADN-Viewer was interfaced with an ad-hoc database that contains only annotations of genes of some organisms. This solution presents two advantages compared to the online banks: the reliability of the data enabling ADN-Viewer to receive sure formatted information from the base, and its opening potentialities towards any existing or future structured databases. ADN-Viewer allows the user to access all the annotated information of the base through an explorer and to visualize them. When a sequence is selected, a level is added in the tree structure of the explorer and the user can access to all corresponding genes.

In the other side, ADN-Viewer will be able to augment the base by providing it pictures of the molecules as well as geometrical and spatial information.

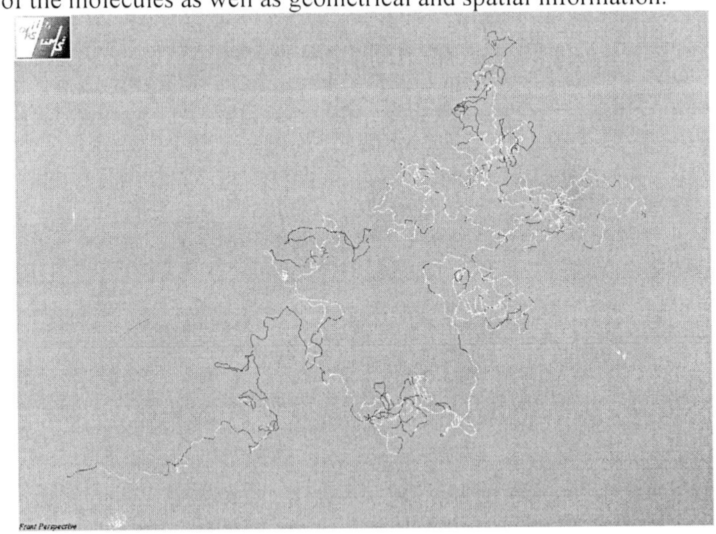

Figure 5. Visualization of genes content for *S. cerevisiae chrI*. Genes regions are white colored and intergenic regions are black colored.

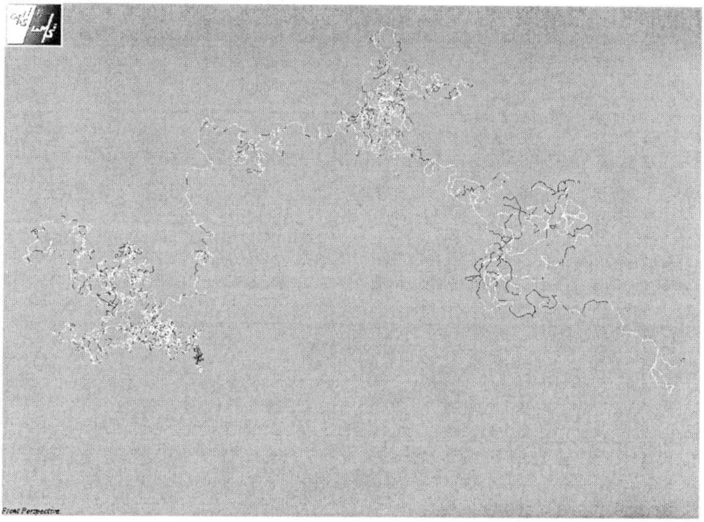

Figure 6. Visualization of genes content for *S. cerevisiae chrIV* (bottom view). Genes regions are white colored and intergenic regions are black colored.

7 Examples of visualization of genomes and biologic elements

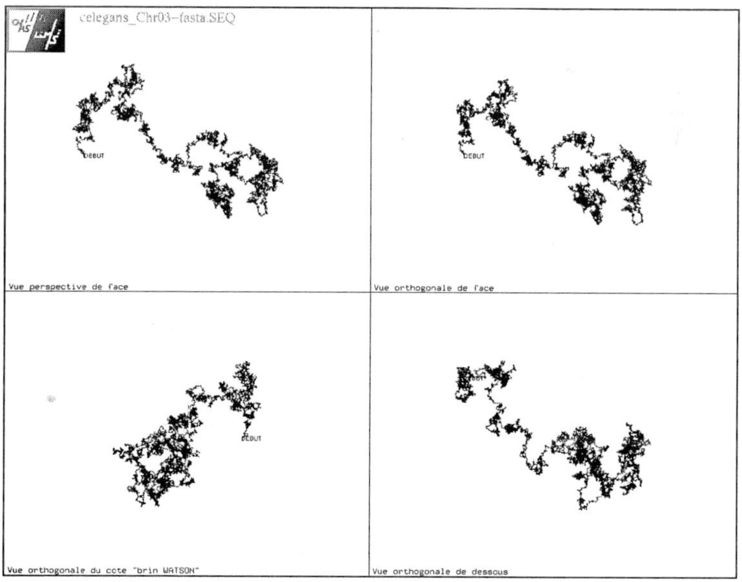

Figure 7. C. elegans_chr3 (11850213 bp).

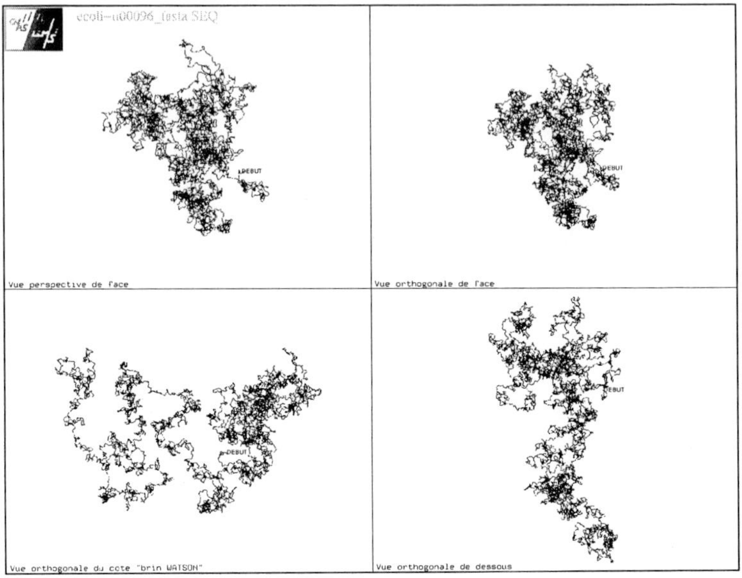

Figure 8. E. coli (4639221 bp).

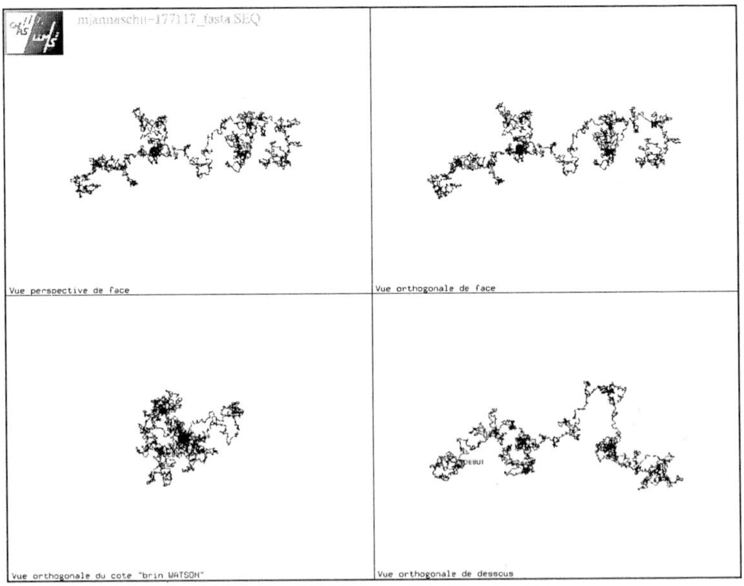

Figure 9. M. jannaschi (1664957 bp).

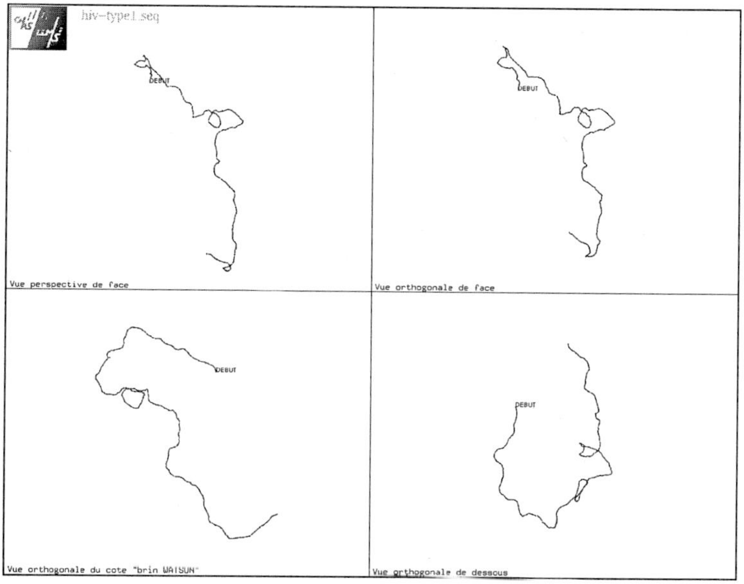

Figure 10. HIV_type1 (9181 bp).

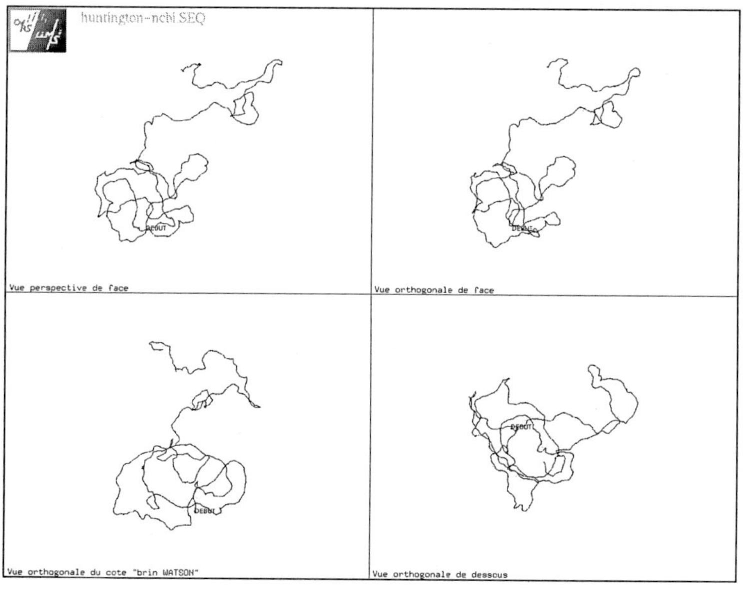

Figure 11. Huntington (55204 bp).

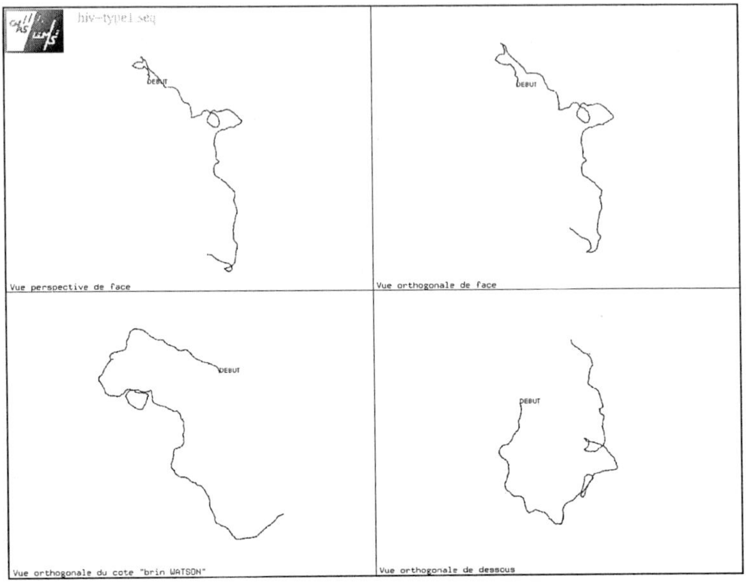

Figure 12. Fotl (1928 bp).

8 Conclusion and future work

The 3D representation and visualization of DNA molecule and interacting with it make it possible to have a global point of view of the sequence, in opposition with the textual format. This brings a new vision and an original approach suitable to launch new bioinformatics studies for the analysis of the genome. Besides, in order to obtain a friendly powerful interactive visualization, a dedicated modeling process is essential and various representations are necessary in order to adapt the visualization according to different analysis cases. In this paper, we carried out work about the spatial architecture of naked DNA, with consideration of both biological and computational constraints. A graphic software tool, named ADN-Viewer, implements this work.

Besides, it was interesting to represent and visualize the annotated biological elements of DNA sequences. To analyze spatial relations between genomic elements, it was possible to display the genes of chromosomes. Nevertheless, this implied to interface *ADN-Viewer* with genomic database and to structure the graphical representations for a better visualization and real-time user interfacing More generally, one can have a global vision of all annotated information in order to study various relations of genes, promoters, terminators, transposons, etc.

Presently and in a future work, various algorithms are in design, dealing with curvature maps computation, geometric distances estimation, compactness cartography elaboration, etc.

We are working now to build a software platform, called **GenoMEDIA**, that integrates the graphical and interactive tool *ADN-Viewer* in its server implementation, a genomic database, and a web server that dialog with the previous both tools. In the close future, a first version of **GenoMEDIA** will be open on the Internet and can be accessible by any user using his own navigator.

For the long-term research, the first work that will be carried out is the validation and comparison of the various conformation models in order to obtain the most reliable prediction. This implies to do different matching and fitting processes between the predicted images and real ones that could be obtained with advanced electronic microscopy.

Acknowledgements

We thank the following academic institutions and people for their financial and advising help: LIMSI-CNRS, Paris-Sud University, the French bioinformatics program inter-EPST CNRS-INRIA-INRA-INSERM, and People from IGM and IBP form Orsay, Monique Marilley from Marseille, Ed. Trifonov from Rehovot university in Israel.

References

1. Rachid Gherbi and Joan Hérisson. ADN_Viewer, a software framework for 3d modeling and stereoscopic visualization of the genome. In Proc. of Graphicon'2000, International conference on Computer Graphics and Vision, Moscow, August-September 2000.
2. Henn C. and Teschner, "Interactive molecular Visualization", Track session at PSB ???
3. Grigoriev A., "Reusable graphical interface to genome information resources", in proc. of PSB conference ???
4. A. Bolshoy, P. McNamara, R.E. Harrington, and E.N. Trifonov. Curved DNA without A-A: Experimental estimation of all 16 DNA wedge angles. *Proc. Natl. Acad. Sci. USA*, 88: 2312-2316, March 1991.
5. Philippe Pasero. *L'organisation du chromosome eucaryote et ses implications dans le contrôle de l'activité génique et la transmission des patrons d'expression.* PhD thesis, University of Aix-Marseille II, Faculty of Sciences of Luminy, December 1993.

6. Marilley M, Pasero P (1996). *Nucleic Acids Res.* **35**:2204-2211
7. Søren Wilken Rasmusen. DNAtools©. http://www.dnatools.org.
8. P. De Santis, A. Palleschi, M. Savino, and A. Scipioni. A theorical model of DNA curvature. *Biophys. Chem.*, 32: 305-317, 1988.
9. Cacchione S, De Santis P, Foti DP, Palleschi A, Savino M (1989). *Biochemistry* **28**, 8706-8713
10. J.D. Watson and F.H.C. Crick. Molecular structure of nucleic acids. *Nature*, 171: 737-738, April 1953.
11. I. Lafontaine and R. Lavery, Curr. Opin. Struct Biol.1999, 9:170-176.
12. Matthews KS (1992). *Microbiology Reviews* **56**:123-136
13. Ponomarenko M.P., Ponomarenko J.V., Kel A.E. and Kolchanov N.A., "Search for DNA conformational features for functional sites. Investigation of the TATA box", in proc. of PSB conference, ?????
14. Perez-Martin J, Rojo F, de Lorenzo V (1994). *Microbiology Reviews* **58**:268-290
15. Suzuki M, Yagi N (1995). *Nucleic Acids Res.* **23**:2083-2091
16. Natale DA, Umek RM, Kowalski D (1993). *Nucleic Acids Res.* **21**:551-560
17. Nickerson CA, Achberger EC (1995). *Journal of Bacteriology* **177**(20): 5756-5761
18. Carmona M, Claverie-Martin F, Magasanik B (1997). *Proc Natl Acad Sci USA* **94**:9568-9572
19. Beloin C, Exley R, Mahé AL, Zouine M, Cubasch S, Le Hégarat F (2000). *Journal of Bacteriology* **182**(16): 4414-4424

---

* LIMSI-CNRS is a laboratory of informatics for mechanics and engineering science, from the French National Centre of Scientific Research, located in a south of Paris in the campus of Paris-South University.

Pairwise RNA Structure Comparison with Stochastic Context-Free Grammars

I.Holmes and G.M.Rubin
Howard Hughes Medical Institute

Abstract

Pairwise stochastic context-free grammars ("Pair SCFGs") are powerful tools for finding conserved RNA structures, but unconstrained alignment to Pair SCFGs is prohibitively expensive. We develop versions of the Pair SCFG dynamic programming algorithms that can be conditioned on precomputed structures, significantly reducing the time complexity of alignment. We have implemented these algorithms for general Pair SCFGs in software that is freely available under the GNU Public License.

1 Introduction

Stochastic Context-Free Grammars (SCFGs) are powerful tools for RNA structure prediction and genefinding[1,2]. However, they are expensive to use. For two sequences of lengths L and M, simultaneous alignment and structure comparison using SCFGs has time complexity $\mathcal{O}(L^3M^3)$; this is cubic compared to $\mathcal{O}(LM)$, the complexity of aligning the same two sequences to a hidden Markov model[3,4]. Such computational power is beyond the reach of most labs.

One way to reduce the time complexity of these algorithms is to constrain the analysis, by supplying either the primary sequence alignment or the secondary structure assignment. Algorithms for the latter task (alignment of supplied structures) have been described[5,6] and implemented in the `Vienna` package[7]. These algorithms require manual specification of the scoring scheme. It is desirable to place such algorithms in a formal framework wherein the scoring scheme can be reliably optimised from a "training set" of trusted structural alignments.

We here present dynamic programming algorithms for two-sequence SCFGs that use devices called "fold envelopes" to restrict the set of base-pairings that the recursion is allowed to consider. These fold envelopes may be based on pre-computed structures. In extreme cases (when the parse trees are deep, e.g. if the structures contain long stem loops) the reduced running time can be as low as $\mathcal{O}(LM)$. One can also easily build "dummy" fold envelopes that reproduce the full, unconstrained algorithm. We describe an implementation of this algorithmic toolkit that works for any SCFG and is freely available under the terms of the GNU Public License[8].

2 Algorithms

SCFGs (sometimes called "Single SCFGs") are flexible models for RNA sequences allowing nested covariation[1]. Pair SCFGs are a generalisation of Single SCFGs, yielding joint probabilities for two sequences at once. In order to reduce the time-complexity of dynamic programming (DP) algorithms for Pair SCFGs, we will start with the analogous treatment for Single SCFGs.

The DP algorithms of interest include the Inside algorithm, the Cocke-Younger-Kasami (CYK) algorithm and the Inside-Outside algorithm[1]. The Inside algorithm calculates the likelihood of the sequence according to the SCFG, summed over all possible parses of the sequence; the CYK algorithm finds the maximum likelihood parse of the sequence; and the Inside-Outside algorithm finds the expected number of times that each grammar production is used, with the expectation taken over the posterior distribution of parses.

These algorithms (for Single SCFGs) work by calculating the partial likelihood of all substrings of the observed sequence, starting with zero-length substrings and working up to the full length. Since the algorithms compute likelihoods for all substrings, their memory usage is $\mathcal{O}(L^2)$ for a sequence of length L. The running time is higher at $\mathcal{O}(L^3)$, since an extra factor of L is incurred in combining adjacent substrings[1].

If the secondary structure of the RNA sequence is already known, a faster approach is to compute *conditional likelihoods* (for the Inside and CYK algorithms) or *conditional expectations* (for the Inside-Outside algorithm), where the given condition is the secondary structure. Rather than iterating over all the substrings, the conditional algorithms only iterate over substrings consistent with the given set of base-pairings.

The situation for Pair SCFGs is directly analogous. Rather than iterating over every *pair* of substrings of the two sequences, we can restrict the algorithms to a limited set consistent with precomputed secondary structures.

We generalise this idea by giving versions of the Inside, CYK and Inside-Outside algorithms for Pair SCFGs that are restricted to any valid sets of substrings of the two RNA sequences. We use the term "fold envelope" to describe such a set of substrings for any one sequence. The problem of calculating likelihoods conditioned on secondary structure is then reduced to one of computing the appropriate fold envelopes. We begin with some notation.

2.1 Notation: Pair SCFG

We deliberately follow the Covariance Model notation of Eddy and Durbin[1]. Although the use of Chomsky normal form would simplify the mathematics, this form is less appropriate for biological sequence analysis.

A Pair Stochastic Context-Free Grammar emits symbols in two sequences, X and Y. The terminals for these symbols are x_i and y_i respectively. For RNA, these take the values '`A`', '`C`', '`G`' and '`U`'. There is also a silent terminal, ϵ, that is only generated by nonterminals of type 'E' (see below).

The grammar has M different nonterminals denoted by $W_1, \ldots, W_M$. Let u and v be indices for states W_u and W_v. There are eighteen different types of state, with properties described in Table 1. These include E (End), N (Null), B (Bifurcation) and fifteen different types of emit state. The emit states each have two-letter identifiers of the form AB, where A denotes the emission in sequence X and B the emission in sequence Y. (For example: states of type 'PL' emit a left-right base pair in sequence X and a single leftwise base in sequence Y, whereas states of type 'NR' emit nothing in sequence X and a single rightwise base in sequence Y.) State W_1 is the "start" state and is always of type N. State W_M is the "end" state and is always of type E; in fact, W_M is the only state that can be of type E. We define s_u to be the *state type* of W_u, taking one of the eighteen values from the first column of Table 1;

The emission and transition probabilities for state u are given by $e_u(\cdot)$ and $t_u(\cdot)$ respectively. We define numbers Δ_u^{XL}, Δ_u^{XR}, Δ_u^{YL} and Δ_u^{YR} which are the number of symbols emitted to left and right in sequences X and Y.

We also define $\mathcal{C}_u$, the *children* of W_u (being the list of indices v for the states W_v that W_u can make a transition to) and $\mathcal{P}_u$, the *parents* of W_u (being the list of indices of states that make a transition to W_u). Bifurcating states (type B) always transit with probability 1 to two Null states, W_{l_u} (left) and W_{r_u} (right). It is possible for l_u to be the same as r_u. The child list $\mathcal{C}_u$ for a bifurcating state is defined to be (l_u, l_r). The parent lists $\mathcal{P}_{l_u}$ and $\mathcal{P}_{r_u}$ do *not* include u; instead, each Null state v has an associated left-parent list $\mathcal{P}_v^L = \{u : s_u = \mathrm{B}, l_u = v\}$ and a right-parent list $\mathcal{P}_v^R = \{u : s_u = \mathrm{B}, r_u = v\}$. This treatment of bifurcations differs slightly from the Covariance Model of Eddy *et al.*

We require that Pair SCFG's have no *null cycles*. That is to say, there is no sequence of productions that starts from an N or B state and returns to the same state without emitting any residues in either X or Y.

For convenience, we define two orderings on the state indices $1 \ldots M$. These are represented by the lists $\mathcal{F}_O$ and $\mathcal{F}_I$. The *outside fill order* $\mathcal{F}_O$ lists all the emitting states, followed by the non-emitting states sorted in topological order (i.e. parents before children). Conversely, the *inside fill order* $\mathcal{F}_I$ lists the emitting states followed by the non-emitting states sorted in *reverse* topological order (i.e. children before parents).

2.2 Notation: Fold Envelope

Let X be a sequence of length L, whose i'th base is x_i (where i starts at zero, i.e. $0 \leq i < L$). Let $X_{i..j}$ be the subsequence of X running from base i to base $j-1$, so that $X_{0..L}$ is the full sequence and $X_{i..i}$ is an empty sequence. In general there are $L+1$ empty subsequences (from $X_{0..0}$ to $X_{L..L}$) and $\frac{1}{2}(L+1)(L+2)$ subsequences in total.

Suppose we have a parse tree of nonterminals aligning X to a SCFG as described in section 2.1 (we discount Y-emissions for the moment, pretending that the grammar is a Single SCFG). Then each node of the parse tree, together with its children, accounts for some subsequence $X_{i..j}$ of X (this subsequence is called the *inside* sequence). Likewise, the parents, siblings and cousins of this node account for the subsequences $X_{0..i}$ and $X_{j..L}$. (the *outside* sequence).

If the nonterminal at this node is W_u, where $s_u \neq \mathrm{B}$, then the subsequence for the immediate child node will be $X_{i+\Delta_u^{\mathrm{XL}}..j-\Delta_u^{\mathrm{XR}}}$. If $s_u = \mathrm{B}$ (i.e. W_u is a bifurcation) then the subsequences for the two child nodes will be $X_{i..k}$ and $X_{k..j}$ for some k, where $i \leq k \leq j$.

In computing the full dynamic programming matrix for a SCFG, one typically considers all subsequences $X_{i..j}$, all emissions $X_{i+\Delta_u^{\mathrm{XL}}..j-\Delta_u^{\mathrm{XR}}}$ and all bifurcations $(X_{i..k}, X_{k..j})$. Our intent is to consider a reduced set of subsequences, rather than the full $\frac{1}{2}(L+1)(L+2)$. In order to manage this, we formally enumerate these allowed subsequences and their associated emissions and bifurcations. The enumeration is called a *fold envelope*.

The fold envelope $\mathcal{E}$ consists of N ordered subsequences. The n'th subsequence is $X_{i_n..j_n}$, where n starts from zero (i.e. $0 \leq n < N$). If a subsequence $X_{i..j}$ is in $\mathcal{E}$, then we define $n[X_{i..j}]$ to be its index within $\mathcal{E}$ (in other words, $n[X_{i_m..j_m}] \equiv m$). For all subsequences $X_{i..j}$, that are *not* in the envelope, $n[X_{i..j}]$ is defined to be the "illegal" value $\emptyset$ (which we typically represent as -1 in actual code).

The ordering is from inside to outside, so that

$$X_{i..j} \in \mathcal{E}, X_{k..l} \in \mathcal{E}, i \geq k, j \leq l \Rightarrow n[X_{i..j}] \leq n[X_{k..l}]$$

For each subsequence $X_{i_n..j_n}$ in the envelope, we precompute the following inward and outward *emission connections*:

$$c_{\mathrm{in}}(n, \Delta^{\mathrm{L}}, \Delta^{\mathrm{R}}) = n\left[X_{i_n+\Delta^{\mathrm{L}}, j_n-\Delta^{\mathrm{R}}}\right]$$

$$c_{\mathrm{out}}(n, \Delta^{\mathrm{L}}, \Delta^{\mathrm{R}}) = n\left[X_{i_n-\Delta^{\mathrm{L}}, j_n+\Delta^{\mathrm{R}}}\right]$$

where Δ^{L} and Δ^{R} can take values from $\{0, +1\}$. Some of these connections may equal $\emptyset$ if the target subsequence is not in the envelope.

We also pre-compute lists of valid inward, outward-left and outward-right *bifurcation connections*:

$$b_{\text{in}}(n) = \{(n_L, n_R) : i_{n_L} = i_n, j_{n_L} = i_{n_R}, j_{n_R} = j_n\}$$

$$b_{\text{outl}}(n) = \{(n_O, n_L) : i_{n_L} = i_{n_O}, j_{n_L} = i_n, j_n = j_{n_O}\}$$

$$b_{\text{outr}}(n) = \{(n_O, n_R) : i_n = i_{n_O}, j_n = i_{n_R}, j_{n_R} = j_{n_O}\}$$

Each element in a bifurcation connection list is a pair of subsequence indices. Together with subsequence n, these subsequences form a bifurcation triplet, i.e. *(outside,inside-left,inside-right)*. In contrast to the emission connections, the subsequence indices in the bifurcation connection lists are guaranteed not to be $\emptyset$.

An envelope is called *global* if it contains (i) all empty subsequences, (ii) the full-length sequence and (iii) at least one parse tree connecting the full-length sequence to one or more empty subsequences via emission or bifurcation connections. Note that by the inside-outside ordering condition, the full-length sequence has to be the last subsequence in a global envelope.

There exists an algorithm to calculate the appropriate fold envelope for a given structure in time that is linear in the length of the fold envelope (Holmes and Rubin, unpublished).

For the pairwise dynamic programming algorithms described below, we need two global envelopes, one for each sequence. We differentiate between these two envelopes by using apostrophes for envelope Y, i.e. the appropriate envelope variables for sequence X are $\{\mathcal{E}, N, L, i, j, n, c, b\}$ and for sequence Y $\{\mathcal{E}', N', L', i', j', n', c', b'\}$.

2.3 The Conditional Inside algorithm

The Conditional Inside algorithm, shown in Figure 1, recursively calculates $\alpha_u(n, n')$, the total likelihood for all joint parses rooted at nonterminal W_u of the two inside subsequences n and n'. The full likelihood of the two sequences is $\alpha_1(N, N')$.

For convenience, we define $\alpha_u(n, \emptyset) = \alpha_u(\emptyset, n') = 0$ for all u, n, n'. Also for convenience, we follow Durbin *et al* in using the notation $e_u(x_i, x_j, y_k, y_l)$ for all emission probabilities, even for nonterminals W_u that emit fewer than four symbols. Thus, for states of type N, $e_u(x_i, x_j, y_k, y_l) \equiv 1$; for states of type LN, $e_u(x_i, x_j, y_k, y_l) \equiv e_u(x_i)$; for states of type PN, $e_u(x_i, x_j, y_k, y_l) \equiv e_u(x_i, x_j)$; for states of type NP, $e_u(x_i, x_j, y_k, y_l) \equiv e_u(y_k, y_l)$ and so on (as per Table 1).

State type (s_u)	Allowed productions	Δ_u^{XL}	Δ_u^{XR}	Δ_u^{YL}	Δ_u^{YR}	Emission probability	Transition probability
N	$W_u \to W_v$	0	0	0	0	1	$t_u(v)$
LN	$W_u \to x_i W_v$	1	0	0	0	$e_u(x_i)$	$t_u(v)$
RN	$W_u \to W_v x_j$	0	1	0	0	$e_u(x_j)$	$t_u(v)$
PN	$W_u \to x_i W_v x_j$	1	1	0	0	$e_u(x_i, x_j)$	$t_u(v)$
NL	$W_u \to y_k W_v$	0	0	1	0	$e_u(y_k)$	$t_u(v)$
LL	$W_u \to x_i y_k W_v$	1	0	1	0	$e_u(x_i, y_k)$	$t_u(v)$
RL	$W_u \to y_k W_v x_j$	0	1	1	0	$e_u(x_j, y_k)$	$t_u(v)$
PL	$W_u \to x_i y_k W_v x_j$	1	1	1	0	$e_u(x_i, x_j, y_k)$	$t_u(v)$
NR	$W_u \to W_v y_l$	0	0	0	1	$e_u(y_l)$	$t_u(v)$
LR	$W_u \to x_i W_v y_l$	1	0	0	1	$e_u(x_i, y_l)$	$t_u(v)$
RR	$W_u \to W_v x_j y_l$	0	1	0	1	$e_u(x_j, y_l)$	$t_u(v)$
PR	$W_u \to x_i W_v x_j y_l$	1	1	0	1	$e_u(x_i, x_j, y_l)$	$t_u(v)$
NP	$W_u \to y_k W_v y_l$	0	0	1	1	$e_u(y_k, y_l)$	$t_u(v)$
LP	$W_u \to x_i y_k W_v y_l$	1	0	1	1	$e_u(x_i, y_k, y_l)$	$t_u(v)$
RP	$W_u \to y_k W_v x_j y_l$	0	1	1	1	$e_u(x_j, y_k, y_l)$	$t_u(v)$
PP	$W_u \to x_i y_k W_v x_j y_l$	1	1	1	1	$e_u(x_i, x_j, y_k, y_l)$	$t_u(v)$
E	$W_u \to \epsilon$	0	0	0	0	1	1
B	$W_u \to W_{l_u} W_{r_u}$	0	0	0	0	1	1

Table 1: The eighteen types of nonterminal in a Pair Stochastic Context-Free Grammar and their associated production rules.

For $n = 0$ to $N-1$, $n' = 0$ to $N'-1$, $u \in \mathcal{F}_I$:

$$\alpha_u(n,n') = \begin{cases} s_u = \text{E:} \quad \delta(i_n - j_n)\ \delta(i'_{n'} - j'_{n'}) \qquad \text{where } \delta(x) = \begin{cases} 1 & \text{if } x = 0 \\ 0 & \text{if } x \neq 0 \end{cases} \\[2ex] s_u = \text{B:} \quad \displaystyle\sum_{\substack{(n_L,n_R) \\ \in\ b_{\text{in}}}} \ \sum_{\substack{(n'_L,n'_R) \\ \in\ b'_{\text{in}}}} \alpha_{l_u}(n_L, n'_L)\ \alpha_{r_u}(n_R, n'_R) \\[2ex] \text{otherwise:} \\ \quad e_u(x_{i_n}, x_{j_n - 1}, y_{i'_{n'}}, y_{j'_{n'}-1}) \displaystyle\sum_{v \in \mathcal{C}_u} t_u(v)\ \alpha_v\left(c_{\text{in}}(n, \Delta_u^{\text{XL}}, \Delta_u^{\text{XR}}),\ c'_{\text{in}}(n', \Delta_u^{\text{YL}}, \Delta_u^{\text{YR}})\right) \end{cases}$$

Figure 1: The Conditional Inside algorithm for Pair SCFGs.

2.4 *The Conditional Outside algorithm*

To estimate counts for production usage conditioned on RNA structure, we need the Conditional Inside-Outside algorithm. The first half of this algorithm (the Conditional Inside algorithm) was already described in section 2.3. We now describe the second half.

The Conditional Outside algorithm shown in Figure 2, recursively calculates $\beta_u(n, n')$, the total likelihood for all joint parses rooted at nonterminal W_u of the outside subsequences n and n'. (Recall that the outside subsequence n of sequence X consists of the two subsequences $X_{0..i_n}$ and $X_{j_n..L}$. Likewise, the outside subsequence n' of sequence Y consists of the two subsequences $Y_{0..i'_{n'}}$ and $Y_{j'_{n'}..L'}$.)

As in section 2.3, we define $\beta_u(n, \emptyset) = \beta_u(\emptyset, n') = 0$ for all u, n, n'. We also use the notation $e_u(x_i, x_j, y_k, y_l)$ for all emission probabilities, as in that section.

We return to the Conditional Outside algorithm in section 2.7.

2.5 *The Conditional CYK algorithm*

The Conditional CYK algorithm, shown in Figure 3, recursively calculates $\gamma_u(n, n')$, the maximum likelihood for a joint parse rooted at nonterminal W_u of the inside subsequences n and n'.

As in section 2.3, we define $\gamma_u(n, \emptyset) = \gamma_u(\emptyset, n') = 0$ for all u, n, n'. We also use the notation $e_u(x_i, x_j, y_k, y_l)$ for all emission probabilities, as in that section.

On its own, the Conditional CYK algorithm may be used for database searching (i.e. to flag homologous structures). Together with the Conditional CYK Traceback algorithm (section 2.6) it can be used to find maximum-likelihood alignments of RNA structures.

2.6 *The Conditional CYK Traceback algorithm*

To recover the conditional maximum likelihood alignment of two RNA sequences following the Conditional CYK algorithm, we need to perform a Conditional CYK Traceback. The algorithm to do this is shown in Figure 4.

This algorithm steps through the $\gamma_u(n, n')$ likelihoods that were calculated by the Conditional CYK algorithm. Whereas the Conditional CYK algorithm goes in the *inside→outside* direction, Conditional CYK Traceback goes *outside→inside*, outputting the maximum likelihood alignment as it goes.

For $n = N-1$ to 0, $n' = N'-1$ to 0, $u \in \mathcal{F}_O$:

$$\beta_u(n,n') = e_u(x_{i_n - 1}, x_{j_n}, y_{i'_{n'}-1}, y_{j'_{n'}}) \left(\sum_{v \in \mathcal{P}_u} t_v(u)\ \beta_v\left(c_{\text{out}}(n, \Delta_u^{\text{XL}}, \Delta_u^{\text{XR}}),\ c'_{\text{out}}(n', \Delta_u^{\text{YL}}, \Delta_u^{\text{YR}})\right) \right.$$

$$+ \sum_{\substack{v \in \mathcal{L}_u,\\ s_v = B}} \ \sum_{\substack{(n_O, n_L)\\ \in\ b_{\text{outl}}}} \ \sum_{\substack{(n'_O, n'_L)\\ \in\ b'_{\text{outl}}}} \beta_v(n_O, n'_O)\ \alpha_{l_v}(n_L, n'_L)$$

$$\left. + \sum_{\substack{v \in \mathcal{R}_u,\\ s_v = B}} \ \sum_{\substack{(n_O, n_R)\\ \in\ b_{\text{outr}}}} \ \sum_{\substack{(n'_O, n'_R)\\ \in\ b'_{\text{outr}}}} \beta_v(n_O, n'_O)\ \alpha_{r_v}(n_R, n'_R) \right)$$

Figure 2: The Conditional Outside algorithm for Pair SCFGs.

For $n = 0$ to $N-1$, $n' = 0$ to $N'-1$, $u \in \mathcal{F}_I$:

$$\gamma_u(n,n') = \begin{cases} s_u = \text{E:} \quad \delta(i_n - j_n)\ \delta(i'_{n'} - j'_{n'}) \qquad \text{where } \delta(x) = \begin{cases} 1 & \text{if } x = 0 \\ 0 & \text{if } x \neq 0 \end{cases} \\[2em] s_u = \text{B:} \quad \max\limits_{\substack{(n_L, n_R)\\ \in\ b_{\text{in}}}} \ \max\limits_{\substack{(n'_L, n'_R)\\ \in\ b'_{\text{in}}}} \ \gamma_{l_u}(n_L, n'_L)\ \gamma_{r_u}(n_R, n'_R) \\ \text{otherwise:} \\ \quad e_u(x_{i_n}, x_{j_n - 1}, y_{i'_{n'}}, y_{j'_{n'}-1}) \ \max\limits_{v \in \mathcal{C}_u} \ t_u(v)\ \gamma_v\left(c_{\text{in}}(n, \Delta_u^{\text{XL}}, \Delta_u^{\text{XR}}),\ c'_{\text{in}}(n', \Delta_u^{\text{YL}}, \Delta_u^{\text{YR}}\right) \end{cases}$$

Figure 3: The Conditional CYK algorithm for Pair SCFGs.

- **Initialisation:**
  - `Let` $u = 1$, $n = N - 1$, $n' = N' - 1$; /* set traceback co-ordinates to start state */
  - `Clear co-ordinates stack;`
- **Main loop:**
  - `Output co-ordinates` (u, n, n');
  - `If` $s_u =$ `E` /* end state? */
    - `If co-ordinates stack is empty then exit;`
    - `Pop co-ordinates` (u, n, n');
    - `Goto` **Main loop**;
  - `else if` $s_u =$ `B` /* bifurcation state? */
    - `Select` (n_L, n_R) `from` $b_{\mathtt{in}}$ `and` (n'_L, n'_R) `from` $b'_{\mathtt{in}}$ `such that` $\gamma_u(n, n') = \gamma_{l_u}(n_L, n'_L)\gamma_{r_u}(n_R, n'_R)$;
    - `Push co-ordinates` (r_u, n_R, n'_R);
    - `Set` (u, n, n') `equal to` (l_u, n_L, n'_L);
    - `Goto` **Main loop**;
  - `else` /* emit or null state */
    - `Let` $(m, m') = (c_{\text{in}}(n, \Delta_u^{\text{XL}}, \Delta_u^{\text{XR}}),\ c'_{\text{in}}(n', \Delta_u^{\text{YL}}, \Delta_u^{\text{YR}}))$;
    - `Select` v `from` $\mathcal{C}_u$ `such that` $\gamma_u(n, n') = t_u(v)\gamma_v(m, m')$;
    - `Set` (u, n, n') `equal to` (v, m, m');
    - `Goto` **Main loop**;

Figure 4: The Conditional CYK Traceback algorithm for Pair SCFGs.

2.7 Estimating conditional counts

To estimate the conditional counts for emission and transition usage following the Conditional Inside and Outside algorithms, we use the following equations:

$$\hat{e}_u(x_i, x_j, y_k, y_l) = \sum_{n \in \mathcal{E}} \sum_{n' \in \mathcal{E}'} \frac{\alpha_u(n, n')\beta_u(n, n')}{e_u(x_i, x_j, y_k, y_l)}$$

$$\hat{t}_u(v) = \sum_{n \in \mathcal{E}} \sum_{n' \in \mathcal{E}'} \alpha_v(n, n')\beta_u(n, n')$$

(The first of these equations strictly is valid only if $e_u(x_i, x_j, y_k, y_l) \neq 0$. In the special case that the emission probability is zero, the estimated count is also zero and this equation does not apply.)

The Conditional version of the expectation-maximization algorithm uses these counts, possibly together with prior distributions such as Dirichlet mixtures, to update the probability parameters for the Pair SCFG. This is repeated until the probability parameters (and the likelihood) do not improve any further.

2.8 Extension to Other Grammars

Dynamic programming recursions for higher-order grammars, such as the pseudoknot-capable grammar described by Rivas and Eddy[9], also involve iterations over subsets of the full sequence. (In the case of Rivas and Eddy's pseudoknot grammar, each subset of the full sequence is a pair of substrings with a "hole" between them.)

It would not present any theoretical difficulty to extend fold envelopes to such higher grammars. The central idea of fold envelopes is to encapsulate the iteration over alignable subsequences. This is just as feasible when the subsequences consist of substring pairs (as with the pseudoknot grammar) as when they consist of single substrings (as with standard Pair SCFG's).

Fold envelopes can also be used to obtain low-complexity recursions for lower-order grammars as special cases of higher-order recursions. For example, HMMs are (formally) a subset of SCFGs, yet one would usually not wish to re-use SCFG algorithms for HMMs, since the HMM algorithms only consider the $\sim L$ substrings of the form $X_{0..j}$, whereas the SCFG algorithms consider all $\sim L^3$ adjacent substring-pairs $X_{i..k}, X_{k..j}$. However, with a fold envelope, the SCFG recursion can be restricted to the substrings used by the HMM recursion.

3 Implementation

We have implemented the algorithms described here in DART, a freely available C++/Unix toolkit available from `www.biowiki.org/dart`.

DART includes implementations of the Conditional Inside, Conditional Outside and Conditional CYK/Traceback algorithms on Pair SCFGs of any topology with states as listed in Table 1. Fold envelopes may be calculated for any structure as well as for the full, unconstrained dynamic programming.

Since Single SCFG's are a subset of Pair SCFG's, DART can also be used for covariance modeling as described by Eddy and Durbin[10]. It can also emulate Single or Pair HMMs.

4 Discussion

In this paper, we have developed algorithms for conditional pairwise dynamic programming to stochastic context-free grammars within the bounds of *fold envelopes* based on precomputed structures. Fold envelopes are efficient and flexible: efficient because they limit the grammar's time complexity, flexible because conditioning on (say) a range of structures (rather than a single structure) merely involves computing an appropriate fold envelope and does not require redesigning the algorithms in Figures 1-4.

With the present investment in large-scale sequencing, computer-assisted identification of conserved RNA structure offers a genomics approach to studies of noncoding RNA genes[2], transcript localisation[11], poly-A stability[12], RNA-RNA duplexes[13] and many other areas. Perhaps the most attractive model from a theoretical viewpoint would be one that described the time-evolution of RNA structure including the effects of natural selection[3]. Such a model may be just around the corner: probabilistic evolutionary models for DNA and protein sequence analysis have made considerable progress in recent years[14,1,15]. In the meantime, we hope the present work may be useful to researchers interested in the evolutionary implications of RNA structure conservation in biology.

4.1 Acknowledgements

We would like to thank Sean Eddy, Elena Rivas, Erwin Frise, Eric Lai and Eliza McKenna for their useful input. This work was supported by the Howard Hughes Medical Institute.

1. R. Durbin, S. Eddy, A. Krogh, and G. Mitchison. *Biological Sequence Analysis: Probabilistic Models of Proteins and Nucleic Acids.* Cambridge University Press, Cambridge, UK, 1998.
2. S. R. Eddy. Noncoding RNA genes. *Current Opinion in Genetics and Development*, 9(6):695–699, 1999.
3. D. Sankoff and R. J. Cedergren. Simultaneous comparison of three or more sequences related by a tree. In D. Sankoff and J. B. Kruskal, editors, *Time Warps, String Edits, and Macromolecules: the Theory and Practice of Sequence Comparison*, chapter 9, pages 253–264. Addison-Wesley, Reading, MA, 1983.
4. V. Bafna, S. Muthukrishnan, and R. Ravi. Similarity between RNA strings. Technical report, Center for Discrete Mathematics and Theoretical Computer Science, 1996.
5. B. A. Shapiro. An algorithm for comparing multiple RNA secondary structures. *Computer Applications in the Biosciences*, 4(3):387–393, 1988.
6. B. A. Shapiro and K. Z. Zhang. Comparing multiple RNA secondary structures using tree comparisons. *Computer Applications in the Biosciences*, 6(4):309–318, 1990.
7. S. Wuchty, W. Fontana, I. L. Hofacker, and P. Schuster. Complete suboptimal folding of RNA and the stability of secondary structures. *Biopolymers*, 49(2):145–165, 1999.
8. The GNU Public License, 2000. Available in full from http://www.fsf.org/copyleft/gpl.html.
9. E. Rivas and S. R. Eddy. The language of RNA: a formal grammar that includes pseudoknots. *Bioinformatics*, 16(4):334–340, 2000.
10. S. R. Eddy and R. Durbin. RNA sequence analysis using covariance models. *Nucleic Acids Research*, 22:2079–2088, 1994.
11. A. Bashirullah, R. L. Cooperstock, and H. D. Lipshitz. RNA localization in development. *Annual Review of Biochemistry*, 67:335–394, 1998.
12. N. Proudfoot. Poly(A) signals. *Cell*, 64(4):671–674, 1991.
13. E. C. Lai and J. W. Posakony. Regulation of Drosophila neurogenesis by RNA:RNA duplexes? *Cell*, 93(7):1103–1104, 1998.
14. J. L. Thorne, H. Kishino, and J. Felsenstein. Inching toward reality: an improved likelihood model of sequence evolution. *Journal of Molecular Evolution*, 34:3–16, 1992.
15. I. Holmes and W. J. Bruno. Evolutionary HMMs: a Bayesian approach to multiple alignment. To appear in Bioinformatics, 2001., 2001.

ESTIMATION OF GENETIC NETWORKS AND FUNCTIONAL STRUCTURES BETWEEN GENES BY USING BAYESIAN NETWORKS AND NONPARAMETRIC REGRESSION

SEIYA IMOTO, TAKAO GOTO and SATORU MIYANO

Human Genome Center, Institute of Medical Science, University of Tokyo, 4-6-1 Shirokanedai, Minato-ku, Tokyo, 108-8639, Japan

We propose a new method for constructing genetic network from gene expression data by using Bayesian networks. We use nonparametric regression for capturing nonlinear relationships between genes and derive a new criterion for choosing the network in general situations. In a theoretical sense, our proposed theory and methodology include previous methods based on Bayes approach. We applied the proposed method to the *S. cerevisiae* cell cycle data and showed the effectiveness of our method by comparing with previous methods.

1 Introduction

The microarray technology provides us enormous amount of valuable gene expression data. The analysis of the relationship among genes has drawn remarkable attention in the field of molecular biology and Bioinformatics. However, due to the cause of dimensionality and complexity of the data, it will be no easy task to find structures, which are buried in noise. To extract the effective information from microarray gene expression data, thus, theory and methodology are expected to be developed from a statistical point of view. Our purpose is to establish a new method for extracting the relationships among genes clearer.

Constructing genetic networks[3,4,5,12,13,19] is one of the hot topics in the analysis of the microarray gene expression data. Bayesian network is an attractive method for constructing genetic networks from a graph-theoretic approach. Friedman and Goldszmidt[12] proposed an interesting method for constructing genetic links by using Bayesian networks. They discretized the expression value and considered to fit the models based on multinomial distributions. However, a problem still remains to be solved in choosing the threshold value for discretizing not only by the experiments. The threshold value assuredly gives essential changes of the results and unsuitable threshold value leads to wrong results. On the other hand, recently, Friedman *et al.*[13] pointed out that discretizing is probably loosing the information. To use the expression data as continuous values, thus, they considered the use of Gaussian models based on linear regression. However this model can only detect linear dependencies and we cannot obtain sufficient results.

In this paper we propose a new method for constructing genetic networks by using Bayesian networks. To capture not only linear dependencies but also nonlinear structures between genes, we use nonparametric regression models with Gaussian noise[11,15,22,23]. Nonparametric regression has been developed in order to explore the complex nonlinear form of the expected responses without the knowledge about the functional relationship in advance. Due to the new structure of the Bayesian networks, a suitable criterion is needed for evaluating models. We derive a new criterion from Bayesian statistics. By using proposed method, we will overcome the defects of previous methods and attain more effective information. In addition, our method includes the previous method[7] as a spacial case. The efficiency of the proposed method is shown by the Monte Calro simulation method. We also demonstrate our proposed method through the analysis of the *S. cerevisiae* cell cycle data[21].

2 Bayesian Network and Nonparametric Regression

Let $\boldsymbol{X} = (X_1, X_2, ..., X_p)^T$ be a p-dimensional random vector and let G be a directed acyclic graph. Under the Bayesian network framework, we look upon a gene as a random variable and decompose the joint probability into the product of conditional probabilities, that is

$$P(X_1, X_2, ..., X_p) = P(X_1|\boldsymbol{P}_1)P(X_2|\boldsymbol{P}_2) \times \cdots \times P(X_p|\boldsymbol{P}_p), \tag{1}$$

where $\boldsymbol{P}_j = (P_1^{(j)}, P_2^{(j)}, ..., P_{q_j}^{(j)})^T$ is a q_j-dimensional vector of parent variables of X_j in the graph G.

Suppose that we have n observations $\boldsymbol{x}_1, ..., \boldsymbol{x}_n$ of the random vector $\boldsymbol{X}$ and the observations of $\boldsymbol{P}_j$ are denoted by $\boldsymbol{p}_{1j}, ..., \boldsymbol{p}_{nj}$, where $\boldsymbol{p}_{ij}$ is a q_j-dimensional vector with k-th element $p_{ik}^{(j)}$, for $k = 1, ..., q_j$. For example, let $\boldsymbol{X}_n$ be an $n \times p$ matrix, where $\boldsymbol{X}_n = (\boldsymbol{x}_1, ..., \boldsymbol{x}_n)^T = (\boldsymbol{x}_{(1)}, ..., \boldsymbol{x}_{(p)}) = (x_{ij})_{i=1,...,n; j=1,...,p}$, $\boldsymbol{x}_i = (x_{i1}, ..., x_{ip})^T$, $\boldsymbol{x}_{(j)} = (x_{1j}, ..., x_{nj})^T$ and $\boldsymbol{x}_i^T$ is the transpose of the vector $\boldsymbol{x}_i$. If X_1 has a parent vector $\boldsymbol{P}_1 = (X_2, X_3)^T$, we obtain $\boldsymbol{p}_{11} = (x_{12}, x_{13})^T, ..., \boldsymbol{p}_{n1} = (x_{n2}, x_{n3})^T$.

It is immediately found that the equation holds when we replace the probability measure P in (1) by densities

$$f(x_{i1}, x_{i2}, ..., x_{ip}) = f_1(x_{i1}|\boldsymbol{p}_{i1})f_2(x_{i2}|\boldsymbol{p}_{i2}) \times \cdots \times f_p(x_{ip}|\boldsymbol{p}_{ip}).$$

Then all we need to do is to consider how to construct the conditional densities $f_j(x_{ij}|\boldsymbol{p}_{ij})$ $(j = 1, ..., p)$.

In this paper, we use nonparametric regression models for capturing the relationship between x_{ij} and $\boldsymbol{p}_{ij} = (p_{i1}^{(j)}, ..., p_{iq_j}^{(j)})^T$ in the form

$$x_{ij} = m_1(p_{i1}^{(j)}) + m_2(p_{i2}^{(j)}) + \cdots + m_{q_j}(p_{iq_j}^{(j)}) + \varepsilon_{ij}, \; i = 1, ..., n; \; j = 1, ..., p,$$

where m_k $(k = 1, ..., q_j)$ are smooth functions from R (a set of the real number) to R and ε_{ij} $(i = 1, ..., n)$ depend independently and normally on mean 0 and variance σ_j^2. For the function m_k, it is assumed that

$$m_k(p_{ik}^{(j)}) = \sum_{m=1}^{M_{jk}} \gamma_{mk}^{(j)} b_{mk}^{(j)}(p_{ik}^{(j)}), \quad i = 1, ..., n; \; k = 1, ..., q_j, \tag{2}$$

where $\{b_{1k}^{(j)}, ..., b_{M_{jk}k}^{(j)}\}$ is a prescribed set of basis functions (such as Fourier series, polynomial bases, regression spline bases, B-spline bases, wavelet bases and so on), the coefficients $\gamma_{1k}^{(j)}, ..., \gamma_{M_{jk}k}^{(j)}$ are unknown parameters and M_{jk} is the number of basis functions.

Then a nonparametric regression model can be written as a probability density function in the form

$$f_j(x_{ij}|\boldsymbol{p}_{ij}; \boldsymbol{\gamma}_j, \sigma_j^2) = \frac{1}{\sqrt{2\pi\sigma_j^2}} \exp\left[-\frac{\{x_{ij} - \sum_{k=1}^{q_j} \sum_{m=1}^{M_{jk}} \gamma_{mk}^{(j)} b_{mk}^{(j)}(p_{ik}^{(j)})\}^2}{2\sigma_j^2} \right], \tag{3}$$

where $\boldsymbol{\gamma}_j = (\boldsymbol{\gamma}_{j1}^T, ..., \boldsymbol{\gamma}_{jq_j}^T)^T$ is a parameter vector with $\boldsymbol{\gamma}_{jk} = (\gamma_{1k}^{(j)}, ..., \gamma_{M_{jk}k}^{(j)})^T$. If a variable X_j has no parent variables, we consider the model based on the normal distributions with mean μ_j and variance σ_j^2.

Finally we have a Bayesian network model based on the nonparametric regression model with Gaussian noise

$$f(\boldsymbol{x}_i; \boldsymbol{\theta}_G) = \prod_{j=1}^{p} f_j(x_{ij}|\boldsymbol{p}_{ij}; \boldsymbol{\theta}_j), \qquad i = 1, ..., n,$$

where $\boldsymbol{\theta}_G = (\boldsymbol{\theta}_1^T, ..., \boldsymbol{\theta}_p^T)^T$ is a parameter vector included in the graph G and $\boldsymbol{\theta}_j$ is a parameter vector in the model f_j, i.e., $\boldsymbol{\theta}_j = (\boldsymbol{\gamma}_j^T, \sigma_j^2)^T$ or $\boldsymbol{\theta}_j = (\mu_j, \sigma_j^2)^T$.

3 Proposed criterion for choosing graph

Let $\pi(\boldsymbol{\theta}_G|\boldsymbol{\lambda})$ be the prior distribution on the unknown parameter vector $\boldsymbol{\theta}_G$ with hyper parameter vector $\boldsymbol{\lambda}$ and let $\log \pi(\boldsymbol{\theta}_G|\boldsymbol{\lambda}) = O(n)$. The marginal probability of the data $\boldsymbol{X}_n$ is obtained by integrating over the parameter space, and we choose a graph G with the largest posterior probability

$$\pi_G \int \prod_{i=1}^{n} f(\boldsymbol{x}_i; \boldsymbol{\theta}_G) \pi(\boldsymbol{\theta}_G|\boldsymbol{\lambda}) d\boldsymbol{\theta}_G, \tag{4}$$

where π_G is a prior probability of G. Friedman and Goldszmidt[12] considered the multinomial distribution as the Bayesian network model $f(\boldsymbol{x}_i; \boldsymbol{\theta}_G)$, and also supposed the Dirichlet prior on the parameter $\boldsymbol{\theta}_G$. In this case, the Dirichlet prior is the conjugate prior and the posterior distribution belongs to the same class of distribution. Then a closed form solution of the integration in (4) is obtained, and they called it BDe score for choosing graph[6,16]. Recall that the BDe score is confined to the multinomial model, and we propose a criterion for choosing graph in more general and various situations.

The essential problem of constructing criteria based on (4) is how to compute the integration. While some methods can be considered for computing the integration such as Markov chain Mote Carlo, we use the Laplace approximation for integrals[7,17,24], because it is not necessary to consider the conjugate prior. The Laplace approximation to the marginal probability of $\boldsymbol{X}_n$ is

$$\begin{aligned}\int \prod_{i=1}^{n} f(\boldsymbol{x}_i; \boldsymbol{\theta}_G)\pi(\boldsymbol{\theta}_G|\boldsymbol{\lambda})d\boldsymbol{\theta}_G &= \int \exp\{nl_\lambda(\boldsymbol{\theta}_G|\boldsymbol{X}_n)\}d\boldsymbol{\theta}_G \\ &= \frac{(2\pi/n)^{r/2}}{|J_\lambda(\hat{\boldsymbol{\theta}}_G)|^{1/2}} \exp\{nl_\lambda(\hat{\boldsymbol{\theta}}_G|\boldsymbol{X}_n)\}\{1 + O_p(n^{-1})\},\end{aligned}$$

where r is the dimension of $\boldsymbol{\theta}_G$,

$$\begin{aligned} l_\lambda(\boldsymbol{\theta}_G|\boldsymbol{X}_n) &= \frac{1}{n}\sum_{i=1}^{n} \log f(\boldsymbol{x}_i; \boldsymbol{\theta}_G) + \frac{1}{n}\log \pi(\boldsymbol{\theta}_G|\boldsymbol{\lambda}), \\ J_\lambda(\boldsymbol{\theta}_G) &= -\partial^2\{l_\lambda(\boldsymbol{\theta}_G|\boldsymbol{X}_n)\}/\partial\boldsymbol{\theta}_G\partial\boldsymbol{\theta}_G^T \end{aligned}$$

and $\hat{\boldsymbol{\theta}}_G$ is the mode of $l_\lambda(\boldsymbol{\theta}_G|\boldsymbol{X}_n)$. Then we have a criterion, BNRC, for selecting graph

$$\begin{aligned}\text{BNRC}(G) &= -2\log\left\{\pi_G \int \prod_{i=1}^{n} f(\boldsymbol{x}_i; \boldsymbol{\theta}_G)\pi(\boldsymbol{\theta}_G|\boldsymbol{\lambda})d\boldsymbol{\theta}_G\right\} \\ &= -2\log \pi_G - r\log(2\pi/n) + \log|J_\lambda(\hat{\boldsymbol{\theta}}_G)| - 2nl_\lambda(\hat{\boldsymbol{\theta}}_G|\boldsymbol{X}_n). \end{aligned} \quad (5)$$

The optimal graph is chosen such that the criterion BNRC (5) is minimal.

This criterion is derived under $\log \pi(\boldsymbol{\theta}_G|\boldsymbol{\lambda}) = O(n)$. If $\log \pi(\boldsymbol{\theta}_G|\boldsymbol{\lambda}) = O(1)$, the mode $\hat{\boldsymbol{\theta}}_G$ is equivalent to the maximum likelihood estimate, MLE, and the criterion is resulted in Bayesian information criterion, known as BIC[20] by removing the higher order terms $O(n^{-j})$ $(j \geq 0)$. Konishi[18] provided a general framework for constructing model selection criteria based on the Kullback-Leibler information and Bayes approach.

It is assumed that the prior density $\pi(\boldsymbol{\theta}_G|\boldsymbol{\lambda})$ is decomposed into the product of the prior densities on $\boldsymbol{\theta}_j$, $\pi_G(\boldsymbol{\theta}_G|\boldsymbol{\lambda}) = \pi_1(\boldsymbol{\theta}_1|\boldsymbol{\lambda}_1)\times\cdots\times\pi_p(\boldsymbol{\theta}_p|\boldsymbol{\lambda}_p)$. Hence $l_\lambda(\hat{\boldsymbol{\theta}}_G|\boldsymbol{X}_n)$ and $\log|J_\lambda(\hat{\boldsymbol{\theta}}_G)|$ in (5) are, respectively,

$$\sum_{j=1}^{p} l_\lambda^{(j)}(\hat{\boldsymbol{\theta}}_j|\boldsymbol{X}_n) \quad \text{and} \quad \sum_{j=1}^{p} \log\left| -\frac{\partial^2 l_\lambda^{(j)}(\boldsymbol{\theta}_j|\boldsymbol{X}_n)}{\partial\boldsymbol{\theta}_j\partial\boldsymbol{\theta}_j^T} \right|,$$

where

$$l_\lambda^{(j)}(\boldsymbol{\theta}_j|\boldsymbol{X}_n) = \frac{1}{n}\sum_{i=1}^{n} \log f_j(x_{ij}|\boldsymbol{p}_{ij};\boldsymbol{\theta}_j) + \frac{1}{n}\log\pi_j(\boldsymbol{\theta}_j|\boldsymbol{\lambda}_j). \tag{6}$$

Thus the BNRC (5) can be obtained by the local scores of graph as follows: We define the local BNRC for the j-th variable X_j by

$$\text{BNRC}_j = -2\log\left\{\pi_{L_j}\int\prod_{i=1}^{n} f_j(x_{ij}|\boldsymbol{p}_{ij};\boldsymbol{\theta}_j)\pi_j(\boldsymbol{\theta}_j|\boldsymbol{\lambda}_j)d\boldsymbol{\theta}_j\right\}, \tag{7}$$

where π_{L_j} is a prior probability of the local structure associated with X_j. We also apply Laplace's method to the BNRC_j and the BNRC is obtained by

$$\text{BNRC} = -2\log\pi_G + \sum_{j=1}^{p}\{\text{BNRC}_j + 2\log\pi_{L_j}\}.$$

Notice that the final graph is selected as a minimizer of the BNRC and it is not necessary minimize each local score BNRC_j, because the graph is constructed as acyclic.

4 Estimating graph and related structures by using BNRC

In this section we express our method in more concrete terms. The key idea of our proposed method is the use of the nonparametric regression and the new criterion for choosing graph from Bayesian statistics.

As for nonparametric regression in Section 2, we use the B-splines[8] as the basis functions in (2). Figure 1 is an example of B-splines of degree 3 with equidistance knots $t_1, ..., t_{10}$. We place the knots dividing the domain $[\min_i(p_{ik}^{(j)}), \max_i(p_{ik}^{(j)})]$ into $M_{jk}-3$ equidistance interval[10] and set M_{jk} B-splines of degree 3.

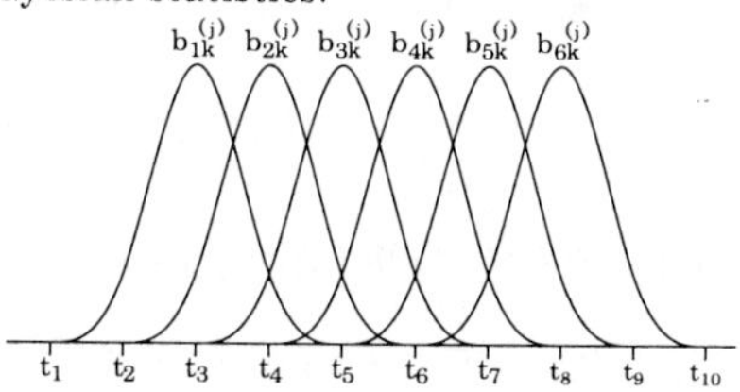

Figrure 1: Example of 6 B-splines of degree 3. $t_1, ..., t_{10}$ are called knots. These knots are equally spaced.

We assume that the prior distribution on the parameter vector $\boldsymbol{\theta}_j$ is

$$\pi_j(\boldsymbol{\theta}_j|\boldsymbol{\lambda}_j) = \prod_{k=1}^{q_j} \pi_{jk}(\boldsymbol{\gamma}_{jk}|\lambda_{jk}).$$

Each prior distribution $\pi_{jk}(\boldsymbol{\gamma}_{jk}|\lambda_{jk})$ is a singular M_{jk} variate normal distribution given by

$$\pi_{jk}(\boldsymbol{\gamma}_{jk}|\lambda_{jk}) = \left(\frac{2\pi}{n\lambda_{jk}}\right)^{-(M_{jk}-2)/2} |K_{jk}|_+^{1/2} \exp\left(-\frac{n\lambda_{jk}}{2}\boldsymbol{\gamma}_{jk}^T K_{jk}\boldsymbol{\gamma}_{jk}\right), \quad (8)$$

where λ_{jk} is a hyper parameter, K_{jk} is an $M_{jk} \times M_{jk}$ matrix, $\boldsymbol{\gamma}_{jk}^T K_{jk}\boldsymbol{\gamma}_{jk} = \sum_{l=3}^{M_{jk}}(\gamma_{lk}^{(j)} - 2\gamma_{l-1,k}^{(j)} + \gamma_{l-2,k}^{(j)})^2$ and $|K_{jk}|_+$ is the product of $M_{jk} - 2$ nonzero eigenvalues of K_{jk}. The score BNRC$_j$ (7) can be obtained as

$$\begin{aligned} \text{BNRC}_j &= -2\log\pi_{L_j} - 2\sum_{i=1}^{n}\log f_j(x_{ij}|\boldsymbol{p}_{ij};\hat{\boldsymbol{\theta}}_j) - 2\sum_{k=1}^{q_j}\log\pi_k(\hat{\boldsymbol{\gamma}}_{jk}|\lambda_{jk}) \\ &+ \log\left|-\frac{\partial^2 l_\lambda^{(j)}(\hat{\boldsymbol{\theta}}_j|\boldsymbol{X}_n)}{\partial\boldsymbol{\theta}_j\boldsymbol{\theta}_j^T}\right| - (\sum_{k=1}^{q_j} M_{jk}+1)\log(2\pi n^{-1}), \end{aligned} \quad (9)$$

where $\hat{\boldsymbol{\theta}}_j = (\hat{\boldsymbol{\gamma}}_j^T, \hat{\sigma}_j^2)^T$ is a mode of $l_\lambda^{(j)}(\boldsymbol{\theta}_j|\boldsymbol{X}_n)$ defined in (6) for fixed λ_{jk}. For computational aspect, we approximate the logarithm of the determinant of the Hessian matrix in (9) by

$$\sum_{k=1}^{q_j}\{\log|B_{jk}^T B_{jk} + n\hat{\sigma}_j^2\lambda_{jk}K_{jk}| - M_{jk}\log(n\hat{\sigma}_j^2)\} - \log(2\hat{\sigma}_j^4),$$

where B_{jk} is an $n \times M_{jk}$ matrix defined by $B_{jk} = (\boldsymbol{b}_{jk}(p_{1k}^{(j)}), \ldots, \boldsymbol{b}_{jk}(p_{nk}^{(j)}))^T$ with $\boldsymbol{b}_{jk}(p_{ik}^{(j)}) = (b_{1k}^{(j)}(p_{ik}^{(j)}), \ldots, b_{M_{jk}k}^{(j)}(p_{ik}^{(j)}))^T$. Hence combining (3), (8) and (9), BNRC$_j$ is resulted in

$$\begin{aligned} \text{BNRC}_j &= C_j + (n - 2q_j - 2)\log\hat{\sigma}_j^2 \\ &+ \sum_{k=1}^{q_j}\left\{\frac{n\beta_{jk}}{\hat{\sigma}_j^2}\hat{\boldsymbol{\gamma}}_{jk}^T K_{jk}\hat{\boldsymbol{\gamma}}_{jk} + \log|\Lambda_{jk}| - (M_{jk}-2)\log\beta_{jk}\right\}, \end{aligned}$$

where $\beta_{jk} = \sigma_j^2\lambda_{jk}$ is a hyper parameter,

$$\begin{aligned} C_j &= -2\log\pi_{L_j} + (n + \bar{M}_{j\cdot} - 2q_j)\log(2\pi) + n - \log 2 \\ &\quad -2(\bar{M}_{j\cdot} - q_j)\log n - \sum_{k=1}^{q_j}\log|K_{jk}|_+, \end{aligned}$$

$$\Lambda_{jk} = B_{jk}^T B_{jk} + n\beta_{jk} K_{jk}, \qquad \bar{M}_{j\cdot} = \sum_{k=1}^{q_j} M_{jk}.$$

By using the backfitting algorithm[15], the modes $\hat{\gamma}_{jk}$ $(k = 1, ..., q_j)$ can be obtained when the values of β_{jk} are given. The backfitting algorithm can be expressed as follow:
Step 1 *Initialize:* $\gamma_{jk} = \mathbf{0}, \quad k = 1, ..., q_j$.
Step 2 *Cycle:* $k = 1, ..., q_j, 1, ..., q_j, 1, ...$

$$\gamma_{jk} = (B_{jk}^T B_{jk} + n\beta_{jk} K_{jk})^{-1} B_{jk}^T (\boldsymbol{x}_{(j)} - \sum_{k' \neq k} B_{jk'} \gamma_{jk'}).$$

Step 3 *Continue Step 2 until a suitable convergence criterion is satisfied.*
The mode $\hat{\sigma}_j^2$ is given by $\hat{\sigma}_j^2 = ||\boldsymbol{x}_{(j)} - \sum_{k=1}^{q_j} B_{jk}\hat{\gamma}_{jk}||^2/n$.

In attention, the modes $\hat{\gamma}_{jk}$ and $\hat{\sigma}_j^2$ depend on the hyper parameters β_{jk} and we have to choose the optimal values of β_{jk}. In the context of our method, it is natural that the optimal β_{jk} are chosen as the minimizers of BNRC_j.

Recall that the B-splines coefficients vectors γ_{jk} are estimated by maximizing (6). The modes of (6) are the same as the penalized likelihood estimates[22,25] and we can look upon the hyper parameters λ_{jk} or β_{jk} as the smoothing parameters in penalized likelihood. Hence, the hyper parameters play an important role for fitting the curve to the data.

5 Computational experiments

Monte Calro simulation

Before analyzing the real data, we used the Monte Carlo simulation method to examine the effectiveness of our method. The data were generated from an artificial graph and structures between variables (Figure 2).

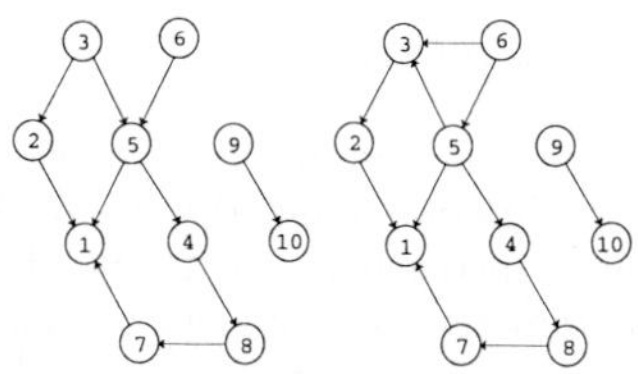

Figure 2: Monte Carlo simulation. (Left) true, (Right) estimate.

$$X_1 = X_2^2 + 2\sin(X_5) - 2X_7 + \varepsilon_1$$
$$X_2 = \{1 + \exp(-4X_3)\}^{-1} + \varepsilon_2$$
$$X_3 = \varepsilon_3, \ X_6 = \varepsilon_6, \ X_9 = \varepsilon_9$$
$$X_4 = X_5^2/3 + \varepsilon_4, \ X_5 = X_3 - X_6^2 + \varepsilon_5$$
$$X_7 = \begin{cases} -1 + \varepsilon_7, & X_8 \le -0.5 \\ X_8 + \varepsilon_7, & -0.5 < X_8 \le 0.5 \\ 1 + \varepsilon_7, & 0.5 < X_8 \end{cases}$$
$$X_8 = \exp(-X_4 - 1)/2 + \varepsilon_8$$
$$X_{10} = \cos(X_9) + \varepsilon_{10}.$$

The results from this Monte Carlo simulation can be summarized as follows: Proposed criterion BNRC can detect linear and nonlinear structures of the data very well. But the BNRC has a tendency toward overgrowth of graph. We then consider the use of Akaike's information criterion known as AIC[1,2] and use both methods. AIC was originally introduced as a criterion for evaluating models estimated by maximum likelihood method. But the estimate by our method is the same as the maximum penalized likelihood estimates and is not MLE. In this case, the modified version of AIC[10] is given by

$$\mathrm{AIC} = -2\sum_{i=1}^{n} \log f_j(x_{ij}|\boldsymbol{p}_{ij}; \hat{\gamma}_j, \hat{\sigma}_j^2) + 2(\sum_{k=1}^{q_j} \mathrm{tr}S_{jk} + 1),$$

where $S_{jk} = B_{jk}(B_{jk}^T B_{jk} + n\hat{\beta}_{jk}K_{jk})^{-1}B_{jk}^T$. The trace of S_{jk} shows the degree of freedom of the fitted curve and is a great help. That is to say, if $\mathrm{tr}S_{jk}$ is nearly 2, the dependency is looked upon linear. We use both BNRC and AIC for decision whether we add up to a parent variable. By using this method, the estimated graph and structures are close to the true model.

Analysis of cell cycle data

We analyze the *S.cerevisiae* cell cycle data discussed by Spellman *et al.*[21] and Friedman *et al.*[13]. The data were collected from 800 genes with 77 experiments.

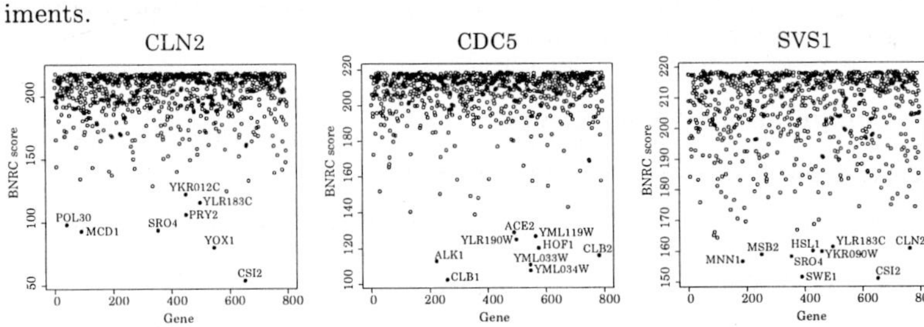

Figure 3: BNRC scores for CNL2, CDC5 and SVS1.

We set the prior probability π_G constant, because we have no reason why the large graph is unacceptable and no information about the size of the true graph. The nonparametric regressors are constructed with 20 B-splines. In fact, the number of B-splines is also a parameter. However, we use somewhat large number of B-splines, the hyper parameters control the smoothness of fitted curve and we cannot visually find differences among fitted curves corresponding to various number of B-splines.

The results of the analysis from the cell cycle data can be summarized as follows: Figure 3 shows BNRC scores when we predict CLN2, CDC5 and SVS1 by one gene. The genes, which give smaller BNRC scores, give a better expression to the target gene. We can observe that which gene is associated with the target gene and we find the set of genes which strongly depend on the target gene. In fact, we can construct a brief network by using this information. We can look upon the optimal graph as a revised version of the brief network by choosing the parent genes and holding the assumption of acyclic. We note that if there is a linear dependency between genes, the score BNRC is also good when the parent-child relationship is reversed. Therefore, the directions of the causal associations in the graph are not strict especially when the dependency is almost linear. Our result basically advocates the result of Friedman *et al.*[13], but, of cause, there are different points in parts. There are some genes that mediate Friedman *et al.*'s result, such as MCD1, CSI2, YOX1 and so on. A large number of the relationships between genes are nearly linear. But we could find some nonlinear dependencies which linear models hardly find. Figure 5 shows the estimated graph associated with genes which were classified their processes into cell cycle and their neighborhood. Here, we omit some branches in Figure 5, but important information is almost shown. As for the networks given by us and Friedman *et al.*[13], we confirmed parent-children relationships and observed that both two networks are similar to each other. Especially, our network includes typical relationships which were reported by Friedman *et al.*[13]. As for the differences between two networks, we paid attention to the parent genes of SVS1. Friedman *et al.*[13] employed CLN2 and CDC5 as the parent genes of SVS1. On the one hand, our result gives CSI2 and YKR090W for SVS1. We check up on the difference of these two results. In the sense of BNRC and AIC, our candidate parent genes are more appropriate than Friedman *et al.*[13]'s. The reason might be the effect of discretizing, because our model suitably fits to both cases in Figure 4. We notice that the range of the fitted curve in Figure 4 (b) is much smaller than other curves. All in all, we conclude that CDC5 gives just weak effects to SVS1 compared with other genes from Spellman *et al.*[21]'s data (see also Figure 3). In fact, as the parent gene of SVS1, the order of BNRC score of CDC5 is 247th. Considering the circumstances mentioned above, our method can provide us valuable information in understandable and useful form.

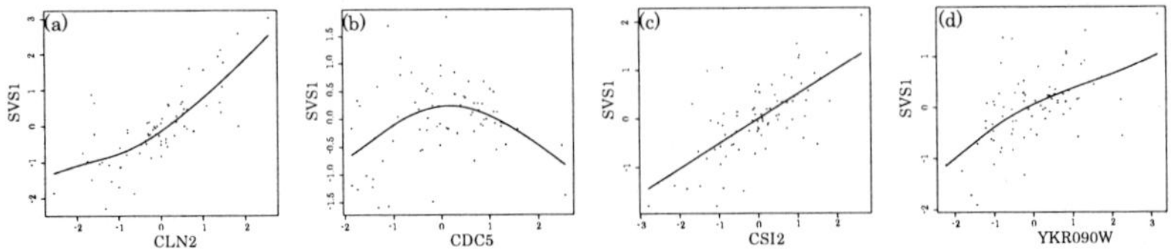

Figure 4: Cell cycle data and smoothed estimates.
(a) and (b) Friedman *et al.*[13], BNRC = 160.45, AIC=167.96;
(c) and (d) Proposed method, BNRC = 135.27, AIC=140.16.

6 Discussion

We proposed the new method for estimating genetic networks from microarray gene expression data by using Bayesian network and nonparametric regression. We derived a new criterion for choosing graph theoretically, and represented its effectiveness through the Monte Calro simulations and the analysis of the cell cycle data. The advantages of our method are mainly as follows: We can use the expression data as continuous values. Not only linear dependencies, we can also detect nonlinear structures and can visualize their functional structures being easily understandable. Fully automatic search can accomplish the creation of optimal graph.

We also pointed out that Friedman *et al.*[13]'s method remained the unknown parameters such as threshold value for discretizing and hyper parameters in the Dirichlet priors which selected by trial and error. These parameters were not optimized in a narrow sense. On the other hand, our proposed method can automatically and appropriately estimate any parameters based on proposed criterion which has a sounder theoretical basis. Besides, our method includes Friedman *et al.*[13]'s as a special case.

We consider the following problems as our future works: (1) We used the statistical models based on Gaussian distribution. However, we derive the criterion BNRC in more general situations. In fact, we can construct the graph selection criterion based on other statistical models. (2) It is a possible case that the outliers cause strange results. Thus, the development of the robust methods and the technique for detecting the outliers are important problems. (3) The intensities of the unions are probably measured by using bootstrap method[9]. We would like to investigate these problems in a future paper.

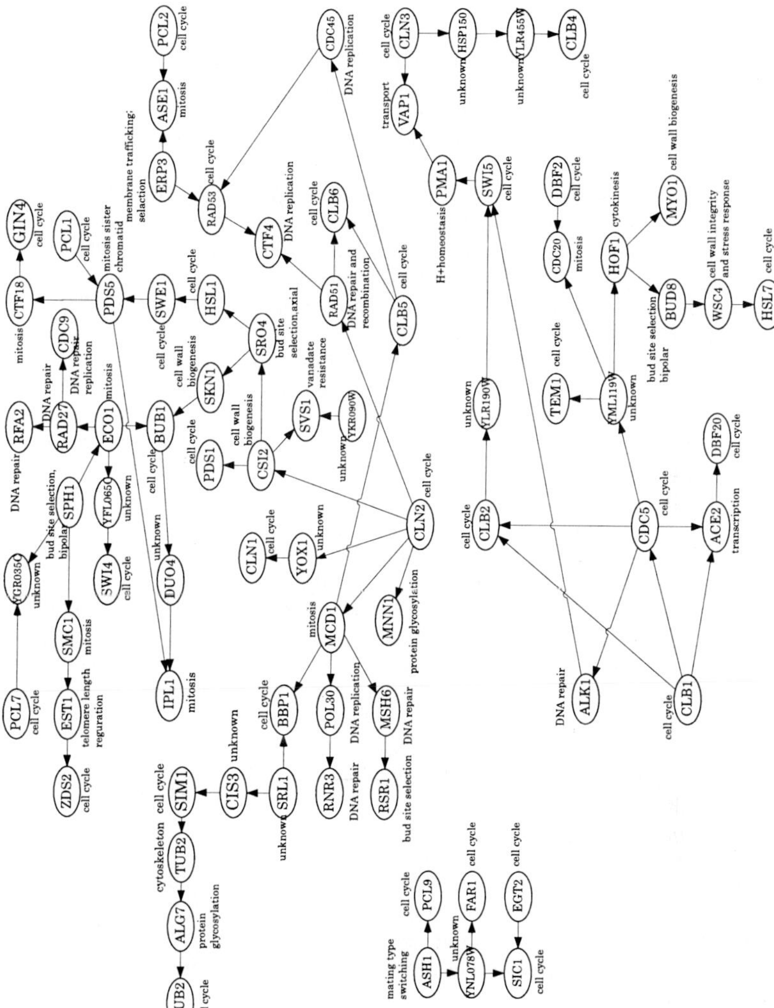

Figure 5: Cellcycle data result.

References

1. H. Akaike, *in* Petrov,B.N. and Csaki,F. eds., *2nd Inter. Symp. on Information Theory*, Akademiai Kiado, Budapest, 267 (1973).
2. H. Akaike, *IEEE Trans. Autom. Contr.*, **AC-19**, 716 (1974).
3. T. Akutsu, S. Miyano and S. Kuhara, *Pacific Symposium on Biocomputing*, 17, (1999).
4. T. Akutsu, S. Miyano and S. Kuhara, *Bioinformatics*, **16**, 727 (2000).
5. T. Akutsu, S. Miyano and S. Kuhara, *J. Comp. Biol.*, **7**, 331 (2000).
6. G. F. Cooper and E. Herskovits, *Machine Learning*, **9**, 309 (1992).
7. A. C. Davison, *Biometrika*, **73**, 323 (1986).
8. C. de Boor, *A Practical Guide to Splines.* Springer, Berlin. (1978).
9. B. Efron, *Ann. Stat.*, **7**, 1 (1979).
10. P. H. C. Eilers and B. Marx, *Statistical Science*, **11**, 89 (1996).
11. R. L. Eubank, *Spline Smoothing and Nonparametric Regression*, Marcel Dekker, New York. (1988).
12. N. Friedman and M. Goldszmidt, *in* M. I. Jordan ed., *Learning in Graphical Models*, Kluwer Academic Publisher. 421 (1998).
13. N. Friedman, M. Linial, I. Nachman and D. Pe'er, *J. Comp. Biol.*, **7**, 601 (2000).
14. P. J. Green and B. W. Silverman, *Nonparametric Regression and Generalized Linear Models.* Chapman & Hall, London. (1994).
15. T. Hastie and R. Tibshirani, *Generalized Additive Models.* Chapman & Hall, London. (1990).
16. D. Heckerman, D. Geiger and D. M. Chickering, *Machine Learning*, **20**, 274 (1995).
17. D. Heckerman, *in* M. I. Jordan ed., *Lerning in Graphical Models* 301, Kluwer Academic Publisher. (1998).
18. S. Konishi, (in Japanese), *Sugaku*, **52**, 128 (2000).
19. D. Pe'er, A. Regev, G. Elidan and N. Friedman, *Bioinformatics*, **17** *Suppl.1*, 215 (ISMB 2001).
20. G. Schwarz, *Ann. Stat.*, **6**, 461 (1978).
21. P. Spellman, G. Sherlock, M. Zhang, V. Iyer, K. Anders, M. Eisen, P. Brown, D. Botstein and B. Futcher, *Mol. Biol. Cell*, **9**, 3273 (1998).
22. B. W. Silverman, *J. R. Stat. Soc. Series* B, **47**, 1 (1985).
23. J. S. Simonoff, *Smoothing Methods in Statistics.* Springer, New York. (1996).
24. L. Tinerey and J. B. Kadane, *J. Amer. Statist. Assoc.*, **81**, 82 (1986).
25. G. Wahba, *J. R. Stat. Soc. Series* B, **40**, 364 (1978).

AUTOMATIC ANNOTATION OF GENOMIC REGULATORY SEQUENCES BY SEARCHING FOR COMPOSITE CLUSTERS

O.V. KEL-MARGOULIS[1,2], T.G. IVANOVA[2], E.WINGENDER[3], A.E. KEL[1,2]

[1] *BIOBASE GmbH, Halchtersche Strasse 33, 38304 Wolfenbuettel, Germany;* [2]*Institute of Cytology & Genetics SB RAN, 10 Lavrentyev pr., 630090, Novosibirsk;* [3]*Research Group Bioinformatics, Gesellschaft für Biotechnologische Forschung mbH, Mascheroder Weg 1, D-38124 Braunschweig, Germany*

A new method was developed for revealing of composite clusters of cis-elements in promoters of eukaryotic genes that are functionally related or coexpressed. A software system "ClusterScan" have been created that enables: (i) to train system on representative samples of promoters to reveal cis-elements that tend to cluster; (ii) to train system on a number of samples of functionally related promoters to identify functionally coupled transcription factors; (iii) to provide tools for searching of this clusters in genomic sequences to identify and functionally characterize regulatory regions in genome. A number of training samples of different functional and structural groups of promoters were analysed. Search for composite clusters in human chromosomes 21 and 22 reveals a number of interesting examples. Finally, a decision tree system was constructed to classify promoters of several functionally related gene groups. The decision tree system enables to identify new promoters and computationally predict their possible function.

1. Introduction

Besides the fact that genomes of eukaryotic organisms contain rather limited number of genes (Ng) [1], the number of different intracellular molecular states (Ns) is enormously huge (Ns >>Ng). In multicellular organisms these are states of cellular ontogenesis in different tissues, organs and cell types, a number of developmental stages and cell cycle phases, the huge amount of influences of different external and internal signals. Every state is characterized and precisely organized by differential expression of specific sets of genes. Therefore it becomes obvious that most of the genes in genome are expressed in various cellular states (gene expression pattern), and it is a combination of active genes that is state specific (gene expression profile). For the majority of genes, transcription regulation plays the most important role in regulation of gene expression. Combinatorial regulation of transcription is organized through binding of a multiplicity of transcription factors (TFs) to their target sites (cis-elements) in regulatory regions. Corresponding TFs interact with each other and with particular components of the basal transcription complex as well as with coactivators/corepressors, histone acetylases/deacetylases, therefore making up function-specific multiprotein complexes. These multi-protein complexes are often referred to as enhncesomes [1]. Functionally related genes involved in the same molecular-genetic, biochemical, or physiological process are often regulated coordinately by specific combinations of transcription factors. On the level of DNA, the blueprint of such common mechanisms of regulation may be seen as specific combinations of TF binding sites located in a close proximity to each other. We call such structures as "composite clusters". We are aiming to reveal a variety of such composite clusters in regulatory regions of eukaryotic genes. Different composite clusters could serve as good benchmarks for identification of new promoters and other regulatory regions in genomes and for functional characterization of the expression of the corresponding new genes as for understanding molecular mechanisms of their regulation. The goal is to find efficient means for automatic annotation of genomic regulatory sequences.

Last years, several computational approaches have appeared addressing the problem of combinatorial regulation of transcription. Specific TF binding site combinations were used for identification of muscle-specific promoters [2,3] for liver-enriched genes [4] and for yeast genes [5]. Recently, we have shown that search for specific combinations of two TF sites - composite elements - is a very effective tool for predicting gene expression patterns. We have demonstrated this approach for promoters of genes highly induced upon immune response [6]. Promoters of genes regulated during cell cycle could be recognized by combination of E2F binding sites with a dozen of oligonucleotide motifs [7]. A number of known examples of composite elements is collected in COMPEL database [8]. This data together with computationally predicted composite structure provide a key for annotation of regulatory regions in genomes.

Annotation of gene regulatory regions requires computational approaches that work with high sensitivity and specificity. One possible way to increase specificity is to develop methods that are trained on groups of co-regulated promoters rather than all promoters. Specific combinations of cis-elements for the vast variety of gene functional groups have to be determined to develop methods for automatic annotation of regulatory genomic sequences.

We have developed a method for revealing of composite clusters of cis-elements in promoters of eukaryotic genes that are functionally related or coexpressed. A software system "ClusterScan" have been created that enables: (i) to train system on representative samples of promoters to reveal cis-elements that tend to cluster; (ii) to train system on a number of samples of functionally related promoters to identify functionally coupled transcription factors; (iii) to provide tools for searching of this clusters in genomic sequences to identify and functionally characterize regulatory regions in genome. A number of training samples of different functional and structural groups of promoters were analysed. Search for composite clusters in human chromosomes 21 and 22 reveals a number of potential cell cycle regulated sequences. Finally, a decision tree system was constructed to classify promoters of several functionally related gene groups. The decision tree system enables to predict an expression pattern of a new potential promoter by classifying it to the one of these promoter groups.

2 Method

2.1 Revealing of composite clusters of TF binding sites.

It is known that most of TF target sites are located in 5' regions of genes. We assume that binding sites for transcription factors that bind together to a regulatory region of a gene tend to be co-localized in a relatively short region inside the 5' regulatory region in order to provide possibility for protein-protein interactions between these factors. Therefore, it is expected that such sites for many different factors will make clusters in 5' regulatory regions that we call: "composite clusters" (CC) . Presence of such composite clusters in genomic sequence might be a good indication of regulatory regions of genes.

We have developed a method for identifying composite clusters of binding sites that are specific for promoter sequences. The method first analyses structure of promoter sequences from a training set of promoters trying to reveal clusters of transcription factor binding sites. For that, the whole library of weight matrices collected in TRANSFAC database [9] were considered. The method is based on genetic algorithm. It selects matrices and optimises cut-off values for every considered matrix in order to maximize the number of clusters in the training set of promoter sequences in contrast to a control set of non-promoter sequences.

Let's M is the set of all weight matrices from TRANSFAC. The following parameters are used for revealing composite clusters: K – a subset of weight matrices selected from the set M; $q^{(k)}_{cut-off}$ ($k \in K$) - cut-off values of the matrix score (a site s considered to be present in a given position of the sequence if the score of the matrix k at this position exceeds the cut-off value $q^{(k)}(s) > q^{(k)}_{cut-off}$); max$d$ – the maximal distance between adjacent binding sites in a cluster. For example, when *mind* = 20bp, the algorithm considers only those clusters where distances between adjacent sites shorter then 20bp. The borders of the clusters are defined by sites that separated from the neighbour sites by the distance longer then max$d$. For a fixed set λ of mentioned above parameters we can search for all clusters in every promoter sequence x. Then, we calculate the following function, that we call "cluster score":

$$CC\_score_{\lambda}(x) = \sum_{i=1,n} number\_of\_sites(i) \times density\_of\_sites(i) \qquad (1)$$

where n – is the number of found clusters in the sequence x; *number_of_sites(i)* – number of sites in the i-th cluster; *density_of_sites(i)= number_of_sites(i)/length_of_cluster(i)* – density of sites in the i-th cluster.

To reveal the best parameter set λ_{best} that exposes clusters in the promoters we apply a *genetic algorithm* (GA-1) that selects the subset K and optimises the values of the parameters *mind* and $q^{(k)}_{cut-off}$. We use the following fitness function:

$$J(\lambda) = \left(\frac{1}{|Y|}\sum_{y\in Y} CC\_score_{\lambda}(y) - \frac{1}{|Z|}\sum_{z\in Z} CC\_score_{\lambda}(z)\right) - R(|K|) \quad (2)$$

In this fitness function we calculate difference between average values of the cluster score of the positive sample (Y) and negative sample (Z). The training set of promoters is the positive sample Y. As the negative sample Z we use a set of exon sequences. *R(m)* – is a function constructed similar to the Akaike Information Criteria [10] that decreases fitness of the models while the number of weight matrices increases. This criterion rescues the model to be over-fitted by getting the high number of free parameters.

2.2 Revealing of functionally coupled transcription factor sets

To analyse in more details the structure of found composite clusters we consider the following basic model of the composite promoter structure. Every eukaryotic promoter (or, more generally, a transcription regulatory region) contains numerous binding sites for different transcription factors that are organized in a number of functionally coupled subsets of factors – "functional sets" (FS). Every FS consists of a group of transcription factors that work together in one regulatory process by synergy or in antagonism through binding to their target sites that located in a close proximity to each other in regulatory regions of genes. Such FSs provide a framework for building up a specific complex of interacting TFs that supply a distinct regulatory function. Many examples of the simplest FSs consisting of two TFs with two adjacent binding sites are collected in the database of composite elements COMPEL [8]. Such FSs may provide gene induction in response to a complex condition, e.g. tissue-specific response to a certain extracellular signals. More complex FSs consisting of several interacting factors and may contain DNA signals of complex origin, such as TATA and GC boxes, Inr element and others. A family of functionally related promoters shares FSs that contain "obligatory" factors with target sites found practically in all promoters of the set and defining the "main" function of these promoters and "facultative" sites that may vary from promoter to promoter and modulate the function in a specific manner. Such FSs being revealed as common for a promoter sample may be good benchmarks for promoter classification.

We describe a FS - μ, characteristic for a group of functionally related promoters, by the following set of parameters: P a set of different TF weight matrices that compose the μ (including "obligatory" and "facultative" matrices). A certain cut-off value $q_{cut-off}^{(p)}$ and importance value $\phi^{(p)}$ are assigned to every weight matrix p ($p \in P$) in μ. For every promoter sequence x we calculate the following function, that we call a "functional score":

$$FS\_score_{\mu}(x) = \sum_{p \in P} \phi^{(p)} \times q_0^{(p)}(x) , \tag{3}$$

where $q_0^{(p)}(x)$ is the score of the best site found in the sequence x by the matrix p ($q_0^{(p)} = 0$, if no sites were found with score $q > q_{cut-off}^{(p)}$).

Optimisation of the parameters of the "functional score" is done by a modification of genetic algorithm (GA-2) similar to the one described in the previous section. In this case, the positive samples (Y) were sets of functionally related promoters. As the negative sample (Z) we use a full set of promoters (EPD database) where Y promoters are excluded.

2.3 Decision tree for classification of promoters

To classify promoters we build a decision tree (T) in the similar way as in [11]. The bottom nodes (i) of the tree (leafs) contain L different promoter classes. The internal nodes (j) of the tree represent different types of FSs - $\mu^{(j)}$. To classify a promoter sequence x the functional score $FS\_score_{\mu^{(i)}}(x)$ is calculated according to the equation (2) at every node as the sequence is passed to the tree. Cut-off values $FS\_score_{cut-off}$ are assigned to every internal node. If $FS\_score_{\mu^{(i)}}(x)$ > $FS\_score_{cut-off}$ the sequence is passed to the left downstream node otherwise to the right downstream node. Finally, the sequence is classified to the one of the L promoter classes.

The decision tree was built by a variant of the *genetic algorithm* (GA-3), that optimizes the structure of the decision tree and cut-off values of the corresponding functions. The algorithm selects the components of $\mu^{(j)}$ at every node of the tree. The fitness function π is calculated on the basis of misclassification rate of decision tree T:

$$\pi(T) = \prod_{i=1,L} \frac{N_{predict}^{(i)}}{N_{real}^{(i)}} \Bigg/ R(\sum m^{(j)}) \tag{4}$$

Here, $N_{real}^{(i)}$ – is the number of promoters of the class (i) in the training set, $N_{predict}^{(i)}$ – is the number of correctly classified promoters of the class (i) , R – is the same function as in (2) calculated on the total number of weight matrices $m^{(j)}$ used in the decision tree.

3 Results

3.1 Composite clusters in promoter sequences of mammalian genes.

We extracted promoter sequences of mammalian genes from EPD database. The considered region was from –500 to +99 relative to the start of transcription. 349 promoters were extracted. We applied GA-1 method and revealed a set of matrices that exposes clusters in the promoters of this group. The following 25 matrices were selected by the algorithm: V$MSX1_01, V$PAX8_B, V$CDXA_01, V$GEN_INI3_B, V$P300_01, V$BARBIE_01, V$E2F_2Q6, V$E2F1_Q6, V$SP1_01, V$AP1_Q6, V$AP1_Q4, V$PAX4_04, V$NFKB_Q6, V$FOXD3_01, VUSF_Q6, VLDSPOLYA_B, V$OCT1_07, V$HNF3B_01, V$STAT_01, V$E2F_Q4, V$ETS1_B, V$OCT1_02, V$MYCMAX_02, V$SRF_C, V$VMAF_01. The maximal distance between adjacent sites *maxd* = 23bp. The average size of the clusters in promoter regions was 2.8 sites per cluster and in exon sequence 0.2 sites per cluster. It means that practically no clusters composed by these sites were observed in exon sequences, whereas in many promoters these sites make clusters of 3–5 sites.

3.2 Functionally coupled transcription factor sets.

Seven sets of promoters were obtained from different sources: promoters for cell-cycle related genes (43 promoters) and brain enriched genes (45 promoters) (collected in this work on the base of literature search), muscle-specific (25 promoters) and immune cell specific genes (24 promoters) [6], erythroid specific genes (10 promoters) (http:/www.bionet.nsc.ru), liver enriched genes (39 promoters) and housekeeping genes (26 promoters) (EPD rel.62). The promoter sequences of the length 600 bp (from –500 to +99 relative start of transcription) were extracted from EMBL database. We have selected these sets since they represent the most distinct functional classes of promoters

Applying the GA-2 method we have revealed functional sets of transcription factors specific for the promoter classes described above (see Table 1).

One can see, that matrices for a number of class-specific factors (such as E2F, NF-AT, MyoD, ..) were taken by the method as "obligatory" (high importance values were assigned). These matrices were included only in one class-specific functional set. Other matrices for some of the ubiquitous factors (such as SP-1, SRF, AP-1...) have been included in many FSs. These factors appeared to play an important role in many types of promoters.

In Fig. 1 we show two distributions of the functional score for cell cycle promoters versus exon sequences. One can see that high values of the score are the characteristic feature of the most cell cycle related promoters.

Table 1. Functional sets of transcription factors specific for different promoter classes. Values of the matrix relative importance are shown in brackets in the front of each TF name.

Promoter class	TF factors selected	$J(\lambda)$ – score
Cell-cycle related	**E2F** (1.00), **TATA** (0.95), **CREB** (0.88), **Sp-1** (0.81)	7.2
Brain enriched	BRLF1 (0.192), ATF (0.038), **CREB (0.450), Sp-1 (0.592), HFH2 (1.00)**	3.8
Muscle-specific	Tal-1 (0.50), **YY-1 (1.0),** Oct-1 (0.40), **MyoD (0.80)**, **SRF (1.0), PAX5 (0.80)**	5.2
Immune cell specific	COMP1 (0.024), STAF (0.017), **NF-kB (1.30), NF-AT (0.957),** Brn-2 (0.059)	6.6
Erythroid specific	**n-myc (0.31)** , GR (0.08), **AP-4 (1.00),** RREB-1 (0.08), v-Maf (.08)	2.0
Liver enriched	**RORalpha1 (1.00),** Sp-1 (0.03), **SREBP-1 (1.00), HNF-1 (0.54),** ER (0.07), GATA-1 (0.03)	2.6
Housekeeping	Egr-2 (0.15), **AhR/Arnt (0.72), ZID (0.94), Elk-1 (0.79),** NRF-2 (0.54), **CREB (.62)**	7.2

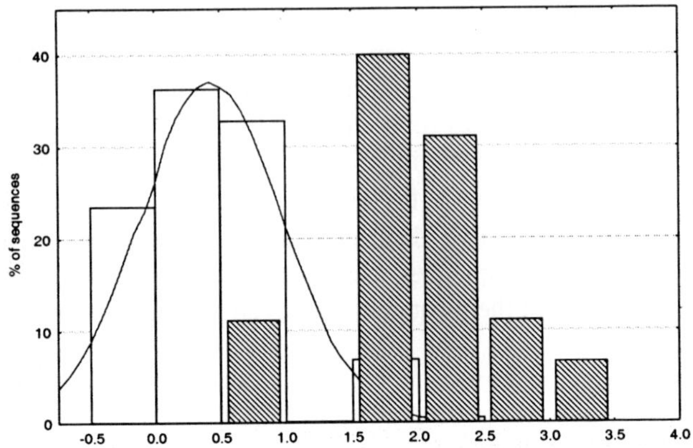

Fig. 1 Histograms of the functional score values in the cell-cycle related promoters (black) and exon sequences (white). The score is given along the x – axis. The functional score is calculated on the bases of set of factors specific for cell cycle promoters that are shown in the Table 1 (first column). In the histogram we show the percentage of the sequences in each set that exhibit the given value of "cell cycle functional score".

3.3 Decision tree classifier of promoters.

A decision tree classifier of the 7 classes of promoters was build by using of the weight matrices found in the previous step. The bottom nodes of the tree contain 7 different promoter classes. The training set of 212 promoters described above was used for optimising the decision tree structure with the help of GA-3. The topology of the one of the decision tree obtained in the analysis is shown in Figure 2.

The following set of TF binding sites appeared to be the most effective for classification of the mentioned sets of promoters: E2F, Oct-1, NF-AT, MyoD, SRF and ER.

Percentage of the correct classification obtained by the tree is shown below each bottom node. One can see that cell cycle related and erythroid specific promoters are classified best (65 – 70% of correct classifications). In contrast, promoters of housekeeping genes and brain-enriched genes are most difficult to classify (34% and 20% of correct classifications correspondingly). It is known that these two classes contain genes with very heterogeneous function and expression. More efforts should be paid for initial grouping of promoters into functionally unified classes.

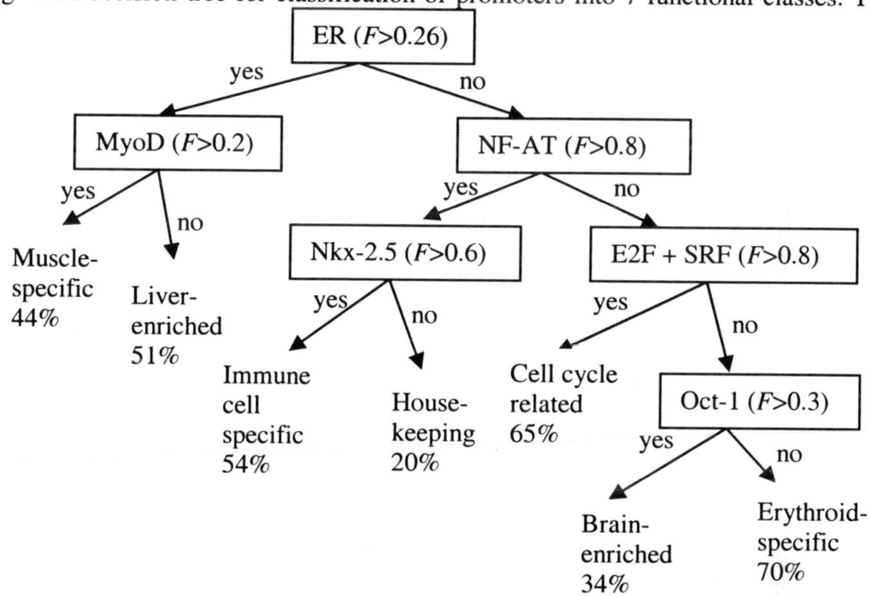

Fig. 2. A decision tree for classification of promoters into 7 functional classes. To classify a new promoter, the sequence (x) is passed down the tree beginning at the top. If the functional score: $F(x) > Fcut\text{-}off$ the sequence is passed down to the left, otherwise to the right. The functions $F(x)$ and cut-offs were optimised by GA-3.

We have applied the developed promoter classifier for identification of new potential cell cycle regulation for a number of known genes retrieved from EMBL. In our previous work [6] we developed a new method for context-specific identification of binding sites for E2F transcription factors – the main regulators of cell cycle progression. This method was applied to reveal new E2F target genes. We scanned EMBL release 6.0, divisions: hum, rod, vrt, and mam. 4611 promoters have been have been retrieved and analysed. As a result, 313 promoters were identified as new potential E2F targets [7].

In the present work, the promoter classifier was applied to the selected promoters to find the most probable target genes. After passing through the decision tree 103 promoters were classified as potential cell cycle regulated promoters.

Some of these promoters were inspected experimentally in our previous work using *in vivo* formaldehyde cross-linking technique to confirm the identity of potential E2F target genes that have been suggested computationally [7]. Using antibodies against various members of the E2F family, the specific E2F binding to several promoters under study in asynchronously growing HeLa cells have been confirmed. The following promoters bearing predicted E2F binding sites were experimentally confirmed to be cell-cycle dependent: *c-fos* and *junB*; the gene encoding TGF-β which acts as an antiproliferative agent to a majority of cell types; ARF locus encoding protein that binds to and stabilizes p53 and thus functions in tumor suppression; *mcm4* and *mcm5* involved in the initiation of DNA replication; von Hippel-Lindau (*VHL*) tumor suppressor gene; and *e2f-1*.

3.4 Search for composite clusters in human chromosomes and identification of new potential cell cycle related genes.

We have applied the ClusterScan system for scanning chromosome sequences of human genome in order to reveal composite clusters benchmarking new potential regulatory regions. As an example, we search for clusters that are specific for cell cycle related genes. We have previously shown that promoters of cell cycle genes are characterized by high frequency of the E2F binding sites [7]. The majority of promoters of cell cycle genes are GC-rich and TATA-less. In some of these promoters E2F binding sites are located just at the transcriptional start site. These data suggest that basal promoters of the cell cycle regulated genes may be characterized by a specific arrangement of known and yet unknown DNA elements resulting in specific composite clusters. In the training set of 29 cell cycle-dependent genes (no orthologs) we have revealed specific basal DNA elements by using Gibbs sampling program [12]. Within region [-45; -16] three motifs were revealed: TATA-like, GC box, and "CCT/ATT" motif. At the start site, [-15;+15], an E2F-like motif, an Inr-like pattern and the motif "CCC/A" were revealed. Downstream of the start site, within [+16; +45], a GAGA-like box was found. For all these motifs positional weight matrices were constructed. All the revealed motifs together with the E2F weight matrix were used for searching composite clusters in the chromosomal sequences.

Analysis of the human chromosome 21 resulted in 20 composite clusters. Of them, 7 clusters are located within annotated repeat families – SINE, LINE and LTR; 1 clusters within CpG islands; 2 within intron sequences of two genes; 4 clusters are found just 5' to the annotated mRNA start of genes with unknown function (see an example of such gene found in the chromosome 22, Fig.3); and 6 clusters do not coincide with any annotation.

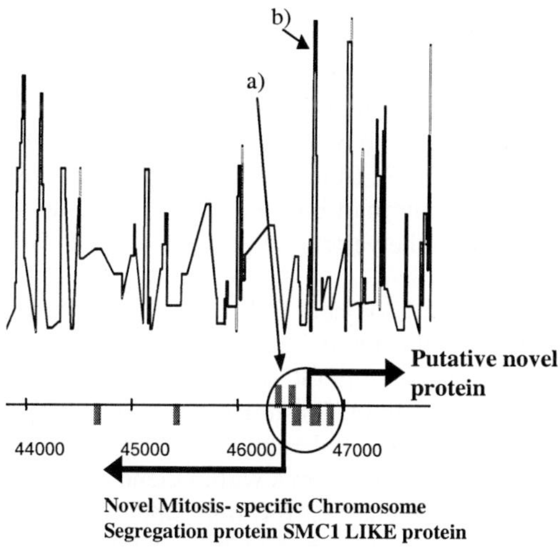

Fig.3 Prediction of a novel cell cycle regulated promoter in a fragment of the chromosome 22. Human DNA sequence from clone RP1-102D24 on chromosome 22 (AC: AL021391) is considered. a) a cluster of E2F sites at the potential starts of transcription of two genes. b) a pick of the composite cluster score (CC_score) comprising basal elements revealed in the training set of cell cycle related genes.

In summary, the computer method presented here allows us to search for clusters of potential cis-regulatory elements and to reveal promoters that belong to several definite functional categories. Experimental verification of some of these promoters confirms the computational predictions. With the advent of the large-scale sequencing projects, it becomes increasingly essential to develop computational methods enabling to analyse transcription regulatory regions of new genes and predict theire regulatory functions.

Acknowledgments

The authors are indebted to Vadim Ratner and Michael Zhang for fruitful discussion of the results. Parts of this work was supported by Siberian Branch of Russian Academy of Sciences, by grant of Volkswagen-Stiftung (I/75941)

References

1. Merika M. and Thanos D. Enhanceosomes. *Curr. Opin. Genet. Dev.* **11,** 205-208 (2001)
2. Wasserman, W. W., Fickett, J. W. Identification of regulatory regions which confer muscle-specific gene expression. *J. Mol. Biol.* **278** , 167-181 (1998)
3. Frech, K., Quandt, K., Werner, T. Muscle actin genes: A first step towards computational classification of tissue specific promoters. *In Silico Biology* **1**, 0005, http://www.bioinfo.de/isb/1998/01/0005/ (1998)
4. Tronche, F., Ringeisen, F., Blumenfeld, M., Yaniv, M. & Pontoglio, M. Analysis of the distribution of binding sites for a tissue-specific transcription factor in the vertebrate genome. *J. Mol. Biol.* **266**, 231-245 (1997)
5. Brazma, A., Vilo, J. & Ukkonen, E. Finding Transcription Factor Binding Site Combinations in the Yeast Genome. In *Proceedings of the German Conference on Bioinformatics* GCB'97, Kloster Irsee, Bavaria, Sept. 22-24, 1997 (H.W.Mewes and D.Frishman eds.), (1997) 57-60
6. Kel, A., Kel-Margoulis, O., Babenko, V., Wingender, E. " Recognition of NFATp/AP-1 Composite Elements within Genes Induced upon the Activation of Immune Cells" *J. Mol. Biol.* **288** , 353-376 (1999)
7. Kel A.E, Kel-Margoulis O.V., Farnham P.J., Bartley S.M., Wingender E., and Zhang M.Q. Computer-assisted identification of cell cycle-related genes - new targets for E2F transcription factors. *J. Mol. Biol.* **309** , 99 – 120 (2001)
8. Kel-Margoulis,O.V., Romaschenko,A.G., Kolchanov,N.A., Wingender,E. and Kel,A.E. TRANSCompel: a database on composite regulatory elements providing combinatorial transcriptional regulation. *Nucleic Acids Res.* **28,** 311-315 (2000)
9. Wingender, E., Chen, X., Hehl, R., Karas, H., Liebich, I., Matys, V., Meinhardt T., Pruss, M., Reuter, I., Schacherer, F. TRANSFAC: an integrated system for gene expression regulation. *Nucleic Acids Res.* **28,** 316-319 (2000)
10. Akaike, H. *IEEE Trans. Autom. Control* **19**, 761−723 (1974)
11. Salzberg, S. Locating protein coding regions in human DNA using a decision tree algorithm *J. Comput. Biol.* **2** , 473-485 (1995)
12. Kel-Margoulis O., Kel A. and Wingender E. Automatic annotation of the regulatory regions of cell cycle related genes on human chromosomes. // Proceedings of the conference, Genome sequencing and biology. Cold Spring Harbor Laboratory, May 9-13, 2001. P.139

EULER-PCR: FINISHING EXPERIMENTS FOR REPEAT RESOLUTION

ZUFAR MULYUKOV, PAVEL A. PEVZNER
Department of Computer Science and Engineering,
University of California, San Diego, CA 92093, USA

Genomic sequencing typically generates a large collection of unordered contigs or scaffolds. Contig ordering (also known as gap closure) is a non-trivial algorithmic and experimental problem since even relatively simple-to-assemble bacterial genomes typically result in large set of contigs. Neighboring contigs maybe separated either by *gaps* in read coverage or by *repeats*. In the later case we say that the contigs are separated by *pseudogaps*, and we emphasize the important difference between gap closure and pseudogap closure. The existing gap closure approaches do not distinguish between gaps and pseudogaps and treat them in the same way. We describe a new fast strategy for closing pseudogaps (repeat resolution). Since in highly repetitive genomes, the number of pseudogaps may exceed the number of gaps by an order of magnitude, this approach provides a significant advantage over the existing gap closure methods.

1 Introduction

Large scale sequencing projects always require a *finishing* phase, i.e., designing and conducting additional experiments for closing gaps and establishing the overall order of contigs. The design of such finishing experiments still requires extensive human intervention using interactive tools, such as sequence editors (Gordon et al., 1998 [1]). A typical DNA sequencing project generates a large collection of unordered contigs or scaffolds. Ordering such contigs is a major effort and often a bottleneck in sequence finishing. Contig ordering is usually done by PCR experiments that correspond to the queries "Are the contigs A and B neighbors?" A naive approach to such "The twenty questions game" requires PCR experiments for every pair of contigs and is very time-consuming. Sorokin et al., 1996,[2] Tettelin et al., 1999,[3] and Beigel et al., 2001 [4] suggested *multiplex PCR* approach that uses pooling strategy to ask more complicated queries "Given sets of contigs $\mathcal{A}$ and $\mathcal{B}$, do they contain contigs $A \in \mathcal{A}$ and $B \in \mathcal{B}$ that are neighbors?".

Contig ordering is closely related to *gap closure*. Neighboring contigs maybe separated either by *gaps* in read coverage or by *repeats*. In the later case of repeat-induced gaps we say that the contigs are separated by *pseudogaps*. For example, in the *Neisseria meningitidis* (NM) project (Parkhill et al., 2000 [5]), Phrap generates 160 contigs, but only half of them are separated by gaps, while the other half is separated by pseudogaps (Pevzner et al.,

2001 [6]). The existing contig ordering algorithms do not distinguish between gaps and pseudogaps and treat them in the same way. This approach is inefficient, since it ignores information available for pseudogaps, such as repeat length and contig sequences adjacent to a particular repeat. Therefore, an algorithm that employs a separate approach to resolving pseudogaps provides a significant advantage over the existing gap closure methods. We describe a new algorithm, EULER-PCR, that significantly reduces the number of finishing experiments for repeat resolution. EULER-PCR software is available by contacting Z.M.

2 Repeat graph

Long repeats present a problem in DNA sequencing since they often lead to multiple solutions of the fragment assembly problem. Figure 1(a) illustrates the "repeat problem" caused by perfect triple repeat that leads to two possible sequence assemblies. The classical "overlap-layout-consensus" approach (Kececioglu and Myers, 1995 [7]) to the assembly problem is based on the notion of the *overlap graph* (Fig. 1(b)). Every read corresponds to a vertex in the overlap graph and two vertices are connected by an edge if the corresponding reads overlap. The DNA sequence corresponds to a path traversing the consecutive reads in this graphs. The fragment assembly problem is thus cast as finding a path in the overlap graph visiting every vertex exactly once, a *Hamiltonian path* problem. However, repeats complicate the overlap graph since repeated regions create edges between non-consecutive reads. The Hamiltonian path problem is NP-complete and the efficient algorithms for solving this problem in large graphs are unknown. This is the reason why fragment assembly of highly repetitive genomes is a notoriously difficult problem. Myers et al., 2000 [8] suggested to mask most of multi-copy repeats, thus breaking the assembly into a large number of contigs. A better approach would be to use the information about repeated regions and try to reduce the number of contigs.

Pevzner et al. 2001, [9] and Pevzner and Tang 2001 [6] developed a new fragment assembly algorithm (EULER) based on the Eulerian path approach. Instead of masking repeats and breaking DNA sequence into a set of contigs, EULER constructs a *repeat graph*, which represents the repeat structure better than the overlap graph does. Given a DNA sequence, the repeat graph can be visualized by glueing together all identical repeated regions (Fig.1(c)). One can see that the repeat graph (Fig.1(c)) is a much simpler representation of repeats than the overlap graph (Fig.1(b)).

To construct the repeat graph from the set of sequencing reads, EULER

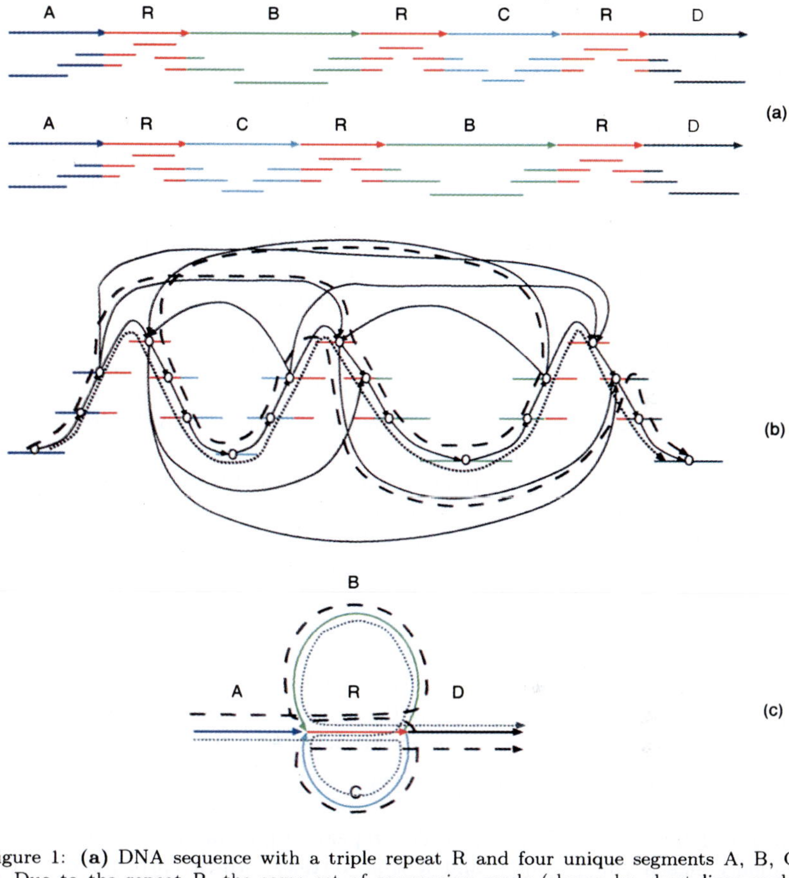

Figure 1: **(a)** DNA sequence with a triple repeat R and four unique segments A, B, C, D. Due to the repeat R, the same set of sequencing reads (shown by short lines under assembled DNAs) can be assembled either as ARBRCRD (upper assembly) and ARCR-BRD (lower assembly), which differ by transposition of B and C. **(b)** Overlap graph for "overlap-layout-consensus" approach. Two Hamiltonian paths, corresponding to two possible fragment assemblies, are shown by dashed (for ARBRCRD) and dotted (for ARCRBRD) lines. **(c)** Repeat graph where three copies of the repeat R are "glued" into a single edge. Every Eulerian path in this graph corresponds to a valid solution of the fragment assembly problem. Two Eulerian paths, corresponding to two possible fragment assemblies, are shown by dashed (for ARBRCRD) and dotted (for ARCRBRD) lines.

breaks the reads into short k-mers (continuous strings of length k). One can view such k-mers as a result of hybridization of reads with a very large virtual DNA chip. These k-mers are represented by edges of the *de Bruijn* graph, while the set all $(k-1)$-mers from set of sequencing reads are represented by vertices of the graph. Two vertices v and w are joined by a directed edge if there is a k-mer in which first $(k-1)$ nucleotides coincide with v, and last $(k-1)$ nucleotides coincide with w (see example in Fig.2). We emphasize that the fragment assembly is now cast as finding a path visiting every *edge* of the graph exactly once, an *Eulerian path* problem. In contrast to the Hamiltonian path problem, the Eulerian path problem is easy to solve even for graphs with millions of vertices since there exist linear-time Eulerian path algorithms. This is the fundamental difference between the EULER algorithm (Pevzner et al., 2001) and the "overlap-layout-consensus" approach. The repeat graph is obtained from de Bruijn graph by collapsing paths in the de Bruijn graph into single edges (see Pevzner et al. 2001 [9] for details).

In this new approach contigs, which would be disconnected if repeats were masked, are represented by edges of a connected graph. We can compute repeat copy numbers by assigning minimal (nonzero) multiplicities to the graph edges that balance in-flow and out-flow on every vertex (Pevzner and Tang, 2001 [6]). For example in Fig. Fig.1(c), repeat edge R has multiplicity 3. Edges with multiplicity higher than one represents a repeat, while edges with unit multiplicity represent conventional contigs. Note, that every repeat corresponds to a single edge in the repeat graph rather than to a collection of vertices in the layout graph. The DNA sequence in Fig.1(a) consists of four unique segments A,B,C,D and one triple repeat R. The corresponding repeat graph (Fig.1(c)) consists of 4+1=5 edges. Two edges X and Y in the repeat graph follow each other if and only if segment X follow segment Y in the DNA sequence. For a repeat edge $e=(v,w)$, edges entering the vertex v are called *entrances* into a repeat, and edges leaving the vertex w are called *exits* from a repeat.

Gaps in read coverage break DNA sequence into a set of Lander-Waterman islands (Lander and Waterman, 1988 [10]) and cause sequencing reads to be assembled into a set of contigs. The repeat graph for a continuous DNA sequence has a single source and a sink vertices, which correspond to the beginning and the end of the DNA sequence. On the other had, the repeat graph for a set of contigs has multiple sources and sinks, corresponding to contig end-points. Fig.3 shows the fragment assembly problem similar to Fig.1 with some reads missing thus leading to two islands in the read coverage. In this case the the repeat graph corresponds to two contigs and there are two possible solutions of the fragment assembly problem: contigs ARB and CRD, or contigs ARD and CRB. A single finishing PCR experiment would resolve

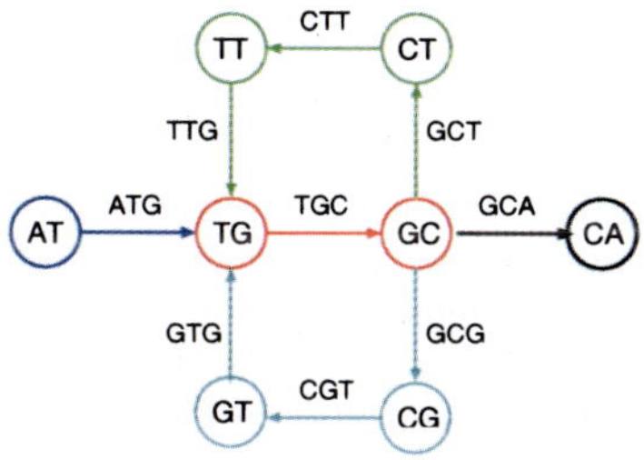

Figure 2: Example of a de Bruijn graph for sequence ATGCTTGCGTGCA, with edges being all 3-mers from this sequence, and vertices being all 2-mers. Due to the repeat TGC there is another sequence ATGCGTGCTTGCA corresponding to another Eulerian path in same de Bruijn graph (compare with Fig.1).

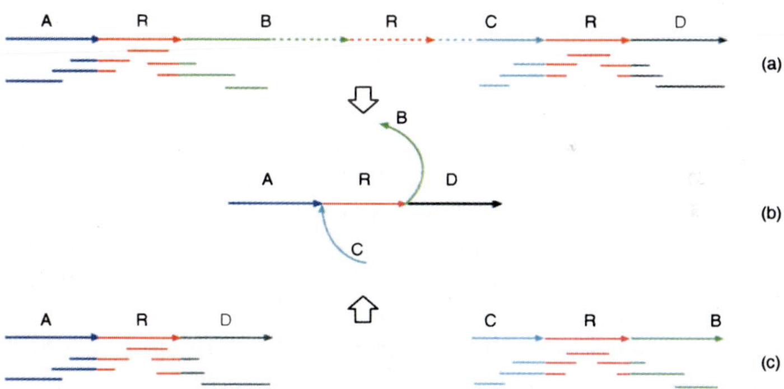

Figure 3: (a) Gap in read coverage breaks DNA sequence into two islands ARB and CRD (B and C correspond to the shortened versions of segments B and C from Fig.1). (b) The repeat graph for these two islands is similar to the repeat graph in Fig.1 with "broken" edges B and C. (c) Another solution of the fragment assembly problem with islands ARD and CRB (instead of ARB and CRD).

the repeat and delineate the correct assembly.

The above examples illustrate that the repeat graphs are valuable tools providing important new insights into the repeat structure, as well as guiding finishing experiments.

3 Repeat resolution

EULER (Pevzner et al., 2001[9]) typically resolves all repeats except long perfect ones that are not contained inside any sequencing read and therefore cannot be resolved without double barreled data. Similarly EULER-DB (Pevzner and Tang, 2001 [6]) typically resolves all repeats that are shorter than the clone length. In a repeat graph such repeats are represented by edges with multiplicities greater than one. Multiplicities of the repeat edges define the repeat copy numbers. Figure 4(a) shows the largest connected component of the repeat graph for the *Neisseria meningitidis* (NM) sequencing project. It is not obvious which edges in this graph correspond to repeats and what are their multiplicities. Pevzner and Tang, 2001[6] described EULER-CN algorithm that find multiplicities of edges in the repeat graph by iteratively balancing the Kirchhoff flow on every vertex.

While Fig. 4(a) looks complicated, it tells us what contigs are possible neighbors and how they are oriented with respect to each other. The traditional sequence assembly algorithms do not output this information, leaving finishers in the dark during the gap closure process. Comparison of figures 4(a) and 4(b) illustrates the advantages of generating a repeat graph instead of a large set of disconnected contigs. For a set of disconnected contigs (Fig. 4(b)), a straightforward way to order them is to conduct PCR experiments for all possible pairs of contigs (*combinatorial PCR*). This results in an extensive finishing effort requiring over 30,000 PCR experiments. The repeat graph in Fig. 4(a) eliminates the need for exhaustive test of every contig pair and suggests conducting PCR experiments only for edges entering and exiting a repeat. Even such simple approach, which tests all possible pairs of edges entering and exiting a repeat one-by-one (*graph-based combinatorial PCR*), requires only 195 PCR experiments to resolve all repeats in the graph in Fig. 4(a) for NM sequencing data.

Tettelin et al., 1999 [3] suggested optimized primer pooling for multiplex PCR to minimize number of experiments for gap closing. However unlike in the case of gap closure, in the case of pseudogap closure the repeat graph provides information about the length of a repeat to be resolved, as well as entrance and exit sequences for the repeat. This information enables us to resolve all pseudogaps in sequencing data in significantly smaller number of

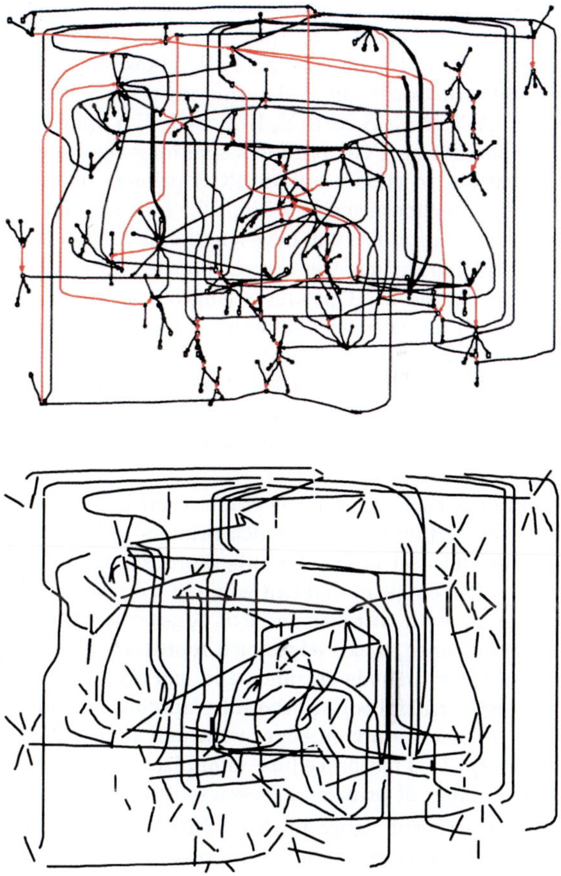

Figure 4: (a) Largest connected component of the repeat graph for *Neisseria meningitidis* project (Parkhill et al., 2001 [5]) as assembled by EULER (Pevzner et al., 2001 [9]). Red edges indicate repeats as determined by EULER-CN. (b) Masking repeats breaks the repeat graph into an unordered set of contigs.

multiplex PCR experiments as compared to resolution of gaps. EULER-PCR uses the repeat graph to select primers pools for multiplex PCR experiments. For NM project, EULER-PCR reduces number of PCR experiments to 21 reactions, which can be run concurrently.

Fig.5 shows an example of a repeat of multiplicity 3, with 3 edges entering and 3 edges leaving the repeat. In this case the sequencing reads do not provide information on which of exits X, Y, Z follow the entrances A, B, C. The sequence reconstruction requires determining 3 correct pairings among 9 possibilities: A-X, A-Y, A-Z, B-X, B-Y, B-Z, C-X, C-Y, and C-Z. This can be accomplished by generating PCR products spanning the repeat. To generate such products, one has to choose unique PCR primers on entrance and exit edges. If we are able to choose such positions for forward and reverse primers, so that *all possible* PCR products will have lengths that are sufficiently different from each other, we can deduce the correct pairings from a single multiplex reaction by measuring the PCR products lengths. Assuming a conservative estimate of the length measurement accuracy for long-range PCR product to be about 10%, the relative pairwise length differences of possible PCR products should be at least 10%. Fig.5 demonstrates that a repeat with 3 entrances and 3 exits can be resolved in a single multiplex PCR experiment using only 3 forward and 3 reverse primers. A single reaction tests all 9 possible pairings between entrance and exit edges.

Repeats can follow each other in the graph, therefore a divide-and-conquer strategy is employed to find edges on which primers to be placed. The repeat graph is partitioned into set of smaller subgraphs which contain one or more repeats as follows: if all entrance and exit edges of a repeat have unit multiplicity, such repeat, along with entrance and exit edges constitutes a single simple subgraph; if some entrance of exit edges or a repeat are repeats on their own, subgraph is expanded to include entrance and exit edges of all repeats in the subgraph until every terminal edge in the subgraph has unit multiplicity. PCR primer pairs will be placed only on edges with unit multiplicity (the terminal edges of the subgraph). This procedure minimizes the number of multiplex PCRs, as well as number of primers, necessary to resolve the repeats.

An example of a repeat subgraph from NM sequencing data is given in Fig.6. The structure of this repeat subgraph reveals that central region of the repeats overlap, while some smaller repeats are completely contained inside larger ones. Thus the repeat graph generated by EULER can provide insights into the history of duplication events in genomes.

After finding the set of repeat subgraphs, EULER-PCR selects such set of primers per reaction tube that will test maximal number of pairings between entrance and exit edges of the repeat subgraphs given the constraint on maxi-

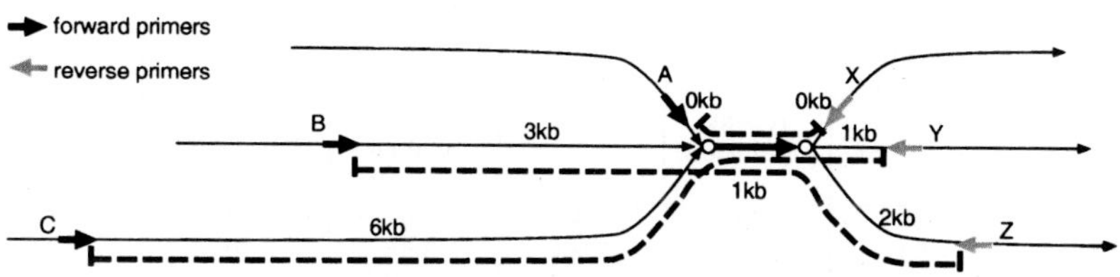

Figure 5: An example of primer placement on edges entering and exiting a repeat. Length of the repeat is 1 kb. Forward primers A, B, and C are placed on distances 0 kb, 3 kb, and 6 kb, respectively, "upstream" from the vertex starting the repeat. Reverse primers X, Y, and Z are placed on distanced 0 kb, 1 kb, and 2 kb, respectively "downstream" from the vertex ending the repeat. With such primer arrangement all nine possible PCR products will have length differing by 10% or more from each other. Lengths of PCR product between primer pairs: A-X, A-Y, A-Z, B-X, B-Y, B-Z, C-X, C-Y, and C-Z varies from 1 kb to 9 kb respectively. One possible outcome of multiplex PCR is shown by dashed lines.

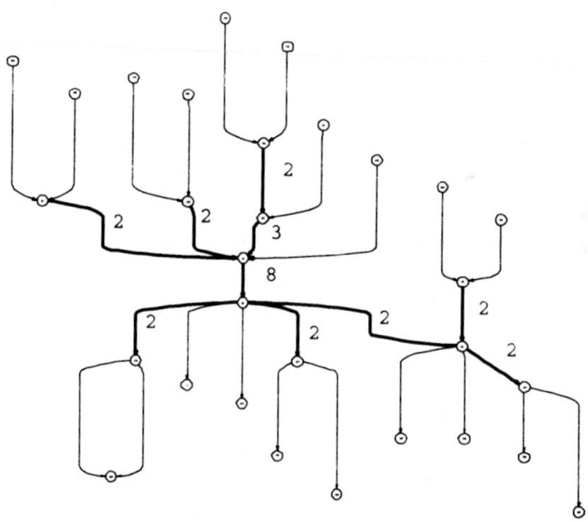

Figure 6: A complex repeat subgraph containing multiple repeats (thick edges). Numbers next to thick edges indicate repeat multiplicities. Thin edges have unit multiplicity.

Table 1: "Combinatorial PCR" and POMP columns show the number of finishing PCR experiments for gap closure. "Graph-based Combinatorial PCR" and EULER-PCR columns show the number of repeat resolution experiments.

Genome	Number of contigs before repeat resolution	Number of contigs after repeat resolution	Number of experiments			
			Combinatorial PCR	Multiplex PCR with primer pooling (POMP)	Graph-based Combinatorial PCR	Multiplex EULER-PCR
CJ	18	15	630	90	16	3
LL	53	8	5,565	490	305	24
NM	123	69	30,135	1800	195	21

mal length of PCR product L_{max}. Primers sequences chosen by EULER-PCR are of length 20 bases or longer. Those sequences are selected in accordance with standard requirements for primer selection (Haas et al., 1998 [11]) regarding uniqueness in the genomic sequence, melting temperature requirements, G or C for 3' base, etc.

Last two columns in the Table 1 show the number of repeat resolution experiments by straightforward graph-based combinatorial PCR and by optimized EULER-PCR. For comparison, though indirect, the table also presents the number of finishing experiments for gap closure by combinatorial PCR and by pipette optimized multiplex PCR (POMP) suggested by Tettelin et al.,1999. [3] The results are presented for *Campylobacter jejuni* (Parkhill et al., 2001 [12]) , *Lactococcus lactis* (Bolotin et al., 2001 [13]), and *Neisseria meningitis* (Parkhill et al., 2000 [5]) sequencing projects. First two methods deal with disconnected contigs and, while closing both gaps and pseudogaps these methods require fairly large number of experiments. For N contigs, combinatorial PCR simply tests all $\binom{2N}{2}$ primer pairs. Optimized multiplex PCR method tests for presence of PCR products between pairs of primer pools, instead of pairs of primers. The number of initial reactions using POMP with pool size $\sqrt{2N}$ if given by $\binom{\sqrt{2N}}{2}$ (Tettelin et al.,1999 [3]). However, up to $2\sqrt{2N}$ additional reactions are needed for each initial PCR product to determine which primers in the pair of pools are mates. Thus, number of POMP reactions is estimated as $\sqrt{2N}\binom{\sqrt{2N}}{2}$. Note, that even straightforward approach to conduct PCRs, using the repeat graph and choosing primers on all possible pairs of edges entering and exiting repeats, requires relatively small number of reactions to resolve repeats. Results for EULER-PCR are generated for long-range PCR setup with relative pairwise difference between PCR products $\epsilon = 0.1$, and maximal PCR product length $L_{max} = 10kb$. Average number of primers per reaction suggested by EULER-PCR is 6 for CJ, 11 for LL, and 9 for NM.

Resolving repeats before gap closure significantly reduces number of finishing experiments. This reduction is especially significant for highly repetitive genomes like *Lactococcus lactis*. Number of contigs after repeat resolution reduces from 53 to only 8 for LL, thus requiring only few gap closing experiments.

Acknowledgments

This paper would not be possible without input, both with data and with discussions, provided by Haixu Tang. We thank David Harper, Julian Parkhill and Alexei Sorokin for providing sequencing data. We also thank Uri Keich and Steffen Heber for feedback and useful discussions. This work was supported by NIH grant 1 R01 HG02366-01 NCHGR.

1. D. Gordon, C. Abajian, and P. Green. Consed: A graphical tool for sequence finishing. *Genome Research*, 8:195–202, 1998.
2. A. Sorokin, A. Lapidus, V. Capuano, N. Galleron, P. Pujic, and S. Ehrlich. A new approach using multiplex long range accurate PCR and yeast artificial chromosomes for bacterial chromosome mapping and sequencing. *Genome*, 6:448–453, 1996.
3. H. Tettelin, D. Radune, S. Kasif, H. Khouri, and S. L. Salzberg. Optimized multiplex PCR: Efficiently closing a whole-genome shotgun sequencing project. *Genomics*, 62:500–507, 1999.
4. R. Beigel, N. Alon, M. Apaydin, L. Fortnow, and S. Kasif. An optimal procedure for gap closing in whole genome shotgun sequencing. In *Proceedings of the Fifth Annual International Conference in Computational Molecular Biology (RECOMB-01)*, Montreal, Canada, April, 2001. ACM Press.
5. J. Parkhill, M. Achtman, K. D. James, S. D. Bentley, C. Churcher, S. R. Klee, G. Morelli, D. Basham, D. Brown, T. Chillingworth, R. M. Davies, P. Davis, K. Devlin, T. Feltwell, N. Hamlin, S. Holroyd, K. Jagels, S. Leather, S. Moule, K. Mungall, M. A. Quail, M. A. Rajandream, K. M. Rutherford, M. Simmonds, J. Skelton, S. Whitehead, B. G. Spratt, and B. G. Barrell. Complete dna sequence of a serogroup a strain of Neisseria meningitidis Z2491. *Nature*, 404:502–506, 2000.
6. P. Pevzner and H. Tang. Fragment assembly with double barreled data. *Bioinformatics*, 17:S225–S233, 2001.
7. J.D. Kececioglu and E.W. Myers. Combinatorial algorithms for DNA sequence assembly. *Algorithmica*, 13:7–51, 1995.
8. E.W. Myers, G.G. Sutton, A.L. Delcher, I.M. Dew, D.P. Fasulo, M.J. Flanigan, S.A. Kravitz, C.M. Mobarry, K.H.J. Reinert, K.A. R., E.L. Anson, R.A. Bolanos, H. Chou, C.M. Jordan, A.L. Halpern, S. Lonardi,

E.M. Beasley, R.C. Brandon, L. Chen, P.J. Dunn, Z. Lai, Y. Liang, D.R. Nusskern, M. Zhan, Q. Zhang, X. Zheng, G.M. Rubin, M.D. Adams, , and J.C. Venter. A whole-genome assembly of drosophila. *Science*, 287:2196–2204, 2000.

9. P. Pevzner, H. Tang, and M. Waterman. An eulerian path approach to DNA fragment assembly. *Proceedings of National Academy of Sciences*, 98:9748–9753, 2001.
10. E.S. Lander and M.S. Waterman. Genomic mapping by fingerprinting random clones: a mathematical analysis. *Genomics*, 2:231–239, 1988.
11. S. Haas, M. Vingron, A. Poustka, and S. Wiemann. Primer design for large scale sequencing. *Nucleic Acids Research*, 26:3006–3012, 1998.
12. J. Parkhill, B.W. Wren, K. Mungall, J. M. Ketley, C. Churcher, D. Basham, T. Chillingworth, R. M. Davies, T. Feltwell, S. Holroyd, K. Jagels, A.V. Karlyshev, S. Moule, M. J. Pallen, C. W. Penn, Q. A. Quail, M. A. Rajandream, K. M. Rutherford, A. H. van Vliet, S. Whitehead, and B. G. Barrell. The genome sequence of the food-borne pathogen Campylobacter jejuni reveals hypervariable sequences. *Nature*, 403:665–668, 2000.
13. A. Bolotin, P. Wincker, S. Mauger, O. Jaillon, K. Malarme, J. Weissenbach, S. D. Ehrlich, and A. Sorokin. The complete genome sequence of the lactic acid bacterium Lactococcus lactis ssp. lactis IL1403. *Genome Research*, 11:731–753, 2001.

The Accuracy of Fast Phylogenetic Methods for Large Datasets

Luay Nakhleh
Dept. of Computer Sciences
University of Texas
Austin, TX 78712

Bernard M.E. Moret
Dept. of Computer Science
University of New Mexico
Albuquerque, NM 87131

Usman Roshan
Dept. of Computer Sciences
University of Texas
Austin, TX 78712

Katherine St. John
Lehman College and
The Graduate Center
City University of New York
New York, NY 10468

Jerry Sun
Dept. of Computer Sciences
University of Texas
Austin, TX 78712

Tandy Warnow
Dept. of Computer Sciences
University of Texas
Austin, TX 78712

Whole-genome phylogenetic studies require various sources of phylogenetic signals to produce an accurate picture of the evolutionary history of a group of genomes. In particular, sequence-based reconstruction will play an important role, especially in resolving more recent events. But using sequences at the level of whole genomes means working with very large amounts of data—large numbers of sequences—as well as large phylogenetic distances, so that reconstruction methods must be both fast and robust as well as accurate. We study the accuracy, convergence rate, and speed of several fast reconstruction methods: neighbor-joining, Weighbor (a weighted version of neighbor-joining), greedy parsimony, and a new phylogenetic reconstruction method based on disk-covering and parsimony search (DCM-NJ+MP). Our study uses extensive simulations based on random birth-death trees, with controlled deviations from ultrametricity. We find that Weighbor, thanks to its sophisticated handling of probabilities, outperforms other methods for short sequences, while our new method is the best choice for sequence lengths above 100. For very large sequence lengths, all four methods have similar accuracy, so that the speed of neighbor-joining and greedy parsimony makes them the two methods of choice.

1 Introduction

Most phylogenetic reconstruction methods are designed to be used on biomolecular (i.e., DNA, RNA, or amino-acid) sequences. With the advent of gene maps for many organisms and complete sequences for smaller genomes, whole-genome approaches to phylogeny reconstruction are now being investigated. In order to produce accurate reconstructions for large collections of taxa, we will most likely need to combine both approaches—each has drawbacks not shared by the other. Because whole genomes will yield large numbers of sequences, the sequence-based algorithms will need to be very fast if they are to run within reasonable time bounds. They will also have to accommodate datasets that include very distant pairs of taxa. Many of the sequence-based reconstruction methods used by biologists (maximum likelihood, parsimony search, or quartet puzzling) are very slow and unlikely to scale up to the size of data generated in whole-genome studies. Faster methods exist (such as the popular neighbor-joining method), but most suffer from accuracy problems, especially for datasets that include distant pairs.

In this paper, we examine in detail the performance of four fast reconstruction methods, one of which we recently proposed (DCM-NJ+MP), and three others that have been used for at least a few years by biologists (neighbor-joining, Weighbor, and greedy parsimony). We ran extensive simulation studies using random birth-death trees (with deviations from ultrametricity), using about three months of computation on nearly 300 processors to conduct a thorough exploration of a rich parameter space. We used four principal parameters: model of evolution (Jukes-Cantor and Kimura 2-Parameter+Gamma), tree diameter (which indirectly captures rate of evolution), sequence length, and number of taxa. We find that Weighbor (for small sequence lengths) and our DCM-NJ+MP method (for longer sequences) are the methods of choice, although each is considerably slower than the other two methods in our study. Our data also enables us to report on the sequence-length requirements of the various methods—an important consideration, since biological sequences are of fixed length.

2 Background

Methods for inferring phylogenies are studied (both theoretically and empirically) with respect to the topological accuracy of the inferred trees. Such studies evaluate the effects of various model conditions (such as the sequence length, the rates of evolution on the tree, and the tree "shape") on the performance of the methods.

The *sequence-length requirement* of a method is the sequence length needed by the method in order to reconstruct the true tree topology with high probability. Earlier studies established analytical upper bounds on the sequence length requirements of various methods (including the popular neighbor-joining [1] method). These studies showed that standard methods, such as neighbor-joining, recover the true tree (with high probability) from sequences of lengths that are exponential in the evolutionary diameter of the true tree. Based upon these studies, we defined a parameterization of model trees in which the longest and shortest edge lengths are fixed [2,3], so that the sequence length requirement of a method can be expressed as a function of the number of taxa, n. This parameterization led us to define *fast-converging* methods, methods that recover the true tree (with high probability) from sequences of lengths bounded by a polynomial in n once f and g, the minimum and maximum edge lengths, are bounded. Several fast-converging methods were developed [4,5,6,7]. We and others analyzed the sequence length requirement of standard methods, such as neighbor-joining (NJ), under the assumptions that f and g are fixed. These studies [8,3] showed that neighbor-joining and many other methods can recover the true tree with high probability when given sequences of lengths bounded by a function that grows exponentially in n.

We recently initiated studies on a different parameterization of the model tree space, where we fix the evolutionary diameter of the tree and let the number of taxa vary [9]. This parameterization, suggested to us by J. Huelsenbeck, allows us to examine the differential performance of methods with respect to "taxon sampling" strategies [10]. In this case, the shortest edges can be arbitrarily short, forcing the method to require unboundedly long sequences in order to recover these shortest edges. Hence, the sequence-length requirements of methods cannot be bounded. However, for a natural class of model trees, which includes random birth-death trees, we can assume $f = \Theta(1/n)$. In this case even simple polynomial-time methods converge to the true tree from sequences whose lengths are bounded by a polynomial in n. Furthermore, the degrees of the polynomials bounding the convergence rates of neighbor-joining and the fast-converging methods are identical—they differ only with respect to the leading constants. Therefore, with respect to this parameterization, there is no significant theoretical advantage between standard methods and the fast-converging methods.

In a previous study [9] we evaluated NJ and DCM-NJ+MP with respect to their performance on simulated data, obtained on random birth-death trees with bounded deviation from ultrametricity. We found that DCM-NJ+MP dominated NJ throughout the parameter space we examined and that the difference increased as the deviation from ultrametricity or the number of taxa increased.

In an unpublished study, Bruno *et al.* [11] compared Weighbor with NJ and BioNJ [12] as a function of the length of the longest edge in the true tree, using random birth-death trees of 50 taxa, deviated from the molecular clock by multiplying each edge length by a random number drawn from an exponential distribution, and using the Jukes-Cantor (JC) model of evolution. They found that Weighbor outperformed the other methods for medium to large values of the longest edge, but was inferior to them for small values—a finding we can confirm only for larger numbers of taxa. At last year's PSB, Bininda-Edmonds et al. [13] presented a study of Greedy Parsimony (which uses a single random sequence of addition and no branch swapping) in which they used very large random birth-death trees (up to 10,000 taxa), deviated from the molecular clock, and with sequences evolved under the Kimura 2-parameter (K2P) model. Unsurprisingly, they found that scaling and accuracy are at odds: the lower the accuracy level, the better the sequence length scaling.

3 Basics

3.1 Model Trees

The first step of every simulation study for phylogenetic reconstruction methods is to generate *model trees*. Sequences are then evolved down these trees, the leaf sequences are fed to the reconstruction methods under study, and the reconstructed trees compared to the original model tree.

In this paper, we use random birth-death trees with n leaves as our underlying distribution. These trees have a natural length assigned to each edge—namely, the time t between the speciation event that began that edge and the event (which could be either speciation or extinction) that ended that edge—and thus are inherently ultrametric. In all of our experiments we modified each edge length to deviate from this restriction, by multiplying each edge by a random number within a range $[1/c, c]$, where we set c, the *deviation factor*, to be 4.

3.2 Models of Evolution

We use two models of sequence evolution: the *Jukes-Cantor* (JC) model [14] and the the *Kimura 2-Parameter+Gamma* (K2P+Gamma) model [15]. In both models, a site evolves down the tree under the Markov assumption; in the JC model, all nucleotide substitutions (that are not the identity) are equally likely, so only one parameter is needed, whereas in the K2P model substitutions are partitioned into two classes (again other than identity): *transitions*, which substitute a purine (adenine or guanine) for a purine or a pyrimidine (cytosine or thymidine) for a pyrimidine; and *transversions*, which substitute a purine for a pyrimidine or vice versa. The K2P model has a parameter which indicates the transition/transversion ratio. We set this ratio to 2 in our experiments. Under either model, each edge of the tree is assigned a value $\lambda(e)$, the expected number of times a random site on this edge will change its nucleotide.

It is often assumed that the sites evolve identically and independently (i.i.d.) down the tree. However, we can also assume that the sites have different rates of evolution, drawn from a known distribution. One popular assumption is that the rates are drawn from a gamma distribution with shape parameter α, which is the inverse of the coefficient of variation of the substitution rate. We use $\alpha = 1$ for our experiments under K2P+Gamma.

3.3 Phylogenetic Reconstruction Methods

Neighbor Joining. Neighbor-Joining (NJ) [1] is one of the most popular distance-based methods. NJ takes a distance matrix as input and outputs a tree. For every two taxa, it determines a score, based on the distance matrix. At each step, the algorithm joins the pair with the minimum score, making a subtree whose root replaces the two chosen taxa in the matrix. The distances are recalculated to this new node, and the "joining" is repeated until only three nodes remain. These are joined to form an unrooted binary tree.

Weighted Neighbor Joining. Weighbor [16], like NJ, joins two taxa in each iteration; the pairs of taxa are chosen based on a criterion that embodies a likelihood function on the distances, which are modeled as correlated Gaussian random variables with different means and variances, computed under a probabilistic model of sequence evolution. Then, the "joining" is repeated until only three nodes remain. These are joined to form an unrooted binary tree.

DCM-NJ+MP. The DCM-NJ+MP method is a variant of a provably fast-converging method that has performed very well in previous studies [17]. In these simulation studies, DCM-NJ+MP outperformed, in terms of topological accuracy, both the provably fast converging DCM*-NJ (of which it is a variant) and NJ. We briefly describe this method now. Let d_{ij} be the distance between taxa i and j.

- *Phase 1:* For each $q \in \{d_{ij}\}$, compute a binary tree T_q, by using the Disk-Covering Method [3], followed by a heuristic for refining the resultant tree into a binary tree. Let $\mathcal{T} = \{T_q : q \in \{d_{ij}\}\}$.
- *Phase 2:* Select the tree from $\mathcal{T}$ which optimizes the parsimony criterion.

If we consider all $\binom{n}{2}$ thresholds in Phase 1, DCM-NJ+MP takes $O(n^6)$ time, but, if we consider only a fixed number p of thresholds, it takes $O(pn^4)$ time. In our experiments, we considered only 10 thresholds, so that the running time of DCM-NJ+MP is $O(n^4)$.

Greedy Maximum Parsimony. The maximum parsimony method that we use in our study (and that was used by Bininda-Edmonds *et al.* [13]) is not, strictly speaking, a parsimony search: for the sake of speed, it uses no branch swapping at all and simply adds taxa to the tree one at a time following one random ordering of the taxa.

3.4 Measures of Accuracy

Since all the inferred trees are binary we use the *Robinson-Foulds* (RF) distance [18] which is defined as follows. Every edge e in a leaf-labeled tree T defines a bipartition π_e on the leaves (induced by the deletion of e), and hence the tree T is uniquely encoded by the set $C(T) = \{\pi_e : e \in E(T)\}$, where $E(T)$ is the set of all internal edges of T. If T is a model tree and T' is the tree obtained by a phylogenetic reconstruction method, then the set of *False Positives* is $C(T') - C(T)$ and the set of *False Negatives* is $C(T) - C(T')$. The RF distance is then the average of the number of false positives and the false negatives. We plot the *RF rates* in our figures, which are obtained by normalizing the RF distance by the number of internal edges in a fully resolved tree for the instance. Thus, the RF rate varies between 0 and 1 (or 0% and 100%). Rates below 5% are quite good, but rates above 20% are unacceptably large.

4 Our Experiments

In order to obtain statistically robust results, we followed the advice of [19,20] and used a number of *runs*, each composed of a number of *trials* (a trial is a single comparison), computed a mean outcome for each run, and studied the mean and standard deviation of these runs. We used 20 runs in our experiments. The standard deviation of the mean outcomes in our studies varied, depending on the number of taxa: the standard deviation of the mean on 10-taxon trees is 0.2 (which is 20%, since the possible values of the outcomes range from 0 to 1), on 25-taxon trees is 0.1 (which is 10%), whereas on 200 and 400-taxon trees the standard deviation ranged from 0.01 to 0.04 (which is between 1% and 4%). We graph the average of the mean outcomes for the runs, but omit the standard deviation from the figures.

We ran our studies on random birth-death trees generated using the r8s [21] software package. These trees have diameter 2 (height 1); in order to obtain trees with other diameters, we multiplied the edge lengths by factors of 0.05, 0.1, 0.25 and 0.5, producing trees with diameters of 0.1, 0.2, 0.5 and 1.0, respectively. To deviate these trees from ultrametricity, we set c, the deviation factor, to 4 (see Section 3). The resulting trees have diameters at most 4 times the original diameters, and have expected diameters of 0.2, 0.4, 1.0 and 2.0. We generated such random model trees with 10, 25, 50, 100, 200, and 400 leaves, 20 trees for each combination of diameter and number of taxa. We then evolved sequences on these trees using two models of evolution, JC and K2P+Gamma (we chose $\alpha = 1$ for the shape parameter and set the transition/transversion ratio to 2). We used a fix factor[22] of 1 for distance correction. The sequence lengths that we studied are 50, 100, 250, 500, 1000 and 2000.

We used the program `Seq-Gen` [23] to generate a DNA sequence for the root and evolve it through the tree under the JC and the K2P+Gamma models of evolution. The software for DCM-NJ was written by Daniel Huson. We used PAUP* 4.0 [24] for the greedy MP method, and the Weighbor 1.2 software package [16].

The experiments were run over a period of three months on about 300 different processors, all Pentiums running Linux, including the 128-processor SCOUT cluster at UT-Austin.

To generate the graphs that depict the scaling of accuracy, we linearly interpolated the sequence lengths required to achieve certain accuracy levels for fixed numbers of taxa, and then, using the interpolation, computed the sequence length, as a function of the number of taxa, that are required to achieve fixed specific accuracy levels (ones that are of interest).

5 Results and Discussion

5.1 Speed

Because we are studying methods that will scale to large datasets (large numbers of taxa and long sequences), speed is of prime importance. Table 1 shows the running time of our various methods on different instances. Note the very high speed and nearly perfect linear scaling of Greedy Parsimony. NJ is known to scale with the cube of the number of taxa; in our experiments, it scales slightly better than that. DCM-

Table 1: The running times of NJ, DCM-NJ+MP, Weighbor, and Greedy MP (in seconds) for fixed sequence length (500) and diameter (0.4)

Taxa	NJ	DCM-NJ+MP	Weighbor	Greedy MP
10	0.01	1.82	0.03	0.01
25	0.02	9.12	0.37	0.02
50	0.06	21.3	3.56	0.05
100	0.37	64.25	44.93	0.10
200	2.6	470.31	352.48	0.25
400	20.13	5432.46	4077.81	0.73

NJ+MP scales exactly as NJ, but runs approximately 200 times more slowly. Finally, Weighbor scales somewhat more poorly—the figures in the table indicate scaling that is supercubic. These figures make it clear that most reasonable datasets (up to a few thousand taxa) can be processed by any of these methods—especially with the help of cluster computing, but also that very large datasets (10,000 taxa or more) will prove too costly for Weighbor and perhaps also DCM-NJ+MP (at least in their current implementations).

5.2 *Sequence-Length Requirements*

We can sort our experimental data in terms of accuracy and, for all datasets on which an accuracy threshold is met, count, for each fixed number of taxa, the number of datasets with a given sequence length, thereby enabling us to plot the average sequence length needed to guarantee a given maximal error rate. We show such plots for two accuracy values in Figure 1: 70% and 85%. Larger values of accuracy cannot

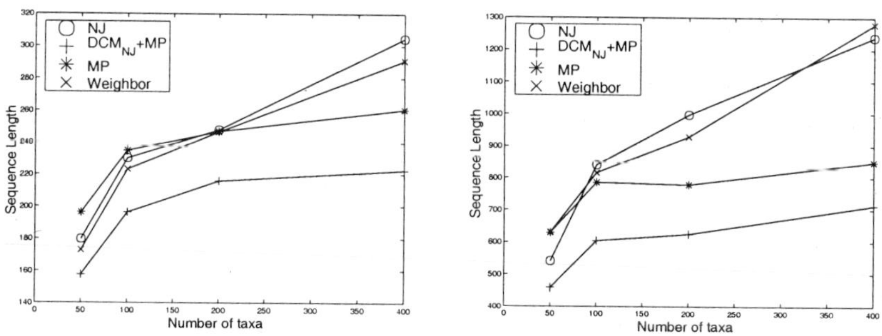

(a) 70% accuracy *(b) 85% accuracy*

Figure 1: Sequence length requirements under the K2P+Gamma model as a function of the number of taxa

be plotted reliably, since they are rarely reached under our challenging experimental conditions. The striking feature in these plots is the difference between the two NJ-based methods (NJ and Weighbor) and the methods using parsimony (DCM-NJ+MP and Greedy Parsimony): as the number of taxa increases, the former require longer and longer sequences, growing linearly or worse, while the latter exhibit only modest growth. The divide-and-conquer strategy of DCM-NJ+MP pays off by letting its NJ component work only on significantly smaller subsets of taxa—effectively shifting the graph to the left—and completing the work with a method (parsimony) that is evidently much less demanding in terms of sequence lengths. Note that the curves are steeper for the higher accuracy requirement: as the accuracy keeps increasing, we expect to see supralinear, indeed possibly exponential, scaling.

5.3 *Accuracy*

We studied accuracy (in terms of the RF rate) as a function of the number of taxa, the sequence length, and the diameter of the model tree, varying one of these parameters at a time. Because accuracy varies drastically as a function of the sequence length and the number of taxa, the plots given in this section have different vertical scales.

For fixed sequence lengths and fixed diameters, we find, unsurprisingly, that the error rate of all methods increases as the number of taxa increases, although the increase is very slow (see Figures 2 and 3, but note the logarithmic scaling on the x-axis). Weighbor indeed outperforms NJ, but DCM-NJ+MP outperforms the other

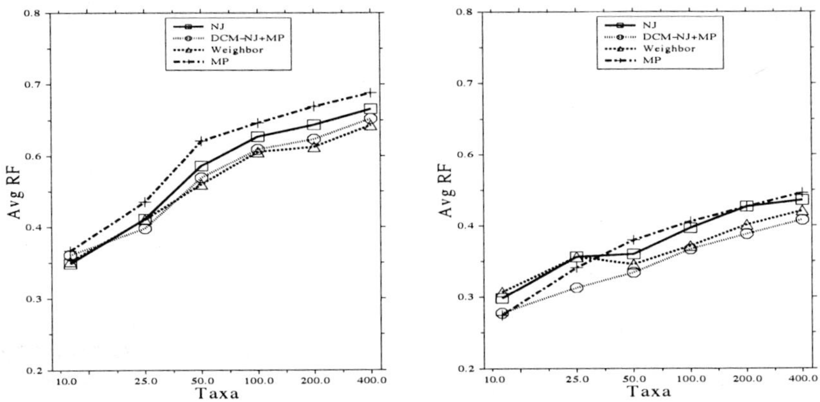

(a) *sequence length 50* (b) *sequence length 100*

Figure 2: Accuracy as a function of the number of taxa under the K2P+Gamma model for expected diameter (0.4) and two sequence lengths

three methods, especially for larger trees—unless the sequences are very short, in which case Weighbor dominates.

If we vary sequence length for a fixed number of taxa and fixed tree diameter, we find that the error rate decreases exponentially with the sequence length (Figure 4). From this perspective as well, DCM-NJ+MP dominates the other methods, more obviously so for larger trees. Interestingly, NJ is the worst method across almost the entire parameter space.

Finally, if we vary the diameter (which varies the rate of evolution) for a fixed number of taxa and a fixed sequence length, we find an initial increase in accuracy (due to the disappearance of zero-length edges), followed by a definite decrease (Figure 5). The decrease in accuracy is steeper with increasing diameter than what we observed with increasing number of taxa—and continually steepens. (At larger diameters—not shown, as we approach saturation, the error rate approaches unity.)

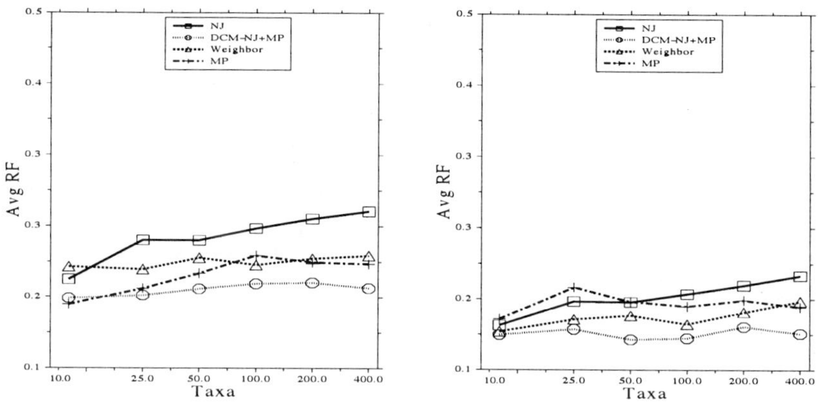

(a) sequence length 500 *(b) sequence length 1000*

Figure 3: Accuracy as a function of the number of taxa under the K2P+Gamma model for expected diameter (2.0) and two sequence lengths

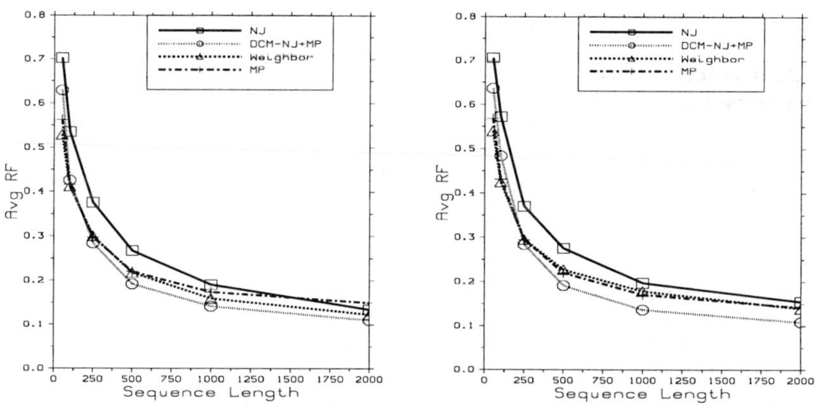

(a) 200 taxa *(a) 400 taxa*

Figure 4: Accuracy as a function of the sequence length under the K2P+Gamma model for expected diameter (2.0) and two numbers of taxa

The dominance of DCM-NJ+MP is once again evident. Comparing NJ and Weighbor, we can see that NJ is actually marginally better than Weighbor at low diameters, but Weighbor clearly dominates it at higher diameters—the two slopes are quite distinct.

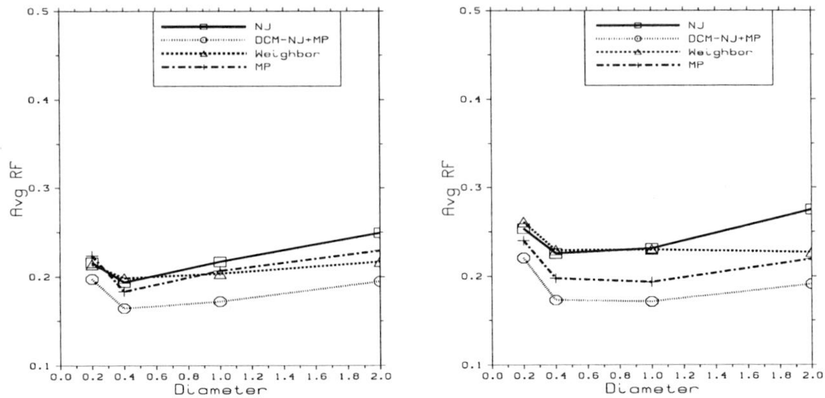

(a) 100 taxa *(a) 400 taxa*

Figure 5: Accuracy as a function of the diameter under the K2P+Gamma model for fixed sequence length (500) and two numbers of taxa

5.4 The Influence of the Model of Sequence Evolution

We reported all results so far under the K2P+Gamma model only, due to space limitations. However, we explored performance under the JC (Jukes-Cantor) model as well. The relative performance of the methods we studied was the same under the JC model as under the K2P+Gamma model. However, throughout the experiments, the error rate of the methods was lower under the JC model (using the JC distance-correction formulas) than under the K2P+Gamma model of evolution (using the K2P+Gamma distance-correction formulas). This might be expected for the Weighbor method, which is optimized for the JC model, but is not as easily explained for the other methods. Figure 6 shows the error rate of NJ on trees of diameter 0.4 under the two models of evolution. NJ clearly does better under the JC model than under the K2P+Gamma model; other methods result in similar curves. Correlating the decrease in performance with specific features in the model is a challenge, but the results clearly indicate that experimentation with various models of evolution (beyond the simple JC model) is an important requirement in any study.

6 Conclusion

In earlier studies we presented the DCM-NJ+MP method and showed that it outperformed the NJ method for random trees drawn from the uniform distribution on tree topologies and branch lengths as well as for trees drawn from a more biologically realistic distribution, in which the trees are birth-death trees with a moderate deviation from ultrametricity. Here we have extended our result to include the Weighbor and

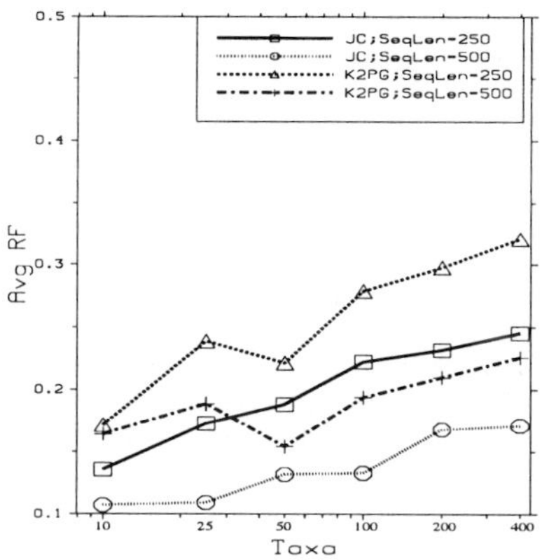

Figure 6: Accuracy of NJ as a function of the number of taxa under JC and K2P+Gamma

Greedy Parsimony methods. Our results confirm that the accuracy of the NJ method may suffer significantly on large datasets. They also indicate that Greedy Parsimony, while very fast, has mediocre to poor accuracy, while Weighbor and DCM-NJ+MP consistently return good trees, with Weighbor doing better on shorter sequences and DCM-NJ+MP doing better on longer sequences. Among interesting questions that arise are: (i) is there a way to conduct a partial parsimony search that scales no worse than quadratically (and might outperform DCM-NJ+MP)? (ii) would a DCM-Weighbor+MP prove a worthwhile tradeoff? (iii) can we make quantitative statements about the accuracy achievable by any method (not just one of those under study) as a function of some of the model parameters?

7 Acknowledgments

This work was supported in part by the National Science Foundation with grants EIA 99-85991 to T. Warnow and ACI 00-81404 to B.M.E. Moret and a POWRE award to K. St. John; by the Texas Institute for Computational and Applied Mathematics and the Center for Computational Biology at UT-Austin (K. St. John); and by the David and Lucile Packard Foundation (T. Warnow).

References

1. N. Saitou and M. Nei. The neighbor-joining method: A new method for reconstructing phylogenetic trees. *Mol. Biol. Evol.*, 4:406–425, 1987.
2. P. L. Erdos, M. Steel, L. Székély, and T. Warnow. A few logs suffice to build almost all trees–I. *Random Structures and Algorithms*, 14:153–184, 1997.
3. P. L. Erdos, M. Steel, L. Székély, and T. Warnow. A few logs suffice to build almost all trees–

II. *Theor. Comp. Sci.*, 221:77–118, 1999.
4. M. Csűrös. Fast recovery of evolutionary trees with thousands of nodes. RECOMB 01, 2001.
5. M. Csűrös and M. Y. Kao. Recovering evolutionary trees through harmonic greedy triplets. *Proc. 10th ACM-SIAM Symp. on Discrete Algorithms (SODA 99)*, pages 261–270, 1999.
6. D. Huson, S. Nettles, and T. Warnow. Disk-covering, a fast-converging method for phylogenetic tree reconstruction. *Comput. Biol.*, 6:369–386, 1999.
7. T. Warnow, B. Moret, and K. St. John. Absolute convergence: true trees from short sequences. *Proc. 12th ACM-SIAM Symp. on Discrete Algorithms (SODA 01)*, pages 186–195, 2001.
8. K. Atteson. The performance of the neighbor-joining methods of phylogenetic reconstruction. *Algorithmica*, 25:251–278, 1999.
9. L. Nakhleh, U. Roshan, K. St. John, J. Sun, and T. Warnow. The performance of phylogenetic methods on trees of bounded diameter. In *Proc. 1st Workshop on Algorithms in Bioinformatics (WABI 01)*, pages 214–226, Aarhus (2001). in *LNCS* 2149.
10. J. Huelsenbeck and D. Hillis. Success of phylogenetic methods in the four-taxon case. *Syst. Biol.*, 42:247–264, 1993.
11. http://www.t10.lanl.gov/billb/weighbor/performance.html.
12. O. Gascuel. BIONJ: an improved version of the NJ algorithm based on a simple model of sequence data. *Mol. Biol. Evol.*, 14:685–695, 1997.
13. O. Bininda-Emonds, S. Brady, J. Kim, and M. Sanderson. Scaling of accuracy in extremely large phylogenetic trees. In *Proc. 6th Pacific Symp. on Biocomputing (PSB01)*, pages 547–557. World Scientific, 2001.
14. T. Jukes and C. Cantor. Evolution of protein molecules. In H.N. Munro, editor, *Mammalian Protein Metabolism*, pages 21–132. Academic Press, NY, 1969.
15. M. Kimura. A simple method for estimating evolutionary rates of base substitutions through comparative studies of nucleotide sequences. *J. Mol. Evol.*, 16:111–120, 1980.
16. W. J. Bruno, N. Socci, and A. L. Halpern. Weighted neighbor joining: A likelihood-based approach to distance-based phylogeny reconstruction. *Mol. Biol. Evol.*, 17(1):189–197, 2000.
17. L. Nakhleh, U. Roshan, K. St. John, J. Sun, and T. Warnow. Designing fast converging phylogenetic methods. In *Proc. 9th Int'l Conf. on Intelligent Systems for Mol. Biol. (ISMB 01)*, 2001. In *Bioinformatics* 17:S190–S198.
18. D. F. Robinson and L. R. Foulds. Comparison of phylogenetic trees. *Mathematical Biosciences*, 53:131–147, 1981.
19. B. M. E. Moret and H. D. Shapiro. Algorithms and experiments: the new (and the old) methodology. *J. Univ. Comput. Sci.*, 7(5):434–446, 2001.
20. C. McGeoch. Analyzing algorithms by simulation: variance reduction techniques and simulation speedups. *ACM Comp. Surveys*, 24:195–212, 1992.
21. M. Sanderson. `r8s` software package. Available from http://loco.ucdavis.edu/r8s/r8s.html.
22. D. Huson, K. A. Smith, and T. Warnow. Correcting large distances for phylogenetic reconstruction. In *Proc. 3rd Workshop on Algorithms Engineering (WAE 99)*, 1999. London, England.
23. A. Rambaut and N. C. Grassly. Seq-gen: An application for the Monte Carlo simulation of dna sequence evolution along phylogenetic trees. *Comp. Appl. Biosci.*, 13:235–238, 1997.
24. D. L. Swofford. PAUP*: Phylogenetic analysis using parsimony (and other methods), 1996. Sinauer Associates, Underland, Massachusetts, Version 4.0.

PRE-mRNA SECONDARY STRUCTURE PREDICTION AIDS SPLICE SITE PREDICTION

DONALD J. PATTERSON, KEN YASUHARA, WALTER L. RUZZO
Computer Science and Engineering
University of Washington, Box 352350
Seattle, WA 98195, USA

Accurate splice site prediction is a critical component of any computational approach to gene prediction in higher organisms. Existing approaches generally use sequence-based models that capture local dependencies among nucleotides in a small window around the splice site. We present evidence that computationally predicted secondary structure of moderate length pre-mRNA subsequences contains information that can be exploited to improve acceptor splice site prediction beyond that possible with conventional sequence-based approaches. Both decision tree and support vector machine classifiers, using folding energy and structure metrics characterizing helix formation near the splice site, achieve a 5–10% reduction in error rate with a human data set. Based on our data, we hypothesize that acceptors preferentially exhibit short helices at the splice site.

1 Introduction

Whole-genome analysis of a single organism or comparison of organisms depends on correct gene annotation. Hence, accurate gene prediction from DNA sequence data is an important goal for bioinformatics, both for purposes of providing "high-throughput annotation" to match high-throughput sequencing, and for the insight it may provide into the underlying biology. Accurate splice site prediction is a critical component of eukaryotic gene prediction. Unfortunately, while current approaches achieve accuracies above 90% with acceptable false negative rates, compounded errors for multi-exonic genes contribute to a substantially higher error rate for full-length gene predictions.[1–3]

Splice site prediction initially depended on very simple models involving consensus sequences in narrow windows around the splice sites.[4] As more data became available, zero-th order Markov models ("Weight Matrix Models" or "Position Specific Scoring Matrices") became possible.[5] With still more data, researchers adopted higher order Markov models ("Weight Array Matrices" or WAMs)[6] and variants such as the "Windowed Weight Array Matrices" of Burge and Karlin,[7] and various kinds of decision trees, such as the "Maximal Dependence Decomposition" model.[7] Despite the increasing sophistication of these models as more training data becomes available, they all basically exploit observed dependencies among *nearby* nucleotides in the vicinity of the splice site.

Much is known about the mechanisms underlying splicing of Group II introns.[8,9] In particular, it is known that certain short RNAs (the U1, U2, U4, U5, U6 snRNAs) hybridize with each other and with complementary segments of the pre-message at the donor, branch point, and acceptor sites. These segments are probably important determinants of the specificity of splicing. The sequence-based models mentioned above are appropriate for characterizing this sequence complementarity. However, it also appears that the information content of these short neighborhoods around the splice sites is not adequate to fully account for the observed high specificity of splicing *in vivo*.

There has long been speculation that secondary structure in pre-mRNAs also plays a role in splicing, and there have been a number of experimentally verified cases where splicing defects have been tied to mutations that alter secondary structure near splice sites.[10–16] However, no clear pattern emerges from these reports, so although secondary structure may play a role in splice site recognition, a single, strongly conserved structure (as found in tRNAs or other functional RNAs) is not expected. Rather, some looser structure or collection of structures might incrementally contribute to the observed specificity of splicing.

For example, it seems plausible that initial hybridization of the spliceosomal snRNAs to the pre-mRNA might be enhanced or inhibited by the presence of short helices in the vicinity of the splice sites, without requiring conservation of a precisely determined structure at an exact position relative to the splice site. This is consistent with observations of Mir and Southern, who examined hybridization of a tRNA to an oligo microarray and reported significant influence of the tRNA's structure on hybridization.[17] In particular, strong hybridization generally seemed to require that the oligo match the entire length of one strand of a helix in the tRNA, together with a few adjacent unpaired bases, and additionally was stabilized by coaxial stacking with another helix.

In this paper, we report positive results from a series of computational experiments designed to discover such correlations between splicing and computationally predicted secondary structure of pre-mRNAs for a sample of human genes. We identified several structure metrics showing subtle but statistically significant correlation to acceptor splice sites (i.e., 3' ends of introns), beyond that already accounted for by a good sequence-based model. Comparable results were obtained with two very different classification methods and hence are unlikely to be simply an artifact of either classifier. Although the net improvement in classifier accuracy was small, approximately a 5–10% reduction in misclassification rate, this could translate into a substantial improvement in the accuracy of full-length gene predictions for genes with 10, 20, or more exons. However, we feel that our most important contribution is not this direct application but rather the evidence that structure does play a role in splicing, and that current structure prediction tools are accurate enough to exploit it. Additionally, structure might play a role in other processes, e.g., ribozyme binding[18] and perhaps mRNA stabilization and degradation. Because computational tools for discovering informative structural features are much less well-developed than tools for features based on primary sequence, we expect the methods outlined here to be of value in other contexts.

In outline, the methodology we employed is as follows. From a test set of annotated, multi-exon human genes, we extracted acceptor splice sites and a representative sample of nearby non-sites matching the acceptor AG dinucleotide consensus. We used Zuker's MFOLD[19,20] to predict foldings for a 100-base window centered on each site/nonsite. Various sequence and structure features, such as per-position dinucleotide frequencies and pairing probabilities, were extracted for use in our classifiers. Each test, using 10-fold cross-validation, examined the change in accuracy between a baseline, sequence-based model and the same model plus one or more of the structure metrics. Tests were performed with two standard machine learning approaches—C4.5 decision trees,[21,22] and support vector machines.[23–26] Details of our methodology are presented in Section 2.

Our results are described in Section 3. To briefly summarize, we obtained statistically significant accuracy improvements with various combinations of three structure metrics. The first was the simplest: energy of the predicted optimal folding. Sequences containing acceptors on average had slightly more stable structures than nonsites. Second, for each position i in a sequence, we computed a "Max Helix" score, roughly an estimate of the probability of a helix within 5 bases of position i. We observed Max Helix scores to be relatively independent of i for nonsites, whereas acceptors showed a dip in Max Helix score roughly 10 bases upstream of the splice site. For our third and most detailed metric, we determined whether each position of a folded sequence was paired and stacked onto the nucleotide preceding it, paired but unstacked, or unpaired, then built a second order Markov model of the resulting ternary sequences. Again, the profiles of acceptors and non-acceptors tended to differ. For example, it appears that acceptors more often have a short helix at the splice site. In all performance comparisons we included the score from a first order Markov model (which we refer to as a weight array model or WAM) trained on the primary sequence near the acceptor site. Some portion of the structural consensus noted above is probably just a reflection of the acceptor sequence consensus. Nevertheless, in our tests, classifiers using one or more of these structural features in addition to WAM score consistently outperformed classifiers using WAM score alone.

2 Methods

2.1 Data Set

For training and testing, we started with 462 unrelated, annotated, multi-exon genes with standard splicing (i.e., excluding cases of alternative or self-splicing) from a data set proposed by Reese *et al.* as a benchmark set for evaluating gene-finding software.[27] Using exon annotations, we extracted a 100-base window centered on each acceptor splice site having sufficient flanking sequence. This formed our collection of positive samples. The non-acceptor, negative sample set was a random sampling of 100-base subsequences centered on an AG dinucleotide that were not annotated as acceptors, but were within 100 bases of an actual acceptor. We imposed these criteria to evaluate how our structure-based methods might enhance gene prediction methods, which must discriminate among several putative acceptor splice sites occurring close to each other. We formed a negative sample set of the same size as our positive set, each with 1,980 subsequences, so that the machine learners gave equal weight to false negative and false positive errors.

We randomly partitioned the 3,960 subsequences into 10 equal-sized groups for cross-validated training and testing, with each group containing an equal number of positive and negative samples. The same groups were used for all tests, allowing comparison of results on a per-group basis, as well as averaged over the 10 groups.

2.2 Sequence-based Metric

Weight Array Model (WAM).[6] A first order WAM models a primary sequence pattern by storing, for each base offset, the probability of observing each base conditioned on the previous base. Given WAMs trained on positive and negative example subsequences and an unclassified subsequence, a log likelihood score can be computed that reflects the likelihood that the subsequence contains a splice site. As in Burge's work,[28] we scored sequences using positions -21 to +3 relative to the putative acceptor site. To ensure that overfitting did not occur, we trained each WAM on 9 groups and scored the remaining group with this model. Cross-validated testing with an optimal threshold classifier confirmed that widening this window by 5 positions on either side did not improve accuracy.

2.3 Pre-mRNA Structure Prediction and Structure Metrics

For each subsequence, we used MFOLD to produce a comprehensive set of foldings, typically hundreds in number, each annotated with a free energy. Low free energy is correlated with folding stability and likelihood. The equilibrium partition function can be used to calculate the probability that a folding will occur in nature, given its free energy and the total free energy of all possible foldings.[29] In computing structure metrics from a given subsequence's many predicted foldings, we used these probabilities to weight each folding's contribution to an aggregate score. More probable foldings (i.e., ones with lower free energy) are accordingly weighted.

For each subsequence, we computed the following structure metrics:

Optimal Folding Energy (OFE). Our simplest metric was the free energy of the optimal folding. This number roughly reflects the stability of the fold and typically is lower with more paired bases.

Max Helix (MH). For each position around the putative splice site, we calculated the probability, according to the equilibrium partition function, of a helix starting or ending at that position. To relax the positional specificity of this metric, for each position, we recorded the maximum probability of a helix start/end within a neighborhood of 5 positions up- and downstream.

Neighbor Pairing Correlation Model (NPCM). A folded structure can be summarized by a string over the three symbol alphabet {S, P, O}, corresponding to whether each position is paired and stacked onto the nucleotide preceding it, paired but unstacked, or unpaired, respectively. The string's length is equal to the length of the original pre-mRNA sequence. For example, a three-base helix flanked by unpaired regions would be represented by ...OPSSO...

Given a set of RNA sequences of equal length, we converted each predicted folding of each sequence into a structure string and a corresponding folding probability. This collection of strings was used to train a second order Markov model, forming an aggregate model of the structure of the collection of sequences.

Although a Markov model can not fully describe the set of structure strings, we believe it can approximate many local features reasonably well. (The extra descriptive power afforded by using a stochastic context-free grammar[30] did not seem warranted at this stage.)

We trained two Markov models as described above—one on acceptor splice site sequences and the other on non-acceptor sequences. We scored the structure string of an unclassified pre-mRNA sequence by computing the posterior probability that each model generated this structure string. We then computed the log of the ratio of the site model probability over the non-site model probability, i.e., the log likelihood ratio.

2.4 Machine Learning Methods

We evaluated our structure metrics by aggregating them into real-valued feature vectors and training two machine learning classifiers, support vector machines (SVMs) and decision trees, on them.

SVMs perform binary classification by partitioning the feature space with a surface implied by a subset of the training vectors near the separating surface called *support vectors*.[24] SVMs are efficient with multi-dimensional data, subsume many other learning methods, and are solidly grounded in statistical theory. (See Hearst *et al.*[26] for a gentle introduction and Burges' tutorial[25] for a more formal, extensive introduction with further references.) In this study, we used Noble's implementation, svm 1.1.[31]

Decision trees are another form of supervised machine learner that classify feature vectors hierarchically. When predicting the class of a vector, a decision tree passes a vector down the tree from the root to a leaf. At each node, the decision tree examines one feature of the vector to determine which branch the vector should recursively travel down. Every leaf on the tree has an associated prediction, which is the classification that is ultimately assigned to the vector.

Decision trees are generated ("trained") by examining a collection of labeled vectors and statistically determining which feature contains the most information relevant to the classification. A node is formed to partition the training vectors into subsets based on this feature. These subsets are independently used to train the next lower level of the tree. When a subset's elements all belong to the same class or the amount of information in the subset is statistically insignificant, a leaf is formed, whose classification is equal to the majority classification of the subset.

We evaluated our feature sets with the C4.5 decision tree software package.[21] C4.5 forms decisions based on axis-parallel hyperplanes, corresponding to threshold tests on one feature at each node of the tree.

2.5 *Testing Methodology*

In all tests, we used accuracy (fraction of samples classified correctly) as our performance metric; observed false negative and false positive rates were roughly equal. We employed slightly different testing methodologies for decision trees and SVMs.

The decision tree methodology began with cross-validated tests of trees trained on WAM score alone, resulting in 10 per-group accuracies, our baseline. For each set of structure metrics, we then repeated the cross-validated tests, allowing the decision tree to train on WAM score in conjunction with combinations of structure metrics. If we observed a significant increase in accuracy relative to the baseline tests, we concluded that the structure metrics contained useful information that the WAM could not capture.

To avoid potential problems comparing performance of SVMs with different dimensionality, we used a *mixed/matched* methodology that only involved comparing results for models trained on data of the same dimensionality. For each of the 10 cross-validations, we trained two models. The *matched* model was trained on feature vectors that contained WAM score and the structure metrics. The *mixed* model was trained on the same data modified by randomly permuting ("mixing") the structure metrics. That is, for each training vector consisting of a WAM score and at least one structure metric, the structure metric components were replaced with those of another training vector, randomly selected (without replacement) independently of the vector's class. The mixed and matched models were then tested on the reserved test set, and significantly lower accuracy with the mixed model evinced useful information in

Table 1: Results of 10-fold cross-validated decision tree testing with Weight Array Model (WAM) and various structure metric combinations: Optimal Folding Energy (OFE), Neighbor Pairing Correlation Model (NPCM), and Max Helix scores (MH). NPCM was scored on positions -50 through +3, and MH scores are taken for positions -10 and +3 only. Mean accuracy (fraction of samples classified correctly), improvement over baseline WAM accuracy ($\Delta \pm$ one standard deviation) and paired Wilcoxon test p-values are shown.

features	*mean acc. (%)*	Δ	p
WAM (baseline)	92.73	-	-
WAM, OFE	93.13	$+0.40 \pm 0.87$	0.066
WAM, OFE, NPCM	93.16	$+0.43 \pm 0.80$	0.022
WAM, OFE, MH	93.21	$+0.48 \pm 0.90$	0.009
WAM, OFE, NPCM, MH	93.13	$+0.40 \pm 0.84$	0.016

the structure metrics not captured in the WAM score.

With no reason to assume the observed accuracy distributions were Gaussian, we conservatively tested statistical significance of accuracy differences using the paired Wilcoxon signed rank test, a nonparametric analogue of the paired t-test. For the paired Wilcoxon test, the p-value is the probability of obtaining test results as extreme as those we observed, assuming the null hypothesis—that the differences between paired accuracies have median zero.

3 Results

We trained decision trees and radial basis kernel SVMs with many combinations of the structure metrics we formulated. Testing with cross-validation as described in Section 2.5, we identified optimal folding energy (OFE), Max Helix, and Neighbor Pairing Correlation Model (NPCM) scores as useful metrics for acceptor recognition.

Decision tree test results in Table 1 show that training on WAM and OFE with each of the remaining structure metrics yielded statistically significant accuracy improvement, relative to training on WAM score alone. Because overfitting causes decision tree performance to degrade with the addition of features with redundant information, we chose only two Max Helix positions (-10 and +3). Adding the combination of OFE and these Max Helix scores yielded a 7% reduction in classification error rate.

We also saw statistically significant accuracy improvements in mixed/matched SVM testing when Max Helix scores for each position from -20 to +3 were combined with OFE. This result independently supports the decision tree results with Max Helix above. To examine the degree of variability of these results due to the randomized mixing step, we repeated the 10-fold cross-validation runs ten times. For each of these ten runs, Table 2 shows the mean accuracy with the mixed models and by how

Table 2: Results of radial kernel SVM mixed/matched testing with Weight Array Model (WAM) score, optimal folding energy (OFE) and Max Helix (MH) scores for each position from -20 to +3. 10-fold cross-validation runs were repeated ten times. For each run, mean mixed model accuracy, improvement with matched model ($\Delta \pm$ one standard deviation) and paired Wilcoxon test p-values are shown. Accuracy with the matched model was 92.90%.

c.v.	*mean accuracy (%)*		
run	*mixed*	Δ	p
1	91.61	+1.29 ±1.18	0.006
2	92.15	+0.76 ±0.76	0.014
3	92.27	+0.63 ±0.62	0.012
4	92.25	+0.66 ±0.71	0.014
5	91.84	+1.06 ±0.99	0.010
6	92.12	+0.78 ±0.73	0.009
7	91.99	+0.91 ±0.74	0.009
8	92.22	+0.68 ±0.71	0.028
9	92.07	+0.84 ±0.47	0.006
10	92.53	+0.38 ±0.63	0.072
mean	92.10	+0.80	-

much it differs from the mean accuracy with the matched models. Properly matching WAM score and structure metrics improved accuracy in all ten of the cross-validation runs, with $p < 0.05$ in nine of the ten runs and with improvements exceeding 1% with $p < 0.01$ in two of them. On average, the improvement was 0.8%, approximately a 10% reduction in misclassification rate.

Figure 1 presents three views of the structure metrics we developed. In each graph, the solid and dotted lines are the mean and standard deviation, respectively, of the metric, calculated across all 10 cross-validation groups. The top graph shows the $\log_{10}$ likelihood ratio of a base pairing (either stacked or unstacked) with any other base within the folding window. From this graph it can be observed that there is an approximately 25% smaller chance of a pair forming at position -5 in an acceptor splice site than in a non-site. This effect is reversed in the region from -2 to +1, where acceptors demonstrate a 25% greater chance of pairing.

The middle graph of Figure 1 further investigates this trend by showing the probability of a stacked pair, S, at a given position, conditioned on the previous two positions being OP, i.e., an unpaired base followed by a paired base. This can be interpreted as the probability that a helix will continue given that it has recently started. From positions -30 to -7 there is a roughly equal chance that such a continuation will occur regardless of whether the sequence is an acceptor or not. Then at positions -6 to -4 the likelihood of a helix continuing drops by approximately 20% in acceptors

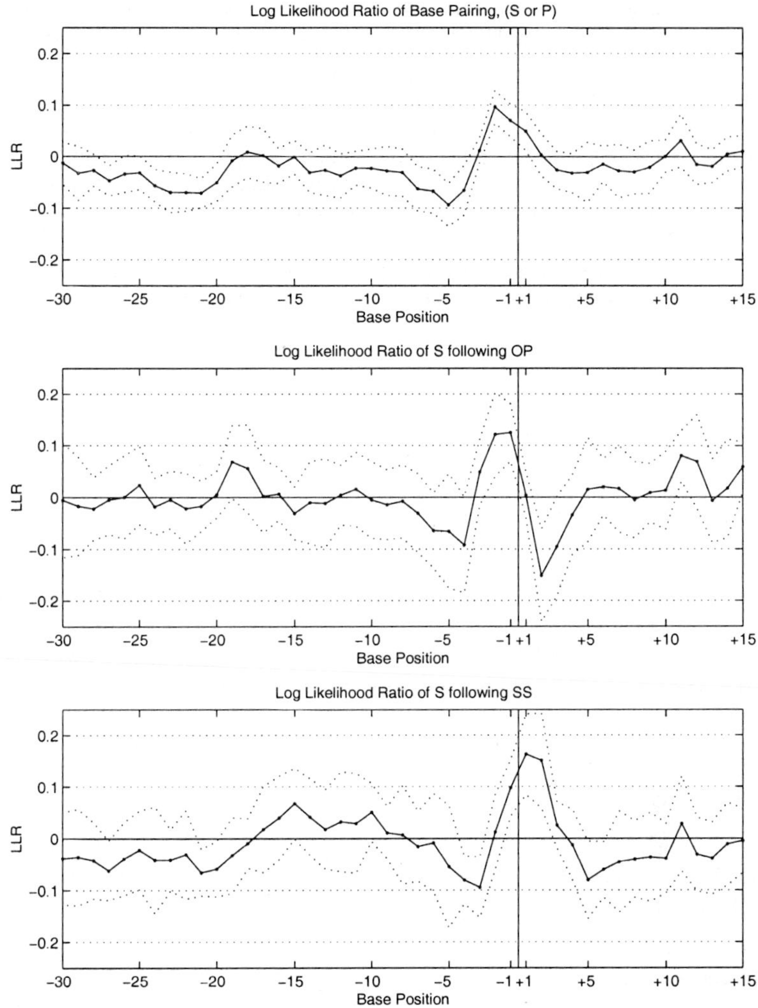

Figure 1: Log likelihood ratios (LLRs) of three structural patterns occurring at different positions relative to the acceptor splice site (vertical line). Top: LLR that a base pairs (either stacked or unstacked) with any other in the folding window. Middle: LLR that a base forms a stacked pair, conditioned on a helix start one position upstream. Bottom: LLR that a base continues a helix that begins three or more positions upstream. Solid and dotted lines show mean $\pm$ one standard deviation across the cross-validation groups.

relative to non-acceptors. Just before the splice site, the trend reverses for the acceptors, suggesting a bias toward helix formation at position -3 and -2. Shortly after the splice site there is a bias away from helix formation.

Finally, the bottom graph shows the conditional probability of a stacked pair, `S`, directly following two other stacked pairs, `SS`. This can be interpreted as the probability that a helix of length 3 or greater will continue. There is a bias at positions -5 to -3 prior to the splice site for termination of helices, but once a helix extends into the splice site region, there is a strong bias toward continuation. After the splice site, the bias is reversed briefly.

Collectively, these three graphs suggest that acceptor sequences are more likely than our non-acceptor sequences to form a short helix at the splice site.

4 Conclusion

We have presented evidence that valuable information can be extracted from predictions of pre-mRNA structure that aid in the location of acceptor splice sites. Multiple machine learners were able to utilize this information to produce statistically significant improvements in accuracy. While specific structural signatures were not detected, general trends toward helix formation in the region of the splice site suggest the possibility of greater exploitation of structural cues by gene finding algorithms.

Similar structural biases were observed at the donor splice site in the same data set, but not with sufficient strength that statistical significance could be ascribed to them. Future research is warranted toward the development of models that capture structural features at the donor splice site, as well as improving upon the acceptor site models we have presented and their biological interpretation.

Acknowledgments

This research was partially supported by grant NSF-DBI 9974498. DJP was supported by a National Defense Science and Engineering Graduate Fellowship, USA. Thanks to Benno Schwikowski and the anonymous referees for helpful comments.

References

1. M. Burset and R. Guigó. Evaluation of gene structure prediction programs. *Genomics*, 34(3):353–367, 1996.
2. R. Guigó, P. Agarwal, J.F. Abril, M. Burset, and J.W. Fickett. An assessment of gene prediction accuracy in large DNA sequences. *Genome Research*, 10(10):1631–1642, 2000.

3. M. Pertea, X. Lin, and S.L. Salzberg. GeneSplicer: a new computational method for splice site prediction. *Nucleic Acids Research*, 29(5):1185–1190, 2001.
4. S. M. Mount. A catalogue of splice junction sequences. *Nucleic Acids Research*, 10(2):459–472, 22 January 1982.
5. R. Staden. Computer methods to locate signals in nucleic acid sequences. *Nucleic Acids Research*, 12:505–519, 1984.
6. M. Q. Zhang and T. G. Marr. A weight array method for splicing signal analysis. *Computer Applications in the Biosciences*, 9(5):499–509, 1993.
7. C. Burge and S. Karlin. Prediction of complete gene structures in human genomic DNA. *Journal of Molecular Biology*, 268:78–94, 1997.
8. Melissa J. Moore, Charles C. Query, and Phillip A. Sharp. Splicing of precursors to mRNA by the spliceosome. In Raymond F. Gesteland and John F. Atkins, editors, *The RNA World: the nature of modern RNA suggests a prebiotic RNA world*, number 24 in Cold Spring Harbor Monograph Series, pages 303–357. Cold Spring Harbor Laboratory Press, 1993.
9. Jonathan P. Staley and Christine Guthrie. Mechanical devices of the spliceosome: Motors, clocks, springs, and things. *Cell*, 92:315–326, 1998.
10. David Solnick. Alternative splicing caused by RNA secondary structure. *Cell*, 43:667–676, December 1985.
11. Domenico Libri, Anna Piseri, and Marc Y. Fiszman. Tissue-specific splicing in vivo of the β-tropomyosin gene: Dependence on an RNA secondary structure. *Science*, 252:1842–1845, June 1991.
12. Béatrice Clouet d'Orval, Yves d'Aubenton Carafa, Joëlle Marie, and Edward Brody. Determination of an RNA structure involved in splicing inhibition of a muscle-specific exon. *Journal of Molecular Biology*, 221:837–856, 1991.
13. Béatrice Clouet d'Orval, Yves d'Aubenton Carafa, Pascal Sirand-Pugnet, Maria Gallego, Edward Brody, and Joelle Marie. RNA secondary structure repression of a muscle-specific exon in HeLa cell nuclear extracts. *Science*, 252:1823–1828, June 1991.
14. James O. Deshler and John J. Rossi. Unexpected point mutations activate cryptic 3' splice sites by perturbing a natural secondary structure within a yeast intron. *Genes & Development*, 5:1252–1263, 1991.
15. Andrés F. Muro, Massimo Caputi, Rajalakshmi Pariyarath, Franco Pagani, Emanuelle Buratti, and Francisco E. Baralle. Regulation of fibronectin EDA exon alternative splicing: Possible role of RNA secondary structure for enhancer display. *Molecular and Cellular Biology*, 19(4):2657–2671, April 1999.
16. Luca Varani, Masato Hasegawa, Maria Grazia Spillantini, Michael J. Smith, Jill R. Murrell, Bernardino Ghetti, Aaron Klug, Michel Goedert, and Gabriele Varani. Structure of tau exon 10 splicing regulatory element RNA and desta-

bilization by mutations of frontotemporal dementia and parkinsonism linked to chromosome 17. *Proceedings of the National Academy of Science USA*, 96:8229–8234, July 1999.

17. K.U. Mir and E.M. Southern. Determining the influence of structure on hybridization using oligonucleotide arrays. *Nature Biotechnology*, 17:788–792, 1999.
18. M. Amarzguioui, G. Brede, E. Babaie, M. Grøtli, B. Sproat, and H. Prydz. Secondary structure prediction and *in vitro* accessibility of mrna as tools in the selection of target sites for ribozymes. *Nucleic Acids Research*, 28(21):4113–4124, 2000.
19. M. Zuker, D.H. Mathews, and D.H. Turner. Algorithms and thermodynamics for RNA secondary structure prediction: A practical guide. In J. Barciszewski and B.F.C. Clark, editors, *RNA Biochemistry and Biotechnology*, NATO ASI Series, pages 11–43. Kluwer Academic Publishers, 1999.
20. D.H. Mathews, J. Sabina, M. Zuker, and D.H. Turner. Expanded sequence dependence of thermodynamic parameters improves prediction of RNA secondary structure. *Journal of Molecular Biology*, 288:911–940, 1999.
21. J. R. Quinlan. *C4.5: Programs for Empirical Learning*. Morgan Kaufmann, 1993.
22. J. R. Quinlan. Bagging, boosting, and C4.5. In *Proceedings of the Thirteenth National Conference on Artificial Intelligence*, pages 725–730, Cambridge, MA, 1996.
23. V. Vapnik. *The Nature of Statistical Learning Theory*. Springer-Verlag, 1995.
24. V. Vapnik. *Statistical Learning Theory*. Wiley, New York, 1998.
25. C.J.C. Burges. A tutorial on support vector machines for pattern recognition. *Data Mining and Knowledge Discovery*, 2(2):121–67, 1998.
26. M. Hearst (Ed.), S.T. Dumais, E. Osuna, J. Platt, and B. Schölkopf. Trends & controversies: Support vector machines. *IEEE Intelligent Systems*, 13(4):18–28, 1998.
27. Martin Reese, David Kulp, Andrew Gentles, and Uwe Ohler. GENIE gene finding data set. http://www.fruitfly.org/sequence/human-datasets.html.
28. C.B. Burge. Modeling dependencies in pre-mRNA splicing signals. In *Computational Methods in Molecular Biology*, pages 129–64. Elsevier Science, 1998.
29. J. McCaskill. The equilibrium partition function and base pair bindings probabilities for RNA secondary structure. *Biopolymers*, 29:1105–1119, 1990.
30. Richard Durbin, Sean R. Eddy, Anders Krogh, and Graeme Mitchison. *Biological Sequence Analysis: Probabilistic models of proteins and nucliec acids*. Cambridge, 1998.
31. W.S. Noble (formerly W.N. Grundy). svm 1.1. http://www.cs.columbia.edu/˜noble/svm/doc/.

FINDING WEAK MOTIFS IN DNA SEQUENCES

S.-H. SZE[1], M.S. GELFAND[2], P.A. PEVZNER[1]
[1] *Department of Computer Science and Engineering, University of California at San Diego, La Jolla, CA 92093-0114.*
[2] *Integrated Genomics — Moscow, P.O. Box 348, 117333, Moscow, Russia.*

Recognition of regulatory sites in unaligned DNA sequences is an old and well-studied problem in computational molecular biology. Recently, large-scale expression studies and comparative genomics brought this problem into a spotlight by generating a large number of samples with unknown regulatory signals. Here we develop algorithms for recognition of signals in corrupted samples (where only a fraction of sequences contain sites) with biased nucleotide composition. We further benchmark these and other algorithms on several bacterial and archaeal sites in a setting specifically designed to imitate the situations arising in comparative genomics studies.

1 Introduction

Large-scale expression analysis and comparative genomics recently generated numerous samples of potentially co-regulated genes whose upstream regions are likely to contain regulatory sites. These samples are often corrupted (with only a fraction of sequences in the sample containing a site) and the corresponding signals may be relatively weak. In difference from previous "gene-by-gene" research efforts, the possibilities of experimental localization of site positions (i.e., via reduction in the length of sequences in the samples by footprinting experiments) in postgenomic era are limited. As a result, computer predictions are often the only realistic way to find regulatory signals in these regions.

The first attempts to find regulatory sites appeared in the early eighties (for reviews, see Gelfand[1], Frech *et al.*[2], or Brazma *et al.*[3]). Current approaches can be roughly subdivided into pattern-driven techniques[4,5,6,7] and profile-based optimization algorithms (greedy search[8], simulated annealing[9], Gibbs sampler[10], and expectation-maximization[11]). Most pattern finding algorithms were developed and tested in a situation when all or most sequences in the analyzed sample contain regulatory sites (mostly single site). This is no longer a valid assumption. Comparative genomics produce samples of genes that are *likely* to be co-regulated, but there is no guarantee that some (or maybe even a majority) of the genes are expressed constitutively or regulated by the same mechanism. Similarly, expression studies often result in the identification of co-expressed genes in response to certain environmental stimuli, but usually do not resolve regulatory cascades and other complex interactions[12].

Several papers reported benchmarking results of signal recognition algorithms. Fickett and Hatzigeorgiou[13] evaluated algorithms for finding eukaryotic promoters. Roulet *et al.*[14] compared predicted affinities of an eukaryotic transcription factor to synthetic oligonucleotides from SELEX data. Frech *et al.*[2] benchmarked programs for finding signals on several prokaryotic and eukaryotic samples. Pevzner and Sze[6] compared several pattern finding algorithms on simulated sequences with implanted signals. Although these results allow one to assess the current state of affairs in controlled situations, they do not provide insight to the behavior of existing programs in real life situations.

Here we are primarily interested in benchmarking with corrupted samples where the signal is present only in a fraction of sequences. This is almost always the case in biological samples. This study models a common situation when it is unclear how to set up the size of the upstream regions and the search parameters (i.e. the length and stringency of the motif). A failure to choose the right parameters may lead to missing the signals. We investigate the effect of using different lengths of upstream sequences and study how the corruption of the sample sequences influences the quality of recognition. We are also interested in how the addition of sequences with a different signal (possibly of different length) to a sample affects recognition. We investigate "how weak" a weak signal should be to become undetectable.

The algorithms WINNOWER and SP-STAR from Pevzner and Sze[6] have been modified to take into account specifics of real biological samples. We compare these programs with a few of the best currently available approaches, including CONSENSUS[15], GibbsDNA[10], and MEME[16]. The choice of the programs is somehow subjective and is limited to those that are the most popular among biologists. We will identify the shortcomings of these approaches and gain insight into what problems future approaches should address.

2 Test Samples

The algorithms were tested on three samples from the *E.coli* genome. Each sample consists of sequences in a $[-1500, 500]$ window with respect to the translational start site annotated with known sites from the Robison *et al.*[17] compilation. The sequences were extracted from GenBank using GenomeExplorer[18]. In our experiments, subsamples within a smaller window of these samples are considered, so that the actual lengths of the sequences used will be smaller.

The first (ARG) sample contains 9 sequences (genes regulated by the arginine repressor ArgR). Each sequence contains a two-part site with the length of each part being 18 nt and separated by 3 nt in all but one sequence where the separation is 2 nt. One of the sequences also has an extra one-part site of length 18 nt. The second (PUR) sample contains 19 sequences with sites of length 16 nt (genes regulated by the purine repressor PurR). Among them, 17 sequences contain 1 site and 2 sequences contain 2 sites. The third CRP sample contains 33 sequences with CRP repressor binding sites of length 22 nt. Among them, 17 sequences contain 1 site, 10 sequences contain 2 sites, 1 sequence contains 3 sites, 1 sequence contains 4 sites, and 4 sequences contain no sites. The sites in the CRP sample are mostly weak. Most sites are found within 200 nt upstream of the translational start site, although a few sites are found up to 400 nt upstream or downstream of the start site.

Define the *majority string* for a collection of strings $\mathcal{W} = \{W_1, \ldots, W_t\}$ as the string W' whose ith letter is the most frequent ith letter in $\mathcal{W}$. We estimate the mutation rate of the signal in each sample by finding the majority string from the set of annotated patterns and computing the average number of substitutions to convert the majority string to each of the annotated patterns. We use the notation of a (l, d)-sample to denote a sample with signals of length l and probability of mutation $p = d/l$ (see Pevzner and Sze[6], the VM mode). The ARG sample contains two-part sites. When only the two-part sites separated by 3 nt are considered with each two-part site treated as one site, the sample corresponds roughly to a (39,11)-sample (i.e., 11 mutations per 39 positions or 28% mismatches on average) and only 8 out of 9 sequences contain a site. When each part is considered a site by itself, the sample is roughly a (18,5.6)-sample (31% mismatches on average) and most sequences contain two sites. The PUR sample corresponds roughly to a (16,3.4)-sample (21% mismatches on average) and most sequences contain one site. The CRP sample corresponds roughly to a (22,9.1)-sample (41% mismatches on average) and most sequences contain one or two sites.

We also study two samples with unknown regulatory sites. The IRON-FACTOR sample contains 12 sequences each of length 250 nt. These sequences are the upstream regions of operons from various gamma-proteobacteria likely to be involved in iron utilization and regulated by homologous repressors other than FUR (E. Panina, personal communication). The PYRO-PURINES sample contains 13 sequences each of length 300 nt (upstream regions of genes involved in the purine metabolism in *Pyrococus horikoshii*). Recently, Gelfand *et al.*[19] made (still unconfirmed) prediction of regulatory sites in this sample.

3 Results

Following Pevzner and Sze[6], we use the performance coefficient $|K \cap P|/|K \cup P|$ to evaluate the performance of signal finding algorithms, where K is the set of known signal positions in a sample and P is the set of predicted positions.

3.1 Samples with a Single Site per Sequence

Most motif finding programs have a performance tradeoff when using a more general model versus a more restricted model. To compare the performance of various approaches fairly, we first assume that the signal length is known as the annotated length and at most one site appears in a sequence. Although the second assumption does not hold for our samples, we assume it here for simplicity and expect that the programs to only be able to pick up the strongest site in a sequence with more than one site. In particular, for the ARG sample, we treat the two-part signal as one signal of length 39 nt and change the annotation to remove the extra one-part site and the exceptional two-part site with separation distance 2 nt. The 3 nt in the separation portion of the two-part signal is considered to be annotated. For the PUR sample, we remove the weaker site from the annotation in each of the two sequences where there are two annotated sites. Since a lot of the sequences in the CRP sample have more than one site, we postpone its test to the later sections when more complicated models are considered. Since there is no convenient way to test GibbsDNA or WINNOWER under the current model (both programs can return more than one site per sequence), we postpone their tests.

We investigate the effect of using different lengths of upstream sequences from 200 nt to 1500 nt. Since all the sites are found upstream of the translational start site, we fix the right end of the sequences to be the position just before the start site and vary the left end. All the programs performed similarly (data not shown). In most cases, the performance was 0.89 on the ARG sample and 0.95 on the PUR sample, independent of the length chosen.

We are interested in how the addition of random sequences to each of the ARG and PUR samples influences the signal recognition. Since most sites can be found within 200 nt upstream of the start site, we fix the upstream sequence length under investigation to be 200 nt. A sample of 666 random fragments of length 200 nt is also given. These sequences contain intergenic regions between convergently transcribed genes which are not expected to contain binding sites for any regulator. An increasing number of these random sequences are added to each sample. Table 1 compares the performance of the various algorithms. We allow each program to return suboptimal solutions in addition to the optimal one and the top-ranked non-overlapping suboptimal solutions are considered. MEME was the best in returning a strong signal as the optimal solution, but sometimes with performance tradeoff since CONSENSUS performed very well in returning an excellent quality result among the top few non-overlapping suboptimal solutions. SP-STAR sometimes performed better than CONSENSUS or MEME but gave inferior results in general.

Table 1: Comparison of the performance of the various algorithms by adding random sequences to the ARG and PUR samples with upstream sequences of length 200 nt under the restricted model where all the programs return at most one site per sequence. Annotations of the samples have been changed to suit the restricted model. For CONSENSUS, the stopping condition is that each sequence has contributed exactly one word to the saved matrices and we consider all the top matrices from each cycle. MEME is run in zoops mode, not allowed to shorten motifs, and is instructed to find three different motifs. SP-STAR is run with local improvements on the top 10% initial signals. The known signal length is 39 nt for the ARG sample and 16 nt for the PUR sample, which is used as an input parameter to all the programs. The top three non-overlapping suboptimal solutions among these results are considered, where each one does not overlap with any of the higher-scored ones, and the one with the highest performance among these non-overlapping solutions is reported along with its suboptimal position in parentheses. Note that sometimes less than three suboptimal solutions are returned from a program.

		number of random sequences added								
sample	program	0	20	40	60	80	100	120	140	160
ARG	CONSENSUS	0.81(1)	1(1)	1(1)	1(1)	1(1)	1(1)	1(2)	1(2)	0.29(3)
	MEME	0.89(1)	0.90(1)	0.90(1)	0.89(1)	0.73(1)	0.34(1)	0.72(2)	0.89(1)	1(1)
	SP-STAR	1(1)	0.81(1)	1(1)	1(1)	0.53(1)	0.53(1)	0.53(2)	0.42(2)	0.53(2)
PUR	CONSENSUS	0.94(1)	1(1)	1(1)	0.94(1)	0.88(1)	0.88(1)	0.88(1)	0.58(1)	0.58(2)
	MEME	0.94(1)	1(1)	0.94(1)	0.94(1)	0.60(1)	0.60(1)	0.60(1)	0.52(1)	0.63(1)
	SP-STAR	0.94(1)	1(1)	1(1)	1(1)	1(1)	1(1)	0.53(1)	0.53(1)	0.53(2)

3.2 Samples with Multiple Sites per Sequence

All the programs in this study can predict multiple sites per sequence. For CONSENSUS, MEME and SP-STAR, the total number of sites in a prediction is restricted to mt, where m is an input parameter to be determined, and t is the number of sequences in a sample. For GibbsDNA, mt is used as the expected number of sites supplied as a parameter to the program. For WINNOWER, all solutions with the total number of sites greater than mt are discarded. Of course, the "mt-restriction" has different implications for different programs, but they represent the closest possible models that these programs offer so that the performances are approximately comparable. We want to set m appropriately so as to obtain the best sensitivity for each program, which means that we have to set m to be as small as possible but should still allow the programs to include most or all sites in a prediction. For the ARG sample, when the signal is considered to be a single (two-part) signal (we do not change the annotation, so there are definitely misses of sites), we can set m to 1. When the signal is considered to be one-part, there are 19 sites in 9 sequences. We can set m to be either 2 or 3. We choose to use $m = 2$ in our experiments since some of the sites will be excluded when smaller window subsamples are considered, which allows a maximum of 18 sites to be predicted with better sensitivity. For the PUR sample, there are 21 sites in 19 sequences, we set m to 1 for similar reasons. For the CRP sample, there are 45 sites in 33 sequences (with two of them overlapping), so we set m to 2. All programs are instructed to return predictions with non-overlapping sites.

The first experiment investigates the effect of the length of upstream sequences. Since almost all the sites are found upstream of the start site, we fix the right end to be the position just before the start site and vary the left end. Table 2 compares the performance of the various algorithms. While CONSENSUS and MEME had good performance in general, GibbsDNA and SP-STAR started to break in some cases when very long upstream sequences are used. WINNOWER only had good performance when short upstream sequences are used (partly due to the fact that we use clique size $k = 2$ instead of $k = 3$ to save computational resources).

The second experiment investigates how the addition of an increasing number of random sequences to the ARG, PUR and CRP samples with upstream sequences of length 200 nt influences the signal recognition. Table 3 compares the performance of the various algorithms, employing the same treatment to allow suboptimal solutions as in Table 1 (excluding GibbsDNA and WINNOWER since the versions we have are not designed for this type of problems). For the ARG samples looking for two-part signals, performance of CONSENSUS and SP-STAR were not bad while MEME returned a good prediction as the top result through a wider range. For the ARG sample (one-part signals) or the CRP sample, SP-STAR had the best performance in returning good solutions among the top results even when a lot of random sequences are added. For the PUR sample, CONSENSUS was the best to return the closest signal as more and more random sequences are added, but it also failed earlier.

In the third experiment we are interested in how the various algorithms perform on samples containing natural but weak sites. We remove sequences successively from the CRP sample (with upstream sequences of length 200 nt) in decreasing order of the strength of the strongest site in a sequence (stronger ones removed first) and investigate when the algorithms break. We compute site strength by the following procedure. Compute the majority string of all the sites and the sum-of-pairs (SP) similarity score of each column of aligned sites. We ignore all positions in the majority string with negative SP column scores and take this string to be the consensus pattern. For the CRP sample, the consensus pattern is found to be `a--tgtga------tcaca-tt`. The site strength is defined to be the similarity score between the site and the consensus pattern computed over retained positions. Table 4 compares the performance of the various algorithms. When not too many CRP sequences are removed, performances of all the programs were comparable except that

Table 2: Comparison of the performance of the various algorithms by using upstream sequences of different lengths from the ARG, PUR and CRP samples allowing multiple sites per sequence. For CONSENSUS, the stopping condition is that the saved matrices contain a maximum of mt words, where m is a parameter and t is the number of sequences, and the first matrix among the list of matrices from each cycle is returned. GibbsDNA is run with the expected number of sites being mt, set to disregard fragmentation and the result with the highest NetMAP score over 100 runs is returned. MEME is run in tcm mode, with the maximum number of sites restricted to mt and not allowed to shorten motifs. WINNOWER is run with clique size $k = 2$ (not tested on the PUR and CRP samples since extensive computation time and resources are needed). SP-STAR is run with local improvements on the top 10% initial signals with the maximum number of sites in a prediction being mt. The known signal length is 39 nt for the ARG sample (2-part) with $m = 1$, 18 nt for the ARG sample (1-part) with $m = 2$, 16 nt for the PUR sample with $m = 1$, and 22 nt for the CRP sample with $m = 2$. All programs return predictions with non-overlapping sites.

sample	program	length of upstream sequences													
		200	300	400	500	600	700	800	900	1000	1100	1200	1300	1400	1500
ARG	CONSENSUS	0.81	0.79	0.71	0.71	0.71	0.71	0.71	0.71	0.71	0.71	0.71	0.71	0.71	0.71
(2-part)	GibbsDNA	0.81	0.79	0.79	0.79	0.79	0.79	0.79	0.79	0.79	0.79	0.79	0.79	0.79	0.79
	MEME	0.73	0.79	0.65	0.65	0.65	0.65	0.65	0.65	0.65	0.65	0.71	0.71	0.71	0.65
	WINNOWER	0.62	0.69	0.69	0.69	0.69	0.69	0.69	0.69	0.69	0.69	0.69	0.69	0.69	0.69
	SP-STAR	0.81	0.79	0.79	0.79	0.79	0.79	0.79	0.79	0.79	0.79	0.79	0.79	0.79	0.00
ARG	CONSENSUS	0.71	0.62	0.69	0.69	0.69	0.69	0.69	0.69	0.69	0.69	0.69	0.62	0.62	0.62
(1-part)	GibbsDNA	0.69	0.68	0.63	0.63	0.63	0.64	0.64	0.67	0.67	0.67	0.67	0.60	0.54	0.58
	MEME	0.48	0.62	0.62	0.80	0.71	0.76	0.80	0.80	0.62	0.62	0.85	0.56	0.85	0.56
	WINNOWER	0.37	0.40	0.36	0.40	0.31	0.36	0.40	0.40	0.40	0.40	0.40	0.14	0.14	0.14
	SP-STAR	0.54	0.83	0.83	0.83	0.55	0.55	0.55	0.55	0.52	0.48	0.48	0.48	0.48	0.48
PUR	CONSENSUS	0.94	0.95	0.95	0.90	0.81	0.85	0.85	0.85	0.85	0.85	0.85	0.76	0.76	0.76
	GibbsDNA	0.89	0.95	0.95	0.90	0.69	0.69	0.55	0.79	0.86	0.86	0.86	0.82	0.82	0.78
	MEME	0.94	0.95	0.95	0.90	0.81	0.81	0.68	0.81	0.81	0.81	0.90	0.81	0.81	0.85
	SP-STAR	0.94	0.94	0.95	0.89	0.95	0.95	0.85	0.85	0.80	0.80	0.80	0.90	0.80	0.48
CRP	CONSENSUS	0.35	0.38	0.44	0.39	0.38	0.37	0.37	0.37	0.32	0.39	0.32	0.31	0.23	0.23
	GibbsDNA	0.47	0.39	0.35	0.35	0.29	0.26	0.33	0.32	0.34	0.25	0.30	0.26	0.00	0.00
	MEME	0.38	0.36	0.45	0.44	0.44	0.44	0.44	0.43	0.43	0.45	0.43	0.41	0.40	0.38
	SP-STAR	0.43	0.32	0.33	0.38	0.33	0.35	0.37	0.25	0.24	0.24	0.24	0.24	0.32	0.32

WINNOWER gave slightly worse results. Since this sample is a very good representative of samples with weak sites, we further investigate in detail the effect of both varying the length of the upstream sequences and the number of sequences removed. We start from the sample with upstream sequences of length 1500 nt and remove sequences in decreasing order of the strength of the strongest site as before. Samples of shorter lengths are obtained by varying the left end. The consensus pattern computed from this larger sample is `aa-tgtga------tcaca-tt`, slightly different than before. Table 5 compares the performance of CONSENSUS, MEME and SP-STAR. While CONSENSUS and MEME had a better performance when the upstream sequence length is

Table 3: Comparison of the performance of the various algorithms by adding random sequences to the ARG, PUR and CRP samples with upstream sequences of length 200 nt allowing multiple sites per sequence. Settings are the same as in Table 2. The treatment of suboptimal solutions and the notations used are the same as in Table 1. When the top three solutions all have performance less than 0.05, we put 0.00 in the entry to emphasize that the run fails completely.

sample	program	number of random sequences added: 0	20	40	60	80	100	120	140	160
ARG (2-part)	CONSENSUS	0.81(1)	0.81(1)	0.81(1)	0.81(2)	0.81(2)	0.81(2)	0.81(2)	0.81(2)	0.28(3)
	MEME	0.73(1)	0.73(1)	0.73(1)	0.56(1)	0.73(1)	0.73(1)	0.73(1)	0.81(1)	0.41(2)
	SP-STAR	0.81(1)	0.66(1)	0.81(1)	0.66(1)	0.66(1)	0.46(1)	0.49(2)	0.46(2)	0.46(3)
ARG (1-part)	CONSENSUS	0.71(1)	0.54(1)	0.64(1)	0.46(1)	0.48(3)	0.45(2)	0.00	0.00	0.00
	MEME	0.48(1)	0.46(1)	0.47(1)	0.43(1)	0.37(1)	0.39(2)	0.00	0.00	0.00
	SP-STAR	0.54(1)	0.56(1)	0.56(1)	0.47(1)	0.43(1)	0.42(1)	0.45(3)	0.12(3)	0.44(3)
PUR	CONSENSUS	0.94(1)	0.94(1)	0.89(1)	0.83(1)	0.83(1)	0.83(1)	0.55(2)	0.55(2)	0.00
	MEME	0.94(1)	0.94(1)	0.89(1)	0.89(1)	0.57(1)	0.52(1)	0.50(1)	0.45(1)	0.50(2)
	SP-STAR	0.94(1)	0.94(1)	0.94(1)	0.94(1)	0.94(2)	0.50(2)	0.50(2)	0.50(2)	0.50(2)
CRP	CONSENSUS	0.35(1)	0.37(1)	0.36(1)	0.31(3)	0.00	0.00	0.00	0.00	0.00
	MEME	0.38(1)	0.40(1)	0.35(1)	0.33(2)	0.27(3)	0.28(3)	0.00	0.25(2)	0.00
	SP-STAR	0.43(1)	0.42(1)	0.35(1)	0.41(1)	0.36(2)	0.37(2)	0.36(3)	0.00	0.00

Table 4: Comparison of the performance of the various algorithms by removing sequences from the CRP sample with upstream sequences of length 200 nt in decreasing order of a sequence's strongest site strength. Settings are the same as in Table 2.

CRP sample	number of CRP sequences removed: 0	3	6	9	12	15	18	21	24
CONSENSUS	0.35	0.39	0.20	0.20	0.10	0.08	0.15	0.09	0.00
GibbsDNA	0.38	0.41	0.36	0.30	0.32	0.16	0.18	0.11	0.00
MEME	0.38	0.42	0.27	0.18	0.12	0.20	0.14	0.15	0.00
WINNOWER	0.19	0.09	0.10	0.13	0.00	0.09	0.00	0.00	0.00
SP-STAR	0.52	0.42	0.34	0.27	0.22	0.23	0.25	0.14	0.00

not too long, SP-STAR was more successful in the difficult cases. The maximum performance achieved was only about 50%, mostly due to the variability of the signal: if we consider the 14 non-degenerate positions in the consensus pattern, about half of the instances are at least 4 mismatches away, which is beyond the limit of the algorithms.

In the fourth experiment we are interested in how the addition of sequences from another sample influences the signal recognition. Similar to before, only upstream sequences of length 200 nt are considered. An increasing number of sequences from the CRP samples sorted in decreasing order of a sequence's strongest site strength are added to each of the ARG and PUR samples (stronger ones added first). Table 6 compares the performance of the

Table 5: Comparison of the performance of the various algorithms by varying both the lengths of the upstream sequences and the number of sequences that are removed from the CRP sample. Settings are the same as in Table 2.

CRP seqs. removed	program	200	300	400	500	600	700	800	900	1000	1100	1200	1300	1400	1500
		CRP sample													
		length of upstream sequences													
0	CONSENSUS	0.35	0.38	0.44	0.39	0.38	0.37	0.34	0.37	0.32	0.39	0.32	0.31	0.23	0.23
	MEME	0.38	0.40	0.47	0.44	0.44	0.42	0.43	0.43	0.43	0.45	0.43	0.41	0.40	0.38
	SP-STAR	0.41	0.31	0.35	0.34	0.33	0.37	0.30	0.30	0.21	0.21	0.20	0.22	0.25	0.19
3	CONSENSUS	0.39	0.40	0.45	0.43	0.42	0.40	0.38	0.38	0.32	0.29	0.28	0.30	0.00	0.00
	MEME	0.38	0.44	0.45	0.44	0.43	0.44	0.44	0.45	0.44	0.45	0.43	0.31	0.40	0.37
	SP-STAR	0.33	0.30	0.39	0.35	0.34	0.30	0.30	0.31	0.23	0.27	0.26	0.23	0.20	0.22
6	CONSENSUS	0.40	0.43	0.39	0.40	0.37	0.37	0.36	0.36	0.34	0.34	0.00	0.37	0.25	0.00
	MEME	0.52	0.44	0.46	0.42	0.43	0.47	0.46	0.43	0.39	0.38	0.39	0.39	0.37	0.00
	SP-STAR	0.33	0.34	0.33	0.30	0.25	0.25	0.30	0.29	0.29	0.31	0.29	0.29	0.22	0.29
9	CONSENSUS	0.41	0.44	0.41	0.41	0.39	0.39	0.39	0.39	0.37	0.37	0.37	0.39	0.41	0.27
	MEME	0.44	0.47	0.46	0.41	0.41	0.46	0.46	0.42	0.44	0.43	0.40	0.40	0.36	0.00
	SP-STAR	0.35	0.38	0.33	0.30	0.28	0.35	0.32	0.31	0.31	0.31	0.31	0.31	0.31	0.31
12	CONSENSUS	0.28	0.41	0.38	0.37	0.35	0.35	0.36	0.27	0.21	0.21	0.21	0.22	0.22	0.00
	MEME	0.38	0.43	0.38	0.36	0.38	0.38	0.35	0.37	0.34	0.31	0.26	0.35	0.31	0.00
	SP-STAR	0.33	0.41	0.40	0.32	0.34	0.31	0.40	0.27	0.27	0.27	0.27	0.27	0.27	0.27
15	CONSENSUS	0.24	0.16	0.35	0.34	0.34	0.34	0.31	0.32	0.02	0.02	0.01	0.18	0.00	0.00
	MEME	0.39	0.34	0.43	0.27	0.39	0.36	0.33	0.37	0.29	0.01	0.27	0.26	0.00	0.26
	SP-STAR	0.29	0.40	0.31	0.29	0.29	0.29	0.33	0.33	0.24	0.24	0.24	0.24	0.24	0.24
18	CONSENSUS	0.12	0.08	0.09	0.09	0.09	0.01	0.01	0.01	0.00	0.00	0.00	0.00	0.00	0.00
	MEME	0.05	0.27	0.26	0.25	0.14	0.16	0.14	0.03	0.00	0.00	0.00	0.00	0.00	0.00
	SP-STAR	0.29	0.32	0.29	0.28	0.28	0.28	0.28	0.28	0.29	0.22	0.26	0.03	0.00	0.00
21	CONSENSUS	0.00	0.00	0.00	0.00	0.00	0.01	0.01	0.01	0.01	0.01	0.00	0.00	0.00	0.01
	MEME	0.10	0.00	0.05	0.00	0.00	0.00	0.01	0.00	0.00	0.01	0.00	0.02	0.00	0.00
	SP-STAR	0.14	0.35	0.23	0.32	0.00	0.01	0.00	0.00	0.00	0.00	0.00	0.00	0.00	0.00
24	CONSENSUS	0.00	0.16	0.04	0.00	0.00	0.00	0.00	0.06	0.00	0.00	0.00	0.00	0.00	0.00
	MEME	0.11	0.09	0.00	0.00	0.00	0.00	0.00	0.00	0.00	0.00	0.00	0.00	0.00	0.00
	SP-STAR	0.11	0.29	0.29	0.29	0.29	0.00	0.00	0.00	0.00	0.00	0.00	0.00	0.00	0.00

various algorithms. Overall, SP-STAR was least affected by the addition of CRP sequences. GibbsDNA and WINNOWER did not have good performance when a lot of CRP sequences are added. For the ARG sample (one-part signals), CONSENSUS and MEME were not very stable when a moderate amount of CRP sequences are added. In fact, excellent solutions were returned as the second (non-overlapping) suboptimal solution in all these cases.

3.3 Samples with Unknown Signals

Table 7(a) shows the results of running SP-STAR on the IRON-FACTOR sample. The consensus shows that the best signal found is highly palindromic

Table 6: Comparison of the performance of the various algorithms by adding CRP sequences in decreasing order of a sequence's strongest site strength to the ARG and PUR samples with upstream sequences of length 200 nt. Settings are the same as in Table 2.

		number of CRP sequences added								
sample	program	0	3	6	9	12	15	18	21	24
ARG (2-part)	CONSENSUS	0.81	0.81	0.81	0.81	0.81	0.81	0.81	0.81	0.81
	GibbsDNA	0.81	0.81	0.81	0.81	0.81	0.81	0.73	0.40	0.38
	MEME	0.73	0.81	0.51	0.81	0.66	0.66	0.73	0.81	0.81
	WINNOWER	0.62	0.62	0.62	0.62	0.62	0.62	0.62	0.62	0.62
	SP-STAR	0.81	0.81	0.81	0.81	0.81	0.81	0.81	0.81	0.81
ARG (1-part)	CONSENSUS	0.71	0.67	0.61	0.61	0.50	0.46	0.00	0.00	0.55
	GibbsDNA	0.69	0.63	0.61	0.53	0.47	0.28	0.13	0.25	0.09
	MEME	0.48	0.48	0.50	0.51	0.53	0.39	0.00	0.51	0.51
	WINNOWER	0.37	0.37	0.37	0.37	0.37	0.00	0.19	0.19	0.19
	SP-STAR	0.54	0.61	0.61	0.61	0.61	0.61	0.61	0.61	0.61
PUR	CONSENSUS	0.94	0.94	0.94	0.89	0.89	0.89	0.85	0.85	0.85
	GibbsDNA	0.89	0.89	0.89	0.85	0.85	0.85	0.74	0.68	0.71
	MEME	0.94	0.94	0.89	0.85	0.85	0.81	0.74	0.74	0.74
	WINNOWER	0.94	0.81	0.89	0.84	0.84	0.84	0.71	0.71	0.71
	SP-STAR	0.94	1	1	0.89	0.89	0.89	0.85	0.85	0.85

which reinforces our belief that it is very likely to be a biological signal. Table 7(b) shows the results of running SP-STAR on the PYRO-PURINES sample. Gelfand *et al.*[19] made a prediction on this sample and we found that our results agree very well with their prediction. We have also run CONSENSUS (using signal lengths found in Table 7 as input parameter) and MEME on the two samples and found that these programs give similar results. If the predictions are assumed to be correct, the IRON-FACTOR sample corresponds to a (29,8.4)-sample (29% mismatches on average), while the PYRO-PURINES sample corresponds to a (22,5.6)-sample (25% mismatches on average).

4 Discussion

We have tested and compared the performance of five programs CONSENSUS, GibbsDNA, MEME, WINNOWER and SP-STAR on several biological samples. All programs perform well on non-corrupted samples when all sequences contain relevant binding sites. This condition is very difficult to satisfy in practice. Indeed, many methods used for sample generation, including clustering of genes with similar expression profiles, analysis of reconstructed metabolic maps, and locating orthologous genes from known regulons in a related species, are very likely to create sequences not belonging to the analyzed regulon. Thus, an important part of the analysis presented here is benchmark-

Table 7: Results on the (a) IRON-FACTOR and (b) PYRO-PURINES samples. Shown is the best solution given by SP-STAR while looking for signals up to 40 nt in length with the maximum total number of sites in a prediction restricted to $2t$, where t is the number of sequences in the sample. The last string shown is the majority string, showing only the positions with positive SP column score.

(a) name	pos	pattern	(b) name	pos	pattern
b_alcA	38	**gagaatagaagtcataattattctcattaa**	*PH0239*	189	**cttttgccagatatatgtctaaaaaa**
b_alcR	166	**ataaaagcgaatgaattgcattatcattaa**	*PH0239*	231	**atttttacataaacatggtgaaatta**
s_foxA	189	**ctaaagggtaataattcttatttacaataa**	*PH0240*	190	**atttcaccatgtttatgtaaaaatca**
v_OM	159	**atatatgcgaatcgttatcatttgtatttt**	*PH0240*	232	**ttttagacatatatctggcaaaagat**
v_reg	189	**aaaaatacaaatgataacgattcgcatata**	*PH0318*	187	**atttaaacatatttatgttaaaaagg**
y_ybtA	103	**attaatgtgaataataaccattatcaataa**	*PH0318*	229	**attttaacatttatacgtcaattagg**
y_ybtP	150	**gttattgataatggttattattcacattaa**	*PH0320*	150	**cgattagcacatatatgtagaaatat**
		ataaatg--aat-atta--att--cattaa	*PH0323*	186	**ttgttaacacgtttatgtaaacaaaa**
			PH0323	229	**attttgacttaaatatggtgatataa**
			PH0438	186	**ctattaacatagccctgtcaaaaggg**
			PH0852	177	**agatttctacaaatatgtcaaaaaca**
			PH0852	220	**attttaccgtgaaaatggtgatataa**
			PH1955	166	**tgattgacatttctttgtcaaaataa**
			PH1955	208	**atttttacatttttctggcaaataag**
					atttt-acatatatatgtcaaaa--a

ing of the programs on corrupted samples. This was modeled in three ways: adding sequences with no sites, removing the strongest sites from a sample, and adding sequences with sites of a different origin to a sample.

In the experiment on addition of random sequences with no sites, MEME outperformed CONSENSUS and SP-STAR on both analyzed samples when at most one site are allowed per sequence. When multiple sites are allowed, SP-STAR performed slightly better than the other programs in the most difficult cases. In the experiment on removal of strong sites, the leaders were MEME and SP-STAR, with GibbsDNA demonstrating comparable or even slightly superior results when not too many sites are removed. In the experiment on addition of sequences from a different sample, GibbsDNA and WINNOWER clearly trailed, with CONSENSUS and SP-STAR being the leaders.

It does not seem possible to recommend a single program for use in all situations. However, this study allows us to make a few practical suggestions. The first one is simple: use all available programs. It seems that the programs are not affected much by varying fragment lengths. As the sites may occur at varying distances from the start site, it is safer to err to the side of using longer fragments. Also, it looks like that asking for at most one site per sequence improves the performance. In this case, additional sites can be found by standard search methods using consensus or positional weight matrix representation.

Acknowledgments

We are grateful to A. Mironov for many helpful discussions and to E. Panina who provided the IRON-FACTOR sample. This work was partially supported by the Russian Fund of Basic Research (99-04-48247 and 00-15-99362), INTAS (99-1476) and the Howard Hughes Medical Institute (55000309).

References

1. M.S. Gelfand. *J. Comp. Biol.*, 2:87–115, 1995.
2. K. Frech, K. Quandt, T. Werner. *Comp. Appl. Biosci.*, 13:89–97, 1997.
3. A. Brazma, I. Jonassen, I. Eidhammer, D. Gilbert. *J. Comp. Biol.*, 5:279–305, 1998.
4. I. Rigoutsos, A. Floratos. *Bioinformatics*, 14:55–67, 1998.
5. J. van Helden, B. Andre, J. Collado-Vides. *J. Mol. Biol.*, 281:827–42, 1998.
6. P.A. Pevzner, S.-H. Sze. *Proc. of the 8th Int. Conf. on Intelligent Systems for Mol. Biol. (ISMB'2000)*, 269–78, 2000.
7. J. Buhler, M. Tompa. *Proc. of the 5th Annual Int. Conf. on Comp. Mol. Biol. (RECOMB'2001)*, 69–76, 2001.
8. G.D. Stormo, G.W. Hartzell. *Proc. Natl. Acad. Sci.*, 86:1183–7, 1989.
9. A.V. Lukashin, J. Engelbrecht, S. Brunak. *Nucleic Acids Res.*, 20:2511–6, 1992.
10. C.E. Lawrence, S.F. Altschul, M.S. Boguski, J.S. Liu, A.F. Neuwald, J.C. Wootton. *Science*, 262:208–14, 1993.
11. T.L. Bailey, C.P. Elkan. *Proc. of the 2nd Int. Conf. on Intelligent Systems for Mol. Biol. (ISMB'1994)*, 28–36, 1994.
12. A.B. Khodursky, B.J. Peter, N.R. Cozzarelli, D. Botstein, P.O. Brown, C. Yanofsky. *Proc. Natl. Acad. Sci.*, 97:12170–5, 2000.
13. J.W. Fickett, A.G. Hatzigeorgiou. *Genome Res.*, 7:861–78, 1997.
14. E. Roulet, I. Fisch, T. Junier, P. Bucher, N. Mermod. *In Silico Biol.*, 1:21–8, 1998.
15. G.Z. Hertz, G.D. Stormo. *Bioinformatics*, 15:563–77, 1999.
16. T.L. Bailey, C.P. Elkan. *Machine Learning*, 21:51–80, 1995.
17. K. Robison, A.M. McGuire, G.M. Church. *J. Mol. Biol.*, 284:241–54, 1998.
18. A.A. Mironov, N.P. Vinokurova, M.S. Gelfand. *Mol. Biol.*, 34:253–62, 2000.
19. M.S. Gelfand, E.V. Koonin, A.A. Mironov. *Nucleic Acids Res.*, 28:695–705, 2000.

EVIDENCE FOR SEQUENCE-INDEPENDENT EVOLUTIONARY TRACES IN GENOMICS DATA

W. VOLKMUTH, N. ALEXANDROV
*Ceres Inc., 3007 Malibu Canyon Road,
Malibu, CA 90265, USA*

Sequence conservation during evolution is the foundation for the functional classification of the ennormous number of new protein sequences being discovered in the current era of genome sequencing. Conventional methods to detect homologous proteins are not always able to distinguish between true homologs and false positive hits in the twilight zone of sequence similarity. Several different approaches have been proposed to improve the sensitivity of these methods. Among the most successful are sequence profiles, multi-linked alignment, and threading. However, evolution might offer up other clues about a protein's ancestry that are sequence independent. Here we report the discovery of two such traces of evolution that could potentially be used to help infer the fold of a protein and hence improve the ability to predict the biochemical function. The first such evolutionary trace is a conservation of fold along the genome, i.e. nearby genes tend to share a fold more often than expected by chance alone—a not unexpected observation, but one which holds true even when no pair of genes being examined share appreciable homology. The second such evolutionary trace is, surprisingly, present in expression data: genes that are correlated in expression are more apt to share a fold than two randomly chosen genes. This result is surprising because correlations in expression have previously only been considered useful for determining *biological* function (e.g. what pathway a particular gene fits into), yet the observed fold enrichment in the expression data permits us to say something about *biochemical* function since fold corresponds strongly with biochemical function. Again, the fold enrichment observed in the expression data is apparent even when no pair of genes being examined share appreciable homology.

1 Introduction

The evolutionary tool of sequence duplication and subsequent sequence divergence is the means used by Nature to create new biochemical and biological function. The full tapestry on which these evolutionary events are played out is becoming available thanks to the multitude of successful whole-genome sequencing projects. Having the entire tapestry allows us to improve our understanding of evolution, and that improved understanding will in turn lead to better prediction of both the biological and biochemical functions of experimentally uncharacterized proteins.

Many experimentally uncharacterized proteins can be identified through sequence similarity to other proteins that *have* been characterized. However, the identification of distant homologs is a fundamental problem in modern computational biology with enormous potential for practical applications. The vast amount of genome sequencing is dramatically increasing the importance of the problem. The function of about 30% of *Arabidopsis thaliana* genes cannot be predicted by sequence similarity search methods[1-5]. About 40% of the identified genes in human chromosomes 21 and 22 do not have detectable homology to known genes[6,7]. Therefore even a small improvement in our ability to identify distant homologs can help us make functional predictions for a large number of newly discovered genes.

To build proteins, Nature draws on what appears to be a more or less fixed repertoire of *folds* developed during the course of evolution. A fold is a structural motif used as a building block in proteins. Various fold classifications exist, for our purposes we use that of the Structural Classification of Proteins (SCOP)[8]. Structural similarity has been recognized as a solid argument for evolutionary relationship between proteins[9,10] and hence, prediction of their function. Success in protein structure prediction by fold recognition is limited by our ability to identify homologous protein with known three-dimensional structure. Progress in fold recognition can be assessed at the regular CASP conferences (http://predictioncenter.llnl.gov/). The recent CASP4 meeting clearly showed that an expert, knowledge-based approach is superior to a purely computational, fully automated approach. The advantage of experts is that they use not only biochemical or biophysical information on protein sequence and structural properties, but also knowledge of protein function derived from biochemical experiments. In support of the utility of expert knowledge, it has in fact already been demonstrated that even something as simple as a key-word comparison of protein descriptions in the database increases the accuracy of fold recognition programs[11]. Clearly then, incorporation of additional data will improve the accuracy of fold prediction

What other kinds of data could be incorporated? The mechanism of duplication often leads to neighboring genes that share a common ancestor. It seems natural then to expect a fold enrichment among neighboring genes, and that is in fact one of the conclusions we report here for both *Arabidopsis* and *Saccharomyces cerevisiae*, consistent with an analysis on a subset of the yeast data analyzed here[12]. However, we can make a stronger statement. Not only do neighboring genes share fold, the distance between pairs of genes in the genome conveys information about structural homology *over and above* the information from the sequence comparison alone.

We will distinguish *biochemical* function of a protein from the gene's *biological* function, where by biological function we mean what the organism accomplishes with the gene and other co-expressed genes, for example a signal transduction cascade. Just as sequence conservation during evolution can be used to infer structural homology and biochemical function, one might wonder if Nature has left a trace of structural information during its evolution of biological (as opposed to biochemical) function.

If a structural trace can be found in the evolution of pathways, that trace could also be helpful in improving the prediction of protein fold. We looked for and found that trace using correlations in expression data, a method used for inferring biological function since the invention of the Northern blot but only recently performed on the genome-wide scale necessary to detect such a trace. As in the case of the fold enrichment seen between nearby genes on the genome, the fold enrichment between pairs of co-expressed genes conveys structural information over and above that from sequence similarity alone. Hence, *biological* function of a gene says something about the gene's corresponding protein fold, and via the fold something about the *biochemical* function of that protein.

In the following, then, we show that both physical distance on the genome and correlation in expression show traces of evolution manifested in structural homology. Those weak signals could in principle be used to improve prediction of structural homology in the twilight zone of sequence similarity.

2 Methods

2.1 Genomes

To investigate the evolutionary trace of fold enrichment along the genome, we used the genomes of the model organisms *Saccharomyces cerevisiae*[13] and *Arabidopsis thaliana*[1-5]. The yeast genome consists of 16 chromosomes, with 6,310 identified ORFs, available from the *Saccharomyces* Genome Database (SGD) at http://genome-www.stanford.edu/Saccharomyces/. The smallest chromosome, chromosome I, is ~0.23 Megabases and has 107 ORFs. The largest chromosome, chromosome IV, is ~1.53 Mb and encodes 819 ORFs. The recently finished Arabidopsis genome consists of five chromosomes, with 25,498 genes predicted. The smallest chromosome is chromosome IV and is ~17.5 Mb in length, containing 3,825 protein encoding genes. The largest chromosome is chromosome I, approximately 29.1 Mb in length, with 6,543 genes. We downloaded Arabidopsis genes from the NCBI web site (http://www.ncbi.nlm.nih.gov).

Only chromosomes II and IV were available as one contig at the time we made our analysis, so we used only 7,852 protein-coding genes from these two chromosomes in our analysis of fold enrichment in the genome neighborhood.

2.2 Microarray Expression Data

Yeast and arabidopsis expression data were downloaded from the Stanford Microarray Database[14]. A subset of non-biological experiments (e.g. assessing the performance of microarrays) was excluded from our analysis. The resulting dataset contained expression data from 345 yeast and 201 Arabidopsis microarray experiments.

The similarity in expression across all experiments between a pair of genes was measured using the Spearman rank correlation coefficient on the normalized ratios[15]. The Spearman r was chosen because it is a robust statistic that will capture any monotonic relationship between a pair of variables, as opposed to the commonly used Pearson correlation coefficient which is suitable for detecting linear relationships between pairs of variables. Missing data points were handled by pairwise deletion of observations from the Spearman *r* calculation and any pair of genes having fewer than 10 experiments in common were ignored. No special attempt was made to account for the redundancy due to experimental replicates or similarities in subsets of experiments. We note, however, that for the purposes of a global correlation analysis at least, it would be more desirable to have a larger number of distinct, diverse experiments than to have experimental replicates since the correlation coefficient implicitly takes inherent experimental variation into account.

Spurious significant correlations might be introduced between a pair of genes that are not actually co-expressed if the two genes are sufficiently similar that cross-hybridization occurs, where "sufficiently similar" is roughly taken to be in the neighborhood of 80% over 50 nucleotides[16-18]. No large scale, systematic experimental study of cross-hybridization on microarrays has been done, so we assessed the degree of cross-hybridization indirectly as follows. Pairwise similarity was measured using Wash-U BLASTN, version 2.0 with M=2 and all other parameters set to their defaults. We compared the overall distribution of correlation coefficients between pairs of genes to the distribution between non-identical chip features showing similarity of >=85% over >=50 nt. There is a clear shift towards one for the similar sequences as seen in Figure 1. The distribution for clones having 70%-85%, >=50 nt similarity is shifted towards 1 as well, but the shift is less pronounced (data not shown). To be conservative and minimize the possibility of cross-hybridization affecting our results, we discarded any chip feature having a sequence with an HSP showing 70% or greater similarity over at least 50 nt to some other gene in the genome. Applying this approach for yeast is straightforward since the microarray features are PCR fragments of the ORFs, the complete sequence of the features is known[19] and there are essentially no introns in yeast. In the case of Arabidopsis, the full sequence of the clones serving as source material for the microarrays is not available, so we mapped ESTs from the clones to the cDNAs from the annotation of the Arabidopsis genome[1-5] and assumed that the entire cDNA sequence was present in the microarray feature, then screened against all other cDNAs in Arabidopsis. While conservative, our approach cannot guarantee complete exclusion of features that might cross-hyb because the genomic annotations often lack full UTRs or have other errors.

The cross-hybridization filtering resulted in 4,280 yeast features and 3,011 Arabidopsis features. The smaller number of filtered features for Arabidopsis was primarily a consequence of feature redundancy on the chip (more than one clone for a given cDNA) and a higher amount of gene duplication in Arabidopsis. A correlation and cluster analysis was performed with the biological sensibility of results conforming to those from the literature, though the Arabidopsis results were less compelling[20-22].

2.3 *Fold Assignment*

For each gene we assigned fold(s) according to the SCOP-1.55 classification of protein structures[8]. Assignment was done by WU-Blastp[23] search against the Astral database of non-redundant SCOP domains at the 95% identity level [24]. We considered all matches with a P-value < .001. At that level of significance approximately 2% of our assignments are wrong[9]. One protein may consist of more than one domain, in those cases multiple folds were assigned to the corresponding gene. Out of 6,310 yeast genes, we assigned folds to 1,839 genes (29%). Out of 27,469 Arabidopsis genes from the TIGR gene index, we assigned folds to 9,147 genes (33%). The distribution of different SCOP folds in the two genomes is shown in Figure 2, with the most frequent folds summarized in Table 1. This is consistent with the most frequent folds in other organisms[25]. More advanced methods of fold assignments, e.g. PSI-BLAST, the profile-profile technique, and threading, increase the coverage, but overall do not change the statistical observations.

2.4 *Non-redundant Set of Proteins*

Since our intent is to determine if we can detect distant homologs, we created a non-redundant set of proteins from the overall set of proteins that had folds assigned to them. To create the non-redundant set, the following procedure was applied: for each protein, beginning with the longest, all shorter proteins were removed from the list if they matched the first protein with a P-value < 1.0e-3.

2.5 *Fold Enrichment Along the Genome*

The relative enrichment of folds along the genome was defined as the ratio of the probability of finding the same fold between pairs of genes a given distance apart in the genome to the probability of finding the same fold between two randomly selected pairs of genes in the genome. The ratio is therefore a function of the distance in nucleotides between gene pairs. At a given distance, a ratio greater than one implies that more similar folds are occurring than one would expect if folds were distributed randomly over that distance. A ratio of one indicates that the folds are distributed randomly.

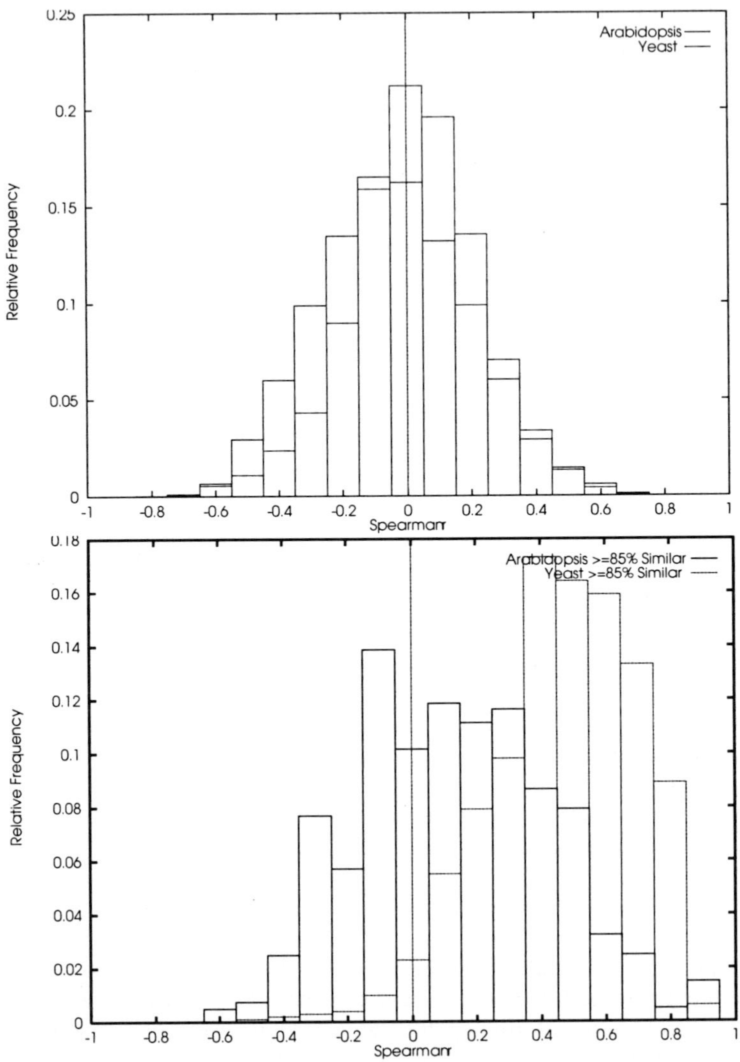

Figure 1. a) Histogram of Spearman *r* distribution for clones showing >=85% similarity over >=50 nt. b) Histogram of Spearman *r* distribution overall for clones included in analysis, i.e. those that show no similarity to any other clone at the 70%, 50 nt level. It should be pointed out that Figure 1a strongly suggests that the Arabidopsis data is of inferior quality to the Yeast data, see comment in the Results section below.

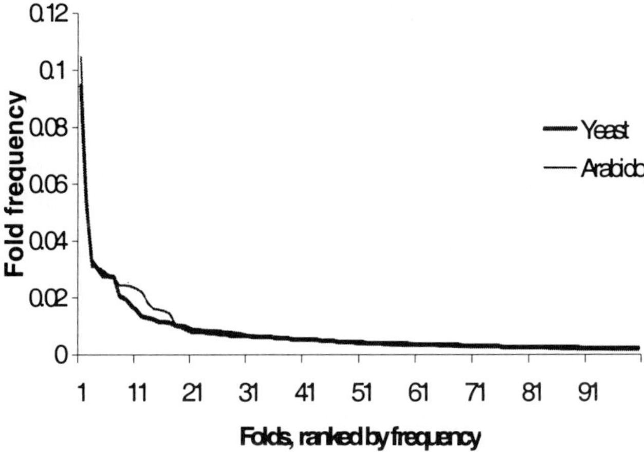

Figure 2. Fold distributions in yeast and Arabidopsis genomes.

SCOP_1.55 fold	Description	Yeast		Arabidopsis	
		rank	frequency	rank	frequency
c.37	P-loop containing nucleotide triphosphate hydrolases	1	0.095	2	0.059
d.144	Protein kinase-like (PK-like)	2	0.054	1	0.105
c.1	TIM beta/alpha-barrel	3	0.033	5	0.030
b.69	7-bladed beta-propeller	4	0.030	10	0.024
a.118	alpha-alpha superhelix	5	0.028	7	0.027
c.2	NAD(P)-binding Rossmann-fold domains	6	0.028	8	0.025
d.58	Ferredoxin-like	7	0.027	3	0.031
g.38	Zn2/Cys6 DNA-binding domain	8	0.021	>279	0
f.2	Membrane all-alpha	9	0.020	18	0.011
c.55	Ribonuclease H-like motif	10	0.018	21	0.009
a.4	DNA/RNA-binding 3-helical bundle	11	0.016	4	0.031
g.44	RING finger domain, C3HC4	22	0.008	6	0.028
a.104	Cytochrome P450	149	0.001	9	0.024

Table1. Ten most frequent folds in the yeast and Arabidopsis genomes. Seven folds belong to the ten most frequent folds in both genomes.

2.6 *Fold Enrichment for Genes with Similar Patterns of Expression*

The relative enrichment for co-expressed genes was defined by calculating the ratio of the probability of having a matching fold at or above a given level of correlation coefficient to that expected by randomly choosing pairs of genes. Fold enrichment for correlated genes is therefore a function of the Spearman *r*, with a ratio greater than one indicating that a pair of correlated genes is more likely to share fold than expected from chance. Error bars were estimated from counting error.

3 Results

3.1 *Fold Enrichment Along the Genome*

One of the most frequent evolutionary events is gene duplication, with the Arabidopsis genome being especially rich in tandemly repeated genes[1-5]. Therefore it is not surprising to see enrichment of homologous genes in the chromosomal neighborhood for both organisms. The effect can, however, still be observed even for the set of non-redundant proteins (Figure 3)

3.2 *Fold Enrichment for Genes with Similar Patterns of Expression*

Figure 4a shows a plot of the fold enrichment in yeast as a function of *r*. Figure 4b shows the corresponding plot for Arabidopsis. Both organisms show enrichment that is significantly elevated from the baseline of 1.0, although the difference is more pronounced for yeast. The enrichment is maintained even when redundant proteins are removed.

In the case of Arabidopsis there is only a weak signal at best, yet we suspect that it is in fact real. We believe that it is weak as compared to yeast at least in part because of dataset size and in part because of overall data quality. A power calculation shows that for the set of yeast data, we can reliably detect correlations down to ~0.4 (significance 0.01, power 80%, Bonferonni correction, power calculation assumed Pearson *r* rather than Spearman *r*). For Arabidopsis the threshold is roughly 0.5. This weaker detection ability is confounded by relatively poor quality data for Arabidopsis. The poorer quality Arabidopsis data means that the threshold for a biologically significant correlation is higher than for the yeast data. The difference in quality is evident from the much smaller overall shift towards a correlation of 1 in the distribution of Figure 1a). With more and better quality data, the peak should presumably become cleaner.

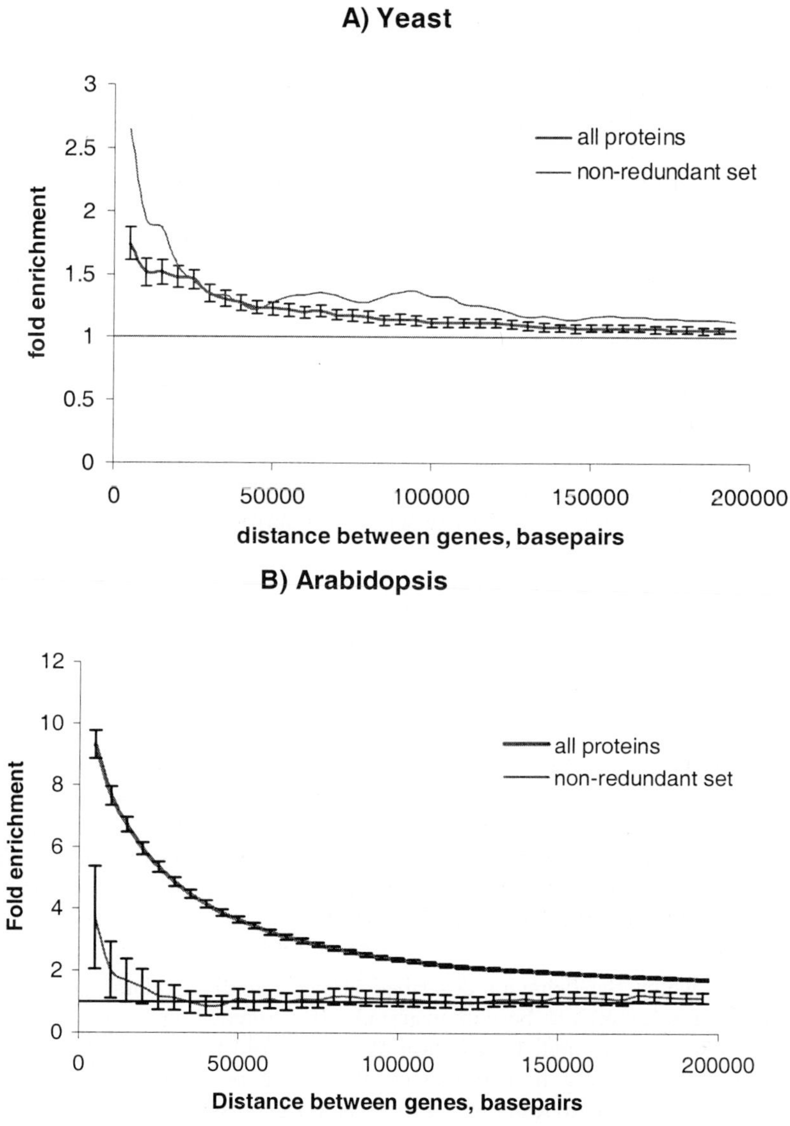

Figure 3. We examined two sets of genes for fold enrichment along the genome. The first set was all genes that had a fold assignment. The second set was a subset of the first consisting of genes whose proteins showed no significant homology to one another. For each set, fold enrichment was measured as the ratio of the frequency of same fold for genes within *d*

nucleotides of each other to the frequency of the same fold in randomly selected genes. The enrichment ratio is then plotted as a function of distance *d*.

4 Summary and Conclusions

Folds among nearby genes in the genome and among co-expressed genes are enriched relative to that expected by chance alone. We examined the distributions of enriched folds and were unable to explain the enrichment through a bias in folds in either case. From this we conclude that the enrichment we see is a more-or-less general feature of folds in organisms; with the large amount of data becoming available for worm[26], human and other organisms we will be able to confirm or rule out this speculation in the near future. Assuming we are correct one should in principle be able to incorporate such information into fold prediction for proteins whose fold is unknown. We are currently evaluating approaches to accomplish that goal.

The mechanism behind the enrichment of folds along the genome seems clear. Gene duplications lead to pairs of genes with similar ancestors, and even after substantial divergence results in no *sequence* homology between nearby genes on the genome, there is still a remnant of structural similarity. It is that remnant which accounts for the observed enrichment.

The mechanism behind the enrichment of folds among co-expressed genes is less clear. One hypothesis is that during the course of evolution of (e.g.) a particular metabolic pathway, a newly duplicated gene is created. For the sake of illustration let us say that the duplicated gene is an enzyme. Since that newly duplicated gene, which includes the promoter region of the original gene, is now redundant, one of two things must happen. Either one of the duplicated genes will disappear or the pair will diverge apart in sequence, with one retaining the original function (by function here we mean both biochemical function as well as biological role), and the other taking on a new function. Since both originally operate on the same substrate there is a structural constraint to how the pair diverge in sequence, and this constraint tends to cause the fold to be maintained.

The extent to which the behavior described above actually explains how Nature evolves pathways remains to be demonstrated. It is interesting to note, however, that we made a correct, blinded prediction of protein fold for a recent target in the CASP4 competition, using in part exactly the above reasoning. The target in question was pectin methylesterase, which is co-expressed with its metabolic pathway neighbor pectate lyase[27]. Both enzymes share exactly the same SCOP fold, the single-stranded right-handed beta-helix.

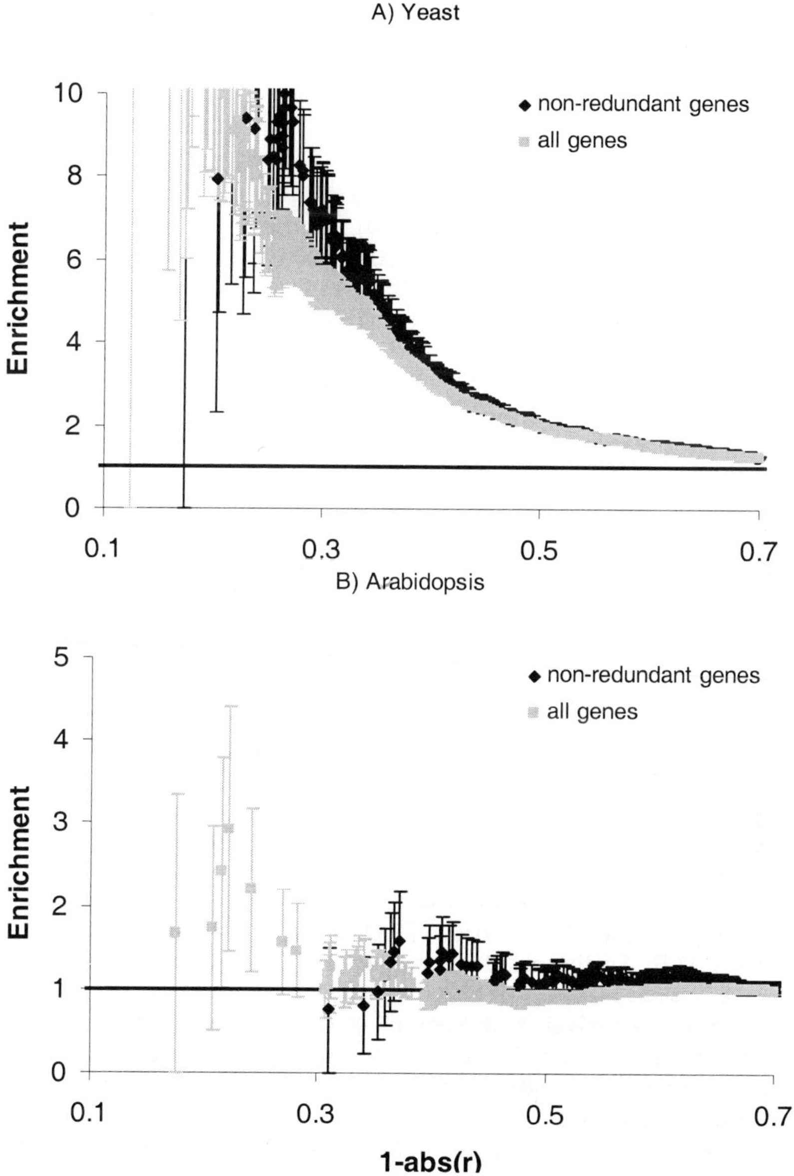

Figure 4. We examined two sets of genes for fold enrichment among co-expressed genes. The first set was all genes that had a fold assignment and showed no significant sequence homology to other genes at the nt level, see methods description for details on selection. The second set was a subset of the first consisting of genes whose proteins showed no significant homology to one another. For each set, fold enrichment was measured as the ratio of the

frequency of same fold for genes correlated at r or better to each other, relative to the frequency of the same fold in randomly selected genes. The enrichment ratio is then plotted as a function of r. Error bars are counting statistics only.

References

1. Theologis, A. *et al. Nature* **408**, 816-20 (2000).
2. Lin, X. *et al. Nature* **402**, 761-8 (1999).
3. Salanoubat, M. *et al. Nature* **408**, 820-2 (2000).
4. Mayer, K. *et al. Nature* **402**, 769-77 (1999).
5. Tabata, S. *et al. Nature* **408**, 823-6 (2000).
6. Dunham, I. *et al. Nature* **402**, 489-95. (1999).
7. Hattori, M. *et al. Nature* **405**, 311-9. (2000).
8. Murzin, A.G., Brenner, S.E., Hubbard, T. & Chothia, C. *J Mol Biol* **247**, 536-40. (1995).
9. Brenner, S.E., Chothia, C. & Hubbard, T.J. *Proc Natl Acad Sci U S A* **95**, 6073-8. (1998).
10. Park, J. *et al. J Mol Biol* **284**, 1201-10. (1998).
11. MacCallum, R.M., Kelley, L.A. & Sternberg, M.J. *Bioinformatics* **16**, 125-9. (2000).
12. Cohen, B.A., Mitra, R.D., Hughes, J.D. & Church, G.M. *Nat Genet* **26**, 183-6. (2000).
13. Goffeau, A. *et al. Science* **274**, 546, 563-7. (1996).
14. Sherlock, G. *et al. Nucleic Acids Res* **29**, 152-5 (2001).
15. Press, W.H., Teukolsky, S.A., Vetterling, W.T. & Flannery, B.R. Numerical Recipes in C. 639-640 (Cambridge University Press, Cambridge Cb2 2RU, UK, 1992).
16. Kane, M.D. *et al. Nucleic Acids Res* **28**, 4552-7 (2000).
17. Girke, T. *et al. Plant Physiol* **124**, 1570-81 (2000).
18. Xu, W. *et al. Gene* **272**, 61-74. (2001).
19. DeRisi, J.L., Iyer, V.R. & Brown, P.O. *Science* **278**, 680-6 (1997).
20. Eisen, M.B., Spellman, P.T., Brown, P.O. & Botstein, D. *Proc Natl Acad Sci U S A* **95**, 14863-8 (1998).
21. Heyer, L.J., Kruglyak, S. & Yooseph, S. *Genome Res* **9**, 1106-15 (1999).
22. Marcotte, E.M., Pellegrini, M., Thompson, M.J., Yeates, T.O. & Eisenberg, D. *Nature* **402**, 83-6 (1999).
23. Gish, W. (1994).
24. Brenner, S.E., Koehl, P. & Levitt, M. *Nucleic Acids Res* **28**, 254-6. (2000).
25. Gerstein, M., Lin, J. & Hegyi, H. *Pac Symp Biocomput* , 30-41. (2000).
26. Kim, S.K. *et al. Science* **293**, 2087-92. (2001).
27. Tierny, Y., Bechet, M., Joncquiert, J.C., Dubourguier, H.C. & Guillaume, J.B. *J Appl Bacteriol* **76**, 592-602. (1994).

Multiple Genome Rearrangement By Reversals

Shiquan Wu and Xun Gu
Center of Bioinformatics and Biological Statistics
Iowa State University
Ames, IA 50011, USA
{sqwu,xgu}@iastate.edu

In this paper, we discuss a multiple genome rearrangement problem: Given a collection of genomes represented by permutations, we generate the collection from some fixed genome, e.g., the identity permutation, in a minimum number of signed reversals. It is NP-hard, so efficient heuristics is important for finding its optimal solution. We at first discuss how to generate two and three genomes from a fixed genome by polynomial algorithms for some special cases. Then based on the polynomial algorithms, we obtain some approximation algorithms for generating two and three genomes in general, respectively. Finally, we apply these approximation algorithms to design a new approximation algorithm for generating more genomes. We also show by some experimental examples that the algorithms are efficient.

1 Introduction

Comparative genomics is one of the most important areas in computational biology and bioinformatics. Sorting by reversal plays a central role in Comparative genomics. The problem was originated in last decade [2,3,4]. Its theme is to determine the evolutionary distances between organisms by using genomic data. Transformations of genomes are widely studied under evolutionary events such as insertion, deletion, point mutation (substitution), reversal, etc [11,12]. Recently, optimal recombination is also discussed[19]. So far, most of the study on comparative genomics has been focused on sorting by reversal[18]. A genome is represented by a permutation and an optimal reversal process is found from any given permutation to the identity permutation.

Sorting by reversal is categorized into two classes: sorting by unsigned and signed reversals, respectively. At first, sorting by unsigned reversals is NP-hard[7,8]. Therefore, only efficient approximation algorithms can be expected to find for the solution of the problem. So far, the best approximation algorithm has been a 1.5-approximation algorithm[10]. It is proved that there exists no polynomial-time 1.0008-approximation algorithm[6].

However, sorting by signed reversals is polynomial-time solvable[13,14]. Many quadratic-time algorithms are widely used for finding the optimal solutions of the problem[5,15,16]. Recently, a linear-time algorithm is found for computing the signed reversal distance between any two signed permutations[1].

Sorting by reversal can be regarded as a problem that generates a permutation from some fixed permutation by a minimum number of reversals. Multiple genome rearrangement by reversals is a generalization of sorting by reversal. It is to generate a given collection of permutations (genomes) from a fixed permutation, e.g., the identity, in a minimum number of reversals. For the unsigned case, the problem is obviously NP-hard (since sorting by unsigned reversals is NP-hard). For the signed case, it is proved that the problem is NP-hard even if two permutations are generated from a permutation[8]. This implies that the problem is extremely hard. Therefore, it is interesting, also our purpose in this paper, to find efficient heuristics, or special cases that are polynomial-time solvable. Heuristics can be combinatorial or experimental algorithms[1,9].

A similar genome rearrangement problem was discussed and an approximation algorithm was given by a local search for the optimal solution on a grid by Sankoff *et al*[17]. In this paper, we discuss a multiple genome rearrangement problem for generating a collection of permutations from some fixed permutation in a minimum number of signed reversals.

The rest of the paper includes five parts. (1) Definitions and models, (2) Related problems, (3) Theorems and algorithms, (4) Experimental applications, (5) Discussion and future work.

2 Mathematical model of multiple genome rearrangement

First of all, we introduce our main definitions and notations. The mathematical model of multiple genome rearrangement problem is also described.

Definition 1 For a signed permutation $p = (p_1 p_2 \cdots p_{|X|})$ on an alphabet X, a signed reversal on the segment $[i, j]$ of p is defined as the following operation from p to $r(p; i, j)$:

$$\begin{aligned} p &= (p_1\ p_2\ \cdots\ p_{i-1}\ \underline{p_i\ \ p_{i+1}\ \ \cdots\ \ p_j}\ p_{j+1} \cdots\ p_{|X|}) \\ r(p; i, j) &= (p_1\ p_2\ \cdots\ p_{i-1}\underline{-p_j \cdots - p_{i+1} - p_i}\ p_{j+1} \cdots\ p_{|X|}) \end{aligned}$$

Definition 2 Let T_0 be a collection of permutations. Define $N(T_0) = \{p | p = r(q; i, j) \text{ for some } q \in T_0, 1 \le i < j \le n\}$, called the reversal neighborhood of T_0. Define $N_1(T_0) = N(T_0)$, $N_2(T_0) = N(N_1(T_0))$, and $N_k(T_0) = N(N_{k-1}(T_0))$, the k-neighborhood of T_0.

Definition 3 A collection T_0 of permutations is called a $k-$Bottleneck family if for any $u, v \in T_0$, the reversal distance between u and v is at most k.

Multiple Genome Rearrangement By Signed Reversal (denoted by $(m, n)-$MGRBSR) Given two collections of permutations $P = \{p_1, p_2, \cdots, p_m\}$ and $Q = \{q_1, q_2, \cdots, q_n\}$ on an alphabet X, we generate Q from P in a minimum number of signed reversals, i.e., find a collection of signed permutations $t_r (1 \le r \le s)$ on X such that (1) any q_j is obtained from some p_i by a series of signed reversals $t_{r_1}, t_{r_2}, \cdots, t_{r_u}$, where each $t_{r_{j+1}}$ is obtained from t_{r_j} by one signed reversal, and (2) s is minimized. Denote $d(P, Q) = s$, the (optimal) signed reversal distance for generating Q from P.

If $m = n = 1$, i.e., $P = \{p\}$ and $Q = \{q\}$, then the problem is reduced to sorting by signed reversal and $d(p, q)$ is the signed reversal distance between p and q. In any optimal reversal process, each q_j is generated from only one p_i. Therefore, it is sufficient for us to consider the case $m = 1$. For $m > 1$, the problem can be split into m $(1, n_j)$–MGRBSR problems and similarly discussed. By rearranging the elements of X, we get $p_1 = 12 \cdots |X|$. Therefore, we discuss how to generate all permutations q_j from p_1.

$(1, n)$–MGRBSR Problem Generate all given permutations $q_j (1 \leq j \leq n)$ from the identity $p = 12 \cdots |X|$ in a minimum number of signed reversals.

We at first consider the $(1, 2)$–, $(1, 3)$–MGRBSR problems and then split the $(1, n)$–MGRBSR problem into some $(1, 2)$–, $(1, 3)$–MGRBSR problems.

3 Related problems

Our $(1, n)$–MGRBSR problem is similar to the genome rearrangement problem discussed by Sankoff *et al*[17] and is closely connected to the following problems[11,12,20].

Multiple alignment Given some sequences, find the alignment with minimum pairwise score.

In our $(1, n)$–MGRBSR problem, we do not consider the pairwise score, but the minimum score of Steiner trees on the given permutations.

Sorting by reversal Given a permutation p, transform it into the identity permutation in a minimum number of signed (or unsigned) reversals.

It generates the identity permutation from a given permutation in a minimum number of signed (or unsigned) reversals. Our $(1, n)$–MGRBSR problem generalizes the problem to generating a collection of permutations.

Star alignment Given some sequences, find one median sequence such that the total alignment score between the median sequence and each given sequence is minimized.

Our $(1, n)$–MGRBSR problem may contain a number of median sequences.

Fixed topology alignment[20] Given some sequences and a topological structure (usually, a tree) T. Each leaf of T is labeled by one given sequence. Assign one sequence to each internal node of T such that the total alignment score for all edges of T is minimized.

Our $(1, n)$–MGRBSR problem is not restricted to a fixed topology and it is a topology-free alignment problem.

4 Theorems and algorithms

In this part, we find algorithms for $(1,n)-$MGRBSR problems. We at first consider $(1,2)-$ and $(1,3)-$MGRBSR problems. If a $(1,2)-$MGRBSR problem contains a pair of close permutations, then we get a polynomial algorithm. If a $(1,3)-$MGRBSR problem consists of two pairs of close permutations, then we also get a polynomial algorithm. Based on these polynomial algorithms, we design approximation algorithms for the general $(1,2)-$ and $(1,3)-$MGRBSR problems, respectively. Next, we discuss a $k-$Bottleneck family for a $(1,n)-$MGRBSR problem. Finally, we split a $(1,n)-$MGRBSR problem into some $(1,2)-$ and $(1,3)-$MGRBSR problems and obtain an approximation algorithm for the general $(1,n)-$MGRBSR problem.

First of all, it is shown that [1]

Theorem 1 The $(1,1)-$MGRBSR problem is solvable in a run time $O(|X|)$.

A linear-time algorithm is presented for computing the reversal distance between two signed permutations [1] (finding the optimal reversal process still costs $O(|X|^2)$). We denote it BMY algorithm and will use it in our algorithms.

We easily have the following approximation algorithm. We at first construct a weighted graph with all given permutations as its vertices. All pairs of the given permutations form its edges. The weight of an edge is defined as the reversal distance of the pair of permutations representing the edge, which is computed by the BMY algorithm. Next we find a minimum weight spanning tree of the graph. Finally, all permutations can be generated from a given permutation along the edges of the spanning tree. With the theorem and the BMY algorithm, the run time is reduced. The steps are stated in the following.

Algorithm A

Input Sequences: $p, q_1, q_2, \cdots, q_n$.

Output Reversal process.

Step 1 Apply the BMY algorithm to construct a graph $G = (V, E, W)$ with $V = \{p, q_1, q_2, \cdots, q_n\}$, $E = \{[u,v] |\ u, v \in V, u \neq v\}$, and $W([u,v]) = d(u,v)$.

Step 2 Find a minimum weight spanning tree T of G.

Step 3 Generate all permutations from p along T.

Theorem 2 Algorithm A finds an approximated solution for any $(1,n)-$MGRBSR problem in a run time $O(n^2|X| + n|X|^2)$.

Proof Step 1 has a run time $O(n^2|X|)$ since it takes $O(|X|)$ time to find each $W([u,v])$ and G has $O(n^2)$ edges. Step 2 has a run time $O(n\ log\ n)$ to find T. Step 3 has a run time $O(n|X|^2)$ since it takes a run time $O(|X|^2)$ to find the optimal reversal process for each edge $[u,v]$ and there are $n-1$ edges T.

It is obvious that the algorithm is a 2-approximation, i.e., the approximated distance is within two times of the optimal one. Furthermore, The number of reversals can be decreased by introducing some median permutations. Suppose we want to generate q_1 and q_2 from p. We at first generate a median permutation q_0 from p, then generate q_1 and q_2 from q_0, respectively. When q_0 is properly chosen, the number of reversals can be improved. The median permutation q_0 is called a Steiner permutation. If we want to generate a collection of permutations, many Steiner permutations will be applied so as to minimize the total reversal distance. These Steiner permutations are called optimal if they minimize the total reversal distance. For a $(1, n)-$MGRBSR problem, there may be $n-1$ optimal Steiner permutations.

In order to find an optimal Steiner permutation q_0 for a $(1,2)-$MGRBSR problem, we can try each permutation on X and finally get it. However, there are $|X|!$ permutations, so the run time is at least $|X|!$. We find some special cases that are polynomial-time solvable.

Theorem 3 (1) Let $V = \{p, q_1, q_2\}$. Assume q_0 is an optimal Steiner permutation. Then $d(q_0, x) \leq d(x, y)$ for any $x, y \in V$.

(2) If $V = \{p, q_1, q_2\}$ contains a pair with a reversal distance at most k, then an optimal Steiner permutation q_0 is found in a run time $O(|X|^{2k+1})$.

(3) If $V = \{p, q_1, q_2, q_3\}$ consists of two pairs with reversal distances at most k, then two optimal Steiner permutations are found in a run time $O(|X|^{4k+1})$.

Proof (1) By contradiction. Suppose that $d(q_0, q_i > d(q_1, q_2)(i = 1, 2)$. Then $d(p, q_0) + d(q_0, q_1) + d(q_0, q_2) > d(p, q_1) + d(q_1, q_2)$, a contradiction.

(2) By (1), the optimal Steiner permutation $q_0 \in N_k(x)$ for some $x \in V$. Since $|N(x)| \leq |X|^2$ and $|N_k(x)| \leq |X|^2|N_{k-1}(x)| \leq |X|^{2k}$. We need a run time $O(|X|)$ to find $d(y, V)(y \in N_k(x))$. Therefore, the run time is $O(|X|^{2k+1})$.

(3) Suppose that p and q_1, and q_2 and q_3 have reversal distances at most k. By (2), for each pair $q_{01} \in N_k(p)$ and $q_{02} \in N_k(q_3)$, compute the total reversal distance from q_{01} and q_{02} to p,q_1,q_2 and q_3. We then get the optimal pair q_{01} and q_{02} as the Steiner permutations. By (2), the run time is $O(|X|^{4k+1})$.

Based on Theorem 3, we can find an optimal Steiner permutation in the k-neighborhood of a permutation in the closest pair for a $(1,2)-$MGRBSR problem. For a $(1,3)-$MGRBSR problem, we can also find two optimal Steiner permutations in the k-neighborhoods of two permutations, each of which corresponds to a closest pair. We have the following algorithms (see Figure 1).

Algorithm B1

Input Sequences: p, q_1, q_2 (with a pair of reversal distance at most k).

Output Optimal reversal process.

Step 1 Find the pair, say p and q_1, with the minimum reversal distance (at most k).

Step 2 Loop over all $u \in N_k(p)$ and update the reversal distance $d = d(u,p) + d(u,q_1) + d(u,q_2)$ and $q_0 = u$ if a better one is found.

Step 3 Generate q_0 from p, q_1 and q_2 from q_0 by BMY algorithm.

Algorithm B2

Input Sequences: p, q_1, q_2, q_3 (two pairs with reversal distances $\leq k$)

Output Optimal reversal process.

Step 1 Find the pairs, say p, q_1 and q_2, q_3, with two minimum reversal distances (at most k).

Step 2 Loop over $u \in N_k(p), v \in N_k(q_3)$. Update the total reversal distance $d = d(u,v) + d(u,p) + d(u,q_1) + d(v,q_2) + d(v,q_3)$ if a better one is found. Also update $q_{01} = u$ and $q_{02} = v$.

Step 3 Generate q_{01} from p, q_1 from q_{01}, q_{02} from q_{01}, q_1 and q_2 from q_{02} by the BMY algorithm.

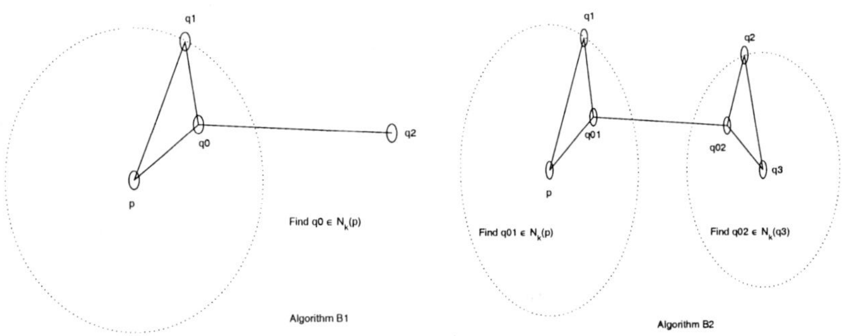

Figure 1: Algorithm B1/B2: Each optimal Steiner permutation is located in some $N_k(x)$

Figure 1 shows that Algorithm B1 (or B2) finds the optimal Steiner permutations in some $N_k(x)$ and terminates within $O(|X|^{2k+1})$ (or $O(|X|^{4k+1})$) run time. Similarly, we generate a collection of close permutations.

Theorem 4 Let $V = \{p, q_1, q_2, \cdots, q_s\}$ be a $k-$Bottleneck family. Then all Steiner permutations can be found in $O(|X|^{2k(s-2)+1})$ run time.

Based on Theorem 4, we can find the optimal reversal process for a small collection of permutations.

Algorithm C

Input Sequences: $p, q_1, q_2, \cdots, q_s$ ($k-$Bottleneck).

Output Steiner permutations and reversal process.

Step 1 Find $N_k(p)$.

Step 2 Loop over all $x_1, x_2 \cdots, x_{s-2} \in N_k(p)$ and update the total reversal distance if a better minimum spanning tree is found for $\{p, q_1, q_2, \cdots, q_s; x_1, x_2 \cdots, x_{s-2}\}$.

By Theorem 4, we find the $s-2$ optimal Steiner permutations in a run time $O(|X|^{2k(s-2)+1})$. For collections that are not $k-$Bottleneck families, we design two approximation algorithms to find their optimal Steiner permutations on the grids constructed from a series of optimal reversal paths.

Algorithm D1

Input Sequences: p, q_1, q_2.

Output Steiner permutations and reversal process.

Step 1 Choose a minimum reversal distance pair, say p, q_1.
Step 2 Find the optimal reversal path P_1 from p to q_1.
Step 3 For $i \geq 2$, find $M_i \in N_k(P_{i-1})(k = 1, 2, 3)$ minimizing $d(M_i, p, q_1, q_2) = d(M_i, p) + d(M_i, q_1) + d(M_i, q_2)$. Find an optimal reversal path P_i from p to M_i, then to q_2.
Step 4 For each u in P_d(the last path in Step 3), find an optimal reversal path W from u to q_2. We get $W_1, W_2, \cdots, W_t$.
Step 5 For each x in all W_i, do a local optimal search to find $q_0 \in N_k(x)$ minimizing $d(q_0, p, q_1, q_2)$.
Step 6 Choose the best q_0 in Step 5 as the global optimal solution.
Step 7 Generate q_0 from p, q_1 and q_2 from q_0 by BMY algorithm.

Algorithm D2

Input Sequences: p, q_1, q_2, q_3.

Output Steiner permutations and reversal process.

Step 1 Choose two minimum reversal distance pairs: $(p, q_1), (q_2, q_3)$.
Step 2 Find optimal reversal paths, $P_1 : p$ to q_1, and $Q_1 : q_2$ to q_3.
Step 3 For $i \geq 2$, find $M_i \in N_k(P_{i-1})$, $N_i \in N_k(Q_{i-1})(k = 1, 2, 3)$ minimizing $d(M_i, N_i)$. Find optimal reversal paths, $P_i : p$ to M_i, then to q_2, and $Q_i : q_2$ to N_i, then to q_3.
Step 4 For each $u \in P_d, v \in Q_c$ (the final paths), find an optimal reversal path W from u to v. We get $W_1, W_2, \cdots, W_t$.
Step 5 For each pair x, y in the grid, do a local optimal search to find $q_{01} \in N_k(x), q_{02} \in N_k(y)$ minimizing $d(q_0, p, q_1, q_2)$.
Step 6 Choose the best q_{01}, q_{02} in Step 5 as the optimal solution.
Step 7 Generate q_{01} from p, q_1 from q_{01}, q_{02} from q_{01}, and q_1 and q_2 from q_{02} by the BMY algorithm.

In fact, in Algorithm D1, we find a series of paths from p to q_1 such that they get closer and closer to q_2. Then construct a grid by using q_2 and the closest path to q_2. Finally, do local optimal searches on the grid.

In Algorithm D2, we find two collections of paths from p to q_1, and from q_2

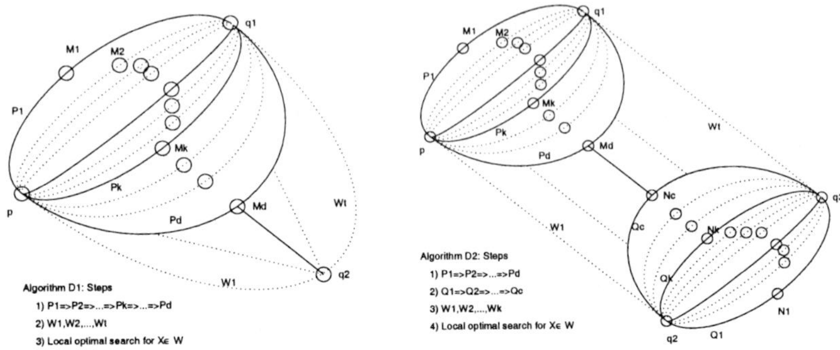

Figure 2: Algorithm D1/D2: Their steps

to q_3, respectively, such that the two collections of paths get closer and closer. Then construct a grid by using the two closest paths. Finally, do local optimal searches on the grid (see Figure 2).

Theorem 5 (1) For any p, q_1, q_2, the approximated Steiner permutation q_0 and reversal process can be found in a run time $O(|X|^{2(k+1)})$.

(2) For any p, q_1, q_2, q_3 the approximated Steiner permutations q_{01} and q_{02} and reversal process can be found in a run time $O(|X|^{2(k+1)})$.

Proof (1) Algorithm D1 at first finds an optimal reversal path from p to q_1. In the next step, it finds another optimal reversal path from p to q_1 that is closer to q_2. After some steps, it constructs a grid by using q_2 and the last optimal reversal path. The algorithm tries each possible permutation x in the grid and then finds an approximated Steiner permutation q_0 from some $N_k(x)$. Each path has length at most $|X|$ and the algorithm goes for at most $|X|$ paths. For each u in the paths, $|N_k(u)| = O(|X|^{2k})$. So the algorithm terminates in $O(|X|^{2(k+1)})$ run time.

(2) Similar to (1).

Based on Algorithm D1/D2 and Theorem 5, we designed an approximation algorithm for finding the Steiner permutations for the $(1, n)-$MGRBSR problem. The main idea is splitting the $(1, n)-$MGRBSR problem into a collection of $(1, 2)-$ and $(1, 3)-$MGRBSR problems. For any given permutations, $p, q_1, q_2, \cdots, q_n$, we at first find a minimum matching $A_i = \{x_i, y_i\}(1 \le i \le c)$ such that (1) $x_i, y_i \in \{p, q_1, q_2, \cdots, q_n\}$, and (2) $\sum_i d(x_i, y_i)$ is minimized. Then we find a minimum matching $w_j = \{u_j, v_j\}(j = 1, 2, \cdots, d)$ such that (1) $u_j, v_j \in \{A_i | i = 1, 2, \cdots, c\}$, (2) $\sum_j d(u_j, v_j)$ is minimized. Next, we apply Al-

gorithm D1/D2 to find two Steiner permutations q_{j1}, q_{j2} for u_j and v_j. Finally, replace all $\{u_j, v_j\}$ by all $\{q_{j1}, q_{j2}\}$ and repeat the process until it terminates.
Algorithm E (See Figure 3)
Input Sequences: $p, q_1, q_2, \cdots, q_n$.
Output Steiner permutations and reversal process.

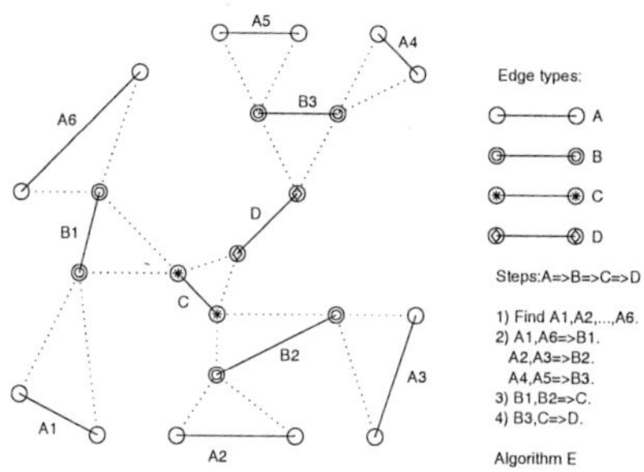

Figure 3: Algorithm E

Theorem 6 Algorithm E approximates the optimal Steiner permutations and the reversal process in a run time $O(|X|^{2(k+1)}n^2)$.

5 Experimental applications

Based on our algorithms, we design a computer program. Applying the program to some specific permutations, we obtain the optimal Steiner permutations for the permutations with different lengths. These examples show that our approximation algorithms are efficient. The three permutations, p, q_1, q_2, are chosen from the genomes of human, sea urchin, and fruit fly, respectively.

$$p = 26\ 13\ 17\ 12 -24\ 15\ 18 -2 -16 -3\ 4 -28\ 7\ 5\ 1\ 10\ 19\ 25\ 22\ 11\ 29\ 14\ 20\ -21\ -8\ 6\ 30\ -23\ 9\ 27.$$

$$q_1 = 26\ 4\ 25\ 22\ 5\ 1 -28\ 19\ 11\ 29\ 20 -21\ 6\ 9\ 27\ 8\ 30\ 23\ -24\ 16\ 14 -2\ 3\ 15 -7\ 10\ 13\ 17\ 12\ 18.$$

$$q_2 = -26 -27\ 12 -24\ 15\ 18 -3\ 4\ 13\ 5\ 7\ 1\ 10\ 19\ 2\ 25\ 16\ 29\ 8\ 9\ -20 -11 -22\ 30\ 23\ 21\ 6\ 28\ 17 -14.$$

By our program, we obtain an optimal Steiner permutation.

$$q_0 = 26 -2 -14 -29 -11 -22 -25 -19 -10 -1 -5\ 13\ 17\ 12 -24\ 15\ 18 -7\ 28 -4\ 3\ 16\ 20 -21 -8\ 6\ 30 -23\ 9\ 27.$$

If we choose the first k genes of p, q_1, q_2 and apply the program for $k = 5$,

10, 15, 20, 25, then we obtain the optimal Steiner permutations $q_0(k)$ from $p(k), q_1(k), q_2(k), q_3(k)$, respectively.

$p(5) =$ -2 -3 4 5 1.
$q_1(5) =$ 4 5 1 -2 3.
$q_2(5) =$ -3 4 5 1 2.
$q_0(5) =$ -2 -1 -5 -4 3.
$p(10) =$ -2 -3 4 7 5 1 10 -8 6 9.
$q_1(10) =$ 4 5 1 6 9 8 -2 3 -7 10.
$q_2(10) =$ -3 4 5 7 1 10 2 8 9 6.
$q_0(10) =$ -10 -1 -5 -7 -4 3 2 -8 6 9.
$p(15) =$ 13 12 15 -2 -3 4 7 5 1 10 11 14 -8 6 9.
$q_1(15) =$ 4 5 1 11 6 9 8 14 -2 3 15 -7 10 13 12.
$q_2(15) =$ 12 15 -3 4 13 5 7 1 10 2 8 9 -11 6 -14.
$q_0(15) =$ 13 12 15 -2 -10 -1 -5 -7 -4 3 11 -9 -6 8 -14.
$p(20) =$ 13 17 12 15 18 -2 -16 -3 4 7 5 1 10 19 11 14 20 -8 6 9.
$q_1(20) =$ 4 5 1 19 11 20 6 9 8 16 14 -2 3 15 -7 10 13 17 12 18.
$q_2(20) =$ 12 15 18 -3 4 13 5 7 1 10 19 2 16 8 9 -20 -11 6 17 -14.
$q_0(20) =$ -20 -14 -11 -19 -10 -1 -5 -7 -4 3 16 8 13 17 12 15 18 -2 6 9.
$p(25) =$ 13 17 12 -24 15 18 -2 -16 -3 4 7 5 1 10 19 25 22 11 14 20 -21 -8 6 -23 9.
$q_1(25) =$ 4 25 22 5 1 19 11 20 -21 6 9 8 23 -24 16 14 -2 3 15 -7 10 13 17 12 18.
$q_2(25) =$ 12 -24 15 18 -3 4 13 5 7 1 10 19 2 25 16 8 9 -20 -11 -22 23 21 6 17 -14.
$q_0(25) =$ 2 -18 -15 24 -12 -17 -13 25 22 11 20 -21 5 1 10 19 -14 -16 -3 4 7 -8 6 -23 9.

n	$d(p, q_1)$	$d(p, q_2)$	$d(q_1, q_2)$	Optimal $d(p, \{q_1, q_2\})$
5	3	2	3	4
10	8	8	9	15
15	12	9	14	19
20	15	13	19	26
25	19	18	24	33
30	21	22	29	40

The optimal reversal distances are computed by our program. They are almost the same as the lengths of the Steiner trees in a metric space. For example, for $n = 10$, we have $(d_1, d_2, d_3) = (8, 8, 9)$. In the Euclidean metric space, we compute the optimal Steiner tree and find that its length is 14.5. The optimal $d(q_0, p) + d(q_0, q_1) + d(q_0, q_2) = 15$. Both are close each other.

Our approximation algorithms can find the optimal solutions for most collections of genomes. In many cases, they are more efficient than the one on local search for optimal solution on a grid[17].

6 Discussion and future work

In this paper, we discuss a $(1, n)-$MGRBSR problem. We design some polynomial algorithms for several special cases and some efficient approximation algorithms for the general problem. The $(1, n)-$MGRBSR problem is one of the most important problems in comparative genomics. We are interested in designing more efficient approximation algorithms for finding optimal solutions for the general $(1, n)-$MGRBSR problem. The problem is very similar to Steiner tree problems in a metric space. With the application of Steiner tree theory, the problem can be solved efficiently. On the other hand, stochastics can also be applied to the discussion of the $(1, n)-$MGRBSR problem. This will be the subject of our future work.

Acknowledgement This work is supported by the NIH grant RO1 GM62118 (to X.G.) and Wu is supported in part by NSF of China (19771025).

References

1. Bader, D.A., Moret, B.M.E. and Yan, M., *A linear-time algorithm for computing inversion distances between signed permutations with an experimental study.* Proc. 7th Workshop on Algorithms and Data Structures (WADS 01), Providence (2001), to appear in Lecture Notes in Computer Science, Springer Verlag.
2. Bafna,V. and Pevzner, P. 1994. *Genome rearrangements and sorting by reversals.* In Proc. 34th IEEE Symp. of the Foundations of Computer Science, 148-157. IEEE Computer Society Press.
3. Bafna, V. and Pevzner, P. 1995. *Sorting permutations by transpositions.* Proceedings of the 6th Annual Symposium on Discrete Algorithms, pages 614-623. ACM Press, January 1995.
4. Bafna, V. and Pevzner, P. 1996. *Genome rearrangements and sorting by reversals.* SIAM Journal on Computing, 25(2):272-289.
5. Berman, P. and Hannenhalli, S. 1996. *Fast sorting by reversal.* Proc. Combinatorial Pattern Matching (CPM), 168-175. Also Lecture Notes in Computer Science.1075.
6. Berman, P., and Karpinski, M. 1998. *On some tighter inapproximability results.* Technical Report TR98-065, ECCC.
7. Caprara, A. 1997. *Sorting by Reversals is Difficult.* Proceedings of the First Annual International Conference on Computational Molecular Bi-

ology (RECOMB'97), ACM Press.

8. Caprara, A. 1999. *Formulations and hardness of multiple sorting by yeversals.* Proceedings of the Third Annual International Conference on Computational Molecular Biology (RECOMB'99), ACM Press.
9. Caprara, A. and Lancia, G. 2000. *Experimental and Statistical Analysis of Sorting by Reversals.* in D. Sankoff and J.H. Nadeau (eds.) Comparative Genomics: Empirical and Analytical Approaches to Gene Order Dynamics, Kluwer Academic Publishers.
10. Christie, D. A. 1998. *A 3/2-approximation algorithm for Sorting by reversals.* Proc. 9th Ann. ACM-SIAM Symp. on Discrete Algorithms, ACM-SIAM, 244-252.
11. Durbin, R., Eddy, S. R., Krogh, A. and Mitchison, G., 1998. Biological sequence analysis: Probabilistic models of proteins and nucleic acids, Cambridge University Press.
12. Gusfield,D., Algorithms on Strings, Trees, and Sequences: Computer Science and Computational Biology. Cambridge University Press, 1997.
13. Hannenhalli, S. and Pevzner, P. 1995. *Transforming men into mice (polynomial algorithm for genomic distance problems.* Proc. IEEE Symp. of the Foundations of Computer Science.
14. Hannenhalli, S. and Pevzner, P. 1995. *Transforming cabbage into turnip (polynomial algorithm for sorting signed permutations by reversals).* Proceedings of the Twenty-Seventh Annual ACM Symposium on Theory of Computing, 178-189.
15. Kaplan, H., Shamir, R. and Tarjan, R. E. 1997. *Faster and simpler algorithm for sorting signed permutations by reversals.* Proc. eighth annual ACM-SIAM Symp. on Discrete Algorithms (SODA 97). ACM Press.
16. Kaplan, H., Shamir, R. and Tarjan, R. E. 2000. *Faster and simpler algorithm for sorting signed permutations by reversals.* SIAM Journal on Computing, 29(3):880-892.
17. Sankoff, D., Sudaram, G. and Kececioglu, J. 1996. *Steiner points in the space of genome rearrangements.* International Journal of the Foundations of Computer Science, 7:1-9.
18. Sankoff,D. and Nadeau,J.H. . 2000. *Comparative Genomics: Empirical and Analytical Approaches to Gene Order Dynamics.* Kluwer Academic Publishers.
19. Wu, S. and Gu. X., 2001. *A greedy algorithm for optimal recombination.* Lecture Notes on Computer Science, 2108:86-90.
20. Wang,L., B. Ma and M. Li, 2000, *Fixed topology alignment with recombination, Discrete Applied Mathematics* 104: 281-300

High Speed Homology Search with FPGAs*

Yoshiki YAMAGUCHI, Tsutomu MARUYAMA
Institute of Engineering Mechanics and Systems, University of Tsukuba, 1-1-1 Ten-ou-dai Tsukuba Ibaraki, 305-8573, JAPAN

Akihiko KONAGAYA
Japan Advanced Institute of Science and Technology, 1-1 Asahidai Tatsunokuchi Ishikawa, 923-1292, JAPAN
Japan Riken Genomic Sciences Center, 1-7-22 Suehiro Tsurumi Yokohama Kanagawa, 230-0045, JAPAN

We will introduce a way how we can achieve high speed homology search by only adding one off-the-shelf PCI board with one Field Programmable Gate Array (FPGA) to a Pentium based computer system in use. FPGA is a reconfigurable device, and any kind of circuits, such as pattern matching program, can be realized in a moment. The performance is almost proportional to the size of FPGA which is used in the system, and FPGAs are becoming larger and larger following Moore's law. We can easily obtain latest/larger FPGAs in the form off-the-shelf PCI boards with FPGAs, at low costs. The result which we obtained is as follows. The performance is most comparable with small to middle class dedicated hardware systems when we use a board with one of the latest FPGAs and the performance can be furthermore accelerated by using more number of FPGA boards. The time for comparing a query sequence of 2,048 elements with a database sequence of 64 million elements by the Smith-Waterman algorithm is about 34 sec, which is about 330 times faster than a desktop computer with a 1GHz PentiumIII. We can also accelerate the performance of a laptop computer using a PC card with one smaller FPGA. The time for comparing a query sequence (1,024) with the database sequence (64 million) is about 185 sec, which is about 30 times faster than the desktop computer.

1 Introduction

In the past several years, there has been a rapid increase in genetic and genomic database, and the pattern matching problems in bioinformatics require huge time for the computations. Many algorithms[4,5,6] and dedicated hardware systems[11,12,13] have been developed. The result obtained there is a trade-off of quality, time and cost. With desktop computer systems, it is unrealistic to check all pattern matching possibilities within a reasonable time. Therefore, simplified (but still very effective) algorithms have been designed and used on the systems. With dedicated hardware systems, the computation time can be

This work was supported by Grant-in-Aid for Scientific Research on Priority Areas (C) "Genome Information Science" from the Ministry of Education, Culture, Sports, Science and Technology of Japan, and Japan Society for the Promotion of Science (JSPS) Research Fellowships for Young Scientists (#5304).

drastically improved, and all the possibilities can be checked, because most of the pattern matching problems have many parallelism in them. However, the cost of the systems are very expensive.

Field Programmable Gate Array (FPGA) is a reconfigurable device designed for rapid prototyping, and any kinds of circuits can be realized on the FPGA in a moment by downloading configuration data from host computers or dedicated memories. The performance is almost proportional to the size of the FPGA because the parallelism of computation is limited by the size. FPGAs are becoming larger and larger following Moore's low (the number of transistors in a fixed size (namely the size of FPGAs) become twice in every 18 months). We can easily obtain the latest/largest FPGAs in the form off-the-shelf PCI boards with FPGAs, which are now being shipped from many companies [7], and we can obtain many kinds of these boards at low cost. FPGAs begin to be used as accelerators in many application areas [8,9,10] and also used in some dedicated hardware systems [11,13] for bioinformatics.

We will show that we can achieve high performance in homology search by only adding one off-the-shelf FPGA board to a Pentium based computer system in use. The performance can be furthermore accelerated by using more number of FPGA boards. In our approach, the search is divided into two phases because the FPGAs do not have enough hardware resources for the pattern matching problems in bioinformatics. Therefore, different configuration data (namely different circuits) are downloaded from the host computer in each phase in order to make up for the limited hardware resources. The configuration data can be easily modified for new FPGA boards, because they are generated from the programs written in hardware description languages without assuming any special hardware resources on the FPGA boards.

This paper is organized as follows. Section 2 describes the overview of our approach, and the details of the approach is given in section 3. Then, experimental results of the approach are shown in section 4. In section 5, current status and future works are given.

2 Overview of the Approach

Our current target problems are homology search problems, and the Smith-Waterman algorithm[1] is used in all comparison between query sequences and database sequences. In this section, we describe the overview of our approach.

2.1 Hardware and Software for our Approach

We need the followings as components of the hardware platform (Figure 1).

1. one off-the-shelf FPGA board (with PCI bus interface), and
2. one host computer (a Pentium based computer, because driver programs for most FPGA boards run only under WINDOWS or LINUX).

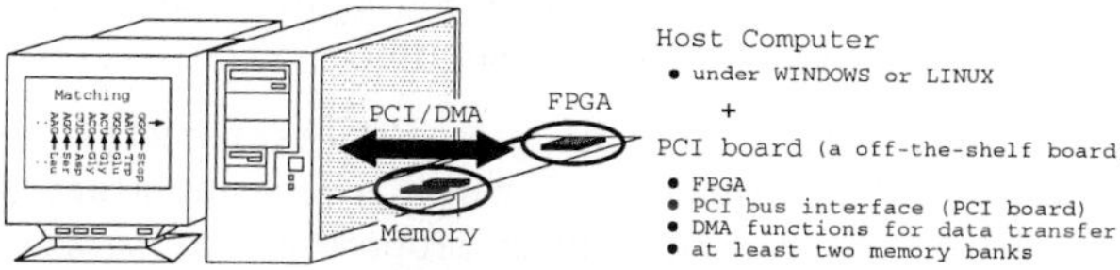

Figure 1: Required Components of Hardware Platform

The softwares which are necessary for our approach are

1. drivers programs to control FPGA boards from the host computer, which are developed by the board maker and attached to the FPGA boards, and
2. CAD tools for the FPGA, only if we need to modify the configuration data for new FPGA boards.

Among the components shown above, what we have developed are

1. the programs for the circuits which are implemented on the FPGA, and
2. interface programs which run on the host computer.

In our programs for the circuits, only the two memory banks to transfer data between the FPGA and the host computer are assumed. Most FPGA boards have at least two memory banks in order to receive data from the host computer while the FPGA is running using another memory bank. Therefore, configuration data for new FPGA boards can be easily generated by only changing some parameters in the programs (FPGA size, memory size in FPGA, I/O pin assignment and so on).

The interface programs need to control FPGA boards using the driver programs. The structure of driver programs depends on the boards, and we need to modify a part of the programs for each board.

2.2 Advantage and Disadvantage of the Approach

Before describing the details of our approach, we would like to summarize the advantage and disadvantage of the approach compared with the dedicated hardware systems.

First, the advantages of our approach are as follows.

1. Many kinds of FPGA boards are shipped from many companies[7], and the costs of the boards are relatively low. We can choose FPGA boards according to our requirements and budgets. For example, the cost of our largest FPGA board is several times the cost of a Pentium based desktop computer system, while the cost of PC card with a smaller FPGA is less than a half of the cost of a laptop computer.
2. It is easy to obtain the boards with latest FPGAs (namely larger FPGAs) as soon as the FPGAs are shipped, which is very important because the performance of the approach is almost proportional to the size of FPGAs.
3. It is possible to replace the FPGA board and the host computer independently.
4. By making the configuration data and the programs for the configuration data open, many users can accelerate their search by only purchasing one off-the-shelf FPGA board.

On the other hand, the disadvantages are as follows.

1. In general, off-the-shelf FPGA boards do not have enough hardware resources for homology search. Especially memory size and memory bandwidth are not sufficient. Furthermore, we assume only two memory banks on the board for data transfer between the host computer and the FPGA, and only the internal memory on the FPGA is used for the homology search in order to maintain portability of our circuits. Because of this limited memory size and memory bandwidth,
   (a) Query sequences can not be compared with long database sequences at once. Therefore, query sequences are always compared with subsequences of the database sequences (automatically divided during the search), and results against only the fragments in the subsequences can be shown (the length of the fragments can be specified by users).
   (b) Some parts of the database sequences are processed twice, and the size of the parts is almost proportional to the length of the query sequences. Therefore, the performance becomes worse as the query sequences becomes longer, though it is negligible if we can use large size FPGAs.
   (c) With smaller FPGAs, the length of query sequences is limited (long query sequences can not be processed). For example, with our PC card with one Virtex XCV300, the maximum length of the query sequence is 1024.

2. Software environment is still poor. Its improvement is one of the our future works.

2.3 *Scalability*

The performance is almost proportional to the number of FPGA boards when

1. the query sequence is compared with many database sequences which are stored in different hard disks, or

2. each database sequence is divided into subsequences which are stored in different hard disks,

because the database sequences or the subsequences can be compared with the query sequence independently. However, the data transfer rate of the PCI bus is limited, and many FPGA boards can not be attached to one host computer. We have not evaluated the performance when we use more than one FPGA board, but according to our estimation, more than two boards should not be attached to one host computer.

By connecting more hardware platforms by Ethernet, we can easily accelerate the performance furthermore. The performance is almost proportional to the number of hardware platforms. Then, the total performance can be comparable with large size dedicated hardware systems.

3 Details of the Approach

In this section, we describe the details of our approach. The features of our approach are :

1. multi-thread computation in order to achieve high performance, and

2. two phase search in order to make up for limited memory bandwidth.

3.1 *Parallel Processing of Dynamic Programming*

Before describing the details, we would like to introduce the dynamic programming algorithm. As shown in Figure 2, a query sequence and a database sequence are compared inserting gaps. Scores for each matching of the elements and inserting gaps are given by score matrices[2,3]. The computation order of pattern matching by dynamic programming is $m \times n$ when the length of the query sequence and database sequence are m and n respectively. Therefore, it is unrealistic to use dynamic programming algorithm against long database sequences on desktop computer systems.

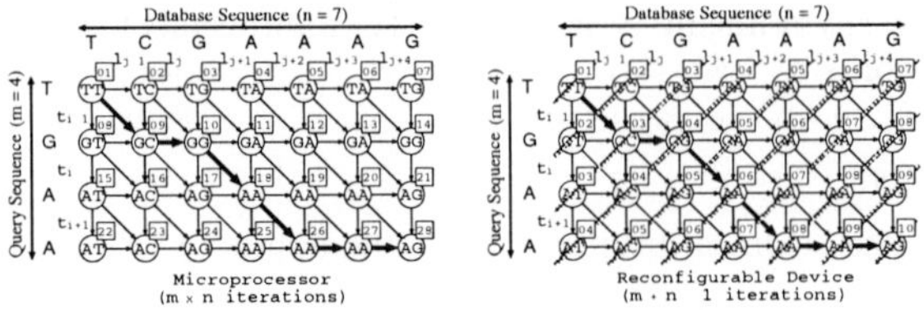

Figure 2: Parallel Processing of Dynamic Programming

With dedicated hardware systems, or reconfigurable devices such as FPGAs, we can process matching of elements in parallel. Figure 2 shows how the matching of the elements is processed in parallel. In the right-hand side part in Figure 2, elements on each diagonal line are processed at once. Therefore, the order of the computation can be reduced to $m+n-1$ from $m \times n$ if m elements can be processed in parallel. If the size of the hardware is not large enough to compare m elements at once, the first p elements (suppose that the hardware process p elements in parallel) of the query sequence are compared with the database sequence at once, and the scores of all p-th elements are stored in temporal memory. Then, the next p elements of the query sequence are compared with the database sequence using the scores stored in the temporal memory.

3.2 Structure of Processing Unit and Multi-thread Computation

Figure 3 shows a structure of our processing unit for dynamic programming. It consists of four stages, and takes four clock cycles to compute scores on each cell on the dynamic programming array (Figure 2). However, by overlapping the computation, we can start to compute the scores of elements on the next diagonal line (Figure 2) in every two clock cycles.

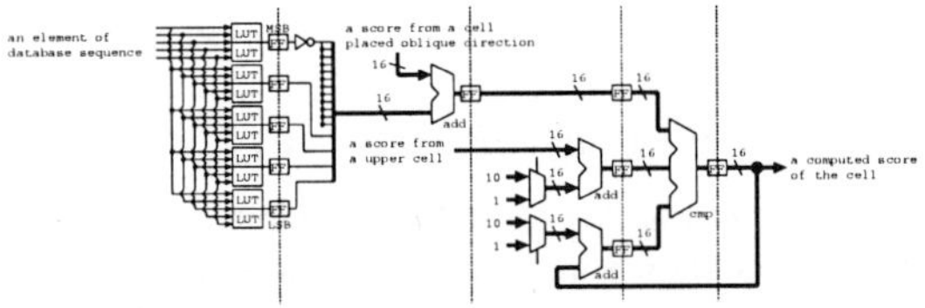

Figure 3: Implementation of a Processing Unit

Figure 4 shows how a database sequence whose length is n, is compared with a query sequence whose length is m. In the figure, each circle represents a processing unit. If the length of the query sequence (m) is not larger than the number of processing elements on the FPGA (p), the query sequence can be processed at once as shown in Figure 4. In this case, it takes $2 \times (m+n-1)$ cycles to compare the two sequences, because the processing units have to wait for one clock cycle to compare elements on the next diagonal lines as described above, which means that the units are idle for one clock cycle in every two clock cycles.

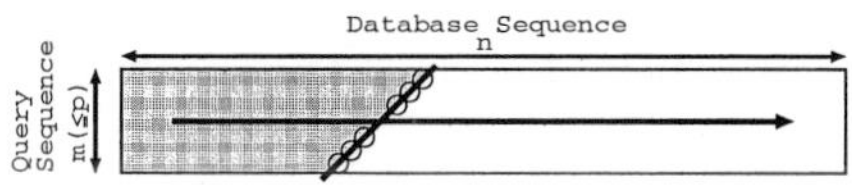

Figure 4: Sequential Execution of Dynamic Programming

Suppose that the length of the query sequence (m) is longer than the number of processing units on the FPGA (p). Then, in the naive approach :

1. first, the first p elements of the query sequence are compared and the intermediate results (all p-th scores on lower edge of upper half in Figure 5(a)) are stored, and
2. then, the next p elements of the query sequence are compared using the intermediate results.

In this case, it takes $2 \times 2 \times (p+n-1)$ cycles to compare the two sequences, and processing units become idle for one clock cycle in every two clock cycles as described above.

We can reduce the computation time by the multi-thread computation method. In the multi-thread computation :

1. first, p elements on the diagonal line in upper half in Figure 5(b) are processed, and the score of p-th element is stored on temporal registers, and
2. then, the next p elements on the diagonal line in lower half are processed without waiting for one clock cycle using the intermediate result.

By interleaving the processing of elements in upper half and lower half, we can eliminate the idle cycles of the processing elements. The clock cycles become $2 \times (p+n-1) + 2 \times p$, which is almost equal to $2 \times n$ because n is much longer than p in most cases.

When the length of the query sequence (m) is longer than twice the number of the processing units ($2p$), the multi-thread computation shown in Figure 5(b) is repeated according to the length of the query sequence. In this case, when the first $2p$ elements in the query sequence are processed, scores of all $2p$-th elements are stored in memories (all n scores are stored in total), and used for the computation of the next $2p$ elements.

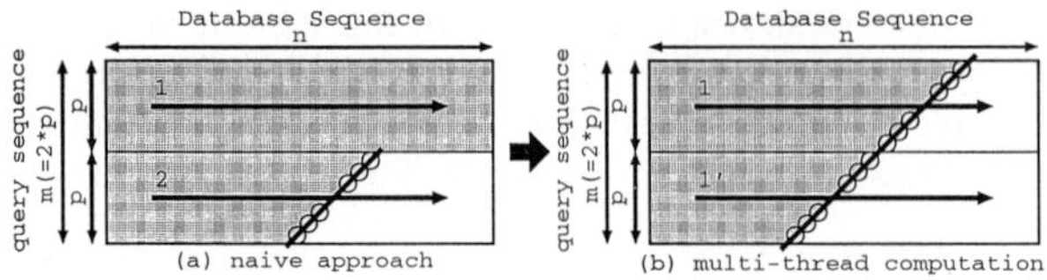

Figure 5: Multi-thread Execution of Dynamic Programming

3.3 Two Phase Search

First Phase

In the first phase, database sequences are divided into sub-sequences, because the size of the intermediate results described above is very large, and can not be stored in the internal memory of the FPGA at once.

Figure 6(a) shows how a long database sequence is compared with the query sequence. The database sequence is divided into sub-sequences of size s (s is decided based on the size of the internal memory of the FPGA). Then, each sub-sequence is compared with the query sequence by the multi-thread method. As shown in Figure 6(a), first, 1 and $1'$ are processed, and then 2 and $2'$ are processed.

In Figure 6(b), in each comparison with the sub-sequences, scores on upper edge (position α_e to α_m) are sampled and compared with scores on lower edge (position β_e to β_m). The score at position α_f is compared with the score at position β_m. The difference of the scores are stored in the two memory banks on the FPGA board, and then sent to the host computer. The host computer sorts the differences, and shows them with the positions on the database sequence. Thus, in our approach, the query sequence is compared with all fragments of the size $k \times l$, and the scores against each fragment by the Smith-Waterman algorithm are shown to the users. The interval of the sampling k and the distance between the two scores $k \times l$ can be specified by users. However, if the k is too small, many data have to be sent to the host computers, and the performance will go down. We assume that the length of the fragment ($k \times l$) is from twice to four times the query sequence.

In this division to the sub-sequences, some parts whose length is $k \times (l-1)$ are overlapped in order to compare the query sequence with all fragments of length $k \times l$ (Figure 6(a)). These parts are compared twice, and become the major overhead in the first phase. The length of the overlapped area ($k \times (l-1)$) is decided based on the length of the query sequence in general. Therefore, this overhead is almost proportional to the length of the query sequence. This overhead becomes larger as the length of the query sequence becomes longer,

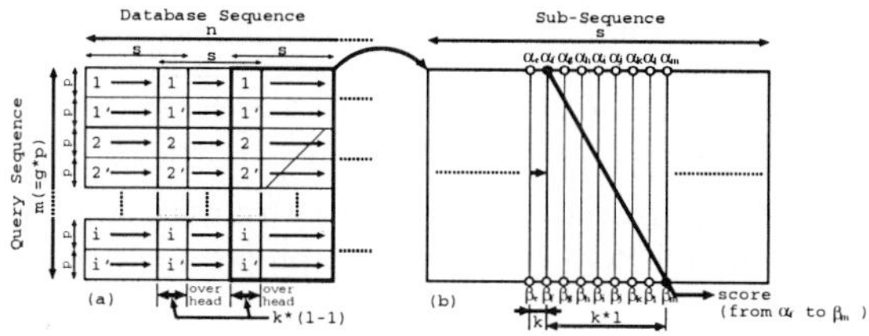

Figure 6: First Phase Execution

and becomes relatively larger as the size of the internal memory of the FPGA becomes smaller (namely the size of the FPGA becomes smaller).

In order to achieve higher performance in the first phase, we need to implement more processing units on the FPGA. The size of the processing unit is proportional to its data width. Therefore, we can implement more units by reducing the data width. However, with narrower data width, the scores may cause overflow or underflow during the comparison with the sub-sequences. In order to avoid the over/underflow, we need to make the size of the sub-sequences smaller too, but this means that more areas have to be overlapped. Therefore, we need to find good balance between the sub-sequence size and the parallelism. In the current FPGAs, the size of the internal memory is not so large, and it gives the best performance when we decide the data width base on only the size of the internal memory size of the FPGA.

Second Phase

In order to display optical alignments, we need to find the path from the upper left position to the lower right position which gives the best score as shown in Figure 2 (in the first phase, only the best score is computed, and all the information about the path is discarded during the computation). We need 2 bits for each cell on the array in Figure 2 to distinguish where the path comes from (from upper, upper left or left). Therefore, the number of elements which can be processed in parallel (namely the performance of the second phase) is decided by the FPGA's data width to the memory banks on the FPGA board, not by the size of the FPGA. If the width is $2 \times p$, p elements can be processed in parallel. In the second phase, only the information about the path is output, because the score is already obtained in the first phase.

However, the number of the fragments that we need to display their alignments, is not so many and the performance of this phase is not so important. If the length of the query sequence is less than a few thousand, we can obtain the optical alignments against one fragment within 1 sec by a desk computer.

4 Experiments

We have tested the performance of our approach under two environments, a desktop computer with a FPGA board and a laptop computer with a PC card.

4.1 Desktop Environment

One FPGA board (RC1000-PP by Celoxica) [14] is used to evaluate the performance on desktop environment. The board has four memory banks, and two of them are used for data transfer between the FPGA board and the host computer. The FPGA (Xilinx XCV2000E) on the board is one of the largest FPGAs that we can obtain now. We could implement 144 processing elements for the first phase of the homology search, and they run at 40 MHz. The size of the internal memory of the FPGA is 640 Kbits, and the length of the subsequence becomes 32768 elements. Therefore, the overhead caused by the overlapped area becomes about 5 - 10% when the length of the query sequence is 2048 and the size of the fragment is several times of the query sequence.

Figure 7 shows the relation between the time of the first phase and the length of the query sequence, when the length of the database sequence is 64 million. The slope of the search time becomes slightly larger as the length of the query sequence becomes larger, because the percentage of the overhead by the overlapped area will gradually increase. The speedup compared with a PentiumIII 1GHz under LINUX (kernel version 2.2.5 and gcc-2.91.66) is 327 times when the length of the query sequence is 2048.

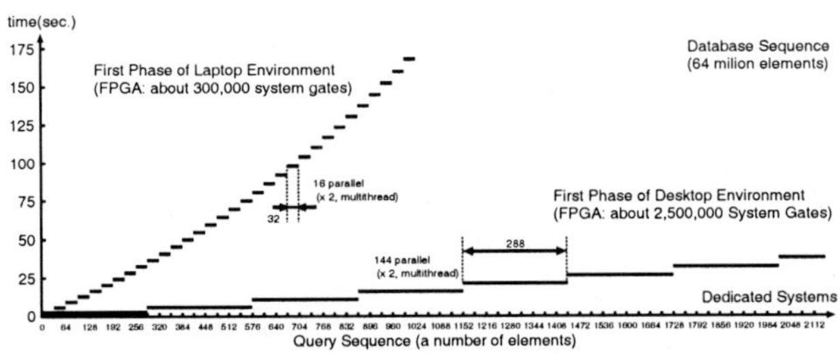

Figure 7: Comparison between Desktop and Laptop Environment

As for the second phase, we can process 32 elements in parallel, because we can write 64 bits to the memory banks on the board at once. The computation time for a query sequence of 2048 elements and a fragment of 8192 elements is about 13 msec. This is about 102 times faster than the PentiumIII 1GHz.

4.2 Laptop Environment

One PC card (Wildcard by Annapolis Micro Systems, Inc.)[15] with one FPGA (XCV300 by XILINX) is used to evaluate the performance on laptop environment. The PC card has two memory banks (32bits width and 256KB per each block), and these banks are used to transfer data between the PC card and the host computer. The size of the FPGA XCV300 is about one seventh of the FPGA XCV2000E.

In this case, we could implement 16 processing units on the FPGA, and they run at 40 MHz. The size of the internal memory of the FPGA is 64 Kbits, and the length of the subsequences becomes 4096 elements. The overhead caused by the overlapped area becomes about 25 to 50% when the length of the query sequence is 1024 and the size of the fragment must be several times of the query sequence. Therefore, we can not compare query sequences longer than 1024.

Figure 7 shows the relation between the time of the first phase and the length of the query sequence, when the length of the database sequence is 64 million. The slope of the search time becomes larger according to the size of the query sequence, because the percentage by the overhead for the overlapped area will increase. The computation time is about 30 times faster than a PentiumIII 1 GHz under LINUX (kernel version 2.2.5 and gcc-2.91.66) when the length of the query sequence is 1024.

As for the second phase, we can process 16 elements in parallel, because we can write 32 bits to the memory bank on the card at once. The computation time for a query sequence of 1024 elements and a fragment of 4096 elements is about 7 msec. This is about 50 times faster than the PentiumIII 1GHz.

5 Current Status and Future Works

We have developed the circuits for homology search, and showed that we can achieve high performance using off-the-shelf FPGA boards. The performance is almost comparable with small to middle class dedicated hardware systems when we use one board with one of the latest FPGAs (Xilinx XCV2000E). The time for comparing a query sequence of 2048 elements with a database sequence of 64 million elements by the Smith-Waterman algorithm is about 34 sec, which is about 330 times faster than a desktop computer with a 1GHz PentiumIII. We can also accelerate the performance of a laptop computer using one PC card with one FPGA (Xilinx XCV300). The time for comparing a query sequence (1024) with database sequence (64 million) is about 185 sec, which is about 30 times faster than the desktop computer.

We are now evaluating the performance for the translated nucleotides. When we need to translate the sequences during the comparison, the size of each unit on the FPGA becomes about 10% larger and the parallelism in the first phase will go down to 120 from 144 (about 20% performance down). We are now improving the circuits of the unit to achieve higher performance.

Some parts of the programs for the homology search are still under development, and we also need to improve other parts. We are also developing softwares for parallel processing of the homology search with more number of pairs of FPGAs and host computers connected by Ethernet.

We are also planning to accelerate other pattern matching problems in bioinformatics with FPGAs.

References

1. Smith T. F. and Waterman M. S.: *Identification of Common Molecular Subsequences*, Journal of Molecular Biology 147, 195-197, (1981).
2. Steven Henikoff and Jorja G. Henikoff: *Amino Acid Substitution Matrices from Protein Blocks*, Proc. Natl. Acad. Sci. 89, 10915-10919, (1992).
3. Jones,D.T., Taylor,W.R. and Thornton,J.M.: *The Rapid Generation of Mutation Data Matrices from Protein Sequences*, CABIOS 8, 275-282, (1992).
4. Altschul,S.F., Gish,W., Miller,W., Myers,E.W. and Lipman,D.J.: *Basic Local Alignment Search Tool*, J. Mol. Biol. 215, 403-410, (1990).
5. Pearson,W.R. and Lipman,D.J.: *FASTA: Improved tools for biological sequence comparison*, Proc. Natl. Acad. Sci. USA 85, 2444-2448, (1988).
6. Altschul,S.F., Madden,T.L., Schaffer,A.A., Zhang,J., Zhang,Z., Miller,W., and Lipman,D.J.: *Gapped BLAST and PSI-BLAST: a new generation of protein database search programs*, Nucl. Aci. Res. 25(17), 3389-3402, (1997).
7. http://www.optimagic.com/boards.html
8. http://www.fccm.org
   (IEEE Symposium on Field-Programmable Custom Computing Machines)
9. http://www.ecs.umass.edu/ece/fpga2002/index.html
   (ACM International Symposium on Field-Programmable Gate Arrays)
10. http://xputers.informatik.uni-kl.de/fpl/index_fpl.html
    (the International Conference on Field-Programmable Logic and Applications)
11. http://www.compugen.com
12. http://www.paracel.com/index.html
13. http://www.timelogic.com
14. http://www.celoxica.com
15. http://www.annapmicro.com
16. http://www.xilinx.com

EXPANDING PROTEOMICS TO GLYCOBIOLOGY: BIOCOMPUTING APPROACHES UNDERSTANDING THE FUNCTION OF SUGAR

C.-W. VON DER LIETH

DKFZ (German Cancer Research Centre), Molecular Modelling, INF 280, 69120 Heidelberg, Germany, {a.bohne,w.vonderlieth}@dkfz.de

The recognition of complex carbohydrates and glycoconjugates as mediators of important biological processes has stimulated investigations into the understanding of the underlying principles. Unfortunately, the rate of generating new information has been slow during the last decade. Carbohydrates differ from the two other classes of biological macromolecules (proteins and DNA/RNA) in two important characteristics: their residues can be connected by many different linkage types and they can form highly branched molecules. As a consequence, carbohydrate chains contain an evolutionary potential of information content, which is several order of magnitude higher in a short sequence than any other oligomer formed by nucleotides or amino acids. This structural variance allows oligosaccharides to encode information for specific molecular recognition and to serve as determinants of protein folding and stability. However, structural complexity is also one of the major barriers for a rapid progress in the field of glycobiology because carbohydrates are laborious to analyze and extremely difficult to synthesize. Knowing that glycosylation of proteins and lipids are the most ubiquitous forms of posttranslational modification, the unexpectedly small number of genes identified in the initial analysis of the human genome sequence provides even more efforts for understanding the biological roles of oligosaccharides.

Whereas the conformational behavior of oligosaccharides has been investigated intensively both by experimental techniques (mainly NMR) as well as various theoretical methods their study (function and structure) has been widely neglected so far in the area of bioinformatics. The utilization of proteomics databases has become indispensable for the daily work of the molecular biologist, but this situation has not been reached for carbohydrate applications yet. Several new and useful applications for the glycosciences (see http://www.dkfz.de/spec/links/glyco_list. html) have appeared on the web during the last few years. Unfortunately, existing data collections are only rarely annotated and cross-linked to other resources. The need to develop and maintain Internet based, databases for carbohydrate structures, carbohydrate binding proteins and carbohydrate active enzymes has recently been emphasized by the Consortium for Functional Glycomics (see http://glycomics.scripps.edu/)

The paper of Cooper, Harrison, Webster, Wilkins and Packer points out the need for the standardization of the entries in glyco-databases. GlycoSuiteDB is an annotated database of glycans structures designed to provide rapid access to information on protein glycosylation. The paper describes how the glycan structure representation is normalized to provide consistency and how to enable different searching criteria. Currently, no generally accepted linear, canonical description for carbohydrates exists. Such a code, which can be processed by computers with ease, will enable efficient automatic cross-linking of distributed carbohydrate data collections by serving as a unique and unambiguous database access key. One of the major points to bring up for discussion during this session will probably be the question how to come to an agreement about a generally accepted linear, canonical description for carbohydrates.

Glycosylation may influence considerably the physicochemical properties and function of a protein. Since the experimental determination of glycosylated sites is difficult to archive - the percentage of annotated glycoprotein entries in SWISS-PROT is still low - the need for developing theoretical approaches to predict glycosylation potential of sequons is obvious. The group of S. Brunak has developed methods to predict glycosylation sites using artificial neural networks that examine correlations in the local amino acid sequence context and their surface accessibility. Here Gupta and Brunak analyze the data available in SWISS-PROT of human proteins with respect to certain functional categories. The proteins are clearly classified highlighting that the N-glycosylation occurs mainly in the 'transport and binding' category of proteins. The authors show that the glycosylation is one of the most important determinants for functional classification of proteins and has definitively taken into account when deciphering protein function and characterizing complete proteomes.

The question if glycans attached to proteins and lipids do show certain secondary or tertiary structural motifs still remains to be answered due to the lack of sufficient crystallographic data. The paper of Bohne and von der Lieth claims that the spatial structure of glycans provides the driving force for many intermolecular interactions and thus predetermines their functions. Moreover, the authors emphasize that flexibility and dynamics of glycans may play a key role in their biological activity and must also be taken into account. The paper describes a new computational method to explore the conformational space of N-glycans. Since the approach is very fast it is well suited for a web-based application.

As already mentioned the treatment of complex carbohydrates and glycoconjugates has so far been no major topic in the area of bioinformatics. The small number of only five papers submitted for this session reflects this situation. Moreover, it turned out to be rather difficult to encourage glycoscientists, who are not used to submit a full paper six months before the conference takes place, to contribute. Nevertheless, I am convinced that the subject is hot and that it will attract more attention in the near future. Since the three selected contributions address scientific questions with emerging interest where much research and development remains to be carried out, they provide an excellent basis for stimulating discussions.

GLYCOSYLATION OF PROTEINS: A COMPUTER BASED METHOD FOR THE RAPID EXPLORATION OF COMFORMATIONAL SPACE OF N-GLYCANS

A BOHNE, C.-W. VON DER LIETH
DKFZ (German Cancer Research Centre), Molecular Modelling, INF 280, 69120 Heidelberg, Germany, {a.bohne,w.vonderlieth}@dkfz.de

Inspection of protein databases suggests that as many as 70% of proteins have potential N-glycosylation sites. Unfortunately glycoproteins often refuse to crystallize and NMR techniques do not allow an unambiguous determination of the complete conformation of the sugar part. Therefore, time-consuming complex simulation methods are often used to explore the conformational space of N-glycans. The generation of a comprehensive data base describing the conformational space of larger fragments of N-glycans taking into account the effects of branching is presented. High-temperature molecular dynamics simulations of essential N-glycan fragments are performed until conformational equilibrium has been reached. Free energy landscapes are calculated for each glycosidic linkage. All possible conformations for each N-glycan fragment are automatically assigned, ranked according to their relative population and stored in a database. These values are recalled for the generation of a complete set of all possible conformations for a given N-glycan topology. The constructed conformations are ranked according to their energy content. Since this approach allows to explore the complete conformational space of a given N-glycan within a few minutes of CPU-time on a standard PC, it is well suited to be used as a Web-Based application.

1 Introduction

Glycosylation is one of the most abundant forms of covalent protein and lipid modification. [1] There are two main types of protein glycosylation: N-glycosylation, in which the oligosaccharide is attached to an asparagine residue, and O-glycosylation, in which the oligosaccharide is attached to a serine or threonine residue. Glycoproteins usually exist as complex mixtures of glycosylated variants (glycoforms). Glycosylation occurs in the endoplasmic reticulum (ER) and Golgi compartments of the cell and involves a complex series of reactions catalyzed by membrane-bound glycosyltransferases and glycosidases. Many of these enzymes are exquisitely sensitive to other events taking place within the cell in which the glycoprotein is expressed. The populations of sugars attached to an individual protein will therefore depend on the cell type in which the glycoprotein is expressed and on the physiological status of the cell, and may be developmentally and disease regulated [2,3].

Inspection of protein databases suggests that as many as 70% of proteins have potential N-glycosylation sites (sequence motif ASN-X-[SER/THR] where X is not PRO) [4] The biological function of glycosylation is still not completely understood [1,5]. However, it is clear that glycoproteins are fundamental to many biological processes including fertilisation, immune defence, viral replications, parasitic infection, cell growth, cell-cell adhesion and inflammation. Glycoproteins and glycolipids are major components of the outer surface of mammalian cells. They represent key structures for the interaction of cells with toxins, viruses, bacteria, antibodies and micro-organisms. The spatial structures of glycans provide the driving force for many intermolecular interactions and thus predetermine their function. Their flexibility and dynamics frequently play a key role in biological activity and must be taken into account.

Glycoproteins often refuse to crystallise [6,7]. The conformational flexibility of the glycan antennae at the surface of the protein obviously hampers crystal growth. In cases where glycoproteins do crystallise, the electron density is effected by high thermal motion of the glycan moiety: the detectable electron density is so low that no defined spatial arrangement can be assigned. For example, despite the existence of a well-resolved X-ray crystal structure of the enzyme ribonuclease B, the poor definition of the electron density associated with the oligosaccharide has prohibited any determination of the N-glycan conformation [1,8]. The current version of the Protein Database [9] contains about 500 glycosylated proteins. Usually only the co-ordinates of the rigid core region of N-glycans are available [7,10]. The question, if complex oligosaccharides do exhibit certain secondary or tertiary structural motifs, still remains to be answered due to the lack of crystallographic data on these molecules.

NMR techniques have been widely used in the conformational analysis of N-glycans [11-14]. However, an unambiguous determination of the complete conformational space is hampered by two effects; First, NMR-derived geometric constraints represent an average value based on conformations for flexible oligosaccharides; Second, only a few (one to three) contacts across a given glycosidic linkage can often be detected. To overcome this drawback, a combination of theoretical and experimental methods is applied to elucidate the dynamic behaviour of flexible N-glycans. This includes complex and time-consuming force field based modeling protocols like molecular dynamics simulations (MD) or Monte Carlo (MC) approaches to explore the conformational space accessible to various glycosidic linkages. Calculations resulting in systematic conformational energy maps are feasible to perform for di- and trisaccharides. A specialised collection of conformational profiles of disaccharides does exists and can be used to construct conformations of complex N-acetyllactosaminic type N-Glycans [15]. Two to seven possible conformations for disaccharides have been described for each glycosidic linkage. An exhaustive generation of all possible conformation of a middle-sized N-glycan possessing eleven residues would easily result in the evaluation of several thousand possible conformations. However, this number would be reduced as it is well known that the extent of branching influences the flexibility of N-Glycans in such a way that the majority of highly-branched glycosidic linkages populate only a restricted conformational area which is generally smaller than that observed in N-linked glycans with smaller numbers of antennae [16].

The aim of this study is a) to derive a comprehensive data base describing the conformational space of larger fragments of N-glycans taking into account the effects of branching, b) to automatically generate a complete set of all possible conformations for a given N-glycan and c) to rank the constructed conformations according to their strain energy. This should allow us to explore the complete conformational space of N-glycans rapidly and add this tool as additional option to our Web-based real-time online application services (see http://www.dkfz.de/spec/)

2 Materials and Methods

The current version of the Complex Carbohydrate Structural Database (known as CarbBank), [17] which is probably the most comprehensive data base of complex carbohydrates, contains about 2000 structurally different N-glycan entries. An analysis revealed that about 225 different tri-, tetra and pentasaccharide fragments are needed to construct all N-glycan structures contained in CarbBank.

The generation of a data base which can be used to enable a rapid exploration of the conformational space accessible to N-glycans can be accomplished in six steps.

Step 1: Extract all N-glycans from CarbBank to create a comprehensive set of all known substructures (CarbBank contains about 2000 N-glycans)
Step 2: Divide all N-glycans into meaningful Tri-, Tetra- and Pentasaccharide substructures. The result of this division is a set of 225 basic fragments which can describe all other N-glycans.
Step 3: Convert the 225 topology data sets of the fragments into 3D structures using the program Sweet II [18].
Step 4: Perform high temperature molecular dynamics simulations until conformational equilibrium has been reached
Step 5: Automatic assignment of conformations for each glycosidic linkage. Rank all possible conformations for each fragment according to their degrees of population.
Step 6: Store Φ,Ψ and ω and their relative populations for each fragment.

2.1 Topology of N-glycans

Complex Carbohydrates differ from the two other classes of biological macromolecules (proteins and DNA) in two important characteristics: their residues (monosaccharide units) can be linked by many different linkage types and they can form highly branched molecules. N-linked glycans, which can be classified structurally into four main types typically consist of 11 to 25 residues exhibiting multiantennary structures (see Figure 1a). Normally, the number of antennae varies between two and five. To take into account the degree of branching we

```
a-D-Manp-(1-2)-a-D-Manp-(1-6)+
```

```
                                     |
                          a-D-Manp-(1-6)+
                                     |       |
           a-D-Manp-(1-2)-a-D-Manp(1-3)+   b-D-Manp-(1-4)-b-D-GlcpNAc-(1-4)-b-D-GlcpNAc-
(1-4)-Asn
                                             |
a-D-Manp-(1-2)-a-D-Manp-(1-2)-a-D-Manp-(1-3)+
```

Figure 1a. The Primary structure of the carbohydrate chain of soybean agglutinin (SBA): a typical oligomannose N-glycan with four antennae. Nomenclature of carbohydrates as used in CarbBank is displayed. (Manp = mannose in its pyranose form , GlcpNAc= Glucosamine). Instead of the commonly used α and β, a and b are used to define the stereochemistry of the anomeric C atom.

```
        [1]-[2]-+
                |
               [3]---+
                |    |
        [5]-[4]-+   [6]-[10]-[11]-[12]
                     |
         [9]-[8]-[7]-+
```

Figure 1b: To simplify its annotation, a more schematic representation of SBA is used throughout the rest of this paper: the residues of SBA are indicated by numbers.

a) Branching points (Tetrasaccharides)

```
a-D-Manp-(1-6)+                                [2]
              |                                  |
     a-D-Manp-(1-6)-b-D-Manp                    [3]-[6]
              |                                 |
a-D-Manp-(1-3)+                                [4]
---------------------------------------------------------------------
---
a-D-Manp-(1-6)+                                [3]
              |                                 |
     b-D-Manp-(1-4)-a-D-GlcpNAc                [6]-[10]
              |                                 |
a-D-Manp-(1-3)+                                [7]
```

b) Linear fragments (Trisaccharides)

```
a-D-Manp-(1-2)-a-D-Manp-(1-6)-a-D-Manp       [1,2,3]
a-D-Manp-(1-2)-a-D-Manp-(1-3)-a-D-Manp       [5,4,3]
a-D-Manp-(1-2)-a-D-Manp-(1-3)-b-D-Manp       [8,7,6]
a-D-Manp-(1-2)-a-D-Manp-(1-2)-a-D-Manp       [9,8,7]
b-D-Manp-(1-4)-b-D-GlcNAc-(1-4)-b-D-GlcpNAc  [6,10,11]
b-D-GlcNAc-(1-4)-b-D-GlcpNAc-(1-4)-Asn       [10,11,12]
```

Figure 1c) Saccharide fragments(CarbBank nomenclature left side, simplified annotation right side) into which the carbohydrate chain of SBA is divided. a) two tetrasaccharides b) five different trisaccharides.

```
          b-D-GlcpNAc-(1-6)+
                           |
   b-D-GlcpNAc-(1-4)-a-D-Manp-(1-6)-a-D-Manp
                           |
              a-GlcpNAc-(1-3)+
```

Figure 1d) Example of a complex branching point (pentasaccharides, not contained in carbohydrate chain of SBA) as can be found in complex N-glycans.

decided to define three topological different types of fragments, a) linear chains (trisaccharides), b) simple branching points (tetrasaccharides) and complex branching points (pentasaccharides) (see Figure 1b). The fragments are assigned in such a way that one residue is always present in two connected fragments.

2.2. The use of Molecular Dynamics Simulation to create free energy landscapes

Computational approaches such as systematic conformational searches, Monte Carlo calculations and molecular dynamics (MD) simulations have been used to explore the conformational space accessible to biological macromolecules [13]. Most of the published methods are based on force-field calculations and produce energy maps which only reflect enthalpy effects and ignore entropic contributions. However, evaluation of both terms including the effects of the aqueous surroundings is necessary to produce reliable free energy landscapes. It is well known, in principle, that with a sufficiently long MD simulation it should be possible to produce a free energy map of a molecular ensemble. However, only recently have technical resources (computation speed, disk storage space) become available which can allow simulations to be continued until equilibrium between all populated conformations is reached. Now that such lengthy simulations are feasible, convenient criteria are required for judging when conformational equilibrium among several conformations has been achieved. Therefore, we have used an empirical algorithm developed by Martin Frank [19] which automatically calculates a parameter called $EQ(t)$ at time point t from the available MD trajectory data. The value of $EQ(t)$ can be used to define if and when any given MD simulation has approached conformational equilibrium (within a desired degree of accuracy).

It has been shown that high temperature MD-simulations (HTMD) of saccharides produce very similar energy landscapes for glycosidic linkages as long-time simulations at room temperature. However, HTMD simulations require significantly shorter time to reach conformational equilibrium than the standard simulations [19]. Such HTMD simulations allowed us to finalise the conformational analysis for all 225 N-glycan fragments on reasonable time scale, resulting thus in important reduction of required computational resources.

The applied simulation protocol was as follows: The SWEET-II environment [18] was used to generate the initial 3D structures of the fragments. The MM3-TINKER force field was used for energy calculation. The influence of the aqueous solvent environment was also considered, based on the GB/SA water model. Additional tethering force was applied during the HTMD simulations in order to avoid puckering of the pyranose ring atoms. The simulation parameters were set to: temperature 1000 K, integration step 10^{-15} s; the lengths of simulations were in the range of several nano-seconds (10^{-9}s) . Two molecular co-ordinates were stored per pico second (10^{-12} s). These structures were subsequently used to perform the population analysis. The TINKER software package [20] was used to perform MD simulations.

2.3 Characterising the conformational space of N-glycan fragments

Glycosidic bonds form the most flexible part of complex carbohydrates (see Figure 2). For the conformational analysis of individual glycosidic linkages most often Ramachandran-type profiles have been used to describe the effects of rotation around the glycosidic bond **Φ,Ψ** co-ordinates. **Φ,Ψ** torsion angles are calculated for each glycosidic linkage and all stored snapshots.

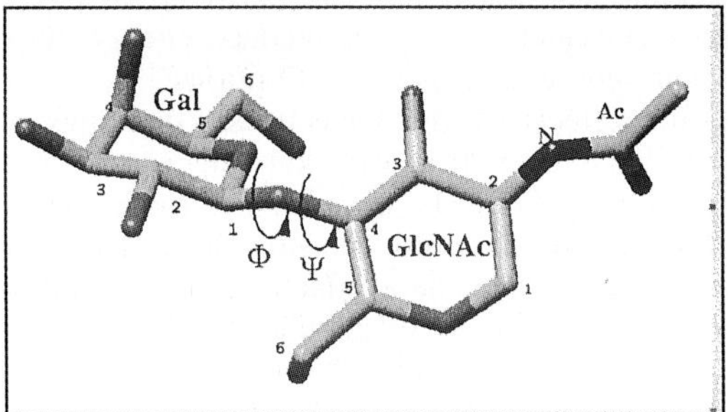

Figure 2: 3D structure of α–D-Galactose (Gal) 1-4 linked to Glucosamine (GlcNAc). Hydrogen atoms are omitted The determination of conformational preferences of oligosaccharides is best approached by describing their preferred conformations of the glycosidic linkage Φ,Ψ torsion angles

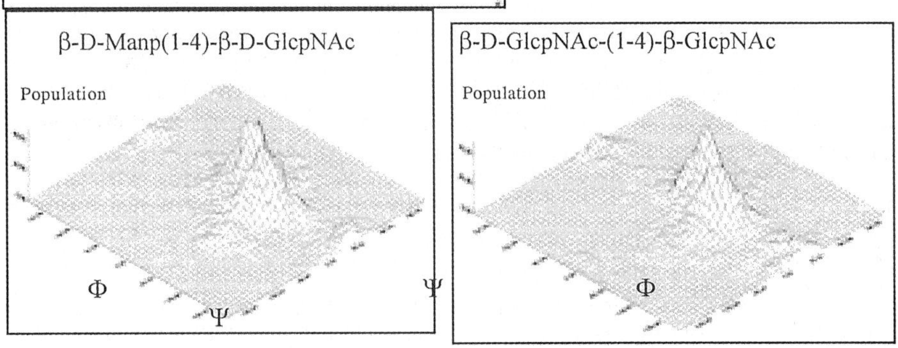

Figure 3: Population maps for both glycosidic linkages of the trisaccharide of β-D-Manp(1-4)-β-D-GlcpNAc-(1-4)-β–GlcpNAc generated from a high temperature molecular dynamics simulations.

In a first step the **Φ,Ψ** torsion angles are analysed separately for each glycosidic linkage. Since we continued HTMD simulations until equilibrium between all populated conformations was reached, we were allowed to assign conformations based on the populations of Φ,Ψ values (Figure 3). This procedure was accomplished in two steps. First, conformations were assigned separately for each individual glycosidic linkage. Therefore, areas showing a maximum in population density had to be identified. Normally, several maxima exist for each glycosidic linkage. Areas belonging to one conformation are encircled around each population maximum. The circled areas are not allowed to overlap in order to maximise the number of assigned values. Figure 4 illustrates how areas defining one conformation were assigned. Each selected conformation is then characterised by the Φ,Ψ values of the centre of the encircled area and its relative population for each glycosidic linkage

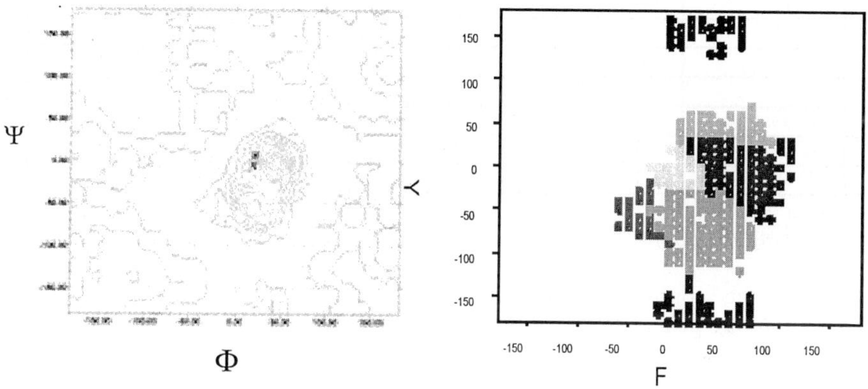

Figure 4: Assignment of areas defining the conformations of the β-D-GlcNAc-(1-4)-β-D-GlcNAc linkage. The analysis is based on the population of the Φ,Ψ space (see Figure 3) derived from the high temperature MD simulation of the trisaccharide β-D-Man-(1-4)-β-D-GlcNAc-(1-4)-β-GlcNAc. The maximum population together with its relative population is stored in the data base for each conformation. The assignment procedure works as follows: starting from maximal values, concentric rings are encircled in such a way that areas are not allowed to overlap and that the number of assigned values becomes maximal.

In a subsequent step the relative population for all existing combinations of conformations taking into account all glycosidic linkages of one fragment (e.g. a pentasaccharide has four glycosidic linkages) is calculated based on assignments for the individual glycosidic linkages. Figure 3 depicts the population of the space for both glycosidic torsion angles of the trisaccharide fragment β-D-Man-(1-4)-β-D-GlcNAc-(1-4)-β-GlcNAc. The automatic assignment procedure found six conformations for the β-D-Man-(1-4)-β-D-GlcNAc glycosidic linkage(Φ/Ψ: 24° /-39°; 38°/4°, 48°/-171°, -42°/-50°, 72°/-155°, 18°/177°) and six for β-D-GlcNAc-(1-4)-β-GlcNAc (Φ/Ψ: 56°/8°; 11°/-39°, 19° /-5°, 33°/173°,-35°/-44° and -56°/50°). The maximum number of possible conformation in this case is 6x6 = 36. The population of all twenty-four conformations is calculated, ranked in descending order and stored. Table 1 shows the combination of values ordered by their relative population as stored in the data base. These data are recalled to exhaustively calculate all possible conformations of larger N-glycan conformations.

Table 1: Relative population and Φ,Ψ value for both glycosidic linkages for the 7 conformations of β-D-Man-(1-4)-β-D-GlcNAc-(1-4)-β-GlcNAc [6-10-11] showing more than 3% population are given.

No	Percent.	Φ [10–11]	Ψ [10–11]	Φ[6–10]	Ψ [6–10]
1	18.43	56	8	24	-39
2	17.76	11	-39	24	-39
3	17.25	56	8	38	4
4	15.80	11	-39	38	4
5	3.53	33	173	38	4
6	3.36	33	173	24	-39
7	3.14	19	-5	24	-39

2.4. Exploring the conformational space of N-glycans

To explore the conformational space of N-glycans an ensemble of possible conformations are generated for a given N-glycan. This task is accomplished in five steps.

Step 1: The structure, which has to be input using the condensed format for oligosaccharides, is divided into its basic structural fragments as described above.

Step 2: An initial 3D structure is generated using SWEET-II. The corresponding Φ,Ψ,ω values and their ranking are recalled form the data base for each fragment.

Step 3: Starting with the largest fragment contained, all possible conformations are constructed setting the glycosidic angles to the recalled Φ,Ψ,ω values one by one according to their ranking. Since the fragments were generated in such a way that always one residue overlaps with the previous fragment, the Φ,Ψ,ω values and the ranking of the already assigned fragments is maintained.

Step 4: The geometry of all generated conformations is relaxed applying a small number of molecular mechanics steps. The resulting force field energy is used to rank the computed structures. A hit list of the low energy structures is displayed.

Step 5: Since the number of generated conformations can be in the order of several thousand, it is not meaningful to perform a complete optimisation for all structures. Therefore, the user is given the possibility to define the number of best hits to be optimised applying a complete optimisation using the MM3 force field.

2.5. Implementation

MD simulations were performed on various high performance computing facilities (IBM-SP2 and HP V class machines) using the MM3 force field implemented in the TINKER software. Analysis of the various MD-trajectories and the generation of the N-glycan structures to explore their conformational space were accomplished using our own software running on a standard PC under Linux. Currently the software is installed on a Pentium 250 MHz PC. Generation of about 1000 conformations and evaluation of the energy content for each conformation takes less than two minutes of CPU time for an N-glycan consisting of eleven residues. Most of the required time is used to relax the large number of generated conformations.

3. Results and Discussion

Low energy conformations as a list of its glycosidic torsion angles Φ,Ψ,ω for some typical N-Glycan can be looked up at http://www.dkfz.de/spec/glydict/ and ordered according to their force field energy content. Here we discuss sugar chain attached to SBA. Overall more than 1000 conformations were generated and ranked according to their energy content. An examination of the glycosidic torsion angles reveals that several energetically similar conformations exist where only the orientation of one linkage has been changed. From X-ray and NMR measurements it is well known that only one conformation dominates the linkages of the core residues close to the protein attachment site. This finding is reflected in our approach for the preferred orientation of the [6-10], [10-11] and [11-12] linkage where only one conformation is present. The predicted torsion angles for the [6-10] (Φ/Ψ =19/-25) and the [10-11] (Φ/Ψ =56/8) linkage have values which are close to experimental data. A statistical analysis of the available X-ray data for glycoproteins as found in the PDB-files showed the following distribution: β-D-GlcNAc-(1-4)-GlcNAc [6-10] (Φ = 47±8; Ψ = -4 ±16). β-D-Man-(1-4) –β-D-GlcNAc [10-11] (Φ = 32±11 ; Ψ = -13 ±20).

A comparison of the preferred orientations of glycosidic linkages found in N-glycan structures as revealed by different experimental and theoretical methods is shown in Table 2. It is well known that complex carbohydrates are rather flexible molecules and therefore often refuse to crystallise. Moreover, in cases when crystals are available, often only for the atoms of the rigid core region have sufficient electron density to be measured making an unambiguous assignment of the conformation of the complete carbohydrate chain impossible. Therefore, only a limited number of X-ray structures and glycosidic linkages exist which can be used for comparison

Table 2: Comparison of generated torsion angles for various glycosidic linkages as found N-glycans with experimental data. Torsion angles are given in degrees.

Method	**systematic search**	**MD und NMR**	**MD and NMR**	**X-Ray**	**Knowledge base**
source	Imberty [7]	Woods [16]	Homans [21-23]	Petrescu[10]	
force field	MM2 Carb	Amber-Glycam	Amber-Homan		MM3 + GB/SA
β-GalNAc (1-4) β-GlcNAc	40/10 35/15 30/-55	36±15/ -33±23	0/-33 54/-2 27(25)/-17/(24) 34(27)/-14(22)	47±8/-4±16	56/8 11/-39 33/173
β-D-Man (1-4) β-GlcNAc	45/10 35/15 30/-55	46±13/-10±16	15/-27 15/-5 13(26)/-16(20) 25(30)/-7(18)	32±11/-13±20	32±11/-13±20
α-D-Man (1-3) α–D-Man	-35/55 -50/-20 30/30	-45/30 -45±10/0±50	-20/30 -20/50 -40(15)/12(27) -38(17)/ 2(23)	-48±11/6±22	-26/47 -39/-25 -55/64 -53/38
α-D-Manp (1-6) α-D-Manp	-50/-170/? -45/100/? -50/-170/? -45/90/?	- 35±15/180±10/? 40+-/10/145+-15/	-2/81/179 8/59/-52 -8(27)/95/31)/?	-55±9/182±5/ -174±10 -54±11/180±15/-55±15 -53±14/109±13//-37±22	-52/-179[1]/-45 -44/-157/-43
α-D-Man (1-2) α–D-Man	-30/60 -50/-20 30/30	-39±9/56±14 -42±9/55±14 -39±9/52±22 -46±8/22±14	-40/-20 -40/0	-58±8/-55±10 -49±8/16±15	35/36 -33/-3 -45/-11 -54/-37 -6/46 -38/-27

with theoretically generated structures. Complex simulation protocols based on force-field calculations like molecular dynamics simulations (MD) in combination with MR-derived constraints are intensively used to explore the conformational space accessible for various glycosidic linkages. All approaches support the fact that complex carbohydrates are rather flexible molecules so that several conformations for each glycosidic linkage can be populated. Nevertheless, the number of conformations found and their exact localisation varies considerably between different methods. An exact correspondence of Φ,Ψ values calculated with different force fields and varying computational approaches cannot be expected. As pointed out by Rasmussen and Fabricius, a minimum in a two-dimensional conformational map may represent an entire family of points in multidimensional space. Therefore, a "difference in Φ,Ψ of 10°, 10° is really no difference at all" [24]. Comparing the Φ,Ψ values generated with our approach with those found in crystals we find at least one conformation which probably populates the same conformation as found in crystal structures. The intention of our approach is not compete with more complex and time consuming simulation techniques which include experimental constraints and explicit solvent molecules. Nevertheless, the ensemble of generated N-glycan conformations represents a realistic description of the conformational space accessible to such molecules.

The presented approach is able to generate a comprehensive set of all possible conformations for a given N-glycan and to rank them according to their energy content. It is very efficient an is therefore well suited to be used as web-based application. We have established an internet interface (http://www.dkfz.de/spec/glydict/), which enables the user to input an N-glycan of interest and to receive an ensemble of possible conformations within a few minutes. The results are distributed via E-mail when the procedure has finished. For each conformation co-ordinates in PDB-format can be downloaded, displayed and analysed locally using 3D visualisation tools like RasMol, Chime or WebMolecule.

The current version of our approach is designed to explore the conformational space for N-glycans. Other types of complex carbohydrate structures like O-glycans or lipopolysaccharides can be as well handled by this approach: only the basic data base has to be expanded with fragments contained in these structures.

The work presented is part of a larger project [25] trying to establish bioinformatics tools for complex carbohydrate structures and to cross-reference sugar structures with existing data collections in the proteomics field.

References

[1] R. Dwek. "Glycobiology: Toward Understanding the Function of Sugar" *Chem. Rev* **96**, 683-720 (1996)

[2] P. Gagneux; A. Varki. "Evolutionary considerations in relating oligosaccharide diversity to biological function" *Glycobiology* **9**, 747-55 (1999)

[3] N. Sharon; H. Lis. "Carbohydrates and Cell Recognition" *Scientific Am.* **268**, 82-89 (1993)

[4] M. Wormald; R. Dwek. "Glycoproteins: glycan presentation and protein-fold stability" *Structure Fold Des.* **15**, R155-60. (1999)

[5] A. Helenius; M. Aebi. "Intracellular Functions of N-Linked Glycans" *Science* **291**, 2364-2369 (2001)

[6] T. Rutherford; D. Neville; S. Homans. "Influence of the extent of branching on solution conformations of complex oligosaccharides: a molecular dynamics and NMR study of a penta-antennary 'bisected' N-glycan" *Biochemistry* **34**, 14131-14137 (1995)

[7] A. Imberty, M. Delange, Y. Bourne et. al.. "Data bank of three-dimensional structures of disaccharides: Part II, N-acetyllactosaminic type N-glycans. Comparison with the crystal structure of a biantennary octasaccharide" *Glycoconjugate J.* **8**, 456-483 (1991)

[8] R. Williams; S. Greene; A. Mc Phearson. "The crystal structure of ribonuclease B at 2.5-A resolution" *J.Biol. Chem.* **262**, 16020-16031 (1987)

[9] H. Berman, J. Westbrook, F. Z et. al.. "The Protein Data Bank" *Nucleic Acids Research,* **28**, 235-242 (2000)

[10] A. Petrescu, S. Petrescu, R. Dwek et. al. "A statistical analysis of N- and O-glycan linkage conformations from crystallographic data." *Glycobiology* **9**, 343-352 (1999)

[11] S. Homans. "Conformation and dynamics of oligosaccharides in solution" *Glycobiology.* **3**, 551-5. (1993)
[12] J. Jimenez-Barbero, J. Asensio, F. Canada et. al.. "Free and protein-bound carbohydrate structures." *Curr Opin Struct Biol.* **9**, 549-55 (1999)
[13] A. Imberty. "Oligosaccharide structures: theory versus experiment" *Curr Opin Struct Biol.* **7**, 617-23 (1997)
[14] C.-W. von der Lieth; T. Kozár; W. Hull . "A (Critical) Survey of Modeling Protocols Used to Explore the Conformational Space of Oligosaccarides" *J. Mol. Struc (Theochem)* **395-396**, 225-244 (1997)
[15] A. Imberty, S. Gerber, V. Tran et. al.. "Data Bank of Three-Dimensional Structure of Disaccharides, A Tool to Build 3D Structure of Oligosaccharide" *Glycoconjugate J* , **7**, 27-54 (1990)
[16] R. Woods, A. Pathiaseril, M. Wormald et. al. "The high degree of internal flexibility observed for an oligomannose oligosaccharide does not alter the overall topology of the molecule" *Eur J Biochem.* **258**, 372-86. (1998)
[17] S. Doubet, K. Bock, D. Smith et. al.. "The Complex Carbohydrate Structure Database" *Trends Biochem. Sci.,* **14**, 475-7 (1989)
[18] A. Bohne; E. Lang; C. von der Lieth. " W3-SWEET: Carbohhydrate Modeling by Internet" *J. Mol. Model* **4**, 33-43 (1998)
[20] Dissertation Martin Frank. "Conformational analysis of oligosaccharides in the free and the bound state" University of Heidelberg (2000), http://www.ub.uni-heidelberg.de/archiv/605/
[20] R. Pappu; R. Hart; J. Ponder. " Analysis and Application of Potential Energy Smoothing for Global Optimization" *J. Phys. Chem. B,* **102**, 25-9742 (1998)
[21] T. Rutherford; S. Homans. "Restrained vs Free Dynamics Simulation of Oligosaccharides: Application to Solution Dynamics of Biantennary and Bisected Biantennary N-Linked Glycans" *Biochemistry* **33**, 9609-9614 (1994)
[22] S. Homans; R. Dwek; T. Rademacher. "Tertiary Structure in N-Linked Oligosaccharides" *Biochem.* **26**, 6553-6560 (1987)
[23] S. Homans, R. Pastore, R. Dwek et. al.. "Structure and Dynamics Oligomannose-Type-Oligosaccharides," *Biochemistry* **26**, 6649-6655 (1987)
[24] K. Rasmussen; J. Fabricius. Optimized Potential Energy Functions in Conformational Analysis of Saccharides" in French , A. D. and Brady, J.W., Ed.; ACS Symposium Series 430: Washington, DC, 1990; Vol. 439, pp 177-190
[25] A. Bohne; T. Wetter; E. Lang et. al.. „Glykowissenschaften, ein neuer Einsatzbereich in der Bioinformatik" in K.Mehlhorn, G. S., Ed.; Springer: Heidelberg, 2000, pp 181-196.

DATA STANDARDISATION IN GLYCOSUITEDB

C.A. COOPER, M.J. HARRISON, J.M. WEBSTER, M.R. WILKINS, N.H. PACKER

Proteome Systems Ltd, Locked Bag 2073,
North Ryde, NSW 1670, Australia

GlycoSuiteDB, a database of glycan structures, has been constructed with an emphasis on quality, consistency and data integrity. Importance has been placed on making the database a reliable and useful resource for all researchers. This database can help researchers to identify what glycan structures are known to be attached to certain glycoproteins, as well as more generally identifying what types of glycan structures are associated with different states, for example, different species, tissues and diseases. To achieve this, a major effort has gone into data standardisation. Many rules and standards have been adopted, especially for representing glycan structure and biological source information. This paper describes some of the challenges faced during the continuous development of GlycoSuiteDB.

1 Introduction

GlycoSuiteDB[1] is a curated and annotated database of glycan structures, available on the Internet at www.glycosuite.com. It was initiated in April 1999 and first made available in September 2000. There are currently more than 6000 entries in GlycoSuiteDB, extracted from approximately 700 references and covering 250 distinct proteins from about 160 different species.

The glycan structures are presented with the biological source from which they were obtained, the literature references in which the glycan structure was described, and the methods used by the researchers to determine the structure. An example entry from GlycoSuiteDB is given in Figure 1. The main aim of GlycoSuiteDB is to store and disseminate information on protein glycosylation in a logical, integrated and searchable way, in order to simplify the study and understanding of glycobiology.

A database of glycosylation faces a different set of complexities from those of nucleic acid or protein sequence databases. An obvious difference is that unlike nucleic and protein sequences, where the individual bases or amino acids are linked together in a linear fashion, glycan structures are branched. Parameters such as anomeric configuration and position of linkage between monomers, contribute to five to six orders of magnitude higher structural diversity than found in proteins. For example, the number of possible structures from six known amino acids is $6! = 720$. The number of possible linear structures from six D-hexose molecules is $6! \times 2^6 \times 4^5 = 47\ 185\ 920$, where the first term is the number of permutations of six linear molecules; the second, the number of possible anomeric configurations; and the third, the position of the linkage. When the ring size (pyranose or furanose) and L-sugars are considered this number increases by 4096. If branching of the chains is considered the number of possibilities increases by more than 100 fold, and naturally occurring substituents such as sulfates and phosphates increase it again. Having said this, the conservative nature of biology dictates that nowhere near this number of possibilities actually occurs, but the discovery of more and more unusual structures emphasizes the need for a consistent, curated catalogue of known glycans.

A glycan database is also complicated by the fact that glycosylation is a finely controlled process dependent upon the availability and activity of the various glycosyltransferases, glycosidases, monosaccharides and precursors[2]. As a result of these factors, one glycosylation site may have many glycan structures, and the same protein expressed at different times of development or from different tissues, can possess different glycan chains. Similarly, in the case of recombinant or viral proteins, the glycosylation machinery of the host organism is the main influence on glycosylation of the protein. More indirectly, levels of various hormones also affect protein glycosylation through a variety of cell type-dependent changes and the production of differentiated phenotypes[3].

GlycoSuiteDB / source / protein / top

entry: 3053-1328

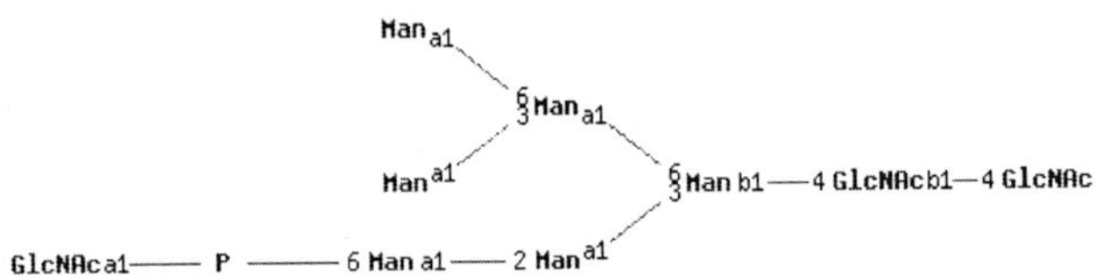

Species	*Homo sapiens* (HUMAN); sample isolated from species *Mesocricetus auratus*
Class	MAMMALIA
Source	UROGENITAL SYSTEM, KIDNEY (cell line BHK-21)
Source notes	NONE
Attached to	ERYTHROPOIETIN (swiss-prot entry P01588); amino-acid ASN-51
Linkage	N-LINKED
Glycosylation sites	N-51, N-65 AND N-110 [NIMTZ ET AL. (1993) EUR. J. BIOCHEM. 213:39-56].
Identified by methods	MALDI-TOF MS, METHYLATION ANALYSIS, MONOSACCHARIDE ANALYSIS, PROTON NMR
References	Nimtz (1995) *Febs Lett.* **365**: 203-208
Glycan structure	GlcNAc(a1-P-6)Man(a1-2)Man(a1-3)[Man(a1-3)[Man(a1-6)]Man (a1-6)]Man(b1-4)GlcNAc(b1-4)GlcNAc
Mass	1679.5319 Da (*monoiso*), 1680.4346 Da (*avg*), total residues: 9
Composition	$Hex_6HexNAc_3P_1$
Release date	04-NOV-00 (*last updated* 10-JAN-01)

Figure 1: An example entry from GlycoSuiteDB

Due to these complexities, all the information in GlycoSuiteDB is manually extracted from the original scientific literature by trained glycobiologists. Whilst doing this, a number of inconsistencies in the way data is represented in these publications became apparent. This paper describes the way in which we have addressed these variations in order to achieve a standardised format in which the entries are presented in GlycoSuiteDB.

2 Glycan Structure Representation

As the glycan structures are the most important data type in the database, it was important to standardise their representation in such a way as to provide consistency and to enable different searching criteria.

2.1 Linear Representation

Since glycans are often branched structures, this poses special challenges for electronic storage. To enable storage and searching of the glycan structures in a textual form, a linear representation or sequence was formulated. The monosaccharide abbreviations[4] and condensed linear form[5] recommended by IUPAC were adopted with additional rules[6] designed to ensure consistent representation of all glycan structures. For example, the IUPAC recommendation states that a branched glycan is represented as a string by placing branches inside square brackets. However, the guidelines for deciding which chain is the parent and which is a branch were limited and not comprehensive enough to ensure constant and precise representation of all glycan structures. In particular, structures are often described in the literature without full assignment of all linkages and anomeric configurations. This complicates converting branched structures to linear form, as the branch linkages are not known.

To address these issues rules were designed[6] and adopted for converting branched glycan structures into linear form:

- Alpha and beta are represented by 'a' and 'b', respectively
- Where the anomeric configuration or linkage point is not known, a question mark is used
- The parent, or primary, chain is defined as the longest chain, all other chains are branches
- If two chains are of the same length, the more branched chain is the primary chain
- If two chains have the same length and degree of branching, then the chain with the lowest alphabetical terminal residue is considered primary,

working from the most terminal residue towards the branch point until a difference is found

- If two chains are still indistinguishable, then the chain with the lowest terminal linkage, working in towards the branch point until a difference is observed, is considered primary

These rules work equally well with *N-,O-* and *C*-linked glycan structures. They also cope with multiply branched structures. For example, structure 1555 (Figure 2) is represented in linear form as: GalNAc(b1-4)[NeuAc(a2-3)]Gal(b1-4)GlcNAc(b1-2)[GalNAc(b1-4)[NeuAc(a2-3)]Gal(b1-4)GlcNAc(b1-4)]Man(a1-3)[GalNAc(b1-4)[NeuAc(a2-3)]Gal(b1-4)GlcNAc(b1-2)[GalNAc(b1-4)[NeuAc(a2-3)]Gal(b1-4)GlcNAc(b1-6)]Man(a1-6)]Man(b1-4)GlcNAc(b1-4)[Fuc(a1-6)]GlcNAc.

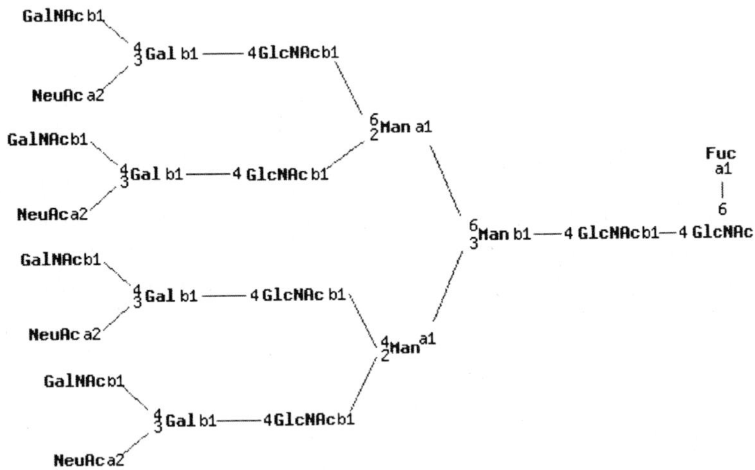

Figure 2: Structure 1555 from GlycoSuiteDB

We have not tried to simplify the terminology of the linear form, as we believe that the 2-D graphical representation that is presented in GlycoSuiteDB (see section 2.3) gives the user a more realistic image of the glycan.

Currently only full glycan structures are entered into GlycoSuiteDB. Fragments of structures, such as lectin recognition patterns, and monosaccharide compositions without any linkage order are not recorded.

2.2 *Searching*

Using the linear format it is possible to search the glycan structure field for structures containing specific residues and linkages. For example, it is possible to perform a search on GlycoSuiteDB for all structures containing the Lewis X motif (Figure 3A).

Figure 3: A) Lewis X motif; and B) Sialyl Lewis X motif

Since the motif is terminal on one branch, the order of the residues in the linear form, following the rules created, is Fuc(a1-3)[Gal(b1-4)]GlcNAc. However, the overall order of this branch within a full linear glycan structure may not be terminal, therefore it is necessary to look in the database for the text 'Fuc(a1-3)[Gal(b1-4)]GlcNAc%' or '%[Fuc(a1-3)[Gal(b1-4)]GlcNAc%' where % is a wild card.

Using this query it was found that there are more than 100 entries in GlycoSuiteDB containing the Lewis X motif. Structures were found where this motif was not terminal in the linear form. For example, structures 1227 and 3803, shown in Figure 4, are represented in linear form as

i)NeuAc(a2-6)Gal(b1-4)GlcNAc(b1-2)Man(a1-3)***[Fuc(a1-3)[Gal(b1-4)]GlcNAc***(b1-2)Man(a1-6)]Man(b1-4)GlcNAc(b1-4)[Fuc(a1-6)]GlcNAc, and

ii)Gal(a1-3)Gal(b1-4)GlcNAc(b1-2)Man(a1-6)***[Fuc(a1-3)[Gal(b1-4)]GlcNAc***(b1-2)Man(a1-3)]Man(b1-4)GlcNAc(b1-4)[Fuc(a1-6)]GlcNAc, respectively.

A similar search was performed looking for Sialyl Lewis X motif (Figure 3B). 32 entries were found to contain this structural feature. Interestingly, Sialyl Lewis X and Lewis X motifs were found on both *N*- and *O*-linked glycans. In addition the search revealed that Lewis X containing structures were found only on native human proteins, whereas Sialyl Lewis X containing structures were reported on glycoproteins isolated from humans, pigs, cattle, mice, rabbit, spotted salamander, tiger salamander and Iberian ribbed newt.

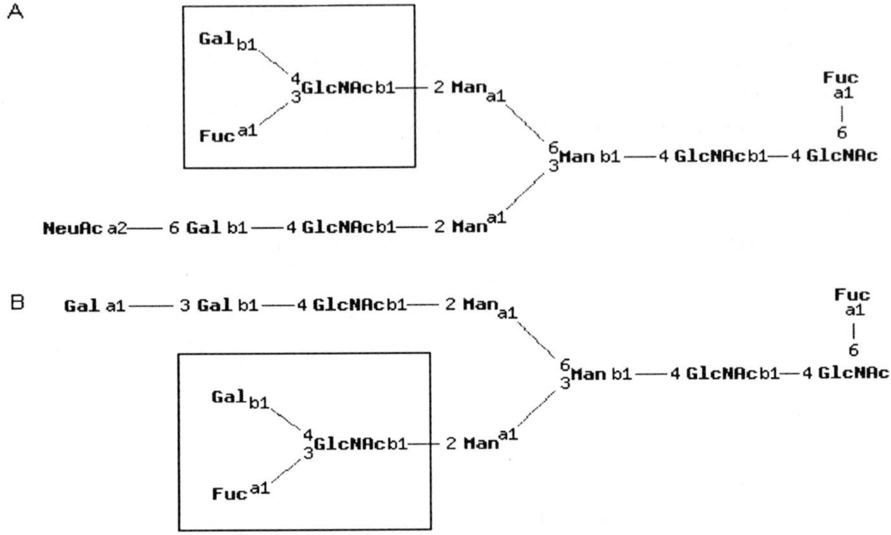

Figure 4: A) Structure 1277 and B) Structure 3803, both containing Lewis X motif (boxed)

2.3 Two-Dimensional Graphical Representation

Standardised linear structures, whilst essential for a searchable database, are not user-friendly because of their visual complexity. GlycoSuiteDB converts the linear string to a more acceptable 2-D graphical representation (as shown in Figures 2, 3 and 4). The use of such software also ensures that the glycan structures are free of syntactical errors and that all glycan structures are valid in terms of linkage. It ensures that monosaccharide residues do not have linkages to positions occupied by other residues, such as other monosaccharides or acetamido groups on GlcNAc.

In addition, based on the linear glycan structure rules given above, we have developed a 'sugarbuilder' tool that allows for the user construction of complex glycans through the incremental addition of monosaccharides. This tool can be used to generate the syntactically correct glycan structure for any given glycan structure. This tool also forms the basis of a user interface that will enable the user to query the glycan structure for full glycan structures, or substructures (such as epitopes), without having to know the linear code rules.

Basic substructure searches, such as those performed in section 2.2, use a pattern-matching technique on the linear code. However, these searches may be limited in that not all structures with a given substructure may be found. Other branches that originate from, or include, the search substructure may be hidden by the presence of nested branch sequences that interrupt the continuous sequence of the search substructure. In the case of GlycoSuiteDB, substructure searches are implemented using mathematical tree-matching algorithms. This type of search algorithm ensures that all glycans that contain the given substructure will be found.

3 Biological Source

Because the glycan structures are very dependent on their biological origin in terms of species, tissue and cell type, each entry in GlycoSuiteDB records these details.

3.1 Taxonomy

Species names are checked against the NCBI taxonomy database (http://www.ncbi.nlm.nih.gov/Taxonomy). Taxonomy class information is also taken from this database. Where the class classification is not given, the closest division is used and the division definition is noted in brackets. For example, Rosidae is the subclass to which *Brassica rapa* (field mustard), *Glycine max* (soybean) and many other plants belong. It is recorded in the database as 'Rosidae (subclass)'.

3.2 Recombinant Proteins

Many glycoproteins under study have been expressed in a recombinant system. This is a common situation if the native protein is not able to be isolated in sufficient quantities to enable the study of the glycan structures. Recombinant glycoproteins are also common due to the increasing demand for glycoprotein drug products. The expression of a protein in different recombinant systems results in different glycosylation patterns compared to the non-recombinant protein and between the different expression systems. In these cases GlycoSuiteDB lists the species name as that from which the DNA encoding the protein originates. The recombinant field then contains the name of the species in which the protein has been expressed. For example, human Interferon omega-1 expressed in *Spodoptera frugipeda*, would have *Homo sapiens* as the species and *Spodoptera frugipeda* in the recombinant field.

Like recombinant proteins, the glycosylation of viral proteins is dependent on the glycosylation machinery of the host species. Thus the species name of a viral glycoprotein is also given as that from which the DNA encoding the protein originates, i.e. the name of the virus, and the recombinant field contains the species from which the viral protein was isolated.

3.3 Tissue and Cell Type

A major source of inconsistency in the literature is that the tissue and cell type source of a glycoprotein is often described in a variety of ways with differing degrees of specificity. For example, mucin samples from the lung are sometimes described as respiratory, bronchial, or tracheobronchial mucins. Since glycosylation is tissue-dependent as well as protein-dependent it is often desirable to be able to search for all types of glycan structures coming from the same tissue source and to have as much information on the biological source as possible. To enable this we developed a standard format for describing the tissue or cell type based on the anatomy categories of the National Library of Medicine's medical subject headings (MeSH).

For each entry in GlycoSuiteDB the tissue or cell type is described in a maximum of four columns: system, division1, division2, division3 and division4 where each is a nested subset of the previous column. For example, Table 1 shows the various subdivisions of the respiratory and hemic systems adopted for GlycoSuiteDB. It is a common problem that tissues or cell types can exist under multiple systems or divisions. Our current position is to standardise our classification so that most of the tissues or cell types are found in one system only. For example, the nose can be a subdivision of the respiratory system or the sensory system. Currently this tissue is classified under the sensory system since the sub-tissue from which the glycoprotein was isolated was the vomeronasal organ, the primary function of which appears to be in sensing pheromones.

Using this information it is possible to search for which tissue/cell types express particular monosaccharides, thus reflecting the activity of glycosyltransferases. For example, a search for where Gal(b1-4) residues are found in *N*-linked glycans derived from human non-recombinant proteins, showed that this particluar residue and linkage has only been isolated to date from *N*-linked glycans in milk, a secretion subdivision of the exocrine system, and urine, an excretion subdivision of the urogenital system. A similar search for Gal(b1-3) shows that this residue is found in *N*-linked glycans from glycoproteins from nearly all human systems.

Table 1: Example of the organisation of tissue and cell type information in GlycoSuiteDB, adapted from the anatomy categories of the National Library of Medicine's medical subject headings (MeSH).

System	Division1	Division2	Division3	Division4
Hemic system	Blood cell	Erythrocyte	Erythrocyte membrane	
		Leukocyte	Mononuclear leukocyte	Lymphocyte
Respiratory system	Lung	Mucosa		
	Pleura	Fluid		

3.4 *Protein Name, Amino Acid Numbering and Glycosylation Sites*

When the protein from which a particular glycan structure has been characterised is known, this information is stored in GlycoSuiteDB. To minimize inconsistencies the protein is searched for in the SWISS-PROT/TrEMBL protein databases and the name used is that preferred in these databases. For example, alpha-1-protease inhibitor and alpha-1-antiproteinase both describe the same protein, for which the preferred name from the SWISS-PROT database is alpha-1-antitrypsin. GlycoSuiteDB entries are cross-linked to the corresponding SWISS-PROT protein entry, and SWISS-PROT links directly to GlycoSuiteDB, where appropriate.

Where known, the amino acids to which an individual glycan structure is linked are also entered, with the numbering of the glycosylated amino acids following the sequence given in SWISS-PROT. If the sequence is not in SWISS-PROT, the numbering follows the sequence numbering given in the relevant literature article.

Not all glycan structures are linked to a particular protein however. Whilst many researchers separate a protein to purity before analyzing its glycans, it is also common for researchers to look at the total glycans present in mixtures of proteins, for example, from a particular tissue or cell line. There are many (greater than 55%) of structures in GlycoSuiteDB therefore which do not have a link to a particular SWISS-PROT protein entry.

3.5 *Disease Names and Cell Lines*

Glycosylation is known to be altered in disease states. For example, alpha fucose linked to the 6 position of reducing terminal GlcNAc in the *N*-linked glycans of alpha-1-antitrypsin is only found when the protein is isolated from patients with hepatocellular carcinoma[7]. In addition different cell lines can result in different glycosylation of the same recombinantly-produced protein, e.g., there is a wide difference in the *N*-linked glycan structures of human interferon-gamma expressed in Chinese hamster ovary cells and Sf9 cells[8].

Disease names have been standardised using the names and definitions National Library of Medicine's medical subject headings (MeSH) and CancerWEB's online medical dictionary (http://www.graylab.ac.uk/omd/). Cell line names have been adopted from American Type Culture Collection (ATCC) (www.atcc.org) and HyperCLDB, the hypertext on cell culture availability extracted from the Cell Line Data Base of the Interlab Project (http://www.biotech.ist.unige.it/cldb/indexes.html).

4 Methods

The quality of GlycoSuiteDB relies on information published in the literature. For glycan structures, the confidence we have that a published structure is correct is dependent on the method, or methods, used in its determination. In our in-house analytical work we use a confidence rating to distinguish data quality based on these methods. Each method used to determine glycan structures has been critically assessed and has been given a confidence value based on the reliability of the method and the value and the extent of the information obtainable from the method. For example, proton NMR is given a value of 10 because it can be used to obtain information on the complete structure of the glycan, e.g., the type and ring form of monosaccharides, relative number of each sugar residue, the linkage positions, anomeric configurations and the sugar sequence. Mass spectrometric methods have been given a value of 5 if they only give the total mass of the glycan from which the possible composition can be predicted[9]. However, if a single structure was fragmented by tandem mass spectrometry, the sugar sequence and linkage position of the glycan can also be deduced. In this case, a method called "fragmentation", with a confidence value of 3, is noted in addition to the mass spectrometry method, to reflect the extra information obtained. Mass spectrometry can, however, only give generic monosaccharide information as many monosaccharides, e.g., glucose and mannose, have the same mass.

The relative values for each method were carefully chosen so that the sum of the confidence values for certain combinations of methods were comparable. For example, a structure determined by proton NMR and methylation analysis is very reliable and has a confidence value of 18. Likewise structures characterized by: i) methylation analysis, monosaccharide analysis and glycosidase treatment; or ii) monosaccharide analysis, glycosidase treatment, mass spectrometry and periodate oxidation; are quite dependable and would have confidence values of 18. A structure that had been determined by its monosaccharide composition and chromatographic elution position compared to a standard would not be considered to be as reliable as this and would have a resultant confidence value of 5 assigned to reflect this.

Although these confidence values are not included in the public version of GlycoSuiteDB because of the controversial nature of the ratings, the user can make their own judgements from the methods field provided.

5 Conclusion

In this article, we have described some of the features of GlycoSuiteDB. There are only 2 other web-based glycan structural databases, CarbBank and Glycominds. Funding for CarbBank was discontinued and the web site now runs unattended, with no new data entries. In its construction, because of its dependence on user entry, there was no standardisation of data procedures implemented, which lead to variability in data formatting and inconsistencies in search output[6]. Glycominds is a new glycan structural database launched on the web in November 2000. Unlike GlycoSuiteDB however, Glycominds advocates the representation of glycan structures in a linear format in order to enable searching for specific epitopes. The data in GlycoSuiteDB is also based on a linear format entry and can be searched in the same way. At present this functionality in both databases is limited, as discussed in Section 2.3 of this paper, and we are addressing this.

GlycoSuiteDB is available on the web (www.glycosuite.com) and there has been considerable focus on data standardisation, which means that it is easily searchable and accurate. Queries can be performed using monosaccharide composition, glycan mass, species, biological tissue/cell type, protein name or any combination of these. GlycoSuiteDB is already extensively linked with the SWISS-PROT protein database and PubMed. Further links to other online databases, such as the Online Mendelian Inheritance in Man (OMIM) database, are planned.

GlycoSuiteDB has been designed to allow researchers to search for precedence and thus to have more confidence in making assumptions on glycan structure. However, in the development of GlycoSuiteDB it has become obvious that there are many variations in glycan structure and that not all assumptions are valid. For example, there are 16 monosaccharide compositions that correspond to more than 10 unique structures each. Moreover, at least 18 unique glycan structures have been isolated with the composition hexose=3, hexNAc=3 and deoxyhexose=1. More than 290 unique *N*-linked and 270 unique *O*-linked glycan structures have been characterised from humans.

An example of a glycan structure that does not conform to precedence is an *N*-linked glycan isolated from chicken ovalbumin[10] (Figure 5). This structure was characterised by FAB-MS and proton NMR, and has three GlcNAc residues all individually linked to the Man6 branch. Using the FAB-MS results only, precedence would probably have predicted that the structure would have only two branches per mannose arm with GlcNAc-GlcNAc units added linearly.

As the above example indicates, we would make the point that precedence does not necessarily mean that a mass equates with a certain structure, and that researchers should be careful as to what structure is assigned. Initially journal articles describing glycan structures focused solely on defining the analytical methods used and on the characterization of major glycans on a particular protein. As advances have been made in the field the focus has shifted to trying to see all the glycan structures present by using more sensitive approaches, such as mass spectrometry, and to try to determine the function of the glycans. This has led to scientists making more assumptions about the glycan structures rather than systematically determining all linkages and anomeric configurations. This is particularly true with N-linked glycans. However, GlycoSuiteDB can assist researchers as a resource to see what is already known about what glycan structures are attached to certain glycoproteins.

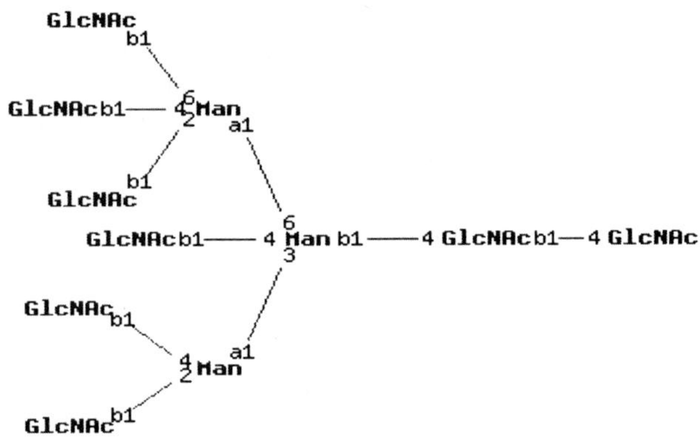

Figure 5: *N*-linked glycan structure isolated from chicken ovalbumin[10]

References

1. C.A. Cooper et al, *Nucleic Acids Res.* **29**, 332 (2001)
2. R. A. Dwek et al, *Annu. Rev. Biochem.* **62**, 65 (1993)
3. C. F. Goochee and T. Monica, *Biotechnology (N. Y.)* **8**, 421 (1990)
4. A. D. McNaught, *Pure Appl. Chem.* **68**, 1919 (1996)
5. N. Sharon, *Eur. J. Biochem.* **159**, 1 (1986)
6. C. A. Cooper et al, *Electrophoresis* **20**, 3589 (1999)
7. A. Saitoh et al, *Arch. Biochem. Biophys.* **303**, 281-287 (1993)

8. D. C. James et al, *Biotechnology (N. Y.)* **13**, 592 (1995)
9. C.A. Cooper et al, *Proteomics* **1**, 340 (2001)
10. M. L. Corradi Da Silva et al, *Arch. Biochem. Biophys.* **318**, 465 (1995)

Prediction of glycosylation across the human proteome and the correlation to protein function

Ramneek Gupta and Søren Brunak
Center for Biological Sequence Analysis, Bldg-208, Bio-Centrum
Technical University of Denmark, DK-2800 Lyngby, Denmark.

1 Introduction

The addition of a carbohydrate moeity to the side-chain of a residue in a protein chain influences the physicochemical properties of the protein. Glycosylation is known to alter proteolytic resistance, protein solubility, stability, local structure, lifetime in circulation and immunogenicity[1,2].

Of the various forms of protein glycosylation found in eukaryotic systems, the most important types are N-linked, O-linked GalNAc (mucin-type) and O-β-linked GlcNAc (intracellular/nuclear) glycosylation. N-linked glycosylation is a co-translational process involving the transfer of the precursor oligosaccharide, $GlcNAc_2Man_9Glc_3$, to asparagine residues in the protein chain. The asparagine usually occurs in a sequon Asn-Xaa-Ser/Thr, where Xaa is not Proline. This is however, not a specific consensus since not all such sequons are modified in the cell. O-linked glycosylation involves the post-translational transfer of an oligosaccharide to a serine or threonine residue. In this case, there is no well-defined motif for the acceptor site other than the near vicinity of proline and valine residues.

We have developed glycosylation site prediction methods for these three types of glycosylation, using artificial neural networks that examine correlations in the local sequence context and surface accessibility. In this paper, we have used glycosylation site information on human proteins to illustrate the contribution of glycosylation to protein function and assess how widespread this modification is across the human proteome.

2 Methods

2.1 Data set

Analysis shown in this paper was derived on a set of human proteins obtained from the SWISS-PROT (rel. 38) database. This consisted of 5,795 well annotated proteins. We chose to work with proteins from a single organism i.e. humans, to restrict the diversity of oligosaccharyltransferase acceptor sites.

Glycosylation in simple organisms, such as yeast, is well studied[3,4], but their glycans are usually high mannosylated structures, and it is not clear how similar their mechanism of glycosylation is to that of humans. Combining data from different organisms would complicate the analysis: possible 'families' of acceptor specificities causing ambiguity in distinguishing acceptor (positive) sites from non-acceptor (negative) sites.

2.2 Functional categories for proteins

Defining protein function is a complicated task, and there are many different ways of describing the roles and functions of a protein in a cell. This is the topic of many on-going ontology projects[5]. Here we chose to use a cellular role descriptor and subcellular location as our categorisations:

15 categories (13 defined, 2 unknown) reflective of the 'cellular role' of the protein in the cell were employed (as shown in Figure 1). The automatic class assignment to sequences was made by an extension of the Euclid system performing a lingustic analysis and clustering of SWISS-PROT keywords[6,7]. Keywords were parsed for the human proteins in SWISS-PROT. For each functional class, the informative weight (Z-score) of each keyword was extracted from a dictionary[6]. Keyword sums gaves scores to all categories for a particular sequence. The central point of the Euclid system is the dictionary. The primary version of this dictionary was generated from an initial set of carefully, hand annotated proteins from different organisms spanning every kingdom of life. From this initial set, a first dictionary was defined which was used to assign all SWISS-PROT proteins and the process of dictionary definition and assignment was reiterated until convergence. This final dictionary obtained was used to assign functional classes to around 5,500 human proteins from SWISS-PROT.

The cellular role categories themselves[8] were derived from an earlier proposed scheme for *Escherichia coli*[9] which was later extended by the TIGR group for other complete genomes. These categories comprise 13 functional classes which are subsets of three superclasses: Energy, Communication and Information. Proteins which do not fit in the 13 categories are assigned to 'Other' (functionally undefined cluster) or to 'Unknown' (sequences which do not contain the relevant keywords needed for classification in the above system).

Subcellular locations of proteins were obtained from SWISS-PROT annotations and PSORT predictions[10] (where no parsable SWISS-PROT annotation was found).

3 Results

3.1 N-Glycosylation

N-linked glycosylation modifies membrane and secreted proteins. This co-translational process occurs in the endoplasmic reticulum and is known to influence protein folding. The modification attributes various functional properties to a protein. To examine if certain categories of proteins were more prone to glycosylation than others, we studied the spread of known glycosylation sites across different categories.

N-glycosylation may also display some positional preferences in the protein chain. Specifically, it has been shown that sites need to be 12-14 residues away from the N-terminus [11] and that glycosylation efficiency is reduced within 60 residues of the C-terminus [12].

In our data set of approximately 6,000 human proteins, only 189 proteins (at 453 *confirmed* sites) were annotated in SWISS-PROT as N-glycosylated (not considering proteins with only POTENTIAL or PROBABLE sites). Figure 1 illustrates the spread of human glycosylation sites along the protein chain and across predicted subcellular locations and keyword based assignment of cellular role categories. Relative positions of sites on proteins were calculated with respect to normalised sequence lengths. The sequence length, divided into tenths is shown along the x-axis, from the N-terminal start on the left to the C-terminal end on the right.

N-glycosylated proteins appeared to almost exclusively belong to the functional category, 'Transport and binding'. This may not be too surprising considering that this category consists largely of membrane and secreted proteins. Only a few proteins belonged to any other cellular role category and most of these appeared involved in central intermediary metabolism. Subcellularly, extracellular proteins were the most favoured and others occurred in membrane proteins and in the endoplasmic reticulum or Golgi.

A clear positional preference for glycosylation sites on protein chains was apparent. The terminal ends of proteins seemed unfavourable and most sites seemed to occur N-terminal to the centre of the protein chain (20 to 40% along the length from the N-terminal start). The frequency of sites smoothly tapered off on both ends from this peak with a longer C-terminal tail. This statistical observation agrees with specific experimental indications of a 12-14 residue distance from the N-terminal and a 60 residue distance from the C-terminal end [11,12]. One peculiar observation from the figure was the C-terminal sites in nuclear proteins. On examination, these turned out to be around 10 proteins which were indeed annotated to be N-glycosylated in the C-terminal. However, this seems to be an anomaly of the sub-cellular prediction by PSORT. For

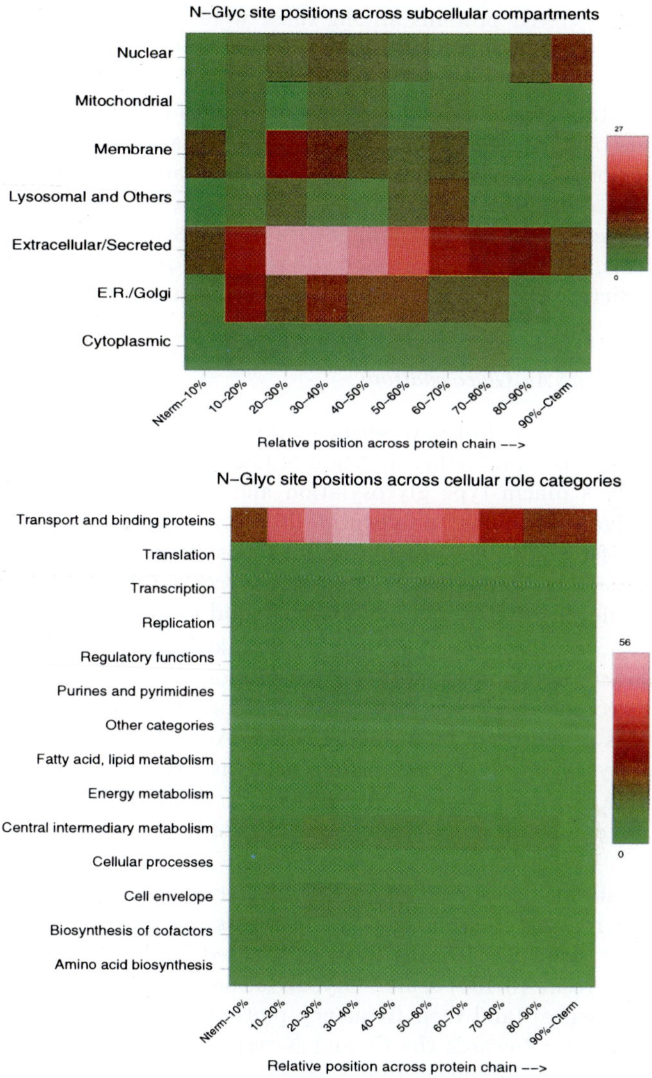

Figure 1: **Categorical distribution of known N-glycosylation sites across the protein chain.** Colour indicates frequency of sites (green to pink in increasing order). Protein chains, normalised in length, are represented across the x-axis from N-terminal to C-terminal. Subcellular locations (top) were predicted using PSORT, and cellular role classification (bottom) by lexical analysis of SWISS-PROT keywords (Alfonso Valencia *et al.*). Most N-glycosylation sites were clustered in the first half of all protein chains, and mainly occurred in extracellular transport and binding proteins.

instance, some secreted proteins among these were Vasopressin-Neurophysin 2-Copeptin precursor, Von Willebrand Factor Precursor and Immunoglobulin Delta Chain C.

Experimental determination of glycosylation sites is difficult to achieve as large amounts of purified protein are needed for the analysis of glycosylation sites. In addition, glycosylation can be an organism- and tissue specific event. Therefore only a few glycoproteins have been characterised so far as reflected in the low percentage of glycoprotein entries in SWISS-PROT (approx. 10% of human proteins, *see also*[13]). This motivates the need for developing theoretical means of predicting the glycosylation potential of sequons.

3.2 O-linked GalNAc Glycosylation

The addition of GalNAc linked to serine or threonine residues of secreted and cell surface proteins, and further addition of Gal/GalNAc/GlcNAc residues [2], is also known as mucin type glycosylation and is catalysed by a family of UDP-N-acetylgalactosamine: polypeptide N-acetylgalactosaminyltransferases (GalNAc-transferases). The modification, a post-translational event, takes place in the cis-Golgi compartment [14] after N-glycosylation and folding of the protein, and affects secreted and membrane bound proteins.

There is no acceptor motif defined for O-linked glycosylation. The only common characteristic among most O-glycosylation sites is that they occur on serine and threonine residues in close vicinity to proline residues, and that the acceptor site is usually in a beta-conformation. A prediction method [15,16] for this type of glycosylation on mammalian proteins has been built earlier and made available as a web server [a]. A database of O-glycosylated sequences is also available[b] and was used in constructing the O-glycosylation site prediction methods [17].

Figure 2 shows the spread of predicted glycosylation sites (O-GalNAc, mucin-type) across different categories and across the protein chain. To construct this plot, sequence lengths were normalised, and relative position expressed on a percent (0-100) scale. Glycosylation sites were binned (10 bins across each sequence), and their frequency plotted across different categories. Sites tend to cluster towards the C- and N-termini of proteins for some categories. This figure also shows that O-glycosylation acceptor sites occur in a wide range of proteins, though glycosylation patterns (frequency, positions across chain) may differ for different types of proteins.

---

[a] http://www.cbs.dtu.dk/services/NetOGlyc/
[b] http://www.cbs.dtu.dk/databases/OGLYCBASE/

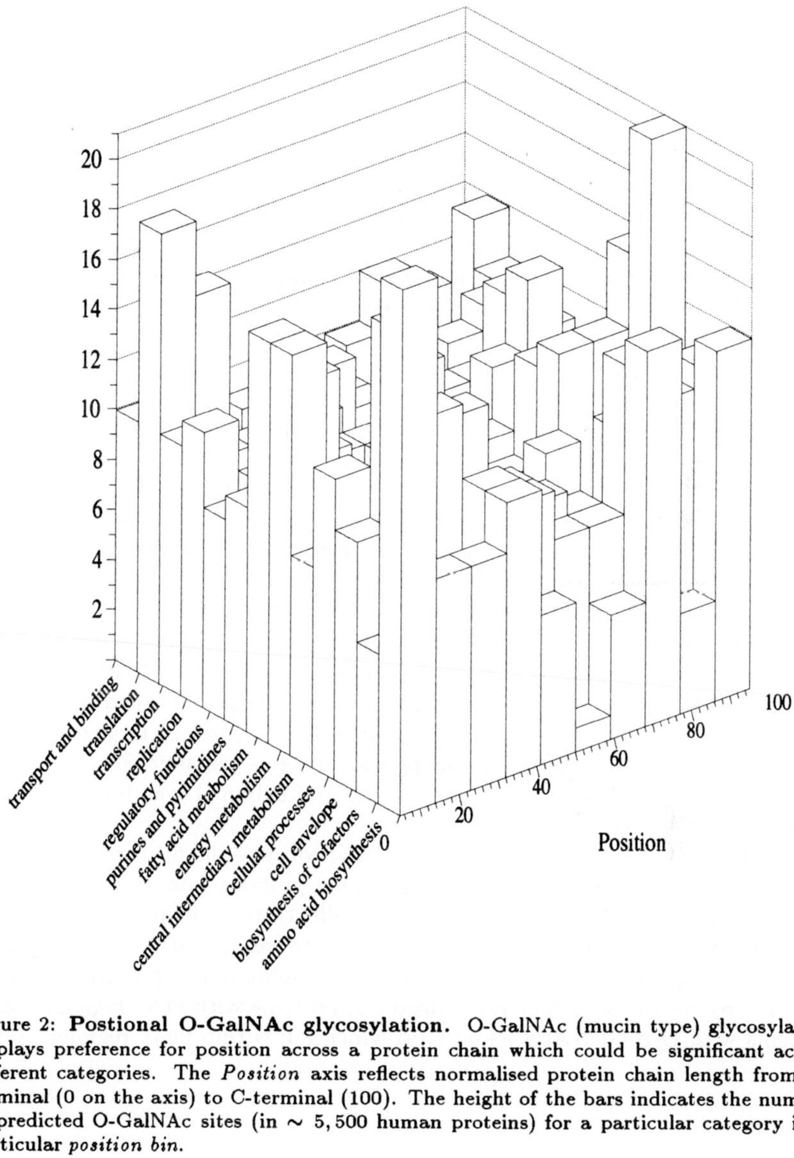

Figure 2: **Postional O-GalNAc glycosylation.** O-GalNAc (mucin type) glycosylation displays preference for position across a protein chain which could be significant across different categories. The *Position* axis reflects normalised protein chain length from N-terminal (0 on the axis) to C-terminal (100). The height of the bars indicates the number of predicted O-GalNAc sites (in $\sim 5,500$ human proteins) for a particular category in a particular *position bin*.

3.3 O-linked GlcNAc Glycosylation

Glycosylation of cytosolic and nuclear proteins by single N-acetylglucosamine (GlcNAc) monosaccharides is known to be highly dynamic and occurs on proteins with wide-ranging functions and cellular roles [18,19]. N-acetylglucosamine, donated by the nucleotide precursor UDP-N-acetylglucosamine, is attached in a beta-anomeric linkage to the hydroxyl group of serine or threonine residues.

So far, all proteins with O-β-GlcNAc linked residues, are also known to be phosphorylated. Evidence suggests that at least in some cases, these two post-translational modification events may share a reciprocal relationship [18,20]. This peculiar behaviour strongly suggests a regulatory role for this modification. Sites which can be both glycosylated and alternatively phosphorylated are also known as 'yin-yang' sites [18].

The acceptor site for O-β-GlcNAc glycosylation does not display a definite consensus sequence, nor are there many annotated sites in public databases. However, the fuzzy motif is marked by the close proximity of Proline and Valine residues, a downstream tract of Serines and an absence of Leucine and Glutamine residues in the near vicinity (data not shown). A prediction method for this type of glycosylation on human proteins has been built and made available[c] as a web server (*in preparation*).

Out of approximately 5,500 human sequences from SWISS-PROT (rel. 38), over 4,600 had at least one predicted O-GlcNAc site. 1,535 of these proteins had at least one high scoring O-GlcNAc site prediction (with 3,154 high scoring Ser/Thr sites). A number of these were DNA-binding proteins and involved in transcriptional regulation. When ranked according to scores, a large fraction at the top of this list were found to be nuclear proteins (as annotated in SWISS-PROT). The O-GlcNAc transferase itself (P100 subunit) was found to have predicted O-GlcNAc sites.

To study if the O-β-GlcNAc modification was specific for certain types of proteins, we classified the potentially modified proteins into cellular role categories and subcellular locations. Figure 3 illustrates the spread of proteins with at least one high-scoring O-β-GlcNAc site, across different categories. Also shown in this figure is the spread of phosphorylated proteins (as predicted [21] by NetPhos[d]), 'Yin-yang' proteins, proteins with PEST regions [22] and proteins with O-β-GlcNAc (+++) sites which fall within PEST regions.

[c] http://www.cbs.dtu.dk/services/YinOYang/
[d] http://www.cbs.dtu.dk/services/NetPhos/

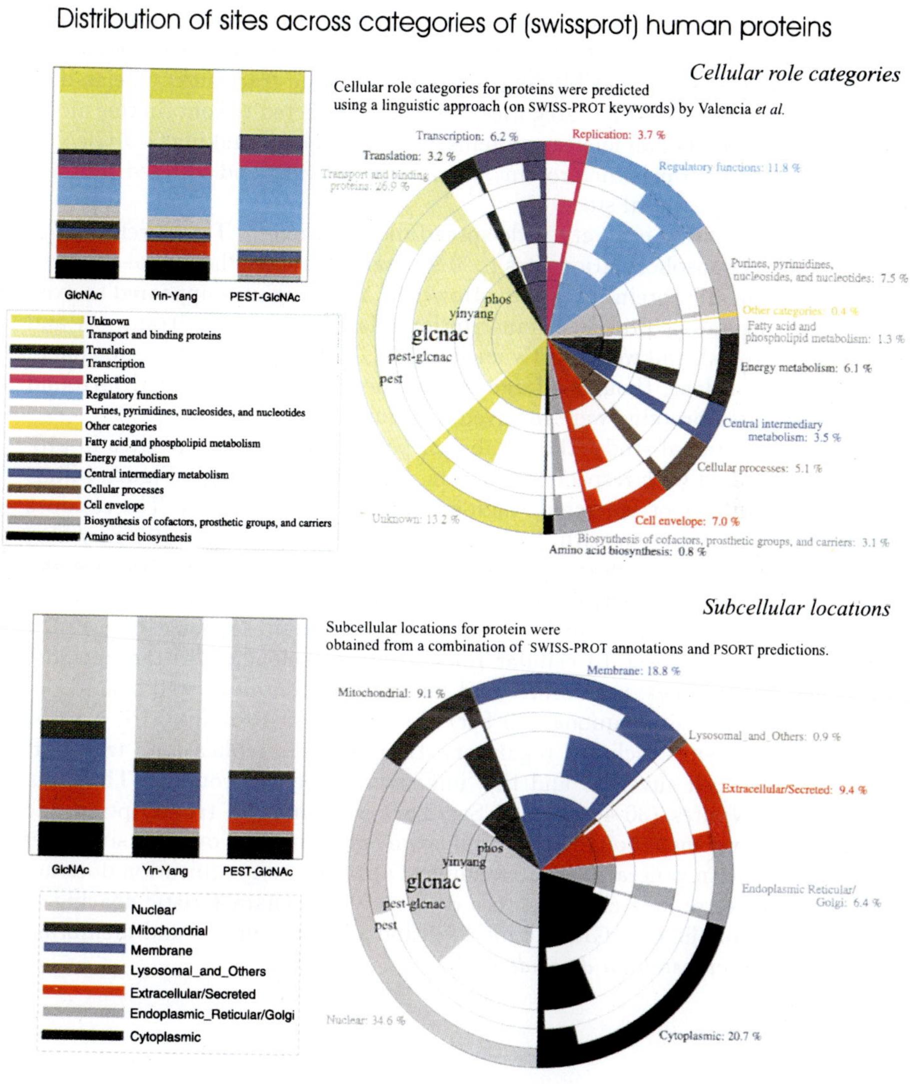

Figure 3: **Predicted O-β-GlcNAc sites across the human proteome.** The two panels (top, bottom) indicate different categorisations of proteins as depicted in the innermost and outmost circles of the pies. Individual rings represent different post-translational modifications and their occurrence in the corresponding category. E.g., phosphorylation occurs widely across all categories of proteins. Potential O-GlcNAc sites occur in half of all nuclear proteins and regulatory proteins. They also occur widely in replication and transcription proteins. Proteins with PEST regions and O-GlcNAc sites are mostly regulatory although PEST regions themselves also occur in other categories.

While the O-β-GlcNAc modification seems to potentially affect almost all types of proteins, most O-GlcNAcylated proteins were either regulatory proteins or 'transport and binding' proteins. A large fraction of unclassified proteins ('unknown' in role categories) were also predicted to contain this modification. Over half of all nuclear proteins contained a high ranking O-β-GlcNAc modified site. Cytoplasmic proteins, membrane proteins and secreted proteins also contained potential sites.

Phosphorylation is a very wide-spread modification[23]. This is reflected in our graphs as phosphorylation sites (> 0.9 potential by NetPhos) appeared well represented in all protein categories. However, Yin-yang sites appeared to exist largely in regulatory proteins, transcription related proteins or 'transport and binding proteins', and were mostly nuclear. O-GlcNAcylated PEST regions were also mostly nuclear, though a large membrane fraction also existed.. Around half of all these proteins were involved in regulatory functions.

In an additional study, the number of potential O-β-GlcNAc sites in proteins was studied with respect to function and cellular location. Figure 4 illustrates the number of predicted (high-scoring) sites per 100 Ser/Thr residues (per protein). Proteins with 1-2 predicted GlcNAc sites (per 100 Ser/Thr) were predominantly nuclear, cytoplasmic or membrane proteins. Nuclear and cytoplasmic proteins carried the highest densities of sites, a few cytoplasmic proteins having as many as 50 high-scoring O-GlcNAc sites among 100 Ser/Thr residues. With respect to cellular roles, proteins belonging to the category 'Purines, pyrimidines, nucleosides and nucleotides' contained well spaced out sites (only a few sites among 100 Ser/Thr residues). Proteins with a wider distribution of sites included regulatory, transcription, replication, 'transport and binding', cell envelope and the 'unknown' category proteins. The highest density of sites (30-40 per 100 Ser/Thr) was found in transcription and regulatory proteins, though some 'unknown' proteins had over 40 sites (per 100 Ser/Thr). In general, the intracellular O-β-GlcNAc modification does not seem to cluster among close residues or display any characteristic spacing as was evident for the O-α-GlcNAc modification affecting surface and membrane proteins of *Dictyostelium discoideum*[24].

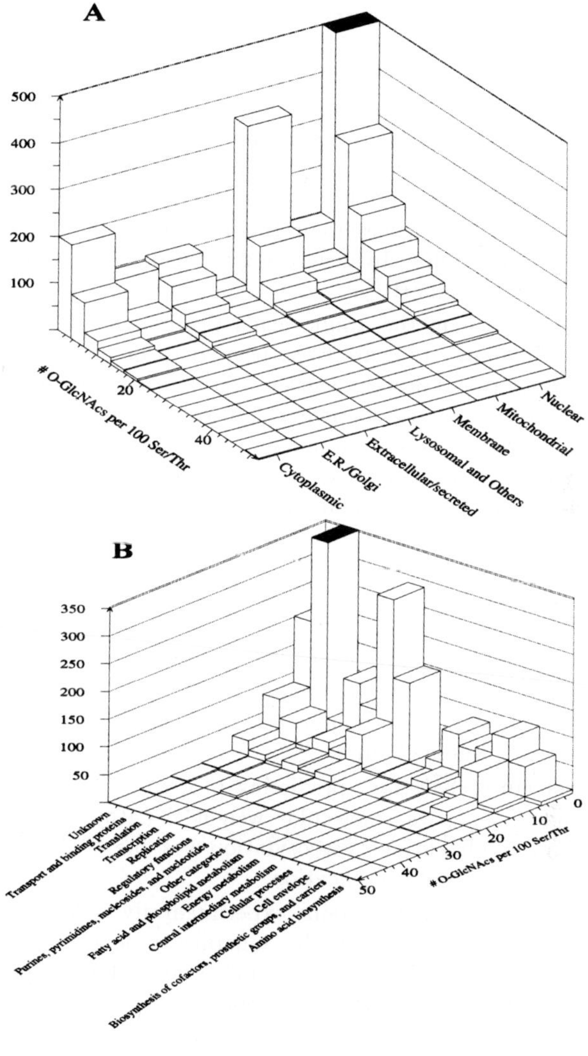

Figure 4: **Number of predicted O-β-GlcNAc sites per 100 Ser/Thr, in different categories of human proteins.** (A) shows proteins in different subcellular locations and (B) indicates cellular role categories. The z-scale (0-500 in A or 0-350 in B) is a frequency count for a particular bin; e.g. 0-2 O-GlcNAcs (per 100 Ser/Thr) occur most frequently for nuclear proteins in (A). These modifications usually do not occur in clusters. Although potential acceptor sites are largely found in nuclear/cytoplasmic proteins (usually regulatory), they also surprisingly occur in membrane proteins (mostly transport and binding proteins).

Human proteome-wide scans revealed that the O-β-GlcNAc acceptor pattern occurs across a wide range of functional categories and subcellular compartments. For humans, the most populated functional categories were regulatory proteins and transport and binding proteins. Nuclear and cytoplasmic proteins were prominent, though membrane and secreted proteins were surprisingly also in high numbers. It is interesting to know that acceptor patterns exist on these proteins too, but the cellular machinery defines protein targetting and consequently influences their modifications. The prediction server guards against this possibility by generating a warning when a potential signal peptide is detected by SignalP[e].

PEST regions, rich in the amino acids Proline (P), Glutamic acid (E), Serine (S) and Threonine (T), are hypothesised to be degradative signals for constitutive of conditional protein degradation[22]. Phosphorylation, a common mechanism to activate the PEST-mediated degradation pathway, may be signalled by deglycosylation in the same region. Our scans revealed that a small fraction of O-GlcNAc sites appeared in PEST regions. Such sites were mostly found in proteins involved in regulatory functions.

4 Final Remarks

Glycosylation is clearly a modification affecting a wide range of proteins, and is now known to affect both intracellular and secreted proteins. Different types of glycosylation have varying site preferences on proteins, and occur in different patterns across the protein chain.

In a project (*in preparation*) predicting protein function solely from protein chain global properties (molecular weight, length, etc.) and potential post-translational modifications, glycosylation was one of the most important determinants for functional classification.

Since characterising glycoproteins experimentally is a tedious and time-consuming task, it is worthwhile at this juncture to develop tools for predicting glycosylation sites. This is essential information for deciphering protein function and characterising complete proteomes.

5 Acknowledgements

The Danish National Research Foundation is acknowledged for support.

---

[e] http://www.cbs.dtu.dk/services/SignalP/

6 References

1. H Lis and N Sharon. Protein glycosylation: Structural and functional aspects. *Cur. J. Biochem.*, 218:1–27, 1993.
2. EF Hounsell, MJ Davies and DV Renouf. O-linked protein glycosylation structure and function. *Glycoconjugate J.*, 13:19–26, 1996.
3. MA Kukuruzinska, ML Bergh and BJ Jackson. Protein glycosylation in yeast. *Annu. Rev. Biochem.*, 56:915–944, 1987.
4. TR Gemmill and RB Trimble. Overview of N- and O-linked oligosaccharide structures found in various yeast species. *Biochim. Biophys. Acta*, 1426:227–237, 1999.
5. M Ashburner, CA Ball, JA Blake, D Botstein, H Butler, JM Cherry, AP Davis, K Dolinski, SS Dwight, JT Eppig, MA Harris, DP Hill, L Issel-Tarver, A Kasarskis, S Lewis, JC Matese, JE Richardson, M Ringwald, GM Rubin and G Sherlock. Gene ontology: tool for the unification of biology. The Gene Ontology Consortium. *Nat. Genet.*, 25:25–29, 2000.
6. J Tamames, C Ouzounis, G Casari, C Sander and A Valencia. EUCLID: automatic classification of proteins in functional classes by their database annotations. *Bioinformatics*, 14:542–543, 1998.
7. C Blaschke, MA Andrade, C Ouzounis and A Valencia. Automatic extraction of biological information from scientific text: protein-protein interactions. In *Proc., Intelligent Systems for Molecular Biology*, pages 60–67, Menlo Park, CA, 1999. AAAI Press.
8. MA Andrade, C Ouzounis, C Sander, J Tamames and A Valencia. Functional classes in the three domains of life. *J. Mol. Evol.*, 49:551–557, 1999.
9. M Riley. Functions of the gene products of Escherichia coli. *Microbiol. Rev.*, 57:862–952, 1993.
10. K Nakai and P Horton. PSORT: a program for detecting sorting signals in proteins and predicting their subcellular localization. *Trends Biochem. Sci.*, 24:34–36, 1999.
11. IM Nilsson and G von Heijne. Determination of the distance between the oligosaccharyltransferase active site and the endoplasmic reticulum membrane. *J. Biol. Chem.*, 268:5798–5801, 1993.
12. I Nilsson and G von Heijne. Glycosylation efficiency of Asn-Xaa-Thr sequons depends both on the distance from the C terminus and on the presence of a downstream transmembrane segment. *J. Biol. Chem.*, 275:17338–17343, 2000.
13. R Apweiler, H Hermjakob and N Sharon. On the frequency of protein glycosylation, as deduced from analysis of the SWISS-PROT database.

Biochim. Biophys. Acta., 1473:4–8, 1999.

14. J Roth, Y Wang, AE Eckhardt and RL Hill. Subcellular localization of the UDP-N-acetyl-D-galactosamine: polypeptide N-acetylgalactosaminyltransferase-mediated O-glycosylation reaction in the submaxillary gland. *Proc. Natl. Acad. Sci. USA*, 91:8935–8939, 1994.
15. JE Hansen, O Lund, J Engelbrecht, H Bohr, JO Nielsen, JES Hansen and S Brunak. Prediction of O-glycosylation of mammalian proteins: specificity patterns of UDP-GalNAc:polypeptide *N*-acetylgalactosaminyltransferase. *Biochem. J.*, 308:801–813, 1995.
16. JE Hansen, O Lund, N Tolstrup, AA Gooley, KL Williams, and S Brunak. NetOglyc: Prediction of mucin type O-glycosylation sites based on sequence context and surface accessibility. *Glycoconjugate J.*, 15:115–130, 1998.
17. R Gupta, H Birch, K Rapacki, S Brunak, and JE Hansen. O-GLYCBASE version 4.0: a revised database of O-glycosylated proteins. *Nucleic Acids Res.*, 27:370–372, 1999.
18. GW Hart, KD Greis, LY Dong, MA Blomberg, TY Chou, MS Jiang, EP Roquemore, DM Snow, LK Kreppel and RN Cole. O-linked N-acetylglucosamine: the 'yin-yang' of Ser/Thr phosphorylation? Nuclear and cytoplasmic glycosylation. *Adv. Exp. Med. Biol.*, 376:115–123, 1995.
19. DM Snow and GW Hart. Nuclear and Cytoplasmic Glycosylation. *Int. Rev. Cytol.*, 181:43–74, 1998.
20. FI Comer and GW Hart. O-Glycosylation of Nuclear and Cytosolic Proteins: Dynamic Interplay Between O-GlcNAc and O-Phosphate. *J. Biol. Chem.*, 275:29179–29182, 2000.
21. N Blom, S Gammeltoft, and S Brunak. Sequence and structure-based prediction of eukaryotic protein phosphorylation sites. *J. Mol. Biol.*, 294:1351–1362, 1999.
22. M Rechsteiner and SW Rogers. PEST sequences and regulation by proteolysis. *Trends Biochem. Sci.*, 21:267–271, 1996.
23. EG Krebs. The growth of research on protein phosphorylation. *Trends Biochem. Sci.*, 19:439, 1994.
24. R Gupta, E Jung, AA Gooley, KL Williams, S Brunak, and J Hansen. Scanning the available *Dictyostelium discoideum* proteome for O-linked GlcNAc glycosylation sites using neural networks. *Glycobiology*, 9:1009–1022, 1999.

Literature Data Mining for Biology

Lynette Hirschman
The MITRE Corporation

Jong C. Park
KAIST

Junichi Tsujii
University of Tokyo

Cathy Wu
Georgetown University

Limsoon Wong
Kent Ridge Digital Labs

Even though the number and the size of sequence databases are growing rapidly, most new information relevant to biology research is still recorded as free text in journal articles and in comment fields of databases like the GenBank feature table annotations. As biomedical research enters the post-genome era, new kinds of databases that contain information beyond simple sequences are needed, for example, information on cellular localization, protein-protein interactions, gene regulation and the context of these interactions. The forerunners of such databases include KEGG [1], DIP [2], BIND [3], among others. Such databases are still small in size and are largely hand curated. A factor that can accelerate their growth is the development of reliable literature data mining technologies.

This year is the third time the Pacific Symposium on Biocomputing has devoted an entire session to natural language processing and information extraction for biology. Compared to the last two years, the field has made tremendous strides. Most of the early work on automated understanding of biomedical papers concentrated on analytical tasks such as identifying protein names [4] or relied on simple techniques such as word co-occurrence[5] and pattern matching[6]. Last year, we began to see work based on more general natural language parsers that could handle considerably more complex sentences [7,8]. This year, we see the emergence of more sophisticated natural language technologies that can handle anaphora, as well as extracting a broader range of information.

Six papers were accepted under peer-review out of a total of seventeen submissions reviewed for this session. We briefly introduce them here:

• The paper by Ding *et al.* examines an issue that is fundamental to literature data mining based on term co-occurrence methods. It systematically compares the impact on recall and precision of mining interaction information when an abstract, a sentence, or a phrase is used as the unit in which to check for term co-occurrence.

• The paper by Hahn *et al.* describes the MEDSYNDIKATE natural language processor designed for acquiring knowledge from medical reports. The system is capable of analysing co-referring sentences and is also capable of extracting new concepts given a set of grammatical constructs.

• The paper by Leroy *et al.* presents the medical parser of the GeneScene system. An interesting aspect of this parser is that it uses prepositions as entry points into phrases in the text, in contrast to earlier approaches which used verbs as entry points. It then fills in a set of basic templates of patterns of prepositions around verbs and nominalized verbs. It also has a set of rules for combining these templates to extract information from more complex sentences.

• The paper by Pustejovsky *et al.* gives us a robust parser for identifying and extracting inhibition relations from biomedical literature. The system is founded on corpus-based linguistics. A particularly interesting feature of this system is its anaphora resolution module. The results reported in this paper focus on *inhibition* relations and demonstrate that it is possible to extract biologically important information from free text with high reliability using a classical approach.

• The paper by Stapley *et al.* is an interesting combination of text processing and machine learning technologies to predict the cellular location of proteins. The performance of the classifier on a benchmark of proteins with known cellular locations is better than a support vector machine trained on amino acid composition and is comparable to an expertly hand-crafted rule-based classifier.[9]

• The paper by Wilbur formalizes the idea of a "theme" in a collection of documents as a subset of the documents and a subset of the indexing terms such that each element of the latter has a high probability of occurring in all elements of the former. An algorithm is then given to produce themes and to cluster documents according to these themes in an optimal way. Results of applying this method to over fifty thousand documents on AIDS are given as an illustration.

The response to the call for papers and the quality of the submitted papers mark this as an emerging field which combines bioinformatics and natural language processing in innovative and productive ways. We find this very encouraging, but we also feel that much research and development remains to be carried out. In particular, the papers in this session illustrate both the promise of literature data mining and the need for challenge evaluations. On the one hand, they show how current language processing approaches can be successfully used to extract and organize information from the literature. On the other, they illustrate the diversity of applications and evaluation metrics. By defining several biologically important challenge problems and by providing the associated infrastructure (annotated data and a common evaluation framework), we can accelerate progress in this field. This will allow us to compare approaches, to scale up the technology to tackle important problems, and to learn what works and what areas still need work. For this purpose, we have organized an additional special session on literature data mining at this Pacific Symposium on Biocomputing to specifically discuss these challenges and benchmarks.

References

1. H. Ogata et al. KEGG: Kyoto Encyclopedia of Genes and Genomes. *Nucleic Acids Res.*, 27(1):29–34, January 1999.
2. I. Xenarios, D.W. Rice, L. Salwinski, M.K. Baron, E.M. Marcotte, and D. Eisenberg. DIP: The database of interacting proteins. *Nucleic Acid Res*, 28(1):289–291, January 2000.
3. G.D. Bader, I. Donaldson, C. Wolting, B.F. Ouellette, T. Pawson, and C.W. Hogue. BIND–the biomolecular interaction network database. *Nucleic Acids Res*, 29(1):242–245, January 2001.
4. K. Fukuda, A. Tamura, T. Tsunoda, and T. Takagi. Toward information extraction: Identifying protein names from biological papers. *Proc. of PSB*, pp. 707–718, Maui, Hawaii, January 1998.
5. B.J. Stapley and G. Benoit. Biobibliometrics: Information retrieval and visualization from co-occurrences of gene names in MEDLINE abstracts. *Proc. of PSB*, pp. 529–540, 2000.
6. S.-K. Ng and M. Wong. Toward routine automatic pathway discovery from on-line scientific text abstracts. *Genome Informatics*, 10:104–112, December 1999.
7. J.C. Park, H.S. Kim, and J.J. Kim. Bidirectional incremental parsing for automatic pathway identification with combinatory categorial grammar. *Proc. of PSB*, pp. 396–407, 2001.
8. A. Yakushiji, Y. Tateisi, Y. Miyao, and J. Tsujii. Event extraction from biomedical papers using a full parser. *Proc. of PSB*, pp. 408–419, 2001.
9. F. Eisenhaber and P. Bork. Evaluation of human-readable annotation in biomolecular sequence databases with biological rule libraries. *Bioinformatics*, 15:528–535, 1999.

MINING MEDLINE: ABSTRACTS, SENTENCES, OR PHRASES?

J. DING[a], D. BERLEANT[a,d], D. NETTLETON[b], AND E. WURTELE[c]
[a]*Department of Electrical and Computer Engineering,*
[b]*Department of Statistics,*
[c]*Department of Botany,*
[d]*berleant@iastate.edu*
Iowa State University, Ames, Iowa 50011, USA

A growing body of works address automated mining of biochemical knowledge from digital repositories of scientific literature, such as MEDLINE. Some of these works use abstracts as the unit of text from which to extract facts. Others use sentences for this purpose, while still others use phrases. Here we compare abstracts, sentences, and phrases in MEDLINE using the standard information retrieval performance measures of recall, precision, and effectiveness, for the task of mining interactions among biochemical terms based on term co-occurrence. Results show statistically significant differences that can impact the choice of text unit.

1 Introduction

The rapid growth of digitally stored scientific literature provides increasingly attractive opportunities for text mining. Concurrently, text mining is becoming an increasingly well-understood alternative to manual information extraction. Most reports on text mining of scientific literature for biochemical interactions have used the MEDLINE repository. Such mining activities have great potential for tasks such as extracting networks of protein interactions as well as for benefiting researchers who need to efficiently sift through the literature to find work relating to small sets of biochemicals of interest. While deep, fully automated literature analysis via natural language understanding (NLU) is an intriguing long-term objective, shallower and human-assisted analysis is both achievable and valuable.

The text processing units from which facts are extracted in MEDLINE mining systems may be the full abstracts, constituent sentences, or phrases. The most basic way to "mine" MEDLINE is simply to use the PUBMED Web interface.[8] The user can submit a query to the database consisting of the AND of two biochemical terms, and abstracts in MEDLINE containing both terms are returned. Such abstracts can be used as monolithic data items in systems that automatically search for interactions among genes based on term co-occurrence within an abstract, as in Stapley and Benoit 2000.[16] A related approach by Shatkay *et al.*[14] infers functional relationships among genes based on similarities among abstracts. Neither of those works identifies the type of interaction (e.g. inhibit, activate, etc.), which is desirable for applications such as automatic construction of networks of interactions. Because an abstract is a relatively large processing unit which contains a great deal of material besides the query terms, it is relatively difficult to automatically determine the type of interaction between the terms without methods that are sensitive to smaller text units such as sentences or phrases.

Easier inference of type of interaction might be expected if retrieval is limited to cases in which the query terms co-occur in the same sentence (Craven and Kumlien 1999,[2] Dickerson *et al.* 2001,[4] Ng and Wong 1999,[6] Rindflesch *et al.* 1999 & 2000,[9,10] Sekimizu *et al.* 1998,[12] Tanabe *et al.* 1999[17]), or in the same phrase (Blaschke *et al.*,[1] Humphreys *et al.*,[5] Ono *et al.*[7]). But such systems will miss interactions that are described over a longer passage, such as this one:

```
...in wild oat aleurone, two genes, alpha-Amy2/A
and alpha-Amy2/D, were isolated. Both were shown
to be positively regulated by gibberellin (GA)
during germination...
```
[21]

The interactions in this example (gibberellin regulates alpha-Amy2/A and alpha-Amy2/D) are described over two sentences, so to extract the interactions in this example a system needs to process text units longer than a sentence. Thus while smaller text units might make it easier to infer many interactions, they will miss others interactions that are expressed over longer passages. Consequently, information retrieval recall must decrease with decreasing text unit size. However a clean qualitative relationship between text unit size and information retrieval precision cannot be inferred from first principles.

Considerations like these revolve around the issue of what the advantages and disadvantages are of different text units, from the standpoint of systems that automatically extract interactions among biochemical terms. This is important when a choice of text processing unit must be made for a text mining system design. Four text units are investigated here: abstracts, adjacent sentence pairs, sentences, and phrases, from the perspective of three standard information retrieval (IR) performance measures: recall, precision, and effectiveness. Recall is the fraction of the relevant items in a test set that are retrieved. Precision is the fraction of retrieved items that are also relevant. Effectiveness is a composite measure combining the recall and the precision. The benefit of the present investigation of the relationships between text unit type and information retrieval performance measures is better understanding of the ability of the different text units to support mining of scientific abstract repositories for interactions among biochemicals.

2 Experimental Procedure: The Data

To compare the merits of different text processing units, a corpus of slightly over three hundred abstracts, termed the Interaction Extraction Performance Assessment (IEPA) corpus, was manually analyzed. The corpus consists of abstracts retrieved from MEDLINE using ten queries (Table 1) to its PUBMED interface.[8] Each query was the AND of two biochemical nouns. The queries were suggested by colleagues who are actively performing research in diverse biological areas, to help make them representative of the kinds of queries users of text mining systems would be interested in. A suggested query was studied only if the number of abstracts retrieved by PUBMED was ten or more to facilitate statistical analysis of results. If more than 100 abstracts conforming to a given query were retrieved, only the most recent abstracts at the time the corpus was defined were studied, enough so that the studied set included approximately forty abstracts describing interaction(s) between the biochemicals in the query, plus those that contained the biochemicals but did not describe interactions between them that were also encountered. Thus the ten queries yielded ten sets of abstracts, with each abstract in a set containing both terms in the query corresponding to that set.

Although each studied abstract contained both biochemical terms in a query, only some of them described interaction(s) between them. An interaction between two terms was defined as a direct or indirect influence of one on the quantity or activity of the other. Examples of interactions between terms `A` and `B` include the following.

- `A increased B.`
- `A activated C, and C activated B.`
- `A-induced increase in B is mediated through C.`
- `Inhibition of C by A can be blocked by an inhibitor of B.`

The following examples do not indicate an interaction between A and B.

- `A increases C, and B also increases C.`
- `C decreases A and B.`

Below are some examples taken from MEDLINE abstracts. Only the smallest text unit containing an interaction is noted, but the interaction is necessarily also present in any larger text unit as well.

`...whereas a combination of` **`gibberellin`** `plus cycloheximide treatment was required to increase alpha-`**`amylase`** `mRNA levels to the same extent.` (PMID is 10198105, query is `gibberellin` and `amylase`, interaction is described within a phrase.)

`...the regulation of hypothalamic` **`NPY`** `mRNA by` **`leptin`** `may be impaired with age.` (PMID is 10868965, query is `leptin` and `NPY`, interaction is described within a phrase.)

`We investigated mechanisms underlying the control of this movement by acetylcholine using an insulinoma cell line, MIN6, in which acetylcholine increases both` **`insulin`** `secretion and granule movement. The peak` activation of movement was observed 3 min after an acetylcholine challenge. The effects were nullified by the muscarinic inhibitor atropine, phospholipase C (**PLC**) inhibitors (D 609 and compound 48/80), and pretreatment with the Ca2+ pump inhibitor, thapsigargin. (PMID is 9792538, query is insulin and PLC, interaction is described within the abstract.)

An abstract was defined to consist of both title and body. A sentence pair was defined as two adjacent sentences. All but the first and last sentence in an abstract therefore appeared in two sentence pairs, once as the first of the pair and once as the second. The text between two successive periods was defined to be a sentence. In addition, the title was defined to be a sentence, as was the body up to the first period. The text between any two successive punctuation marks {. : , ;} was defined as a phrase. The title up to its first punctuation mark was also defined as a phrase, as was a complete title containing no punctuation mark, and also the body of the abstract up to the first punctuation mark.

While both members of the query occurred in each abstract, in only some of the abstracts did both terms or their synonyms occur within adjacent sentences. In only some of these sentence pairs did both occur within just one sentence of the pair. Finally, in only some of those sentences did both occur in the same phrase.

3 Experimental Procedure: Measuring Information Retrieval Quality

Recall and precision measure the completeness and correctness of information retrieval, respectively. Effectiveness assesses overall performance by combining both recall and precision,[15] while a generalized form of effectiveness includes the relative weights of recall and precision as a parameter in the calculation.[19]

In the present case, recall is the fraction of all those interactions between two biochemical terms in the corresponding set of abstracts that are stated within a sentence, phrase, or other text unit under consideration:

$$\text{recall} = \frac{\text{\# of interactions between A and B occurring within a type of text unit}}{\text{\# of interactions between A and B occurring within abstracts}}$$

where A and B are query terms or their synonyms.

Intuitively, recall here measures the capacity of a given text unit to contain the interactions present in MEDLINE abstracts. Any interaction described within a particular text unit is also described within all larger text units. Therefore, since the largest unit considered here is the abstract the recall for abstracts is exactly 1.

Precision refers to the fraction of abstracts, sentences, phrases, etc. containing both biochemical terms that also describe an interaction between them:

$$\text{precision} = \frac{\#\,\text{of interactions between A and B occurring within a type of text unit}}{\#\,\text{of times A and B co-occur in that type of text unit}}$$

where A and B are query terms or their synonyms. Intuitively, precision here measures the richness of a given text unit as "ore" from which to mine biochemical interactions from term co-occurrences.

Effectiveness combines recall and precision with the harmonic mean (the reciprocal of the arithmetic mean of the reciprocals, appropriate e.g. for calculating average travel speed for a trip):

$$\text{effectiveness} = \frac{1}{\frac{1}{2}\cdot\frac{1}{\text{recall}} + \frac{1}{2}\cdot\frac{1}{\text{precision}}} = \frac{2\times\text{recall}\times\text{precision}}{\text{recall}+\text{precision}}$$

Generalized effectiveness (G) parameterizes effectiveness with a weight coefficient w specifying the relative weights given to recall and precision:

$$G = \frac{1}{w\cdot\frac{1}{\text{recall}} + (1-w)\cdot\frac{1}{\text{precision}}} = \frac{\text{recall}\times\text{precision}}{w\times\text{precision} + (1-w)\times\text{recall}}, \quad 0\le w\le 1.$$

Generalized effectiveness can account for differences among applications and users in their needs for recall compared to precision.

4 Data Analysis

Information retrieval performances for abstracts, sentence pairs, sentences, and phrases were assessed by tabulating, for each query and each text unit, term co-occurrences and the subset of co-occurrences describing interactions. The recall, precision, and effectiveness of each were then tabulated (Tables 1 and 2). Because preliminary study showed that often an interaction is described using a synonym of a query term rather than the query term itself, occurrences of synonyms were treated as occurrences of query terms.

Table 1. Queries and the recall, precision, and effectiveness for each, given abstracts (Ab), sentences (Se), and phrases (Ph) as text units from which to extract interactions between the query terms or their synonyms, in MEDLINE abstracts containing both query terms. (The last query, an outlier, is discussed further in Appendix A.)

Query terms	Recall			Precision			Effectiveness		
	Ab	Se	Ph	Ab	Se	Ph	Ab	Se	Ph
insulin & PLC	1	.80	.54	.38	.58	.69	.55	.68	.61
leptin & NPY	1	.88	.53	.52	.46	.53	.69	.60	.53
AVP & PKC	1	.85	.60	.83	.65	.78	.91	.74	.68
Beta-amyloid & PLC	1	.86	.71	.67	.83	.89	.80	.85	.79
prion & kinase	1	.79	.71	.70	.79	.77	.82	.79	.74
UCP & leptin	1	.96	.69	.53	.57	.73	.69	.71	.71
insulin & oxytoxin	1	.89	.65	.45	.63	.73	.62	.74	.69
gibberellin & amylase	1	.89	.71	.95	.94	.96	.97	.92	.82
oxytoxin & IP	1	.98	.80	.68	.73	.77	.81	.83	.79
flavonoid & cholesterol	1	.25	.10	.55	.50	.50	.71	.33	.17

Table 2. Information retrieval measures for different types of text units. Recall and precision figures are means over the relevant figures for each query (shown in Table 1 for all text unit types except sentence pairs). Each figure was appropriately weighted, by the number of abstracts in the set associated with that query (in the case of precision of abstracts), the number of co-occurrences for that query within the text unit under consideration (in the case of precision of sentence pairs, sentences, and phrases), or by the number of interactions described for that query within the associated set of abstracts (for recall).

TEXT UNIT → ↓ IR MEASURE	Abstracts	Sentence pairs	Sentences	Phrases
Recall	1	0.916	0.849	0.621
Precision	0.571	0.345	0.638	0.743
Effectiveness	0.727	0.501	0.729	0.677

Table 2 suggests a trend of increasing precision for smaller text units, except for sentence pairs which rated poorly overall. Phrases, the smallest unit, had the highest precision. Precision differences were significant at the 0.05 level except in the case of abstracts vs. sentences (Appendix B).

With respect to effectiveness, sentences were significantly better than phrases at the 0.05 level, indicating that the advantage of phrases over sentences in precision is outweighed by the disadvantage in recall. Abstracts measured about equal to sentences in effectiveness. The measured effectiveness advantage of abstracts over phrases did not reach significance (p=0.17 two-tailed). Abstracts, sentences, and phrases all rated significantly higher than sentence pairs.

Application of the generalized effectiveness formula to the figures in Table 2 rates abstracts as most effective when recall is of overriding concern, phrases as most effective when precision is of overriding concern, and sentences as most effective over an intermediate range of weightings (Table 3).

Table 3. Ranges of weight parameter w for which each text unit measured as best in generalized effectiveness (w can range between 0 and 1).

TEXT UNIT →	Abstract	Sentence pair	Sentence	Phrase
$w \rightarrow$	$w>0.511$	–	$0.339<w<0.510$	$w<0.338$

5 Discussion and Conclusion

In view of the results reported here it is not surprising that researchers have reported interesting results for text mining in MEDLINE based on abstracts, sentences, and phrases. Tables 2 and 3 and the statistical significance summary in the preceding section indicate that each of these units has advantages and disadvantages compared to the others.

Sentence pairs fared so poorly in precision that an analysis was undertaken to understand why. Although considering pairs of sentences nearly doubled (99%) the number of distinct co-occurrences found compared to limiting consideration to sentences, the number of distinct interactions went up by only 8%. In fact, the dominant contributor to the already low precision of sentence pairs is interactions that are actually described in a single sentence within the pair. For the remaining interactions, those for which each term was in a different sentence, the precision was a mere 0.05. This in turn suggests that compared to the effort it would take to build a system to extract biochemical interactions from sentences, it might not be worth much additional effort to deal with sentence pairs as well. Even large expenditures of computation time or system development effort to achieve quality anaphora resolution across adjacent sentences would result in only modest benefit.

Regardless of the text unit chosen for a system that extracts biochemical interactions from MEDLINE, interactions contained in an abstract were often described using a synonym of the query term. Thus we counted synonyms as query term instances in deriving the retrieval performance measures reported here.

Increasing the sophistication of text processing can raise precision without degrading recall, thereby raising effectiveness, as suggested by Craven and Kumlein's[2] Figure 2 and accompanying discussion, and by Thomas *et al.*'s[18] Table 5. Sophisticated text processing techniques seem likely to benefit smaller text units more than larger ones because their generally shorter lengths, simpler structures, and higher proximity of relevant verbs and biochemical nouns make their processing more tractable. For example, appropriate verbs ("bind," "activate," etc.) in close proximity to biochemical terms are likely to be better indicators of an interaction than more distant verbs. However, ease of analysis would not be an issue if complete automatic natural language understanding were available, which would in principle enable precisions of 1 for all text units. This would swing the advantage back to longer text units because the principle of decreasing recall for smaller text units, in conjunction with the theoretical possibility of the same precision, 1, for all text units implies potential superiority of longer text units. However, complete automatic natural language understanding is currently not possible nor is it likely to be for some time. Effectiveness figures for the current state of the art for biochemical interaction extraction using sophisticated text processing were derivable from two reports, summarized in table 4.

Table 4. Effectiveness of sophisticated text processing techniques is higher than the baseline figures in Table 2 above for both the sentence and phrase text units. For phrases, sophisticated techniques led to an effectiveness higher than that of any entry in Table 1 above. (However comparisons across reports should be interpreted with caution.)

Report	Comment	Text unit	Best effectiveness	Baseline average	Baseline range
Rindflesch *et al.*[9]	"RESULTS" section	Sentence	0.75	0.72	0.33-0.92
Ono *et al.*[7]	Their Table 3	Phrase	0.89	0.65	0.17-0.82

Sophistication in text processing techniques can be important for reasons other than improving IR performance. For example, automatic construction of signal transduction pathways is an application that requires accounting for verbs.

Another application that clearly favors small text units is the simultaneous display of targeted passages from the often unwieldy body of scientific literature. It is better for this purpose to display sets of relevant sentences or phrases taken from numerous abstracts on a screen than it is to display one or two entire abstracts with occasional embedded relevant passages, particularly if it is convenient to move from a short relevant passage to its containing abstract, such as by clicking.

In summary, abstracts, sentences, and phrases are all competitive for automatic extraction of interactions among biochemicals from MEDLINE. Not surprisingly, sophisticated text processing appears to increase IR performance relative to more basic text processing. However, a very large range of choices is possible in designing systems with advanced text processing capabilities. For example, just defining a set of verbs that indicate interactions will be difficult to characterize definitively. To provide a relatively clean baseline we avoided verb analysis, although a suitable accounting of verbs might be expected to increase precision particularly for smaller text units.

Appendix A: An Outlier Query

It is interesting to consider an outlier from among our ten queries. For the query "cholesterol AND flavonoid," smaller text units fared more poorly than for other queries (Table 1). Closer inspection of these abstracts showed that flavonoid is a large family of chemicals, and the name of a specific flavonoid is usually stated in the first sentence of an abstract. In the rest of the abstract, the name of the specific flavonoid is used instead of the general term "flavonoid." Therefore the term "flavonoid" tends to be distant from the term "cholesterol" in the abstracts, leading to relatively low recall, precision, and hence effectiveness for sentence pairs, sentences, and phrases. This factor should be considered in the context of particularly general chemical terms.

Appendix B: Statistical Procedure

We conducted separate analyses for precision and effectiveness. The structure of the data suggests an analysis based on the usual linear model for a block design, where each query serves as a block. The model often used for such data is

$$Y_{ij} = \mu + \alpha_i + \beta_j + e_{ij}$$

where Y_{ij} denotes a measure of information retrieval quality (recall, precision, or effectiveness) for the method using the ith processing unit (i=1, 2, 3, or 4 corresponding to abstract, sentence pair, single sentence, or phrase, respectively) on the jth set of abstracts. The e_{ij} represent independent random errors with mean zero and variance σ^2/w_j, where w_j is a weight equal to the number of abstracts used in the determination of Y_{ij}. We assume that the distribution of $d_{ii'j}=e_{ij} - e_{i'j}$ is symmetric for all $i \neq i'$ and j=1,...,10. The parameters $\alpha_1,\ldots,\alpha_4$ represent the statistical effects associated with the processing units. These are the quantities of interest. The α_i are typically constrained to sum to zero for easier interpretation, and the μ parameter is introduced as an intercept. Thus α_i greater (less) than zero implies above (below) average performance for the ith method relative to the others for any particular chemical pair. The $\beta_1,\ldots\beta_{10}$ quantities are the statistical effects associated with each of the 10 sets of abstracts corresponding to the 10 queries.

For the IR performance measures of precision and effectiveness, we are interested in testing for differences among pairs of text units. For two different text units indexed by i and i', we may formally write our null and alternative hypotheses as $H_{ii'}: \alpha_i = \alpha_{i'}$ and $K_{ii'}: \alpha_i \neq \alpha_{i'}$, respectively. To test $H_{ii'}$ against $K_{ii'}$ we will compute the usual weighted t-statistic using the differences $D_{ii'j} = Y_{ij} - Y_{i'j}$ with weights $w_j (j = 1,\ldots,10)$. The formula for the weighted t-statistic is

$$t_{ii'} = \overline{D}_{ii'} / s(\overline{D}_{ii'}), \text{ where}$$

$$\overline{D}_{ii'} = \sum_{j=1}^{10} w_j D_{ii'j} / \sum_{j=1}^{10} w_j \text{ and}$$

$$s(\bar{D}_{ii'}) = \sqrt{\sum_{j=1}^{10} w_j (D_{ii'j} - \bar{D}_{ii'})^2 / \{(10-1)\sum_{j=1}^{10} w_j\}}.$$

To assess the significance of an observed value of $t_{ii'}$, we condition on the magnitudes of the observed differences and note that under the null hypothesis the probability of a positive difference is equal to the probability of a negative difference. This follows from the fact that $D_{ii'j}=Y_{ij}-Y_{i'j}=\alpha_i-a_{i'}+d_{ii'j}=d_{ii'j}$ when the null hypothesis is true. Now, under the null hypothesis, all 2^{10} possible assignments of signs to $|D_{ii'1}|, \ldots |D_{ii'10}|$ are equally likely assuming $d_{ii'j} = e_{ij} - e_{i'j}$ are independent for $i \neq i'$. Thus the conditional null distribution of $t_{ii'}$ places probability mass $1/2^{10}$ on each of the 2^{10} values obtained by computing $t_{ii'}$ for the 2^{10} possible assignments of signs to $|D_{ii'1}|, \ldots |D_{ii'10}|$. The relevant two-tailed p-value is obtained by counting the proportion of those 2^{10} values whose magnitudes match or exceed the observed value of $|t_{ii'}|$. This is essentially the randomization test for matched pairs described, for example, in Section 5.11 of Conover.[3] We have augmented this slightly by using the number of abstracts as weights in our test statistic to account for variation in the number of abstracts used to compute the measures of performance.

To illustrate the testing procedure we used, consider testing for a difference between the effectiveness of sentence pairs and single sentences. The relevant differences (one for each query) are -0.19, -0.23, -0.28, -0.18, -0.17, -0.28, -0.24, -0.22, -0.25, and +0.14. The preponderance of negative signs immediately suggests greater effectiveness for the single sentence method. The weighted t-statistic is $t_{23} = -5.97$. If we were to randomly assign signs to the observed differences, the chance of getting a weighted t-statistic as far from zero as -5.97 is only 6/1024 $\approx$ 0.0059. This is the p-value of the test, and it can be computed by calculating that there are only 6 sign configurations (among the 1024 possible configurations) that yield a t-statistic, weighted to reflect the number of examined abstracts associated with each query, as far from zero as -5.97.

Because it is so unlikely (probability 0.0059) to see a value of the test statistic as extreme as -5.97 when the null hypothesis is true, we reject the null hypothesis and conclude that single sentences are significantly more effective than sentence pairs. Other results for effectiveness, and results for precision, are shown in Table 5.

Two columns of Table 5 contain p-values that have been adjusted for multiple testing using the restricted step-down method,[13] for which a clear description is provided in Section 2.7 of Westfall and Young.[20] The use of adjusted p-values is conservative and reduces the chance of errantly rejecting a true null hypothesis simply because many hypotheses are being tested. Motivation for the use of adjusted p-values may be found in several statistical texts on the subject of simultaneous inference.

Table 5. Tests of null hypotheses of no difference between text units. Sentences and phrases are significantly different. Precision of phrases is significantly different from that of abstracts, while other cells do not reach significance. (Ab=abstract, Se=sentence, and Ph=phrase.)

Comparison	Precision			Effectiveness		
	Weighted t-statistic	P-value	Adjusted p-value	Weighted t-statistic	P-value	Adjusted p-value
Ab vs. Se	-1.34	0.3516	0.3516	0.25	0.8398	0.8398
Ab vs. Ph	-3.00	0.0488	0.0488	1.36	0.1719	0.1719
Se vs. Ph	-5.14	0.0078	0.0234	5.26	0.0039	0.0117

References

1. C. Blaschke, M. Andrade, C. Ouzounis, and A. Valencia, "Automatic extraction of biological information from scientific text: protein-protein interactions" *AAAI Conference on Intelligent Systems in Molecular Biology*, 60-67 (1999).
2. M. Craven and J. Kumlien, "Constructing biological knowledge based by extracting information from text sources" *7th International Conference on Intelligent Systems for Molecular Biology (ISMB-99).*
3. W. Conover, *Practical Nonparametric Statistics, 2nd Ed.* (Wiley, NY, 1980).
4. J. Dickerson, D. Berleant, Z.Cox, W. Qi, D. Ashlock, and E. Wurtele, "Creating metabolic network models using text mining and expert knowledge" *Atlantic Symposium on Computational Biology and Genome Information Systems & Technology (CBGIST 2001)*, 26-30.
5. K. Humphreys, G. Demetriou, and R. Gaizauskas, "Two applications of information extraction to biological science journal articles: enzyme interactions and protein structures" *Pacific Symposium on Biocomputing 5,* 502-513 (2000).
6. S.-K. Ng and M. Wong, "Toward routine automatic pathway discovery from on-line scientific text abstracts" *Genome Informatics* **10**, 104-112 (1999).
7. T. Ono, H. Hishigaki, A. Tanigami, and T. Takagi, "Automated extraction of information on protein-protein interactions from the biological literature" *Bioinformatics* **17**, 155-161 (2001).
8. PUBMED interface to MEDLINE, U.S. National Library of Medicine, http://www.ncbi.nlm.nih.gov/entrez/query.fcgi?db=PubMed.
9. T. Rindflesch, L. Hunter, and A. Aronson, "Mining molecular binding terminology from biomedical text" *Proceedings of the AMIA '99 Annual Symposium.*
10. T. Rindflesch, L. Tanabe, J. Weinstein, L. Hunter, "EDGAR: extraction of drugs, genes and relations from the biomedical literature" *Pacific Symposium on Biocomputing 5,* 514-525 (2000).

11. W. Salamonsen, K. Mok, P. Kolatkar, and S. Subbiah, "BioJAKE: a tool for the creation, visualization and manipulation of metabolic pathways" *Pacific Symposium on Biocomputing 4,* 392-400 (1999).
12. T. Sekimizu, H. Park, and J. Tsujii, "Identifying the interaction between genes and gene products based on frequently seen verbs in MEDLINE abstracts" *Genome Informatics* (Universal Academy Press, Inc., 1998).
13. J. Shaffer, "Modified sequentially rejective multiple test procedures" *Journal of the American Statistical Association* **81**, 826-831 (1986).
14. H. Shatkay, S. Edwards, W. Wilbur, and M. Boguski, "Genes, themes, and microarrays: using information retrieval for large-scale gene analysis" *8th Int. Conf. on Intelligent Systems for Mol. Bio. (ISMB 2000)*, La Jolla, Aug. 19-23.
15. W. Shaw, R. Burgin, and P. Howell, "Performance standards and evaluations in IR test collections: cluster-based retrieval models" *Information Processing and Management* **33** (1), 1-14 (1997).
16. B. Stapley, and G. Benoit, "Biobibliometrics: information retrieval and visualization from co-occurrences of gene names in medline abstracts" *Pacific Symposium on Biocomputing 5*, 529-540 (2000).
17. L. Tanabe, U. Scherf, L. Smith, J. Lee, L. Hunter, and J. Weinstein, "MedMiner: an internet text-mining tool for biomedical information, with application to gene expression profiling" *BioTechniques* **27**, 1210-1217 (1999).
18. J. Thomas, D. Milward, C. Ouzounis, S. Pulman, and M. Carroll, "Automatic extraction of protein interactions from scientific abstracts" *Pacific Symposium on Biocomputing* 5, 538-549 (2000).
19. C. Van Rijsbergen, *Information Retrieval*, Butterworths (1979).
20. P. Westfall, and S. Young, *Resampling-Based Multiple Testing: Examples and Methods for P-Value Adjustment* (Wiley, New York, 1993).
21. R. Willmott, P. Rushton, R. Hooley, and C. Lazarus, "DNase1 footprints suggest the involvement of at least three types of transcription factors in the regulation of alpha-Amy2/A by gibberellin" *Plant Molecular Biology* **38** (5), 817-825 (1998).
22. L. Wong, "A protein interaction extraction system" *Pacific Symposium on Biocomputing* 6, (2001).

CREATING KNOWLEDGE REPOSITORIES FROM BIOMEDICAL REPORTS: THE MEDSYNDIKATE TEXT MINING SYSTEM

UDO HAHN MARTIN ROMACKER

Text Knowledge Engineering Lab, Freiburg University, D-79098 Freiburg, Germany

http://www.coling.uni-freiburg.de

STEFAN SCHULZ

Department of Medical Informatics, Freiburg University Hospital, D-79104 Freiburg

http://www.imbi.uni-freiburg.de/medinf

MEDSYNDIKATE is a natural language processor for automatically acquiring knowledge from medical finding reports. The content of these documents is transferred to formal representation structures which constitute a corresponding text knowledge base. The system architecture integrates requirements from the analysis of single sentences, as well as those of referentially linked sentences forming cohesive texts. The strong demands MEDSYNDIKATE poses to the availability of expressive knowledge sources are accounted for by two alternative approaches to (semi)automatic ontology engineering. We also present data for the knowledge extraction performance of MEDSYNDIKATE for three major syntactic patterns in medical documents.

1 Introduction

The application of methods from the field of natural language processing to biological data has long been restricted to the parsing of molecular structures such as DNA [1,2]. More recently, however, efforts have also been directed to capturing content from biological documents (research reports, journal articles, etc.), either dealing with restricted information extraction problems such as name recognition for proteins or gene products [3,4,5], or more sophisticated ones which aim at the acquisition of knowledge relating to protein or enzyme interactions, molecular binding behavior, etc. [6,7,8,9].

Current information extraction (IE) systems, however, suffer from various weaknesses. First, their range of understanding is bounded by rather limited domain knowledge. The templates these systems are supplied with allow only factual information about particular, a priori chosen entities (cell type, virus type, protein group, etc.) to be assembled from the analyzed documents. Also, these knowledge sources are considered to be entirely static. Accordingly, when the focus of interest of a user shifts to (facets of) a topic not considered so far, new templates must be supplied or existing ones must be updated manually. In any case, for a modified set of templates the analysis has to be rerun for the entire document collection. Templates also provide either no or severely limited inferencing capabilities to reason about the template fillers – hence, their understanding depth is low. Finally, the potential of IE systems for dealing with textual phenomena is rather weak, if it is available at all. Reference relations spanning over several sentences, however, may cause invalid knowledge base structures to emerge so that incorrect information may be retrieved or inferred.

With the SYNDIKATE system family, we are addressing these shortcomings and aim at a more sophisticated level of knowledge acquisition from real-world texts. The source documents we deal with are currently taken from two domains, *viz.* test reports from the information technology domain (ITSYNDIKATE [10]) and medical finding reports, the framework of the MEDSYNDIKATE system [11]. MEDSYNDIKATE is designed to acquire from each input text a maximum number of simple facts ("The *findings correspond to* an *adenocarcinoma.*"), complex propositions ("*All* mucosa layers show an inflammatory infiltration that mainly consists of lymphocytes."), and evaluative assertions ("The findings correspond to a *severe chronical* gastritis."). Hence, our primary goal is to extract conceptually deeper and inferentially richer forms of relational information than that found by state-of-the-art IE systems. Also, rather than restricting natural language processing intentionally to few templates, we here present an open system architecture for knowledge extraction where text understanding is constrained only by the unpredictable limits of available knowledge sources, the domain ontology, in particular.

To achieve this goal, several requirements with respect to language processing proper have to be fulfilled. As most of the IE systems, we require our parser to be robust to underspecification and ill-formed input (cf. the protocols in [12]). Unlike almost all of them, our parsing system is particularly sensitive to the treatment of textual reference relations as established by various forms of anaphora [13]. Furthermore, since SYNDIKATE systems rely on a knowledge-rich infrastructure, particular care is taken to provide expressive knowledge repositories on a larger scale. We are currently exploring two approaches. First, we automatically enhance the set of already given knowledge templates through incremental concept learning routines [14]. Our second approach makes use of the large body of knowledge that has already been assembled in medical taxonomies and terminologies (e.g., the UMLS). That knowledge is automatically transformed into a description logics format and, after interactive debugging and refinement, integrated into a comprehensive medical knowledge base [15].

2 System Architecture

In the following, major design issues for MEDSYNDIKATE are discussed, with focus on the distinction between sentence-level and text-level analysis. We will then turn to two alternative ontology engineering methodologies satisfying the need for the (semi)automatic supply of large amounts of background knowledge.

The overall architecture of SYNDIKATE is summarized in Figure 1. The general task of any SYNDIKATE system consists of mapping each incoming text, T_i, into a

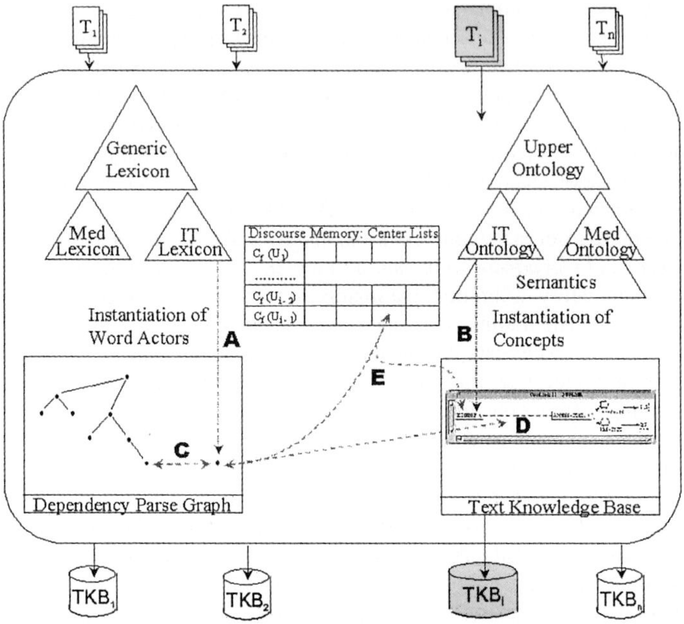

Figure 1: System Architecture of SYNDIKATE

corresponding *text knowledge base*, TKB_i, which contains a formal representation of T_i's content. This knowledge will be exploited by various information services, such as inferentially supported fact retrieval or text summarization.

2.1 Sentence-Level Understanding

Grammatical knowledge for syntactic analysis resides in a fully lexicalized dependency grammar (cf. [12] for details), we refer to as *Lexicon* in Figure 1. Basic word forms (lexemes) constitute the leaf nodes of the lexicon tree, while grammatical generalizations from lexemes appear as lexeme class specifications at different levels of abstraction. The *Generic Lexicon* in Figure 1 contains entries which are domain-independent (such as *move*, *with*, or *month*), while domain-specific extensions are kept in specialized lexicons serving the needs of particular subdomains, e.g., IT (*notebook*, *hard disk*, etc.) or medicine (*adenocarcinoma*, *gastric mucosa*, etc.).

Conceptual knowledge is expressed in a KL-ONE-like representation language (cf.[11] for details). These languages support the definition of complex concept descriptions by means of conceptual roles and corresponding role filler constraints which introduce type restrictions on possible fillers. Taxonomic reasoning can be defined as being primitive (following explicit links), or it can be realized by letting a classifier engine compute subsumption relations between complex conceptual descriptions. A distinction is made between concept classes (types) and instances (representing concrete real-world entities). Most lexemes (except, e.g., pronouns, prepositions) are directly associated with one (or, in case of polysemy, several) concept type(s). Accordingly, when a new lexical item is read from the input text, a dedicated process (word actor) is created for lexical parsing (step A in Figure 1), together with an instance of the lexeme's concept type (step B). Each word actor then negotiates dependency relations by taking syntactic constraints from the already generated dependency tree into account (step C), as well as conceptual constraints supplied by the associated instance in the domain knowledge (step D) [12]. As with the *Lexicon*, the ontologies we provide are split up between one that serves all applications, the *Upper Ontology*, while specialized ontologies account for the conceptual structure of particular domains, e.g., information technology (NOTEBOOK, HARD-DISK, etc.), or medicine (ADENOCARCINOMA, GASTRIC-MUCOSA, etc.).

Semantic knowledge is concerned with determining relations between instances of concept classes based on the interpretation of so-called minimal *semantically interpretable subgraphs* of the dependency graph. Such a subgraph is bounded by two content words (nouns, verbs, adjectives) which may be directly linked by a single dependency relation or indirectly by a sequence of dependency relations linking non-content words only (e.g., prepositions, auxiliaries). Hence, a conceptual relation may either be constrained by dependency relations (e.g., the *subject:* relation may only be interpreted conceptually in terms of AGENT or PATIENT roles), by intervening non-content words (e.g., some prepositions impose special role constraints, such as *"with"* does in terms of HAS-PART or INSTRUMENT roles), or it may only be constrained by conceptual compatibility between the concepts involved (e.g., for genitives) [16]. The specification of semantic knowledge shares many commonalities with domain knowledge. Hence, the overlap in Figure 1.

2.2 Text-Level Understanding

The proper analysis of textual phenomena prevents inadequate text knowledge representation structures to emerge in the course of sentence-centered analysis [17]. Consider the following text fragment:

(1) Der Befund entspricht einem hochdifferenzierten *Adenokarzinom.*
(The findings correspond to a highly differentiated *adenocarcinoma.*)

(2) *Der Tumor* hat einen Durchmesser von 2 cm.
(*The tumor* has a diameter of 2 cm.)

With purely sentence-oriented analyses, *invalid* knowledge bases are likely to emerge, when each entity which has a different denotation at the text surface is treated as a formally distinct item at the symbol level of knowledge representation, although different denotations refer literally to the same conceptual entity. This is the case for *nominal anaphora*, an example of which is given by the reference relation between the noun phrase *"Der Tumor" (the tumor)* in Sentence (2) and *"Adenokarzinom" (adenocarcinoma)* in Sentence (1). A false referential description appears in Figure 2, where TUMOR.2-05 is introduced as a new representational entity, whereas Figure 3 depicts the adequate, intended meaning at the conceptual representation level, *viz.* maintaining ADENOCARCINOMA.6-04 as the proper referent.

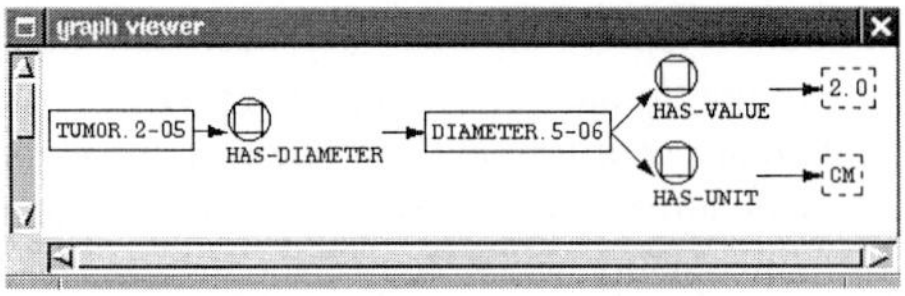

Figure 2: Unresolved Nominal Anaphora

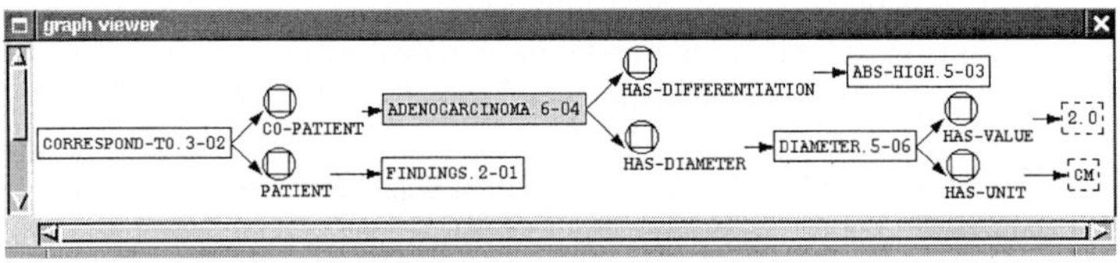

Figure 3: Resolved Nominal Anaphora

The methodological framework for tracking such reference relations at the text level is provided by *center* lists [13] (cf. step E in Figure 1). The ordering of their elements indicates that the most highly ranked element is the most likely antecedent of an anaphoric expression in the subsequent utterance, while the remaining elements are ordered according to decreasing preference for establishing referential links.

S_1	[FINDINGS.2-01: Befund, ADENOCARCINOMA.6-04: Adenokarzinom]
S_2	[ADENOCARCINOMA.6-04: Tumor, DIAMETER.5-06: Durchmesser, CM: cm]

Table 1: Center Lists for Sentences (1) and (2)

In Table 1, the tuple notation takes the conceptual correlate of each noun in the text knowledge base in the first place, while the lexical surface form appears in second place. Using the center list of Sentence (1) for the interpretation of Sentence (2) results in a series of queries whether FINDINGS is conceptually more special than TUMOR (answer: No) or ADENOCARCINOMA is more special than TUMOR (answer:

Yes). As the second center list item for S_1 fulfils all required constraints (mainly the one that ADENOCARCINOMA IS-A TUMOR), in the conceptual representation structure of Sentence (2), TUMOR.2-05, the literal instance (cf. Figure 2), is replaced by ADENOCARCINOMA.6-04, the referentially valid identifier (cf. Figure 3). As a consequence, instead of having two unlinked sentence graphs for Sentence (1) and (2) (e.g., cf. Figure 2) the reference resolution for nominal anaphora leads to joining them in a single coherent and valid text knowledge graph in Figure 3.

Given a fact retrieval application, the validity of text knowledge bases becomes a crucial issue. Disregarding textual phenomena will cause dysfunctional system behavior in terms of incorrect answers. This can be illustrated by a query Q such as

```
Q :  (retrieve ?x (Tumor ?x))
A-:  (Tumor.2-05, Adenocarcinoma.6-04)
A+:  (Adenocarcinoma.6-04)
```

which triggers a search for all instances in the text knowledge base that are of type TUMOR. Given an invalid knowledge base (cf. Figure 2), the incorrect answer (A-) contains two entities, *viz.* TUMOR.2-05 and ADENOCARCINOMA.6-04, since both are in the extension of the concept TUMOR. If, however, a valid text knowledge base such as the one in Figure 3 is given, only the correct answer, ADENOCARCINOMA.6-04, is inferred (A+).

2.3 Ontology Engineering

MEDSYNDIKATE requires a knowledge-rich infrastructure both in terms of grammar and domain knowledge, which can hardly be maintained by human efforts alone. Rather a significant amount of knowledge should be generated automatically. For SYNDIKATE systems, we have chosen a dual strategy. One focuses on the incremental learning of new concepts while understanding the texts, the other is based on the reuse of available comprehensive (though semantically weak) knowledge sources.

Concept Learning from Text. Extending a given core ontology by new concepts as a by-product of the text understanding process builds on two different sources of evidence — the already given domain knowledge, and the grammatical constructions in which unknown lexical items occur in the source document.

The parser yields information from the grammatical constructions in which lexical items occur in terms of the labellings in the dependency graph. The kinds of syntactic constructions in which unknown words appear are recorded and later assessed relative to the credit they lend to a particular hypothesis. Typical linguistic indicators that can be exploited for taxonomic integration are, e.g., appositions *('the symptom @A@'*, with '@A@' denoting the unknown word) or exemplification phrases *('symptoms like @A@')*, These constructions almost unequivocally determine '@A@' when considered as a medical concept to denote an instance of a SYMPTOM.

The conceptual interpretation of parse trees involving unknown words in the text knowledge base leads to the derivation of concept hypotheses, which are further enriched by conceptual annotations. These reflect structural patterns of consistency, mutual justification, analogy, etc. relative to already available concept descriptions in the ontology or other concept hypotheses. Grammatical and conceptual evidence of this kind, in particular their predictive "goodness" for the learning task, are represented by corresponding sets of linguistic and conceptual quality labels. Multiple concept hypotheses for each unknown lexical item are organized in terms of hypothesis spaces, each of which holds alternative or further specialized conceptual readings. An inference engine coupled with the classifier, the so-called quality machine, estimates the overall credibility of single concept hypotheses by taking the available set of quality labels for each hypothesis into account (cf. [14] for details).

Reengineering Medical Terminologies. The second approach makes use of the large body of knowledge that has already been assembled in comprehensive medical terminologies such as the UMLS [18]. The knowledge they contain, however, cannot be applied directly to MEDSYNDIKATE, because it is characterized by inconsistencies, circular definitions, insufficient depth, gaps, etc., and the lack of an inference engine.

The methodology for reusing weak medical knowledge consists of four steps [15]. First, we create automatically KL-ONE-style logical expressions by feeding a generator with data directly from the UMLS, i.e., the concepts and the semantic links between concept pairs. In a second step, the imported concepts, already in a logical format, are submitted to the classifier of the knowledge representation system (in our case, LOOM) in order to check whether the terminological definitions are consistent and non-circular. For those elements which are inconsistent, their validity is restituted and definitional circles are removed manually by a medical domain expert. In the final step the knowledge base which has emerged so far is manually rectified and refined (e.g., by checking the adequacy of taxonomic and partonomic hierarchies).

3 Evaluating Knowledge Extraction Performance

3.1 Evaluation Framework

In quantitative terms, SYNDIKATE is neither a toy system nor a monster. The *Generic Lexicon* currently includes 5,000 entries, while the *MED Lexicon* contributes 3,000 entries. Similarly, the *Upper Ontology* contains 1,500 concepts and roles, to which the *MED Ontology* adds 2,500 concepts and roles. However, recent experiments with reengineering the UMLS have resulted in a very large medical knowledge base with 164,000 concepts and 76,000 relations [15] that is currently under validation.

We extracted the text collection from the hospital information system of the University Hospital in Freiburg (Germany). All finding reports in histopathology from the first quarter of 1999 were initially included, altogether 4,973 documents. However, for the time being MEDSYNDIKATE covers especially the subdomain of gastro-intestinal diseases. Thus, 752 texts out of these 4,973 were extracted semi-automatically in order to guarantee a sufficient coverage of domain knowledge. From this collection, a random sample of 90 texts was taken and divided into two sets. 60 of them served as the training set which was used for parameter tuning of the system. The remaining 30 texts were then used to measure the performance of the MEDSYNDIKATE system with unseen data. The configuration of the system was frozen prior to analyzing the test set.

In the empirical study proper, three basic settings of dependency graphs were evaluated, *viz.* ones containing genitives, prepositional phrases, as well as constructions including modal verbs or auxiliaries. Genitives and prepositional phrases relate fundamental biomedical concepts via associated roles at the conceptual level. Modal and auxiliary verbs create a complex syntactic environment for the interpretation of verbs, and, hence, the conceptual representation of medical processes and events. For each instance of these configurations semantic interpretations were automatically computed the result of which was judged for accuracy by two skilled raters.

Still, the way how a (gold) standard for semantic interpretation can be set up is an issue of hot debates [19]. In fact, conceptually annotated medical text corpora do not exist at all, at least for the German language. At this level, the ontology we have developed eases judgements, since it is based on a fine-grained relation hierarchy with clear sortal restrictions for role fillers. In anatomy, e.g., we use relations such as ANATOMICAL-PART-OF, which is itself a subrelation of PHYSICAL-PART-OF and PART-OF, and specialize it in order to account for subtle PART-OF relationships. A very specific relation such as ANATOMICAL-PART-OF-MUCOSA refers to a precise subset of entities to be related by the interpretation process. Therefore, relating BRAIN to MUCOSA by ANATOMICAL-PART-OF-MUCOSA obviously would be considered as incorrect, whereas relating LAMINA-PROPRIA-MUCOSAE would be considered a reasonable interpretation.

3.2 Quantitative Analysis

The following tables contain data for both the training and the test set indicating the quality of knowledge extraction as obtained for the three different syntactic settings. Besides providing data for recall and precision, the tables are divided into two assessment layers: *"without interpretation"* means that the system was not able to produce an interpretation because of specification gaps, i.e., at least one of the two content words in a minimal dependency graph under consideration was not specified. Note that even for the training set which was intended to generate optimal results we were unable to formulate reasonable and generally valid concept definitions for some of the content words we encountered (e.g., for fuzzy expressions of locations: *"In der Tiefe der Schleimhaut"* (*"In the depth of the mucosa"*)). The second group *"with interpretation"* is divided into four categories. The label *correct (non-ambiguous)* qualifies, if just a single and correct conceptual relation was computed by the semantic interpretation process. However, if the result was correct but yielded more than one conceptual relation, the label *correct (ambiguous)* was assigned. An interpretation was considered *incorrect* when the conceptual relation was inappropriate. Finally, *NIL* was used to indicate that an interpretation was performed (both concepts for the content words were specified) but no conceptual relation could be computed.

Genitives. In the medical domain, as indicated by Table 2 the recall and precision values for the interpretation of genitives are very encouraging both for the training set (92% and 93%) and the test set (93% and 93%), respectively.[a] However, since genitives, in general, provide no additional constraints how the conceptual correlates of the two content words involved can be related, the number of ambiguous interpretations amounts to 13% and 36%, respectively.

	Training Set	Test Set
Recall	92%	93%
Precision	93%	93%
# occurrences ...	168	91
... **with interpretation**	158 (94%)	86 (95%)
[confidence intervals]	[89%-97%]	[90%-98%]
...... correct (non-ambiguous)	125.5 (75%)	48.5 (53%)
...... correct (ambiguous)	22 (13%)	33 (36%)
...... incorrect	6.5	3.5
...... NIL	4	1
... **without interpretation**	10 (6%)	5 (5%)

Table 2: Evaluation of Genitives

Auxiliaries and Modals. Table 3 contains the results for modal verbs or auxiliaries. A semantic interpretation of modal/auxiliary verb complexes relates a content-bearing verb with the conceptual correlate of the syntactic subject. In case of a passive construction the direct-object-to-subject normalization has to be carried out.

Recall and precision for the training set are high (94% and 98%, respectively) and, therefore, indicate that semantic interpretation can cope with almost all occurrences given an optimal degree of specification. The values for recall and precision dropped to 80% and 84%, respectively, in the test set. The increase of *NIL* results reveals that the granularity of the underlying domain model is insufficient as far as conceptual relations are concerned. Although the corresponding concepts are modelled, no conceptual relation between them could be determined.

[a] Confidence intervals for .95 probability are given in square brackets.

	Training Set	Test Set
Recall	94%	80%
Precision	98%	84%
# occurrences ...	131	55
... **with interpretation**	125 (95%)	52 (95%)
[confidence intervals]	[92%-99%]	[84%-99%]
...... correct (non-ambiguous)	122 (93%)	43,5 (79%)
...... correct (ambiguous)	1	0
...... incorrect	0	0,5
...... NIL	2	8 (15%)
... **without interpretation**	6 (5%)	3 (5%)

Table 3: Evaluation of Modal Verbs and Auxiliaries

Prepositional phrases (PPs) are crucial for the semantic interpretation of a text, since they introduce a wide variety of conceptual relations, such as spatial, temporal, causal, or instrumental ones. The importance of PPs is reflected by their relative frequency. In the training set and the test set, we encountered 1,108 prepositions, which is a little bit less than 10% of the words in both sets (approximately 11,300).[b] Provided also that the preposition's syntactic head and its modifier participate in the interpretation, at the phrase level, more than 25% of the texts' contents is encoded by PPs (certainly, this data also reflects a considerable degree of genre dependency).

	Training Set	Test Set
Recall	85%	85%
Precision	79%	81%
# occurrences ...	562	278
... **with interpretation**	548 (98%)	253 (91%)
[confidence intervals]	[96%-99%]	[86%-93%]
...... correct (non-ambiguous)	401,5 (71%)	167 (60%)
...... correct (ambiguous)	32,5 (6%)	37,5 (13%)
...... incorrect	43 (8%)	30,5 (11%)
...... NIL	71 (13%)	18 (6%)
... **without interpretation**	14 (2%)	25 (9%)

Table 4: Evaluation of Prepositional Phrases

Considering the results for semantic interpretation of PPs (cf. Table 4), the values for recall and precision are almost the same for the training set and the test set. Recall climaxed at 85% for both the training set and the test set, whereas precision reached 79% for the training set and 81% for the test set. Getting almost the same performance

[b]Only 940 of these 1,108 were included in the empirical analysis, since 168 did not form a minimal subgraph. Phrases like *"zum Teil"* (*"partly"*) map to a single meaning — as evidenced by the English translation correlate — and were therefore excluded.

for both sets also reveals a stable level of semantic interpretation of PPs.[c]

4 Conclusions

We have introduced MEDSYNDIKATE, a system for mining knowledge from biomedical reports. Emphasis was put on the role of various knowledge sources required for 'deep' text understanding. When turning from sentence-level to text-level analysis, we considered representational inadequacies when text phenomena were not properly accounted for and, hence, proposed a solution based on centering mechanisms.

The enormous knowledge requirements posed by our approach can only be reasonably met when knowledge engineering does not rely on human efforts only. Hence, a second major issue we have focused on concerns alternative ways to support knowledge acquisition and guarantee, this way, a reasonable chance for scalability of the system. We made two proposals. The first one deals with an automatic concept learning methodology that is fully embedded in the text understanding process, the other one exploits the vast amounts of medical knowledge assembled in various knowledge repositories such as the UMLS. We, finally, provided empirical data which characterizes the knowledge extraction performance of MEDSYNDIKATE in terms of three major syntactic structures, *viz.* genitives, modals and auxiliaries, and prepositional phrases. These reflect, at the linguistic level, fundamental categories of biomedical ontologies — states, processes, and actions.

References

1. D. B. Searls. String variable grammar: A logic grammar formalism for the biological language of DNA. *Journal of Logic Programming*, 24(1/2):73–102, 1995.
2. S. Leung, C. Mellish, and D. Robertson. Basic gene grammars and DNA-ChartParser for language processing of *Escherichia coli* promoter DNA sequences. *Bioinformatics*, 17(3):226–236, 2001.
3. F. Fukuda, T. Tsunoda, A. Tamura, and T. Takagi. Toward information extraction: Identifying protein names from biological papers. In *PSB 98 – Proceedings of the 3rd Pacific Symposium on Biocomputing*, pages 705–716. Maui, Hawaii, 4-9 January, 1998.
4. N. Collier, C. Nobata, and J. Tsujii. Extracting the names of genes and gene products with a hidden Markov model. In *COLING 2000 – Proceedings of the 18th International Conference on Computational Linguistics*, pages 201–207. Saarbrücken, Germany, 31 July - 4 August, 2000.

---

[c] Corresponding data for an alternative test scenario, knowledge extraction from information technology (IT) test reports, is not as favorable as for the medical domain. For PPs, 70% / 64% recall and 77% / 66% precision were measured for the training set and the test set, respectively. A reasonable argument why we achieved better results in the medical domain than in the IT world might be that in the medical texts a considerably lower degree of linguistic variation is encountered.

5. T. Ono, H. Hishigaki, A. Tanigami, and T. Takagi. Automated extraction of information on protein-protein interactions from the biological literature. *Bioinformatics*, 17(2):155–161, 2001.
6. M. Craven and J. Kumlien. Constructing biological knowledge bases by extracting information from text sources. In *Proc. of the 7th Intl. Conference on Intelligent Systems for Molecular Biology*, pages 77–86. Heidelberg, Germany, August 6-10, 1999.
7. C. Blaschke, M. A. Andrade, C. Ouzounis, and A. Valencia. Automatic extraction of biological information from scientific text: Protein-protein interactions. *Intelligent Systems for Molecular Biology*, 7:60–67, 1999.
8. K. Humphreys, G. Demetriou, and R. Gaizauskas. Two applications of information extraction to biological science journal articles: Enzyme interactions and protein structures. In *PSB 2000 – Proceedings of the 5th Pacific Symposium on Biocomputing*, pages 502–513. Honolulu, Hawaii, USA, 4-9 January, 2000.
9. T. Rindflesch, J. Rajan, and L. Hunter. Extracting molecular binding relationships from biomedical text. In *ANLP 2000 – Proceedings of the 6th Conference on Applied Natural Language Processing*, pages 188–195. Seattle, WA, USA, April 29 - May 4, 2000.
10. U. Hahn and M. Romacker. Content management in the SYNDIKATE system: How technical documents are automatically transformed to text knowledge bases. *Data & Knowledge Engineering*, 35(2):137–159, 2000.
11. U. Hahn, M. Romacker, and S. Schulz. How knowledge drives understanding: Matching medical ontologies with the needs of medical language processing. *Artificial Intelligence in Medicine*, 15(1):25–51, 1999.
12. U. Hahn, N. Bröker, and P. Neuhaus. Let's PARSETALK: Message-passing protocols for object-oriented parsing. In H. Bunt and A. Nijholt, editors, *Advances in Probabilistic and Other Parsing Technologies*, pages 177–201. Dordrecht, Boston: Kluwer, 2000.
13. M. Strube and U. Hahn. Functional centering: Grounding referential coherence in information structure. *Computational Linguistics*, 25(3):309–344, 1999.
14. U. Hahn and K. Schnattinger. Towards text knowledge engineering. In *AAAI'98 – Proceedings of the 15th National Conference on Artificial Intelligence*, pages 524–531. Madison, Wisconsin, July 26-30, 1998.
15. S. Schulz and U. Hahn. Knowledge engineering by large-scale knowledge reuse: Experience from the medical domain. In *Proc. 7th Intl. Conf. on Principles of Knowledge Representation and Reasoning*, pages 601–610. Breckenridge, CO, April 12-15, 2000.
16. M. Romacker, S. Schulz, and U. Hahn. Streamlining semantic interpretation for medical narratives. In *AMIA'99 – Proceedings of the Annual Symposium of the American Medical Informatics Association*, pages 925–929. Washington, D.C., Nov. 6-10, 1999.
17. U. Hahn, M. Romacker, and S. Schulz. Discourse structures in medical reports – watch out! The generation of referentially coherent and valid text knowledge bases in the MED-SYNDIKATE system. *International Journal of Medical Informatics*, 53:1–28, 1999.
18. *Unified Medical Language System*. Bethesda, MD: National Library of Medicine, 2001.
19. P. Zweigenbaum, J. Bouaud, B. Bachimont, J. Charlet, and J.-F. Boisvieux. Evaluating a normalized conceptual representation produced from natural language patient discharge summaries. In *AMIA'97 – Proceedings of the 1997 AMIA Annual Fall Symposium (formerly SCAMC).*, pages 590–594. Nashville, TN, October 25-29, 1997.

FILLING PREPOSITION-BASED TEMPLATES TO CAPTURE INFORMATION FROM MEDICAL ABSTRACTS

G. LEROY, H. CHEN

Department of Management Information Systems, University of Arizona, 1030 E. Helen St,
Tucson, AZ 85721, USA

Due to the recent explosion of information in the biomedical field, it is hard for a single researcher to review the complex network involving genes, proteins, and interactions. We are currently building GeneScene, a toolkit that will assist researchers in reviewing existing literature, and report on the first phase in our development effort: extracting the relevant information from medical abstracts. We are developing a medical parser that extracts information, fills basic prepositional-based templates, and combines the templates to capture the underlying sentence logic. We tested our parser on 50 unseen abstracts and found that it extracted 246 templates with a precision of 70%. In comparison with many other techniques, more information was extracted without sacrificing precision. Future improvement in precision will be achieved by correcting three categories of errors.

1 Introduction

The explosion of information in the biomedical field provides researcher with great opportunities to study cell growth, differentiation and death, and the associated regulating processes. The biochemical pathways seem to be interconnected and consequently form a complex network involving numerous genes and proteins. The enormous amount of information available on individual pathways and their potential connections makes it hard for a single researcher to investigate and formulate relationships, especially in a new or unfamiliar domain. We believe researchers would benefit from a toolkit to assist them in summarizing and reviewing the existing literature.

We are currently building such a toolkit for the biomedical field, called GeneScene. GeneScene will derive information from the relevant journals and assist in reviewing existing literature, identifying gaps in existing knowledge, and as such help lead the way to new and interesting hypotheses and field research. The complete toolkit will contain four components: 1. the extracted, stored, and integrated gene pathway analysis data from abstracts from several journals, 2. a visualization component that will allow researchers to browse and search for information, get an overview of the collection, retrieve particular abstracts, and modify the representational map, 3. personalization and collaboration options for the researchers, and 4. the possibility to map microarray data onto the literature-based data. GeneScene will be developed in three consecutive phases (see Figure 1).

Initially, information will be extracted from individual sentences and put into preposition-based templates. Then, the sentence-based information will be combined with information from existing knowledge sources, allowing additional checking. At this point, meta-information such as the publication date will also be extracted. Finally, all information will be made available to researchers in a software toolkit allowing revision, modification, and information sharing.

Development Phase	Processing	Tools
Phase 1: Sentence Level Processing	1. Templates with biomedical information 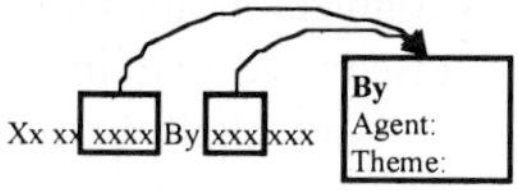 2. Combine Templates	English Syntax Prepositional Templates WordNet 1.6 UMLS SPECIALIST Lexicon AZ Noun Phraser Rewrite Rules for Templates
Phase 2: Abstract Level Processing	1. Extract high-level Information: date, authors, affiliation 2. Label Information in templates: gene, disease, etc 3. Combine Sentence Information 4. Database storage	NCBI UMLS Metathesaurus English Syntax (anaphora) SQL Server
Phase 3: Collection Level Processing	1. Combine data from abstracts 2. Develop software toolkit: - information visualization Import & map microarray data 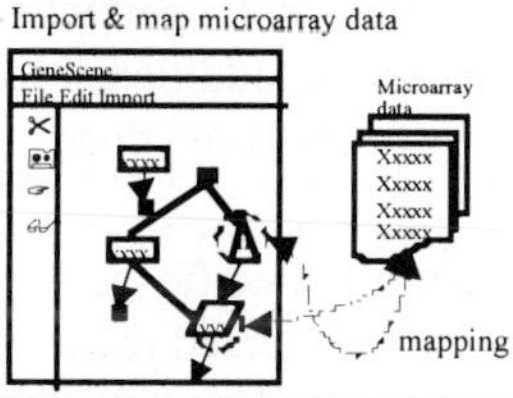	Java C++ SQL

Figure 1: GeneScene Development Overview.

This paper discusses our approach to and initial results for the first phase of the project: extracting the relevant information from individual sentences in medical abstracts. Careful review of the literature and our own strengths and weaknesses led us to a new approach to this problem: a preposition-based medical parser. Our approach is new since we do not focus on pre-specified genes and interactions; additionally, we do not try to parse the complete sentence structure. Instead, we use basic templates as building blocks. These templates are based on English closed word classes, such as prepositions and conjunctions. We use rewrite rules to combine the basic templates and rewrite them into more complex patterns that reflect the underlying sentence logic, which is necessary to correctly represent the information. In the following sections we describe previous research, followed by our own approach and evaluation, and a discussion of future work.

2 Background

The approaches currently described in the literature range from general-purpose parsers to pre-specified extraction of particular information. The general-purpose parsers are based on sound linguistic principles and aim to detect the complete structure of a sentence. The complexity of the medical language used instigates parsing errors and problems with overall processing speed. Yakushiji et al.[1] built a full parser and increased its speed with two preprocessors to reduce the workload of the full parser. The first preprocessor recognizes noun chunks; the second reduces parts-of-speech ambiguity. The authors discovered that medical abstracts use more complicated sentence structures than the ones their parser was based on. They reported that 53% of their test bed's structures was not extracted. Park et al.[2] used a slightly more specific approach with a bi-directional incremental parser based on a combinatory categorical grammar. With this grammar, verbs are expected to be surrounded by a particular sentence structure. For example, "inhibits" expects a noun phrase to its left and to its right. The authors focused their parser on a few verbs of interest. Their approach resulted in high precision (80%) and somewhat lower recall (48%) of protein-protein relations. We believe that a perfect medical parser would be invaluable; however, it would still need an additional logic module since, as Rindflesh et al.[3] point out, a linguistic analysis does not provide a semantic interpretation.

Several approaches focus on extracting specific gene, protein and interaction information from abstracts. Sekimizu et al.[4] collected the most frequently used verbs in their collection of abstracts. They used partial and shallow parsing techniques to extract noun phrases from sentences and developed rules to find the subject and object of the high-frequency verbs. They estimated their precision at 73%. Thomas et al.[5] used a statistical parser to fill templates with information on proteins and their interactions. They concentrated on three verbs (interact with, associate with, bind to) for which they developed templates. They calculated recall and precision in four different manners for three samples of abstracts. Recall ranged from 24% to 63%, and precision from 60% to 81%. BioNLP[6] uses three components, two of which are BioKleisli[7] to query multiple medical databases and BioJAKE[8] to visualize and manipulate metabolic pathways. The third component is of interest here; it extracts gene names and their relations from free text based on an existing thesaurus, together with additional rules to identify existing and new gene names. The relations are limited to a predefined set of verbs. Once the genes are found, the sentences are matched against predefined syntactic structures and the verb thesaurus to identify the nature of the relation between the genes. Unfortunately, there was no evaluation data and the authors indicated that their pattern matching was not sophisticated enough to handle all sentences. The rules used by BioNLP are based on work by Fukuda et al.,[9] who achieved very high precision extracting proteins (95% to 98%).

Other specific approaches extract information about a subset of genes and interactions. The PIES project,[10] requires users to submit key terms, such as "calyculin," and searches Medline for abstracts containing these terms. From the matching abstracts, "inhibit" and "activate" interactions are considered. The authors use BioNLP to extract the relevant information from the sentences, and the Graphviz software package (available online at http://www.research.att.com/) to visually display the results. An interesting addition to their system is that users can save and update the retrieved information. Unfortunately, no evaluation was provided. Blaschke et al.[11] used a comparable approach and asked users to provide the protein names to retrieve abstracts. They focused on the sentences containing the protein names and one of 14 pre-defined words representing actions. No systematic evaluation was reported. Stephens et al.[12] started from thesauri containing gene names and possible relations. They represented documents as vectors with a dimension equal to the size of the thesaurus and calculated the association between the genes based on the similarity of the vectors. When related genes were found, they retrieved the verb in that sentence. If it was found in their relation thesaurus, they accepted it as the relation between the two genes. The information is represented in a representational graph where distance represents similarity.

3 GeneScene

3.1 Selecting Abstracts and Sentences

GeneScene will ultimately integrate gene pathway information from thousands of abstracts. We will not require researchers to pre-specify genes or interactions. Instead, to extract the relevant information with sufficiently high precision, we plan to filter at three levels: the journal, the abstract, and the sentence level. Filtering at the "journal level" will be straightforward: we will initially concentrate on journals with a high impact factor, as defined by the Institute for Scientific Information (ISI, http://www.isinet.com/isi/index.html), that are also indicated as top journals by the biomedical researchers advising us in this project. The journal impact factor measures the frequency with which the "average article" in a journal has been cited in a particular year. It indicates a journal's relative importance in the field. At the "abstract level" we plan to focus on general abstracts. For abstracts describing clinical studies we plan to extract information only from the conclusion and discussion sections. Finally, at the "sentence level," we will evaluate individual sentences based on WordNet information, to ensure that actual information and not e.g. the hypothesis is extracted. WordNet is a general English ontology (http://www.cogsci.princeton.edu/~wn/). We are currently building a WordNet-based thesaurus of catch phrases that will help us classify sentences. Sentences containing phrases such as "we show," "we demonstrate," "we established," "we hypothesized," "we expect" can be mapped to WordNet and its verb hierarchies. For example, "hypothesize" and "speculate" are both more specific ways (hypernyms) of "expect" and as such belong to the "expect" hierarchy. We will identify hierarchies containing phrases that indicate sentences to be included and other hierarchies that indicate sentences to be excluded. The classification system's main contribution will be to exclude sentences discussing expectations and hypotheses instead of results from GeneScene.

3.2 Preposition-based Parsing

There are two major phases our parser works through when processing a sentence. During the first phase, the extraction phase, the basic templates are identified. Prepositions form the entry point in a sentence. We then retrieve the main verb, adverbs, negation, and noun phrases around the preposition to fill the templates. Classification of words into word classes is currently based on WordNet 1.6. However, we noticed that the WordNet vocabulary will be insufficient to process a large collection of medical abstracts. Terms that necessarily need to be recognized as verbs, adjectives or adverbs, are not always found in WordNet 1.6. Fortunately, they are part of the SPECIALIST Lexicon, a component of the Unified Medical Language System (UMLS). For example, the verbs "phosphorylate," "overexpress," and "dysregulate," and the adjectives "oncogenic," "mitogenic," and "transcriptional" are not part of WordNet 1.6 but can be found in the SPECIALIST Lexicon. Lack of time prevented us from integrating the SPECIALIST Lexicon as a component in our parser for this evaluation. Instead, a small lexicon was added for the terms we discovered so far that are not found in WordNet.

We believe that by building templates around prepositions, we are able to capture more information than when looking for particular genes. We capture genes and proteins, but also e.g. diseases, cell phases, gene locations. In addition, we believe that precision will be high because, while we cover all possible sentence structures, we only extract the information that fits our templates. Although we intend to cover most prepositions, we report here on initial results of the templates developed for the two prepositions: "by" and "of." These were chosen because they frequently appear in medical abstracts and are representative of the complexity involved in processing medical abstracts. Additionally, our choice of prepositions allows us to demonstrate the second phase of parsing, the recombination phase, where we rewrite the basic templates into combined templates that capture the underlying logic of the abstract. In the following, we discuss both phases in detail.

During the extraction phase, we focus on filling basic templates with phrases surrounding the preposition. We first retrieve the main verb close to the preposition. Then, we search for noun phrases to the left and right of the verb and preposition. Noun phrase detection is currently based on a variant of stop word phrasing: punctuation, auxiliaries, verbs, and closed class words are used as indicators of the start and end of phrases. For example, in the sentence "Remarkably, despite the inhibition **of** cell proliferation and apoptosis, the degeneration **of** lens fibers and aberrant expression **of** filensin were only ..." we can extract three templates (described later) surrounding "of." For example, for the second template the closest boundaries are a comma on the left and a conjunction on the right. We also use a stop word list to cleanse the strings. For example, auxiliaries should not be part of an agent or theme. We keep track of begin and end indices of the template in the sentence. This information will be necessary to take overlapping arguments into account when combining templates. We employ additional selectional restriction to limit the phrases that can be agents or themes. A determiner, adjective, adverb, closed class word, a number, or a phrase containing a percentage cannot be the agent or theme. For example, in the sentence "..., JNK activity was increased by 150%," the "150%" is not the agent of the activity. It is restricted from this function.

In the following, we provide an overview of the templates currently being tested. A first template is built around the preposition "by." For this template we capture two main sentence structures (Structure 1 and 2) used to fill the by-template:

Structure 1: String1 – [modifier | negation] – main verb – **by** – String2
Structure 2: String1 – [modifier | negation] – nominalized verb – **by** – String2
Rule: agent = String2, action = verb, theme = String1
By-template: agent – [modifier | negation] action - theme

Both structures are used to fill the template as follows: *action* is the main verb or the underlying verb form of the nominalized verb and can be modified by a negation or a modifier. String2 is the *agent* of the action and String1 is the *theme* of the action. Auxiliaries can appear in the structure but they will not be part of the final template. If no verb is found, then only an agent is searched for; otherwise, both agent and theme are searched for. A modifier can be an adjective, an adverb, or a verb in the past tense. For example, the sentence "Apoptosis induced **by** the p53 tumor suppressor can attenuate cancer growth in preclinical animal models," results in the following template: (p53 tumor suppressor – induce – apoptosis).

A second template is built around the preposition "of." We capture two similar structures as with the by-template. However, with a nominalized verb, an agent is not searched for and "null" is inserted instead. The theme is found after the preposition in the sentence. For example, the sentence "This effect was accompanied by an increased expression **of** the cyclin-dependent kinase inhibitor p21(WAF1/CIP1) and a decreased expression **of** cyclin A," results in the following of-templates: (null – [increased] express – cyclin-dependent kinase inhibitor p21(WAF1/CIP1)) and (null – [decreased] express – cyclin A). The nulls in templates are important for the rewrite rules of the second recombination phase. Negation is also captured, for example the sentence "However, E2F is not a general regulator **of** oxidative phosphorylation genes since ...," results in the following template: (E2F – [not][general] regulate – oxidative phosphorylation genes).

We do not only capture genes and proteins, but all information. For example the sentence "This arrest response appeared independent **of** p53/p21cip1/waf-1 function," results in the following template: (arrest response – [independent] appear – p53/p21cip1/waf-1 function). Other approaches miss this information. Labeling the content of the templates, e.g. "gene" or "bacteria," will follow in a later phase by mapping to data from the UMLS and the National Center for Biotechnology Information (NCBI) database.

During the recombination phase, templates are combined and rewritten. A first set of rewrite rules looks at specific prepositional combinations. In the following, we describe the individual templates that need to be extracted from a sentence, as described above, and the resulting combined template. We use the "*" notation to indicate a pointer to another template.

Prepositional Combination 1:

Of-template:	null – [modifier| negation] action1 – theme1
By-template:	agent2 – null – null
Rule:	no other by- or of-template can be found in between
Combined:	agent2 – [modifier| negation] action1 – theme1

For example, "Inactivation **of** the pRb proteins in mouse brain epithelium **by** the T121 oncogene induces aberrant proliferation and ....," resulted in the following combined template (T121 oncogene – inactivate – pRb proteins).

Prepositional Combination 2:

Of-template:	null – [modifier| negation] action1 – theme1
By-template:	agent2 – [modifier| negation] action2 – theme2
Rule:	theme1 = theme2
Combined:	agent2 – action2 – *of-template

For example, the sentence "... suggests the existence **of** cell type-specific inhibitory pathways induced **by** these signals," results in the combined template (signals – induce – (NULL – exist – cell type-specific inhibitory pathways))

Prepositional Combination 3:

Of-template1:	null – [modifier| negation] action1 – theme1
Of-template2:	null – [modifier| negation] action2 – theme2
Rule:	action2 = verb form of theme1
Combined:	null – [modifier| negation] action1 – *of-template2

An example of this third combination is the following: "...distribution through the modulation **of** the expression **of** cell cycle-related genes ..." which results in the template (null – modulate – (null – express – cell cycle-related genes)).

Prepositional Combination 4:

By-template:	agent1 - action1 - null
Of-template:	null – [modifier] action2 - theme2
Rule:	[modifier] + verb form of agent 1 = [modifier] action2
Combined:	*of-template – action1 – null

An example of this combination is the sentence "...that are activated **by** severe depletion **of** cell energy stores." The by-template (severe depletion - activate – null) and the of-template (null – [severe] deplete – cell energy stores) are combined into ((null – [severe] deplete – cell energy stores) – activate – null).

A second set of rewrite rules focuses on conjunctions. Two non-overlapping templates based on the same preposition and connected by "and" are combined. The missing element in the second template (following the "and") is copied from the first template. Currently, we only test for missing themes.

Conjunctional Combination:

X-template1:	agent1 – [modifier| negation] action1 – theme1
X-template2:	agent2 – [modifier| negation] action2 – null
Rule:	conjunction "and," no overlap between templates, prepositions in both templates have to be identical
X-template1:	agent1 – [modifier| negation] action1 – theme1
X-template2:	agent2 – [modifier| negation] action2 – theme1

For example, from the sentence "Given that E2F1 activity is stimulated **by** p300/CBP acetylase and repressed **by** an RB-associated deacetylase, we ...," the following templates are extracted: (p300/CBP acetylase – stimulate – E2F1 activity) and (RB-associated deacetylase – repress – null). These are connected by "and," and the rewrite rule changes the second template to (RB-associated deacetylase – repress – E2F1 activity).

3.3 Evaluation

Following a tuning-phase, we used the keyword "E2F1" to retrieve 50 new abstracts. Both titles and the actual abstracts were processed, resulting in a total of 474 sentences and 246 templates. Table 1 provides an overview of the results. We only consider templates that contained at least two non-null elements. For example, when an agent name is captured, but no other information, the resulting template (e.g. pRb – null – null) is currently not considered for evaluation. A template was scored as correct when all noun phrases were complete, when no modifier or negation was missing, and when the template correctly represented that subpart of the sentence.

To calculate recall, we counted the instances where templates could have been built. For the of-template this meant all occurrences of the preposition except when it was used in expressions such as "some of which," numeric expressions such as "5 of 7," or noun phrases without action words such as "B-subunits of replicative DNA polymerases." For the by-template this meant all occurrences of the preposition except when it was used in expressions such as "by which," or as the first word in a sentence. For the combination templates, there were no exceptions. Precision, recall, and F-measure were calculated according to the following formulas:

Precision = total correct templates / total extracted templates

Recall = total correct templates / total possible templates

F-measure = (2 * recall * precision) / (recall + precision).

Table1: Performance Analysis

	Total	Average Per abstract	Average Precision (%)	Average Recall (%)	F-measure
General Analysis:					
Abstracts:	50	-	-	-	-
Sentences:	474	9.5	-	-	-
Templates built:	246	4.9	70	47	56
Template Specific Analysis:					
Of-Templates built:	189	3.8	74	52	61
By-Templates built:	58	1.2	72	43	54
Combo-Templates built:	22	0.5	45	38	42

The average precision was 70% for all templates combined. It was slightly higher for of-templates (74%) and by-templates (72%) separately. Since combined templates can only be correct if the two underlying templates are correct, this precision is lower (45%). Recall was 47% in general, 52% for of-templates, and

43% for by-templates. As with precision, recall of combined templates depends on the other two templates being recalled and, as such, was lower (38%). If we had taken a less general approach and concentrated on only those relations that contain the term "E2F1," then we would have extracted a maximum of 110 templates. Many approaches take an even more specific approach and require not one, but two genes to be present in a sentence. In that case, fewer relations would have been extracted. Although it is possible to test all possible combinations of known genes, our approach does not depend on any pre-specified name list. Additionally, we also extract information elements that are not genes or proteins.

In Table2, we provide an overview of the distribution of errors that shows there are major categories of errors that can be systematically addressed.

Table 2: General Error Analysis

Error Type:	Fraction (%)
Template not yet developed:	24
Agent/Theme overextension:	28
Modifier incomplete:	9
Agent/Theme incomplete:	4
Agent/Theme contains rubbish terms:	15
Error in Combinations:	4
Error due to WordNet limitation:	1
Other:	14

A closer look reveals that almost 70% of the errors belong to just three categories. The first category accounts for 24% of the errors. These were incorrect because combinational templates not yet designed were not incorporated, resulting in a misrepresentation of the information. For example, the sentence "... for the induction **of** the p21 promoter **by** activated Ras, ..." resulted in the templates (NULL – induct – p21 promoter) and (activated Ras – promote – p21). Since the "activated Ras" does not promote "p21" but the "induction of the p21 promoter," this is a missed "of-by" combination resulting in an erroneous second template. These errors will be corrected with additional combination rules. Although it is a challenge to add more template combinations without introducing new errors, correcting this category of errors would increase precision significantly.

The errors due to overextension of the agent and theme phrases form a second main error category, representing 28% of the total errors. In almost all cases, these errors were due to a word not being recognized as a conjugated verb. For example, in the sentence "We show that the E2Fs control the expression of several genes that are involved in cell proliferation," the word "control" was not recognized as the conjugated verb, resulting in an erroneous agent "E2Fs control." To address this second category of errors, we will try and implement proven noun phrasing techniques based on our experience with the Arizona Noun Phraser.[13]

A final major error category contains the agents or themes with rubbish terms. For example, from the sentence "Increased expression of neutrophins (e.g. NGF,

BDNF) and …," the "(e.g." became part of the theme. We expect improvements by processing more abstracts since that will make our stop word list, which is used to filter and cleanse this irrelevant information from the templates, more complete.

We want to remark on our decision to convert nominalized verbs to their base verb form. This was done to increase the compilation powers of GeneScene when we combine all information. In some cases, the transformation of nominalized verbs to their base verb form might seem unsuitable. However, by transforming e.g. "the expression of CDK4" and "CDK4 is expressed" to the same form "null – express – CDK4" the relation is strengthened. This will provide researchers with important clues since a frequently found relation often indicates consistent findings. A very rarely found relation can be an erroneous finding stated by an author, an error in the processing of the abstracts, or a very interesting and rare finding. Furthermore, this process will allow us to represent more information visually in the same manner, making the overall picture less demanding to understand. For example, name labels in "green" ink to indicate "expression," or a colored arrow from the agent to the theme indicating that the agent is responsible for the expression of the theme.

4 Conclusion

We feel that our approach has a lot of potential for different reasons. First of all, we achieved an average precision of 70% without focusing on a subset of the available information. We expect to improve this precision by correcting the main error categories discussed earlier. Most approaches to automated extraction of biomedical information report precision between 60% and 80%,[2,3,5] depending on the different definitions of precision used and also on the diversity of the extracted information. It can be expected that systems focusing on a very specific subset of the information will be more precise than general system. However, we do not focus on certain types of information. The agent and themes do not have to be proteins or genes; the action does not need to belong to a pre-specified set of interaction verbs. We also use a liberal definition of modifiers for verbs, allowing us to capture details about the relation. Furthermore, by focusing on the prepositions and their particular combinations, we are able to capture the underlying sentence logic. The combination of e.g. a by-template followed by an of-template is different from an of-template followed by a by-template. Finally, we want to note that the development of our parser is a continuing effort. We expect to improve its precision, and to process larger sets of abstracts in the near future.

Acknowledgments

We would like to thank Ann Lally and Jesse Martinez for their inspiring comments and suggestions. We feel grateful to the National Library of Medicine and Princeton University for making the UMLS and WordNet available to researchers. This research was sponsored in part by the following grants:

NSF Digital Library Initiative-2, "High-performance Digital Library Systems: From Information Retrieval to Knowledge Management," April 1999-March 2002.

National Institute of Health and National Library of Medicine, "UMLS Enhanced Dynamic Agents to Manage Medical Knowledge," February 2001-February 2004.

References

1 A. Yakushiji, Y. Tateisi, Y. Miyao, and J. Tsujii, *Pacific Symposium on Biocomputing*, 408 (2001)

2 J.C. Park, H.S. Kim, and J.J. Kim, *Pacific Symposium on Biocomputing*, 369 (2001)

3 T.C. Rindflesch, L. Hunter, and A.R. Aronson, *Proc AMIA Symp*: 127 (1999)

4 T. Sekimisu, H.S. Park, and J.i. Tsujii, *Genome Informatics*: 62 (1998)

5 J. Thomas, D. Milward, C. Ouzounis, S. Pulman, and M. Carroll, *Pacific Symposium on Biocomputing* **5**: 538 (2000)

6 S.-K. Ng, and M. Wong, *Genome Informatics* **10**: 104 (1999)

7 S.B. Davidson, C. Overton, V. Tannen, and L. Wong, *International Journal on Digital Libraries*: (1996)

8 W. Salamonsen, K.Y.C. Mok, and P. Kolatkar, *International Journal of Digital Libraries* **1**(1): 36 (1997)

9 K. Fukuda, T. Tsunoda, A. Tamura, and T. Takagi, *Pacific Symposium on Biocomputing*, 705 (1998)

10 L. Wong, PIES, *Pacific Symposium on Biocomputing*, 520 (2001)

11 C. Blaschke, M.A. Andrade, C. Ouzounis, and A. Valencia, *ISMB*: 60 (1999)

12 M. Stephens, M. Palakal, S. Mukhopadhyay, R. Raje, and J. Mostafa, *Pacific Symposium on Biocomputing*, 483 (2001)

13 K.M. Tolle, and H. Chen, *Journal of the American Society of Information Systems* **51**(4): 352 (2000)

Robust Relational Parsing over Biomedical Literature: Extracting Inhibit Relations[a]

J. Pustejovsky, J. Castaño, J. Zhang
Department of Computer Science, Brandeis University
415 South St., Waltham, MA 02454, U.S.A.

M. Kotecki, B. Cochran
Department of Physiology, Tufts University, 136 Harrison Ave., Boston, MA, U.S.A.

We describe the design of a robust parser for identifying and extracting biomolecular relations from the biomedical literature. Separate automata over distinct syntactic domains were developed for extraction of nominal-based relational information versus verbal-based relations. This allowed us to optimize the grammars separately for each module, regardless of any specific relation resulting in significantly better performance. A unique feature of this system is the use of text-based anaphora resolution to enhance the results of argument binding in relational extraction. We demonstrate the performance of our system on *inhibition*-relations, and present our initial results measured against an annotated text used as a gold standard for evaluation purposes. The results represent a significant improvement over previously published results on extracting such relations from Medline: Precision was 90 %, Recall 57 %, and Partial Recall 22%. These results demonstrate the effectiveness of a corpus-based linguistic approach to information extraction over Medline.

1 Introduction

A vast amount of new biological information is made available in electronic form on a regular basis. Medline contains over 10 million abstracts, and approximately 40,000 new abstracts are added each month. Although there are growing numbers of sequence databases and other hand-constructed databases, most new information is unstructured text in Medline and full-text journals. This information, which is coming to be referred to as the "biobibliome", is a repository of biomedical knowledge that is larger and faster growing than the human genome sequence itself (Stapley and Benoit[22]). In this age of genomics and proteomics, the ability to process this natural language based information computationally is becoming increasingly important. It is now not uncommon for biologists to study protein complexes and pathways composed of dozens of dynamically interacting proteins. With the recent advent of high sensitivity methods to rapidly identify components of multiprotein complexes (Link

[a]This work was supported by NIH grant R01-LM06649 to Prof. Pustejovsky at Brandeis and Prof. Cochran at Tufts. Direct all correspondences to `jamesp@cs.brandeis.edu`.

et al.[10]), the extent of this complexity is likely to grow exponentially in the next few years. For this reason, the automatic extraction of information from Medline articles and abstracts will play an increasingly critical role in aiding in research and speeding up the discovery process.

To begin addressing this problem computationally, we have begun developing advanced natural language tools for the automated extraction of structured information from biomedical texts as part of a project we call MEDSTRACT (www.medstract.org). Previously we have reported a strategy for the automatic extraction and compilation of biomedical acronyms call *Acromed.* Here we utilize this and other NLP techniques to extract reported relationships between biological entities using the inhibit relation as an example.

The use of computational linguistic techniques for automatically extracting information from biological texts (in particular from Medline) has received increasing attention lately (e.g., Tagaki et al.[23], Sekimizu et al.[21], Hishiki et al.[6], Andrade et al.[1], Blasche et al.[2], Craven et al.[4], Rindfleisch et al. [20], Pustejovsky et al.[17]). Much of the work reported on thus far has focused on specific protein-protein interactions, and in particular, on predicates implicated in binding activities (cf. Sekimizu et al.[21], Blasche et al.[2], and Rindfleisch et al. [20]). Craven et al.[4] use a relational learning algorithm to induce pattern-matching rules on shallow parsed trees for protein-location relations. Although the precision is quite high (92%), their recall is quite low: (21%). The data set they examine is the YPD corpus (Yeast Protein Database). Rindfleisch et al.[20] use shallow parsing combined with UMLS semantic types to extract binding relations from Medline. Their results gave precision of 73% and recall of 51%. Proux et al.[14] also use shallow parsing and domain knowledge (gene type identification) to extract gene-gene interactions from the Flybase corpus. This work is the first we know to pose the problem of retrieving partial x relation information (only one argument of the relation.) Their results were: Precision 81%, Recall 44% and Partial recall (they call it *weak interaction*) 26%. Finally, Sekimizu et al.[21] apply shallow parsing using a general purpose parser (EngCC) to retrieve assertions corresponding to the most frequent set of verbs from Medline abstracts. Their average estimated precision was 73%, for identifying the right subject and object in the relation. No recall is given because no gold standard was created. Partial projected precision for some relations considered in other works mentioned here are: (*interact*: 83.3%, *bind*: 72%, *inhibit*: 83.3%). The results from these experiments are summarized in Table 1 below.

Within information extraction (IE) tasks, entity extraction is typically viewed as a procedure distinct from relation extraction. For example, in enterprise IE systems, products, dates, and company names are easily distinguished from ventures, buy-outs, and product release relations. For most ordinary us-

Lab	Relation	Type Constraints	Data Set	Precision	Recall	P. Recall
Crav.[4]	Location	Protein	YPD	92%	21%	–
Rind.[20]	binding	UMLS	MEDLINE	73%	51%	–
Proux[14]	interact	gene	Flybase	81%	44%	26%
Seik.[21]	several	–	MEDLINE	73%	?	–

Table 1: Previous Relation Extraction Results

age of language, however, and for Medline in particular, the syntax of the sublanguage maps imperfectly to basic semantic distinctions, such as entities and relations. That is, not all entity-looking phrases are entity types; specifically, relations may be expressed as nominalizations (*phosphorylation of GAP by the PDGF receptor*) as well as verbal predications (*X inhibits/phosphorylates Y*). Things become even more complicated for IE when true entities embed relational information by virtue of their semantics, such as the relational entities *the Ron receptor* and *Tissue inhibitors of metalloproteinases.* The difficulty in this example is that such entities are proteins and also incorporate relational semantic information; "x inhibits metalloproteinase". Such considerations demand more sophisticated linguistic processing than is typically employed for IE tasks in enterprise deployments, and certainly richer than the statistical techniques that have received attention in the bioinformatics community recently (cf. Jenssen et al.[8], Marcotte et al.[12]).

2 Design and Methodology

In this paper, we address the problems mentioned above by exploiting a combination of lexical semantic techniques and corpus analytics (Pustejovsky et al[15,16]). In the section below, we briefly describe this methodology as employed in the Medline domain for targeted information extraction tasks.

Semantic Automata: We begin by constructing simple *semantic automata* from the UMLS database for the relations we are interested in targeting (cf. Humphreys et al.[7]). For example, for *inhibit*-relations and *regulate*-relations, there are four basic selectional patterns (or frames), corresponding to the two options available for each of the two arguments to the relation. These frames are summarized in the table below.

The family of syntactic forms for a lexical item and the mappings to semantic values are part of the typing information encoded within a word, as seeded by UMLS and stored in the lexicon. It should be noted that because the syntactic typing of *inhibit* and *regulate* is transitive, additional semantic automata corresponding to the syntactic passive forms are automatically

ARG-TYPES	Obj = Bio-entity	Obj = Process
Subj = Bio-entity	*(entity,entity)*	*(entity,process)*
Subj = Process	*(process,entity)*	*(process,process)*

Table 2: Selection Patterns

generated. Furthermore, nominal and verbal predicative forms have distinct syntactic distributions and different semantic bindings; hence, they map to different semantic automata, as we see in Section 3.3 below.

Corpus Analytics: We then apply corpus analytics over a subset of Medline corresponding to the target relations, e.g., *inhibit.* Corpus analytics involves several steps (cf. Pustejovsky and Hanks[18]):

i. Create concordances over the predicates (verbal or nominal) associated with the semantic automata;

ii. Automatically cluster complementation patterns of the relation over the concordances, to propose grammar patterns;

iii. Semi-automatically verify and amend grammar rules to ensure correctness and completeness of the patterns for the automata;

For this experiment, we focused on only expressions corresponding to the *inhibit*-relation. The methodology used and the subsequent grammar developed, however, applies to any binary relation with similar semantic typing restrictions. By limiting our case study to this relation set, we were able to create a gold standard corpus with which to evaluate our algorithm. Examples of the concordances used to derive grammar patterns are shown below.

1. `A peptide representing the carboxyl-terminal tail of the met receptor` inhibits `kinase activity.`
2. Whereas `phosphorylation of the IRK by ATP` is inhibited by `the nonhydrolyzable competitor adenylyl-imidodiphosphate.`
3. `The Met tail peptide` inhibits `the closely related Ron receptor` but does not affect ...

A set of 2,000 abstracts was first selected from the Medline database, identified from the concordance constructed around verbal forms and nominal forms of the stem *inhibit.* This set of abstracts was used as a development corpus to optimize each of the components of our system. ¿From this development corpus, a subset of 500 abstracts was used as a core for manual mark-up by domain-expert biologists. In the development-test cycle, a complete month of Medline abstracts was used for robustness test.[b] We then preprocessed and tagged 500 abstracts of the development corpus , with the following results: There were 497 verbal instances of *inhibit* (*inhibited*, *inhibit* and *inhibits*); There were 342 instances of nominal forms (187 correspond to *inhibition*, and 155 instances to *inhibitor/s*. All the other forms were either gerundives instances (13 of *inhibiting*) or instances of compound forms, e.g., *bisphosphonate-inhibited* (7 instances). Given this distribution and some independent linguistic considerations to be mentioned later, we focused primarily on the development of the grammar of verbal predication.

3 A Description of the system

3.1 General Architecture

As mentioned above, the task of recognizing and extracting *inhibit*-relations between biological entities and processes is part of a much larger research effort underway at Brandeis and Tufts, called MEDSTRACT. The goal of MEDSTRACT is to provide tools and resources to biomedical researchers for better search, retrieval, and navigation of new facts and products within the biological literatures. An illustration of the relevant portion of the architecture is shown below in Figure 1.

3.2 Preprocessing

After identifying the corresponding fields of the Medline documents, titles and abstracts are tokenized. Tokens are then tagged, using a Brill-like rule-based decision procedure. A lexicon with single or multiple tags for each word is used. If the word in question has multiple tags in the lexicon, then it is tested to match a set of disambiguation rules. If it matches any, the corresponding tag is assigned. Otherwise the most probable tag is assigned. The source lexicons used were the lexicon produced by Brill's tagger and the corresponding lexicon for the UMLS Thesaurus (Humphreys et al[7]), with its corresponding syntactic information. The tagged elements were then stemmed with a version of the Porter stemmer. The information corresponding to the *string*, its syntactic tag, and the corresponding stem is stored in a *preterminal* object.

[b]We are currently developing gold standards for other *inhibit*-relations, and testing the robustness of the algorithm on these sets; e.g., *block*, *regulate*, *stimulate*, etc.

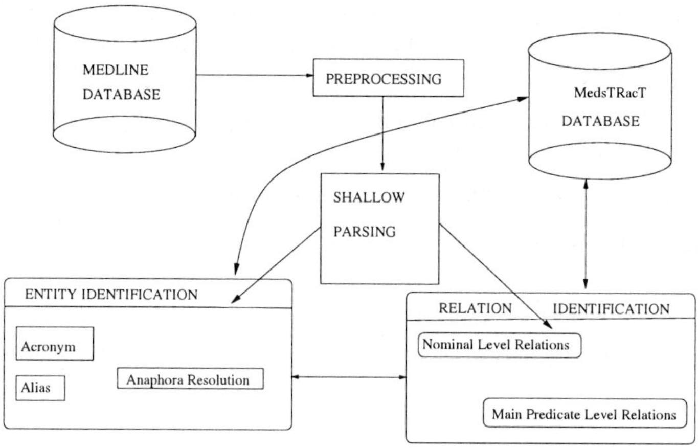

Figure 1: System Architecture

3.3 Type Identification

Our system uses one of two resources for dynamic semantic typing of the input: (a) the UMLS Thesaurus can be exploited to assign types to nouns or noun phrases according to the UMLS type ontology; (b) the GO ontology is also available as a type resource for specific genomic data. For the present experiment, however, neither resource was used, since we were focused primarily on evaluating the construction and deployment of syntactic patterns from semantic automata. Integration of semantic tags into the parsing procedure is under development. Furthermore, we wanted to test the robustness of syntactic techniques independently of typing information. The UMLS types were however used in the anaphora resolution task, as one of the parameters in ranking the possible antecedents list.

3.4 Shallow Parsing Module

The construction of shallow parse trees involves a cascade of five separate automata, each focusing on a distinct family of grammatical constructions. This is very much in the spirit of Hindle[5], McDonald[11] and Pustejovsky et al.[16]. These can be distinguished as follows:

Level I: Noun chunking, groups Proper Nouns and common nouns. It also groups some double prepositions, and compound relational terms.

Level II: Creates noun phrase chunks without prepositional phrases (including adjectives and determiners). It also creates relational terms chunks (verbal chunks, including some adjectival and adverbial terms).

Level III: Creates chunks for coordinated nouns or noun chunks and coordinated verbal chunks or verbs.

Level IV: Creates chunks of noun phrases with *of*-prepositional phrase.

Level V: Identification of subordinate clauses chunks.

3.5 Relation Identification Module

As mentioned briefly above, the concordances derived for *inhibit*-relations distinguished the verbal forms from the nominal forms. Because of their distinct argument binding and complementation behaviors, we decided to develop separate automata for each form, and then merge the results in a subsequent database population phase. In fact, however, there is reason to believe that keeping the results extracted from the two modules separate is actually desirable for database purposes as well; this is due in large part to the degree of relevance associated with 'given' versus 'new' information as presented in documents (cf. Pustejovsky et al,[19]).
The relation identification module was built independent of the specifics of how the verb *inhibit* and associated nominals behave in Medline. Rather, this module was defined and designed to work on the output of the shallow parsing module to identify argument and relational chunks, independently of any specific lexical item. The extraction of a particular relation (e.g. *inhibit* or *regulate*), is accomplished by specifying stems that denote the required relation.

Sentence-level parsing identifies the following constructions:

SENTENCE-LEVEL RELATION IDENTIFICATION

1. Main predicate relational chunk in the sentence.
2. Subject nominal chunk (Nominal chunks at 4th level above)
3. Object nominal chunks.
4. Subordinate clauses (identifying also antecedents of relative clauses, and main predicates of object clauses).
5. Sentential coordination.

It has also the capability of identifying:

1. Preverbal adjuncts.
2. Post Object target adjuncts (ambiguous between adjuncts and nominal modifiers, PP attatchment ambiguity)

In the nominal domain, head nouns may typically carry relational semanctis; for example the noun *inhibitor* can refer to both the relation as well as the biological entity itself, the parsing decisions involved for these forms are distinct from the verbal form. The constructions and relations identified by the nominal-level module are given below:

NOMINAL-LEVEL RELATION IDENTIFICATION

1. Nominal chunks of Level IV.
2. prepositional relational chunks.

Note that relations inside Level IV are decomposed first, i.e., *of*-prepositional relations. Our next step will be to add reduced relative clauses and gerundive relations to this parser module.

3.6 Anaphora Resolution Module

Identifying the arguments of the relations may not be enough for identifying the actual entities involved in the relation. Quite often anaphors (e.g., *it, they*) and sortal anaphoric noun phrases (e.g. *the protein, both enzymes*) are the actual arguments to a relation, but unfortunately are not specific enough to establish a unique reference to an entity or process. Although the use of anaphoric terms seems to be relatively infrequent in Medline abstracts, the use of sortal anaphors is quite prevalent. This module focuses on the resolution of biologically relevant sortal terms (i.e., proteins, genes, and bio-processes), as well as pronominal anaphors, including third person pronouns and reflexive pronouns. The initial data source for this resolution algorithm is the preprocessed Medline text (shallow parsed), where each noun phrase (NP) has been identified and annotated with a syntactic tag and semantic tag(s). The anaphora resolution algorithm examines the text sequentially and represents each sentence as a "frame environment". Every NP within a sentence is a potential referent and is made into an entity with a unique ID and syntactic/semantic tags, and added to the sentence environment in which it occurs. If an NP is identified as an anaphor, then the resolution algorithm will attempt to resolve it by traversing through the sentence environments from the most recent (which contains the anaphor), back to the first sentence of the abstract, and selecting the NP among the sentences that has the highest compatibility with the anaphor as the antecedent (cf. Kennedy and Boguraev[9]). The choice of antecedent is determined by matching syntactic and semantic features of the candidate NP with that of the anaphor, which includes person/number agreement, semantic type, as well as physical string comparisons. In the case that more than one NP is found to be equally compatible, preference is given to the one that is most adjacent to the anaphor in the text. If an anaphor requires multiple antecedents (e.g., the anaphor *both enzymes*) then the resolution algorithm will continue in the sentence environment where the first antecedent is found, and then select the subsequent antecedent which is most compatible with both the anaphor and the first antecedent.

Each resolved anaphor retains its assigned antecedent(s) in memory so as to enable cascading anaphoric links of coreference between an anaphor and a previous discourse referent which could be another anaphor. In addition, special filters are used to exclude the resolution of expletives as well as restricting antecedents of reflexive pronouns to occur in the same sentence as the anaphor.

4 The Evaluation Test

4.1 The mark-up

A new data set of abstracts was collected using a different search, using the strings *protein* and *inhibit*. We ensured that there was no overlap between the training set and the evaluation set. This data set consisted of 56 abstracts, which was manually annotated in XML format, as described at `www.medstract.org`. Those instances which had an argument which referred to an antecedent were annotated as were those corresponding strings for the relations. If the instance in question was particularly difficult to annotate, the comments of the annotator were included. The corresponding entities were annotated with the appropriate semantic type; however, the type information was not used or processed in this experiment. Below is an example of parsed output showing types and bindings of entities in a relation, together with an anaphoric binding

```
<Entity id="83" Type="small molecule"> Cyanide</Entity>,
<Entity id="84" Type="small molecule">azide</Entity>,
<Entity id="85" Type="small molecule">p-hydroxymercuribenzoate</Entity>,
<Entity id="86" Type="small molecule">iodoacetamide</Entity>, and
<Entity id="87" Type="small molecule">oxygen </Entity>
<InhibitRelation id="88" Inhibitor="83, 84, 85, 86, 87"
Inhibitee="82">inhibit </InhibitRelation>
<Entity id="82" Antecedent="81">the enzyme</Entity>
```

The antecedent of the string "the enzyme" corresponds to a previous occurrence of:

```
<Entity id="81" Type="Protein">Formate dehydrogenase</Entity>
```

If a particular instance of a relation did not have an argument which could be interpreted from the document, then the argument value was annotated as *unspecified*.

4.2 Results

There were 95 instances of the *inhibit*-relation annotated in the 56 articles. Our system identified 84 of these instances: 56 were correct instances: (57% Recall) There were 21 instances in which one argument was identified correctly, but the second was not identified, and there was no False Positive argument: (22% Partial Recall). There were 8 False Positive (incorrect) answers: (Precision 90%). We understand it is important to consider the partial information retrieved. If the two arguments are absolutely necessary for any retrieval purpose, those instances which have only one argument specified can be easily filtered out.

Relation	Type Constraints	Data Set	Precision	Recall	Partial Recall
inhibit	No	MEDLINE	90.4%	58.9%	22%

Table 3: Summary of our Results

These results show a marked improvement over previously reported techniques from the literature. It is interesting to analyze the results corresponding to each submodule. In the Sentence-level (main predicate) module, 45 instances were returned of which 36 were correct instances. Only 4 were Partial correct answers and 5 were False Positives: (Precision 88.8%). In the Nominal-level module, 39 instances were returned: 20 of which were correct; 17 were partial correct answers and 2 were False Positives: (Precision 94.8%).

Module	Precision	Recall	Partial Recall	False Negatives
Sentence level	88.8%	37.8% %	4.2%	9.5%
Nominal level	94.8%	21%	17.8%	9.5%

Table 4: Results Per Module

We observed that there was a marked difference in precision between the sentence-level module and the nominal-level module. There was also a difference between the answers with one argument (partial correct answers), a difference which is also reflected in the corpus. Six instances of the markup had *unspecified* arguments, all of which were nominal instances (e.g., *A lupus inhibitor*); five instances of the markup had arguments which had no string representation, and the argument was deduced by the annotator from a preceding instance of the same relation in the context (these instances were marked as anaphoric). 17 instances were reiterations of a previously specified relation, of which 11 were nominal and 6 were verbal. This is summarized in table 5 below.

Relation Type	Anaphoric Instances	Reiterations	Redundancy
Main Pred	0%	6.3%	6.3
Nominal Rel	5.2%	11.5%	17.8%

Table 5: Corpus Redundancy Information

This supports the view that nominal instances of the relation tend to be more redundant, as a total of 23 instances were redundant (24.2%). We also counted how many arguments were anaphoric in nature (e.g. Entity 82 above). Eleven such instances were anaphoric. We applied our Anaphora Resolution module, resulting in the recognition of 10 anaphors from the 11 in the mark-up, with 8 correct and 2 incorrect results.[c]

5 Discussion and Conclusion

In this paper, we presented the results of our initial experiments on identifying and extracting biomolecular relations from the biomedical literature. Our performance represents a significant improvement over previously published results on comparable relation extraction from Medline. We attribute this performance to the integration of lexical semantic techniques, intensive corpus analytics over the corpus and the design of general automata over syntactic chunks. The results of this integration indicated that two separate modules would be most appropriate for relational parsing, allowing us to optimize verbal-based relations separately from the nominal-based cases. We are currently testing it over gold standards for new relational classes and extending the coverage of the grammar to improve the recall.

6 Bibliography

1. Andrade, Miguel A. and Valencia, Alfonso. Automatic annotation for biological sequences by extraction of keywords from MEDLINE abstracts. Development of a prototype system. AAAI, 1997.
2. Blasche, Christian; Andrade, Miguel A.; Ouzounis, Christos and Valencia, Alfonso. Automatic extraction of biological information from scientific text: protein-preotein interactions. AAAI,1999.
3. Buckley, C. Implemetation of the SMART Information Retrieval System. Technical Report 85-686, Cornell University, Computer Science. 1985.
4. Craven, Mark and Kumlien, Johan. Constructing Biological Knowledge Bases by Extracting information from Text Sources. In *In Proceedings of the 7th International Conference on Intelligent Systems for Molecular Biology* 1999.

[c]Due to space limitations, we refer the reader to `www.medstract.org` for an analysis and discussion of the false positive results from the present experiment.

5. Hindle, D. "Deterministic Parsing of Syntactic non-fluencies", *Proceedings of the 21st Annual Meeting of the Association for Computational Linguistics*, 1983.
6. Hishiki, T.; Collier, N.; Nobata, C.; Okazaki-Ohta, T.; Ogata, N.; Sekimizu, T.; Steiner, R.; Park, H.S. and Tsujii, J. Developing NLP Tools for Genome Informatics: An Information Extraction Perspective. In *In Proc. of Genome Informatics* pp81-90, Tokyo, Japan, 1998.
7. Humphreys, B. L., Lindberg, D. A. B., Schoolman H. M., and Barnett G. O. "The Unified Medical Language System: An informatics research collaboration", *Journal of the American Medical Informatics Association* 5:1, 1998.
8. Jenssen, T. K., Laegreid, A., Komorowski, J. and Hovig, E. A literature network of human genes for high-throughput analysis of gene expression. Nat Genet 28, 21-8, (2001).
9. Kennedy, C. and B. Boguraev Anaphora for Everyone: Pronominal Anaphora Resolution without a Parser. In: Proceedings of the 16th International Conference on Computational Linguistics (COLING), Vol. I, August 1996, Kopenhagen, 113-118.
10. Link, A. J., Eng, J., Schieltz, D. M., Carmack, E., Mize, G. J., Morris, D. R., Garvik, B. M. and Yates, J. R., 3rd. (1999). Direct analysis of protein complexes using mass spectrometry. Nature Biotechnology 17, 676-82.
11. McDonald, D. D. "Robust Partial Parsing through incremental multi-algorithm processing", in *Text-based Intelligent Systems*, P. Jacobs, ed. 1992.
12. Marcotte, E. M., Xenarios, I. and Eisenberg, D. Mining literature for protein-protein interactions. Bioinformatics 17, 359-63, (2001).
13. Ohta, Yoshihiro; Yamamoto, Yasunori; Okazaki, Tomoko; Uchiyama, Ikuo and Takagi, Toshihisa. Automatic Construction of Knowledge Base from Biological Papers. AAAI, 1997.
14. Proux, D. and Rechenmann, F. and Laurent, J. A Pragmatic Information Extraction Strategy for gathering Data on Genetic Interactions. In *ISMB 2000* 279-285, 2000.
15. Pustejovsky, J., S. Bergler, and P. Anick. (1993) "Semantic Methods for Corpus Analysis," *Computational Linguistics*, 19.2.
16. Pustejovsky, J., B. Boguraev, M. Verhagen, P. Buitelaar, M. Johnston, (1997) Semantic Indexing and Typed Hyperlinking AAAI Symposium on Language and the Web, Stanford, CA.
17. Pustejovsky, J.; Castaño, J.; Cochran, B.; Kotecki, M.; Morrell, M. Automatic Extraction of Acronym-meaning pairs from MEDLINE databases. In *Proceedings of Medinfo*, 2001.
18. Pustejovsky, J. and P. Hanks *Very Large Lexical Databasees: A Tutorial Primer* Association for Computational Linguistics, Toulouse, July, 2001.
19. Pustejovsky, J.; Castaño, J.; Cochran, B.; Kotecki, M. Exploiting Given versus New Information for Information Extraction Tasks *in preparation.*
20. Rindflesch, Thomas C; Rajan, Jayant V. and Hunter, Lawrence. Extracting Molecular Binding Relationships from Biomedical Text. In *Proceedings of the ANLP-NAACL 2000*, pages 188–195 Association for Computational Linguistics, 2000.
21. Sekimizu, T.; Park, H. S. and Tsujii, J. Identifying the Interaction between Genes and Gene Products based on Frequently Seen Verbs in Medline Abstracts. In *Proc. of Genome Informatics*, pp62-71, Tokyo, Japan, 1998.
22. Stapley, B. J. and Benoit, G. (2000). Biobibliometrics: information retrieval and visualization from co- occurrences of gene names in Medline abstracts. Pacific Symposium on Biocomputing 529-40.
23. Fukuda,K; Tsunoda,T.; Tamura, A. and Takagi, T. Toward Information Extraction: Identifying protein names from biological papers. In *PSB*, 707-718, 1998.

PREDICTING THE SUB-CELLULAR LOCATION OF PROTEINS FROM TEXT USING SUPPORT VECTOR MACHINES

B.J. STAPLEY [a], L.A. KELLEY, M.J.E. STERNBERG [b]
Biomolecular Modelling Laboratory, Imperial Cancer Research Fund, 44 Lincoln's Inn Field, London. WC2A 3PX, United Kingdom.
(b.stapley, l.kelley, m.sternberg)@icrf.icnet.uk

We present an automatic method to classify the sub-cellular location of proteins based on the text of relevant medline abstracts. For each protein, a vector of terms is generated from medline abstracts in which the protein/gene's name or synonym occurs. A Support Vector Machine (SVM) is used to automatically partition the term space and to thus discriminate the textual features that define sub-cellular location. The method is benchmarked on a set of proteins of known sub-cellular location from *S.cerevisiae.* No prior knowledge of the problem domain nor any natural language processing is used at any stage. The method out-performs support vector machines trained on amino acid composition and has comparable performance to rule-based text classifiers. Combining text with protein amino-acid composition improves recall for some sub-cellular locations. We discuss the generality of the method and its potential application to a variety of biological classification problems.

1 Introduction

The sub-cellular localisation of a protein is a key element in understanding its function. In order to carry out its physiological role, a protein must be often be proximal to other components involved in that process; thus knowledge of sub-cellular localization can restrict the number of possible pocesses with which a protein can be involved. Location can also alter the experimental approach to characterising a protein - e.g. purification.

Despite the importance of a protein's sub-cellular localisation, automatic prediction or extraction of this property has proved a suprisingly difficult task[1]. It has been know for sometime that the amino acid composition of protein can be an indicator of its sub-cellular location [2]. It is also clear that many cellular compartments have proteins assigned to them according to targeting signals within the protein sequences; however, such signals are not universal or necessarily clearly defined.

[a]present address; Biomolecular Sciences, University of Manchester Institute of Science Technology, PO Box 88, Manchester, UK, M60 1QD

[b]present address; Department of Biological Sciences, Imperial College of Science, Technology and Medicine, London, SW7 2AY, United Kingdom

An alternative approach, pioneered by Eisenhaber and Bork is to use the existing textual information relevant to a protein to classify it to a particular sub-cellular location [3]. This is achieved using a set of manually generated biological rules and the SWISS-PROT annotations of the proteins [4]. After tokenizing the annotations the rules are applied and a sub-cellular location extracted. They named this method Meta-Annotator. The authors report that 88% of SWISS-PROT entries can been assigned to a cellular compartment by this method. This compares very favourably with the 22% that is achieved by simple matching of relevant keywords within the documents.

Despite the success of this technique, it has two inherent weaknesses: first, a set of rules must be generated - this is obviously less intensive than manually classifying the documents, but is subjective and costly in time; second, in order to tokenize the documents they must already be structured - free text cannot be treated in such a manner without recourse to natural language parsing (NLP). NLP is beginning to show great promise within the field of biological informatics and has been successfully applied to extracting protein-protein interactions [5], metabolic pathways [6], and drug/gene relationships [7] from biological text. Although NLP often achieves very good precision, recall is often disappointing - problems of synonymy and polysemy are very difficult to overcome. In this work we investigate whether a simpler approach to the problem can be successfully applied.

The method described in this paper is to treat the protein as a vector of terms from relevant Medline documents. This approach derives from the vector-based model common in information retrieval [8]. The term weights of a vector are a functions of their frequencies within the document collection as a whole and the frequency within the relevant documents. Given a set of protein term-vectors the task is to find some function that partitions the space according to the localisation of the protein. For this task we employ support vector machines (SVM) [9].

Support vector machines are a mathematical method for performing simultaneous dimension reduction and binary classification [9]. SVMs have been applied to the problems of pattern recognition [10], regression estimation [10] and information retrieval [11,?]. Because SVMs cope well with high dimensionality and are very fast to train, they are particularly suited to problems in text data-mining/information retrieval. Kwok studied the use of SVMs in text catagorization of Reuters newswire documents [12]. In this paper, we apply an analogous approach to Medline/SWISS-PROT documents.

We evaluate the performance of SVMs in classifying a set of proteins of known sub-cellular locations from *S. cerevisiae*. Text relevant to these proteins is obtained from Medline by key-word matching of the gene naming terms.

SVMs trained on the resulting term vectors classify the proteins with good precision and recall. We also show that SVMs trained on amino acid compositions are out-performed by our SVMs trained to text data and that combining amino acid composition and term vectors can enhance classification for some sub-cellular locations.

2 Methods

2.1 Document and term processing

To obtain term vector representations of *cerevisiae* proteins we employed the following procedure. First, we scanned 22517 Medline documents for occurrences of yeast gene naming terms. These terms and synonyms were obtained from the Saccharomyces Genome Database gene registry[13 c]. For each protein, any document that contained an occurrence of the gene name or aliases of that gene was considered relevant. This resulted in a collection of 12596 documents. We employed stop word removal, stemming and removed stemmed terms that occurred in fewer than five documents. The term representation of a gene is a function of the number of relevant Medline documents and the occurrence statistics of the terms. We employed a variant of inverse document frequency (IDF) that takes account of the number of Medline documents relevant to a particular gene. The weight of term i for gene k is given by :

$$\log(1 + \sum_j f_j(w_i)) - \log N(w_i) - \log(1 + R_k) \tag{1}$$

where $f_j(w_i)$ is the frequency of term i in document j, $N(w_i)$ is the number of documents containing term i, and R_k is the number of medline documents relevant to gene k. Cooley suggests that the specific nature of term weighting may not not be crucial to the performance of SVMs in text classification[11]

2.2 Classification

The assignment of yeast proteins to sub-cellular compartments was obtained from the MIPS web site [d]. According to MIPS, 2233 proteins have known locations in one or more of 16 categories. We limit our test and training data to these proteins. The locations and numbers of proteins at each location is shown in 1. For each location class, our training set consisted of half the number of genes that fall into this category plus half the of the remaining

[c] http://genome-www.stanford.edu/Saccharomyces/registry.html

[d] available from http://mips.gsf.de/proj/yeast/catalogues/subcell/index.html

negative examples. The test set consists of the remaining proteins - positive and negative cases.

Table 1: Number of positive examples in training and test sets for sub-cellular location

Role/location	+ve in training set	+ve in test set
organisation of plasma membrane	67	63
organisation of cytoplasm	279	245
organisation of cytoskeleton	47	52
organisation of endoplasmatic reticulum	68	80
organisation of Golgi	44	33
nuclear organisation	267	341
organisation of chromosome structure	19	18
mitochondrial organisation	174	155
peroxisomal organisation	19	12
vacuolar and lysosomal organisation	27	16
extracellular/secretion proteins	10	5

2.3 *Training of SVM's*

We used the support vector machine program SVM Light package v3.50 [14] [e]. We trained a SVM for each classification using a linear kernel function with C calculated as $\frac{1}{mean(x \cdot x)}$.

2.4 *Evaluation*

We evaluate the classification performance using a variety of methods. For traditional text retrieval evaluation measures based on precision/recall have been widely used [15]. We use precision/recall plots calculated on the distance of each test vector from the SVM decision boundary; however, comparison of performance between them is difficult because the classes contain different numbers of positive examples. To assess the global performance of classification methods we employed micro- and macro- averaging of the precision/recall data. Micro-averaging determines precision and recall of a set of binary classifiers averaged over the number of documents; this equates to evaluating average

[e] available from http://ais.gmd.de/thorsten/svm_light

performance a document selected randomly from the test collection. In macro-averaging, the recall/precision are averaged over the number of classes. Macro-averaging estimates the expected performance of an SVM trained on a new class; whereas micro-averaging estimates the performance of the system with new documents. For our purposes, micro-averaging is more useful.

We also use the F1 measure proposed by van Rijsbergen [?]. F1 is given by $\frac{2rp}{r+p}$ where p and r are precision and recall respectively. We determine the maximal value of F1 for the performance of each system on a particular classification.

3 Results

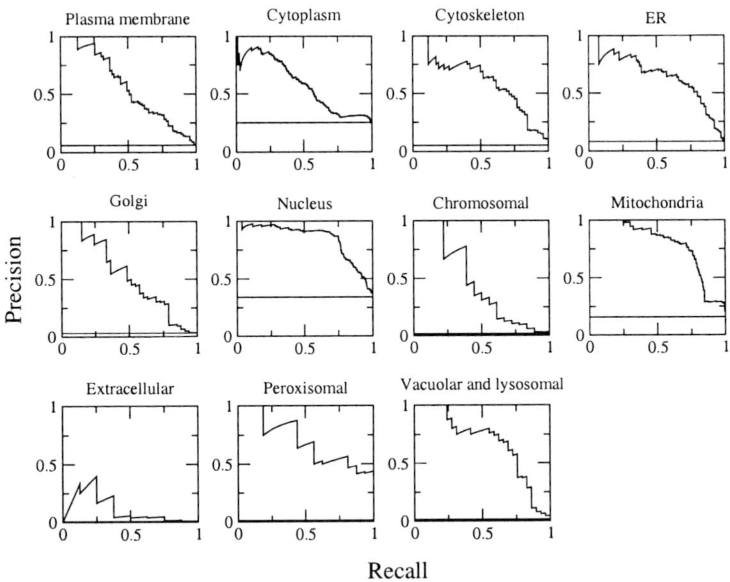

Figure 1: Precision/recall plots for location classifiers trained on term vectors. Horizontal lines indicate the performance of a random classifier

Precision/recall graphs for the various classifications are shown in figure 1. The performance of a random classifier is shown as a horizontal line in each plot. At low levels of recall, the precision is generally very high (95%+).

Classes with a large number of positive examples - nuclear, cytoplasmic, and mitochondrial - are better predicted than the rarer classifications. This is reflected in averaged precision/recall shown in figure 2. The better apparent performance from micro-averaging is a result of better prediction of bigger classes. The values of $Max(F1)$ are shown in table 2 and are much greater in all cases than a random classifier.

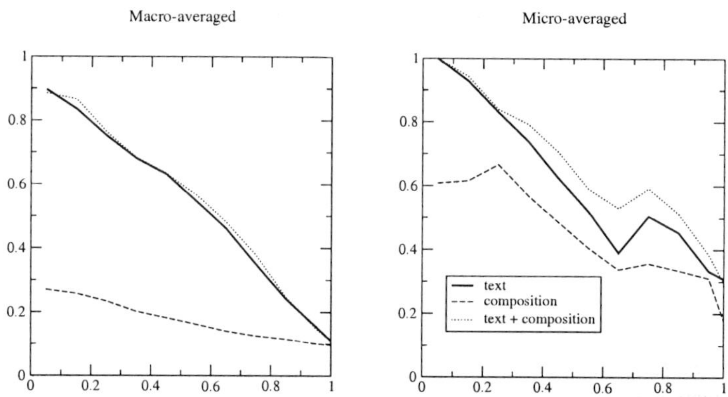

Figure 2: Micro and macro averaging of classification to 11 locational categories.

3.1 Sequence and text together improve classification

It has been known for some time that the amino acid composition of a protein can be used as an indicator of its sub-cellular localisation [16,2]. In particular, membrane associated proteins tend towards hydrophobicity, while intracellular proteins tend to be low in cysteine and rich in aliphatic and charged amino acids. Nuclear proteins generally contain disproportionally more charged and polar residues.

Figure 3 illustrates the performance of support vector machines in discriminating protein localisation based of their fractional composition of the twenty amino acids. It can been seen that composition is a poor predictor of ER, cytoskeleton, golgi, peroxisomal and vacuolar proteins, but good at

Table 2: Maximum F1-value for classifications

Role/location	Max F1			
	text alone	text + composition	composition alone	random
organisation of plasma membrane	0.54	0.56	0.47	0.12
organisation of cytoplasm	0.55	0.60	0.48	0.39
organisation of cytoskeleton	0.62	0.61	0.13	0.10
organisation of endoplasmatic reticulum	0.65	0.66	0.10	0.14
organisation of Golgi	0.54	0.53	0.10	0.14
nuclear organisation	0.80	0.82	0.61	0.51
organisation of chromosome structure	0.52	0.52	0.20	0.04
mitochondrial organisation	0.75	0.75	0.36	0.27
peroxisomal organisation	0.67	0.65	0.03	0.01
vacuolar and lysosomal organisation	0.69	0.69	0.06	0.02
extracellular/secretion proteins	0.31	0.33	0.12	0.01

predicting cytoplasmic, membrane and nuclear proteins. Composition also contains limited information on mitochondrial and chromosomal proteins. For extra-cellular proteins the scarcity of data makes assessment difficult, but the composition of these proteins gives better than random predictions.

When the test and training vectors derived from terms within Medline are extended to include the amino acid composition, performance in classifying proteins to the cytosol and nucleus is enhanced (table 2). In particular recall is improved. This may reflect improved performance on those proteins which have relatively few citations in the literature.

3.2 Detecting errors in annotation

Any manual method of gene annotation is liable to errors of omission and mis-classification. We checked apparent false negatives and positives to assess whether they were genuine by inspection of the relevant Medline documents.

For the cytosolic classification, the top scoring 'false' positive is cdc42, a Rho-type GTPase involved in bud site assembly and cell polarity. cdc42 contains a CAAX motif for geranylgeranyl modification and is likely to be associated with cell membranes. Ziman *et al.*,[17] determined that cdc42 exists in both a soluble form and membrane associated form within the cell; thus cdc42 should be included in cytosolic classification. A similar situation exists with ypt1 which is a GTP-binding protein required for vesicle transport from ER to Golgi and within the Golgi stack. It also undergoes geranylgeranyl modification, but the abundance and significance of any cytosolic form of the protein is not clear.

There are several proteins which MIPS assigns to the nucleus that our

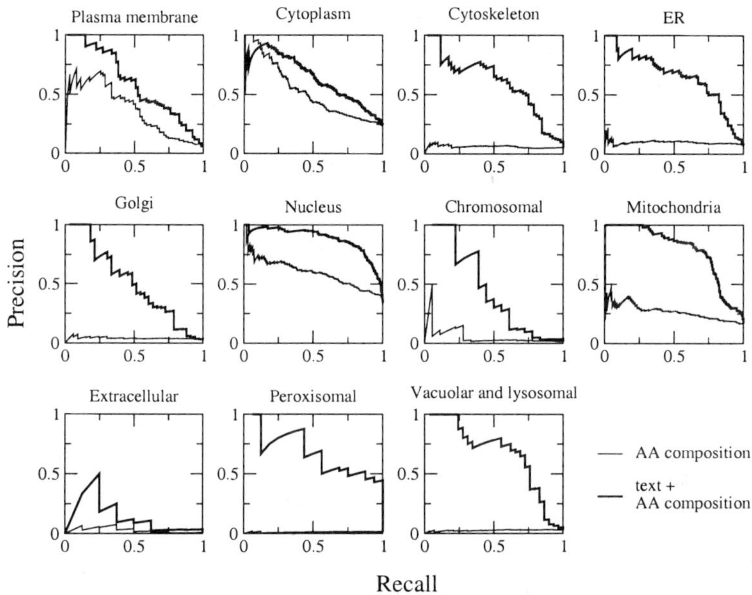

Figure 3: Precision/recall plots for location classifiers trained on term and/or amino acid composition vectors.

method correctly flags as being non-nuclear. These include: UBC6 - a ubiquitin-conjugating enzyme, anchored in the ER membrane with the catalytically active domain in cytoplasm [18]; hts1 - a histidyl-tRNA synthetase which is located exclusively in the mitochondria and cytosol [19]; and SMI1 protein involved in beta-1,3-glucan synthesis which has been shown to localise in patches at bud sites [20].

4 Discussion

4.1 Functional classification using SVMs and text

Here we show that functional classification of genes can be facilitated by text analysis of documents relevant to a gene. Other than a list of gene naming terms and synonyms, our method uses no prior knowledge of the problem domain nor any information from previously compiled sequence databases.

Automatic functional assignment of proteins can be used to improve manual assignment by spotting errors and increasing recall. Such errors may be simple mistakes, or the result of partial or incorrect information or understanding on the part of the human classifier. Even in the absence of such errors, assessments of what constitutes a correct assignment of documents into a classification will vary from user to user; thus there is a theoretical limit to the precision of an automatic classifier. For nuclear and mitochondrial proteins, automatic classification may be approaching this limit.

Although amino acid composition is generally a poor indicator of subcellular location, for some locations sequence provides a strong signal. In such cases, combining text and composition features can enhance recall.

4.2 Comparison with other methods

To compare our classification methods to that of Eisenhaber and Bork, we tested their algorithm (Meta-Annotator) on a subset of our original data that is present in SWISS-PROT [4]. Meta-Annotator is outstandingly good at predicting mitochondrial proteins and very good at predicting nuclear proteins.

Because Meta-Annotator joins the golgi and endoplasmic reticulum (ER) into a single class, we modified our treatment of this locational class. A single SVM trained to distinguish golgi or ER from others performed very poorly, probably because the intersection of these two sets is very small (8 cases) according to the MIPS classification. We therefore used the max(F1) value from micro-averaging of two SVMs trained on the ER and golgi proteins independently. Text classification using SVMs out-performs Meta-Annotator for cytoplasmic and golgi/ER proteins.

It should be borne in mind when comparing the two approaches that Meta-Annotator involves a large ammount of manual intervention. Not only is the method only applicable to a previously manually curated protein database (Swiss-Prot), but it also has encoded into more than 1000 logical rules derived from a human expert. Our approach requires no human input other than a list of gene names and synonyms. Given these facts it is little wonder than Meta-Annotator can generally out-perform our method. It is encouraging that a generic automatic approach can perform so well. With a larger set of training documents the SVM approach may be improved.

4.3 Combining features for functional classification of proteins

In this paper we have demonstrated that combining disparate features of a protein can aid in the functional classification of that protein. With the advent of many high-throughput studies of genes and proteins, many more features can

Table 3: Comparison of text SVMs and Meta-Annotator

Role/location	Meta_A precsion/recall	$Meta_A F1$	max(F1) for text
organisation of cytoplasm	49/32	0.38	0.54
organisation of Golgi/ER	75/48	0.58	0.62
nuclear organisation	87/86	0.86	0.80
mitochondrial organisation	90/93	0.91	0.75

be used as training data for binary classifiers. These include protein interaction data, features of the protein or DNA sequence and expression array data. The inclusion of a variety of independent or semi-independent features should improve recall since data for every protein may not be available from every experiment. For example, our method can be applied to proteins/genes of unknown sequence or conversely, sequence information can be used to infer function in the absence of any text relevant to the protein/gene.

Support vector machines are well suited to classification tasks of high dimensionality in which many features may be noisy or irrelevant. There is no doubt that data from expression array and protein interaction experiments can yield insights into gene function, but the quality of such data is hard to determine. The method presented here may ameliorate some of these problems.

Finally, entities other than proteins and genes can be represented as high dimensional vectors of text terms; these include whole organisms, protein complexes, protein domains or motifs, small molecules, cells and arbitrary text documents. In short, any entity which contains text, or for which relevant texts can be retrieved can be placed within a classification scheme.

5 References

1. F. Eisenhaber and P. Bork. Wanted: subcellular localization of proteins based on sequence. *Trends Cell Biol*, 8(4):169–70, Apr 1998.
2. K. Nishikawa and T. Ooi. Correlation of the amino acid composition of a protein to its structural and biological characters. *J Biochem (Tokyo)*, 91(5):1821–4, May 1982.
3. F. Eisenhaber and P. Bork. Evaluation of human-readable annotation in biomolecular sequence databases with biological rule libraries. *Bioinformatics*, 15(7-8):528–35, Jul-Aug 1999.
4. A. Bairoch and R. Apweiler. The protein sequence data bank and its supplement trembl in 1999. *Nucleic Acids Res*, 27(1):49–54, Jan 1 1999.
5. L. Wong. PIES, a protein interaction extraction system. In *Pac Symp Biocomput*, pages 520–31, Hawaii, 2001.

6. S. K. Ng and M. Wong. Toward routine automatic pathway discovery from on-line scientific text abstracts. In *Genome Inform Ser Workshop Genome*, volume 10, pages 104–112., 1999.
7. T. C. Rindflesch, L. Tanabe, J. N. Weinstein, and L. Hunter. Edgar: extraction of drugs, genes and relations from the biomedical literature. In *Pac Symp Biocomput*, pages 517–28., Hawaii, 2000.
8. C. J. van Rijsbergen. *Information Retrieval.* Butterworths, 2nd edition, 1979.
9. V.N. Vapnik. *The nature of statistical learning theory.* Springer Verlag, Heidelberg, DE, 1995.
10. V. Vapnik, S. Golowich, and A. Smola. Support vector method for function approximation, regression estimation, and signal processing. In M.C. Mozer, M.I. Jordan, and T. Petsche, editors, *Advances in Neural Information Processing Systems*, volume 9, pages 281–287, 1997.
11. R. Cooley. Classication of news stories using support vector machines. In *International Joint Conference on Articial Intelligence Text Mining Workshop*, 1999.
12. J. T-Y. Kwok. Automated text categorization using support vector machine. In *Proceedings of the International Conference on Neural Information Processing (ICONIP)*, pages 347–351, Kitakyushu, Japan, 1999.
13. J. M. Cherry, C. Ball, K. Dolinski, S. Dwight, M. Harris, J. C. Matese, G. Sherlock, G. Binkleyand H. Jin, S. Weng, and D. Botstein. Saccharomyces genome database. ftp://genome-ftp.stanford.edu/pub/yeast/SacchDB/, 2000.
14. T. Joachims. Text categorization with support vector machines: learning with many relevant features. In *10th European Conference on Machine Learning*, pages 137–142, Chemnitz, DE, 1998. Springer Verlag, Heidelberg, DE.
15. R. Baeza-Yates and B. Ribeiro-Neto. *Modern information retrieval.* Addison-Wesley, Harlow, England, 1999.
16. J. Cedano, P. Aloy, J. A. Perez-Pons, and E. Querol. Relation between amino acid composition and cellular location of proteins. *J Mol Biol*, 266(3):594–600., Feb 28 1997.
17. M. Ziman, D. Preuss, J. Mulholland O., J. M. 'Brien, D. Botstein, and D. I. Johnson. Subcellular localization of cdc42p, a saccharomyces cerevisiae-binding protein involved in the control of cell polarity. *Mol Biol Cell*, 4(12):1307–16., Dec 1993.
18. U. Lenk and T. Sommer. Ubiquitin-mediated proteolysis of a short-lived regulatory protein depends on its cellular localization. *J Biol Chem*, 275(50):39403–10., Dec 15 2000.

19. M. I. Chiu, T. L. Mason, and G. R. Fink. Hts1 encodes both the cytoplasmic and mitochondrial histidyl-trna synthetase of saccharomyces cerevisiae: mutations alter the specificity of compartmentation. *Genetics*, 132(4):987–1001., Dec 1992.
20. H. Martin, A. Dagkessamanskaia, G. Satchanska, N. Dallies, and J. Francois. Knr4, a suppressor of saccharomyces cerevisiae cwh mutants, is involved in the transcriptional control of chitin synthase genes. *Microbiology*, 145:249–58, Jan 1999.

A THEMATIC ANALYSIS OF THE AIDS LITERATURE

W. JOHN WILBUR
National Center for Biotechnology Information, National Library of Medicine, National Institutes of Health, Bethesda, MD, U.S.A.

Faced with the need for human comprehension of any large collection of objects, a time honored approach has been to cluster the objects into groups of closely related objects. Individual groups are then summarized in some convenient manner to provide a more manageable view of the data. Such methods have been applied to document collections with mixed results. If a hard clustering of the data into mutually exclusive clusters is performed then documents are frequently forced into one cluster when they may contain important information that would also appropriately make them candidates for other clusters. If a soft clustering is used there still remains the problem of how to provide a useful summary of the data in a cluster. Here we introduce a new algorithm to produce a soft clustering of document collections that is based on the concept of a theme. A theme is conceptually a subject area that is discussed by multiple documents in the database. A theme has two potential representations that may be viewed as dual to each other. First it is represented by the set of documents that discuss the subject or theme and second it is also represented by the set of key terms that are typically used to discuss the theme. Our algorithm is an EM algorithm in which the term representation and the document representation are explicit components and each is used to refine the other in an alternating fashion. Upon convergence the term representation provides a natural summary of the document representation (the cluster). We describe how to optimize the themes produced by this process and give the results of applying the method to a database of over fifty thousand PubMed documents dealing with the subject of AIDS. How themes may improve access to a document collection is also discussed.

1 Introduction

There are at least two reasons for interest in clustering a set of documents. One is to improve retrieval efficiency and the other is to improve human understanding of the data in the collection. The first of these goals proved elusive historically because the quality of the retrieval degraded due to the clustering.[1, 2] With the much greater speed and memory of current computers the interest in clustering for efficiency has waned. However, the need for improved human understanding of large data sets has reached critical proportions with the advent of the Internet as well as the many large databases of documents that are now becoming available in different specialty areas.

Improved human understanding of data through clustering may consist of graphical aids in visualizing the data[3-5] as well as methods of examining textual summaries of cluster content. Given a predefined set of clusters there are many machine learning methods for inducing representations for the clusters.[6-9] These methods play an important role in the human comprehension of information from a variety of points of view. However, our interest is somewhat different from these in that we desire to find a rich representation of the topics or themes that occur in a database and view document clusters as a means to this end. For example, in the AIDS data that we study here "blood transfusion" is an important theme. This theme is described by a rich terminology and it is this terminology that we refer to in using the word theme. There is also the cluster of documents in the database that discuss this theme and this cluster plays an essential role in discovering the theme, but the theme and the cluster are treated as of equal importance each helping to define the other.

A number of methods have been proposed whereby topical groupings of terms are derived from a document collection in an effort to improve the representation of the documents. Such methods may or may not involve a clustering of the documents, but they are of interest since they address the problem of theme generation that is our interest. There are information bottleneck methods,[10, 11] probabilistic latent semantic indexing,[12] and mixture models.[13-15] The bottleneck approach produces term groupings that maximize the information relative to the document collection. This groups terms with a high mutual co-occurrence, but seems unsuited to produce the natural themes that occur in text. Theme generation as we conceive it will be almost certain to reduce the information relative to the documents because of the large number of terms that are grouped together in a theme. We believe this is simply the wrong paradigm for theme generation. Probabilistic latent semantic indexing and most mixture models assume in principle that a document arises from a single source even if that source is not determined. Again this is theoretically unsuitable for our purposes as a document does not arise from a single theme. Rather a document often contains multiple themes. The one approach in the literature that seems theoretically most consistent with our goal is the Multiple Cause Mixture Model.[15] While this approach solves the one document-multiple theme problem, it along with the other methods mentioned here has another unfortunate property. It requires that all terms that occur in documents must be forced into some topical word group even if they are function words, etc.

Aside from the theoretical problems mentioned here there is the practical problem that previous methods require the whole database to be processed before any result is obtained. This is very computationally expensive and out of reach for large collections (even the AIDS data we study here). Our approach is unique in that it produces one theme and one document cluster at a time. Because of this simplicity our method is readily applied to very large collections (even millions of documents). Indeed a very large collection may yet take a long time for complete analysis, but one can produce useable collections of themes without a complete analysis. The approach we present here is related to earlier work,[16, 17] but the model is different in its explicit treatment of themes, is simpler, and allows a much more efficient computation.

2. Preliminaries

Let D be a database of documents and let the set of all index terms that appear in at least one of the members of D be denoted by T. Let R denote the "occurs in" relationship between elements of T and D. Then it is customary to represent R as a subset of the product set of T and D, i.e., $R \subseteq T \times D$. The set R is known as a relation and if $(t,d) \in R$ we say t occurs in d and may also write tRd. If $U \subseteq T$ and $V \subseteq D$ it is standard usage to define

$$\begin{aligned} R[U] &= \{d \in D \mid \exists t \in U \ni (t,d) \in R\} \\ R^{-1}[V] &= \{t \in T \mid \exists d \in V \ni (t,d) \in R\}. \end{aligned} \tag{2.1}$$

For a single point t we write $R[t] = R[\{t\}]$ and for a single point d likewise $R^{-1}[d] = R^{-1}[\{d\}]$.

By a theme we mean a particular subject area that is discussed by some subset of the documents in the database D. Interestingly such a subject area is generally also characterized by a particular subset of index terms T that are used to describe that subject area. Intuitively then a theme means nonempty sets $U \subseteq T$ and $V \subseteq D$ with the property that all the elements of U have a high probability of occurring in all the elements of V. We require not only that this be true, but that it be true in some optimal sense which we will make explicit.

3. The Theme Generation Algorithm

In order to apply the EM algorithm we will follow the notation of Little and Rubin.[18] Our description will be in terms of the sets U and V which we have used to outline the concept of a theme. There is observed and missing data.

$$\begin{aligned} Y_{obs} &= R \\ Y_{miss} &= \{z_d\}_{d \in D} \end{aligned} \tag{3.1}$$

The observed data is the relation R. The missing data is a set of indicator variables that are defined by

$$z_d = \begin{cases} 1, & d \in V \\ 0, & d \notin V. \end{cases} \tag{3.2}$$

The parameters are

$$\Theta = U\left(\|U\| = n_U\right),\ \{p_t, q_t\}_{t \in U},\ \{r_t\}_{t \in T}. \tag{3.3}$$

Here n_U is a constant positive integer and the size of the set U (number of elements). For any $t \in U$, p_t is the probability that for any $d \in V$, tRd, and q_t is the probability that for any $d \in D-V$, tRd . For any $t \in T$, r_t is the probability that for any $d \in D$, tRd . Constants in the process in addition to the integer n_U are the set of prior probabilities $\{pr_d\}_{d\in D}$ that are the prior probabilities that the elements d belong to V .

In order to develop the EM algorithm approach we will need to make an independence assumption about the statistical properties of R. This kind of assumption is common in many contexts for the purpose of facilitating the mathematical analysis of complicated data:

Independence Assumption. Within $T \times V \cap R$ all the atomic events tRd are independent of each other and likewise for $T \times (D-V) \cap R$.

Finally to facilitate the writing of mathematical formulas we will use the indicator variables $\{u_t\}_{t\in T}$ defined by

$$u_t = \begin{cases} 1, & t \in U \\ 0, & t \notin U \end{cases} \tag{3.4}$$

and the delta notation

$$\delta_{td} = \begin{cases} 1, & tRd \\ 0, & \neg tRd. \end{cases} \tag{3.5}$$

We must work with the quantity

$$p(R,\{z_d\}|\Theta) = p(R|\{z_d\},\Theta)\,p(\{z_d\}|\Theta). \tag{3.6}$$

Computing from the right side we obtain

$$p(\{z_d\}|\Theta) = \prod\nolimits_{d\in D} pr_d^{z_d}\,(1-pr_d)^{1-z_d} \tag{3.7}$$

$$p(R|\{z_d\},\Theta) = \prod\nolimits_{t,d}\left\{\left[\left(p_t^{\delta_{td}}\,(1-p_t)^{1-\delta_{td}}\right)^{u_t}\left(q_t^{\delta_{td}}\,(1-q_t)^{1-\delta_{td}}\right)^{1-u_t}\right]^{z_d}\left(r_t^{\delta_{td}}\,(1-r_t)^{1-\delta_{td}}\right)^{1-z_d}\right\} \tag{3.8}$$

It is next necessary to take the expectation of the log of (3.6) over the distribution $p(\{z_d\}|R,\Theta)$. In this we may ignore (3.7) because it will yield a constant and have no influence on the subsequent maximization. Thus we may compute

$$\begin{aligned} E(\ln P(R|\{z_d\},\Theta)) = {} & \sum\nolimits_t u_t \sum\nolimits_d pz_d\left(\delta_{td}\ln p_t + (1-\delta_{td})\ln(1-p_t)\right) + \\ & \sum\nolimits_t u_t \sum\nolimits_d (1-pz_d)\left(\delta_{td}\ln q_t + (1-\delta_{td})\ln(1-q_t)\right) + \\ & \sum\nolimits_t (1-u_t)\sum\nolimits_d \delta_{td}\ln r_t + (1-\delta_{td})\ln(1-r_t) \end{aligned} \tag{3.9}$$

In order to complete this calculation it is necessary to compute pz_d based on R and Θ . This we do from Bayes' theorem

$$pz_d = p(z_d = 1 | R, \Theta) = p\left(z_d = 1 | \{\delta_{td}\}_{t\in T}, \Theta\right)$$
$$= \frac{p\left(\{\delta_{td}\}_{t\in T} | z_d = 1, \Theta\right) pr_d}{p\left(\{\delta_{td}\}_{t\in T} | z_d = 1, \Theta\right) pr_d + p\left(\{\delta_{td}\}_{t\in T} | z_d = 0, \Theta\right)(1 - pr_d)} \tag{3.10}$$

Individual probabilities on the right side are given by

$$p\left(\{\delta_{td}\}_{t\in T} | z_d = 1, \Theta\right) = \prod_t \left(p_t^{\delta_{td}} (1-p_t)^{1-\delta_{td}}\right)^{u_t} \left(r_t^{\delta_{td}} (1-r_t)^{1-\delta_{td}}\right)^{1-u_t}$$
$$p\left(\{\delta_{td}\}_{t\in T} | z_d = 0, \Theta\right) = \prod_t \left(q_t^{\delta_{td}} (1-q_t)^{1-\delta_{td}}\right)^{u_t} \left(r_t^{\delta_{td}} (1-r_t)^{1-\delta_{td}}\right)^{1-u_t} \tag{3.11}$$

Because of the common factor in these expressions it is convenient to write

$$pz_d = \frac{1}{1 + exp(-score_d + C)} \tag{3.12}$$

where

$$C = \sum_{t\in U} ln\left(\frac{1-p_t}{1-q_t}\right)$$
$$score_d = \sum_{t\in U} \delta_{td}\, ln\left(\frac{p_t(1-q_t)}{q_t(1-p_t)}\right) + ln\left(\frac{pr_d}{1-pr_d}\right). \tag{3.13}$$

The final step is to carry out the maximization of (3.9) over Θ. Too accomplish this we note that we may begin by choosing the values of p_t, q_t, and r_t so that the individual sums on the right in (3.9) are maximal if in the case of p_t and q_t, $u_t = 1$ and if in the case of r_t, $u_t = 0$. This is straightforward and yields

$$p_t = \sum_d \delta_{td} pz_d / \sum_d pz_d$$
$$q_t = \sum_d \delta_{td} (1 - pz_d) / \sum_d (1 - pz_d) \tag{3.14}$$
$$r_t = n_t / N$$

Here we have defined

$$n_t = \|R[t]\| = \sum_d \delta_{td}$$
$$N = \|D\|. \tag{3.15}$$

Now for each t we define a quantity which is the difference between the contribution coming from t in the sum (3.9) depending on whether $u_t = 1$ or $u_t = 0$.

$$\alpha_t = n_{st}\, ln\left(\frac{p_t}{q_t}\right) + \left(n_s - n_{st}\right) ln\left(\frac{1-p_t}{1-q_t}\right) +$$
$$\left(n_t - n_{st}\right) ln\left(\frac{q_t}{r_t}\right) + \left(N - n_t - n_s + n_{st}\right) ln\left(\frac{1-q_t}{1-r_t}\right) \tag{3.16}$$

In addition to (3.15) we here employ the definitions

$$n_s = \sum_d pz_d$$
$$n_{st} = \sum_d \delta_{td}\, pz_d. \tag{3.17}$$

The maximization is completed by choosing the n_U largest α_t's and setting $u_t = 1$ for each of them and $u_t = 0$ for all others. If there is ambiguity due to equal α_t's choices are made arbitrarily to obtain the number n_U.

4. A Practical Algorithm

Our interest is in a practical algorithm for applications. With that objective we will outline here our approach first as a series of steps and then give more detail in how to begin the computation and how to control it.

Input: *R*, the number n_U, and the set of prior probabilities $\{pr_d\}_{d\in D}$.
Step 1: Compute the probabilities $\{pz_d\}_{d\in D}$ through the use of (3.13).
Step 2: Compute p_t, q_t, and r_t, all $t \in T$ from (3.14) and (3.15).
Step 3: Compute the α_t, all $t \in T$ from (3.16) and (3.17).
Step 4: Select the n_U points $t \in T$ for which α_t is the greatest to define the set U and the indicator values $\{u_t\}_{t\in T}$.
Step 5: Test for convergence and if not converged return to Step 1.

By examining the steps listed it is evident that if we can obtain the probabilities $\{pz_d\}_{d\in D}$ in Step 1, the remaining steps are relatively straightforward to perform (we will discuss convergence below). As a general approach we have found it quite satisfactory to restrict the values pz_d to either 0 or 1. This is simply accomplished by setting a cutoff value and using (3.13) to compute

$$pz_d = \begin{cases} 1, & score_d > cutoff \\ 0, & score_d \le cutoff \end{cases}. \tag{4.1}$$

In practice we find that the number of *d*'s for which $pz_d = 1$ can have a large variation from one iteration to the next if we use a fixed cutoff. We have found improved stability by defining an integer we term the *stringency*. The stringency must be positive and not greater than n_U. We then set

$$cutoff = \frac{stringency}{n_U} \sum_{t\in U} ln\left(\frac{p_t\left(1-q_t\right)}{q_t\left(1-p_t\right)}\right). \tag{4.2}$$

When (4.1) and (4.2) are used to implement Step 1 we will refer to the result as the *binary* form of the algorithm. The algorithm begins by assigning the values pz_d to be 0 or 1 depending on some preliminary guess as to what V might be. This allows the first iteration through Steps 1-5. On the second and all subsequent iterations the equations (4.1) and (4.2) are used in Step 1. In Step 5 convergence is tested by observing when all quantities become fixed. In practice this is easily ascertained by observing when the value of C in (3.13) takes the same identical value on successive iterations.

Control of the algorithm is important in that it generally has the potential to converge to a local maximum in many different ways. Such control could be exerted through the choice of the values $\{pr_d\}_{d\in D}$. However our approach is generally to set these values all to 0.5 so that they have no influence in (3.13). We only exert control by the initial choice of the $\{pz_d\}_{d\in D}$ as binary values reflecting some estimate of V. Occasionally this is not satisfactory and we wish to force the algorithm to converge with certain d included in V. Then we set the values of the corresponding pr_d close to 1 or equivalently the values of $ln\left(pr_d/(1-pr_d)\right)$ large so that these particular d become locked into V.

5. Focusing a Theme

The binary form of the theme generation algorithm described in the foregoing works well in that it is successful in producing a large number of different themes on a database. The difficulty is that one must decide on the value n_U prior to generating a theme and this may not be optimal. If it is too small it will not allow the full theme to develop and if it is too large it will allow extraneous material to be pulled in to be part of the theme. In order to deal with this problem we have developed a method of focusing a theme to the optimal size. The method works by starting with a value of n_U that is too small and running the algorithm to stability or near stability and then increasing the size of n_U by a small amount to $n_{U'}$ and again running the algorithm close to stability. Let U and U' denote the two term sets corresponding to the two themes obtained at these two successive points. At each such step we check two things. First, are the two themes close together? To measure closeness let α_t represent the value from (3.16) corresponding to a $t\in U$ and likewise α_t' for $t\in U'$. We may then define a Dice coefficient of similarity between the two themes by

$$Dice(U,U')=\left(\sum_{t\in U\cap U'}\alpha_t+\alpha_t'\right)/\left(\sum_{t\in U}\alpha_t+\sum_{t\in U'}\alpha_t'\right). \qquad (5.1)$$

We require that

$$Dice(U,U')>0.9 \qquad (5.2)$$

at each successive increment of n_U to $n_{U'}$. This is a continuity condition that is necessary because during expansion a theme my become unstable and suddenly in a single step metamorphose into a completely different theme or into one that is only distantly related to the theme from the previous step. If such a sudden change in the theme takes place we halt the process of focusing at the previous step. Our second concern is that the theme actually improves at each step. In order to measure improvement we require a fixed integer smaller than the number of terms in the theme. We will call this number the *focal size* of the theme and denote it by f. Then we define the *focus* of a theme U to be the average of the f largest α_t, $t \in U$. We denote the focus of a theme U by $\bar{\alpha}(U,f)$. Then we consider a step in focusing to be an improvement provided

$$\bar{\alpha}(U,f) \leq \bar{\alpha}(U',f). \qquad (5.3)$$

Thus if the process of focusing the theme does not end because of a violation of (5.2) it will eventually end because of a violation of (5.3) when we have reached at least a local maximum in the focus possible for that theme.

6. Themes from the AIDS data

In February of 2001 we extracted all documents in PubMed that had assigned the MeSH® term "Human Immunodeficiency Syndrome". This comprised 52,970 documents consisting of title, abstract (when present), and MeSH terms. This set is the database D for the thematic analysis presented here. We used as the index term set T all MeSH terms (with and without qualifiers assigned in the documents and with and without stars) as well as terms from the titles and abstracts. Title and abstract were broken into single word and two word terms and any term containing a stop word was discarded. No stemming was performed and no punctuation is allowed in the terms.

Our first step was to generate a set of themes with stringency 10 and n_U equal to 30. These parameter settings tend to give undersize themes. We attempted to generate such an initial theme for each document in D. For each $d \in D$ we used a vector document retrieval algorithm to obtain the 100 documents, $\{d_i\}_{i=1}^{100}$, in D most similar to d. We then set pz_d to 1 for all the $\{d_i\}_{i=1}^{100}$ and 0 for all other documents. With this initialization we then attempt to generate a theme. We succeeded in generating a theme in 42,395 of the 52,970 attempts. In some cases the set of documents $\{d_i\}_{i=1}^{100}$ has insufficient similarity within the set to produce a theme. When duplicates were removed the 42,395 themes resulted in a set of 7,311 unique themes. These themes provided the seeds for the focusing process described in the previous section.

For the focusing process we chose a focal size f of 10. Beginning with each of the 7,311 seed themes we carried out the focusing process and obtained a set of 5,236 unique focused themes. In a significant number of cases different seed themes produced the same focused theme. While the 5,236 themes are unique, there are many pairs of themes that are closely related to each other. We processed all pairs of the 5,236 themes and marked a pair (U,U') as equivalent if they satisfied

$$Dice(U,U') \geq 0.9 . \qquad (6.1)$$

We generated the equivalence relation based on the marked pairs and chose one of the largest themes from each class. This yielded a set of 1164 unique themes with a certain distance between any two themes in the set. These 1164 themes provide a picture of the AIDS literature in PubMed.

While one can view each of the 1164 themes, this is still a relatively large number of themes to examine. In order to facilitate browsing the data we performed single link clustering of the 1164 themes with different thresholds according to

$$Dice(U,U') \geq threshold \qquad (6.2)$$

to obtain clusters of themes that could be examined by a human. We performed the clustering at five levels beyond the baseline of 0.9 and obtained the numbers of clusters shown in Table 1.

Table 1. An analysis of the 1164 themes by single link clustering. As the threshold decreases there are fewer clusters of larger size.

Level	Threshold	Clusters
1	0.9	1164
2	0.8	772
3	0.7	477
4	0.6	287
5	0.5	171
6	0.4	92

We have developed a web interface to allow browsing of the 1164 themes. At level 1 this allows access to the individual themes. At higher levels one views the individual themes grouped into clusters and may still view an individual theme or may select a cluster and view its summary or differences between its members.

7. An Example theme

Here we give a partial listing of the term set for the theme developed on pneumocystis pneumonia in AIDS. This theme has 70 terms associated with it. This is of intermediate size. While some themes have only 30 terms many have over 70 and a few over 200 terms.

Table 2. The "pneumocystis pneumonia" theme in the AIDS database. The top twenty and the bottom twenty terms are listed. Terms ending with "!!t" or "!!p" are from the text (title or abstract). Terms ending with "!!T" or "!!P" are from the title. All other terms are MeSH terms.

Γ_t	weight	term
4203.16	7.67756	pneumocystis!!t
4131.06	7.45661	carinii!!t
4044.71	7.2371	pneumocystis carinii!!p
3451.11	6.22396	pneumonia, pneumocystis carinii!
3189.29	5.88545	pneumonia!!t
3148.41	8.89602	pneumocystis!!T
3128.89	6.16155	carinii pneumonia!!p
3076.95	11.3618	carinii!!T
3019.83	11.3184	pneumocystis carinii!!P
2056.77	6.119	pneumonia!!T
2034.68	10.5684	carinii pneumonia!!P
1334.72	4.80077	pneumonia, pneumocystis carinii!complications
1196.67	6.04187	pcp!!t
1166.19	7.33732	pneumonia pcp!!p
874.62	4.7418	pneumonia, pneumocystis carinii!drug therapy
860.536	2.73083	acquired immunodeficiency syndrome!complications
765.718	4.90711	pentamidine!
747.283	4.9847	pentamidine!!t
687.268	4.81072	pneumonia, pneumocystis carinii!diagnosis
644.147	4.36415	pneumonia, pneumocystis carinii!etiology
....	....	....
288.126	5.02338	pneumocystis pneumonia!!p
279.602	5.41492	pneumocystis carinii!isolation & purification
279.038	3.85202	bronchoalveolar lavage fluid!
276.319	4.35901	trimethoprim sulfamethoxazole!!p
269.075	4.23562	bronchoscopy!
266.897	6.37945	pentamidine!administration & dosage*
258.448	4.98364	pneumonia, pneumocystis carinii!mortality
239.531	4.00696	trimethoprim-sulfamethoxazole combination!
239.136	1.70665	diagnosis!!t
236.596	7.98609	carinii infection!!P
231.826	4.30286	trimethoprim!
221.698	4.05579	prophylaxis!!T
220.914	2.85947	respiratory!!t
218.138	4.89353	trimethoprim!therapeutic use
217.447	4.23204	pentamidine!adverse effects
214.698	4.48168	transbronchial!!t
208.269	4.20944	trimethoprim-sulfamethoxazole combination!therapeutic use
199.971	4.20475	sulfamethoxazole!
193.587	4.89464	sulfamethoxazole!therapeutic use
190.416	3.63024	alveolar!!t

8. Future Plans

Our immediate plan is to extract the literature from MEDLINE® that deals with genetics (over one million documents) and produce a set of themes for this subject area. It is unclear whether such a large set of themes will lend itself to browsing, though we plan to experiment with browsing. We are more optimistic regarding a different strategy. We plan to treat the individual themes as documents and make them accessible through Boolean querying much as for documents. Because the terms in themes are rated by their associated α_t values, these values may be used to produce ranked retrieval. This is straightforward in the case of a single query term and for Booleans could make use of some kind of extended Boolean[19] or fuzzy logic. Once a user has selected a theme consistent with his interests he has the option of using it to produce ranked retrieval of the documents in the database. This is based on the weights associated with the terms in a theme (see Table 2) and is defined by (3.13). Another potential application of themes is to the problem of term disambiguation. If a term occurs in multiple themes and the term occurs in a document then one may compare the themes with the document to see which theme best fits the context and interpret the term accordingly. We hope to investigate the usefulness of this approach in future work.

References

1. P. Willett, "Recent trends in hierarchical document clustering: A critical review" *Information Processing & Management* **24**, 577-597(1988).
2. E.M. Voorhees, The cluster hypothesis revisited. Proceedings of the Eighth Annual International ACM SIGIR Conference on Research and Development in Information Retrieval, (Association for Computing Machinery, New York, 1985) 188-196.
3. M.A. Hearst, C. Karadi, Cat-a-Cone: An interactive interface for specifiying searches and viewing retrieval results using a large category hierarchy. in N.J. Belkin, A.D. Narasimhalu, P. Willett, eds. ACM SIGIR'97, (ACM Press, Philadelphia, Pennsylvania, 1997)
4. J.A. Wise, "The ecological approach to text visualization" *Journal of the American Society for Information Science* **50**, 1224-1233(1999).
5. P. Au, M. Carey, S. Sweraz, Y. Gua, S.M. Ruger, New paradigms in information visualization. in N.J. Belkin, P. Ingwersen, M.-K. Leong, eds. ACM SIGIR2000, (ACM Press, Athens, Greece, 2000) 307-309.
6. P. Langley. *Elements of Machine Learning* (Morgan Kaufmann Publishers, Inc., San Francisco, 1996).
7. T.M. Mitchell. *Machine Learning* (WCB/McGraw-Hill, Boston, 1997).
8. A.D. Gordon. *Classification* (Chapman & Hall/CRC, New York, 1999).

9. R.O. Duda, P.E. Hart, D.G. Stork. *Pattern Classification* (Second ed.) (John Wiley & Sons, Inc., New York, 2000).

10. N. Slonim, N. Tishby, Document clustering using word clusters via the infomation bottleneck method. in N. Belkin, P. Ingwersen, M.-K. Leong, eds. ACM SIGIR2000, (ACM Press, Athens, Greece, 2000) 208-215.

11. L.D. Baker, A.K. McCallum, Distibutional clustering of words for text classification. in W.B. Croft, A. Moffat, C.J. van Rijsbergen, R. Wilkinson, J. Zobel, eds. ACM SIGIR'98, (ACM Press, Melbourne, Australia, 1998) 96-103.

12. T. Hofmann, Probabilistic latent semantic indexing. Twenty-second Annual International SIGIR Conference on Research and Development in Information Retrieval, (1999)

13. T. Hofmann, Learning and representing topic. Conference for Automated Learning & Discovery: Workshop on Learning from Text and the Web, (CMU, 1998)

14. H. Li, K. Yamanishi, Document classification using a finite mixture model. Conference of the Association for Computational Linguistics, (Madrid, Spain, 1997) 39-47.

15. M. Sahami, M. Hearst, E. Saund, Applying the multiple cause mixture model to text categorization. in L. Saitta, ed. Machine Learning: Proc. of the Thirteenth International converence, (Morgan Kaufmann, San Francisco, California, 1996) 435-443.

16. H. Shatkay, W.J. Wilbur, Finding themes in MEDLINE documents: Statistical similarity search. IEEE ADL2000, (Bethesda, Maryland, 2000) 183-192.

17. H. Shatkay, W.J. Wilbur, Genes, themes, and microarrays. ISMB2000, (San Diego, California, 2000) 317-328.

18. R.J.A. Little, D.B. Rubin. *Statistical Analysis with Missing Data* (John Wiley & Sons, New York, 1987).

19. G. Salton, E.A. Fox, H. Wu, "Extended boolean information retrieval" *Communications of the ACM* **26**, 1022-1036(1983).

GENOME, PATHWAY AND INTERACTIONS BIOINFORMATICS

PETER KARP, PEDRO R. ROMERO
Bioinformatics Research Group, SRI International
333 Ravenswood Ave., Menlo Park, CA 94025, USA
{pkarp,promero }@ai.sri.com

ERIC NEUMANN
Beyond Genomics
40 Bear Hill Road, Waltham, MA
Eneumann@BeyondGenomics.com

The completion of major metazoan genomes, such as the human genome, has propelled life science research into a new era. Now that the "Book of Life", as some have called it, has been sequenced, and the vocabulary (genes) is being catalogued, the task now at hand is to identify the syntax and semantics of the book, and make sense out of what currently looks to us more like Lewis Carrol's "Jabberwocky" than Shakespeare's "Hamlet". The research and development for the next generation of bioinformatics tools for this task is on our critical path to unlocking the secrets of the human genome.

At the heart of this new challenge is the understanding of the interplay of genes and protein products. From various kinds of interactions (protein-gene, protein-protein), causal, regulated networks of biological pathways arise. Such networks are responsible for the development, maintenance, and responsiveness of all living systems. The collection and organization of pathway information is critical and still needs to be effectively addressed. Assimilating such information and turning it into knowledge of how living systems function (or function aberrantly in disease states) will become increasingly important for both basic research and drug discovery.

A key driving factor in our ability to understand how biological systems function is the emergence of new high-throughput functional-genomics technologies. Data from various kinds of experiments that have a bearing on pathways are being created at an increasing rate. The large ensemble of information they produce contains patterns that are a reflection of pathway dynamics, and therefore can be used to deduce pathway causal structures. These technologies and the information they generate include micro-array technologies that produce gene expression profiles, and 2-hybrid systems that produce information about protein-protein interactions.

Key to advancing our knowledge of biochemical pathways and networks is the intelligent analysis and mining of functional-genomics data in order to infer pathways and their regulation. For example, gene-expression profiles are dependent on the actual pathways that are in place within the target tissue, and can themselves be used to determine gene regulatory mechanisms as well as signal transduction cascades. Expression data is a source of pathway "causal" information.

In order to help elucidate these functional relationships, researchers are applying a wide range of approaches to analyzing micro-array data. Methodologies such as Boolean networks, Bayesian networks, genetic algorithms, and simulation analyses are being used to help build new pathways or extend existing ones. These approaches each have their own advantages and disadvantages, and must be compared using a well defined set of expression benchmark problems. This session attempts to address the benchmark problem in particular. Only in this way will researchers be able to objectively evaluate different kinds of approaches for analyzing such diverse information.

Information collected from protein interaction sources (both experimental and literature-based) will also yield information about pathways, but from a physicalassociation perspective. Evidence from various kinds of protein-protein interaction experiments, such as 2-hybrid, hybrid-competitive, and 3-hybrid systems, will suggest the outlines of protein-binding cascades. Conclusive proof is not forthcoming because the partial protein domains that are created in such experimental systems may often give rise to false-positive and false-negative signals. It is from the careful analysis of such data, and its corroboration with other information, that the actual pathways will emerge.

Once pathways are elucidated, a new round of challenges present themselves. How do we compare and align individual pathways, and entire biochemical networks? Can we infer network properties when the immense number of quantitative parameters that govern their behavior is not known? Can we use estimation methods to infer the values of those parameters? How accurately can pathway behavior be simulated when parameter values are known?

Pathway bioinformatics still includes many challenges, and holds many promises for our basic understanding of biological systems, for drug discovery, and biotechnology.

PATHWAY LOGIC: SYMBOLIC ANALYSIS OF BIOLOGICAL SIGNALING

STEVEN EKER, MERRILL KNAPP, KEITH LADEROUTE, PATRICK LINCOLN, JOSE MESEGUER, AND KEMAL SONMEZ

SRI International 333 Ravenswood Ave, Menlo Park, CA, 94025
E-mail: lincoln@csl.sri.com

The genomic sequencing of hundreds of organisms including *homo sapiens*, and the exponential growth in gene expression and proteomic data for many species has revolutionized research in biology. However, the computational analysis of these burgeoning datasets has been hampered by the sparse successes in combinations of data sources, representations, and algorithms. Here we propose the application of symbolic toolsets from the formal methods community to problems of biological interest, particularly signaling pathways, and more specifically mammalian mitogenic and stress responsive pathways. The results of formal symbolic analysis with extremely efficient representations of biological networks provide insights with potential biological impact. In particular, novel hypotheses may be generated which could lead to wet lab validation of new signaling possibilities. We demonstrate the graphic representation of the results of formal analysis of pathways, including navigational abilities, and describe the logical underpinnings of the approach. In summary, we propose and provide an initial description of an algebra and logic of signaling pathways and biologically plausible abstractions that provide the foundation for the application of high-powered tools such as model checkers to problems of biological interest.

1 Introduction

Biological Signaling Pathways. The tremendous growth of genomic sequence information combined with technological advances in the analysis of global gene expression has revolutionized research in biology and biomedicine[1]. However, the vast amounts of experimental data and associated analyses now being produced have created an urgent need for new ways of integrating this information into theoretical models of cellular processes for guiding hypothesis creation and testing. Investigation of mammalian signaling processes, the molecular pathways by which cells detect, convert, and internally transmit information from their environment to intracellular targets such as the genome, would greatly benefit from the availability of such predictive models. Although signaling pathways are complex, fundamental concepts have emerged from contemporary research indicating that they are amenable to analysis by computational methods. For example, most signaling pathways involve the hierarchical assembly in space and time of multi-protein complexes or modules that regulate the flow of information according to logical rules[2,3]. Moreover, these pathways are embedded in networks having stimulatory, inhibitory, cooperative, and other connections to ensure that a signal will be interpreted appropriately in a particular cell or tissue[4,5].

Modeling Cellular Signaling Networks. Various models for the computational analysis of cellular signaling networks have been proposed involving approaches that incorporate rate and/or concentration information[6,7]. However, these critical approaches are currently limited by the great difficulty of obtaining true intracellular rate or concentration information. Moreover, they could be limited by the potentially stochastic features of cellular scale populations of signaling molecules[8]. Because of these problems, we have chosen to focus exclusively on an abstract level involving the logic of signal. Previous work at a similar level of abstraction includes EcoCyc, the pathway/genome and metabolic reaction database for *E. coli*[9,10]. Initial steps allowing simulation (Section 2.1) of biological pathways include work animating the EcoCyc database[11], and the use of π-calculus to represent and forward-simulate a small signaling pathway[12]. Here we describe an approach to the development of logical models based on the application of formal methods tools to mammalian signaling pathways[13,14]. Specifically, we describe the application of rewriting logic to the symbolic representation of a major receptor-mediated pathway in mammalian cells: receptor tyrosine kinase (RTK) signaling through the epidermal growth factor receptor (EGFR) leading ultimately to activation of an autocrine loop[15] (Figure 1).

1.1 Mathematical Models of the Cell and Levels of Abstraction

At the continuous level of abstraction, natural processes are described by detailed approaches drawn from the physical sciences involving continuous mathematics and analyzed using sophisticated numerical computation packages. However, while chemical or molecular events ultimately constitute biological processes, the complexity of these processes severely limits their accurate and effective description in terms of purely physical/chemical phenomena. This problem can be resolved at the discrete level of abstraction, where natural processes are described by purely symbolic expressions. Although this highly abstracted approach is an established means to analyze physical systems such as computer designs[16,17], it is also applicable to less predictable phenomena such as biological signaling processes. Indeed, biologists routinely reason about these processes at the discrete level, although this reasoning consists of informal notations and potentially ambiguous representations of important concepts like pathways, cycles, and feedback loops, with poor tool support. New rigorous but abstract models are needed for biology that: (i) accommodate conventional types of discrete reasoning based on experimentation, (ii) formally define a model and allowable reasoning steps, and (iii) provide predictive power for generating testable hypotheses.

Consider an analogy with algebraic analysis, such as the task of accurately and efficiently computing the polynomial x^2-y^2, given values for x and y. One implementation of this task is based on the expression $(x+y)\times(x-y)$. Because this latter expression consists of two additions and only one multiplication it is inherently faster on many hardware platforms than the original expression, which requires one addition and two multiplications. To proceed with the implementation of this example, it would be necessary to decide whether the two expressions are equivalent. However, it is not apparent how many tests of equivalence would be sufficient to make this determination. A symbolic or formal methods approach to this task would be to assume a set of symbolic rewrite or inference rules and to reason algebraically. For example, starting from $(x+y)\times(x-y)$, it could be reasoned by the distributivity law that this expression is equal to $(x+y)\times x$ - $(x+y)\times y$. Again by distributivity it could be reasoned that the expression is equal to $x\times x + y\times x - x\times y - y\times y$. Using associativity, commutativity, and laws of subtraction, the $x\times y$ terms cancel, and it could be shown that the expression is equal to $x\times x$ - $y\times y$, and by the definition of exponentiation it could be shown that this is equal to x^2-y^2. Thus, subject to the validity of the axioms, it could be demonstrated that these polynomial expressions are symbolically equivalent **for all** x and **for all** y. This type of reasoning is categorically different from numeric testing, but can be computationally challenging.

Major breakthroughs in efficient symbolic reasoning have occurred in the last decade of research in computer science. The framework of model checking[18,19], exponentially more efficient representations of Boolean and other functions[20], decision procedures, and efficient implementation of rewriting[21] represent quantum jumps in the ability to reason about symbolic systems, even when those systems may potentially have more states than there are atoms in the universe.

1.2 Pathway Logic

Here we propose and describe *Pathway Logic*, an algebraic structure enabling the symbolic analysis of biological signaling pathways analogous to the standard definitions and laws for polynomials referred to above. We use the EGFR pathway as an example[15] in this discussion, but Pathway Logic could be applied to pathways regulating very diverse biological processes.

We use the Maude system to express the algebraic structure of Pathway Logic[13]. Maude is a high-performance reflective multiparadigm language and system which supports a wide range of applications. Maude implements rewriting logic, a logic of state and concurrent computation, and supports efficient logical reflection. This makes Maude remarkably extensible and powerful, and enables the creation of executable environments for different logics, languages, and models of computation, including abstract models of discrete biological computation.

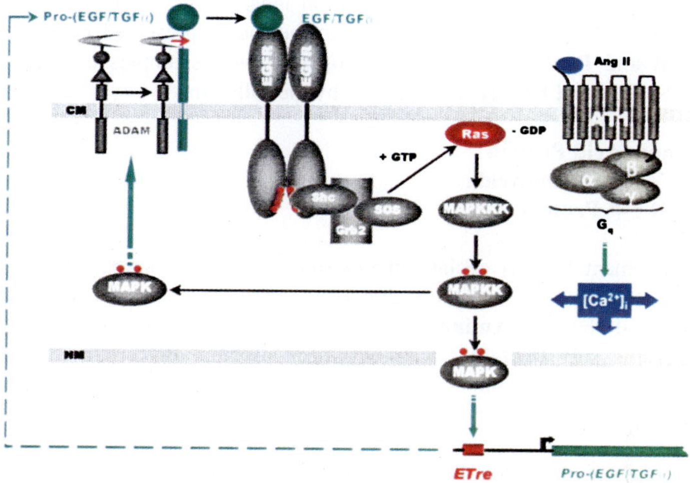

Figure 1. Fragment of the mammalian EGFR system illustrating activation of a downstream mitogenic signaling pathway involving the gene for the autocrine EGFR ligand TGFα. Also shown is a potential mechanism for cross-communication between the EGFR and a G protein-coupled receptor (AT1). Adapted from Gschwind et al.[15].

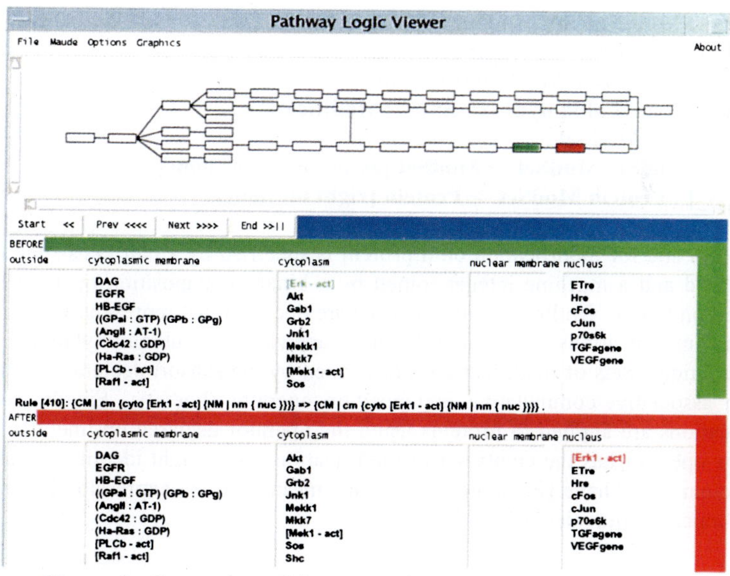

Figure 2. Screenshot of the current Pathway Logic viewer.

In what follows, numbered lines and bold font is used to highlight text written in the Maude executable specification language[13,14] to define the Pathway Logic algebra. The most basic sorts within the model presented here are termed **AminoAcid** and **Protein**. These sorts together with examples of their specific members, **S**, **T**, **Y** and **EGFR**, are declared by the following:

```
1 sorts AminoAcid Protein .
2 ops S T Y : -> AminoAcid .
3 op EGFR : -> Protein .
```

Statement 1simply declares the existence of the two sorts, **AminoAcid** and **Protein**. Statement 2 declares the amino acids serine **S**, threonine **T**, and tyrosine **Y** as constants of the sort **AminoAcid**, and statement 3 declares the **EGFR** as a constant of the sort **Protein**. The keywords **op** and **ops** are used to declare *operators* from a list of domain sorts into a range sort. In both of the examples above, the list of domain sorts is empty, indicating that the declared operators are *constants*.

Important ideas for the approach described here are explained using the **EGFR** pathway (Figure 1). These ideas are posttranslational protein modification, protein association, and cellular compartmentalization.

Protein Modification. We specify an algebra of protein modifications as follows:

```
4 sort Site Modification ModSet .
5 subsort Modification < ModSet .
6 op _ _ : AminoAcid MachineInt -> Site .
7 ops phos acetyl ubiq hydrox : Site -> Modification .
8 op none : -> ModSet .
9 op _ _ : ModSet ModSet -> ModSet [assoc comm id: none] .
10 op [_-_] : Protein ModSet -> Protein [right id: none] .
```

A site for modification on a protein is specified by a pair consisting of an amino acid and a machine integer joined by a binary juxtaposition operator (_ _) declared on line 6. On line 7 four operators are declared that represent the common protein modifications phosphorylation, acetylation, ubiquitinylation, and hydroxylation. Sets of modifications are formed by the subsort declaration (line 5) and an associative-commutative juxtaposition operator (line 9). Finally, sets of modifications are applied to proteins using the operator declared on line 10. Note that this operator has the empty set of modifications as its right identity. Thus, for any protein **P,** we have **[P - none] = P,** which means that the expression **[P - none]** is algebraically equivalent to **P**.

Like many signaling proteins, the EGFR is posttranslationally modified by phosphorylation[5]. Using this algebra, a common phosphorylation state of the EGFR can be modeled as follows:

11 **[EGFR - phos(Y 1092) phos(Y 1110) phos(Y 1172) phos(Y 1197)]**

This expression indicates that the modified EGFR is phosphorylated at tyrosines 1092, 1110, 1172, and 1197 (Y1092, Y1110, Y1172, and Y1197, respectively) after activation by EGF[22]. However, because there is only one known activation state of the EGFR, this expression can be simplified to:

12 **[FR – act]**

Protein Association. Signaling proteins commonly associate to form functional complexes[22]. This important phenomenon is algebraically represented by the following declarations:

13 **sort Complex .**
14 **subsort Protein < Complex .**
15 **op _:_ : Complex Complex -> Complex [comm] .**

That is, each protein is a singleton complex and two such complexes could be associated by the ":" operator on line 16 to obtain a multiprotein complex or module. Notice that this ":" operator has been declared commutative, but it is not assumed associative. Therefore, parentheses must be used to describe complexes formed from other complexes, such as the association shown on line 16 below between **EGF** and the **EGFR**.

16 **(EGF:([EGFR - act] : [EGFR - act]))**

Protein Compartmentalization. In the eukaryotic cell, proteins and other molecules exist in complex mixtures that are compartmentalized[2]. These compartmentalized mixtures (here termed "Soup" for convenience) are algebraically represented by the following declarations:

17 **sorts Soup Enclosure MemType.**
18 **subsort Complex < Soup .**
19 **op empty : Soup .**
20 **op _ _ : Soup Soup -> Soup [assoc comm id: empty] .**
21 **ops CM NM MM : -> MemType .**

```
22 op {_|_{_}} : MemType Soup Soup -> Enclosure .
```

An **Enclosure** is defined as a cellular membrane plus its **Soup**. A **MemType** denotes a specific membrane such as the cell membrane (CM) or the nuclear membrane (NM) (Figure 1). As with individual protein complexes (line 15), soups can also be combined as shown on line 20 by means of the binary soup union operator (with juxtaposition syntax). This union operator models the presumed fluid or dynamic nature of some subcellular compartments by specifying associative and commutative laws, so that no parentheses are needed and the order in which molecules exist in the soups does not matter.

1.3 Rewriting Logic: Symbolic Modeling of Biochemical Reactions

The algebraic structures or models described in Section 1.2 provide symbolic representations of protein modifications and eukaryotic cellular organization by means of an algebraic specification **S** involving sorts, subsorts, operators, and equational laws. To symbolize biochemical events such as signaling processes, we use theories in rewriting logic[23]. A rewrite theory is a pair **(S, R)** with **S** being an algebraic specification and **R** being a collection of rewrite rules. Each rewrite rule is of the form **l : t ⇒ t'**, with **l** being a label, and **t** and **t'** being algebraic expressions in the algebra specified by **S**. Each rewrite rule specifies a local change or reaction that can occur in the system modeled by the theory **(S, R)**. These rewrite rules can precisely express biochemical processes or reactions involving single or multiple subcellular compartments. For example, the following text describes the first step in the activation of the EGFR signaling pathway, the binding of EGF to the EGFR[24] (Figure 1): **"Activated MAPK$^{erk1/2}$ is rapidly translocated to the nucleus where it is functionally sequestered and can regulate the activity of nuclear proteins including transcription factors."** In Maude syntax, this signaling process is described by the following rewrite rules:

```
23 rl [410]: {CM | cm {cyto [Erk1 – act] {NM | nm {nuc}}}} =>
                {CM | cm {cyto {NM | nm {nuc [Erk1 – act]}}}} .
24 rl [436B]: [Erk1 – act] ETre => [Erk1 – act] [ETre – act] .
```

In Figure 1, MAPK represents ERK1 and ETre is a consensus DNA binding site for transcription factors that activate expression of the *TGFα* gene. Notice that in the new state of the system represented by the right hand side of rule 410, activated ERK1 is present in the nucleus following translocation. The right hand side of rule 436B indicates that activated ERK1 has induced transcription of the *TGFα* gene through the ETre element in its 5′-regulatory region. These rewrite rules describe a local change that could occur when an instance of the left-hand side of each rule exists in a cell. Mathematically, the rules in **R** are applied modulo the equivalence between expressions defined by algebraic laws in **S**.

In a system specified by a rewrite theory **(S, R)**, rewriting logic allows reasoning about *possible* complex changes given the basic changes specified in **R**, such that any such change is possible if and only if it can be proved to be derived using the rules in **R**[14]. These complex changes can be concurrent; that is, different parts of a compartment can change simultaneously and independently. Furthermore, the changes can be multi-step, allowing reasoning about *possible future* states of the system. Thus, under very reasonable assumptions rewrite theories can be executed in Maude to describe a biological signaling process over time according to a symbolic model, and can be formally analyzed to reason about properties of the states reachable from an initial state.

2 Analysis Techniques

Given a formal symbolic model of some part of the signaling pathways in a cell as a rewrite theory, several kinds of automatic analysis can be performed. We note that each biochemical reaction that is represented as a rewrite rule conserves proteins. This conclusion has two important consequences for the term rewriting system that simplify its analysis. First, from a given initial state the set of reachable states is finite. Second, each Soup variable occurring in the left-hand side must also occur in the right-hand side, as otherwise there would be destruction of arbitrary numbers of molecules bound to the variable.

2.1 Static Analysis

As a prelude to analyzing the dynamic behavior of a model, we can first perform static analysis. As one example, just considering the simplest algebraic part **S** of the model (without the rewrite rules **R**), we can identify sorts that are inhabited only by declared constants. Typically, these sorts will capture the notion of a family of related proteins (the constants). A rule containing variables ranging over such sorts is (for the purposes of ground rewriting) equivalent to a set of rules obtained by instantiating such variables with the constants inhabiting their sorts in all possible ways.

2.2 Forward and Backward Search

The simplest form of analysis of the dynamic behavior of a model is to *run* the model from a given initial state by applying the rules in an arbitrary order for some fixed number n of rewrites or until no more rules are applicable. In Maude, this process is done with the *rewrite* command. Since most models will be nondeterministic and the future states reachable from a given initial state form a graph where paths diverge, converge, or cycle back on themselves, the *search* command in Maude supports a breadth-first search through this transition graph looking for states that match some pattern, possibly with a side condition. Using search, all possible outcomes can be identified from a given initial state. When the rules are unconditional and each variable that occurs in the left-hand side occurs in the right-hand side it is possible to flip the rules over and run the model backwards – either as a simulation or as a search. Using such reversed rules, we can ask questions of the form "from what initial state(s) can we get to some desired (or known) final state?"

2.3 Explicit state Model Checking

The search command allows us to examine the transition graph produced from a given initial state for states satisfying some static property, such as the existence of a protein in some particular phosphorylation state. We may want to ask more complex queries about the paths in the transition graph, such as "if we reach a state that satisfies property **P**, is it true that we must eventually reach a state that satisfies property **Q**?" We may also want to restrict our attention to the subset of paths in the transition graph that satisfy some *fairness* criterion such as "if reaction **R** is always possible, then eventually reaction **R** happens".

A language suitable for framing such queries is propositional linear temporal logic (LTL). Here the propositions correspond to properties that can be statically checked for each state. The familiar propositional calculus with its operators such as $\wedge$ ("and"), $\rightarrow$ ("implies"), and $\neg$ ("not") is extended with temporal operators such as $\Box$ ("always") and $\Diamond$ ("eventually"). Standard techniques based on Büchi automata can be used to check if the transition graph produced from a given initial state satisfies an LTL formula[17].

2.4 Meta-analysis

In practice there may be uncertainty regarding the experimental evidence for certain reactions. Thus, we consider a parameterized specification that describes a finite family of models. Rather than a fixed set **R** of reactions, we have some base set $\mathbf{R}_{base}$ of reactions about which we are confident, together with a set $\mathbf{P} = \{P_1, \ldots, P_n\}$ of parameters (often n will be 1) and for each parameter P_i, a set $\{A_{i,1}, \ldots, A_{i,m_i}\}$ of alternative instantiations. Here each $A_{i,j}$ is a set of reactions. Such a specification describes a family of $\prod_{i=1}^{n} m_i$ distinct models, which are obtained by choosing different combinations of instantiations for $P_1, \ldots, P_n$ and adjoining them to $\mathbf{R}_{base}$.

Given a LTL formula and an initial state we can now search this family of models to find those models for which an LTL formula is true in the initial state.

2.5 Key Benefit: Expressive Questions

The application of model checking and other analysis techniques is important for the expressiveness of the questions for which answers can be effectively computed. In particular, providing a complete search of the space of all possible executions of an abstraction of system has been found in other domains to be more useful than the forward simulation (testing) of just some possibilities for that system. The complete symbolic exploration of all reaction interactions can provide useful insights, and can directly enable a biologist to ask questions such as "If EGF is not present to stimulate it's pathway, but angiotensin II is, is the ERK signal silent?" (Figure 1). This kind of expressive question can be directly encoded in temporal logic as follows: $\Box(\Box(\text{AngII} \wedge \neg\text{EGF}) \rightarrow \neg\Diamond\text{ERK1})$. The answer to such queries (including traces demonstrating counterexamples) can be effectively computed using the techniques described above. Thus far, we have encoded hundreds of reactions relating to signaling pathways in mammalian cell cycle control, and we have computed all possible outcomes from certain interesting states using the Maude search command.

3. Graphical Representation for Pathway Logic

To make Maude-generated models easily accessible to a user we have developed a graphical viewer tool. The representation is a directed transition graph, with nodes representing the states of proteins within a model and transitions representing rewriting with respect to applicable rules. The state of the system is specified by the contents of each of its compartments. One possible realization of the viewer is shown in Figure 2. In the top part of the graphical user interface is a canvas displaying the directed transition graph constructed from Maude generated paths or traces. Circles depict the states, and arrows indicate the state transitions. Every transition is associated with the rewriting operation with respect to a certain rule, and the rules are shown below the arrows. Below the canvas is a set of boxes that specify the Soups before and after a rewriting transition. In this example the Maude output from a simple search operation shows potential crosstalk between known pathways. The user has selected a step in the pathway to examine in detail, with the green before state and red after state displayed in detail, and highlighted in the pathway overview.

4. Summary

Our major hypothesis is that useful computational analysis can be performed on biological signaling networks at a very high level of abstraction.

We have presented an approach of applying formal methods to the analysis of biological pathways. Most previous related work has focused on continuous models of such systems, an approach restricted by lack of detailed *in vivo* rate and concentration data and the computational complexity of simulations at that level of abstraction. Some previous work such as that on EcoCyc and PiFPC has begun to show the benefits of higher level abstractions[26,12]. The PiFPC work is progressing toward stochastic and rate-based models within a formal framework. Here we propose a formal framework and the application of modern model checking and other symbolic techniques to signaling networks at the higher level of abstraction, but enabling the answers to queries of a different nature than simple forward simulation.

Future Work. In the near future, we will explore the automated connection between different levels of abstraction of biological modeling. Using automated symbolic abstraction methods we will be able to simultaneously represent low-level molecular details, and higher-level protein module-at-a-time or compartment-at-a-time structure, thus enabling scalable computational analysis of extremely large systems. We will also begin to symbolically represent delays, which will begin to allow reasoning about circadian and other cellular rhythms, without reliance on detailed *in vivo* rate data. We will also experiment with perturbations of the pathways, computing possible outcomes of induced signal or network changes. Finally, we will use temporal logic specifications to study the property-sensitive differences between related pathways.

Acknowledgements

We thank Tom Garvey, Peter Karp, Pedro Romero, Raymonde Guindon, Andrea Lincoln, Marianna Yanovsky, Natarajan Shankar for helpful discussions, and the anonymous reviewers for their comments.

References

1. O.G. Vukmirovic and S.M. Tilghman. Exploring genome space. *Nature*, 405:820-822, 2000.
2. J.D. Jordan, E.Landau, and R. Iyengar. Signaling networks: The origins of cellular multitasking. *Cell*, 103:193-200, 2000.

3. K.W. Kohn. Molecular interaction map of the mammalian cell cycle control and DNA repair systems. *Mol Biol Cell*, 10:2703-2734, 1999.
4. D. Fambrough, K. McClure, A. Kazlauskas, and E.S. Lander Diverse signaling pathways activated by growth factor receptors induce broadly overlapping, rather than independent, sets of genes. *Cell*, 97:727-741, 1999.
5. T. Pawson and T.M. Saxton. Signaling networks--do all roads lead to the same genes? *Cell*, 97:675-678, 1999.
6. K.W. Kohn. Functional capabilities of molecular network components controlling the mammalian G1/S cell cycle phase transition. *Oncogene*. 16:1065-75, 1998.
7. G. Weng, U.S. Bhalla, and R. Iyengar. Complexity in biological signaling systems. *Science*, 284:92-96, 1999.
8. H.H. McAdams and A. Arkin. It's a noisy business! Genetic regulation at the nanomolar scale. Trends Genet, 15:65-69, 1999.
9. P.D. Karp, M. Riley, S.M. Paley, A. Pellegrii-Toole, and M. Krummenacker. Ecocyc: Encyclopedia of Escherichia coli genes and metabolism. *Nucleic Acids Res*, 26(1):50-53, 1998.
10. C.A. Ouzounis and P.D. Karp. Global properties of the metabolic map of Escherichia coli. *Genome Res*, 10(4):568-576, 2000.
11. P.R. Romero and P. Karp. Nutrient-related analysis of pathway/genome databases. In R.B. Altman et al., editor, Pacific Symposium on Biocomputing 2001, pages 471-482. World Scientific, 2001.
12. Aviv Regev, William Silverman, and Ehud Shapiro. Representation and simulation of biochemical processes using the π-calculus process algebra. In R.B. Altman et al., editor, Pacific Symposium on Biocomputing 2001, pages 459-470. World Scientific, 2001.
13. Manuel Clavel, Francisco Duran, Steven Eker, Patrick Lincoln, Narciso Marti-Oliet, Jose Meseguer, and Jose Quesada. A tutorial on Maude. SRI International, March 2000, http://maude.csl.sri.com.
14. Manuel Clavel, Francisco Duran, Steven Eker, Patrick Lincoln, Narciso Marti-Oliet, Jose Meseguer, and Jose Quesada. Towards Maude 2.0. In K.Futatsugi, editor, Proc. 3rd. Intl. Workshop on Rewriting Logic and its Applications, volume 36 of ENTCS Elsevier, 2000.
15. A. Gschwind, E. Zwick, N. Prenzel, M. Leserer, and A. Ullrich. Cell communication networks: epidermal growth factor receptor transactivation as the paradigm for interreceptor signal transmission. *Oncogene,* 20:1594-1600, 2001.
16. Steven P. Miller and Mandayam Srivas. Formal verification of the AAMP5 microprocessor: A case study in the industrial use of formal methods. In WIFT '95: Workshop on Industrial-Strength Formal Specification Techniques, pages 2-16, Boca Raton, FL, 1995. IEEE Computer Society.

17. Ed Clarke, Orna Grumberg, and Doron Peled. Model Checking. MIT Press, 1999.
18. J.P. Queille and J. Sifakis. Specification and verification of concurrent systems in Cesar. In Proceedings of the 5th International Symposium on Programming, volume 137 of Lecture Notes in Computer Science, pages 337-351, Turin, Italy, April 1982. Springer-Verlag.
19. E.M. Clarke, E.A. Emerson, and A.P. Sistla.Automatic verification of finite-state concurrent systems using temporal logic specifications. *ACM Transactions on Programming Languages and Systems*, 8(2):244-263, April 1986.
20. R.E. Bryant. Graph-based algorithms for Boolean function manipulation. *IEEE Transactions on Computers*, C-35(8):677-691, August 1986.
21. Manuel Clavel, Francisco Duran, Steven Eker, Patrick Lincoln, Narciso Marti-Oliet, Jose Meseguer, and Jose Quesada. The Maude System. In P. Narendran and M.Rusinowitch, editors, Procs. 10th Intl. Conference on Rewriting Techniques and Applications (RTA-99), volume 1632 of Lecture Notes in Computer Science, pages 240-243. Springer-Verlag,1999.
22. W. Kolch. Meaningful relationships: The regulation of the Ras/Raf/MEK/ERK pathway by protein interactions. *Biochem J.*, 351:289-305, 2000.
23. Jose Meseguer. Conditional rewriting logic as a unified model of concurrency. *Theoretical Computer Science*, 96(1):73-155, 1992.
24. J. Schlessinger. Cell signaling by receptor tyrosine kinases. *Cell*, 103:211-225, 2000.
25. Peter D. Karp. Pathway Databases: A Case Study in Computational Symbolic Theories. *Science* 293(5537): 2040-2044, Sept 2001.

TOWARDS THE PREDICTION OF COMPLETE PROTEIN—PROTEIN INTERACTION NETWORKS

SHAWN M. GOMEZ[1] and ANDREY RZHETSKY[1,2]

Columbia Genome Center[1], and Department of Medical Informatics[2], Columbia University, 1150 St. Nicholas Avenue, Unit 109, New York, NY 10032, USA

{smg42, ar345@columbia.edu}

We present a statistical method for the prediction of protein—protein interactions within an organism. This approach is based on the treatment of proteins as collections of conserved domains, where each domain is responsible for a specific interaction with another domain. By characterizing the frequency with which specific domain—domain interactions occur within known interactions, our model can assign a probability to an arbitrary interaction between any two proteins with defined domains. Domain interaction data is complemented with information on the topology of a network and is incorporated into the model by assigning greater probabilities to networks displaying more biologically realistic topologies. We use Markov chain Monte Carlo techniques for the prediction of posterior probabilities of interaction between a set of proteins; allowing its application to large data sets. In this work we attempt to predict interactions in a set of 40 human proteins, known to form a connected network, and discuss methods for future improvement.

1. Introduction

Increases in the number of sequenced genomes have led to rapid growth in the number of biological systems with characterized molecular components. Understanding of how these individual components are integrated together into a complete system, however, has lagged. Part of the difficulty in this undertaking originates in the fact that experimental data as to the existence of interactions between any two molecules are extremely sparse. While advances have also been made with, for example, high—throughput two—hybrid studies and complementary interaction databases, a comprehensive view of these molecular interaction networks is still lacking.

Networks consisting of proteins, DNA, RNA, and various small molecules, are formed due to one molecule's propensity to bind or otherwise influence another and hence alter system function. In this article, functional areas that provide this ability for one molecule to interact with another are referred to as domains. For example, subsequences of DNA where specific proteins bind are one class of domain (as are the amino acid subsequence responsible for binding activity within the protein). Interactions between proteins are of particular interest, as they are responsible for the majority of "active" biological function. To date, protein—protein interactions are also the predominant type of interaction with significant quantities of supporting experimental data sets. As a result of these two factors, the work described here is focused on protein interaction networks, and more specifically, the *a priori* prediction of interactions between proteins as well as the prediction of whole networks.

Here we describe a statistical model capable of predicting protein—protein interactions which can be extended to other classes of molecules. While some previous methods have focused on using gene fusion events for the prediction of interactions [1, 2], our approach is more general in that any type of experimental evidence supporting an interaction can be used in prediction. Based upon experimentally verified interactions and estimates of network topology, this approach generates posterior probabilities conditioned on data for all possible interactions. The work described here is an extension of earlier work [3], which described the fundamentals of this model in some detail and presented small examples of its application. In this paper we describe the results of a more challenging application of the model and discuss methods for its improvement. A primary goal of this work is to provide a method for generating predictions that would be useful to the experimental biology community. In particular we feel that the prediction of molecular interactions, along with the ability to assign a probability to a given interaction, could be of significant benefit in the generation of new hypotheses and the prioritizing of appropriate (and perhaps more focused) experiments.

2. Model description

We start by representing a network as an oriented graph, $G = <V, E>$, where the vertices, V, of the graph are connected to each other through the edges, E. In this paper, edges represent a physical binding between corresponding proteins. Each vertex represents a protein, although extension of the model to handle other types of molecules (e.g. DNA) is rather straightforward. Each protein can be broken into smaller subunits consisting of one or more domains. We treat domains as evolutionarily conserved, elementary units of function. We assume that the domains are responsible for the generation of edges within the network; a simple example being a phosphorylation site and a kinase domain capable of phosphorylating that site. The upstream kinase domain is where the edge originates, and the edge terminates at the phosphorylation site. Domains themselves are found through the use of current tools and databases capable of assigning domains to proteins (e.g. Pfam)[4]. In addition to interaction data we use an additional parameter, characterizing the network topology, in the prediction of a network. While we describe the model in some detail here, greater description of certain aspects of the model can be found elsewhere [3].

2.1 Assigning probabilities to edges

Assigning a probability to a given network consists of two independent steps. The first step consists of assigning a probability p_{ij} to the existence of an edge connecting the proteins i and j or not connecting them $(1 - p_{ij})$. This process of assigning an edge can be thought of as the toss of biased coins (one coin per edge) for all possible edges, $|V|^2$ edges in all. The coin may be biased by prior information, assigning probabilities greater than 0.5 to vertices likely to be "attracted" to each other and form an edge. Probabilities of less than 0.5 can be assigned between proteins that "repel" each other and are thus unlikely to interact. Then for a protein network with a fixed number of vertices and a particular set of edges E between them, the probability of this network becomes

$$P(E) = \prod_{e_{ij} \in E} p_{ij} \prod_{e_{kl} \notin E} (1 - p_{kl}) \, .$$

How then do we define these individual edge probabilities p_{ij}?

We treat each protein as a collection of domains, and each of these domains has a tendency to attract or repel other domains between distinct proteins. Specifically, we define a probability of attraction $p(d_m, d_n)$ that exists for each upstream and downstream domain (as defined by moving with the "flow" or temporal sequence of the pathway), d_m and d_n, respectively. If the orientation is unknown, $p(d_m, d_n) = p(d_n, d_m)$, and the edge is undirected and both directed edges are present. Identical to edge probabilities between proteins, probabilities greater than 0.5 represent attraction while those less than 0.5, repulsion. For a pair of multidomain proteins i and j, where v_i and v_j are the set of unique protein domains for each, the probability of an edge forming between the two is

$$p_{ij} = \sum_{d_m \in v_i} \sum_{d_n \in v_j} \frac{p(d_m, d_n)}{|v_i||v_j|} \, .$$

Thus the probability of an edge forming between a pair of proteins is dependent on the relative attraction and repulsion of each protein's complement of domains, taken over all upstream—downstream pairwise combinations. This expression is a reasonable assumption as long as the number of edges incoming to or outgoing from a vertex is independent of the number of domains per protein; we have verified this assumption previously [3].

We determine the probability between a pair of domains, $p(d_m, d_n)$, by observing the frequency with which domain d_m appears upstream of domain d_n within experimental protein—protein interaction data. Specifically, we use

$$p(d_m, d_n) = \frac{1}{2}\left(1 + \frac{k_{mn}}{k_m k_n + \Psi}\right),$$

where Ψ is a positive real-valued pseudocount, k_{mn} is the number of edges in the training set that contain at least one domain d_m at the vertex of edge origin and at least one domain d_n at the vertex of edge destination, k_m is the number of distinct vertices that contain at least one domain d_m, and k_n is the number of distinct vertices that contain at least one domain d_n. This expression generates domain attraction probabilities greater than or equal to 0.5. As discussed later, probabilities of less than 0.5 are reserved for future modeling of repulsive interactions between domains, as observed, for example, in domain combination studies [5]. In this work, Ψ was assigned a value of 1. We assume that data supporting the existence of a particular interaction is usually backed by several experiments, while experiments showing the absence of an interaction are generally underrepresented by having either failed (and these failures not reported) or have not been performed. Thus this expression does not "penalize" for lack of an interaction, but assumes it to be the lack of supporting data. In the absence of any supporting data, all interactions between domains (and hence proteins) are equally likely.

In summary, we observe the frequency with which domain X lies immediately upstream or downstream of domain Y within experimental protein—protein interaction data. For an arbitrary pair of proteins, each with their own set of domains, we are then able to assign a probability to the likelihood of an edge forming between them. A complete network with a defined set of edges can similarly be assigned a probability; networks with many favorable edges will have a higher probability than a network with many unlikely edges.

2.2 Assigning probabilities to network topologies

The second part of our model deals with a global property of the network, that being its topology. The topology of a network is defined here as the distribution of edges going into and out of each vertex of the network. The number of edges going into a vertex is termed the *indegree*, and the number outgoing, the *outdegree*. In this model, we sort networks into a finite number of bins each representing a specific topology, where biologically realistic topologies have greater probabilities. Since multiple networks may be characterized by the same topology, each bin represents a collection of networks each with the same topological probability. For each network we compute the number of vertices that have outdegree zero, n_0^{out}, one, n_1^{out}, two, n_2^{out}, and so on to n_N^{out}. The vertices of a particular indegree are similarly computed. Networks with identical sets $\{n_x^{in}\}$ and $\{n_y^{out}\}$ are then grouped into a single bin. The probability of this bin is defined as

$$P(\{n_x^{in}\};\{\pi_x^{in}\},|V|) * P(\{n_y^{out}\};\{\pi_y^{out}\},|V|),$$

where

$$P(\{n_z\};\{\pi_z\},|V|)=\frac{|V|!}{n_0!\ldots n_N!}\prod_{z=0}^{N}\pi_z^{n_z}.$$

The probability distributions π_x^{in} and π_y^{out} give the probability of a network having x incoming and y outgoing edges respectively (described in more detail below).

It is easy to see that the probability of a network is simply the product of this distribution and $P(E)$ described in section *2.1*:

$$P(E)\times P(\{n_x^{in}\};\{\pi_x^{in}\},|V|)\times P(\{n_y^{out}\};\{\pi_y^{out}\},|V|).$$

Networks with favorable edge sets and favorable topologies will be more likely to be selected under our model.

The probability distributions π_x^{in} and π_y^{out} were estimated from yeast data taken from the Database of Interacting Proteins (http://dip.doe-mbi.ucla.edu/) [6, 7] and were found to follow a power—law distribution [3]; generally associated with scale—free behavior. This property was observed independently for the yeast protein network by Jeong and colleagues [8] and is also typical of a variety of other systems, both biological and man—made [5, 9, 10]. In this case, the probability of a vertex having k incoming or outgoing edges is

$$\pi_k = ck^{-\gamma},$$

with the values of c and γ different for each. For this work fits for each distribution gave $c = 0.30$ and $\gamma = 1.97$ for outgoing edges, and $c = 0.56$, $\gamma = 2.80$ for incoming. For π_0^{in} and π_0^{out} we used

$$\pi_0^{in}=1-\sum_{k=1}^{\infty}\pi_k^{in},\quad \pi_0^{out}=1-\sum_{k=1}^{\infty}\pi_k^{out}.$$

We used these distributions in the predictions described here, however, their use in other distributions is also possible. See the discussion section for further detail.

3. Methods

For training data, we used a combined dataset of protein—protein interaction data for both *Saccharomyces cereviseae* and *Homo sapiens*. We used the Pfam database (Pfam6.2; 2773 domains) and the HMMER package to determine the domains within each proteins (0.01 significance threshold). For the yeast data, we used a comprehensive list of interactions downloaded from Stanley Field's lab home page (http://depts.washington.edu/sfields/). This data included interactions from a number of sources [7, 11, 12]. We analyzed a total of 708 protein—protein interactions from yeast, all of which had at least 1 domain. For human data, we used a set of 778 interactions downloaded from the Myriad Genetics Pronet Online web site (http://www.myriad-pronet.com/).

In this study, we attempted to predict interactions between a set of 40 human proteins known to form a fully connected network. Proteins were chosen from Pronet, with some of the proteins involved in the process of apoptosis. These interactions were not included as part of the original training set. Except for the requirement that all proteins of the network must be defined by at least one domain, this network was chosen at random. Proteins used in this analysis, and their indices in all figures, are given in Figure 1.

4. Results

4.1 Predictions of human protein—protein interactions

Edge probabilities based on domain—domain interaction data alone indicated that 97 edges had probabilities > 0.5 (see Figure 1). Note that we assumed that edges were not directed and thus the matrix shown here is symmetric. A total of 44 edges were in the original data set. Of these 44 edges, 8 are observed (18%) in the predicted 97 with probabilities > 0.5. Three out of eight interactions were involved in the heat shock pathway (read as (Y-axis, X-axis) on the figure); CHIP (12, 12) self—interaction, HSPA8—MRJ (24, 27), and HSPA8—PLCG1 (24, 30). The remaining 5 included FLN1—KSR1 (16, 25), PS2—CIB (32, 13), GDI2—RAB6 (20, 37), RAB6—GAPCenA (37, 18), and RAB6—RAB6KIFL (37, 38).

To see if any of the remaining 89 predicted edges represent known edges, we attempted a brief literature search. While often requiring significant expertise in a given pathway to adequately evaluate these results, we were still able to find obvious successes. The predictions of GDI1 (Guanine Nucleotide Dissociation Inhibitor, vertex 19) interacting with Rab11A, Rab3A, Rab5A, and Rab6 (vertices 34, 35, 36, 37 respectively) are in fact correct, and again not in the original data [13-15].

The prediction of CHIP interacting with TTC1 (tetratricopeptide repeat domain 1)(12, 40) is also understandable (though likely not a correct prediction, it may also be questionable in the original data) as the tetratricopeptide domain is a common protein—protein interaction motif, and a number of TPR containing proteins are known to interact with members of the heat shock protein family [16]. While purely speculative, the interaction of CIB (calcium and integrin binding protein with FLN1 (filamin) is interesting, as filamin has recently been shown to be a scaffold protein that interacts with calcium receptor and other cell signaling proteins [17]. While the prediction of only 8 known edges is disappointing, it is not unexpected due to limitations in the training data, and so it is quite possible that most of the predicted interactions are simply "noise." The valid prediction of the GDI-Rab interactions, however, were encouraging. Limitations of the data and methods for improving the model are presented in greater detail in the Discussion.

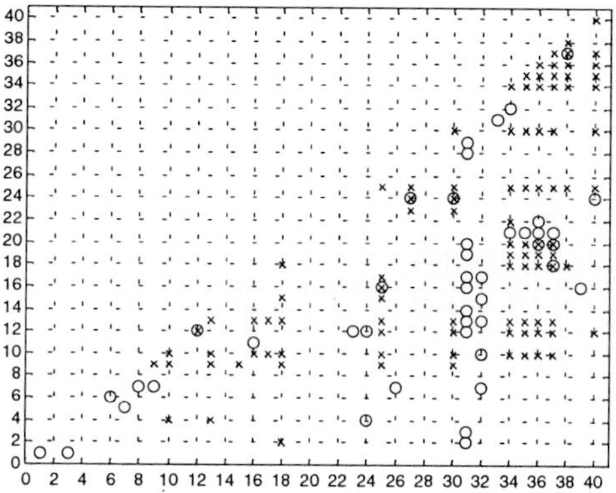

Figure 1. Known and predicted edges. Known edges are shown as open circles, while predicted edges are displayed as an "x." Proteins and their indices in all figures are: 1) ANT2, 2) APP (695), 3) B-CAT, 4) BAG3, 5) BAK, 6) Bax-beta, 7)Bcl-xL, 8) BCL2A1, 9)Bcl2-alpha, 10) Calsenilen, 11) CAV1, 12) CHIP, 13) CIB, 14) D-CAT, 15) DRAL, 16) FLN1, 17) FLNB, 18) GAPCenA, 19) GDI1, 20) GDI2, 21) GGTB, 22) GTPBP1, 23) HSPA4, 24) HSPA8, 25) KSR1, 26) MCL1, 27) MRJ, 28) PSAP, 29) PKP4, 30) PLCG1, 31) PS1 (467), 32) PS2 (448), 33) QM, 34) RAB11A, 35) RAB3A, 36) RAB5A, 37) RAB6, 38) RAB6KIFL, 39) TF, 40) TTC1. Values given in parentheses for proteins 2, 31, and 32 refer to alternative splice forms.

4.2 Markov chain Monte Carlo (MCMC) simulations

We used a MCMC simulation approach for computing the posterior probabilities of all edges within the network [18, 19]. This approach, particularly useful in generating posteriors from complicated distributions, allowed us to adequately sample from the astronomically large number of possible network configurations (for $|V|$ vertices there are $2^{|V||V|}$ possible networks). In our approach we used a uniform prior distribution over all networks, as we had no prior information that would cause us to prefer one network over another. Starting with an arbitrary network, and using a reversible—jump methodology [20], edges were both added and removed at each iteration of the algorithm. Addition and removal of edges moves the network from the current state X to a proposed state Y. Using a symmetric proposal distribution, the new state is accepted with probability

$$\alpha(x, y) = \min\left\{1, \frac{L(Y)}{L(X)}\right\},$$

where $L(.)$ is the likelihood of the network. If the proposed state is accepted, it becomes the current state. This method thus samples networks from the space of all possible networks while keeping each edge occupied, or unoccupied over time, in proportion to its posterior probability.

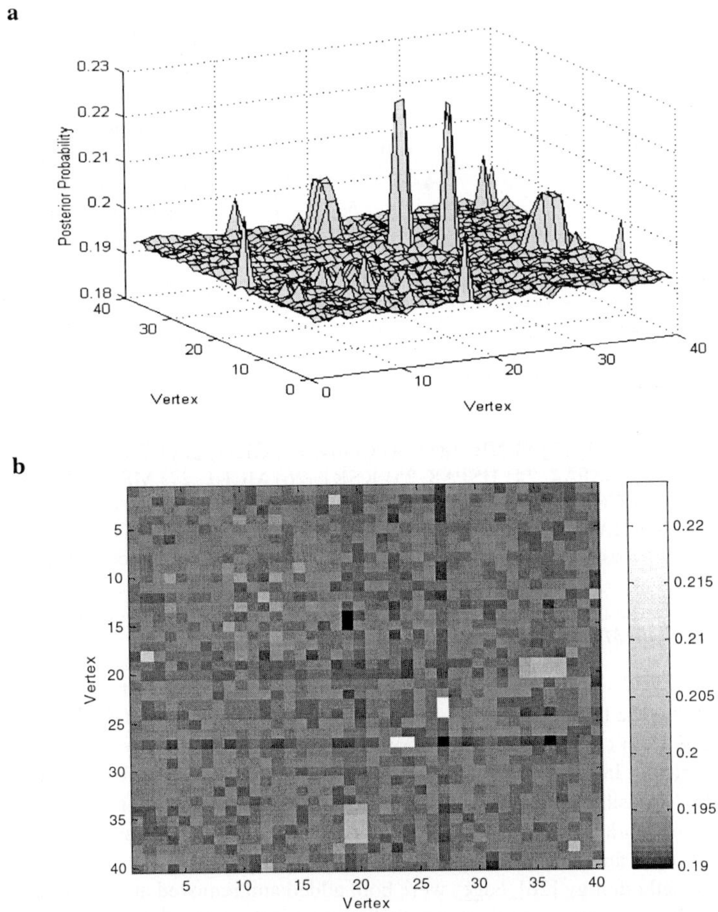

Figure 2. Network posterior probabilities. See text for further details.

The posterior distribution generated from approximately 10^7 samples is shown in Figure 2a and b. In 2a it can be seen that a few edges are readily apparent; rising well above the surrounding background. The two tallest peaks are of the HSPA8—MRJ interaction. Edges such as these show up rapidly in our simulations, while low—

probability edges can take considerably greater amounts of sampling to distinguish them from background. Figure 2b shows the posterior probabilities for each edge of the network. The lower probability (darker) "lines" running horizontally at vertices 20 and 27 and vertically along vertex 27 show the influence of the nonsymmetrical edge distributions. For example, since vertex 27 has a relatively high probability connection, the edge distribution tends to suppress the addition of new edges to the same vertex. Of course any vertex can have multiple incoming and outgoing edges, however due to the scale—free property of these networks, highly connected vertices are relatively rare.

5. Discussion

Obviously, we would have preferred to have better accuracy in our predictions, though the prediction of a few edges not in the data set and that were able to be confirmed through a literature search was very encouraging. However due to inadequacies in the training data and current limitations of the model, a large number of potential errors is unavoidable at this time. In our previous analysis, we used cross-validation to measure the effectiveness of the model. By starting with a large network (642 edges), and either adding or removing a single edge and determining whether the network probability increased or decreased as a result, we estimated 7% false negative and 10% false positive error rates [3]. The study described here, however, was significantly more challenging. We should also note that it is extremely difficult to evaluate the true accuracy of these results. Of the edges that could not be matched to known edges, it is quite possible that some of these are also correctly predicted. In fact, a primary goal of this effort is to generate just such predictions of currently unknown interactions.

Our approach, however, is primarily limited by the use of a rather limited set of previously defined domains. For example, of the 6202 proteins within yeast, nearly 40% were unable to be assigned any type of domain. Furthermore, of the 2238 edges used here, only 708 originated and terminated at proteins each with at least one domain. Similar limitations are seen within the human data. In addition, while yeast proteins tend to be characterized by a single domain, multidomain proteins are closer to the norm in humans, and thus the limited amount of training data is again a factor. Obviously, the lack of adequate domain coverage presents serious difficulties, as our model requires at least 1 instance of a particular domain—domain interaction in the training set to predict it in "real" data. To address this issue, we are in the process of developing a method that should be capable of providing 100% coverage.

As discussed in the model description, we currently use a multinomial distribution to characterize the distribution of edges going into and out of each vertex of the network, with the bin probabilities taken from fits to yeast data. While not optimal, the use of yeast parameters seemed an acceptable first—pass attempt as, for example, edge distributions from metabolic networks (which also follow power—law behavior) have been shown to be very similar across species [21]. While we plan to acquire distributions for a number of species, it appears that the lack of reasonably large data sets could be a hindrance, with improper edge distributions perhaps masking interactions that would otherwise be apparent, particularly in interspecies predictions. In the interim, we plan to use parameters from a well—characterized system (e.g. yeast) in a distribution with identical mean but with greater variance. This requirement can be fulfilled with the incorporation of the negative multinomial distribution (instead of the multinomial distribution) into our simulations, defined as

$$P(n_1, n_2, \ldots n_k) = \frac{\Gamma\left(N + \sum_{i=1}^{k} n_i\right)}{\left(\prod_{i=1}^{k} n_i!\right)\Gamma(N)} Q^{-N} \prod_{i=1}^{k} \left(\frac{P_i}{Q}\right)^{n_i}; \quad (n_j \geq 0) \cdot$$

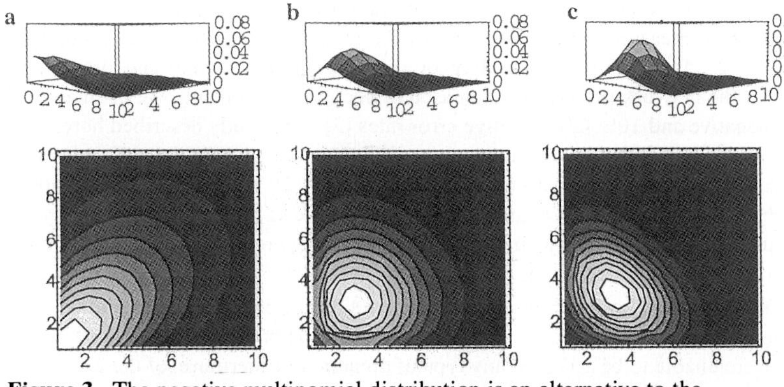

Figure 3. The negative multinomial distribution is an alternative to the multinomial. Parts (a) & (b) show the negative multinomial (suface plot above its corresponding contour plot) in comaprison to the multinomial distribution in (c). For the multinomial, $P_i = 0.25$ and N = 14. For the neg. multinomial, P_i was set equal to 0.25 times a constant, with *NP* held constant. For part (a) the const = 4, while for part (b), const = $1\text{x}10^{-6}$. See text for further details.

In Figure 3a and b, we show the negative multinomial with different parameters P_i, while Figure 3c shows a multinomial distribution. It can be seen that by increasing P_i we are able to increase the variance of the distribution while keeping the expected value identical to the multinomial distribution shown in part c. Note that while we can match the expected value, we can only generate a variance that is greater than, but not equal to, the multinomial's. This is because a negative binomal distribution tends to a Poisson distribution as the variance decreases, and the Poisson distribution has typically larger variance than a multinomial distribution with the same mean.

From an implementation standpoint, this approach, while capable of handling large networks, benefits significantly from the use of appropriate computational resources. We ran a C programming language implementation of our method, which proved to be significantly more rapid than our previous implementation in Matlab. In addition, we have the benefit of a 5—node Beowulf cluster running Linux, with each node having 2, 1GHz CPUs. The availability of appropriate hardware and software was invaluable, as it can take a considerable amount of time to establish a stationary distribution (1-2 days in this case) and to generate the posterior (many days to generate a posterior with resolution of low—probability edges).

In the future, we plan on implementing "repulsive" interactions between domains. This can be achieved by assigning domain—domain interaction probabilities of < 0.5 to interactions that are never present. While requiring careful normalization and balancing with "attractive" probabilities, this feature should provide enhanced resolution of predicted interactions (bigger peaks and deeper valleys in the posterior plots). While having its own set of favorable and unfavorable properties, two—hybrid data should prove particularly valuable for this approach.

6. Conclusion

This work has attempted to describe a probabilistic approach to the prediction of protein—protein interactions *a priori*. Other approaches to predicting protein interactions are also being developed. Recently, work by Bock and Gough [22] described a Support Vector Machine approach for this prediction. This approach was based on primary structural data (the protein sequence) and utilized the DIP database for training data. A benefit of the approach described here is that we can assign probabilities to both edges and to complete networks (or subgraphs). Given a target protein(s), a ranking of most likely interaction candidates can be generated directly, providing some direct measure as to how confident one is as to the existence of a given interaction.

Our use of Markov chain Monte Carlo techniques provides a computationally feasible way to calculate the posterior probability of a network given data as:

$$P(network_i \mid data) = \frac{P(data \mid network_i)P(network_i)}{\sum_{all\ networks} P(data \mid network_j)P(network_j)} .$$

While we have assumed a uniform prior distribution over all possible networks, the model does not require this. This framework allows new information (in the form of priors) to be added into the calculation as it becomes available.

In summary, while requiring further improvement, we feel that this approach holds significant potential. Its Bayesian basis allows the integration of disparate types of data into a single prediction. The discussed improvements should allow for more accurate predictions of both known and unknown interactions and will hopefully provide predictions of some value to the biological community.

References

[1] E. M. Marcotte, M. Pellegrini, et al., *Science* **285**, 751 (1999)
[2] A. J. Enright, I. Iliopoulos, et al., *Nature* **402**, 86 (1999)
[3] S. M. Gomez, S.-H. Lo, et al., *Genetics* (to appear, 2000)
[4] A. Bateman, E. Birney, et al., *Nucleic Acids Res* **28**, 263 (2000)
[5] G. Apic, J. Gough, et al., J. Mol. Biol. **310**, 311 (2001)
[6] I. Xenarios, E. Fernandez, et al., *Nucleic Acids Res* **29**, 239 (2001)
[7] I. Xenarios, D. W. Rice, et al., *Nucleic Acids Res* **28**, 289 (2000)
[8] H. Jeong, S. P. Mason, et al., *Nature* **411**, 41 (2001)
[9] A. L. Barabasi and R. Albert, *Science* **286**, 509 (1999)
[10] R. Albert, H. Jeong, et al., *Nature* **406**, 378 (2000)
[11] T. Ito, K. Tashiro, et al., *Proc Natl Acad Sci U S A* **97**, 1143 (2000)
[12] P. Uetz, L. Giot, et al., *Nature* **403**, 623 (2000)
[13] D. M. Hutt, L. F. Da-Silva, et al., *J Biol Chem* **275**, 18511 (2000)
[14] S.-K. Wu, P. Luan, et al., *J. Biol. Chem.* **273**, 26931 (1998)
[15] O. Ullrich, H. Stenmark, et al., *J. Biol. Chem.* **268**, 18143 (1993)
[16] C. A. Ballinger, P. Connell, et al., *Mol. Cell. Biol.* **19**, 4535 (1999)
[17] H. Awata, C. Huang, et al., *J. Biol. Chem.* **4**, 4 (2001)
[18] W. R. Gilks, S. Richardson, et al., "Markov chain Monte Carlo in practice." (Chapman & Hall/CRC, New York, 1996)
[19] W. K. Hastings, *Biometrika* **57**, 97 (1970)
[20] P. J. Green, *Biometrika* **82**, 711 (1995)
[21] H. Jeong, B. Tombor, et al., *Nature* **407**, 651 (2000)
[22] J. R. Bock and D. A. Gough, *Bioinformatics* **17**, 455 (2001)

IDENTIFYING MUSCLE REGULATORY ELEMENTS AND GENES IN THE NEMATODE *CAENORHABDITIS ELEGANS*

D. GUHATHAKURTA, L.A. SCHRIEFER, M.C. HRESKO
R.H. WATERSTON, G.D. STORMO
Department of Genetics, Washington University School of Medicine
4566 Scott Avenue, Campus Box: 8232, St. Louis. MO 63110. USA
Phone: (314)747–5535; Fax: (314)362–7855
Email: {dg,larry,coutu,stormo}@genetics.wustl.edu; bwaterst@watson.wustl.edu

We report the identification of several putative muscle–specific regulatory elements, and genes which are expressed preferentially in the muscle of the nematode *Caenorhabditis elegans*. We used computational pattern finding methods to identify *cis*–regulatory motifs from promoter regions of a set of genes known to express preferentially in muscle; each motif describes the potential binding sites for an unknown regulatory factor. The significance and specificity of the identified motifs were evaluated using several different control sequence sets. Using the motifs, we searched the entire *C. elegans* genome for genes whose promoter regions have a high probability of being bound by the putative regulatory factors. Genes that met this criterion and were not included in our initial set were predicted to be good candidates for muscle expression. Some of these candidates are additional, known muscle expressed genes and several others are shown here to be preferentially expressed in muscle cells by using GFP (green fluorescent protein) constructs. The methods described here can be used to predict the spatial expression pattern of many uncharacterized genes.

1 Introduction

Establishing where and when a gene is expressed and understanding the underlying regulatory network which guides its expression are critical in understanding gene function in a multicellular organism. The transcription regulatory apparatus which directs temporo–spatial expression of genes is encoded in the DNA, in the form of organized arrays of transcription factor (TF) binding sites[1,2]. These *cis*–regulatory sites are recognized sequence–specifically by cognate TFs which control and guide the expression pattern of genes.

We are interested in identifying muscle–specific regulatory elements, and genes expressed in muscle as tools to study muscle development. Out of the thousands of genes which express in a particular tissue, the most interesting genes to study for understanding its development, differentiation, function and structure, are the ones which are *preferentially* expressed in that tissue. Preferential expression can be either *selective* (expression in a *subset* of tissues or cell types in an organism) or *specific* (expression in *only one* tissue or cell type). Experimentally elucidating novel, additional genes which are expressed specifically or selectively in a tissue, or finding new *cis*–regulatory elements which function only in one tissue type is time consuming as well as challenging. Hence, computational methods which can accurately predict tissue–specific genes and regulatory elements are of great value. Here we describe computational approaches for the identification of muscle–specific *cis*–regulatory elements and genes which are expressed preferentially in the *C. elegans* muscle.

We obtained a list of 35 genes known to be selectively or specifically expressed in the muscle of *C. elegans*. We used a subset of these genes as the training set and the remaining genes as the test set. From the promoter regions of the training set genes we identified several conserved motifs using two different computational methods. We evaluated the significance and specificity of these motifs for muscle–expressed genes using the test and several control sets. The identified motifs describe potential target binding sites for novel transcription factors. Using these motifs, we searched the *C. elegans* genome for genes whose promoter regions have a high probability of being bound by the regulatory factors. These genes were considered as potential candidates for muscle expression. Several identified candidates were known muscle genes (present in both training as well as test sets). We have tested the expression pattern of some of the other candidates using GFP technology and found that several of these genes are indeed expressed preferentially in *C. elegans* muscle.

The methods described here can be used in identifying regulatory elements and genes in other tissues and cell types in *C. elegans* and other eukaryotic organisms.

2 Data

Training set: Upstream regions (–2000 to –1, relative to the translation start) of 19 genes known to be expressed selectively or specifically in *C. elegans* muscle[3,4,5].
Control set 1: Upstream regions of a completely *different* set of 16 genes known to be expressed selectively or specifically in the *C. elegans* muscle[3,4,5].
Control set 2: Upstream regions of 500 genes, randomly selected from the *C. elegans* genome.
Control set 3: Upstream regions of 19 genes known to be expressed selectively or specifically in the *C. elegans* intestine[6,7] (J.D. McGhee, personal communication). None of these genes have any known expression in muscle.

Complete gene lists (with additional references) are available at http://ural.wustl.edu/~dg/PSB02.html. All sequences were downloaded from the WormBase anonymous ftp server: ftp://ftp.wormbase.org/pub/wormbase/.

3 Methods

3.1 *Identification of regulatory motifs*

Two local multiple sequence alignment methods, Consensus[8] and ANN–Spec[9], were run on the training set sequences to identify conserved motifs. Both Consensus and ANN–Spec use weight matrix models to represent un–gapped sequence motifs. Since the TF binding sites in a set of similarly regulated sequences are expected to be conserved to a certain extent, conserved motifs identified by these programs represent potential regulatory elements.

Consensus: Consensus[8] uses a greedy algorithm and searches for a matrix with a low probability of occurring by chance, or, equivalently, having a high information content (I.C.)[10]. Version 6.c of Consensus was used. The top scoring results were reported from different runs. Different pattern lengths were tested, and both strands of the DNA were searched for motifs since TFs can bind to either strand. Patterns with high I.C. and the lowest expected frequency were considered.

ANN–Spec: ANN–Spec[9] uses a simple artificial neural network and Gibbs sampling[11] method to define DNA binding site patterns. The program searches for the parameters of a simple perceptron network (weight matrix) which maximize the specificity for binding a positive sequence set (or training set) compared to a background sequence set. Binding sites in the positive data set are found with the resulting weight matrix and these sites are then used to define a local multiple sequence alignment. ANN–Spec Version 1.0 was used. A comparison of ANN–Spec and other related programs has shown that ANN–Spec is able to identify patterns of higher specificity when training with background sequences (C.T. Workman and G.D. Stormo, unpublished observation). Hence, for ANN–Spec, a background sequence set of upstream regions from 3000 randomly picked genes was used. Different motif lengths were tried and both strands of the DNA were searched for motifs. Due to the non–deterministic nature of the algorithm, multiple training runs are performed (100), with each run iterating 2000 times. The results were sorted by their best attained objective function values. Weight matrices corresponding to the ten highest scoring runs were observed. If >5 of these top scoring ten runs give a motif with one consistent pattern consensus, that pattern is considered significant.

3.2 Searching for "sites"in sequences

The Patser program (G.Z. Hertz and G.D. Stormo, unpublished) allows one to score the words of a sequence against a weight matrix. Once the weight matrices for regulatory motifs are obtained by Consensus or ANN–Spec, the matrices can be used as input for Patser to identify high scoring sub–sequences (or "sites") in a given set of sequences. Patser calçulates the p–value (or probability) of observing a particular score or higher at a particular sequence position[12]. A "cutoff" score for eliminating low scoring sub–sequences is also calculated numerically. From an alignment of sites in a binding site pattern, the program calculates the cutoff score as follows. The "true" information content of an alignment of sites is given by:

$$I_{sites} = -\sum_b \sum_k f(b,k) \ln \frac{f(b,k)}{p(b)} \qquad \text{....... (1)}$$

where, f(b,k) is the frequency of observing a base, b, at a particular position, k, in a binding site, p(b) is the prior probability for base b in the genome, k sums over all l positions of the pattern (l being the length of the pattern), and b sums over all four DNA bases. *ln(probability)* of observing binding sites in a random sequence is related to this true information content[10]: $ln(probability) \leq I_{sites}$. The "sample size adjusted" information content of an alignment is the true information content *minus* the average information content expected from an arbitrary alignment of random sites. Patser approximates the target *ln(probability)* of the cutoff score (*i.e.* the probability of observing a score greater or equal to the cutoff score) as −(sample size adjusted information content); the cutoff score can then be calculated from this *ln(probability)* value.

3.3 Establishing spatial expression pattern of genes

A major advance in the attempts to localize gene expression and proteins is the recent advent of green fluorescent protein (GFP) as a reporter molecule in living organisms[13]. GFP is a protein from jellyfish that emits green fluorescence when excited by blue light, even when expressed in heterologous organisms. Here, the promoters (−6000 to −1) of the genes which are predicted to be expressed in muscle are fused with the GFP−coding sequence using genetic recombination, so that the GFP is under the regulatory control of the promoter. Suitable DNA constructs for promoter::GFP are injected into the gonad of the hermaphrodite worms. A portion of the progeny segregating from the injected animals express GFP under the control of the promoter of interest. Green fluorescence from the GFP is observed in the different cells and tissues of these progeny. (Detailed description of the method is available at: http://ural.wustl.edu/~dg/PSB02.html.)

4 Results

4.1 Identification of regulatory motifs

One very strong motif with the consensus CCCGCGGGAGCCCG (Motif 1, Figure 1) was obtained using both Consensus and ANN−Spec. Some shorter motifs were also found which appeared to be parts of the above motif and were ignored. Instances of this motif (sub−sequences scoring above the Patser cutoff value) were identified in the training set. These sites were then deleted from the sequences and Consensus and ANN−Spec programs were re−run which resulted in identification of several other motifs (motifs 2 through 5, Figure 1). We checked to see if the motifs found in our analysis were previously reported. Motifs 4 and 5 are very similar to the G−rich binding sites of the ubiquitous, Sp−1 like, transcription factor which has been shown to regulate the expression of many different classes of genes including housekeeping and muscle genes[14]. Since our objective was to identify muscle–specific regulatory elements, we did not consider motifs 4 and 5 any further. Motifs 1, 2, and 3 were novel since they did not have any matches to any known sites in the TRANSFAC database[15].

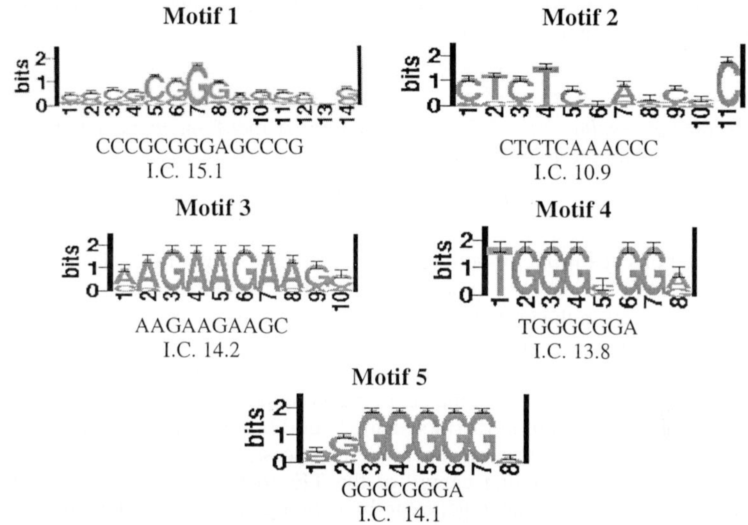

Figure 1: Motifs identified using Consensus and ANN–Spec programs. The motif consensus, information content of the motifs in bits and sequence logos[16] are given.

4.2 DNA binding probability and significance of identified motifs

A "site" in a sequence is simply a high scoring sub–sequence which is obtained by the Patser program using the motif weight matrix as an input. A term which is proportional to the probability of a TF molecule binding to its sites in a sequence can be obtained from the "site scores" calculated by the Patser program. The free energy of binding ($-\Delta G$) of a TF molecule to a DNA site is proportional to the score (s) of the site[17,18], $\Delta G = -RTs$, where, R is the universal gas constant and T is the absolute temperature. Let us assume that occupancy of DNA sites by a TF follows the Boltzmann distribution under the condition of thermodynamic equilibrium. Then, the probability that a TF molecule occupies a site with score s, is given by:

$$P(s) \;\alpha\; e^{-\Delta G/RT} \qquad \text{..... (2a)}$$

$$\text{or, } P(s) \;\alpha\; e^{s} \qquad \text{..... (2b)}$$

This is called the probability proportionality value (*pp–value*). Since we will use these *pp–values* only for the purpose of comparison between different sequences (see below), the proportionality constant in equation 2 is of no consequence to us, and we assume that the constant is equal to 1. For any motif, *m*, there can be multiple sites in a given sequence scoring above the Patser cutoff; the *pp−value* for binding of a TF molecule to *any* of its several binding sites in a sequence is given by the sum of individual terms:

$$P_m^{seq} = \sum_{Sites} e^{s} \qquad \text{.... (3)}$$

The average *pp−value* for a TF, corresponding to motif *m*, to be bound to a sequence in a given set of N sequences is:

$$< P_m^{seq} > = \frac{1}{N} \sum_{Seqs} \sum_{Sites} e^{s} \qquad \text{..... (4)}$$

We calculate this *pp−value* for both the training and the random sets (Table 1).

For efficient gene regulation TFs need to bind effectively to the regulatory elements in the promoter region i.e. the binding energy (−ΔG) of the TF to the promoter region of the regulated gene should be higher compared to other (background) sequences. In other words, the probability of binding to the regulated sequences should be higher compared to background sequences, assuming the components of the cells are in thermodynamic equilibrium and the binding events follow the Boltzmann distribution. There are two possible ways in which this may be achieved. First, the binding energy of the TF to one individual site in the upstream region of the regulated gene can be very high (*Mode 1*); alternatively, in the absence of a very strong binding site, there can be multiple weaker sites, with lower binding energies (individually), but the combined effect of these sites may result in high binding probability (*Mode* 2) (see equation 3). A combination of both strategies is also possible. We do not know which mode is more suitable for describing gene regulation by the putative TFs for the identified DNA binding site motifs. Therefore, we determined several relevant parameters for both modes of binding (Table 1) for the training set as well as the control set 2 (random set). First, using Patser, potential binding sites for each motif were determined in both sets. We then calculated: average number of sites per sequence, average score of the binding sites, average of the maximum scoring sites from each sequence, and a measure of the probability of binding of a TF to its sites in a particular sequence. Parameters were determined from (a) only the highest scoring sites in each sequence (*Mode 1*) (b) all sites scoring above the respective Patser cutoff scores (*Mode* 2). For the purpose of comparison, we have also shown the parameters for an unrelated pattern, ACTGATA ("GATA" in Table 1), which is obtained by Consensus and ANN−Spec from the promoters of a list of genes expressed in the *C. elegans* intestine (D. GuhaThakurta, J.D. McGhee and G.D. Stormo, unpublished observation). Sites corresponding to this motif have been shown to be important for intestine−specific expression of the *ges−1* gene in *C. elegans*[6].

Table 1: Potential TF binding site parameters for training and random set. Column 1: average number of sites per sequence above the Patser calculated cutoff, Column 2: average score per site, Column 3: *pp–value* (equation 4) calculated from all sites scoring above the Patser calculated cutoff value. For columns 4 and 5, the highest scoring sites in each sequence were determined using Patser (highest scoring sites in some sequences may have scores below the cut–off values). Average of the highest scores from each sequence is given in column 4. Column 5 shows the *pp–value* based on only the highest scoring sites in each sequence. The values in columns 3 and 5 in both sets are to be multiplied by a factor of 10^4.

	Training Set					Random Set						
Motif Index	*Avg. no. of sites per seq.*	*Avg. score per site*	*pp–val all sites > cutoff*	*Avg. highest score per seq.*	*pp–val with highest scoring sites*	*Avg. no. of sites per seq.*	*Avg. score per site*	*pp–val all sites > cutoff*	*Avg. highest score per seq.*	*pp–val with highest scoring sites*	*Ratio* C^T_3 / C^R_3	*Ratio* C^T_5 / C^R_5
	C^T_1	C^T_2	C^T_3	C^T_4	C^T_5	C^R_1	C^R_2	C^R_3	C^R_4	C^R_5	R_C	R_H
1	3.45	10.5	61.9	11.79	47.8	0.69	9.51	2.55	7.1	1.92	24.3	24.9
2	5.15	7.17	4.19	9.36	3.16	2.68	6.84	0.74	7.04	0.49	5.7	6.4
3	1.6	10.32	10.1	9.3	2.68	0.66	9.15	1.60	7.25	0.51	6.3	5.25
GATA	1.6	6.12	1.1	6.1	0.73	2.15	6.38	2.8	6.54	2.19	0.39	0.33

Ratios, R_C and R_H, are called *discrimination factors* or *R–factors.* Given two sequences, one from the training set and another from the random set, the discrimination factors show how likely it is for the cognate TF to bind a training set sequence as opposed to a random sequence. The R–factors are nearly identical using all sites above the cutoff (R_C) or using only the **h**ighest scoring sites (R_H). For the cognate TF corresponding to motif 1, it is about 24–25 times more likely that the TF will bind to a training set sequence. The R–factor for motif 1 is much higher than that of motifs 2 or 3, which might explain why the motif appeared to be the most significant one in our first round of Consensus and ANN–Spec runs. In eukaryotic gene regulation, it is common for multiple TFs to act together and bind DNA in a cooperative fashion[1,2,14,19,20]. If this is the case here, then the combined effect of multiple TFs binding to the sites could be dramatic even though the individual discrimination factors for motifs 2 and 3 are on the order of 5–6. The R–factor for the unrelated GATA motif is less than 0.4.

4.3 Specificity of identified motifs for muscle genes

The *combined pp–value* for multiple motifs is calculated for the upstream sequence of each gene in the *C. elegans* genome. For lack of more specific information regarding the mode of TF binding and interaction at this point, we assume that for selective or specific expression of a gene in the muscle context: *(1) all* relevant TFs (corresponding to the motifs 1, 2 and 3) need to bind to the upstream sequence, and *(2)* if there are multiple sites scoring above the Patser cutoff for a particular motif, *any one* of those binding sites may to be occupied by the corresponding TF. For a particular upstream sequence, the combined *pp−value* for multiple motifs is calculated by taking a product of individual *pp−values* (from equation 3) for the different motifs:

$$P^{seq} = \prod_{m=1}^{3} P_m^{seq} \qquad ..... (5)$$

All (19,804) upstream sequences were sorted according to the log of the combined *pp−value,* $\ln(P^{seq})$ (equation 5). Two sorted lists were generated, viz. list 1, where the combined *pp−values* were calculated for each sequence using only the highest scoring sites corresponding to the three motifs (*Mode 1*); and list 2, where the *pp−values* for each sequence was calculated using all sites for the three motifs scoring above the respective Patser cutoffs (*Mode* 2). Based on the position of the genes in a sorted list, a "*specificity score*" can be calculated for a given sequence set. Suppose the positions of N genes of a given set in the sorted list are: x_1, .. x_n, .. x_N. The probability that a particular gene is at position x_n or higher in the sorted list is given by: (x_n/19,804). The joint probability of observing N genes at positions x_1 ... x_N or higher in the list is given by the product of individual probabilities. We consider the log of this probability, and to have a measure independent of the number of sequences, we define the specificity score as:

$$\text{Specificity Score} = -\frac{1}{N}\left[\sum_n \ln\left(\frac{x_n}{19{,}804}\right)\right] \qquad ...... (6)$$

We calculate the specificity score for the training and the three control sets. Using list 1, the specificity score for the training set, and control sets 1, 2 and 3 were 4.76, 2.63, 1.04 and 0.63 respectively. Using list 2 the specificity scores for the training set and three control sets (in the same order as above) were 4.22, 2.5, 1.01 and 0.8. The specificity scores for the second muscle gene set (control set 1) are not as high as that for the training set, but still substantially higher than the random (control set 2) or the intestine (control set 3) gene sets. The higher specificity scores for the training set and the control set 1, show that the identified motifs are specific (or, at least selective) for the muscle genes.

4.4 Selection of candidates for testing muscle expression by GFP

We considered the two sorted lists from section 4.3. Several known muscle genes were placed high on the lists (within top 25, Table 2); for reference the highest scoring intestine gene was placed at only 2029 in list 1 and 3943 in list 2. To select a few genes for GFP−expression testing, we took the top−scoring 25 genes from list 1, since the specificity score of muscle genes was higher using this list compared to list 2. To minimize the false positive rate, we checked for the presence these genes in the top scoring 50 genes in the list 2. Genes which score high in *both* lists were considered good candidates for muscle expression. Several of these candidates were known muscle genes (training set or control set 1). The remaining ones were selected for GFP–expression testing (Table 2, Figure 2).

Table 2: the list of top scoring 25 genes from list 1 (see text, section 4.4). genes which were in the training set are in bold. genes which were in control set 1 are in italics. all previously known muscle genes are indicated in column 5 and candidates which have been verified here to have muscle–specific or selective expression are indicated in column 6.

Pos. in List 1	*Pos. in List 2*	*Gene ID*	*Gene name or putative product*	*Previously known muscle expression*	*GFP verified expression in muscle*
1	**1**	**F09F7.2**	**mlc–3**	**Yes**	
2	**4**	**ZK617.1b**	**unc–22**	**Yes**	
3	*13*	*K10B3.8*	*gpd–2*	*Yes*	
4	**9**	**Y105E8B.c**	**tmy–1**	**Yes**	
5	7	F55C7.2	unknown		
6	8	Y44A6D.3	unknown		
7	*12*	*F08B6.4b*	*unc–87*	*Yes*	
8	**25**	**F11C3.3**	**unc–54**	**Yes**	
9	2	C49A1.6	unknown		
10	6	F58F6.3	unknown		
11	11	B0513.1	gei–5		Yes
12	15	F47F6.1	unknown		Yes
13	46	F02E9.2b	lin–28		
14	10	R08B4.2	transcription factor		No
15	19	F14B8.5a	unknown		
16	44	F41E7.6	carnitine octanoyltransferase		Yes
17	14	T22C1.7	unknown		Yes
18	18	W06F12.1a	ser/thr kinase		
19	41	F29C12.1	unknown		
20	*40*	*T22E5.5*	*mup–2*	*Yes*	
21	*31*	*C16C2.2*	*eat–16*	*Yes*	
22	51	C05D11.4	let–756		
23	64	C09D8.2	ptp–3		
24	**36**	**K12F2.1**	**myo–3**	**Yes**	
25	**31**	**D1081.2**	**unc–120**	**Yes**	

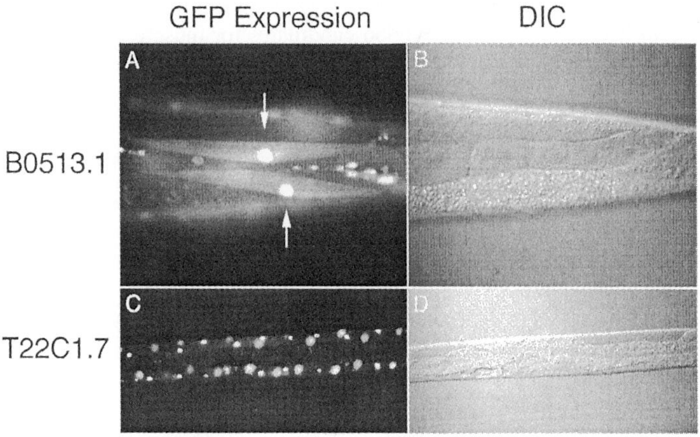

Figure 2: Experimental verification of candidate genes; expression pattern of two genes are shown (additional figures are available at http://ural.wustl.edu/~dg/PSB02.html/). Fluorescence (A, C) and corresponding DIC (differential interference contrast) (B, D) images of transgenic worms expressing GFP under the control of 6 Kb region upstream of B0513.1 and T22C1.7. **A**. GFP–dependent fluorescence is detected in the nuclei (arrows) and in the cytoplasm of bodywall muscle cells. **C**. GFP dependent fluorescence is detected in nuclei of the body wall muscle cells close to the outer edge of the animal. This focal plane shows in–focus nuclei from two quadrants and out–of–focus nuclei are from the other two quadrants.

4.5 Identifying spatial expression of candidate genes

Accurately establishing that a gene is expressed in *only one* cell type using the currently available techniques (e.g. *in–situ* hybridization, GFP) is difficult. This is in part because the observance of expression of in only one cell type does not rule out low or transient expression in other tissues. In addition, depending on the technique used, detection of expression in certain tissues can be problematic. Hence some of the genes which are described in the literature as muscle specific, may actually be expressed in a few other tissues.

We determined the spatial expression of candidate genes in adult and larval worms using GFP–reporter constructs, in which GFP is under the control of the promoter regions of the candidate genes. Complete identification of the spatial GFP–expression pattern is difficult and still ongoing, however, general statements concerning localization can be made. T22C1.7 is expressed predominately in the bodywall muscle (Fig. 2c) and in cells tentatively identified as pharyngeal, vulva and intestinal muscle. Thus, this gene could be muscle specific, but we need to critically identify its expression in other cells before we can make this claim. B0513.1 is clearly expressed in the bodywall (Fig. 2a) and vulval muscle. Its GFP expression also includes a limited set of non–muscle tissue including intestine, neurons and, probably, hypodermis. F41E7.6 has GFP expression in intestinal muscle, sphincter muscle and anal depressor muscle. It is also expressed in non–muscle tissue of the pharyngeal–intestinal valve and a small number of neurons. Muscle expression of the F47F6.1 gene was observed under a GFP–dissecting microscope in the initial progeny (F1) of the injected adult worms. However, no transmitting lines for this gene were established, and therefore further detailed investigation of this gene was not done. R08B4.2 is expressed predominately in neuronal tissue. However, its expression in muscle tissue cannot be ruled out for reasons mentioned above. Transient expression of this gene in muscle during embryogenesis is possible and we are in the process of characterizing the expression pattern during development. There was no observed GFP expression from the promoter regions of genes F58F6.3 and C09D8.2. Possible reasons for this are: experimental error, low level or transient GFP expression, or these genes could be pseudogenes that are not expressed. We are continuing with experiments to determine the expression for the remainder of the genes in Table 2.

5 Discussion

Using computational pattern recognition methods we identified several potential muscle–specific regulatory elements. These putative regulatory elements were then used to predict other genes which might be preferentially expressed in the muscle tissue. Out of the top 25 genes in list 1, 23 score highly (within the top 50) in list 2. Out of these 23, 6 were from the training set, 4 were known muscle genes which were not included in our training set, and 4 more have been experimentally shown to have muscle–selective expression. Thus, checking for consistency in the two lists gives a high true positive rate for identification of muscle–specific or selective genes. We are in the process of checking the expression of some lower scoring genes (e.g. genes at positions 25 through 100). We believe some of these genes will also show muscle–selective or specific expression.

A number of additional considerations are likely to increase the efficiency of identification of muscle genes. Here, we started with a partial set of muscle genes for the purpose of cross validation and evaluation against an independent test set; including all known muscle genes in our training set could increase the quality and specificity of the regulatory motif weight matrices and lead to more efficient detection of other muscle genes. A more thorough computational study should also be helpful; e.g. a study of the distance distribution and orientation of the sites can illustrate the possible modes by which the TFs interact with the DNA sites and with each other. This should lead to building better models for the TF–DNA interaction and allow us to identify muscle genes with higher specificity[14,19,20,21]. In addition we need to initiate experiments to test whether the predicted regulatory elements are functional and guide muscle–selective or specific expression. These studies will not only help in more efficient identification of muscle–specific genes but facilitate our understanding of muscle–specific regulatory elements and mechanisms which guide gene expression in this tissue. Clearly, the studies described here can be helpful in understanding the underlying regulatory mechanism, and in identifying new genes which are expressed in other spatial contexts and tissues not only in *C. elegans* but also other eukaryotic organisms. This knowledge may also find applications in gene therapy, where using tissue–specific regulatory elements, one can design promoters for the purpose of gene delivery to specific tissues[22].

Acknowledgments

John Spieth is acknowledged for helpful discussions and advice on our work. Jim McGhee is thanked for providing a list of intestine genes. The work was supported by grant (HG00249) from National Institutes of Health to GDS.

References

1. Arnone, M.I. and Davidson, E.H. *Development*, **124**, 1857, (1997)
2. Yuh, C–H., Bolouri, H., and Davidson, E.H. *Science*, **279**, 1896, (1998)
3. Moerman, D.G., and Fire, A. In Riddle, D.L., Blumenthal, T., Meyer, B.J., Priess, J.R. (eds.) "*C. elegans II*", Cold Spring Harbor Laboratory Press, 417, (1997)
4. Waterston, R.H., In Wood, W.B. (ed.) "The nematode *C. elegans*", Cold Spring Harbor Laboratory Press, 281 (1988)
5. The *C. elegans* consortium, *Science*, **282**, 2012, (1998)
6. Egan, C.R. *et.al., Development,* **170**, 397, (1995).
7. Mochii, M., *et.al. Proc. Natl. Acad. Sci. USA.*, **96**, 15020–15025 (1999)
8. Hertz, G.Z. and Stormo, G.D. *Bioinformatics*, **15**, 563, (1999)
9. Workman, C.T. and Stormo, G.D. *Pacific Symp. Biocomp.*, **5**, 464, (2000)
10. Schneider, T.D., *et.al. J. Mol. Biol.*, **188**, 415, (1986).
11. Lawrence, C.E., *et.al. Science*, **262**, 208, (1993)
12. Staden, R. *Comp. App. Biosci.*, **5**, 89, (1989)
13. Chalfie, M., *et.al. Science,* **263**, 802, (1994)
14. Wasserman, W.W. and Fickett, J.W. *J. Mol. Biol.*, 278, 167, (1998)
15. Wingender, E. *et.al. Nucleic Acids Res.*, **29**, 281, (2001)
16. Schneider, T.D. and Stephens, R.M. *Nucleic Acids Res.* **18**, 6097, (1990)
17. Stormo, G.D. *J. Theo. Biol.*, **195**, 135, (1998)
18. Stormo, G.D., and Fields, D.S. *Trends Biochm. Sci.*, **23**, 109, (1998)
19. Fickett, J.W. *Gene*, 172, GC19, (1996)
20. Wagner, A. *Bioinformatics*, **15**, 776, (1999)
21 Klingenhoff, A., *et.al. Bioinformatics*, **15**, 180 (1999)
22. Nettelbeck, D.K., *et.al. Trends Genet.*, **16**, 174, (2000)

COMBINING LOCATION AND EXPRESSION DATA FOR PRINCIPLED DISCOVERY OF GENETIC REGULATORY NETWORK MODELS

ALEXANDER J. HARTEMINK

Duke University Department of Computer Science
Box 90129, Durham, NC 27708-0129

DAVID K. GIFFORD, TOMMI S. JAAKKOLA

MIT Artificial Intelligence Laboratory
200 Technology Square, Cambridge, MA 02139

RICHARD A. YOUNG

Whitehead Institute for Biomedical Research
Nine Cambridge Center, Cambridge, MA 02142

We develop principled methods for the automatic induction (discovery) of genetic regulatory network models from multiple data sources and data modalities. Models of regulatory networks are represented as Bayesian networks, allowing the models to compactly and robustly capture probabilistic multivariate statistical dependencies between the various cellular factors in these networks. We build on previous Bayesian network validation results by extending the validation framework to the context of model induction, leveraging heuristic simulated annealing search algorithms and posterior model averaging. Using expression data in isolation yields results inconsistent with location data so we incorporate genomic location data to guide the model induction process. We combine these two data modalities by allowing location data to influence the model prior and expression data to influence the model likelihood. We demonstrate the utility of this approach by discovering genetic regulatory models of thirty-three variables involved in *S. cerevisiae* pheromone response. The models we automatically generate are consistent with the current understanding regarding this regulatory network, but also suggest new directions for future experimental investigation.

1 Introduction

While genomic expression data has proven tremendously useful in providing insights into cellular regulatory networks, other valuable sources of data are increasingly becoming available to aid in this process. The wide range of data modalities presents a significant challenge, but also an opportunity since principled fusion of these diverse information sources will help reveal synergistic insights not readily apparent when sources are examined individually. We approach the information fusion challenge by developing principled methods for the automatic induction (discovery) of genetic regulatory network models from both genomic location and expression data. In our modeling framework, models of regulatory networks are represented as Bayesian networks, allowing the models to compactly and robustly capture probabilistic multivariate statistical dependencies between the various cellular factors in these networks.[1,2,3,4] Here we extend our previously developed model validation framework based on Bayesian networks[1] to the context of model induction. We discover models by using a heuristic search algorithm based on simulated annealing to visit high-scoring regions of the model posterior and then using posterior model averaging to compute likely statistical dependencies between model variables. We combine genomic location and expression data to guide the model induction process by permitting the former to influence the model prior and the latter the model likelihood.

In this paper, we apply our methodology to examine the regulatory network responsible for controlling the expression of various genes that code for proteins involved in *Saccharomyces cerevisiae* pheromone response pathways. The protein Ste12 is the ultimate target of the pheromone response signaling pathway and binds DNA as a transcriptional activator for a number of other genes. Data from genomic location analysis indicates which intergenic regions in the yeast genome are bound by Ste12, both in the presence and absence of pheromone.[5] Because pheromone response and mating pathways play an essential role in the sexual reproduction of yeast and because we have access to location data regarding the binding locations of Ste12 within the yeast genome, this is a natural choice of regulatory network to examine.

We begin in Section 2 by considering various model induction methodologies. In Section 3 we discuss the collection and preparation of data for model induction in the context of pheromone response. We present various results of our model induction approach in Section 4, including the impact of using data from genomic location analysis to add edge constraints representing prior information. We conclude in Section 5 with a discussion of the results presented in this paper and offer some directions for future work.

2 Model induction

Methods for the induction of Bayesian network models from observational data generally fall into two classes: constructive methods based on the examination of conditional independence constraints that hold over the empirical probability distributions on the variables represented in the data, and search methods that seek to maximize some scoring function that describes the ability of the network to explain the observed data. We concentrate here exclusively on the latter although recent work[6] suggests that the two methods are equivalent under reasonable assumptions. In a search context, the Bayesian scoring metric (BSM) is an especially common choice for the scoring function, although other choices can be made if the BSM is difficult to compute exactly.

We consider heuristic rather than exhaustive search strategies, since the identification of the highest-scoring model under the BSM for a given set of data is known to be NP-complete.[7] Commonly used local heuristic search algorithms include greedy hill-climbing, greedy random, Metropolis,[8] and simulated annealing; the last three are successive generalizations of one another. We have implemented these search algorithms and have observed in this particular context that simulated annealing consistently finds the highest scoring models among these algorithms. For reasons of limited space and simplified exposition, we therefore concentrate here only on the simulated annealing algorithm and results generated through its use.

The simulated annealing algorithm is so named because it operates in a manner analogous to the physical process of annealing. During the search process, the Metropolis algorithm is run as a subroutine at various temperatures T. The prevailing temperature and the score difference between graphs determine the transition probability within Metropolis, with higher temperatures indicating more permissive transitions. Initially, the temperature is set very high (allowing almost all changes to be made), but is gradually reduced according to some schedule until it reaches zero, at which point the Metropolis subroutine is equivalent to the greedy random algorithm. The schedule that the temperature is constrained to follow can be varied to produce different kinds of search algorithms. The schedule we employ allows for "reannealing" after the temperature becomes sufficiently low.

We extend our simulated annealing algorithm to search for models with constraints specifying which edges are required to be present and which are required to be absent. This allows for the incorporation of prior information about edges in the graph since this kind of constrained search algorithm is equivalent to an unconstrained search algorithm with a nonuniform prior over structures that gives zero weight to models that either include edges required to be absent or do not include edges required to be present. In this way, data from other sources (such as location data) can be easily incorporated.

We do not use our algorithm to isolate a single model because model selection tends to over-fit the data by selecting the single maximum a posteriori model and ignoring completely other models that score nearly as well. A more principled Bayesian approach is to compute probabilities of features of interest by averaging over the posterior model distribution. For example, if we are interested in determining whether the data D supports the inclusion of an edge in graph S between two variables X and Y, we compute:

$$\mathrm{p}(E_{XY}|D) = \sum_S \mathrm{p}(E_{XY}|D,S)\cdot \mathrm{p}(S|D) \tag{1}$$

$$= \sum_S 1_{XY}(S)\cdot \mathrm{e}^{\mathsf{BSM}(S)} \tag{2}$$

where E_{XY} represents an edge from variable X to variable Y, $1_{XY}(S)$ is an indicator function that is 1 if and only if graph S includes E_{XY} as an edge, and $\mathsf{BSM}(S)$ is the Bayesian scoring metric for graph S. However, this sum is difficult to compute because the space of graphs S is enormous. Fortunately, it is possible to approximate this sum since the vast bulk of its mass lies among the highest scoring models.[a] For example, if we restrict our attention to the N highest scoring models, and index these by the variable i, we have:

$$\mathrm{p}(E_{XY}|D) \approx \frac{\sum_{i=1}^{N} 1_{XY}(S_i)\cdot \mathrm{e}^{\mathsf{BSM}(S_i)}}{\sum_{i=1}^{N} \mathrm{e}^{\mathsf{BSM}(S_i)}} \tag{3}$$

Using model averaging in this way reduces the risk of over-fitting the data by considering a multitude of models when computing the probabilities of features of interest.

3 Data preparation

3.1 Expression data

A set of 320 samples of unsynchronized *Saccharomyces cerevisiae* populations of varying genotype were observed under a diversity of experimental conditions. The set of samples ranges widely but consists primarily of observations of various wild-type and mutant strains made under a variety of environmental conditions including exposure to different nutritive media as well as exposure to stresses like heat, oxidative species, excessive acidity, and excessive alkalinity. Whole-genome expression data for each of these 320 observations was collected using four low-density 50-micron Affymetrix Ye6100 GeneChips per observation (roughly a quarter of the genome can be measured on each chip).

[a] The exponential factor in the sum has the effect of drowning out all but the highest scoring models, even though these highest scoring models are relatively infrequent.

The reported "average difference" values from these 1280 Affymetrix GeneChips were normalized using maximum a posteriori normalization methods based on exogenous spiked controls.[9] The output of this process was a 6135×320 matrix of normalized log expression values, one row for each gene in the yeast genome and one column for each experimental observation.

From the 6135 genes of the *S. cerevisiae* genome, 32 were selected either on the basis of their participation in the pheromone response signaling cascade or as being known to affect other aspects of the mating response in yeast. Descriptions of the roles of the genes and proteins that were selected are presented in Table 1 and compiled from information from a variety of sources.[10,11]

Table 1 Descriptions of the 32 genes selected for model induction. The color mnemonics are used later in Figure 2: genes expressed only in MATa cells are magenta, genes expressed only in MATα cells are red, genes whose promoters are bound by Ste12 are blue, genes coding for components of the G-protein complex are green, genes coding for core components of the signaling cascade complex are yellow (except FUS3 which is already blue), genes coding for auxiliary components of the signaling cascade are orange, and genes coding for components of the SWI-SNF complex are aqua.

Gene	Color Mnemonic	Function of Corresponding Protein
STE2	magenta	transmembrane receptor peptide (present only in MATa strains)
STE3	red	transmembrane receptor peptide (present only in MATα strains)
GPA1	green	component of the heterotrimeric G-protein (Gα)
STE4	green	component of the heterotrimeric G-protein (Gβ)
STE18	green	component of the heterotrimeric G-protein (Gγ)
FUS3	blue	mitogen-activated protein kinase (MAPK)
STE7	yellow	MAPK kinase (MAPKK)
STE11	yellow	MAPKK kinase (MAPKKK)
STE5	yellow	scaffolding peptide holding together Fus3, Ste7, and Ste11 in a large complex
STE12	blue	transcriptional activator
KSS1	orange	alternative MAPK for pheromone response (in some dispute)
STE20	orange	p21-activated protein kinase (PAK)
STE50	orange	unknown function but necessary for proper function of Ste11
MFA1	magenta	a-factor mating pheromone (present only in MATa strains)
MFA2	magenta	a-factor mating pheromone (present only in MATa strains)
MFALPHA1	red	α-factor mating pheromone (present only in MATα strains)
MFALPHA2	red	α-factor mating pheromone (present only in MATα strains)
STE6	magenta	responsible for the export of a-factor from MATa cells (present only in MATa strains)
FAR1	blue	substrate of Fus3 that leads to G1 arrest; known to bind to STE4 as part of complex of proteins necessary for establishing cell polarity required for shmoo formation after mating signal has been received
FUS1	blue	required for cell fusion during mating
AGA1	blue	anchor subunit of a-agglutinin complex; mediates attachment of Aga2 to cell surface
AGA2	magenta	binding subunit of a-agglutinin complex; involved in cell-cell adhesion during mating by binding Sag1 (present only in MATa strains)
SAG1	red	binding subunit of α-agglutinin complex; involved in cell-cell adhesion during mating by binding Aga2 (present only in MATα strains; also known as Agα1)
BAR1	magenta	protease degrading α-factor (present only in MATa strains)
SST2		involved in desensitization to mating pheromone exposure
KAR3		essential for nuclear migration step of karyogamy
TEC1		transcriptional activator believed to bind cooperatively with Ste12 (more active during induction of filamentous or invasive growth response)
MCM1		transcription factor believed to bind cooperatively with Ste12 (more active during induction of pheromone response)
SIN3		implicated in induction or repression of numerous genes in pheromone response pathway
TUP1		implicated in repression of numerous genes in pheromone response pathway
SNF2	aqua	implicated in induction of numerous genes in pheromone response pathway (component of SWI-SNF global transcription activator complex)
SWI1	aqua	implicated in induction of numerous genes in pheromone response pathway (component of SWI-SNF global transcription activator complex)

The normalized levels of expression for these 32 genes were extracted from the 6135 $\times$ 320 normalization output matrix to yield a matrix of data with 32 rows and 320 columns, one row for each gene and one column for each observation. This data was then log-transformed and discretized using discretization level coalescence methods which incrementally reduce the number of discretization levels for each gene while preserving as much total mutual information between genes as possible.[2] In this case, each gene was discretized to have four levels of discretization while preserving over 98% of the original total mutual information between pairs of genes.

In addition to the 32 variables representing levels of gene expression, an additional variable named `mating_type` was considered. The variable `mating_type` represents the mating type of the various haploid strains of yeast used in the 320 observations and can take one of two values, corresponding to the MATa and MATα mating types of yeast. The inclusion of this variable is necessary because, *e.g.*, the MFA1 and MFA2 genes responsible for producing the mating pheromone a-factor are expressed only in MATa strains of yeast. The data used as input for model induction was thus a matrix of 33 rows and 320 columns, 32 rows representing the discretized levels of log expression for 32 genes involved in pheromone response and one row representing the mating type of the strain in each experiment, either MATa or MATα.

3.2 Location data

Data from genomic location analysis, gathered using a chromatin immunoprecipitation assay, revealed the genes in the yeast genome whose upstream regions were bound by Ste12 under both presence and absence of pheromone. Of the 32 pheromone response genes in this paper, STE12, FUS1, FUS3, AGA1, and FAR1 promoters are all bound by Ste12, the first three being bound significantly both before and after the addition of pheromone, and the latter two being bound significantly only after the addition of pheromone. A description of the assay and a more detailed presentation of the results can be found in the paper by Ren, *et al.*[5]

4 Model averaging results

The implementation of our search algorithm is written in C and is capable of searching about 200,000-250,000 (not necessarily unique) models per minute on a 400MHz Pentium II Linux workstation. Although the code keeps a small hash-table of the scores of recently visited models, it is not especially optimized and could likely be sped up.

We used our search implementation to visit high-scoring regions of the model posterior and present the results of two of those runs here. In the first run, we traversed the model space without constraints on the graph edges. In the second run, we incorporated the available location data by requiring edges from STE12 to FUS1, FUS3, AGA1, and FAR1. The top left and right histograms in Figure 1 show the distributions of scores for all models visited during the unconstrained and constrained simulated annealing runs, respectively. For comparison, the bottom histogram in Figure 1 shows the distribution of scores for all models visited when we perform a lengthy random walk through the space of models, accepting every proposed local change (equivalent to infinite-temperature Metropolis). From this figure, we see that the simulated annealing algorithm is quite effective in gradually concentrating its efforts on extremely high scoring models.

After gathering the five hundred highest scoring models that were visited during each run of the search algorithm, we computed the probability of edges being present by using the weighted average approximation shown in Equation 3 (with $N = 500$). Results of this computation for the unconstrained and constrained searches are presented in Tables 2 and 3, respectively. The estimated probability of an edge can be exactly 1 if (and only if) the edge appears in all 500 highest scoring models.

We then compiled a composite network for each of these that consists of all edges with estimated posterior probability over 0.5. These networks are shown in Figure 2. Graph nodes have been augmented with color information to indicate the different groups of variables with known relationships in the literature, as indicated in Table 1 and below. Graph edges have also been augmented with color information: solid black edges have posterior probability of 1, solid blue edges have probability between 1 and 0.99, dashed blue edges have probability between 0.99 and 0.75, and dotted blue edges have probability between 0.75 and 0.5. The strength of an edge does not indicate how *significantly* a parent node contributes to the ability to explain the child node but rather an approximate measure of how *likely* a parent node is to contribute to the ability to explain the child node.

In both of the networks presented in Figures 2, we observe a number of interesting properties. In each case, the `mating_type` variable is at the root of the graph, and contributes to the ability to predict the state of a large number of variables, which is to be expected. The links are generally quite strong indicating that their presence was fairly consistent among the 500 highest scoring models. Almost all the links between `mating_type` and genes known to be expressed only in MATa or MATα strains occur with posterior probability above 0.99. Moreover, in both networks there exists a

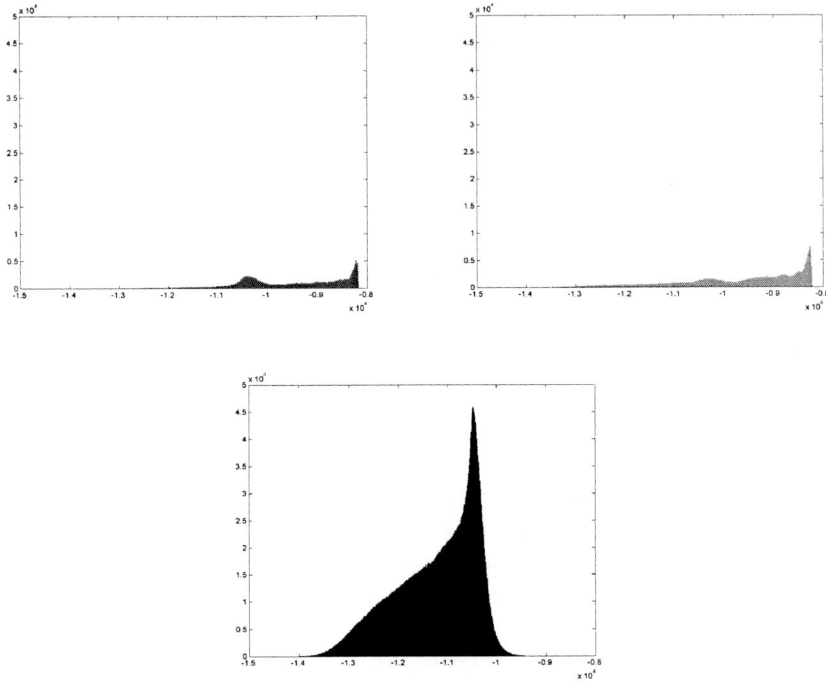

Figure 1. Histograms of scores for all models visited during simulated annealing runs (frequency versus log posterior probability of model). The top left and right histograms are for the unconstrained and constrained simulated annealing runs, respectively. For comparison, the bottom histogram was generated by a random walk through the space of models, accepting every proposed local change.

directly-connected subgraph consisting of genes expressed only in MATa cells (magenta) and a directly-connected subgraph consisting of genes expressed only in MATα cells (red). In each case the subgraph has the `mating_type` variable as a direct ancestor with strong predictive power, as expected.

The heterotrimeric G-protein complex components GPA1, STE4, and STE18 (green) form a directly-connected component in the constrained graph but only GPA1 and STE18 are connected in the unconstrained graph. Indeed, even the link between GPA1 and STE4 in the constrained graph is fairly weak. On the other hand, SWI1 and SNF2 (aqua) are weakly adjacent in the unconstrained graph, but not adjacent in the constrained graph, though in both cases they are close descendants of TUP1. STE11 and STE5, two of

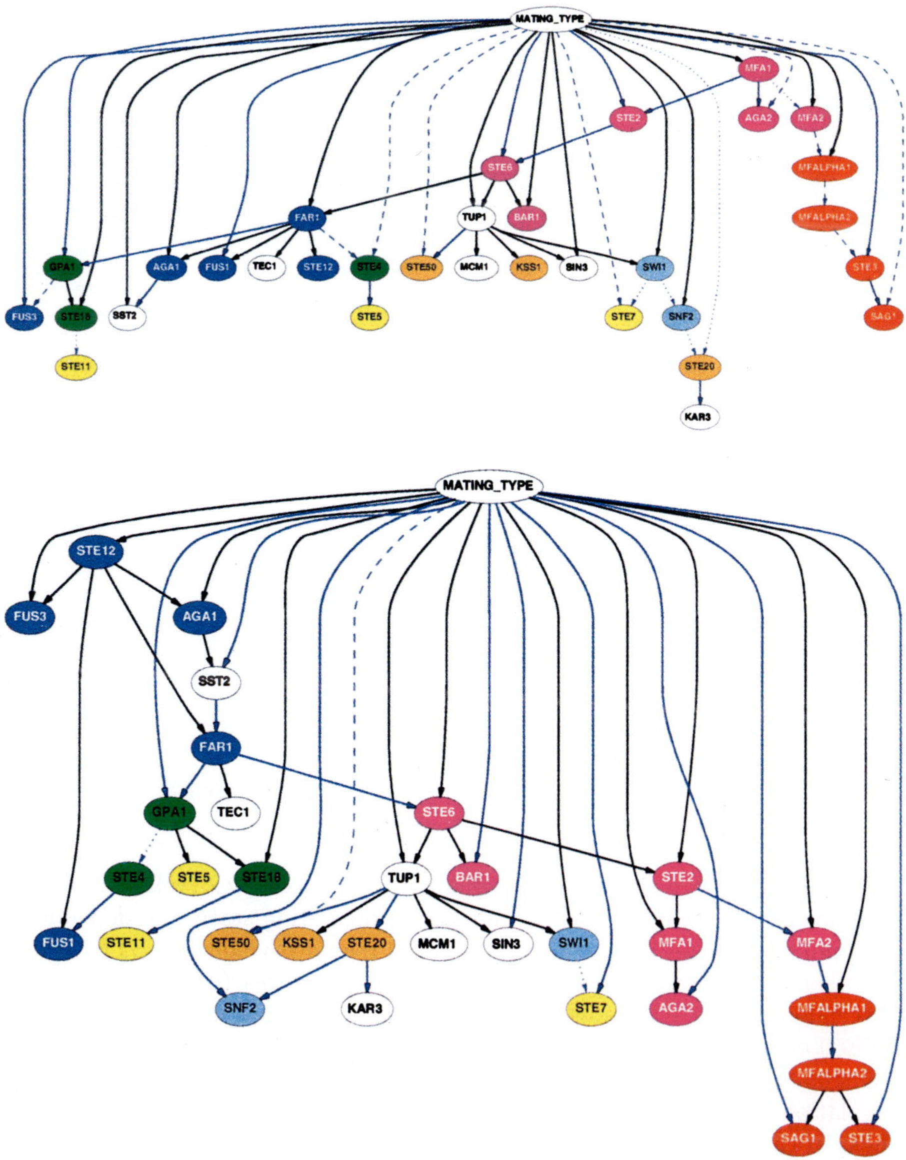

Figure 2. Bayesian network models learned by model averaging over the 500 highest scoring models visited during the unconstrained and constrained simulated annealing search runs, respectively. Edges are included in the figure if and only if their posterior probability exceeds 0.5. Node and edge color descriptions are included in the text.

Table 2. Posterior probabilities of edges being present in the unconstrained search as estimated by a weighted average over the 500 highest scoring models.

From	To	Posterior Probability	From	To	Posterior Probability
MATING_TYPE	MFA1	1.0000000	MATING_TYPE	STE4	0.9310940
STE6	TUP1	1.0000000	MATING_TYPE	SAG1	0.9191640
MATING_TYPE	TUP1	1.0000000	MATING_TYPE	STE50	0.8662340
FAR1	FUS1	1.0000000	MFA1	MFA2	0.6304400
TUP1	SWI1	1.0000000	SWI1	STE7	0.6292930
MATING_TYPE	SWI1	1.0000000	STE18	STE11	0.6159960
MATING_TYPE	MFALPHA1	1.0000000	SWI1	SNF2	0.5659030
MATING_TYPE	SNF2	1.0000000	MATING_TYPE	STE20	0.5551610
TUP1	SIN3	1.0000000	SNF2	STE20	0.5533950
MATING_TYPE	SIN3	1.0000000	STE20	SNF2	0.4340970
FAR1	AGA1	1.0000000	TUP1	STE20	0.4154630
MATING_TYPE	AGA1	1.0000000	STE50	STE11	0.3824810
MATING_TYPE	MFA2	1.0000000	STE2	MFA2	0.3695600
TUP1	MCM1	1.0000000	MATING_TYPE	STE12	0.2741810
MATING_TYPE	SST2	1.0000000	MATING_TYPE	KSS1	0.1964160
FAR1	TEC1	1.0000000	STE5	STE7	0.1710230
GPA1	STE18	1.0000000	STE50	STE7	0.1708470
MATING_TYPE	STE18	1.0000000	MATING_TYPE	MFALPHA2	0.0708669
STE6	FAR1	1.0000000	MFALPHA2	MFALPHA1	0.0596000
MATING_TYPE	FAR1	1.0000000	STE3	MFALPHA2	0.0596000
STE6	BAR1	1.0000000	GPA1	STE4	0.0594723
MATING_TYPE	BAR1	1.0000000	MFALPHA1	MFA1	0.0569844
FAR1	STE12	1.0000000	MFA1	FUS3	0.0339687
TUP1	KSS1	1.0000000	SWI1	STE20	0.0311423
MFA1	STE2	0.9998720	FUS3	MFA1	0.0255037
STE4	STE5	0.9998720	STE11	STE7	0.0090948
AGA1	SST2	0.9998720	MATING_TYPE	STE11	0.0017818
STE2	STE6	0.9998720	STE2	MFALPHA1	0.0016013
STE3	SAG1	0.9998720	SNF2	STE11	0.0015236
FAR1	GPA1	0.9998720	MFA1	MFALPHA1	0.0005666
MFA1	AGA2	0.9998720	AGA2	MFALPHA1	0.0003554
TUP1	STE50	0.9997690	STE11	KAR3	0.0002888
STE20	KAR3	0.9997110	STE6	STE2	0.0001277
MATING_TYPE	STE3	0.9995500	AGA2	MFA1	0.0001277
MATING_TYPE	GPA1	0.9994340	GPA1	STE5	0.0001277
MATING_TYPE	STE2	0.9979080	FAR1	SST2	0.0001277
MATING_TYPE	FUS3	0.9965830	SAG1	STE3	0.0001277
MATING_TYPE	STE6	0.9955280	SST2	SAG1	0.0001277
MATING_TYPE	FUS1	0.9937100	STE4	GPA1	0.0001277
MATING_TYPE	AGA2	0.9850980	STE2	AGA2	0.0001277
MATING_TYPE	STE7	0.9753010	AGA2	FUS3	0.0001277
FAR1	STE4	0.9405280	STE20	STE50	0.0001235
MFALPHA2	STE3	0.9404000	STE12	FUS3	0.0000620
MFALPHA1	MFALPHA2	0.9404000	STE18	STE50	0.0000619
GPA1	FUS3	0.9404000	MATING_TYPE	TEC1	0.0000519
MFA2	MFALPHA1	0.9349580			

the core elements of the primary signaling cascade complex (yellow), are seen as descendants of G-protein complex genes, indicating statistical dependence that may be the result of common or serial regulatory control. STE7 occurs elsewhere, however. Auxiliary signaling cascade genes (orange) are always descendants of TUP1, sometimes directly and sometimes more indirectly, but STE50 and KSS1 are siblings in both cases. In general, the auxiliary cascade elements do not tend to cluster with the core elements, suggesting that the regulation of their transcript levels may occur by a different mechanism than those of the genes in the core signal transduction complex.

In both networks, TUP1 appears with a large number of children, consistent with its role as a general repressor of RNA polymerase II transcription. Both networks have MCM1 and SIN3 as children of TUP1; Tup1 and Mcm1

Table 3. Posterior probabilities of edges being present in the constrained search as estimated by a weighted average over the 500 highest scoring models. As the four edges required by location analysis appear in all visited graphs, their posterior probability is 1 by definition.

From	To	Posterior Probability	From	To	Posterior Probability
STE6	STE2	1.0000000	MATING_TYPE	GPA1	0.9979470
MATING_TYPE	STE2	1.0000000	MATING_TYPE	STE7	0.9959610
STE2	MFA1	1.0000000	SST2	FAR1	0.9956230
MATING_TYPE	MFA1	1.0000000	MATING_TYPE	AGA2	0.9915840
GPA1	STE5	1.0000000	MFA2	MFALPHA1	0.9911710
STE6	TUP1	1.0000000	MATING_TYPE	STE50	0.8627090
MATING_TYPE	TUP1	1.0000000	GPA1	STE4	0.7190930
STE12	FUS1	1.0000000	SWI1	STE7	0.6980530
TUP1	SWI1	1.0000000	MFA1	FUS3	0.3288450
MATING_TYPE	SWI1	1.0000000	FAR1	STE4	0.2809070
MATING_TYPE	MFALPHA1	1.0000000	MATING_TYPE	STE4	0.2809060
TUP1	SIN3	1.0000000	MATING_TYPE	KSS1	0.1904580
STE12	AGA1	1.0000000	STE50	STE7	0.1808790
MATING_TYPE	AGA1	1.0000000	AGA2	FUS3	0.1517050
MATING_TYPE	MFA2	1.0000000	STE5	STE7	0.0452417
TUP1	MCM1	1.0000000	STE11	STE7	0.0114721
AGA1	SST2	1.0000000	MATING_TYPE	MFALPHA2	0.0081962
MFALPHA2	STE3	1.0000000	MFA1	MFALPHA1	0.0044375
MATING_TYPE	STE6	1.0000000	MATING_TYPE	FAR1	0.0043766
FAR1	TEC1	1.0000000	STE2	MFALPHA1	0.0024661
GPA1	STE18	1.0000000	MATING_TYPE	STE5	0.0008114
MATING_TYPE	STE18	1.0000000	STE18	STE50	0.0004177
MFALPHA2	SAG1	1.0000000	AGA2	MFALPHA1	0.0003914
STE12	FAR1	1.0000000	MATING_TYPE	KAR3	0.0003614
STE6	BAR1	1.0000000	STE20	STE50	0.0003008
MATING_TYPE	STE12	1.0000000	SWI1	SNF2	0.0002817
TUP1	KSS1	1.0000000	SNF2	STE20	0.0002817
MFA1	AGA2	1.0000000	MATING_TYPE	STE20	0.0001019
STE12	FUS3	1.0000000	STE11	KAR3	0.0000791
MATING_TYPE	FUS3	1.0000000	STE6	MFALPHA1	0.0000523
MATING_TYPE	SNF2	0.9999970	STE4	GPA1	0.0000420
FAR1	STE6	0.9999950	MFA1	MFA2	0.0000371
MATING_TYPE	BAR1	0.9999950	STE7	STE50	0.0000239
MFALPHA1	MFALPHA2	0.9999950	MATING_TYPE	STE11	0.0000221
STE4	FUS1	0.9999860	MATING_TYPE	TEC1	0.0000164
MATING_TYPE	SIN3	0.9999840	STE50	STE11	0.0000148
STE18	STE11	0.9999790	FAR1	FUS1	0.0000138
STE2	MFA2	0.9999630	SNF2	STE7	0.0000085
FAR1	GPA1	0.9999580	MFALPHA2	MFALPHA1	0.0000050
MATING_TYPE	STE3	0.9998550	GPA1	STE6	0.0000050
STE20	SNF2	0.9997180	MFA2	MFALPHA2	0.0000050
TUP1	STE20	0.9997180	STE18	KAR3	0.0000038
STE20	KAR3	0.9995980	STE11	STE50	0.0000024
MATING_TYPE	SAG1	0.9994460	SNF2	STE11	0.0000020
TUP1	STE50	0.9983870	MFALPHA2	STE50	0.0000018
MATING_TYPE	SST2	0.9979600			

are known to interact in the cell[12] and this result that the level of Tup1 is helpful in predicting the level of Mcm1 suggests a possible regulatory relationship between the two. FAR1 is a parent of TEC1 and GPA1 in both networks. Far1, Tec1, and Gpa1 are all known to be cell-cycle regulated and all three are classified as being transcribed during early G_1 phase.[13] This result suggests that Far1 may play a role in regulating the expression of Tec1 and Gpa1, providing a possible mechanism for their previously observed G_1 phase co-expression.

Though it is produced at higher levels in MATa cells, it is known that Aga1 is produced in both MATa and MATα cells.[14] The graphs are each consistent with this knowledge, including a frequent predictive edge from `mating_type` to AGA1, but not clustering AGA1 with other mating type specific genes (magenta and red) as it is likely regulated differently. In both graphs, AGA1 and SST2 are adjacent, consistent with the fact that the two are expressed very similarly, both peaking at the M/G_1 phase of the cell-cycle.[15]

5 Discussion

When we interpret automatically generated Bayesian networks, it should be remembered that edges indicate a statistical dependence between the transcript levels of genes, but do not necessarily specify the form or presence of a physical dependence. For example, a variable X can seem to be influencing a variable Z if a critical intermediating variable Y remains unmodeled. As another example, in both networks in Figure 2, a link appears between MFA2 and MFALPHA1. even though these mating factors are never both expressed in haploid *S. cerevisiae* strains. However, cells expressing one are less likely, statistically, to be expressing the other; hence the link. The weakness of the link indicates that other variables such as `mating_type` are frequently successful in explaining away this statistical dependence. In general, multiple biological mechanisms may map to the same set of statistical dependencies and thus be hard to distinguish on the basis of statistical tests alone. Moreover, if there is not sufficient data to observe a system in a numberof different configurations, we may not be able to uncover certain dependencies at all.

The composite network resulting from unconstrained search based only on genomic expression data has a few apparent limitations. Most strikingly, the search method is unable from expression data alone to learn the correct regulatory relationships between Ste12 and its targets. By fusing expression data with location data, the constrained search is able to consider statistical dependencies in the expression data that are consistent with the physical relationships already identified in the location data. In this way, location data proves to be quite complementary to expression data: since it can help identify network edges directly, location data dramatically decreases the amount of expression data needed to discover regulatory networks by statistical methods.

When genomic location data suggests that particular edges should be present, our algorithms currently modify the model prior so that graphs lacking these suggested edges have zero weight. However, we know that location data can be noisy. We can relax our assumption of zero weight, and instead modify the model prior so that graphs lacking these suggested edges have small but positive weight. This is permissible within our framework but adds the extra complication that the relative weight of models lacking suggested edges needs to be specified (presumably based on the degree of confidence in the location data). Values for this weight that are too high or low lead to the under- or over-inclusion of suggested edges, respectively.

Additionally, it is possible that a protein may bind DNA but have no impact on downstream gene regulation. Location data provides information about physical relationships while expression data provides information about statistical relationships; the two are not guaranteed to agree.

There remain a number of ways to extend this work in the future. Among these are the use of search algorithms that more frequently visit high scoring regions of the model search space, incorporation of data from other sources besides expression and location data, leveraging time-series data and dynamic Bayesian networks to model feedback processes, leveraging interventional data to uncover causal processes, and adding the ability to discover annotated network edges refining the type of relationship learned between model variables.[1]

Acknowledgments

The authors wish to thank Tomi Silander for access to B-Course source code and anonymous reviewers for their helpful comments. In addition, Hartemink gratefully acknowledges support through the Merck/MIT Graduate Fellowship in Informatics.

References

1. A. J. Hartemink, *et al.* In *Pac. Symp. Biocomp.*, 6:422–433, 2001.
2. A. J. Hartemink. *Principled Computational Methods for the Validation and Discovery of Genetic Regulatory Networks.* PhD thesis, MIT, 2001.
3. N. Friedman, *et al.* In *RECOMB 2000.* ACM-SIGACT, Apr. 2000.
4. D. Pe'er, *et al.* In *ISMB 2001.* ISCB, Jul. 2001.
5. B. Ren, *et al.* *Science*, 290(5500):2306–2309, Dec. 2000.
6. R. G. Cowell. In *UAI 2001.* Morgan Kauffman, Jul. 2001.
7. D. M. Chickering. In D. Fisher and H.-J. Lenz, eds, *Learning from Data: AI and Statistics V*, chap. 12, 121–130. Springer-Verlag, 1996.
8. N. Metropolis, *et al.* *J. Chem. Phys.*, 21:1087–1091, 1953.
9. A. J. Hartemink, *et al.* In *BiOS 2001*, 132–140. SPIE, Jan. 2001.
10. E. A. Elion. *Curr. Opin. Microbiology*, 3(6):573–581, Dec. 2000.
11. M. Costanzo, *et al.* *Nuc. Acids Res.*, 29(1):75–79, 2001.
12. I. Gavin, M. Kladde, and R. Simpson. *Embo J.*, 19:5875–5883, 2000.
13. R. J. Cho, *et al.* *Mol. Cell*, 2:65–73, Jul. 1998.
14. A. Roy, *et al.* *Mol. Cell. Biol.*, 11(8):4196–4206, Aug. 1991.
15. P. T. Spellman, *et al.* *Mol. Biol. Cell*, 9:3273–3297, 1998.

THE ERATO SYSTEMS BIOLOGY WORKBENCH: ENABLING INTERACTION AND EXCHANGE BETWEEN SOFTWARE TOOLS FOR COMPUTATIONAL BIOLOGY

M. HUCKA[1,2], A. FINNEY[1,2], H. M. SAURO[1,2] H. BOLOURI[1,2,3,4], J. DOYLE[1,2], H. KITANO[1,2,5]

[1] *ERATO Kitano Symbiotic Systems Project, M-31 Suite 6A 6-31-15 Jingumae Shibuya-ku, Tokyo 150-0001, Japan*

[2] *Control and Dynamical Systems 107-81, California Institute of Technology, CA 91125, USA*

[3] *Science and Technology Research Centre, University of Hertfordshire, AL10 9AB, UK*

[4] *Division of Biology 216-76, California Institute of Technology, CA91125, USA*

[5] *Sony Computer Science Laboratories, 3-14-13 Higashi-gotanda Shinagawa-ku, Tokyo 141-0022, Japan*

Researchers in computational biology today make use of a large number of different software packages for modeling, analysis, and data manipulation and visualization. In this paper, we describe the ERATO Systems Biology Workbench (SBW), a software framework that allows these heterogeneous application components—written in diverse programming languages and running on different platforms—to communicate and use each others' data and algorithmic capabilities. Our goal is to create a simple, open-source software infrastructure which is effective, easy to implement and easy to understand. SBW uses a broker-based architecture and enables applications (potentially running on separate, distributed computers) to communicate via a simple network protocol. The interfaces to the system are encapsulated in client-side libraries that we provide for different programming languages. We describe the SBW architecture and the current set of modules, as well as alternative implementation technologies.

1 Introduction

The ERATO Systems Biology Workbench (SBW) is a framework for allowing both legacy and new application resources to share data and algorithmic capabilities. Our target audience is the computational biology community whose interest lies in simulation and numerical analysis of biological systems. Our work has been motivated by the desire to achieve interoperability between a set of tools developed by our collaborators: *BioSpice* [1], *DBsolve* [2], *E-Cell* [3], *Gepasi* [4], *ProMoT/DIVA* [5], *Jarnac* [6], *StochSim* [7], and *Virtual Cell* [8]. Since these applications are written in a variety of languages and run on a variety of platforms, it was essential not to limit integration capabilities to resources implemented in a single language or platform. SBW allows communication between processes potentially located across a network on different hardware and operating systems. SBW currently has bindings to C, C++, Java, Delphi and Python, with more planned for in the future, and it is portable to both Windows and Linux.

We are aware that our target community is largely not composed of professional software programmers. Any software development carried out in this community tends to be secondary to the main research effort. As a result, we have endeavored to make integration of software components into SBW as straightforward as possible. SBW is also an open-source framework, to allow the community to evolve and grow SBW with their changing needs. In addition, many laboratories have budgetary constraints. Unlike the licensing terms of a number of other frameworks, our use of an open-source license (GNU Lesser General Public License, LGPL [9]) guarantees that SBW will remain available at no cost indefinitely, while simultaneously allowing developers the freedom to release closed-source modules that work with SBW.

SBW does not attempt to be more than a mechanism to enable the integration of applications. The architecture of SBW does not exclude its integration with other frameworks and integration technologies. In fact, we hope to integrate other frameworks to extend the functionality available to users and developers. In many configurations, SBW will be a small component in a larger system (size being quantified as CPU, disk and memory usage).

We begin by describing SBW from the user's perspective in Section 2, then from the developer's perspective in Section 3. In Section 4, we go on to compare its features to those of other tools for building interoperable software. Finally, in Section 5 we describe the various modules made available in the initial beta release of SBW in November 2001.

2 SBW from the User's Perspective

When an application has been modified to interact with SBW, we describe it as being *SBW-enabled*. This means the application can interact with other SBW-enabled applications. The kinds of possible interactions depend on the facilities that have been exposed to SBW by the applications' programmers. Typical SBW-enabled applications also provide for ways of exchanging models and data using an XML-based common representation format, the Systems Biology Markup Language (SBML) [10]. SBW is not a controller in the system—the flow of control is entirely determined by what the individual modules and the user do. SBW doesn't define any particular type of user interaction with modules: the user can control modules from either script language interpreter, GUIs or some hybrid of GUI and interpreter. The interpreter approach is inevitably more flexible enabling access to all modules in the SBW environment in a single application environment. In the remainder of this section, we present a scenario where the user is controlling events using via GUIs only.

A user will typically start up the first SBW-enabled application as they would any other program. The user doesn't need to do anything specific to start SBW itself. Figure 1 shows an example of using a collection of SBW-enabled software modules. The upper left-hand area in the figure (partly covered by other windows) shows an SBW-enabled version of JDesigner[11], a visual biochemical network layout tool. This module's appearance is nearly identical to that of its original non-SBW-enabled counterpart, except for the presence of a new item in the menu bar called **SBW**. This is typical of SBW-enabled programs: the SBW approach strives to be minimally intrusive. In this example, the user has created a network model in JDesigner, then has decided to run a time-series simulation of the model. To do this, the user has pulled down the **SBW** menu and selected one of the options listed, *Jarnac Analysis*, to invoke the SBW-enabled simulation program *Jarnac*[6]. This has brought forth a control GUI, shown underneath the plot window in the lower right-hand area of Figure 1; the user has then input the necessary parameters into the control GUI to set up the time-series simulation, and has finally clicked the **Run** button in the GUI to start the simulation.

In this example, the control GUI used SBW calls to instruct the simulation module (Jarnac) to run with the given parameters and send the results back to the controlling GUI module, which then sent the results to a plotting module. This example scenario illustrates the interactions involved in using SBW and four modules: the visual JDesigner, the computational module Jarnac, a time-series simulation control GUI, and a plotting module.

3 SBW from the Developer's Perspective

SBW uses a *broker-based, message-passing architecture* that allows dynamic extensibility and configurability. As mentioned above, software modules in SBW can interact with each other as peers in the overall framework. Modules are started on demand through user requests or program commands. Modules are executables which have their own event loops. All remote calls run in their own threads. As shown in Fig. 2, interactions are mediated through the *SBW Broker*, a small program running on a user's computer; the Broker enables locating and starting other modules and establishing communications links

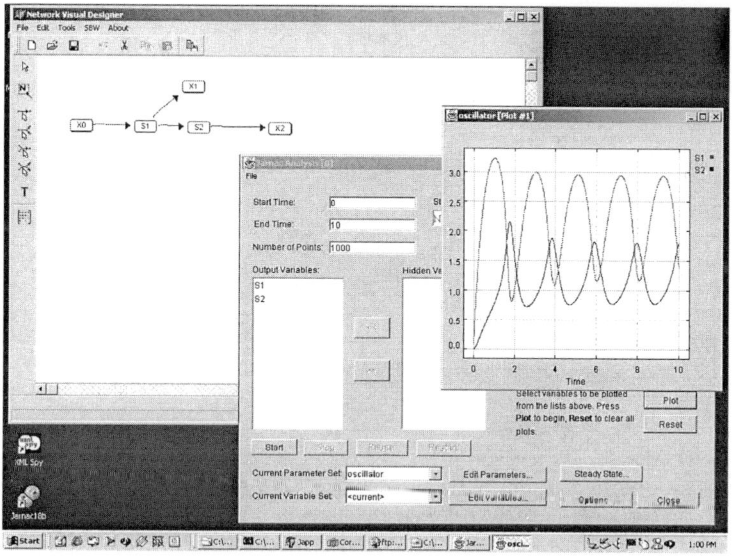

Figure 1. Example of applications interacting through SBW.

between them. Communications are implemented using a fast, lightweight system with a straightforward programming interface.

Broker-based architectures are a common software pattern [12]. They are a means of structuring a distributed software system with decoupled components that interact by remote service invocations. In SBW, the remote service invocations are implemented using *message passing*, another tried and proven approach [13,14]. Because interactions in a message-passing framework are defined at the level of messages and protocols for their exchange, it is easier to make the framework neutral with respect to implementation languages and platforms: modules can be written in any language, as long as they can send, receive and process appropriately-structured messages using agreed-upon conventions. The *dynamic extensibility and configurability* quality of SBW is that components—i.e., SBW modules—can be easily exchanged, added or removed, even at run-time, under user or program control.

From the application programmer's point of view, it is preferable to isolate communications details from application details. For this reason, we provide an

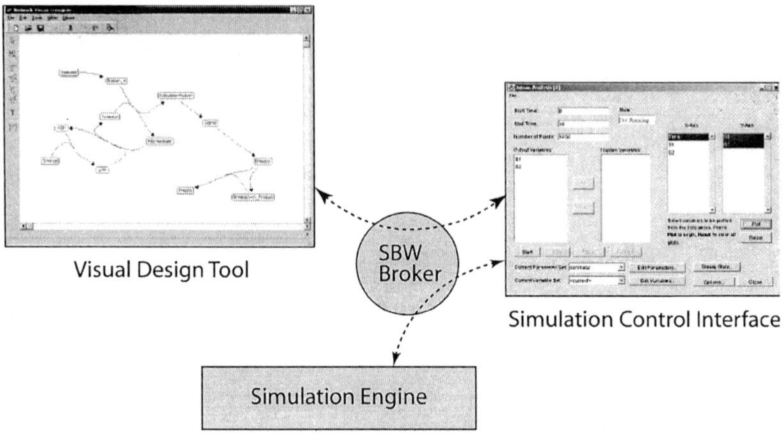

Figure 2. Illustration of the relationship between the SBW Broker and SBW modules. At the API level, communications appear to be direct (i.e., module-to-module); however, the implementation passes messages back-and-forth through the SBW Broker.

Application Programming Interface (API)[15] that hides the details of constructing and sending messages and provides ways for methods in an application to be "hooked into" the messaging framework.

We strove to develop an API for SBW that provides a natural and easy-to-use interface in each of the different languages for which we have implemented libraries. By "natural", we mean that it uses a style and features that programmers accustomed to that language would find familiar. For example, in Java, the high-level API is oriented around providing SBW clients with proxy objects whose methods implement the operations that another application exposes through SBW.

An SBW module provides one or more interfaces or *services.* Each service provides one or more methods. Modules register the services they provide with the SBW Broker. The module optionally places each service it provides into a *category.* By convention, a *category* is a group of services from one or more modules that have a common set of methods.

As an example of how simple the high-level API is to use in practice, the following is Java code demonstrating how one might invoke a simulator from a hypothetical module:

```
// Define the interface for the Java compiler.
interface Simulator
{
  void loadSBML(string);
  void setTimeStart(double);
  void setTimeEnd(double);
  void setNumPoints(integer);
  double[] simulate();
}

double[] runSimulation(String modelDefinition, double startTime,
                       double endTime, integer numPoints)
{
  try
  {
    // Start a new instance of the simulator module.
    Module module
      = SBW.getModuleInstance("edu.caltech.simulator");

    // Locate the service we want to call in the module.
    Service srv = module.findServiceByName("simulation");
    Simulator simulator
      = (Simulator) srv.getServiceObject(Simulator.class);

    // Send the model to the simulator and set parameters.
    simulator.loadSBML(modelDefinition);
    simulator.setTimeStart(startTime);
    simulator.setTimeEnd(endTime);
    simulator.setNumPoints(numPoints);

    // Run the simulation and return the result.
    return simulator.simulate()

  } catch (SBWException e) {
    // Handle problems here.
  }
}
```

As the example above shows, using an SBW-enabled resource involves getting a reference to the module that implements a desired service and invoking methods on that service.

4 Comparison to Related Efforts

The idea of creating a framework that enables the integration of disparate software packages is not new. When we began this project, we considered using an existing framework and simply augmenting it with additional facilities. But after examining a number of other options, we were forced to conclude that none of the existing systems provided an adequate combination of simplicity, support for major programming and scripting languages, support for dynamically querying modules for services they offer, support for distributed computing on Windows and Linux (with a clear ability to be ported to other platforms), and free availability of open-source implementations for Windows and Linux.

4.1 Frameworks for Computational Biology

One of the projects most similar to SBW is ISYS [16]. This system provides a generalized platform into which components may be added in whatever combination the user desires. The system provides a bus-based communications framework that allows components to interoperate without direct knowledge of each other, by using a publish-and-subscribe approach in which components place data on the bus and other components can listen for and extract the data when it appears. ISYS components include graphical visualization tools and database access interfaces. This style of interoperability is an alternative to the more direct communications in SBW, but could be used to the same ends.

The main drawbacks of ISYS for our goals is that it is not freely available and distributable. Moreover, it is largely Java-based and does not offer direct support for components written in other languages.

4.2 General-Purpose High-Level Frameworks

In terms of communications frameworks, SBW has many similarities to Java RMI [17] and CORBA [18]. Both of the latter technologies enable a programmer to tie together separate applications potentially running on different computers, and both offer directory services so that modules can dynamically query and discover the services being made available by other modules. Unfortunately, Java RMI is only truly practical when all applications are written in Java, conflicting with our goal of supporting as many languages as possible. Although RMI-over-IIOP [19] is an option, this simply means that the non-Java components would have to use CORBA.

CORBA is the industry standard for broker-based application object integration. We decided against using CORBA as the basis of SBW primarily because of issues of standards compliance, complexity and maintenance. CORBA is a large and complicated standard and has a steep learning curve. We felt it would have been too much to ask of most researchers, whose time is limited and main goals are in developing domain-specific applications, to acquire CORBA development skills. Further, there are no open-source, standard-compliant implementations of CORBA that support sufficiently many languages in the same implementation. The implication is that SBW modules written in different languages would have to interact with CORBA implementations from different open-source projects. We were concerned about the difficulties of managing not only compatibility of different CORBA packages, but also the installation process, user documentation, and long-term maintenance.

Notwithstanding these issues, we are not in principle opposed to using CORBA. Indeed, we plan to design an interface that will provide a CORBA bridge to SBW for those developers who prefer to use this technology.

4.3 Low-Level Communications Frameworks

SBW uses a custom message-passing communications layer with a simple tagged data representation format and a specialized protocol layered on top of TCP/IP sockets. We examined several alternatives before implementing the scheme used in SBW.

Two attractive, recent alternatives were SOAP [20] and XML-RPC [21]. The latter is essentially a much-simplified version of the former; both provide remote procedure calling facilities that use HTTP as the protocol and XML as the message encoding. We performed an in-depth comparison of XML-RPC and SBW's messaging protocols [22] and concluded that XML-RPC and SOAP would not work for the goals of SBW. The HTTP and XML layers impose a performance penalty not present in SBW's simpler protocol and encoding scheme. Further, the HTTP protocol is not bidirectional: HTTP is oriented towards client-server applications in which a client initiates a connection to a server listening on a designated TCP/IP port. The implication of using XML-RPC for SBW is that each module would have to listen on a different TCP/IP port. This would add needless complexity to SBW.

Another alternative for the message-passing functionality in SBW is MPI [14]. We declined using MPI primarily because at this time there does not appear to be a standard Java interface, and because MPI is considerably more complex than the simple message-passing scheme used in SBW. However, MPI remains an option for reimplementing the communications facility in SBW if it proves useful to do so in the future.

5 SBW Modules

In this section, we describe a variety of different modules that we have implemented and released with the SBW beta release in November 2001.

5.1 Inspector Module

The inspector module is a GUI based tool which allows a user to explore the SBW environment. It enables other modules and their services and methods to be inspected. In the future we hope to extend the inspector module to enable individual methods of a module service to be executed, in which case the inspector will provide an excellent tool for testing new modules.

5.2 JDesigner

JDesigner [11], developed by Herbert Sauro, allows users to draw biochemical networks on screen. It can save models in SBML [10] format. We provide an SBW interface to JDesigner which allows other modules connected to SBW to gain access to the functionality of JDesigner. In particular, it is possible for a remote module to request SBML code from JDesigner. In addition, we also provide an interface which allows remote modules to control many details of JDesigner, for example providing the ability to rearrange the network on-screen.

JDesigner has a menu option "SBW" which lists the services registered with the SBW Broker in the "Analysis" category (e.g. the MATLAB Model Generator, described below). JDesigner passes the SBML representing the drawn model to the selected service.

5.3 Network Object Model

The most frequently requested module is some means of parsing and interpreting SBML. SBML is defined in terms of XML and for many developers in our community it is a non-trivial task to code a parser for SBML. We have therefore written a module, called the Network Object Model or NOM with methods that load and generate SBML as well as methods for accessing and modifying the object model constructed from the loaded SBML. The NOM module can be used as a SBML clipboard for moving data between applications.

5.4 Optimization Module

A frequently need in modeling is the ability to fit parameters to a model. This problem is recast as the minimization of some predefined fitting function by adjusting model parameters. We have collaborated with Pedro Mendes to allow user access to the extensive optimization algorithms in Gepasi.

5.5 *Plotting Module*

This module provides a 2D graph plotting service.

5.6 *MATLAB Model Generator*

This translation module creates either ODE or Simulink models for MATLAB[23] from SBML. This modules provides services in the "Analysis" category and thus can be invoked from JDesigner or similar modules. We anticipate integrating the MATLAB application itself as a module in later releases enabling SBW modules to be invoked from the MATLAB command line and scripts.

5.7 *Simulation Control GUI*

The simulation control GUI is a non-scripting interface to a simulation server such as Jarnac. The simulation control GUI service is in the "Analysis" category. The GUI enables users to set up simulation runs, edit parameters or variables and plot the resulting runs. In addition the GUI can also be used to compute the steady state and carry out metabolic control analysis. Any simulator in the "Simulation" category can be controlled from this GUI interface. The Gillespie, Gibson and Jarnac modules (see below) provide services in that category.

5.8 *Gillespie Stochastic Simulator*

The stochastic simulator module is based on the Gillespie algorithm [24]. The code which forms the basis of this module was provided by Baltazar Aguda[25]. Once a model is loaded, the module allows a user to change parameters and variables in addition to graphing of results and collection of run data.

5.9 *Gibson Stochastic Simulator*

This stochastic simulator uses an algorithm[26], developed by Gibson and Bruck, based on the Gillespie algorithm which includes optimizations to reduce simulation run times.

5.10 *Jarnac Simulator*

Jarnac [6] is an ODE-based biochemical network simulator. Simulations are controlled via a scripting language. Services supported by Jarnac include matrix manipulation, time-course simulation, steady-state analysis and metabolic control analysis. Adding an SBW interface to Jarnac permits two types of interaction. In one mode, Jarnac can act as a server for carrying out simulations. This allows users access to the capabilities of Jarnac without having to interact with a scripting interface.

The second mode of operation is from the scripting interface itself. In this mode, the user is able to explore and use SBW modules from a command line. Interaction is achieved by requesting Jarnac to create a Jarnac object interface to the desired module. This allows a user to use a module as if it were part of Jarnac itself.

6 Summary

The SBW is a very flexible and straightforward system to integrate a range of heterogeneous software components written in a variety of languages and running on a variety of platforms. At the time of this writing, we have completed the implementation of the SBW Broker and the libraries that implement the SBW protocol in Delphi, C, C++, and Java. Full documentation of the SBW design is available from the project web site [27]. A beta release of the SBW software and several sample modules was made in November, 2001, and is available from the project web site.

Acknowledgments

This work has been funded by the Japan Science and Technology Corporation under the ERATO Kitano Systems Biology Project. The Systems Biology Workbench has benefitted from the input of many people. We wish to acknowledge in particular the authors of BioSpice, DBsolve, Cellerator, E-Cell, Gepasi, ProMoT/DIVA, StochSim, and Virtual Cell, and the members of the `sysbio` mailing list. We also thank Mark Borisuk, Mineo Morohashi and Tau-Mu Yi for support, comments and advice.

References

1. A. P. Arkin, *Simulac* and *Deduce*, `http://gobi.lbl.gov/~aparkin/Stuff/Software.html` (2001).
2. I. Goryanin, T. C. Hodgman, and E. Selkov, *Bioinformatics* **15**(9), 749–758 (1999).
3. M. Tomita et al., *Bioinformatics* **15**(1), 72–84 (1999).
4. P. Mendes, *Trends in Biochemical Sciences* **22**, 361–363 (1997).
5. M. Ginkel et al., in *Proceedings of the 3rd MATHMOD*, ed. I. Troch and F. Breitenecker, 525–528 (2000).

6. H. M. Sauro in *Animating the Cellular Map*, ed. J-H. S. Hofmeyr, J. M. Rohwer, and J. L. Snoep (Stellenbosch University Press, 2000).
7. D. Bray, C. Firth, N. Le Novère, and T. Shimizu, *StochSim*, `http://www.zoo.cam.ac.uk/comp-cell/StochSim.html` (2001).
8. J. Schaff, B. Slepchenko, and L. M. Loew, in *Methods in Enzymology*, ed. M. Johnson and L. Brand, **321**, 1–23 (Academic Press, 2000).
9. Free Software Foundation, GNU Lesser General Public License, `http://www.gnu.org/copyleft/lesser.html` (1991).
10. M. Hucka, A. Finney, H. M. Sauro, and H. Bolouri, Systems Biology Markup Language (SBML) Level 1, `http://www.cds.caltech.edu/erato` (2001).
11. Herbert S. Sauro, JDesigner: A simple biochemical network designer, `http://members.tripod.co.uk/sauro/biotech.htm` (2001).
12. F. Buschmann et al., *Pattern-Oriented Software Architecture: A System of Patterns* (John Wiley & Sons, 1996).
13. Jim Farley, *Java Distributed Computing* (O'Reilly & Associates, 1998).
14. W. Gropp, E. Lusk, and A. Skjellum, *Using MPI: Portable Parallel Programming with the Message Passing Interface* (MIT Press, 1999).
15. A. Finney, H. Sauro, M. Hucka, and H. Bolouri, Programmer's manual for the Systems Biology Workbench (SBW), `http://www.cds.caltech.edu/erato/sbw/docs/api/` (2001).
16. A. Siepel et al., *Bioinformatics* **17**(1), 83–94 (2001).
17. M. Hughes, M. Shoffner, and D. Hamner, *Java Network Programming* (Manning Publications Co., 1999).
18. OMG, *CORBA* Specification, `http://www.omg.org` (2001).
19. Sun Microsystems, RMI over IIOP, `http://java.sun.com/products/rmi-iiop/` (2001).
20. D. Box et al., Simple Object Access Protocol (SOAP) 1.1: W3C note 08 May 2000, `http://www.w3.org/TR/SOAP/` (2000).
21. D. Winer, XML-RPC, `http://www.xmlrpc.com/spec/` (2001).
22. M. Hucka, A. Finney, H. M. Sauro, and H. Bolouri, A comparison of two alternative message-passing approaches for SBW, `http://www.cds.caltech.edu/erato/sbw/docs/xml-rpc-comparison/` (2001).
23. Mathworks, Matlab, `http://www.mathworks.com/products/matlab/` (2001).
24. D. Gillespie, *J. Comput. Phys.* **22**, 403–434 (1976).
25. Baltazar Aguda, personal communication.
26. M. Gibson and J. Bruck, *J. Phys. Chem.* **104**, 1876–1889 (2000).
27. The ERATO Systems Biology Workbench Development Group, `http://www.cds.caltech.edu/erato/` (2001).

GENOME-WIDE PATHWAY ANALYSIS AND VISUALIZATION USING GENE EXPRESSION DATA

M. P. KURHEKAR, S. ADAK
Bioinformatics Group,
IBM India Research Lab, Block 1, Indian Institute of Technology, Hauz Khas, New Delhi, India 110016
{mkurhekar, asudeshn} @in.ibm.com

S. JHUNJHUNWALA
Department of Biochemical Engineering and Biotechnology, Indian Institute of Technology, Hauz Khas, New Delhi, India 110016

K. RAGHUPATHY
Department of Electrical Engineering, Indian Institute of Technology, Chennai, India

Visualization of results for high performance computing pose special problems due to the complexity and the volume of data these systems manipulate. We present an approach for visualization of c-DNA microarray gene expression data in metabolic and regulatory pathways using multi-resolution animation at different levels of detail. We describe three scoring functions to characterize pathways at the transcriptional level based on gene expression, coregulation and cascade effect. We also assess the significance of each pathway score on the basis of their biological connotation.

1 Introduction

1.1. Microarrays and Gene Expression

DNA and other microarray technologies are on their way to becoming standard tools of modern life sciences research[6]. A DNA microarray experiment shows the expression levels of thousands of genes at a single time point. This has given scientists the ability to get a snapshot view of genome-wide expression patterns. The vast quantity of data generated by genomic expression arrays affords researchers a significant opportunity to transform biology, medicine, and pharmacology using systematic computational methods. The availability of genomic (and eventually proteomic) expression data promises to have a profound impact on the understanding of basic cellular processes, the diagnosis and treatment of disease, and the efficacy of designing and delivering targeted therapeutics. Particularly relevant to these objectives is the development of a deeper understanding of the various mechanisms by which cells control and regulate the transcription of their genes.

1.2. Pathways

Living organisms behave as complex systems that are flexible and adaptive to their surroundings. At the cellular level, organisms function through intricate networks of chemical reactions and interacting molecules. These networks or biochemical pathways may be considered as the wiring diagrams[12] for the complete biological system of an organism. The best characterized among them are metabolic pathways, the biological networks that involve enzymatic reactions of chemical compounds. Regulatory pathways are another class of pathways that represent protein-protein interactions. Pathways are the key to understanding how an organism reacts to perturbations in its environment (e.g. heat shock, chemical or hormone stimulus) or internal changes (e.g. disease, development, etc.)

1.3. Pathways and Gene Expression

With the advent of microarrays, it is hoped that knowledge of genome-wide expression levels will speed up the understanding of biological systems. Microarrays can serve as an efficient tool to study the expression levels of genes involved in pathways. The effect of induction and repression of target genes on metabolic and regulatory pathways is particularly important in drug development. In this paper, we present a method for using genomic expression data to elucidate and visualize the effect of different stimuli on these genetic networks.

Typical automatic analysis of microarray expression data is performed by clustering the expression profiles: using pair-wise measures such as correlation[1,7] and mutual information[2]; using more multivariate methods like principal components[18] and Fourier analysis[20]. Clustering methods are based on the microarray expression data and subsequent efforts are made to correlate clusters with pathways[22]. Several authors have suggested methods for synthesizing pathways using gene expression data: linear models[5], Boolean networks[16] and Bayesian networks[9]. However, it is difficult to evaluate such reversed engineered pathways in view of known metabolic and regulatory pathways. This has led to efforts being made to map reconstructed pathways onto known pathways[8, 22].

In this paper, we present a method for scoring of putative pathways. The scores are defined to measure the impact of gene expression levels from a series of microarray experiments on metabolic and regulatory pathways. We also present an animated visualization technique that allows the user to observe the complex changes that occur in pathways as tracked by the changing expression levels in a series of microarray experiments. All methods have been implemented in a stand-alone JAVA application.

2 Methods

2.1 Expression Data

In this paper, we consider series of microarray experiments, which measure genome-wide expression levels, as observed over time or over increasing levels of different stimuli like temperature, radiation, drug dosage etc. Well-known examples of such series of microarray experiments include:

- *Yeast* – diauxic shift microarray data[4], yeast sporulation data[3], yeast response to various environmental changes[7], yeast cell cycle data[20]
- *E.coli* – heat shock microarray time series[19]
- *Human* – response of human fibroblasts to serum[11]

For each microarray experiment series, let G be the set of genes investigated in a series of T experiments. For each gene $g\chi\ G$, we regard the expression data as a mapping from g to an ordered series of numbers, $X_{t,g}$ ($t=1,2,\ldots,T$). In this, $X_{t,g}$ denotes the expression ratio of the gene g in the t-th microarray experiment of the series. The expression ratio is calculated as:

$$X_{t,g} = L_{t,g} / L_{0,g} \tag{1}$$

where $L_{t,g}$ is the actual expression level for the gene g in the t-th microarray experiment and $L_{0,g}$ is the expression level of the gene g in the reference sample. That means $L_{0,g}$ is the unperturbed case.

2.2 KEGG: The Pathway Database

The KEGG (Kyoto Encyclopedia of Genes and Genomes) database provides a catalog of metabolic and regulatory pathways that may be considered wiring-diagrams of genes and molecules[18]. In addition, it provides up-to-date links from the gene catalogs generated by genome sequencing projects. More than diagrams, the KEGG database also provides direct links from the genes to the gene products (enzymes and other proteins) involved in the biochemical pathways. This feature of KEGG is particularly useful in mapping gene expression data to known metabolic and regulatory pathways.

In the case of a series of microarray experiments, visualizing the course of a pathway is highly informative and essential in understanding how the pathway is affected over the sequence of experiments. While visualization is essential to understanding each individual pathway, it is also necessary to provide the user some indication of the relative importance of the more than 100 different pathways in the KEGG database. In this paper, we describe three kinds of pathway scores, which are based on "activity", "coregulation" and "cascade" effects in pathways.

Method Outline

The methods described in this paper allow scoring and visualization of the putative pathways in the KEGG database according to the gene expression levels in a microarray experiment series. The method can be summarized as follows:

- Given the input
  - Gene expression data from a microarray experiment series
  - Putative pathways of the KEGG database
- Answer the questions
  - Which pathways are most affected during the course of the experiments?
  - What is the nature of the effect? (Details such as which genes in a pathway are most affected, are the genes over-expressed or under-expressed, which reactions are disrupted etc.)
- By providing the output
  - Pathway scores – these quantify "activity", "coregulation", and "cascade" effects in pathways as measured by the gene expression levels from the microarray experimental data.
  - Pathway animated view – these show the effects on individual pathways over the course of a microarray experiment series.

3 Pathway Scoring

The pathway scoring methods described below measure the changes in metabolic and regulatory pathways as indicated by genome-wide gene expression levels. A high level of gene expression indicates that the cell required the particular protein coded by the gene and hence the expression of the gene has been induced. Thus, significant induction in the genes of a pathway shows that the pathway is being used more extensively than at the reference time point. Similarly, significant repression in the genes involved in a pathway shows that the pathway has been de-activated. By measuring the gene expression through a series of microarray experiments, it is thus possible to measure the effect on biochemical pathways as the cell is subjected to different stimuli. In this paper, we describe three kinds of pathway scores which progressively try to capture the complexity of biochemical pathways in living cells:

- The *Activity score* for a pathway gives a summary measure of the extent to which a pathway is perturbed from the reference state. This score will rank those pathways higher in which more genes were over-expressed or under-expressed with reference to reference state.
- The *Coregulation score* gives an indication of co-expression of the genes in a pathway under the given experimental conditions. It assigns higher scores to pathways whose genes show similar patterns of expression.
- The *Cascade score* takes into account the structure of a pathway as well as measuring *activity* and *coregulation.* It gives a measure of the extent to which a metabolic pathway is affected by analyzing the microarray data

along reaction chains. If the first enzyme in a series of reactions is, say, over produced, this should be accompanied by an increase in production of the subsequent enzymes in the reaction chain. A high score is given to such over-expressed or under-expressed chains of reactions.

Since it is important to assess the relative importance of pathways rather than the absolute scores, each type of score is further normalized on a scale of 0-100 as follows: Relative Score = [Score/Max. Score]x100, where Max. Score is the maximum score (of the same type) among all putative pathways in KEGG.

Another important normalization required for the scoring functions is based on the number of enzymes in a pathway for a given organism. For example, in case of a pathway like prostaglandin and leukotriene metabolism is probably defunct in yeast as only two of the enzymes in this pathway are known to be present in yeast and they are entirely disconnected. Such defunct pathways should be differentiated from valid pathways while scoring, and their score should not be given importance. For a metabolic pathway *P*, the "validity factor normalization" with respect to the organism under investigation is defined as follows:

$$VF_{org}(P) = 1, \text{ if } P_{org}/P_{ref} >= 0.3 \qquad (2)$$
$$= P_{org}/P_{ref}, \text{ if } P_{org}/P_{ref} < 0.3, \quad \text{where}$$

P_{org} : number of enzymatic reactions in the organism specific version of *P*

P_{ref} : number of enzymatic reactions present in the reference version of *P* as provided by the KEGG database, it is the unperturbed case.

(Enzymatic reactions are uniquely identified by the substrate-product-enzyme combination). Thus, if only a few enzymes in a particular metabolic pathway are known to exist in an organism, the pathway will be given a low score by discounting the original score by the "validity factor". The threshold of 0.3 was used for P_{org}/P_{ref} in defining the validity factor, as it was found to be empirically suitable.

3.1 Activity Score

Consider a pathway *P* and let the set of genes involved in the pathway be denoted by G_P. The activity score for the pathway *P* with respect to a user-defined threshold η is defined as follows using (1) and (2):

$$\text{Activity Score}(P,\eta) = \text{VF} \wedge \sum_{g \in G_p} \sum_{t=1}^{T} I(g,t) \text{ , where} \qquad (3)$$

$$I(g,t) = 1 \text{ if } X_{t,g} > \eta_1 \text{ or } X_{t,g} < 1/\eta_2\text{; 0 otherwise.}$$

Thus, according to activity score, pathways will be scored higher if there are more genes that are over-expressed above a given threshold value η_1 or under-expressed below a given threshold value $1/\eta_2$. The thresholds represent the minimum fold-deviation, with respect to the reference sample, that is considered meaningful. Similar activity scores have been previously used[21] where the threshold is determined based on the data. However, due to inherent noise in the experimental data, it is difficult to validate and interpret the resulting scores when the threshold is data dependent.

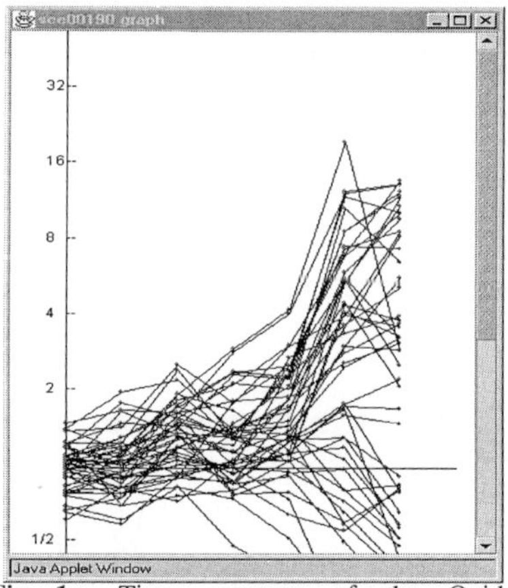

Fig 1a. Time progress of the Oxidative Phosphorylation pathway showing high activity is verified by the activity score. (Diauxic shift)

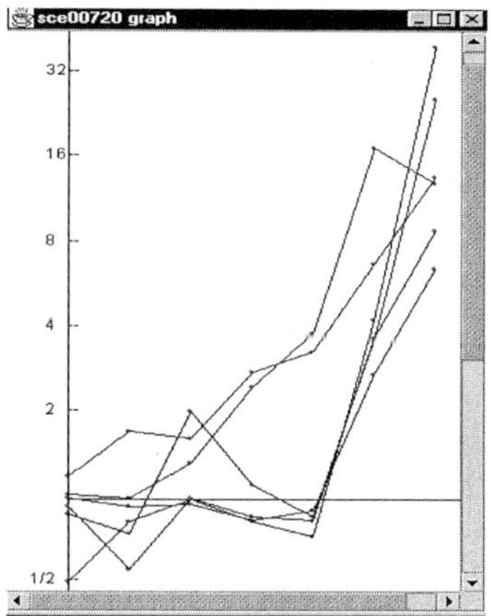

Fig 1b. Time progress of CO_2 fixation pathway shows high coregulation is verified by the coregulation score. (Diauxic Shift)

3.2 Coregulation Score

The activity score merely considers the number of genes that are over-expressed or under-expressed in a pathway, but it does not capture similarity in expression patterns among the genes of a pathway. Coregulation score ranks those pathways higher in which genes show greater similarity in their expression pattern.

Coregulation scores have been previously used[21] which were based on the averages of all pairwise correlations among all genes in a pathway. However, the coregulation score as defined in [21] did not perform very well with experimental data. This is probably because pairwise correlations fail to capture the simultaneous co-expression of all genes in a pathway. We define a slope coregulation score that captures simultaneous coregulation by looking at the variation in the "slopes" among all genes in a pathway.

Consider a pathway P and let the set of genes involved in the pathway be denoted by G_P and $N_P = |G_P|$. The slope coregulation score for a pathway P is defined as follows using (1) and (2):

$$\text{Slope Coregulation Score}(P) = \text{VF} \times \sum_{t=1}^{T} \text{SlopeScore}_{t,p} \text{ , where}$$

$$\text{SlopeScore}_{t,P} = N_p \Bigg/ \left[\sum_{g \in G_p} \left(\text{Trend}(g,t) - \text{Mean}\left[\text{Trend}(g,t)\right]\right)^2 \right],$$

$$\text{where Trend}(g,t) = \left(\log_2 \left(X_{t,g}\right) - \log_2 \left(X_{t-1,g}\right)\right)\left(A_t - A_{t-1}\right) \text{ and}$$

$A_t - A_{t-1}$ represents the fold change in the experimental condition for the microarray experiment series. For example, in case of microarray time series data, this represents the time lapse between experiments t and t-1; in case of microarray experiments performed over increasing temperature levels, this represents the change in temperature between experiments t and t-1, etc.

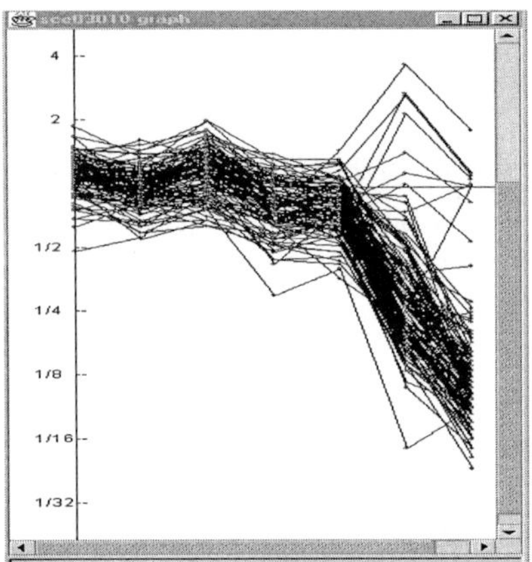

Fig 2a. Time progress of Ribosomal proteins which have a coregulation score of 100 - highest among all pathways. (Diauxic shift).

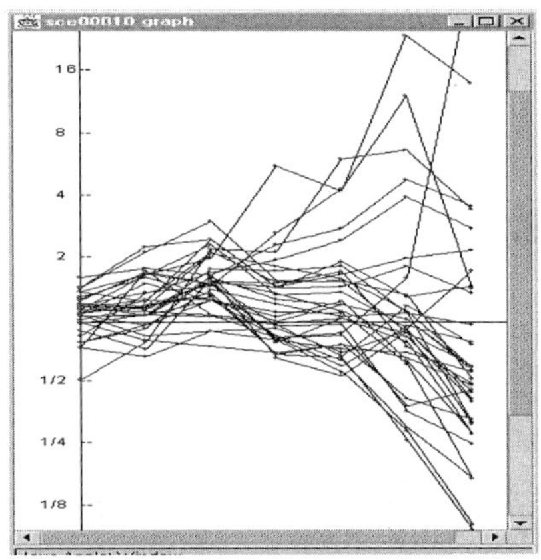

Fig2b. Time progress of the glycolysis pathway which have a coregulation score of 25 - highest among metabolic pathways (Diauxic shift).

3.3 Cascade Score

Using coregulation scores, pathways in which the genes have similar expression patterns will be ranked higher. However, this does not account for a) genes whose expression levels do not show much deviation from the expression level at the reference time point; b) the structure of the pathway. The first problem can be resolved by combining activity and coregulation scores, as was done in a combined scoring function described in [21]. The combined scoring function still does not take into account the structure and ordering of reactions in metabolic pathways.

Here, we define a cascade score that accounts for both these features in scoring pathways. This method of scoring is particularly useful to find out in which pathway a reaction chain is active or shutdown for the particular experiment. The cascade score is valid only for metabolic pathways, as it requires a network of linked reactions. It is computed as follows:

- Step 1: Create a list of all enzymatic reactions in the pathway. Discard all reactions in which the gene (corresponding to the enzyme) does not show any significant fold-deviations from the reference expression level.
- Step 2: Form all possible reaction chains (paths in a graph). Score each chain based on the coregulation of enzyme pairs as they occur in the chain. The edge weight W for an edge between two genes h and g is given by:

  $W(g \rightarrow h) = \#(h=1|g=1) / \#(g=1)$, where $g=1$ means g is active.
- Step 3: Find the chain with the highest score – assign the score for the chain as the cascade score for the pathway.

Details of the method for calculating cascade score are given in [15].

The table below gives the results for the three scoring methods for different yeast microarray experiment series.

Table 1. Top pathways based on scoring functions :

Microarray Experiment Series	Activity Score		Coregulation Score		Cascade Score
	Regulatory	Metabolic	Regulatory	Metabolic	
Diauxic Shift[4]	Ribosome	Oxidative Phosphoryl-ation	Electron Transport System - II	Reductive Carboxylate cycle	TCA cycle
Alpha Factor	Cell cycle	Riboflavin	Cell cycle	Riboflavin	Purine Metabolism
Elutriation	Cell cycle	Pentose Phosphate	Cell cycle	Porphyrin and chrolophyll metabolism	Glyoxylate and dicarbonate metabolism
Sporulation	Ribosome	Oxidative Phosphoryl-ation	Proteasome	Terpenoid Biosynthesis	Vitamin metabolism
Heat Shock	Ribosome	Purine Metabolism	Proteasome	Galactose	Fatty acid biosynthesis

The above table gives an idea of how the different analysis can give different results and how the scores point the affected pathways that are related with the experiment data. Consider the diauxic shift experiment, in [4] it is shown to be related with ribosome and TCA cycle pathways. Diauxic shift is known to be related with Oxidative Phosphorylation and Electron Transport System-II. Co-regulation of the Electron Transport System-II becomes evident due to significant activity of the Oxidative Phosphorylation pathway. Figure 1b clearly shows reductive carboxylate (CO_2 fixation) pathway getting activated during diauxic shift response.

4 Multiresolution, Animated Visualization of Pathways

As mentioned earlier, KEGG contains information on a large number of putative pathways. The pathway scores are useful in directing the user to the "right" pathway in the context of a microarray experiment series. However, visualization of the pathways is necessary to show a user the details of pathway effects as measured by changes in gene expression levels in response to stimuli. The visualization technique of [14] requires a single absolute level for each member of the set of genes, which is a severe limitation. Also identification of affected pathways based on color rather than numerical value is prone to errors. Our technique removes these inadequacies.

4.1 Pointing the User in the Right Direction – Multiresolution Viewing

The metabolic pathways in KEGG are classified hierarchically at three levels of detail i.e. three resolutions. Resolution 1 is a coarse grained representation of the complete network of metabolic pathways. Resolution 2 provides medium grade resolution is terms of functionality like carbohydrate metabolism, nucleotide metabolism, etc. and contains pathways related to that function. The finest resolution is resolution 3, which shows the reaction network as well as the compounds and the enzymes involved. KEGG also organizes its regulatory pathways into groups at resolution 2 based on broad functionality.

The (activity, coregulation, cascade) score for a pathway group at the resolution 2 level is simply the average of the corresponding scores of the pathways belonging to that group. The scores at the resolution 2 level are normalized on a scale of 0-100, with the highest pathway being given a score of 100.

The user is directed using the relevant summary pathway scores at each of the coarser resolutions (resolution 1 and 2). Clickable maps allow the user to navigate easily through the pathways. Example of this multiresolution view is shown in figure 3 and figure 4. Figure 3 shows the resolution 2 view for energy metabolism that had the highest activity score among all pathway groups in resolution 1.

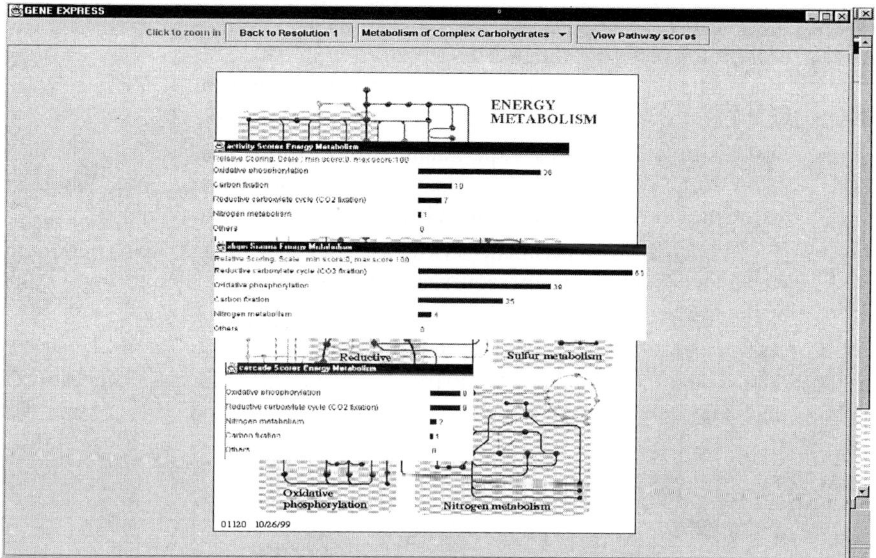

Fig 3. Second Resolution grouping of pathways belonging to Nucleotide Metabolism. The activity, co-regulation and cascade effect scores are also shown

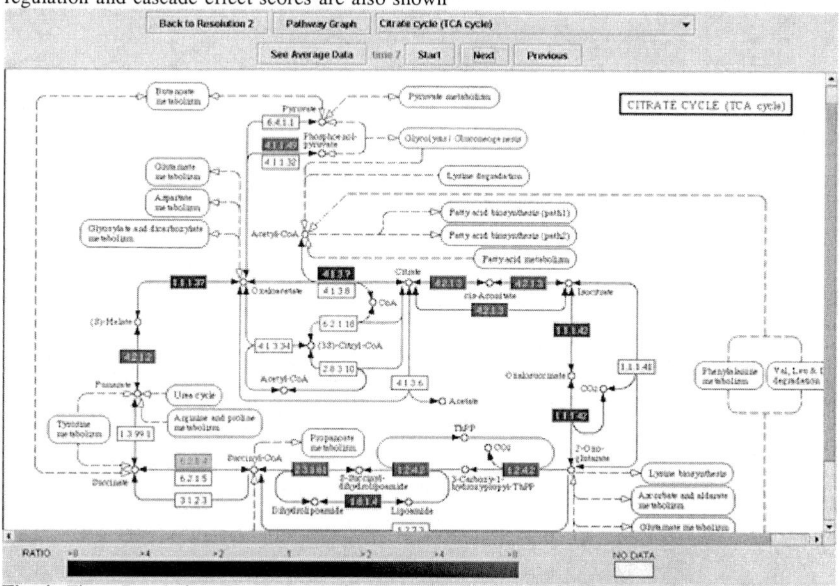

Fig 4. Finest Resolution (Resolution 3) for Citrate Cycle. The enzymes are colored according to their activity at time instant 7 of the Diauxic shift experiment [1].

4.2 Directing the User to Impact - Animated Visualization

At any resolution 3 pathway, the user is presented with several choices for viewing the expression data for all the genes involved in the pathway. One such choice is an animated view:

- For a single microarray experiment, the organism specific pathway map from KEGG is colored. The enzyme boxes are colored based on their expression level (red indicating induction and blue indicating repression).
- For a microarray experiment series, the user can use a "next" button to view the experiments in sequence, allowing a visual monitoring of the pathway changes.

Fig 4 shows the resolution 3 view for the citrate cycle pathway at the last time point of the diauxic shift microarray experiment series[4]. The user can use the previous and next buttons to observe this pathway at each time point.

5 Results and Discussion

While the potential utility of expression data is immense, some obstacles will need to be overcome before significant progress can be realized. First, data from expression arrays is inherently noisy. Second, gene expression is regulated in a complex and seemingly combinatorial manner. Third, our knowledge regarding genetic regulatory networks is extremely limited. Never the less, gene expression data from microarrays are very useful for understanding biochemical pathways, their progress with time and their response to experimental stimuli.

The scoring and visualization methods used here give a natural way for using genome-wide expression data in understanding how biological systems function. However, current methods can be improved if protein microarrays become widely available. This is because current DNA microarray technology measures mRNA expression levels, which are only an indication of the level of activity of the final protein, also the mapping from gene to proteins/enzymes is many to many, which can result in misleading scores. Directions for future work include analysis and visualization for microarray experimental data that corresponds to two or more classes[1, 21].

References

1. A.A. Alizadeh et al, "Distinct types of diffuse large B-cell lymphoma identified by gene expression profiling" *Nature* **403**, 503 (2000)
2. A.J. Butte et al, "Mutual information relevance networks: Functional genomic clustering using pairwise entropy measurements" *PSB* (2000)
3. S. Chu et al, "The transcriptional program of sporulation in budding yeast" *Science* **282**, 699 (1998)

4. J.L. DeRisi et al, "Exploring the metabolic and genetic control of gene expression on a genomic scale" *Science* **278**, 680 (1997)
5. P. D'haeseleer et al, "Linear modeling of mRNA expression levels during CNS development and injury" *PSB* **4**, 41 (1999)
6. D.J. Duggan et al, "Expression profiling using cDNA microarrays". *Nature genetics supplement* **21**, 10 (1999)
7. M.B. Eisen et al. "Cluster analysis and display of genome wide expression patterns" *PNAS* **95**, 14863 (1998).
8. M. Fellenberg et al, "Interpreting clusters of gene expression profiles in term of metabolic pathways" *German Conf. On Bioinformatics* Poster (1999)
9. N. Friedman et al, "Using bayesian networks to analyze expression data" *Proc. RECOMB* 127 (2000)
10. Goto et al, "Organizing and computing metabolic pathway data in terms of binary relations" *PSB* (1997)
11. V.R. Iyer et al, "The transcriptional program in the response of human fibroblasts to serum" *Science* **283**, 83 (1999)
12. M. Kanehisa in *Bioinfomatics: Databases and Systems*, "KEGG: From genes to biochemical pathways" Ed. S Letovsky (1999)
13. P.D. Karp et al, "Integrated access to metabolic and genomic data" Journal of Computational Biology (1996)
14. P.D. Karp et al, "Integrated pathway/genome databases and their role in drug discovery" Trends in Biotechnology (1999)
15. M.P. Kurhekar et al, "Analysis of pathways using cascade scores" *Technical Report*, IBM India Research Lab. (Draft under submission)
16. S. Liang et al, "REVEAL: A general reverse engineering algorithm for inference of genetic networks" *PSB* **3**, 18 (1998)
17. Nakao et al, "Genome-scale Gene Expression Analysis and Pathway Reconstruction in KEGG" *Genome Informatics* **10**, 94 (1999)
18. Raychauduri et al, "Prinicipal component analysis to summarize microarray experiments: application to sporulation time series" *PSB* (2000)
19. Richmond et al, "Genome-wide expression profiling in Escherichia coli K-12" *Nucleic Acids Research* **27**, 3821 (1999)
20. P.T. Spellman et al, "Comprehensive identification of cell cycle regulated genes of the yeast Saccharomyces cerevisiae by microarray hybridization" *Mol. Biol. Cell* **9**, 3273 (1998)
21. P. Tamayo et al, "Interpreting patterns of gene expression with self-organizing maps: Methods and application to hematopoietic differentiation" *PNAS* **96** 2907 (1999)
22. A. zien et al, "Analysis of gene expression data with pathway scores" *Proc. ISMB'00* (2000)

EXPLORING GENE EXPRESSION DATA WITH CLASS SCORES

PAUL PAVLIDIS
Columbia Genome Center, Columbia University, pp175@columbia.edu

DARRIN P. LEWIS and WILLIAM STAFFORD NOBLE[a]
Department of Computer Science, Columbia University, {dplewis,noble}@cs.columbia.edu

Abstract

We address a commonly asked question about gene expression data sets: "What functional classes of genes are most interesting in the data?" In the methods we present, expression data is partitioned into classes based on existing annotation schemes. Each class is then given three separately derived "interest" scores. The first score is based on an assessment of the statistical significance of gene expression changes experienced by members of the class, in the context of the experimental design. The second is based on the co-expression of genes in the class. The third score is based on the learnability of the classification. We show that all three methods reveal significant classes in each of three different gene expression data sets. Many classes are identified by one method but not the others, indicating that the methods are complementary. The classes identified are in many cases of clear relevance to the experiment. Our results suggest that these class scoring methods are useful tools for exploring gene expression data.

1 Introduction

Researchers interested in discovering meaningful patterns in gene expression data often ask, "What functional categories of genes are most interesting in the data?" This question is usually answered by indirect means, by making use of two fundamentally different general methodologies: "supervised" and "unsupervised."[1] [b] In this paper, we describe an analysis method which we term "semi-supervised," which combines elements of supervised and unsupervised approaches using existing classifications. This approach attempts to circumvent some of the limitations of the supervised and unsupervised methods, and is designed to directly identify "interesting" gene classes.

[a] Formerly William Noble Grundy: see www.cs.columbia.edu/~noble/name-change.html

[b] The supervised and unsupervised methodologies can be used to seek information about the genes on the arrays, or the samples from which the RNA was extracted; here we focus on the analysis of genes.

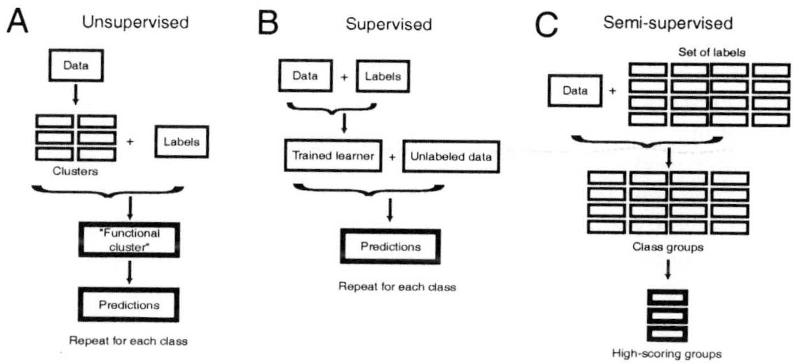

Figure 1: **Schematics of three general methodologies for analyzing gene expression data.** Boxes represent sets of genes or descriptions of gene classifications ('labels'), and arrows represent transitions between analysis stages. Thick outlines indicate the main outputs of each method. **A. Unsupervised.** The data is divided into clusters based on profile similarity. A *post hoc* analysis using classification labels can be used to identify "functional clusters" which contain many genes from the same class. Unannotated genes in these classes are predicted to have a related function. This *post hoc* analysis can then be repeated for multiple classes. **B. Supervised.** Data together with classification labels are used to train a learning algorithm, which can then be used to make predictions about unannotated genes. The learner must be retrained to recognized each class. **C. Semi-supervised.** Data together with a constellation of classification labels are used simultaneously to partition the data into class groups, based entirely on the labels. Scoring methods (the topic of this paper) are then used to identify groups that have particular characteristics.

An important concept for our discussion is that of a "gene class." We define a gene class as a group of genes with related functions, or which are otherwise grouped together based on biologically relevant information. For example, a class could represent a signal transduction or metabolic pathway, the members of a protein complex, or an enzymatic activity. A gene can be a member of any number of classes, and hundreds if not thousands of gene classes can be defined[2]. The goal of computational analysis is often to identify new class members, but here we are primarily concerned with making direct use of the existing gene classifications.

Before describing semi-supervised gene expression analysis, it is useful to describe the approaches from which it is derived. The three general methods (supervised, unsupervised, and semi-supervised) are depicted schematically in Figure 1. The unsupervised approach is perhaps the most familiar to gene expression researchers, who often use clustering algorithms to identify genes with similar expression patterns.[3,4] Clustering is unsupervised because the only input is the expression data, without any additional use of prior knowledge about the genes (Figure 1A). Genes with similar expression patterns are grouped together by clustering, without any knowledge of the genes' functions. Using clustering as a functional genomics technique thus requires *post hoc* interpretation of clusters in terms of the functions of the genes in the clusters. For example, if in a given cluster many genes are found to be in the same class, the experimenter might hypothesize that other genes in the cluster have related functions, and that the function is relevant in some way to the biological process under investigation. Another way to use unsupervised methods is for class discovery: genes which cluster together are hypothesized to have some functional or regulatory relationship. A strength of the cluster-driven approach is that it encourages an exploratory approach to the data, but it does not automatically generate hypotheses about the functions of the genes in the clusters.

In constrast, prior knowledge about genes is directly exploited by supervised methods, such as support vector machine classification [5,6] (Figure 1B). This type of algorithm is often referred to as a "learner." The learner is trained by a "teacher" to identify a particular gene class. The learner can predict the classifications of previously unanotated genes.[5] Thus, the supervised approach can be used to do the same job as the unsupervised method by complementary means. Supervised methods can yield superior performance in grouping genes of particular functions together,[5] but require identification of the class of interest ahead of time.

The semi-supervised approach is intermediate between the supervised and unsupervised approaches (Figure 1C). A score is assigned to each of a large number of predefined gene classes, and classes with high scores are considered potentially more interesting than classes with low scores. Thus in the semi-supervised approach, a large collection of teachers is available, but only some of the teachers provide "true" classifications. The goal of the learner in this case is to select the true classes from among this large collection of candidate classes. In comparison, in unsupervised learning, there is no teacher to provide the true classifications, while in supervised learning there is only a single teacher from which to learn the classifications.

To implement semi-supervised learning, we consider three methods for scoring classes: the tendency of genes in the class to be co-expressed, the significance of the expression profiles in the context of the experimental design, and the learnability of the gene class. The first method measures how well the genes in the class cluster together, that is, how similar their expression profiles are. The method we apply uses the average pairwise correlation between the expression profiles in a class [7]. Although this co-expression measure is a powerful means for scoring classes, profile similarity alone is too limiting as a metric for class importance. This is because while it may sometimes be true that genes which cluster together have related functions, it is **certainly** not always the case that genes with related functions cluster together.

The second scoring method measures the statistical significance of the expression pattern of each gene with respect to the experimental design. Using statistical methods such as analysis of variance (ANOVA), each gene can be assigned a p-value corresponding to the probability that the variations in gene expression across the conditions could have been observed by chance. Such analysis of each gene is commonly conducted in expression studies to assess which genes changed expression level during the experiment. Although such scores cannot be used as a means of identifying new members of a class, or in class discovery, we show here that the scores for the genes in a class can be meaningfully combined to provide a score for a class as a whole.

The third scoring method measures the learnability of a candidate gene class. The particular score we use here is a p-value derived from the total hold-one-out cross-validated error rate of a k-nearest neighbor classifier. This metric measures the distinctness of genes within the class relative to genes outside of the class.

Some previous work suggests that the semi-supervised approach is likely to be fruitful. Gerstein and Jansen (2000) have shown how classes can be ranked by coexpression,[7] while Hakak *et al.* [8] used the statistical signficance of individual genes to assess the significance of one class of interest. Mirnics *et al.*[9] compared the distributions of expression ratios in gene classes to that of the bulk data. Zien et al.[10] report the use of "conspicuousness" (related to the statistical significance approach we describe) and "synchrony" (essentially the same as expression pattern similarity) alone and in combination, as a means of identifying biologically relevant biochemical pathways among sets of hypothetical pathways. Ben-Dor *et al.*[11] discuss tissue classification and class discovery based on "surprise scores" that are similar to the statistical measures we describe here. However, the use of these methods as a general means for identifying gene classes of interest in a data set does not appear to have been fully explored.

In this paper, we apply the similarity-based, statistical-significance-based, and learnability-based methods to three previously published gene expression data sets, using publically-available gene classifications. In all three cases we show how to calculate p-values that can be used to accurately assess the significance of a particular class score. All three types of scores identify interesting classes of genes in all three data sets. Importantly, we show that the methods are to a large extent complementary, with each giving high scores to classes that the other does not.

Data set	Type	Arrays	Cond.	Genes	Classes	Reference
Yeast	Spot	79	79	2465	145 (MIPS)	Eisen *et al.*, 1998
Brain	Affy	24	6×2	5552	581 (GO)	Sandberg *et al.*, 2000
Cancer	Affy	38	3	5092	397 (GO)	Golub *et al.*, 1999

Table 1: **Summary of the three gene expression data sets.** The type of array, either spotted cDNA[14] (spot) or Affymetrix olignucleotide[15] (Affy), is listed, together with the number of arrays, conditions (Cond), genes and classes present. Genes were counted only if they were a member of at least one class. In the brain data, six brain regions are examined in two mouse strains. In the "classes" columns, MIPS and GO refer to the classification scheme (MIPS functional catalog,[16] or Gene Ontology,[2] respectively).

2 Methods

We used three publically available gene expression data sets to evaluate our methods. The data sets were chosen to represent a range of situations where the methods we describe might be useful. The first ("yeast") is from Eisen *et al.*,[3] and consists of 79 experiments in a variety of conditions. The conditions include different time points during the cell cycle and during the responses to various stresses (heat, cold, etc.). There is only one array per condition. The second ("brain") is from the work of Sandberg *et al.*,[12] and consists of replicate analysis of six brain regions in two mouse strains, for a total of 24 arrays. The last ("cancer") is from the work of Golub et al.,[13] who performed microarray analysis of acute leukemias. Each sample is from an individual patient, and was identified by Golub *et al.* as belonging to one of three groups of tumor type. We used the "training" data set from their work. The data sets are summarized in Table 1.

Our classification schemes were drawn from publically available databases. For the yeast data, we used the MIPS functional catalog[16] (www.mips.biochem.mpg.de). For the brain data and the cancer data, we used the publically available Gene Ontology[2] classifications (www.geneontology.org). Both schemes are hierarchical; that is, they consist of nested descriptions of genes that increase in detail as one descends down the hierarchy. Thus we expect a certain amount of redundancy in our results, as similar classes will receive similar scores. While we have not attempted to address this replication issue directly, we did find that restricting the classes to a particular size range to be useful for reducing the complexity of the results. Thus we limited our analysis to classes that had between 5 and 200 members. The number of classes meeting our criteria for each data set are listed in Table 1.

In our first scoring approach, we wished to calculate a measure that represents how coregulated the genes in a class are. The measure we used is the average of the Pearson correlation coefficient for the pairwise comparisons of genes in the class, omitting comparisons of genes to themselves.[7] If the expression vectors for the genes in a class are correlated, then the average correlation between the genes will be high.

Some genes (or more precisely, UniGene clusters) were represented more than once on the Affymetrix arrays, and this replication can skew the average score for a class. To deal with this issue, we gave each member of a set of n replicates a weight of $1/n$ in calculation of the average, and comparisons between replicates were not included in the average correlation. We note that this correction is crude; not all replicates are equivalent because the various "replicates" can come from different sequences representing different splice variants, or probe sets which are of varying sensitivity, and thus should not truly be considered replicates. We apply this simple correction to ameliorate the problems caused by giving the replicates the same weight as unreplicated genes, but leave a more thorough treatment as a topic for further study.

To convert the raw average corrrelations into p-values, the background distribution of scores expected under the null hypothesis was determined empirically by generating scores for 500,000 randomly selected sets of genes. Separate distributions were calculated for each class size for each expression data set. For small classes this distribution is quite broad, while for large classes it is narrower (not shown). The p-value for a class was then calculated as the fraction of simulated classes of the same size which had higher scores than the real class. The smallest p-value we could thus directly measure is thus 1/500000 (2×10^{-6}). Classes with p-values less than this value were provisionally set to 1×10^{-6}. This p-value is the "correlation score" for a class, and is calculated for all classes.

Our second measure applies statistical measures of significance of the expression pattern with respect to the experimental design. For the brain and cancer data sets, we used ANOVA[17] to obtain a separate significance score for each gene, in the form of a p-value. ANOVA is a standard statistical method for testing hypotheses about multiple means. In this context, genes with low p-values show more significant changes in expression between groups. For the brain data, we focused on genes showing differences among the six brain regions in two mouse strains in a two-way ANOVA, while for the cancer data we generated p-values for differences among the three tumor types (ALL-Bcell, ALL-Tcell and AML) in a one-way ANOVA. We used the $-\log_{10}(p\text{-value})$ as the score for each gene in our subsequent calculations. For the yeast data, which had no replication across the 79 conditions, we used the standard deviation of the expression Zien *et al.*[10]

The average of the log transformed p-values for the genes in a class forms the basis of the class score. This summation is equivalent to calculating the joint probability of the genes in the class under the null hypothesis, assuming independence of the genes (an assumption which is certainly not correct, but our results indicate that this simplification is acceptable). These average values can be converted to p-values in a manner identical to that we used for the correlation scores, by calculating the average $-\log_{10}(p\text{-value})$ for 500,000 randomly chosen sets of genes to generate a background distribution, with a separate distribution calculated for each class size. To deal with replicated genes, we used a $1/n$ weighting scheme analogous to that described for the classification scores. Because these scores take the experimental design into consideration, we refer to it as the "Experiment score."

The third method we tested measures the learnability of the class by a simple supervised learning algorithm, yielding a "learnability score." In order for a class to be learnable, the genes must not only cluster together in space (i.e., be co-expressed), but also be sufficiently distinct from other genes in the data set to be distinguishable as a class. The degree to which this is possible using the k-nearest-neighbor (KNN) algorithm forms the basis of our third method. The KNN classifier predicts the label of an unclassified example as the label belonging to the majority of the k closest examples in Euclidean space. Because KNN is unique among supervised learning algorithms in that there is no training step, we can efficiently compute hold-one-out cross-validation error rates. These rates form the basis for the scoring scheme. In this work we set k to one. The use of a different learning algorithm might yield different results than those we report here.

To convert these raw scores into p-values, the null distribution can be calculated analytically, instead of empirically as for the correlation and experiment scores. The calculation is based on the observation that, for randomly labeled data, the probability of KNN misclassifying a randomly selected data point X depends only on the size $\mathcal{P}$ of the gene class and the size D of the entire data set. Say that example X belongs to the positive class $\mathcal{P}$. To encounter an error on this example, KNN must place X into the negative class $\mathcal{N}$, which can only occur if fewer than $\lfloor \frac{k}{2} \rfloor$ of the next k points, chosen at random, have labels $\mathcal{P}$. This outcome is expressed in the following conditional probability: $\mathbf{Pr}(X_{\mathcal{N}}|X \in \mathcal{P}) = \left(\sum_{i=0}^{\lfloor \frac{k}{2} \rfloor} \binom{P-1}{i}\binom{D-P}{k-i}\right) / \binom{D-1}{k}$, where $X_{\mathcal{N}}$ denotes example X being classified in class $\mathcal{N}$ by KNN. This probability, along with prior probabilities derived from the class sizes, yields the overall probability of a false positive or false negative misclassification $\mathbf{Pr}(X_{\mathcal{N}}|X \in \mathcal{P})\mathbf{Pr}(X \in \mathcal{P}) + \mathbf{Pr}(X_{\mathcal{P}}|X \in \mathcal{N})\mathbf{Pr}(X \in \mathcal{N})$, which can be used to compute a binomial cumulative distribution. In this way, a p-value can be obtained for any KNN cross-validated total error.

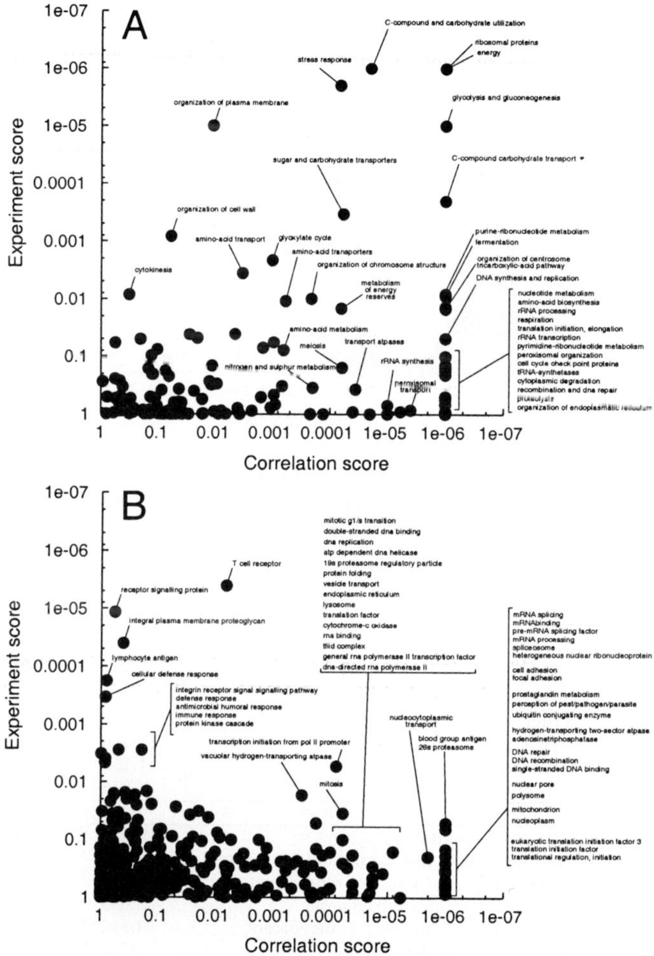

Figure 2: **Summary of "experiment" and "correlation" results.** See next page for legend.

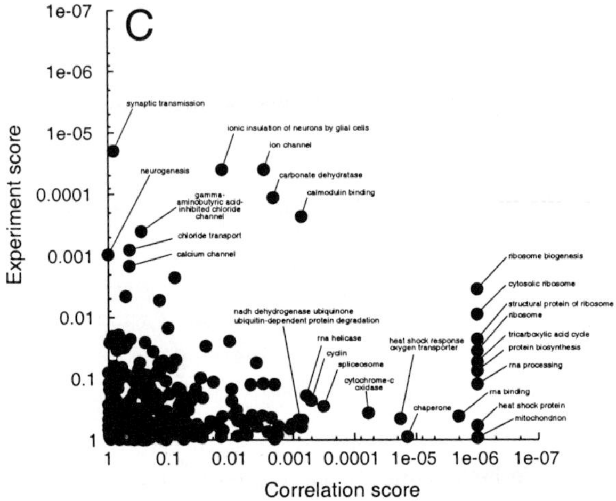

Figure 2: **(Continued) Summary of "experiment" and "correlation" results. A.** Yeast data. **B.** Cancer data. **C.** Brain data. In all three panels, each point represents a single gene class. The correlation score is plotted on the horizontal axis while the experiment score is plotted on the vertical axis. Text labels indicate the identities of some individual high-scoring classes and groups of classes.

Class	p-value
Brain	
histogenesis and organogenesis	3.861^{-4}
Cancer	
structural protein of ribosome	1.969^{-23}
protein biosynthesis	5.160^{-7}
RNA binding	1.158^{-6}
cell motility	1.876^{-4}
Yeast	
Transport facilitation	1.550^{-10}
lipid fatty-acid and isoprenoid biosynthesis	2.682^{-4}
lipid fatty-acid and isoprenoid metabolism	3.869^{-4}
glycolysis and gluconeogenesis	3.058^{-4}
sugar and carbohydrate transporters	3.816^{-4}
tricarboxylic-acid pathway	5.00^{-4}
rRNA processing	5.807^{-4}

Table 2: **Summary of the KNN results.** Only classes which were not given p-values less than 10^{-3} by another method are listed. Closely related classes are indented. The total number of significant classes identified by this method were: Brain: 1; Yeast: 22; Cancer: 4. Most of these were identified by the correlcation score method.

3 Results

We measured correlation, experiment, and learnability scores for the yeast, cancer, and brain data sets. The results for the correlation and experiment scores are summarized in Figure 2. The learnability method yielded fewer significant classes than the other methods, so its results are summarized in Table 2.

As predicted (see Methods), due to the hierarchical nature of the classifications some of the high-scoring classes shown in Figure 2 are closely related to each other. For example, in Figure 2B, multiple classes closely corresponding to mRNA splicing factors (mRNA splicing, spliceosome, etc.) are given high correlation scores. This redundancy makes it somewhat difficult to make an accurate count of how many classes are given high scores. However, some important trends are discerned by inspecting the data.

Of the three methods, the learnability measure yielded the fewest "interesting" classes. However, some of the classes it identifies are different than the ones identified by the other methods (Table 2). Thus it forms a useful complement to the other two methods, and has in addition the advantage of computational speed.

We observed that several classes consisting of "housekeeping" genes, such as the ribosomal proteins and "RNA processing," are given high correlation scores in all three data sets, but not necessarily high experiment scores. The appearance of these classes in three disparate data sets suggests that such housekeeping genes show a very high coordination of expression levels that is not dependent on the experimental context.

In contrast to the correlation scores, high experiment scores tended to be given to classes that are highly specific to the experimental design. For example, the highest experiment-scoring class in the cancer data (Figure 2B) is "T-cell receptor," which is appropriate considering that the tumors studied fell into groups depending on whether they were derived from T-cells or B-cells.[13] Similarly, the highest scoring classes in the brain data set were "synaptic transmission," "ion channels" and "ionic insulation of neurons by glial cells," all of which might be relevant to functional differences among the brain regions studied.[12] In the yeast data set, fewer classes received high experiment scores without also receiving high correlation or learnability scores. The major exceptions are "organization of plasma membrane" and possibly "stress response." The former class consists primarily of permeases for sugars and other small molecules. The latter class consists of genes that change expression level in response to stress. These classes are relevant because the yeast data was gathered during various stressful conditions and metabolic states.[3]

4 Discussion

Our contributions in this work are three-fold. First, we provided an explicit description of the class scoring problem, formulating it as intermediate between supervised and unsupervised approaches. Second, we described three methods for semi-supervised analysis, which capture different features of the data. Finally, we demonstrated the use of these methods on real data, showing they reveal interesting biologically relevant features of the data. In our view, one of the chief appeals of the semi-supervised method is that it uses prior knowledge in ways that unsupervised methods cannot, while maintaining a flexibility that supervised methods lack.

Interestingly, the three methods we used often give different classes high scores; that is to say, they are complementary in the kinds of information they provide. This result is particularly apparent in the comparison of experiment scores to correlation scores for the cancer and brain data sets. The learnability score yields only a small number of additional high-scoring classes. Of the three methods, the experiment scores appear to be the most specific for each data set, while the correlation scores, and to some extent learnability scores, tended to focus on "housekeeping" classes. It will be interesting to see if this trend is evident as we examine additional data sets.

There are several issues we encountered during our experiments that suggest avenues for future research and improvements to the methods. Most obviously, we are at the mercy of the existing annotations. A major reason for this limitation is the current incompleteness of annotations based on the Gene Ontology. Thus our methods should prove to be even more useful as database annotations improve. Because some classes are redundant, for our purposes some simplification of the classification schemes would also be desirable.

Another issue is our assumption, for the experiment-score analysis, that the ANOVA p-values for each gene are independent. This is clearly not the case. At one extreme, some genes are represented more than once in a data set. In general, the correlation structure of the data will affect the statistical significance of a given gene pattern. There are many methods for correcting p-values in such a situation,[18] but we have not attempted to apply them here and leave this as an issue for future study.

A final issue is the requirement for a computationally intensive determination of the background distribution of experiment and correlation scores. It is possible that this computation can be avoided by estimating the distributions.[7] Even if such estimates are not exact, they are likely to provide a reasonable calibration of the scores for the effect of class size. We note that we can also probably afford to sacrifice some precision in p-value computation, because as long as the method provides guidance through the hundreds of gene classes, we consider it a success.

Acknowledgments

This work was supported by an Award in Bioinformatics from the PhRMA Foundation, and by National Science Foundation grants DBI-0078523 and ISI-0093302.

5 References

1. A. Brazma and J. Vilo. *FEBS Letters*, 23893:1–8, 2000.
2. M. Ashburner, C. A. Ball, et al. *Nat Genet*, 25(1):25–9., 2000.
3. M. B. Eisen, P. T. Spellman, et al. *Proc Natl Acad Sci U S A*, 95(25):14863–8., 1998.
4. P. Tamayo, D. Slonim, et al. *Proc Natl Acad Sci U S A*, 96:2907–2912, 1999.
5. M. P. Brown, W. N. Grundy, et al. *Proc Natl Acad Sci U S A*, 97(1):262–7., 2000.
6. T. R. Hvidsten, J. Komorowski, et al. In *Proceedings of the Pacific Symposium on Biocomputing*, pages
7. M. Gerstein and R. Jansen. *Curr Opin Struct Biol*, 10(5):574–84., 2000.
8. Y. Hakak, J. R. Walker, et al. *Proc Natl Acad Sci U S A*, 98(8):4746–51., 2001.
9. K. Mirnics, F.A. Middleton, et al. *Neuron*, 28:53–67, 2000.
10. A. Zien, R. Kuffner, et al. In *Proc Int Conf Intell Syst Mol Biol*, pages 407–417, 2000.
11. A. Ben-Dor, N. Friedman, et al. In *Proc Int Conf Intell Syst Mol Biol*, pages 31–38, 2000.
12. R. Sandberg, R. Yasuda, et al. *Proc Natl Acad Sci U S A*, 97(20):11038–43., 2000.
13. T. R. Golub, D. K. Slonim, et al. *Science*, 286(5439):531–7., 1999.
14. J. DeRisi, L. Penland, et al. *Nat Genet*, 14(4):457–60., 1996.
15. D. J. Lockhart, H. Dong, et al. *Nat Biotechnol*, 14(13):1675–80., 1996.
16. H. W. Mewes, D. Frishman, et al. *Nucleic Acids Res*, 28(1):37–40., 2000.
17. J. H. Zar. Prentice Hall, 1998.
18. P.H. Westfall and S.S. Young. John Wiley & Sons, Inc., New York, 1993.

GUIDING REVISION OF REGULATORY MODELS WITH EXPRESSION DATA

JEFF SHRAGER[a] and PAT LANGLEY
Institute for the Study of Learning and Expertise
2164 Staunton Court, Palo Alto, CA 94306

ANDREW POHORILLE
Center for Computational Astrobiology and Fundamental Biology
NASA Ames Research Center, M/S 239-4, Moffett Field, CA 94035

BioLingua is a computational system designed to support biologists' efforts to construct models, make predictions, and interpret data. In this paper, we focus on the specific task of revising an initial model of gene regulation based on expression levels from gene microarrays. We describe BioLingua's formalism for representing process models, its method for predicting qualitative correlations from such models, and its use of data to constrain search through the space of revised models. We also report experimental results on revising a model of photosynthetic regulation in Cyanobacteria to better fit expression data for both wild and mutant strains, along with model mutilation studies designed to test our method's robustness. In closing, we discuss related work on representing, discovering, and revising biological models, after which we propose some directions for future research.

1 Introduction and Motivation

There is general agreement that scientists need computational tools to assist in analyzing the rapidly increasing amount of biological data. Unfortunately, most existing software makes only limited contact with the methods that practicing biologists use in formulating and evaluating their models. In particular, most computational tools in biology have focused on knowledge-lean methods for data analysis, such as clustering, whereas biologists typically reason in a knowledge-rich manner using models of biological processes.

In this paper, we describe BioLingua, a suite of computational tools designed to assist working biologists in building and reasoning about their process models. Our goal in developing the system has been to match the ways in which biologists think about explanatory models, rather than to apply existing algorithms to available data in ways seldom pursued by biologists themselves. Working biologists, like other scientists, use data and models interactively, utilizing their models to interpret new experimental results and in turn revising these models in response to observations.

[a] Also affiliated with Department of Plant Biology, Carnegie Institution of Washington. Email: jshrager@andrew2.stanford.edu

In the sections that follow, we describe our initial version of BIOLINGUA, which supports data intepretation and model revision in the arena of regulatory models. We start by defining the task of revising an intitial model given expression data and then report on BIOLINGUA's approach to representing models, using them to make predictions, and carrying out heuristic search through the space of candidate models. After this, we discuss related work on representing knowledge about biological processes and discovering models that encode them. In closing, we note some limitations of our system and suggest directions for future research on computational discovery aides for biologists.

2 The Task of Revising Regulatory Models

One important facet of biological theory concerns the regulation of gene expression. Although scientists understand the basic mechanisms through which DNA produces proteins and thus biochemical behavior, they have yet to determine most of the regulatory networks that control the degree to which each gene is expressed. However, for particular organisms under certain conditions, biologists have developed partial models of gene regulation. The measurement and analysis of gene expression levels, either through Northern blots or cDNA microarrays, has played a central role in the elucidation of regulatory models, as both measures quantify gene activity in terms of RNA concentration.[b]

There are two typical ways in which expression data are used to extend knowledge about regulatory mechanisms. The most common computational approach involves the use of clustering to infer which genes occur in coregulated classes. This knowledge-lean approach lets one reduce the high dimensionality of microarray data to a manageable level, but the result is typically descriptive rather than explanatory in nature. A second paradigm, more commonly used by practicing biologists, uses data about expression levels to test specific pathway hypotheses. This knowledge-rich approach lets one evaluate proposed explanations, but it generally does not move beyond these hypotheses to suggest improved regulatory models.

We have designed BIOLINGUA to combine the best aspects of these two approaches to regulatory model discovery. We can state the task in semi-formal terms as:

- *Given*: a partial model of gene regulation for some organism;
- *Given*: data about the expression levels of relevant genes;
- *Given*: knowledge of biological processes that regulate gene expression;
- *Find*: an improved regulatory model that explains the data better.

[b]The distance between these measures and actual biochemical activity is considerable, but they still provide valuable information to biologists.

Computational tools that support this task will let biologists use microarray data both to test their regulatory models and to revise them in response to relevant observations.

3 An Approach to Regulatory Model Revision

Now that we have stated the task of revising an initial regulatory model based on microarray data, we can describe the approach that BIOLINGUA takes to this discovery problem.

3.1 Representing Models of Gene Regulation

Before we can develop algorithms to improve regulatory models, we must select some representation for those models. Most work in machine learning and data mining, including that in biological domains, draws on representational formalisms like decision trees, logical rules, or Bayesian networks that were designed by artificial intelligence researchers themselves. These formalisms are often adequate for representing complex regularities and making accurate predictions, but they make little or no contact with notations commonly used by practicing scientists.

In contrast, we are committed to representing biological models in terms that are familiar to biologists themselves. In biology talks and publications, these models are often depicted graphically. Figure 1 presents one such model, which we obtained from a plankton biologist, that aims to explain why Cyanobacteria bleaches when exposed to high light conditions and how this protects the organism. Each node in the model corresponds to some observable or theoretical variable; each link stands for some biological process through which one variable influences another. Solid lines in the figure denote internal processes, while dashes indicate processes connected to the environment.

The model states that changes in light level modulate the activity of DFR, a protein hypothesized to serve as a sensor. This in turn activates NBLR, which then reduces the number of phycobilisome (PBS) rods that absorb light, which is measurable photometrically as the organism's greenness. The reduction in PBS serves to protect the organism because the reduced PBS array absorbs less light, which can be damaging at high levels. The organism's health under high light conditions can be measured in terms of the culture density. The model also posits that DFR impacts health through a second pathway, by influencing an unknown response regulator RR, which in turn down regulates expression of the gene products psbA1, psbA2, and cpcB. The first two positively influence the level of photosynthetic activity (Photo), which, if left unaltered, would also damage the organism.

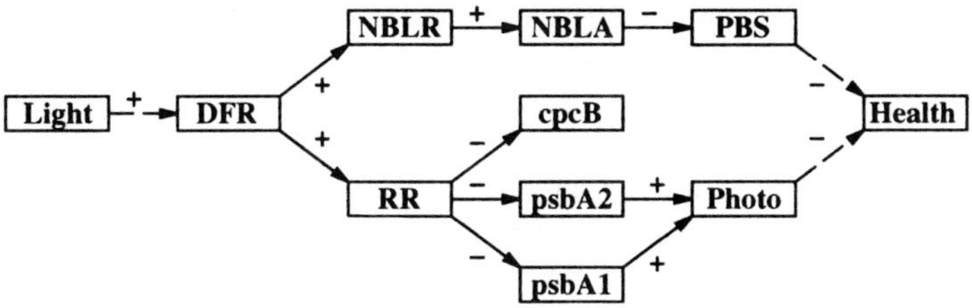

Figure 1: An initial model for regulation of photosynthesis in Cyanobacteria.

Note that this model, although incorporating quantitative variables, is qualitative in that it specifies directions of influence but not their degree. For instance, one causal link indicates that increases in NBLR will increase NBLA, but it does not state whether that relation obeys a linear or some other law, nor does it specify any parameters. We have focused on qualitative models not because quantitative ones are undesirable, but because biologists usually operate on the former, and we want our computational tools to support their typical reasoning styles.

Another characteristic of the model is that it is both partial and abstract. The biologist who proposed this model made no claim about its completeness, and clearly viewed it as a working hypothesis to which additional genes and processes should be added as indicated by new data. Some processes are abstract in the sense that they denote entire chains of subprocesses. For instance, the link from DFR to NBLR stands for a signaling pathway, the details of which hold little relevance at this level of analysis. The model also includes abstract variables like RR, which denotes an unspecified gene, or possibly a set of genes, that acts as an intermediary controller. BioLingua's formalism lets it express such partial, abstract, and qualitative models of the type that biologists propose and reason about.

3.2 Microarray Data on Gene Regulation

Like many other researchers, we are excited about the potential of cDNA microarrays for elucidating biochemical processes. Briefly, these devices measure the expression level for hundreds to thousands of an organism's genes, as reflected by the concentration of mRNA for each gene relative to that in a control condition. One can collect such measurements under different environmental conditions (e.g., clean vs. polluted water), for different organisms (e.g., healthy vs. cancerous tissue), or for different points in time.

We have access to such microarray data for several strains of Cyanobacteria under high light conditions that cause the organism to bleach and reduce its photosynthetic activity over a period of hours. These data include measurements of the expression levels for about 300 genes believed to play a role in photosynthesis, although we have focused on those genes mentioned in the model. We have array data collected at 0, 30, 60, 120, and 360 minutes after high light was introduced, with four replicated measurements at each time point. The dimensionality of these data, and thus the number of parameters required in a numeric model, is much higher than the number of observations, providing another reason to favor qualitative models over quantitative ones.

3.3 Making Predictions and Evaluating Models

BioLingua needs some procedure to map a biological model like that in Figure 1 onto the microarray data we have available. Since its models are qualitative, they cannot directly predict the continuous expression levels, but they can predict which variables should be correlated and the direction of those correlations. For each pair of variables (nodes) in a model, the system enumerates the paths that connect those variables. BioLingua transforms each such path into a predicted correlation by multiplying the signs on its links and, when the predictions for all paths between two nodes agree, predicting that correlation.[c]

However, when the correlations predicted by two or more paths disagree, BioLingua must resolve the ambiguity in some manner. In a quantitative model, each path would have its own degree of influence, and one could sum their effects to determine the outcome. Lacking such quantitative information, the system can still annotate the model to indicate that the positive (or negative) paths are dominant, and thus predict a positive (or negative) correlation. This extended formalism lets any qualitative model predict a positive or negative correlation for each pair of observed variables, provided one is willing to pay the cost of adding assumptions about dominance. For example, the model in Figure 1 has three paths between the expression levels for DFR and Health. The product of signs on each path is positive, meaning that they each predict a positive correlation between the two variables. However, if the link from NBLA to PBS were positive, this path would make a different prediction and the model would need a dominance annotation to resolve the ambiguity.

This procedure lets BioLingua generate qualitative correlations between pairs of variables in a given model. Naturally, the system can compare these

[c]Note that some paths pass through unobservable variables like RR; although we cannot measure such terms' values, that does not keep BioLingua from utilizing them in predictive paths between observable variables like DFR and psbA1.

predictions to the observed correlations, which it computes from corresponding expression levels in the arrays across different time steps. BioLingua treats any correlation that fails a significance test, in this case $p < 0.05$, as zero. The system incorporates these matches against the data in its evaluation metric for models. However, it also includes a measure of model complexity which favors simpler models and a term which favors models that make more predictions (i.e., a Popperian bias toward hypotheses that are easier to reject), which we found necessary to guard against degenerate models. The specific function used to evaluate candidates is

$$E = B(variables) + B(links) + B(annotations) + B(errors) - B(predictions) ,$$

where $B(X)$ denotes the total number of X (e.g., links or errors) times the number of bits needed to encode X. In this scheme, each variable and each link requires 4 bits, each disambiguation annotation requires 0.1 bit, and each prediction error and each prediction requires 3 bits. The resulting measure, which is similar to minimum description length, gives the overall quality for each model.

3.4 Revising Regulatory Models to Explain Microarray Data

As with most research on computational knowledge discovery, one can view the revision of biological models in terms of heuristic search through a space of candidate models. This framework requires one to make a number of design decisions, including the state from which to initiate the search, the operators used to generate new states, the knowledge used to constrain these operators' application, the evaluation metric used to select among competing states, the overall scheme for search control, and the criterion used to halt the search.

Biologists often have some abstract qualitative model in mind at each stage of their research. BioLingua takes such a model as the starting point for its search process. Some natural operators for revising such a model include adding a signed link, removing a link, and reversing the sign on a link. In the current implementation, BioLingua's evaluation function for selecting among models is simply the measure of model quality E described earlier. The control scheme that utilizes this function is greedy search through the model space, with failure to improve on the evaluation metric as the halting criterion.

For example, to generate an improved regulatory model for the photosynthetic process in Cyanobacteria under high light, BioLingua starts from the model in Figure 1. This model's 11 variables and 12 causal links lead to some 350 one-step revisions that produce distinct models, resulting from link rever-

sals, link additions, and link deletions. The system generates each of these candidates, calculates their E scores given the expression data, and selects the best one as the current model. It then repeats this process, continuing until further changes fail to yield improvements in the evaluation metric.

4 Experimental Results on Photosynthetic Regulation

Ultimately, BIOLINGUA's success as a discovery system will depend on whether it can use expression data to improve biological models. Here we report initial experiments designed to test the program's abilities on this dimension.

4.1 Improving Models of Wild and Mutant Cyanobacteria

We have already described an initial model, shown in Figure 1, of bleaching in Cyanobacteria that we obtained from biologists, along with expression data on the genes that regulate this process over time. The data lead to 18 positive correlations and 10 negative correlations among the observed expression levels.

When given this initial model and these qualitative data, BIOLINGUA's revision module carries out its greedy search through the model space, taking eight steps and examining 2382 candidates along the way. Additional revisions lead to no improvement in the evaluation function, causing the system to halt. Figure 2 shows the final revised model that results from this search process, which matches the observed expression levels better than the starting model and has a better evaluation score ($E = -46$ rather than $E = 12.2$).

This model differs from the initial one in some important ways. These include deletion of the links from DFR to NBLR, from psbA1 to Photo, from RR to psbA2, and from RR to cpcB. The revised model also contains three new links, indicating a positive influence from cpcB to NBLR and negative influences from psbA1 to psbA2 and from psbA2 to cpcB. The revision process has also changed signs on the links from RR to psbA1, from PBS to Health, and from Photo to Health.

In addition to proposing regulatory models for wild strains of an organism, biologists also desire to model mutant strains. We have access to array data for a nonbleaching mutant of Cyanobacteria under the same high light conditions as for the wild strain. Because such a mutant presumably differs genetically from the wild organism in at most a few ways, it seems natural to utilize BIOLINGUA's revision module to formulate a model of the mutant's regulatory processes. In this case, the system considers 2270 candidates while taking nine steps through the model space. Figure 3 presents the resulting model, which has a better score ($E = -24.6$) than the initial one ($E = 12.2$).

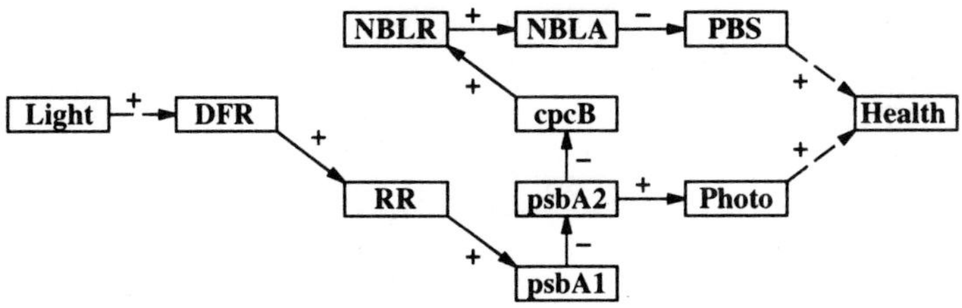

Figure 2: A revised model for regulation of photosynthesis in wild Cyanobacteria.

There are a number of differences between the revised model for the mutant strain and the initial model. These include deletion of the links from DFR to RR, from RR to psbA2, from RR to cpcB, and from psbA1 to Photo. The mutant model also specifies three new links, indicating positive influences from psbA1 to cpcB and from cpcB to psbA2, along with a negative influence from NBLA to RR. The revision mechanism has also changed signs on the links from psbA2 to Photo and from Photo to Health.

These revised models have some biological plausibility, but they also have problematic aspects. Generally speaking, it seems plausible that DFR influences photosynthetic activity through NBLR (in the wild strain) or a psbA1 cascade (in the mutant strain), and additional experiments could test these proposals. On the other hand, in both cases the revision process produced models with cascades whereas the initial model had separate influences, specifically from RR. Although such chains are not impossible, there is no reason to prefer such structures. Additional knowledge, either in the form of biological constraints or an improved evaluation metric, could resolve this ambiguity.

4.2 Robustness of the Approach

Although the previous runs demonstrate BioLingua's relevance to problems in model revision that arise among practicing biologists, they do not provide evidence of its robustness. To evaluate BioLingua's revision module along this dimension, we designed an experiment to determine whether the quality of the final revised model degrades gracefully with decreasing correctness of the initial model. Thus, we took the revised model from Figure 2 as our target T and generated different initial models by taking random steps through the model space. In this manner, we generated ten distinct models that differed from T by one step, another ten that differed by two steps, and so forth, halting at five steps from the target. We then ran the revision algorithm on each initial model with the expression data that produced the model in Figure 2.

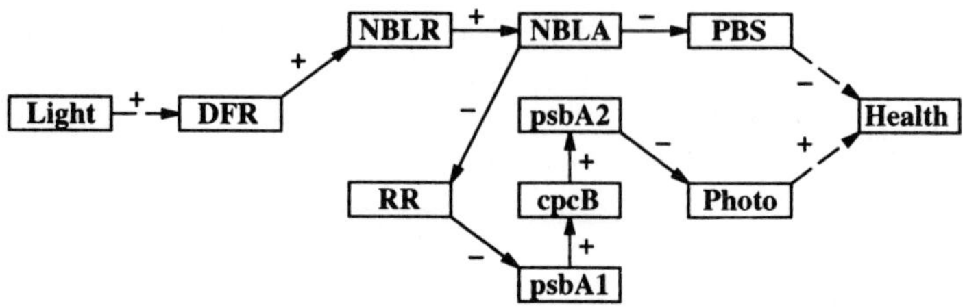

Figure 3: A revised model for regulation of photosynthesis in mutant Cyanobacteria.

We measured two dependent variables as a function of distance from the target model. The first involved the revised model's accuracy at predicting qualitative correlations, specifically the number of correctly predicted correlations or non-correlations over the total number of possible correlations. The second was simply the distance (number of steps in the search space) between the revised model and the target model T. We hypothesized that both measures would get worse, on average, with distance between the initial and target models, but that this degradation would be graceful.

The results were generally consistent with our expectations. The predictive accuracy of the target model on the expression data was 94 percent, whereas the revised models from runs starting one, two, three, four, and five steps from the target had average accuracies of 84, 79, 78, 65, and 63 percent, respectively. Similarly, the average distance of these revised models from the target, in terms of steps through the model space, was 3.5, 3.5, 5.9, 4.4, and 5.0, respectively. Thus, the method's behavior degraded as the revision task became more difficult, but this occurred in a graceful manner.

5 Related Research on Computational Discovery

Our approach to computational biological discovery builds on three previous lines of research. The first framework has focused on the explicit representation of knowledge about biological pathways. For instance, Karp et al.'s EcoCyc[1] encodes most established pathways for E. Coli and lets users display this knowledge graphically. Kanehisa[2] reports another effort that has produced KEGG, which codifies similar knowledge about a range of organisms. The knowledge stored in these systems is impressive, including information about metabolic pathways, regulatory pathways, and molecular assemblies, but their ability to reason over this knowledge remains limited. Tomita et al.[3] describe another framework, E-Cell, which stores similar knowledge and includes mechanisms for predicting behavior, but even E-Cell lacks the ability to revise its models in response to observations, which is BIOLINGUA's central feature.

A second framework has focused explicitly on the discovery of biological knowledge from data. We have already contrasted our approach with the more common technique of clustering microarray data in a knowledge-lean manner, but there exists some other work on constructing process explanations from such data. For example, Koza et al.[4] use heuristic search methods to estimate, from time-series data about concentrations, the structure and parameters of a metabolic model. Zupan et al.[5] describe GENEPATH, a system that comes somewhat closer to our approach in that it combines biological knowledge and data about the effects of mutations to propose qualitative genetic networks. Hartemink et al.,[6] although not focused on discovery, propose a similar notation for encoding regulatory models and another evaluation metric that could direct search through the model space.

A third research framework has focused not on constructing models from scratch but rather on revising existing theories to improve their fit to data. For example, Ourston and Mooney[7] present a method that uses data to revise models stated as sets of propositional Horn clauses, whereas Towell[8] reports a related approach transforms such models into multilayer neural networks, then uses backpropagation to improve their fit to observations. Our technique comes closer to Karp's HYPGENE,[9] which uses qualitative phenomena to revise a model cast in biological terms, but which differs considerably in its formalism and reasoning mechanisms. This framework has emphasized supervised rather than unsupervised data, but it shares the notion of revising an initial model.

Each of these frameworks has clear merits. Our research is novel in that it combines these three themes into a single system for the computational discovery of biological knowledge.

6 Concluding Remarks

BIOLINGUA is a computational tool kit designed to assist biologists in stating process models, using those models to make predictions, interpreting observations in light of those predictions, and improving their models in response. Our initial work has focused on revising a given regulatory model to better fit observed expression levels, an approach that differs considerably from the knowledge-lean methods typically applied to such data.

We illustrated BIOLINGUA's application to this task in the context of a particular model of photosynthetic regulation in Cyanobacteria and expression data collected for that organism. We presented the system's formal representation for biological process models, a method that uses such models to predict qualitative correlations between expression levels, and an algorithm that carries out heuristic search through the space of regulatory models, guided by data and a bias toward simpler models. In addition, we demonstrated the system's revision of an initial model of photosynthetic regulation, given expression data for wild and mutant Cyanobacteria. We also studied BIOLINGUA's ability to recover a model's structure after mutilating it to varying degrees, and the system exhibited reasonable robustness on this task.

Although our results to date are encouraging, we must extend BIOLINGUA in a number of directions before it can become a useful tool for biologists. For example, the current system can add, remove, and reverse causal links to the initial model, but it cannot introduce new variables that correspond to observed expression levels for known genes, which seems desirable. Achieving this functionality means adding a new revision operator and thus enlarging the space of candidate models, which in turn will require an improved search mechanism. This expanded search process would benefit from interaction with biologists, who could help to guide the decision process in cases where different models have similar scores.

Future versions of the system should support link types that correspond to additional biological concepts. For example, BIOLINGUA should distinguish between metabolic processes, which are effectively instantaneous, and regulatory processes, which typically take place over time. This distinction will also mean extending our formalism and prediction mechanism to support time-delayed effects. One response to this challenge comes from qualitative physics, which describes dynamic systems in terms of qualitative differential equations. This approach is consistent with our bias toward qualitative models.

A more fundamental issue concerns BIOLINGUA's current modeling formalism. Although biologists state some models in terms of measurable statistical variables, such as gene expression levels, they often describe an organism's behavior in terms of mechanical processes that operate on individual molecules. Karp's work[9] on modeling the Tryptophan operon provides one approach to representing such mechanisms. Future versions of BIOLINGUA should support the ability to make statistical predictions from such mechanical models, and thus make better contact with biologists' conceptual repertoire.

In the longer term, we envision BIOLINGUA developing into an interactive discovery aide that lets a biologist specify initial models, focus the system's attention on particular data and parts of those models it should attempt to improve, select among candidate models with similar scores, and generally control high-level aspects of the discovery process. Combined with other planned extensions, this facility should make BIOLINGUA a more valuable tool for practicing biologists.

Acknowledgements

This work was supported by the NASA Ames Director's Discretionary Fund, by the NASA Biomolecular Systems Research Program, and by NTT Communication Science Laboratories, Nippon Telegraph and Telephone Corporation. We thank Arthur Grossman and C. J. Tu for the initial model, for microarray data, and for advice on biological plausibility.

References

1. P. D. Karp, M. Riley, M. Saier, I. T. Paulsen, S. Paley, and A. Pellegrini-Toole, "EcoCyc: Electronic Encyclopedia of E. coli genes and metabolism." *Nucleic Acids Research*, **28**, 56 (2000).
2. M. Kanehisa, "A database for post-genome analysis." *Trends in Genetics*, **13**, 375–376 (1997).
3. M. Tomita, K. Hashimoto, K. Takahashi, T. Shimizu, Y. Matsuzaki, F. Miyoshi, K. Saito, S. Tanida, K. Yugi, J. C. Venter, and C. Hutchison, "E-Cell: Software environment for whole cell simulation." *Bioinformatics*, **15**, 72–84 (1999).
4. J. R. Koza, W. Mydlowec, G. Lanza, J. Yu, and M. A. Keane, "Reverse engineering and automatic synthesis of metabolic pathways from observed data using genetic programming." *Pacific Symposium on Biocomputing*, **6**, 434–445 (2001).
5. B. Zupan, I. Bratko, J. Demsar, J. R. Beck, A. Kuspa, and G. Shaulsky, "Abductive inference of genetic networks." *Proceedings of the Eighth European Conference on Artificial Intelligence in Medicine* (Cascais, Portugal, 2001).
6. A. J. Hartemink, D. K. Gifford, T. S. Jaakkola, and R. A. Young, "Using graphical models and genomic expression data to statistically validate models of genetic regulatory networks." *Pacific Symposium on Biocomputing*, **6**, 422–433 (2001).
7. D. Ourston and R. Mooney, "Changing the rules: A comprehensive approach to theory refinement." *Proceedings of the Eighth National Conference on Artificial Intelligence*, 815–820 (AAAI Press, Boston, 1990).
8. G. Towell, *Symbolic Knowledge and Neural Networks: Insertion, Refinement, and Extraction.* Doctoral dissertation, Computer Sciences Department, University of Wisconsin, Madison (1991).
9. P. D. Karp, "Hypothesis formation as design." In *Computational Models of Scientific Discovery and Theory Formation*, Ed. J. Shrager and P. Langley (Morgan Kaufmann, San Francisco, 1990).

DISCOVERY OF CAUSAL RELATIONSHIPS IN A GENE-REGULATION PATHWAY FROM A MIXTURE OF EXPERIMENTAL AND OBSERVATIONAL DNA MICROARRAY DATA

C. YOO*, V. THORSSON♠, and G.F. COOPER*
*Center for Biomedical Informatics, University of Pittsburgh
8084 Forbes Tower, 200 Lothrop St., Pittsburgh PA 15213
♠The Institute for Systems Biology
4225 Roosevelt Way NE, Suite 200, Seattle Washington 98105

This paper reports the methods and results of a computer-based search for causal relationships in the gene-regulation pathway of galactose metabolism in the yeast *Saccharomyces cerevisiae*. The search uses recently published data from cDNA microarray experiments. A Bayesian method was applied to learn causal networks from a mixture of observational and experimental gene-expression data. The observational data were gene-expression levels obtained from unmanipulated "wild-type" cells. The experimental data were produced by deleting ("knocking out") genes and observing the expression levels of other genes. Causal relations predicted from the analysis on 36 galactose gene pairs are reported and compared with the known galactose pathway. Additional exploratory analyses are also reported.

1 Introduction

Causal knowledge makes up much of what we know and want to know in science. Thus, causal modeling and discovery are central to science. Experimental studies, such as biological interventions with corresponding controls, often provide the most trustworthy methods we have for establishing causal relationships from data. In such an experimental study, one or more variables are manipulated and the effects on other variables are measured. On the other hand, observational data result from passive (i.e., non-interventional) measurement of some system, such as a cell. In general, both observational and experimental data may exist on a set of variables of interest. For example, in biology, there is a growing abundance of observational gene expression data. In addition, for selected variables of high biological interest, there are data from experiments, such as the controlled alteration of the expression of a given gene.

Microarray technology has opened a new era in the study of gene regulation. It allows a relatively quick and easy way to assess the mRNA expression levels of many different genes. Large time-series datasets generated by microarray experiments can be informative about gene regulation. Microarray data have been analyzed using classification or clustering methods[1,2] and gene pathway (network) methods[3,4,5,6,7]. Dutilh[8] gives a short review of genetic networks. A more thorough review of genetic networks based on biological context was published by Smolen, et al.[9]. Wessels, et al.[10] conducted a limited comparison study of selected continuous genetic network models[3,11,12]. Unlike these previous methods, we introduce a method that models experimental interventions explicitly when evaluating hypotheses about causal relationships. Independently, Pe'er, et al.[13] have similarly modeled interventions but they did not model latent variables.

This paper reports the results from the analysis of a gene-expression dataset that was gathered by experimentation on galactose genes in the yeast *Saccharomyces cerevisiae*[14]. Our analysis focuses on the discovery of pairs of genes (X, Y) in which the expression of gene X has a causal influence on the expression of gene Y. As a representation of causation, we use causal Bayesian networks that include measured gene expression levels as well as possible latent causes that are not measured, such as the cellular level of proteins and small molecules[15]. The results of our causal analyses are compared with the known pathway. We also report novel causal relationships found in the analysis, which we believe deserve additional study.

2 Modeling Methods

A causal Bayesian network (or causal network for short) is a Bayesian network in which each arc is interpreted as a direct causal influence between a parent node (variable) and a child node, relative to the other nodes in the network[16]. Figure 1 illustrates the structure of a hypothetical causal Bayesian network structure containing five nodes. The probabilities associated with this causal network structure are not shown.

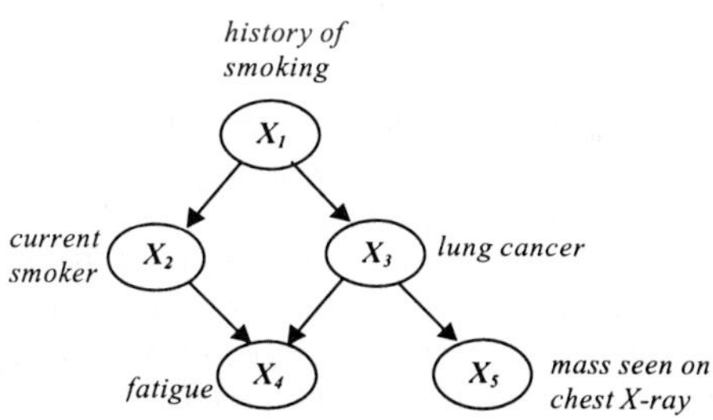

Figure 1. A hypothetical causal Bayesian network structure

The causal network structure in Figure 1 indicates, for example, that a *history of smoking* can causally influence whether *lung cancer* is present, which in turn can causally influence whether a patient experiences *fatigue*. The causal Markov condition gives the conditional independence relationships specified by a causal Bayesian network: *A node is independent of its non-descendants (i.e., non-effects) given its parents (i.e., its direct causes).*

The causal Markov condition permits the joint probability distribution of the n variables in a causal Bayesian network to be factored as follows[16]:

$$P(x_1, x_2, ..., x_n \mid K) = \prod_{i=1}^{n} P(x_i \mid \pi_i, K) \quad (1)$$

where x_i denotes a state of variable X_i, π_i denotes a joint state of the parents of X_i, and K denotes background knowledge.

Discovery of causal networks is an active field of research in which numerous advances have been—and continue to be—made in areas that include causal representation, model assessment and scoring, and model search[17,18,19,20].

2.1 The Modeling of Experimental Interventions

In this section, we briefly describe how to represent experimental intervention in a causal Bayesian network. First, consider that we have a Bayesian network S that represents the causal relationships among a set of genes (in terms of the regulation of expression). We need to augment this network to represent the experimental interventions (manipulations) that were performed to obtain the microarray data at hand. To do so, let M_{X_i} be a variable that represents the value k (from 1 to r_i) to which the experimenter deterministically manipulated gene X_i (e.g., a "knock out" of X_i). To represent this deterministic manipulation, we augment S so that (1) variable M_{X_i} is a parent of X_i in S, and (2) for all the joint states of π_i' we have that $P(X_i = k \mid M_{X_i} = k, \pi_i') = 1$, where π_i' are the parents of X_i other than M_{X_i}. Details about this representation are discussed by Cooper and Yoo[21].

For given microarray data D, we are interested in deriving the posterior probability of a causal network hypothesis S given data D and background knowledge (priors) K; that is, we want to know $P(S \mid D, K)$. In particular, we would like to find causal networks with posterior probabilities that are relatively high. A key step in the Bayesian derivation of $P(S \mid D, K)$ is to derive the marginal likelihood, namely $P(D \mid S, K)$. Specifically, $P(S \mid D, K)$ is proportional to $P(D \mid S, K) \times P(S \mid K)$, where $P(S \mid K)$ denotes prior belief (perhaps from background biological knowledge) that S is a valid causal hypothesis.

If D contains only passively observed data (no interventions), then Equation 2 provides a method for deriving the marginal likelihood, where r_i is the number of states that X_i can have, q_i denotes the number of joint states that the parents of X_i can have, N_{ijk} is the number of cases in D in which node X_i is observed to have state k when its parents have the states that are indexed by j, Γ is the gamma function, $N_{ij} = \sum_{k=1}^{ri} N_{ijk}$, α_{ijk} and α_{ij} express parameters of the Dirichlet prior distributions, and $\alpha_{ij} = \sum_{k=1}^{ri} \alpha_{ijk}$. The derivation and detailed explanation of Equation 2 are given in Cooper and Herskovits[22] and Heckerman, et al.[23]. Briefly, the equation assumes (1) discrete variables, (2) causal mechanisms that are local and independent (e.g., belief about the causes of gene X_i do not influence belief about the causes of gene X_j), (3) data exchangeability (i.e., the order in which the experiments were performed is irrelevant), (4) a particular representation of parameter prior probabilities that is based on Dirichlet probability density functions, and (5) no missing data or latent variables. The marginal likelihood given by Equation 2 is sometimes called the BDe metric[23].

$$P(D \mid S,K) = \prod_{i=1}^{n} \prod_{j=1}^{q_i} \frac{\Gamma(\alpha_{ij})}{\Gamma(\alpha_{ij} + N_{ij})} \prod_{k=1}^{r_i} \frac{\Gamma(\alpha_{ijk} + N_{ijk})}{\Gamma(\alpha_{ijk})} \quad (2)$$

Consider microarray data D, some of which was obtained under intervention and some of which was obtained by passive observation (i.e., data on "wild types"). As proven by Cooper and Yoo[21], augmenting S to contain manipulation variables of the type M_{X_i}, as described above, is equivalent to having the terms N_{ijk} in Equation 2 denote just those cases in which X_i was *passively observed* (e.g., not manipulated) to have value k when its parents were in state j. We used Equation 2 under this modification to derive $P(D \mid S, K)$ when D contains a mixture of data obtained under manipulation and under passive observation. For parameter priors, we assumed that $\alpha_{ijk} = \frac{1}{r_i q_i}$, which for the BDe metric is a commonly used non-informative parameter prior.

3 Scoring Methods of Structures with a Latent Variable

As mentioned in Section 2, Equation 2 assumes no latent variables. If we wish to model the possibility of a latent causal influence (which indeed we do), we need to extend Equation 2. Exact extensions involve applying Equation 2 an exponential number of times, one application for each possible state of the latent variables[24]. Such an exact method is not computationally feasible. Thus, in practice, we need to use approximation methods to evaluate $P(S \mid D, K)$ when S contains latent variables. In the remainder of this section, we explain the approximation method that we applied in this paper.

3.1 Causal Hypotheses Being Modeled

Figure 2 displays six local causal hypotheses (E_1 through E_6) that we model, shown as causal network structures. Variable X is the expression level of a given gene. Variable Y is the expression level of another gene. H is a latent (hidden) variable.

We denote an arbitrary pair of nodes in a given causal network S as (X, Y). If there is at least one directed causal path from X to Y or from Y to X, we say that X and Y are *causally related* in S. If X and Y share a common ancestor, we say that X and Y are *confounded* in S.

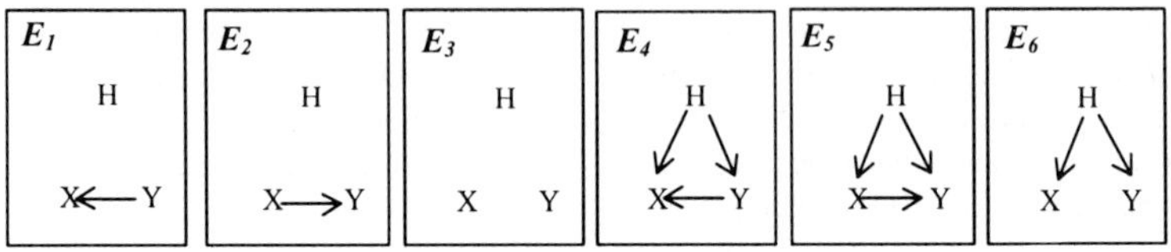

Figure 2. Six causal hypotheses on a pair (X, Y) of measured variables.

To derive the marginal likelihood (i.e., *score*) of causal structures E_1, E_2, and E_3, we can use Equation 2, since the hidden variable H does not influence either X or Y, and thus, can be ignored[21]. For structures E_4, E_5, and E_6, for which H *is* a confounding influence of X and Y, we use the scoring method discussed in the next section.

3.2 Implicit Latent Variable Scoring (ILVS) Method

Since explicit scoring of latent-variable models needs exponential time (in the number of database samples), approximation methods have been introduced in the literature, including methods based on stochastic simulation and on expectation maximization[23]. Unfortunately, these methods often require long computation times before producing acceptable approximations. Therefore, we developed a new method called the Implicit Latent Variable Scoring (ILVS) method[15].

The basic idea underlying ILVS is to (1) transform the scoring of a latent model S (e.g., model E_5 in Figure 2) into the scoring of multiple non-latent variable models, (2) score those non-latent models efficiently using Equation 2, and then (3) combine the results of those scores to derive an overall score (i.e., marginal likelihood). For instance, consider scoring E_5 with two types of samples. One type is data for which X and Y were passively observed. We can derive the marginal likelihood of this data using the causal network in Figure 3(a), which contains no latent variable. Let $P(D_o \mid E_5, K)$ denote this marginal likelihood. The other type of samples is data for which X was manipulated and Y was observed. We use the causal network in Figure 3(b) to derive the marginal likelihood of this data, namely $P(D_m \mid E_5, K)$. The different appearance of the arcs in Figures 3(a) and 3(b) signifies that these arcs are representing different distributions of X and Y. Figure 3(a) represents a situation in which X and Y are dependent based on a combination of direct causal influence of X on Y and on the confounding of X and Y by hidden process H. For the situation modeled by Figure 3(b), the experimental manipulation of X removes all causal influence of H on X. Therefore, Figure 3(b) represents a situation in which X and Y are dependent based only on the direct causal influence of X on Y – there is no additional confounding influence. Continuing the Bayesian analysis, if (as in ILVS) we assume our beliefs about the distribution of X and Y in the Figure 3(a) situation are independent of the beliefs about their distribution in the Figure 3(b) situation, then the overall marginal likelihood of all the data (the passively observed data and the experimentally manipulated data) is $P(D \mid E_5, K) = P(D_o \mid E_5, K) \times P(D_m \mid E_5, K)$. It is straightforward to extend the analysis to also include data in which Y was manipulated and X was passively observed.

In deriving the marginal likelihood of E_4 and E_6, ILVS uses a technique similar to the one just described for E_5. Yoo and Cooper[15] provide algorithmic details of ILVS and a proof of its convergence to the correct generating structure in the large sample limit.

$$X \Rightarrow Y \qquad\qquad X \rightarrow Y$$
(a) (b)

Figure 3. Two non-latent variable structures used to score a latent-variable structure.

4 Gene Expression Dataset Analyses

We applied the ILVS algorithm to a gene-expression dataset to produce putative causal relationships among the genes. This section briefly describes the dataset and summarizes the steps we followed in preparing the data for analysis.

4.1 Dataset Description

The cDNA microarray data we analyzed were obtained from experiments that focused on the galactose utilization pathway in the yeast *Saccharomyces cerevisiae* as reported by Ideker, et al.[14]. The experiments included single gene deletion involving nine of the key genes[1)] that participate in yeast galactose metabolism. All microarray experiments were repeated four times. For each experiment, one of the nine genes was deleted, or alternatively, the experiment used a *wild-type* cell wherein no genes were deleted. For each of those 10 experimental conditions, galactose was available extracellularly in one set of experiments and absent in another set. Thus, there were a total of 20 different experimental conditions. Since each of those 20 experiments was repeated four times, the overall dataset contains results from 80 experiments. In each

---

1) Nine galactose genes are *Gal1, Gal2, Gal3, Gal4, Gal5(PGM2), Gal6(LAP3), Gal7, Gal10,* and *Gal80.*

experiment, 5,936 gene expression levels were measured, corresponding to almost the entire *Saccharomyces cerevisiae* genome.

4.2 Dataset Preparation

This section describes the five data preparation steps that we applied to the data. We tried different methods of discretization (i.e., forcing all knock out genes to have expression level of 0; discretization with clustering) but found no significant difference in the overall predicted performances.

1) Genes that had expression levels missing in all four repetitions of a given experiment were excluded from the analysis (n = 195 genes were excluded).
2) If the expression level of a gene was missing in some experiment, its value was estimated as the mean value of the available measurements for that gene.
3) Negative intensities were assumed to be 0.
4) Let X_i^* denote the intensity for gene X_i, which serves as an indicator of the expression level of X_i in an experiment in which some gene (not necessarily X_i) was knocked out. Similarly, let X_i^r denote the intensity, which is an indicator of the expression level of X_i when no genes were manipulated (wild type). The relative intensity for gene X_i was calculated as $\log(X_i^*/X_i^r)$. If X_i^* was 0, we used the minimum $\log(X_i^*/X_i^r)$ value for gene X_i.over all 80 experiments for gene X_i. If X_i^r was 0, we used maximum $\log(X_i^*/X_i^r)$ value for gene X_i over all 80 experiments. If both X_i^* and X_i^r were 0, $\log(X_i^*/X_i^r)$ was set to 0.
5) Discretization was performed based on each gene's expression level mean m and standard deviation δ over all 80 samples. All genes were assigned three states: 0 was assigned to any value less than $m-\delta$, 1 was assigned to any value greater than or equal to $m-\delta$ and less than $m+\delta$, and 2 was assigned to any value greater than or equal to $m+\delta$.

4.3 Analyses

We applied ILVS to every pair of the 5,936 yeast genes that includes one or both genes from the nine galactose genes. For each gene pair, we used the method in Section 3.2 to derive a posterior probability for each of the six causal hypotheses in Figure 2. We analyzed our results in two main parts.

The first part focused on just the nine galactose genes that were manipulated. Since the causal relationships among these nine genes are understood relatively well, we assume that these generally accepted relationships are correct and can serve as a standard against which to compare the output of ILVS. The ILVS method was considered to label a given gene pair as having causal relationship R (among the six possibilities in Figure 2) if $P(R \mid D, K)$ was greater than 0.5. For each gene pair (*X*, *Y*), there were 8 cases in each *X* and *Y* (and all other genes for that matter) were passively observed, 8 cases in which *X* was knocked out and *Y* observed, and 8 cases in which *Y* was knocked out and *X* observed.

The second part of our analysis was more exploratory than evaluative. In particular, we examined all 9 × 5,732 = 51,588 gene pairs consisting of one gene from the nine galactose genes and one gene from outside that set. Let *X* denote one of the nine galactose genes and let *Y* denote one of the other 5,732 genes studied. For each gene pair (*X*, *Y*), there were 8 cases in which *X* and *Y* were observed, 8 cases in which *X* was knocked out and *Y* observed, and *no cases* in which *Y* was knocked out and *X* observed. We identified every gene pair for which one of the six causal hypotheses in Figure 2 was greater than probability 0.9. Each pair represents a hypothesis about the nature of a causal relationship that has yet to be characterized.

5 An Investigation of ILVS: Results

This section first presents the results of exploring relationships just among the nine galactose genes, and then between those nine genes and all other genes in the dataset.

5.1 Galactose Genes

The first column in Table 1 shows pairwise causal relationships that represent generally accepted biological knowledge about the galactose gene-regulation pathway, as summarized in Ideker, et al.[14]. The table also shows the results of applying ILVS to the 36 pairs of galactose genes. Only the causal relationships that had a posterior probability greater than 0.5 are shaded. No relationship had a causal hypothesis with a probability higher than 0.9, possibly due to the small number of samples in the dataset.

Upon comparing the ILVS predictions with the known galactose metabolic pathway, we show the results in Table 1. Each shaded row in reversed font represents ILVS predictions that agree with accepted biological knowledge. The shaded rows in bold font denote that for reference structure T, there exists uncertainty about its validity. Other shaded rows represent predictions that are inconsistent with accepted biological knowledge.

The errors in Table 1(a) are due to ILVS assuming genes are independent, when biological knowledge indicates an expected dependence. We label these as false negatives. There are at least two plausible reasons for these errors. First, the sample size (24 samples per pair) is small, and thus, unless the dependence is strong, that dependence may not be apparent from the data. Second, the biological knowledge we used as a reference standard expresses general patterns of causal dependency among the galactose genes; not all of those patterns were necessarily revealed by the experiments performed in creating the dataset that was given as input to ILVS.

Table 1. The most probable causal hypotheses predicted by ILVS as representing relationships among the nine manipulated galactose genes under study

Reference Structure T	ILVS Predicted Structure: Structure S	$P(S\|D,K)$	$P(T\|D,K)^{\S}$
Gal6---Gal7	Gal6 Gal7	0.80	0.02
Gal2⇒Gal7	**Gal2 Gal7**	**0.80**	**0.01**
Gal1---Gal5	Gal1 Gal5	0.80	0.03
Gal2⇒Gal5	**Gal2 Gal5**	**0.75**	**0.01**
Gal2⇒Gal10	**Gal2 Gal10**	**0.74**	**0.01**
Gal1---Gal6	Gal1 Gal6	0.73	0.03
Gal6---Gal10	Gal6 Gal10	0.73	0.02
Gal80⇒Gal5	Gal5 Gal80	0.70	0.04
Gal2⇒Gal1	**Gal1 Gal2**	**0.57**	**0.16**
Gal2⇒Gal6	**Gal2 Gal6**	**0.56**	**0.01**
Gal5---Gal7	Gal5 Gal7	0.55	0.02
Gal5---Gal10	Gal5 Gal10	0.54	0.07
Gal3⇒Gal6	Gal3 Gal6	0.51	0.10
Gal3⇒Gal5	*Gal3 Gal5*	0.45	0.08
Gal3⇒Gal10	*Gal3 Gal10*	0.43	0.25
Gal4→Gal6	*Gal4 Gal6*	0.41	0.08
Gal3---Gal80	*Gal3 Gal80*	0.33	0.11
Gal4→Gal5	*Gal4 Gal5*	0.28	0.08
Gal5---Gal6	*Gal5 Gal6*	0.28	0.09

(a) E_3

Reference Structure T	ILVS Predicted Structure: Structure S	$P(S\|D,K)$	$P(T\|D,K)^{\S}$
Gal80⇒Gal1	Gal80→Gal1	0.56	0.16
Gal3⇒Gal1	*Gal3→Gal1*	0.33	0.33
Gal3⇒Gal7	*Gal3→Gal7*	0.31	0.31
Gal80⇒Gal10	*Gal80→Gal10*	0.30	0.17
Gal80⇒Gal7	*Gal7→Gal80*	0.25	0.11

(b) E_1 and E_2

Reference Structure T	ILVS Predicted Structure: Structure S	$P(S\|D,K)$	$P(T\|D,K)^{\S}$
Gal4→Gal7	**Gal4⇒Gal7**	**0.83**	**0.01**
Gal10---Gal7	**Gal10⇒Gal7**	**0.81**	**0.004**
Gal4→Gal80	**Gal4⇒Gal80**	**0.58**	**0.10**
Gal4→Gal1	*Gal4⇒Gal1*	0.47	0.04
Gal2⇔Gal3	*Gal2⇒Gal3*	0.42	0.45*
Gal4→Gal10	*Gal4⇒Gal10*	0.41	0.003
Gal2↔Gal4	*Gal2⇒Gal4*	0.40	0.33†
Gal3⇒Gal1	*Gal3⇒Gal1*	0.33	0.33
Gal3⇒Gal7	*Gal3⇒Gal7*	0.31	0.31

(c) E_4 and E_5

Reference Structure T	ILVS Predicted Structure: Structure S	$P(S\|D,K)$	$P(T\|D,K)^{\S}$
Gal80⇒Gal6	Gal6---Gal80	0.63	0.06
Gal1---Gal7	***Gal1---Gal7***	***0.60***	***0.60***
Gal1---Gal10	***Gal1---Gal10***	***0.57***	***0.57***
Gal2⇔Gal80	**Gal2---Gal80**	**0.57**	**0.35***
Gal4→Gal3	*Gal3---Gal4*	0.49	0.29

(d) E_6

Notation: The symbol → represents the relationship given by causal structure E_1 and E_2 (from Figure 2). Likewise, a blank space is used for E_3, a ⇒ for E_4 and E_5, and a --- for E_6. In column 1, a birectional arc indicates that there is a known feedback pathway between the two genes (e.g., *Gal2⇔Gal3*). The symbol † indicates summing the posterior probabilities for E_1 and E_2. A * indicates summing the probabilities for E_4 and E_5. The column labeled $P(T|D,K)^{\S}$ gives the probability that ILVS assigned to the reference structure in column 1. See the text for an explanation of the shaded results.

The errors in Tables 1(b), 1(c), and 1(d) result from the most probable hypothesis (according to ILVS) being inconsistent with assumed biological knowledge. There are only two pairs (the shaded ones in Table 1(d)) for which

ILVS obtains exactly the correct hypothesis. Consider, however, the following relaxation: A hypothesis is correct if it indicates that there is a causal pathway from X to Y (with or without confounding) and according to existing biological knowledge there is indeed a causal pathway from X to Y (with or without confounding). For example, under this interpretation *Gal80*→*Gal1* would be correct, since the reference structure is *Gal80*⇒*Gal1*, which includes a causal path from *Gal80* to *Gal1*. Under this relaxation of correctness, 12 of the 17 unique relationships (71%) in Tables 1(b), 1(c), and 1(d) are correct.

5.2 Galactose Genes and Other Genes

In this section, we report the results of an exploratory analysis. The purpose of this section is to illustrate an initial step in using computer-based, data-intensive methods to hypothesize causal relationships.

Table 2. Types of highly probable (>0.9) gene pairs predicted by ILVS from 51,588 considered pairs.

E_1, E_4, and E_6	E_2	E_3	E_5
1,329 (2.58%)	4 (0.008%)	586 (1.14%)	1,113 (2.16%)

Table 3. Conditional distributions of four genes that are reported by ILVS in Table 2 to be highly probable (>0.9) effects of *Gal1* or *Gal2*. For example, Table 3(a) presents *P*(*YBR096W* | *Gal2*).

(a)

YBR096W	*Gal2*: low	no ch	high
low	0.021	0.301	0.083
no change	0.021	0.688	0.833
high	0.958	0.011	0.083

(b)

YMR086W	*Gal2*: low	no ch	high
low	0.021	0.301	0.083
no change	0.021	0.688	0.833
high	0.958	0.011	0.083

(c)

SSU1	*Gal2*: low	no ch	high
low	0.958	0.011	0.083
no change	0.021	0.688	0.083
high	0.021	0.301	0.833

(d)

SER3	*Gal1*: low	no ch	high
low	0.026	0.493	0.026
no change	0.949	0.493	0.026
high	0.026	0.013	0.949

In an analysis of 51,588 gene pairs, ILVS scored 5.9% of the node pairs as having a high probability (>0.9) causal hypothesis (see Table 2). Since the 5,732 genes were not experimentally deleted, ILVS could not distinguish among E_1, E_4, and E_6 when one of the nine galactose genes were treated as variable X in Figure 2; therefore, E_1, E_4, and E_6 are grouped in Table 2. The four unconfounded causal relationships are *Gal2*→*YBR096W*, *Gal2*→*YMR086W*, *Gal2*→*SSU1*, and *Gal1*→

SER3. Conditional distributions in Table 3 suggest that *Gal2* is acting as a relative inhibitor for *YBR096W* and *YMR086W*, and as an activator of *SSU1*. Table 3(d) suggests that *Gal1* is acting as an activator for *SER3*. Interestingly *SER3* is one of only two proteins that are known to bind with *Gal1* and its regulatory role is unknown[25].

To evaluate all 51,588 gene pairs, ILVS required about four and a half hours of CPU time on a Pentium 500MHz Linux machine with 384M RAM.

6 Summary and Discussion

ILVS is a novel, efficient causal discovery algorithm that can model causal hypotheses with (and without) latent variables. The method attains its efficiency by modeling one pair of variables at a time and by evaluating latent-variable models implicitly, rather than explicitly. In previous work, the ILVS algorithm has been shown to be asymptotically correct in the large sample limit[15]. Thus, with enough valid data, it is guaranteed to find the correct causal relationship between each pair of variables in a dataset. The ILVS method can use data obtained from passive observation and from active experimental manipulation. Since much gene-expression data is of both types, the ILVS method is of particular relevance to work on discovery of gene-regulation pathways from gene-expression data.

We applied the ILVS method to an available dataset containing gene expression levels from experiments that focused on galactose metabolism. These early results are promising, but in need of improvement. The error rates in re-discovering the known galactose gene-regulation pathway were high. Possible reasons include a small set of samples and limited experimental conditions and variation; the influence of these two limitations is supported by the fact that ILVS did not give a high probability (> 0.9) to any of the galactose causal relationships that it hypothesized. For the false positives output by ILVS, some may simply be wrong, while others may represent unknown causal relationships within galactose gene regulation.

An exploratory analysis of the galactose dataset yielded approximately 3,000 causal relationships (out of 51,588) that appear highly probably (according to ILVS). Those hypothesized relationships have not yet been investigated in the laboratory. In future work, we intend to work with yeast biologists to begin to explore some of the most interesting and promising of these hypotheses.

Algorithmic improvements that we are pursuing include performing Bayesian model averaging (rather than model selection), modeling continuous data directly (rather than discretizing the data), developing and using informative models of the types of measurement noise that exists in microarray experiments, modeling covariates (e.g., galactose levels) in evaluating the causal relationships among a pair of genes, modeling gene-regulation feedback, and incorporating into our Bayesian analyses informative structure and parameter prior probabilities.

Acknowledgments

We thank the Computational Systems Biology Group at the University of Pittsburgh and Carnegie Mellon University for helpful discussions. This research was supported by NSF grant IIS-9812021 and NASA grant NRA2-37143.

References

1. Spellman, P.T., G. Sherlock, M.Zhang, V.Iyer, K.Anders, M.Eisen, P. O. Brown, D. Botstein, and B. Futcher (1998). *Mol. Bio. Cell* 9, p3273–3297.
2. Getz, G., E. Levine, and E. Domany (2000). *Proc. Na. Ac. Sc.*, 97(22).
3 Chen, T., V. Filkov, and S. S. Skiena (1999). *RECOMB.*
4. D'haeseleer, P., S.Liang, and R.Somogyi (2000). *Bioinfo.*, 16(8):707-726.
5. Friedman, N., M. Linial, I. Nachman, and D Pe'er (2000). *J. Comp. Bio.*
6. Hartemink, A.J., D.K. Gifford, T.S. Jaakkola, and R.A. Young (2001). *PSB.*
7. Maki, Y., D. Tominaga, M. Okamoto, S. Watanabe,Y. Eguchi (2001) *PSB.*
8. Dutilh, B. (1999). Gene Networks from Microarray Data. Unpublished paper.
9. Smolen, P., D.Baxter, and J.Byrne (2000). *Bul. Math. Bio.* 62, p247–292.
10. Wessels, L.F.A., E.P. Van Someren, and M.J.T. Reinders (2001). *PSB.*
11. Arkin, A., P. Shen, and J. Ross (1997). *Science* 277, p1275–1279.
12. Weaver, D., C.Workman, and G. Stormo (1999). *PSB* 4, p112–123.
13. Pe'er,D., A.Regev, G.Elidan, and N.Friedman (2001). *ISMB.*
14. Ideker, T., V.Thorsson,J. Ranish, C.Rowan, J.Buhler, J.Eng, R.Bumgarner, D.Goodlett, R.Aebersold, L. Hood (2001). *Science* 292: 929–934. Dataset is available at: http://www.sciencemag.org/cgi/content/full/292/5518/929/DC1
15. Yoo, C. and G. F. Cooper (2001). *CBMI research report* CBMI-173.
16. Pearl, J. (1988). Probablistic Reasoning in Intelligent Systems. *Morgan Kaufmann*, San Mateo, CA.
17. Glymour,C., G.Cooper (1999). Computation, Causation, Discovery. *MIT Pr.*
18. Spirtes, P., C.Glymour, R.Scheines (2000). Causation, Prediction, and Search. *MIT Press.*
19. Pearl, J. (2000) Causality. *Cambridge University Press.*
20. UAI 2001: http://robotics.stanford.edu/~uai01/pastconferences.htm.
21. Cooper, G.F., and C. Yoo (1999). *UAI '99*, Stockholm, Sweden.
22. Cooper, G.F., and E. Herskovits (1992).*Mach. Learning, 9*, 309-347.
23. Heckerman,D., D.Geiger, D.Chickering (1995). *Mach. Learning 20* 197-243.
24. Cooper, G.F. (1995). *Journal of Intelligent Information Systems, 4*, p71-88.
25. Ito,T.,Chiba,Ozawa,Yoshida,Hattori,Sakaki (2001) *Proc. Natl. Acad. Sci.* 10.

PHYLOGENETIC GENOMICS AND GENOMIC PHYLOGENETICS

SCOTT STANLEY, BENJAMIN A. SALISBURY
Genaissance Pharmaceuticals, 5 Science Park,
New Haven, CT 06511, USA

Phylogenetics has a long history, with many contributions to the field actually predating Darwin. Genomics, in contrast, is barely a decade old. While these two fields may be disparate in their ages, each has much to contribute to the further development of the other. Phylogenetics provides a rich source of methodologies able to facilitate and enhance genomic research. Genome structure and function analyses can be strengthened through examining evolutionary history, whereby patterns and rates of genomic evolution can be inferred. The analysis of evolutionary history is similarly empowered by a genomic perspective, providing both a wealth of characters ideally suited for phylogenetic analysis as well as interesting subject matter in the form of comparative genomics.

This session includes three examples of research that address issues important to both phylogenetics and genomics. Zilversmit et al. provide a case study where high throughput sequencing enables phylogenetics at a genomic scale. Wang et al. propose novel phylogenetic algorithms and compare their performance to other phylogenetic algorithms designed to analyze gene order. Page and Cotton address the problem in genomic evolution imposed by gene duplication. Together, these studies address several important areas in the new field of phylogenomics.

The genomic revolution has been both enabled and defined by high throughput DNA sequencing. Advances in DNA sequencing technologies have allowed for the generation of extremely large DNA sequence data sets, many including entire genomes. Zilversmit et al. demonstrate how high throughput sequencing efforts can be applied to phylogenetic questions, using the insect family Drosophilidae as an example. They outline a method that can be applied to other non-model organisms, enabling other researchers to apply high throughput techniques to phylogenetics. They discuss how the size of data sets affects the results of phylogenetic analyses, particularly as data sets grow to include many regions from throughout the genome. In this way, Zilversmit et al. present a study that is truly phylogenetic and genomic.

Wang et al. present a novel phylogenetic method designed for gene order data, allowing researchers to conduct phylogenetic analyses using gene order data from whole or partial genomes. Gene order represents an attractive alternative to DNA sequence data, particularly when rates of DNA sequence evolution are high, or when divergence time is great. While DNA sequence data have provided an important source of characters for phylogenetic analyses in recent years, limitations associated with high rates of evolution together with the limited number of character states available have forced researchers to look to other sources of molecular data to overcome these problems. This is particularly true when ancient divergences are involved. Gene order is one such example, and Wang et al. have added another tool to the phylogeneticists' toolbox for working on difficult phylogenetic problems. Wang et al. compare distance-based and parsimony-based methods of analyzing gene order phylogenetically. While studies of gene order have generally been restricted to genomes from organelles, it is possible to apply the methods presented by Wang et al. to nuclear genomes as well. Wang et al. found that their novel method based on a coding scheme proposed by Bryant outperformed the distance- based methods as well as the other parsimony method. Perhaps more importantly, this paper highlights the need for continued development of alternative phylogenetic methods that can be applied to something other than DNA sequence data.

Page and Cotton provide a beautiful example of the complementarity of phylogenetics and genomics. They demonstrate the use of gene tree reconciliation and a new method of mapping duplications to elucidate vertebrate phylogeny and genome evolution. Gene phylogenies may differ from each other and from species phylogenies because of complex histories of gene duplication, lineage sorting and gene loss. Page and Cotton use genomic-scale data (118 gene families) for phylogeny estimation and then map inferred gene duplications onto the resulting phylogeny to test for hypothesized genome duplications. Remarkably, they find that minimizing gene duplications produces a tree highly concordant with traditional views (where individual gene trees would resemble that phylogeny estimate poorly). Two commonly hypothesized genome duplications early in vertebrate history, however, were not evident in this analysis.

Our hope is that these papers can help further the dialogue between phylogeneticists and those in the field of genomics. Many researchers in systematics remain unaware of the importance of genomics to their own work, and vice versa. We feel that each has much to offer the other. This session is intended to spotlight those synergies and establish "phylogenomics" as a bona fide and important new field of study.

Acknowledgments

We are grateful to all the authors who submitted papers to this session and to the generous individuals who carefully reviewed them.

SHALLOW GENOMICS, PHYLOGENETICS, AND EVOLUTION IN THE FAMILY DROSOPHILIDAE

M. ZILVERSMIT, P. O'GRADY, R. DESALLE

American Museum of Natural History, Department of Invertebrate Zoology, Central Park West @ 79th Street, New York, NY 10024, USA

The effects of the genomic revolution are beginning to be felt in all disciplines of the biological sciences. Evolutionary biology in general, and phylogenetic systematics in particular, are being revolutionized by these advances. The advent of rapid nucleotide sequencing techniques have provided phylogenetic biologists with the tools required to quickly and efficiently generate large amounts of character information. We use family Drosophilidae as a model system to study phylogenetics and genome evolution by combining high throughput sequencing methods from the field genomics and standard phylogenetic methodology. This paper presents preliminary results from this work. Separate data partitions, based on either gene function or linkage group, are compared to a combined analysis of all the data to assess support on phylogenetic trees.

1 Introduction.

The traditional goal of molecular systematics has been to use the character information found in one or a few well-characterized loci to infer the evolutionary history of a selected group of taxa. The assumptions here are twofold; first, that the gene history will accurately represent the history of the taxa sampled and second, that the evolutionary rate of the gene selected will provide ample characters to resolve each node in the phylogeny. Many authors have rightly pointed out that the history of a single gene sequence sampled for a species is not necessarily reflective of the history of that species.[1] Reasons for this discrepancy include horizontal gene transfer, lineage sorting of ancestral polymorphisms, and other phenomena.[2] Furthermore, depending on the divergence times of the taxa being examined, one or a few genes may not contain a sufficient number of characters to robustly reconstruct all nodes in the phylogeny.

A variety of genome projects have successfully employed high throughput methods to rapidly and reliably generate large numbers of sequences.[3,4] As a result, a wealth of characters are now readily available to the molecular systematist. Many systematists have begun to take a "phylogenomic" approach by using genomic information to infer phylogenetic relationships. However, other than prokaryotic and organellar genomes, it is unlikely that most systematists will ever be able perform research on a truly genomic scale. Using a large number of loci, sampled throughout the genome, is a far more powerful way to infer phylogenetic relationships than currently being employed. Such an approach not only yields a more accurate view of how the genome as a whole is evolving, but will also increase the likelihood that loci evolving at many different rates are sampled, thereby increasing the numbers of characters supporting each node. Even though the sequences of entire genomes are not being compared, this genomic sample approach to systematics can be referred to as shallow genomics.

We are currently taking a similar approach to address phylogenetic relationships within the family Drosophilidae. We have developed a relatively simple and cost-effective system for high throughput gene sequencing. High throughput sequencing is not new, as it is exactly this sort of approach that enabled the existence of the genome projects. The methods used for those projects and in commercial laboratories are beyond the reach of most academic laboratories because of the tremendous cost of reagents, equipment, and labor. We have developed a system that is inexpensive enough to be used by several researchers in the same laboratory while being simple enough to be completed by a single individual. This was accomplished by pooling existing methods and protocols (from publications, Web sites and personal communication) and then combining and modifying them.[5,6] The improvements are enhanced by using a capillary sequencer, a system that greatly increases ease and efficiency of data generation relative to slab gel systems.

1.1 Drosophilidae as a shallow genomic model system

This family is an excellent model system for genomic studies because the entire sequence of *Drosophila melanogaster* is now completed and that of a second drosophilid, *D. pseudoobscura*, is nearing completion. Coupling this impressive amount of sequence data with gene location data derived from in situ hybridization to the polytene chromosomes, we can insure that the entire genome is sampled.

Phylogenetic relationships among drosophilid genera based on molecular and some morphological analyses are incongruent.[7,8] The major conflict is in the placement of the endemic Hawaiian Drosophilidae, a lineage consisting of two genera, *Drosophila* and *Scaptomyza*. Grimaldi[9] proposed that the Hawaiian *Drosophila* was actually closely related to a clade of mycophagous genera, the *Hirtodrosophila* genus complex. The Hawaiian *Scaptomyza* formed a clade with the continental *Scaptomyza* and were actually more closely related to the subgenus *Drosophila* than the Hawaiian *Drosophila*. In contrast, molecular data,[7] as well as previous morphological work,[10] suggest that the Hawaiian *Drosophila* and all *Scaptomyza* (Hawaiian and continental) form a clade. This group is closely related to the subgenus *Drosophila*, not the *Hirtodrosophila* genus complex.

Our goals are to (1) test the monophyly of the Hawaiian Drosophilidae and (2) determine the sister group relationships of the Hawaiian *Drosophila* and the genus

Scaptomyza. These analyses will examine congruence between partitions generated from different chromosomes and different functional classes of genes (i.e., transcription factors and enzymes). Although the explicit intention of using shallow genomic methods in our lab is for phylogentics, we expect that the data gathered will reveal information about genome evolution in a number of drosophilid species. Our methods allow us to gather data from unstudied genes and gene regions to examine not only organismal evolution, but molecular and genome evolution as well.

2 Methods

1.1 Primer design and testing

Our primer design takes advantage of the large number of known sequences from *Drosophila*, humans, and other organisms to design large numbers of primers that are functional across large phylogenetic distances. Like the CATS system,[5] we have systematically scanned GenBank for genes and gene regions that might be appropriate for primer design and use. Oligonucleotide primers homologous to *D. melanogaster* genes are designed from each chromosomal segment via GADFLY (Genome Annotation Database of *D. melanogaster*). These sequences, and those from two or more homologous genes from other species, are then input to the CODEHOP web site.[6] CODEHOP produces ortholog blocks of sequence and a series of potential primers using the genetic code and codon bias tables. Optimal primers are synthesized based on the following criteria:

(1) The primers should amplify a fragment between 300-800 bp. Shorter fragments are better if template DNA has been fragmented, while longer targets are more economical, as 800 bp is the current maximum fragment size that can be sequenced with a single primer in both directions.
(2) Target sequences should be variable as to the level of sequence conservation between taxa, representing as complete a range as possible.
(3) Primers are chosen with low to intermediate degeneracy, (less than or equal to 32-fold).

In order to automate the sequencing step, we have incorporated the sequences for the T3 and T7 universal sequencing primers into our oligonucleotides. These universal primer sites are also located on the vector (TOPO pCR4; Invitrogen) used for cloning PCR products should that be necessary. With this addition only this single set of primers is needed to sequence any number of different genes, either directly or ligated into a vector, and enables significantly greater throughput in less time for virtually no extra cost.

Primers are initially screened for success in PCR, assessing both for effectiveness in amplifying gene fragments from divergent taxa (e.g., multiple diptera, plus sturgeon, teleost and mammal samples) and the number of products produced. Testing on such a broad range of taxa is done to indicate both how well a primer will work with members of the family Drosophilidae, for our immediate test study, and its amplification success among a variety of other taxa for later work. In order to reduce problems due to secondary amplification and paralogy, only those primers producing single bands are used for sequencing.

1.2 Gene fragment amplification, processing and sequencing

PCR products are then purified either using a standard isopropanol precipitation method adapted for microtitre plates or via filtration using SOPE resin (Edge Biosystems) with lab-made resin columns in a microtitre-style column rack. The former method has a significant cost effectiveness advantage and the latter permits superior speed and ease at low cost for a commercial product. The purified PCR products are then used as the template for dye terminator (BigDye; ABI) sequencing reactions using 0.5ul dye terminator mixed with an equal part dye terminator extender buffer (Tris-HCl, pH9) per reaction with only universal sequencing primers. Using such a low volume of dye terminator allows for some of the most significant cost reduction for a product that cannot easily be replaced. The sequencing reactions are also purified by an adapted alcohol precipitation method and resuspended in 5ul ABI deionized formamide and sequenced on an ABI Prism 3700 DNA analyzer.

The Sequencher software package (Gene Codes) was used to edit raw sequences and create consensus sequences from several clones. Sequences were aligned either by eye or by using a multiple alignment program.[11] PAUP, version 4.0[12] was then used to generate phylogenetic trees using parsimony, likelihood, and distance methods. Other sequence analyses were performed in MacClade.[13]

2.3 Phylogenetic Analyses

PAUP 4.0[13] was used to perform all analyses in this study. Most parsimonious trees (MPTs) were found using an exhaustive search algorithm. Tree characteristics, shown in Figures 1 and 2, are number of MPTs, number of steps on the shortest tree, number of parsimony informative characters (PICs), and the consistency index (CI). Bootstrap proportions[14] and decay indices[15] were used as measures of support on the phylogeny. Partitioned branch support[16] was calculated using TreeRot[17]. Best estimate data removal indices[18] were also calculated.

3. Results

3.1 Primer Design and Amplification.

To date, we have designed 500 primers, amplifying a range of different functional classes of genes, for loci located on each major linkage group in *Drosophila*

Table 1. Characteristics of the loci used in this study.

Gene/Locus	Size (bp)	PICs	% PIC[1]	Functional Class
16S*	908	66	7.2 / 3.5	mitochondrial ribosomal RNA
28S*	812	29	3.6 / 1.5	nuclear ribosomal RNA
Adh*	771	198	25.7 / 10.5	enzyme
amy	186	4	2.2 / 0.2	enzyme
BcDNA	630	119	18.9 / 6.3	peptide transporter
boss	225	31	13.8 / 1.7	signal transduction
COII*	688	127	18.5 / 6.8	mitochondrial protein coding
dpp	436	59	13.5 / 3.1	TGF-B receptor ligand
dsh	468	20	4.3 / 1.1	segment polarity
esc	374	55	14.7 / 2.9	gene silencing
fkh	201	35	17.4 / 1.9	RNA polymerase II trans. fact.
Gpdh*	770	119	15.5 / 6.3	enzyme
glass	763	48	6.3 / 2.6	RNA polymerase II trans. fact.
kuz	544	2	0.4 / 0.01	enzyme (metalloendopeptidase)
mago	360	9	2.5 / 0.5	germ plasm assembly
pdm2	599	4	0.7 / 0.02	RNA polymerase II trans. fact.
Sod*	439	131	29.8 / 7.0	enzyme
sia	448	56	12.5 / 3.0	zinc finger domain
snf	295	116	39.3 / 6.2	small nuclear ribonucleoprotein
Xdh*	2078	449	21.6 / 23.9	enzyme
PSAP	448	74	16.5 / 3.9	enzyme
26S prot	431	101	23.4 / 5.4	26S proteosome regulatory subunit
CG3869	268	26	9.7 / 1.4	unknown
Total	**13124**	**1878**		

1. Percent parsimony informative characters in each locus / percent parsimony informative characters in entire data matrix. * Primers used prior to this project.

melanogaster. Of these primer pairs 179 have been synthesized and tested. Thus far, we have been able to successfully amplify 125 of the 179 primer pairs we have tested. Primer pairs that fail to give any amplification product are immediately rejected. Of the 125 primer pairs that gave a PCR product, about 2/3 give strong, single band amplification products that we anticipate little difficulty sequencing 16 of which are at a level of completion to be used in the preliminary data matrix (Table 1). These loci are found on the X, second and third chromosomes of *Drosophila melanogaster* and represent a broad range of functional classes of genes.

3.2 Combined Phylogenetic Analysis.

Our phylogenetic results indicate that the endemic Hawaiian Drosophilidae (Hawaiian *Drosophila* plus *Scaptomyza*) are monophyletic. *Scaptomyza* is more

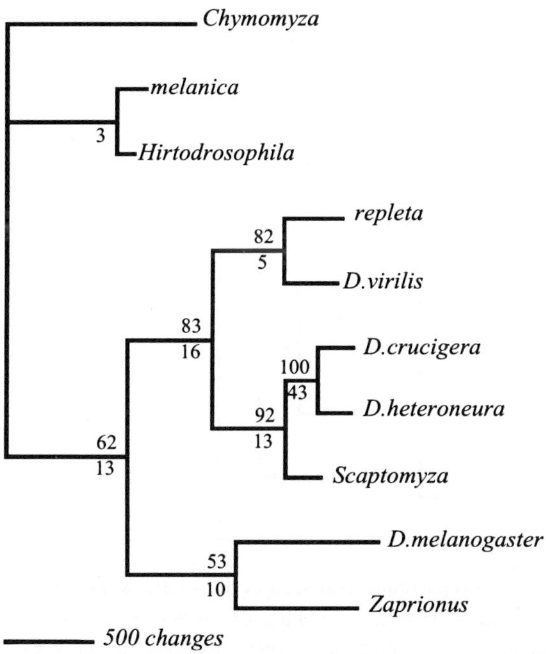

Figure 1. Phylogram showing relationships based on maximum parsimony analysis of 23 molecular partitions (Table 1). 1MPT, 7591 steps, 1879 PICs. Support is indicated above (bootstrap proportions) and below (decay indices) each node.

closely related to the Hawaiian *Drosophila* species *D. crucigera* and *D. heteroneura*, suggesting that the genus *Scaptomyza* may have originated on Hawai'i and subsequently colonized the rest of the world. The Hawaiian Drosophilidae clade is the sister group of two subgenus *Drosophila* species groups, *virilis*

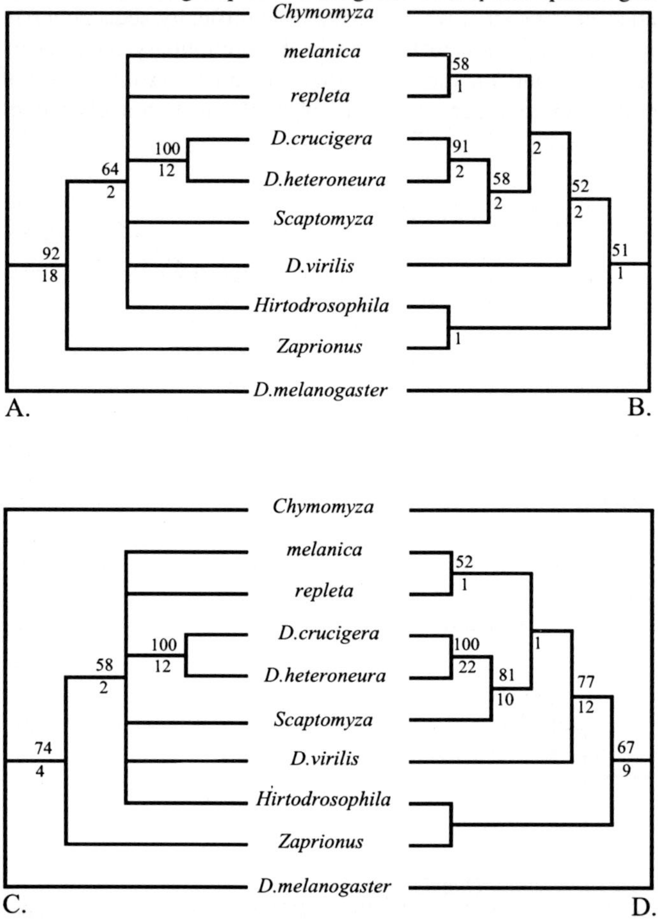

Figure 2A-D. Phylogenetic analysis of several partitions examined in this study. A. Enzyme tree (Adh, amy, Gpdh, kuz, PSAP, Sod, Xdh). 6MPTs, 3539 steps, 977 PICs, CI – 0.760. **B.** Transcription factor tree (BcDNA, boss, dpp, dsh, fkh, glass, mago, pdm2, sia). 1MPT, 2106 steps, 381 PICs, CI = 0.872. **C.** Chromosome 2 tree (26S, dpp, Gpdh, esc, pdm2, kuz, Adh, amy, mago). 10 MPTs, 2247 steps, 551 PICs, CI = 0.823. **D.** Chromosome 3 tree (PSAP, Sod, sia, Xdh, glass, boss, fkh). 1 MPT, 3320 steps, 824 PICs, CI = 0.782. Support is indicated above (bootstrap proportions) and below (decay indices) each node.

(represented by *D. virilis*) and *repleta*. These data suggest that the genus *Drosophila* is not monophyletic with respect to the genera *Scaptomyza* and *Zaprionus*. Interestingly, the subgenus *Sophophora*, represented by *D. melanogaster,* is quite basal within the drosophilid taxa we sampled here. This group may constitute a lineage separate from the genus *Drosophila*, as some previous studies have suggested.[19] Also of interest is that though other forms of statistical support are relatively strong here, best estimate data removal indices[18] are relatively low (1-3 for all nodes with the exception of the *Scaptomyza* -Hawaiian clade, estimated at 5), indicating that still more data is needed to produce the strongest phylogeny.

The 23 molecular loci were analyzed in several partitions. First, we divided the genes based on functional class, either enzymes (Fig. 2A) or transcription factors (Fig. 2B). The enzyme tree (Fig. 2A) is based on the coding regions of seven loci, which contain a total of 977 parsimony informative characters, and is quite unresolved. It does, however, support the sister group relationship of the two Hawaiian *Drosophila* species sampled, *D. crucigera* and *D. heteroneura*, as well as the placement of genera *Hirtodrosophila* and *Zaprionus* as close relatives of the subgenus *Drosophila*. The transcription factor tree (Fig. 2B), which is based on nine loci containing a total of 381 parsimony informative characters, is most similar to the combined analysis tree, although support on this phylogeny is generally not high.

Next we examined the three major linkage groups in the *Drosophila melanogaster* genome. Loci located on the first chromosome yielded a completely unresolved phylogeny (data not shown). Those on the second (Fig. 2C) and third (Fig. 2D) chromosomes, however, were quite resolved. The third chromosome tree (Fig. 2D) was most similar to the combined analysis tree (Fig. 1). Bootstrap proportions at four nodes on this tree were near or above 70%. The major difference was in the placement of several subgenus *Drosophila* species groups, *repleta*, *virilis*, and *melanica*.

4. Discussion

Using a genome-based approach is necessary for the studies of phylogenetics and molecular evolution to reach their full potential. Studying these elements of biology by using a few, carefully selected genes has been a practice born out of the limitations of technologies and techniques as much as out of intellectual paradigms. Now that it is clear that many of these restrictions no longer apply it is time to benefit from genomic data.

4.1 Drosophilidae Phylogeny: Molecular Evidence

Note that seven of the loci used (16s, 28s, Adh, COII, Gpdh, Sod and Xdh) in this preliminary analysis (Table 1) were not developed for this project and have been used in previous studies of drosophilid systematics and evolution. However, these seven loci do not represent a broad range of chromosomal positions or functional classes (they are all ribosomal or mitochondrial genes or code for enzymes). This sampling is biased toward enzyme coding genes, which evolve slowly relative to other types of sequences, such as transcription factors or introns.

We have examined a number of partitions, defined based on linkage group or functional class, in addition to the combined data matrix. Such comparisons highlight the benefits of shallow genomics and combined phylogenetic analysis by showing that a tree based on a single gene, gene class, or linkage group does not generate a phylogeny as resolved and robust as the entire data matrix analyzed simultaneously. A genomic-based sample may be more effective at reconstructing the phylogenetic relationships than an analysis producing a tree based on the congruencies or conflicts of just a few genes.

We divided the genes based on functional class and analyzed two partitions, enzymes (Fig. 2A) and transcription factors (Fig. 2B). While the enzyme tree is mostly unresolved at the tips, it is well supported at the base. In contrast, the transcription factor tree shows the most support for nodes at the tips. This is consistent with the notion that transcription factors are more rapidly evolving relative to enzyme coding genes. These results suggest that, when approaching a specific phylogenetic question, it may be most effective to select a gene based on function. Evaluation by linkage group (Fig. 2C-D) shows a similar example of a group of slow versus fast genes. Thus using a combination, and representative sampling, of all gene aids in producing a tree that has both better resolution and support, and is based more closely on the overall evolution of the taxa.

4.2 Future Directions: Interpreting and Evaluating Conclusions from Shallow Genomic Data

One of the limitations to generating a large number of gene sequences via shallow genomic methods is that there are not many algorithms that allow one to easily analyze these disparate data fully. For example, the program TreeRot.2b[17] which is widely available can only calculate the partitioned branch support for up to 12 data sets. Partitioned branch support data was only efficiently produced for the shallow genomic data after the program was altered by the creator to process up to 30 partitions. Note also that the data removal indices mentioned above are only listed as best estimates. It is necessary to list them tentatively because the methods and algorithms for properly calculating this measure of support between 23 data sets does not exist. These are just a few examples of how technical and algorithmic systems available are inadequate for large data sets. Now that this volume of data can be produced, it is up to those in the field of informatics and computer science to meet the challenge presented by it.

4.3 Additional Data on Genome Evolution

It has been mentioned above that we expected our methods to reveal multiple indications and artifacts of the evolution of the genome studied. Although our work is still in its preliminary stages we have already come across a number of interesting phenomena using previously unstudied genes and gene regions. Each of these is a paper unto itself, but we will summarize three examples to highlight the information that can be revealed with the survey methods we describe above.

The first example concerns the evolution of CAG repeats. The glass gene encodes a specific RNA polymerase II transcription factor which has been implicated in photoreceptor determination in *Drosophila melanogaster*.[20] We have examined a portion of exon four in several members of the genus *Drosophila* as part of the genus level project described above. The glass gene of *D. micromelanica*, a species in the *melanica* group, contains a large polyglutamine (CAG) repeat region not been found in any other species yet examined. This region is interesting in that it may be used as a model to better understand the molecular basis of how repeat regions form. Such work has applications beyond genome evolution in *Drosophila*, and can be used in the study of a number of diseases since polyglutamine repeats, particularly CAG repeats, have been implicated in at least eight human neurological disorders. These diseases include Huntington's Disease for which *Drosophila* is now one of the emerging model systems.[21,22]

In the course of examining our loci we have also come across an interesting case of horizontal gene transfer. In the 26s proteosome encoding region there is a 60 base pair insertion in three of the taxa in our analysis. BLAST[23] searches revealed that these inserts are possibly viral in origin. Whether these represent excised viruses, some sort of coevolving polydnavirus, or a third option remains to be seen and is a very interesting question under study in our laboratory.

The third element of interest found was an unusually long non-coding region in the CG3869 gene fragment, with significant phylogenetic uses beyond our higher level study. This gene was discovered only during the genome project and its function is still unknown. Finding introns in *Drosophila* genes and designing primers for them is not especially difficult, yet it is not generally worth the effort as *Drosophila* introns tend to be rather small. The CG3869 non-coding region, by contrast, is roughly 500bp and shows a surprising amount of variation, including a number of indels and microsatellites. This region has already proven to be a useful in species and population level studies done in our lab and the primers are in high demand from those in other labs who want to do the same.

It is the accumulation of data on individual changes like those above that will allow people using the shallow genome techniques to examine the finer aspects of genome evolution, beyond the coarser level of just gene placement, duplication or elimination

5 Conclusion

High throughput sequencing has had a significant impact on large scale genome sequencing projects.[24] Shallow genome sequencing techniques target a subset of the total genome, making genomics much more tractable for evolutionary and population genetic studies. Determination of sequences from non-model system taxa (or even studies of polymorphism within model systems) will not only prove effective in addressing the questions above, but will also lead to a better understanding of the genes involved in human development and disease.[25] A phylogenetic approach allows for a more complete understanding of the architecture of genomic change that occurs over evolutionary time. The genomic information we collect in the next few decades will not only aid us in reconstructing phylogeny, but will also address a wide range of questions pertinent to how genomes evolve.

Acknowledgements

The first and second authors contributed equally to the content of this paper. Additional help came from Jessica Chen, Jeremy Lynch, Julian Stark, Ilya Temkin and Jake Wintermute. Dr. Michael Sorenson graciously wrote a new version of TreeRot.2 (version "c") to be compatible with our large data matrix. We also would like to thank two anonymous reviewers for their comments and suggestions.

References

1. D. R. Maddison, *Syst. Zool.* **40**:315-340 (1991).
2. J.B. Clark, W.P. Maddison and M.G. Kidwell. *Mol. Biol. Evol.* **11**:40-50 (1994).
3. G. M. Rubin, *et al., Science* **287**: 2222-2224 (2000).
4 S. A. Chervitz, *et al., Science* **282**:2022-2027 (1998).
5. L. A. Lyons, M. M. Raymond, and S. J. O'Brien, *Animal Biotechnology* **5**:103-111 (1997).
6. T. M. Rose, *et al., Nucleic Acids Res.* **26**:1628-1635 (1998).
7. R. DeSalle and D. Grimaldi, *Ann. Rev. Ecol. Syst.* **22**:447-475 (1991).

8. R. DeSalle and D. Grimaldi, *J. Hered.* **83**:182-188 (1992).
9. D. Grimaldi, *Bull. Am. Mus. Nat. Hist.* **197**:1-139 (1990).
10. L. H. Throckmorton, *Univ. Tex. Publs.* **6615**:335-396. (1966)
11. W. C. Wheeler and D. Gladstein, American Museum of Natural History, New York (1993).
12. D. L. Swofford, Sinauer Press, Washington (2000).
13. W. P. Maddison and D. R. Maddison, Sinauer Press, Washington (1992).
14. J. Felsenstein, *Evolution* **39**: 783-791 (1985).
15. K. Bremer, *Evolution* **42**:795-803 (1988)
16. R. Baker and R. DeSalle, *Syst. Biol.* **46**:654-673 (1997).
17. M .D. Sorenson, TreeRot, version 2, Boston University, Boston MA (1999).
18. J. Gatesy, P.M. O'Grady, and R.H. Baker, *Cladistics*, **15**:271-313 (1999)
19. L. H. Throckmorton, in *Handbook of Genetics: Invertebrates of Genetic Interest,* R. C. King, Ed. (Plenum, New York, 1975).
20. K. Moses , M.C. Ellis and G.M. Rubin., *Nature* **340**:531-536 (1989)
21. G. R Jackson, *et al., Neuron* **21**, 633-642 (1998).
22. Mitchell, A. , *Nature* **395**, 841 (1998)
23. S. F. Altschul, *et al., Nucleic Acids Res.* **25**:3389-3402 (1997)
24. J. C. Venter, *et al., Science* **280**:1540-1542 (1998).
25. A. R. Templeton, *Ann. Rev. Ecol. Syst.* **30**:23-49 (1999).

Fast Phylogenetic Methods for the Analysis of Genome Rearrangement Data: An Empirical Study

Li-San Wang
Dept. of Computer Sciences
University of Texas, Austin, TX 78712

Robert K. Jansen
Section of Integrative Biology
University of Texas, Austin, TX 78712

Bernard M.E. Moret
Dept. of Computer Science
University of New Mexico
Albuquerque, NM 87131

Linda A. Raubeson
Dept. of Biological Sciences
Central Washington University
Ellensburg, WA 98926

Tandy Warnow
Dept. of Computer Sciences
University of Texas
Austin, TX 78712

Evolution operates on whole genomes through mutations that change the order and strandedness of genes within the genomes. Thus analyses of gene-order data present new opportunities for discoveries about deep evolutionary events, provided that sufficiently accurate methods can be developed to reconstruct evolutionary trees. In this paper we present two new methods of character coding for parsimony-based analysis of genomic rearrangements: one called MPBE-2, and a new parsimony-based method which we call MPME (based on an encoding of Bryant), both variants of the MPBE method. We then conduct computer simulations to compare this class of methods to distance-based methods (NJ under various distance measures). Our empirical results show that two of our new methods return highly accurate estimates of the true tree, outperforming the other methods significantly, especially when close to saturation.

1 Introduction

1.1 Gene Orders as a Source of Phylogenetic Data

While DNA sequences have greatly improved our understanding of evolutionary relationships, they have also left open many crucial phylogenetic questions. The research community has thus sought other sources of phylogenetic signal, looking for characters that evolve slowly or have a large number of states, since such characters generally have a higher signal-to-noise ratio than DNA sequences. One source of such characters is the category of "rare genomic changes" [1]. Rare genomic changes are defined as large-scale mutational events in genomes; among many possibilities are *genomic rearrangements*, which include both gene duplications [2] and changes in gene order [3]. The relative rarity of genomic rearrangements makes these characters very attractive as sources of phylogenetic data. Although it has been suggested that there are not enough genomic rearrangements to provide sufficient numbers of characters for resolving phylogenetic relationships in most groups (e.g., chloroplast genomes [4]), increased genome sequencing efforts are uncovering many new genome rearrangements for use in phylogeny reconstruction. For example, gene-order comparisons for two ascomycete fungal nuclear genomes (*Saccharomyces cervisiae* and *Candida albicans*) estimated that there have been approximately 1100 single-gene inversions since the divergence of these species [5].

1.2 Genome rearrangement evolution

Some organisms have a single chromosome or contain single-chromosome organelles (such as mitochondria [6,7] or chloroplasts [3,4]), whose evolution is largely independent of the evolution of the nuclear genome for these organisms. Whole-genome sequencing projects are providing us with information about the ordering and orientation of the genes, enabling us to represent the chromosome by an ordering (linear or circular) of signed genes (where the sign of the gene indicates its orientation). Evolutionary processes on the chromosome can thus be seen as transformations of signed orderings of genes. With a number assigned to the same gene in each genome, a genome can be represented by a *signed permutation* of $\{1, \ldots, n\}$—a permutation in which each number is given a sign; if the genome is circular, so is the permutation.

An *inversion* lifts a contiguous subpiece of the permutation, reverses its order and the orientation of every gene within it, then puts the resulting piece back in the same location; for it to happen requires two concurrent breaks in the DNA. A *transposition* lifts a contiguous subpiece of the permutation and puts it back unchanged between two contiguous permutation elements not in the subpiece; it requires three DNA breaks. An *inverted transposition* is a transposition that also reverses the order of the subpiece and the orientation of every gene within it.

The *Generalized Nadeau-Taylor (GNT)* model [8,9] describes the process responsible for the change in gene order along the edges of a given phylogeny. The model includes the three types of rearrangement events just described; within each type, all events have equal probability (e.g., any inversion is as likely as any other), but the model includes two parameters to indicate the probabilities of each type of event: α and β are the probabilities that an event is a transposition or an inverted transposition, respectively—and thus $(1-\alpha-\beta)$ is the probability that an event is an inversion. Each edge e of the tree has an associated parameter λ_e, which is the expected value of a Poisson distribution for the number of events taking place along this edge. The process that this model describes, when given a rooted binary tree with an ancestral gene order at the root, as well as the values of the various parameters, produces a set of signed gene orders at the leaves of the model tree.

1.3 Approaches for Reconstructing Phylogenies from Gene Order Data

We now describe two basic classes of methods for reconstructing phylogenies from whole genomes. In the first class of *distance-based methods*, we use a method such as the Neighbor Joining Method [10] in conjunction with an estimator of evolutionary distances, to infer an edge-weighted tree whose matrix of leaf-to-leaf distances approximates the estimated distance matrix. Thus, the estimation of evolutionary distances is an important component of distance-based estimation of phylogenies.

The *true evolutionary distance* (*t.e.d.*) between two leaves in the true tree is simply the length (in terms of actual numbers of rearrangements) of the unique simple path between these two leaves in the tree. Theory has established that if we can estimate all t.e.d.s sufficiently accurately, we can reconstruct the tree T using even very simple methods [11,12]. Estimates of pairwise distances that are close to the t.e.d.s will in general be more useful for evolutionary tree reconstruction than edit distances, because edit distances usually *underestimate* t.e.d.s, by an amount that can be very significant as the number of rearrangements increases [13,14].

The other basic approach we examine are called "maximum parsimony" type approaches, since they are similar to the maximum parsimony problem for biomolecular sequences. Given a set R of allowed rearrangement events, the *length* of a tree T with all nodes labeled by genomes is defined as the sum of the edit distances with respect to R over all edges in T. The *parsimony score* of T with respect to R is the minimum length over all possible labelings of the internal nodes. The *Maximum Parsimony on Rearranged Genomes* problem asks for the tree topology T that has minimum parsimony score with respect to R. The problem is difficult even when R is very restricted: the time complexity is unknown (but believed to be NP-hard) when R is the set of all transpositions and is NP-hard when R is the set of all inversions.

Sankoff *et al.* [15] proposed a different optimization problem for phylogeny reconstruction from gene-order data: seek the tree with the minimum number of breakpoints rather than that with the minimum number of evolutionary events. The resulting tree is called the *breakpoint phylogeny*. When the breakpoint distance is linearly correlated with the t.e.d., minimizing the number of breakpoints also minimizes the total number of evolutionary events; Blanchette *et al.* [6] observed such a relationship in a group of metazoan mitochondrial genomes. Computing the breakpoint phylogeny is NP-hard for just three genomes [16], a special case known as the *Median Problem for Breakpoints (MPB)*. Blanchette *et al.* reduced the MPB to the travelling salesman problem and developed the software suite `BPAnalysis` to approximate the breakpoint phylogeny; this approach was subsequently refined and enormously accelerated by Moret *et al.* with the `GRAPPA` software suite [17]. However, these approaches fail on large datasets—a 16-taxon problem gives rise to several quadrillion trees!

Our experiments have shown that selecting trees with smaller total edge length (under any of these measures) leads to more accurate reconstructions [18,19].

1.4 Our Contribution

This paper provides the first thorough empirical study of fast phylogenetic reconstruction methods for gene-order data, using both distance-based and parsimony-based approaches. It also introduces two new analysis methods based on encodings of gene orders as sequences of state characters. In Section 2 we describe the various methods tested in our experiments; in Section 3 we discuss the experimental setup; and in Section 4 we present our results, in terms of efficiency and of topological accuracy. We find that the NJ method, used with our t.e.d. estimator *EDE*, and the MPME method, are both highly accurate, outperforming all of the other methods in this study.

2 Phylogenetic Methods Under Study

The methods used in our experiments can be grouped under *distance-based methods*, which use various distance estimators to recover *true evolutionary distances*, and *parsimony-based methods*, which convert the gene-order data into character codings and use conventional parsimony algorithms to reconstruct the phylogeny.

2.1 Distance-Based Methods

Our basic distances are the breakpoint (BP) and the inversion (INV) distances. The first measures the number of adjacencies that are disrupted in moving from one ordering into the other, while the second measures the minimum number of inversions required to transform one ordering to the other. Both are computable in linear time (the second through the method of Bader *et al.* [20]). Using these distances, we have three t.e.d. estimators, all of which can be computed in low-order polynomial time.

The a-IEBP (Approximately Inverting the Expected BreakPoint distance) method [9] approximates, with known error bound, the expected breakpoint distance obtained after k random events in the GNT model, for any setting of the parameters α and β. Consequently, given two genomes, we can estimate the true evolutionary distance (t.e.d.) between them by selecting the number of events most likely to have created the observed breakpoint distance. Simulations [9] show that the method is robust even under wrong assumptions about model parameters.

The e-IEBP (Exactly Inverting...) method [21] improves the accuracy of a-IEBP by providing an exact calculation of the expected breakpoint distance, at the cost of increased running time. In our simulations [21], e-IEBP produces more accurate trees than a-IEBP when used with NJ.

The EDE (Empirically Derived Estimator) method [18,19] estimates the t.e.d. by inverting the expected inversion distance. We derived the estimator through a nonlinear regression on simulation data. The evolutionary model in the simulations uses only inversions, but NJ using EDE distances shows high accuracy in simulations [21,18] even when transpositions and inverted transpositions are present.

2.2 Parsimony-Based Methods

All methods discussed in this section are based on character-encodings generated from the signed permutation. These character matrices are then subjected to parsimony searches—for which good implementations have long been available.

The *Maximum Parsimony on Binary Encodings (MPBE)* [22,23] has running time exponential in the number of genomes, but runs very fast in practice. In MPBE, each gene ordering is translated into a binary sequence, where each site from the binary sequence corresponds to a pair of genes. (The ordering of the sites is immaterial in this encoding.) For the pair (g_i, g_j), the sequence has a 1 at the corresponding site if g_i is immediately followed by g_j in the gene ordering and a 0 otherwise (note that g_i and g_j can be negative and that, since (g_i, g_j) and $(-g_j, -g_i)$ denote the same adjacency, w need only one site for both). There are $\binom{n}{2}$ pairs, where n is the number of genes in each genome, but we drop the sites where every sequence has the same value.

Our first new encoding, *MPBE-2*, is a subset of an MPBE encoding designed to eliminate any character denoting the ancestral condition (the identity permutation in our simulations). For example, if the adjacency 1–2 is scored as one character for MPBE, but we can safely assume it is also the state of the common ancestor of the taxa, then we will not include this character in the MPBE-2 encoding. That is, we attempt in MBE-2 to be true to the cladistic goal of using only shared derived mutations to support sister-group relationships. This also has the consequence of reducing dependencies among characters, although it cannot fully eliminate these dependencies.

Our second new encoding builds on this observation by developing an ordered multistate encoding that avoids multiple encodings for each position. This new encoding, which we call *Maximum Parsimony on Multistate Encodings* (MPME). is inspired by a proposal of Bryant's [24], itself based on an earlier characterization approach of Sankoff and Blanchette. Let n be the number of genes in each genome; then each gene order is translated into a sequence of length $2n$. For every i, $1 \leq i \leq n$, site i takes the value of the gene immediately following gene i and site $n + i$ takes the value of the gene immediately following gene -i. For example, the circular gene ordering (1,-4,-3,-2) corresponds to the MPME sequence of $(\text{-}4, 3, 4, \text{-}1, 2, 1, \text{-}2, \text{-}3)$. Each site can take up to $2(n-1)$ different values; the unbounded number of states per characters is a drawback in practical implementations, which usually assume that this number is bounded by a small constant. For example, the bound is 32 in PAUP* 4.0 [25]; even after remapping the set of successor values into a consecutive set of symbols, the number of symbols often exceeds the PAUP bound for larger problems.

3 Design of the Experiments

The goal of our experiments is to compare the tradeoffs (time vs. accuracy) offered by NJ with those offered by the parsimony-based methods; thus we present results for both running time and accuracy.

3.1 Quantifying Accuracy

Given an inferred tree, we assess its *topological accuracy* by computing the *normalized false negative (FN) rate* with respect to the *true tree*. The true tree may not be the model tree itself: the evolutionary process may cause no changes on some edges of the model tree, in which case we define the true tree to be the result of *contracting* those edges in the model tree.

For every tree there is a natural association between every edge e and the bipartition on the leaf set induced by deleting e from the tree. Let T be the true tree and let T' be the inferred tree. An edge of T is *missing* in T' if T' does not contain an edge defining the same bipartition; such an edge is then called a *false negative* (FN). We normalize these values by dividing the number of false negatives by the number of internal edges in the true tree, thus producing a value between 0 and 1 (which we express in terms of percentages).

3.2 *The Experiments*

For each setting of the parameters (number of leaves, number of genes, probability of each type of rearrangement, and edge lengths), we generate 60 *runs*. In each run, we generate a model tree, and a set of genomes at the leaves as follows. First, we generate a random leaf-labeled tree (from the uniform distribution on topologies); the leaf-labeled tree and the parameter settings thus define a model tree in the GNT model. We run the GNT simulator on the model tree and produce a set of genomes at the leaves. The numbers of genes in each genome are 37 (typical of genes in animal mitochondrial genomes [6]) and 120 (typical of genes in plant chloroplast genomes [23]).

Our GNT simulator[9,18] takes as input a rooted leaf-labeled tree and the associated parameters (edge lengths and the relative probabilities of inversions, transpositions, and inverted transpositions). On each edge, it applies random rearrangement events to the genome at the ancestral node according to the model with given parameters α and β. We use `tgen` (from D. Huson) to generate random trees. These trees have topologies drawn from the uniform distribution, and edge lengths drawn from the discrete uniform distribution on intervals $[a, b]$, where we specify a and b. Table 1 summarizes the settings. We then compute NJ trees on each of the five distance matrices (`BP`, `INV`, `a-IEBP`, `e-IEBP`, and `EDE`) and the most parsimonious trees from the heuristic search using the three encodings (MPBE, MPBE-2, and MPME). When the parsimony search returns more than one tree, we use the majority-rule consensus (generally not a fully resolved tree) for comparison to the true tree. We use PAUP* 4.0b8 [25] for NJ, to compute the false negative rate between two trees, and for the parsimony search using the three encodings. The upper bound for the running time is 240 mins., the heuristic search uses Tree-Bisection-Reconnection (TBR) operations to generate new trees, at any time we hold the 5 trees having the lowest parsimony score, and we use the NJ trees with our five distances as the starting trees. All experiments were conducted on Pentium-class machines.

Table 1: Settings For The Empirical Study.

Parameter	Value ($*$ for 120 genes only)
# genes	37, 120
# leaves	40, 80*, and 160*
expected # events/edge	uniform within $[1,3]$, $[1,5]$, $[1,10]$, $[3,5]^*$, $[3,10]^*$, and $[5,10]^*$
probability settings: (α, β)	$(0,0)$, $(1,0)$, $(0,1)$, $(\frac{1}{2},\frac{1}{2})$, $(0,\frac{1}{2})$, $(\frac{1}{2},0)$, $(\frac{1}{3},\frac{1}{3})$
datasets per setting	60

4 Results of the Experiments

As mentioned, MPME will exceed 32 states per character for large problems. The problem worsens with increasing rate of evolution; for runs with 120 genes, 160 taxa, and edge lengths in $[5, 10]$, PAUP *always* rejects the MPME data matrix. We ignore all MPME datasets rejected by PAUP—thereby introducing an unknown, but undoubtedly favorable bias in our accuracy results for MPME on large problems.

Figure 1 shows histograms of the running times of the parsimony-based methods for two sizes of problems; on smaller problems (40 taxa), the parsimony search ran quickly (20 mins.), but larger numbers of taxa caused sharp increases in running times—to the point where MPME generally reached the time limit. In comparison, the NJ-based methods ran faster—typically in 8 minutes or less (this time includes calculation of pairwise distances, which can be computationally expensive for the `IEBP` methods), with no variation among runs using a particular estimator.

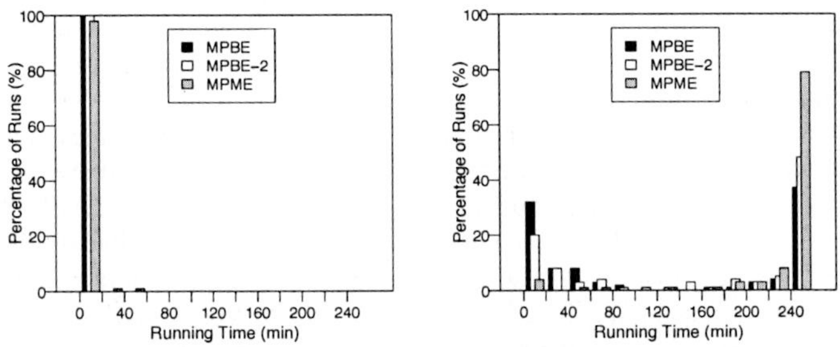

120 genes, 40 taxa — 120 genes, 160 taxa

Figure 1: PAUP running times for the three parsimony-based methods. The vertical bars right of 240 mins. represent the runs that exceeded the parsimony search limit and were cut off.

Due to space limitations, we present in Figures 2, 3, and 4 only a sample of our results. We show three different problem sizes, which we can think of as small, medium, and large. For 37 genes, both distance- and parsimony-based methods (except MPME) yield FN of at least 10%—the low number of genes reduces the amount of phylogenetic information. For 120 genes, trees produced by parsimony-based methods and NJ using `a-IEBP`, `e-IEBP`, and `EDE` have FN at most 20% (10% for higher rate and 40 taxa), and outperform NJ(INV) and NJ(BP) by a large margin when the amount of evolution is high. While MPME usually produces the most accurate trees among the parsimony-based methods, it is considerably slower than MPBE; indeed, we expect its performance on larger datasets is time-limited—had we given it more time to run, it would have surpassed the other MP-based methods easily. With 37 genes, increasing the rate of evolution improves the accuracy of MPME, but wors-

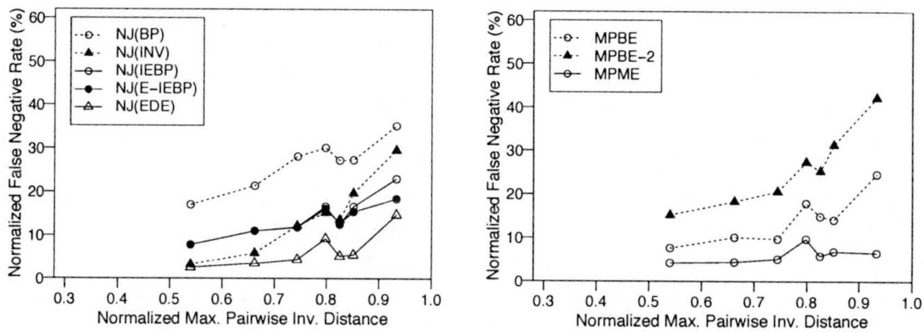

$(\alpha, \beta) = (0, 0)$

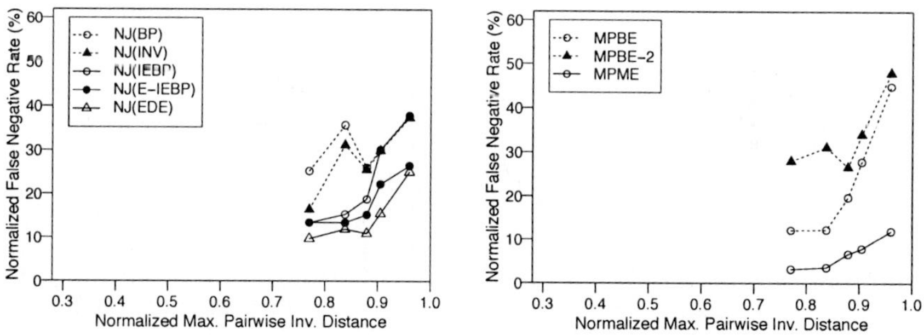

$(\alpha, \beta) = (\frac{1}{2}, \frac{1}{2})$

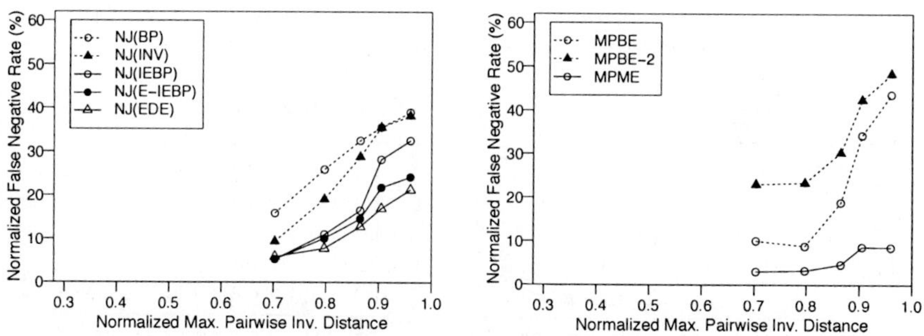

$(\alpha, \beta) = (\frac{1}{3}, \frac{1}{3})$

Figure 2: Topological accuracy of phylogenetic methods on problems with 37 genes and 40 taxa. The x-axis is normalized by the number of genes, the highest inversion distance two gene orders can have. Our plots result from binning the values into range of evolutionary distances (maximum pairwise inversion distance in the dataset) and plotting the average value for each bin. See Section 1.2 for the definition of the model weights (α, β).

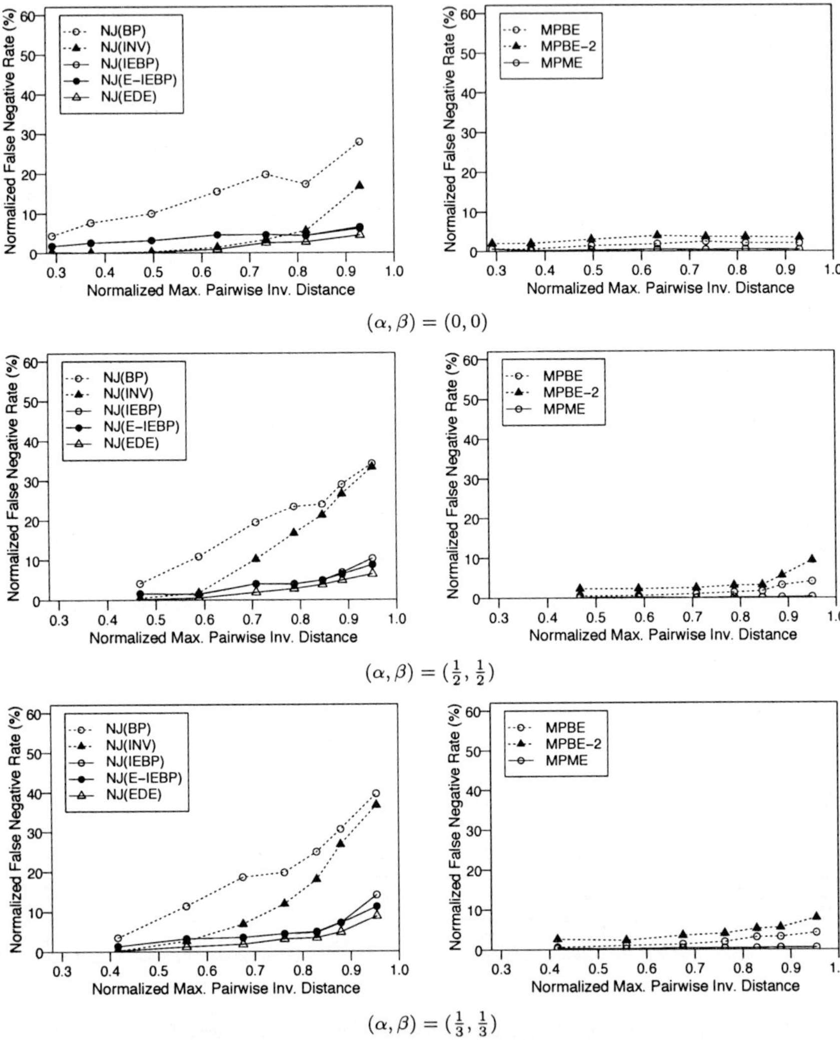

Figure 3: Topological accuracy of phylogenetic methods on problems with 120 genes and 40 taxa. See Section 1.2 for the definition of the model weights (α, β).

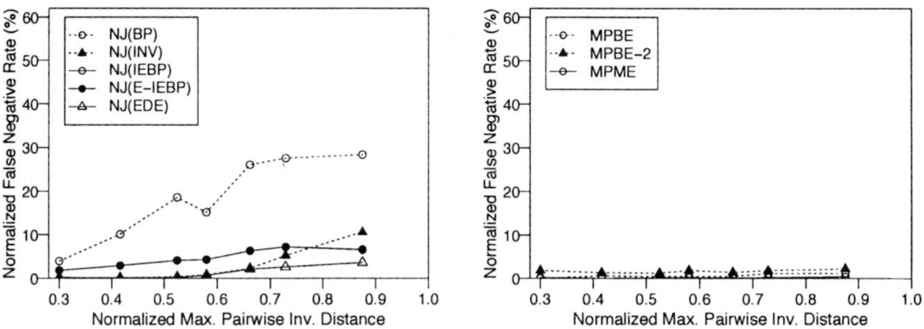

$(\alpha, \beta) = (0, 0)$

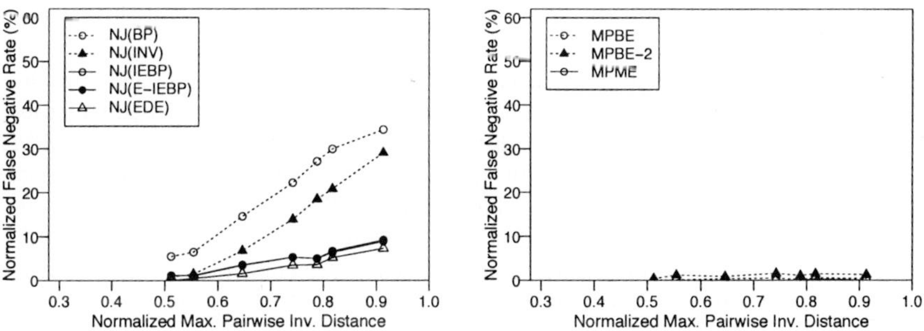

$(\alpha, \beta) = (\frac{1}{2}, \frac{1}{2})$

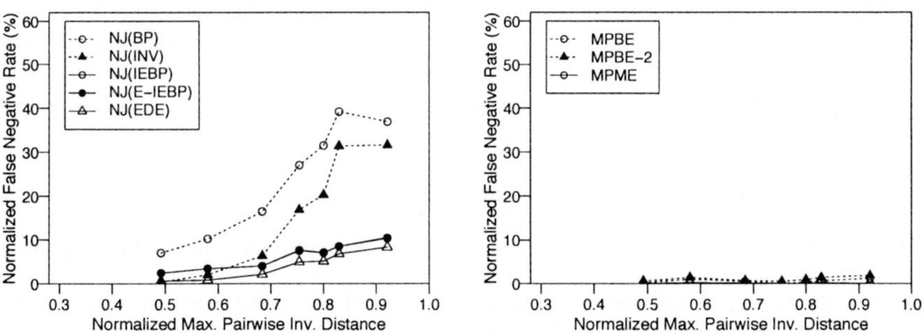

$(\alpha, \beta) = (\frac{1}{3}, \frac{1}{3})$

Figure 4: Topological accuracy of phylogenetic methods on problems with 120 genes and 160 taxa. See Section 1.2 for the definition of the model weights (α, β).

ens that of MPBE and MPBE-2, whereas all three methods improve in accuracy for larger evolutionary rates with 120 genes.

NJ(EDE) is clearly the most accurate distance-based method, and this difference is most noticeable as the dataset gets close to saturation. Furthermore, it is also one of the fastest (calculating both `a-IEBP` and `e-IEBP` distances is more expensive). Also, NJ(EDE) is competitive with both MPBE and MPBE2, although mostly not as accurate as MPME (except for the inversion only scenario, and even then only for datasets which are far from saturated). Of all the methods we studied, MPME is the most accurate: it behaves well at all rates and is much better at high rates. Our results suggest that using an encoding that attempts to capture more details about the gene order (like MPME) preserves useful phylogenetic information that a parsimony-based search (with sufficient time) can put to good use. The choice between the best two methods (NJ(EDE) and MPME) may thus be dictated by running time rather than accuracy concerns: while NJ(EDE) is very fast and thus always usable, MPME will be too computationally expensive to use for some datasets.

5 Conclusion

We have introduced two new encoding methods for gene-order data and compared them to a previous encoding method (MPBE) and to NJ analyses based on various estimates of the true evolutionary distance (`EDE`, `a-IEBP`, and `e-IEBP`). MPME and NJ(EDE) are clearly the best two choices in our study, returning much more accurate trees than the other methods. Furthermore, MPME almost always outperforms NJ(EDE). However, while the advantage gained by MPME is significant, MPME is also the slowest of the methods we studied. An important direction for future research is thus to develop new heuristics that are as accurate as MPME, yet easy to implement for practical use.

Acknowledgement

This work was supported in part by the National Science Foundation under grants to R.K. Jansen (DEB 99-82092), to B.M.E. Moret (ACI 00-81404), and to L.A. Raubeson (DEB 00-75700), and by the David and Lucile Packard Foundation to T. Warnow.

References

1. A. Rokas and P. W. H. Holland. Rare genomic changes as a tool for phylogenetics. *Trends in Ecology and Evolution*, 15:454–459, 2000.
2. S. Mathews and M. J. Donoghue. The root of angiosperm phylogeny inferred from duplicate phytochrome genes. *Science*, 286:947–950, 1999.
3. L.A. Raubeson and R.K. Jansen. Chloroplast DNA evidence on the ancient evolutionary split in vascular land plants. *Science*, 255:1697–1699, 1992.
4. R.G. Olmstead and J.D. Palmer. Chloroplast DNA systematics: a review of methods and data analysis. *Amer. J. Bot.*, 81:1205–1224, 1994.
5. C. Seoighe et al. Prevalence of small inversions in yeast gene order evolution. *Proc. Natl.*

Acad. Sci. USA, 97:14433–14437, 2000.

6. M. Blanchette, M. Kunisawa, and D. Sankoff. Gene order breakpoint evidence in animal mitochondrial phylogeny. *J. Mol. Evol.*, 49:193–203, 1999.
7. J.D. Palmer. Chloroplast and mitochondrial genome evolution in land plants. In R. Herrmann, editor, *Cell Organelles*, pages 99–133. Wein, 1992.
8. J.H. Nadeau and B.A. Taylor. Lengths of chromosome segments conserved since divergence of man and mouse. *Proc. Nat'l Acad. Sci. USA*, 81:814–818, 1984.
9. L.-S. Wang and T. Warnow. Estimating true evolutionary distances between genomes. In *Proc. 33th Annual ACM Symp. on Theory of Comp. (STOC 2001)*. ACM Press, 2001.
10. N. Saitou and M. Nei. The neighbor-joining method: A new method for reconstructing phylogenetic trees. *Mol. Biol. & Evol.*, 4:406–425, 1987.
11. T. Warnow. Some combinatorial problems in phylogenetics. In *Proc. Int'l Colloq. Combinatorics & Graph Theory*, Balatonlelle, Hungary, 1996.
12. K. Atteson. The performance of the neighbor-joining methods of phylogenetic reconstruction. *Algorithmica*, 25(2/3):251–278, 1999.
13. D. Huson, S. Nettles, K. Rice, T. Warnow, and S. Yooseph. The hybrid tree reconstruction method. *J. Experimental Algorithmics*, 4:178–189, 1999. http://www.jea.acm.org/.
14. D. Swofford, G. Olson, P. Waddell, and D. Hillis. Phylogenetic inference. In D. Hillis, C. Moritz, and B. Mable, editors, *Molecular Systematics*. Sinauer Assoc. Inc., 1996.
15. D. Sankoff and M. Blanchette. Multiple genome rearrangement and breakpoint phylogeny. *J. Comp. Biol.*, 5:555–570, 1998.
16. I. Pe'er and R. Shamir. The median problems for breakpoints are NP-complete. *Elec. Colloq. on Comput. Complexity*, 71, 1998.
17. B.M.E. Moret, S.K. Wyman, D.A. Bader, T. Warnow, and M. Yan. A new implementation and detailed study of breakpoint analysis. In *Proc. 6th Pacific Symp. Biocomputing PSB 2001*, pages 583–594. World Scientific Pub., 2001.
18. B.M.E. Moret, L.-S. Wang, T. Warnow, and S. Wyman. New approaches for reconstructing phylogenies based on gene order. In *Proc. 9th Intl. Conf. on Intel. Sys. for Mol. Bio. (ISMB 2001)*, pages S165–S173, 2001. In *Bioinformatics* 17.
19. B.M.E. Moret, J. Tang, L.-S. Wang, T. Warnow, and S. Wyman. New approaches for reconstructing phylogenies based on gene order. *J. Comput. Syst. Sci.*, 2001. to appear.
20. D.A. Bader, B.M.E. Moret, and M. Yan. A fast linear-time algorithm for inversion distance with an experimental comparison. *J. Comput. Biol.*, 8(5):483–491, 2001.
21. L.-S. Wang. Improving the accuracy of evolutionary distances between genomes. In *Proc. 1st Workshop Algs. in Bioinformatics WABI'01*, pages 176–190. Springer Verlag. LNCS 2149.
22. D. Sankoff and J.H. Nadeau, editors. *Comparative Genomics : Empirical and Analytical Approaches to Gene Order Dynamics, Map Alignment and the Evolution of Gene Families*. Kluwer Academic Pubs., 2000.
23. M.E. Cosner, R.K. Jansen, B.M.E. Moret, L.A. Raubeson, L. Wang, T. Warnow, and S.K. Wyman. A new fast heuristic for computing the breakpoint phylogeny and experimental phylogenetic analyses of real and synthetic data. In *Proc. 8th Int'l Conf. on Intelligent Systems for Mol. Biol. ISMB-2000*, pages 104–115, 2000.
24. D. Bryant. A lower bound for the breakpoint phylogeny problem. In *Proc. 11th Ann. Symp. Combinatorial Pattern Matching CPM'00*, pages 235–247. Springer Verlag, 2000.
25. D. Swofford. *PAUP* 4.0*. Sinauer Associates Inc, 2001.

VERTEBRATE PHYLOGENOMICS: RECONCILED TREES AND GENE DUPLICATIONS

R.D.M. PAGE, J.A. COTTON

Division of Environmental and Evolutionary Biology, IBLS,
Graham Kerr Building, University of Glasgow, Glasgow G12 8QQ, UK
E-mail: r.page@bio.gla.ac.uk

Ancient gene duplication events have left many traces in vertebrate genomes. Reconciled trees represent the differences between gene family trees and the species phylogeny those genes are sampled from, allowing us to both infer gene duplication events and estimate a species phylogeny from a sample of gene families. We show that analysis of 118 gene families yields a phylogeny of vertebrates largely in agreement with other data. We formulate the problem of locating episodes of gene duplication as a set cover problem: given a species tree in which each node has a set of gene duplications associated with it, the smallest set of species nodes whose union includes all gene duplications specifies the locations of gene duplication episodes. By generating a unique mapping from this cover set we can determine the minimal number of such episodes at each location. When applied to our data, this method reveals a complex history of gene duplications in vertebrate evolution that does not conform to the "2R" hypothesis.

1 Introduction

Most genes belong to large gene families, so the analysis of the gene family evolution represents a considerable challenge for the study of genome evolution. Within vertebrates, paralogy (the relationship between genes within a family) is pervasive, and gene duplication has clearly been particularly common [1], but a broadly similar pattern is found in prokaryotes. The timing and frequency of gene duplications is of particular interest, given that gene (and genome) duplication has been posited as a major factor in the evolution of complexity in vertebrates [2]. A popular – and controversial [3,4] – hypothesis of vertebrate genome evolution postulates two successive genome duplications early in vertebrate evolution (the "2R" hypothesis). Understanding the evolution of vertebrate genomes requires a well supported phylogenetic framework for vertebrates, and methods for locating episodes of gene duplication. In this paper we explore the use of reconciled trees [5,6] to address the latter question.

1.1 Reconciled trees

Conventional phylogenetic methods use molecular sequences as characters of organisms, which conflates organismal and gene phylogenies. However, gene phylogenies are not species phylogenies - processes such as gene duplication, gene loss, and lineage sorting can introduce important differences between the correct phylogenetic tree for a set of genes and the correct tree for the corresponding species. An alternative is to investigate the relationship between gene trees and species trees using reconciled trees. A reconciled tree [5,6] is a map between a gene tree and a given species tree, with gene duplications and losses being postulated to explain any incongruence between the two trees. If the species tree is unknown then the most parsimonious estimate of the species tree is that minimizing the number of gene duplications required on a gene tree [7,8]. We can extend the method to many genes, so the most parsimonious species tree is that which implies the minimum number of gene duplication (or duplication and loss) events over the set of gene families ("gene tree parsimony" [7]). The map between a gene tree and a species can be computed in linear time [9], making reconciled trees practicable for very large analyses, and potentially even for genome-wide comparisons.

1.2 Vertebrate phylogeny

To test the performance of gene tree parsimony on a real dataset, we constructed a data set of 118 vertebrate gene families [a] based on data from the HOVERGEN database [10]. The higher-level phlyogeny and ancient evolution of the vertebrate in many ways represents an ideal test-case for these methods, because there has been considerable recent interest in both their phylogeny and in evolution by gene duplication in the group. A fairly robust consensus on the main relationships within the group had emerged, based on morphological evidence from both fossil and extant taxa [11], but analyses of whole mitochondrial genomes have produced unorthodox and controversial phylogenies, provoking new debate [12].

The species tree we obtained using gene tree parsimony (Fig. 1) differs little from a conventional view of vertebrate phylogeny [11], in marked contrast to the unorthodox trees obtained from mitochondrial genomes [12]. This result confirms preliminary findings [13] that reconciled tree methods can reconstruct phylogeny accurately in the face of gene duplication and loss.

1.3 Genome duplications

The timing and location of gene duplications is a key problem in understanding the evolution of gene families and genomes. Existing techniques for mapping gene trees onto species trees can identify gene duplications, but do not necessarily locate them precisely on the species tree. Furthermore, gene duplication

[a] The GENETREE file and individual alignments and gene trees are available from http://kimura.zoology.gla.ac.uk/vertebrate_data.

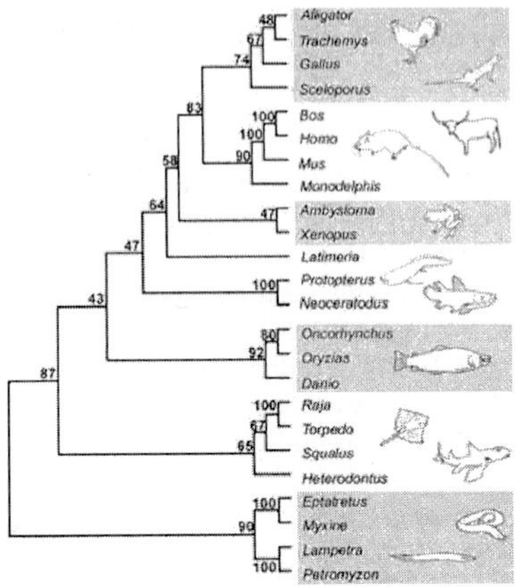

Figure 1: Phylogeny of vertebrates reconstructed using gene tree parsimony in GENETREE[15] on a set of 118 nuclear genes. Bands of shading identify higher taxonomic groups of vertebrates. This is the majority-rule consensus of 100 species trees, generate from 100 bootstrap trees for each gene tree. Figures on nodes are bootstrap percentages.

events can occur on any scale, from small pieces of DNA carrying fragments of genes right up to polyploidisation events due to hybridisation or incorrect division, so duplications on individual gene trees could be correlated, occurring as a result of the same molecular events. Identifying these events is complicated by the fact that most gene families are known from only some species, so there can be considerable uncertainty in where particular duplications occurred on the species tree. We need techniques that can identify these "duplication episodes" by clustering individual gene duplications [16,17]. We now present a method for achieving this and apply the technique to our vertebrate data set.

2 Locating gene duplications

2.1 Terminology

We will restrict ourselves to rooted trees. The immediate ancestor of a node in a tree is its *parent*, and the immediate descendants of a node are its *children*. A node with no children is a *leaf*. Let G be a rooted tree for m genes obtained from $n \leq m$ species (a *gene tree*), and S be a rooted tree for the species (a *species tree*). For each node in S the set of nodes that are its descendants form that nodes *cluster*. The cluster of the root is $\{1, \ldots, n\}$, the clusters of the leaves are $\{1\}, \{2\}, \ldots, \{n\}$. Following Margush and McMorris [14], we use the shorthand of treating the node and its cluster as synonymous. Hence, for any pair of nodes x and y in S, if $x \subset y$ then x is a descendant of y. For any node $g \in G$, let $\eta(g)$ be the set of species in which occur the extant genes descendant from g (if g is a leaf then $\eta(g)$ is the species from which gene g was obtained). For any $g \in G$, let $M(g)$ be the node in S with the smallest cluster satisfying $\eta(g) \subseteq M(g)$. A map from G into S associates each node $g \in G$ with a node $M(g) \in S$, and can be visualized using a reconciled tree [6]. Let l and r be the left and right children of a node $g \in G$. If either l or r (or both) map onto $M(g)$ (i.e., $M(l) = M(g)$ and/or $M(r) = M(g)$) then we infer that g is a gene duplication [5].

2.2 The problem

The problem of locating gene duplications using reconciled trees was first addressed by Guigó *et al.* [16], who noted that the map between gene tree and species tree puts bounds on the location of a given duplication, rather than necessarily locating the duplication precisely. Whereas the map between gene and species tree associates each node g in the gene tree with a single node $M(g) = s$ in the species tree, the actual gene duplication may have occurred anywhere along the path between $M(g)$ and $M(parent(g))$ [b]. Given this ambiguity, our task is to find the optimal placement of the duplications required to reconcile a set of gene trees $G_1, G_2, \ldots, G_k$ with a species tree S. It is important to clearly distinguish between episodes of gene duplication and *genome* duplication. Guigó *et al.* refer to *any* clustering of gene duplications as a "genome duplication," regardless of whether the whole genome or only a part of it duplicated. Here we use the term "episode" as the generic term for two or more duplications in different gene families that can be explained by a single event.

2.3 Guigó et al.'s algorithm for placing duplications

Guigó *et al.* partition gene duplications into three categories:

free: if g is the root of G.
locked: if g is not the root of G.
absolutely locked: if g is locked and $M(parent(g)) = parent(M(g))$.

[b] Note that moving a duplication down the species tree towards the root will require additional losses to be postulated. However, given that many apparent "losses" in reconciled trees may be due to lack of knowledge (such as poor taxonomic or genomic sampling), rather than actual gene loss, invoking additional losses does not seem unreasonable.

Examples of these three categories can be seen in Figure 2b. Guigó *et al.* sketched an algorithm to cluster gene duplications into the minimum number of locations on the species tree. First we identify the set of allowed locations A_g in the species tree for a duplication g. If g is the root of the gene tree then $A_g = \{s \in S : M(g) \subseteq s\}$ (the set of all nodes in the species tree from $M(g)$ down to the root). If g is not the root of the gene tree then $A_g = \{s \in S : M(g) \subseteq s \subseteq M(parent(g))\}$ (the set of nodes in the species tree from $M(g)$ down to, but not including, the node into which the parent of g is mapped). Duplications are placed as follows:

> Step 1: Place on the species tree S all absolutely locked duplications (for which $A_g = M(g)$). The set of locations of absolutely locked duplications is $D_{absolute}$.
>
> Step 2: For all locked duplications g_l for which $A_{g_l} \cap D_{absolute} \neq \emptyset$ find the absolutely locked duplication(s) ($g_a : A_{g_l} \cap A_{g_a} \neq \emptyset$). If $|A_{g_l} \cap A_{g_a}| > 1$ place g_L at the $s \in A_{g_l} \cap A_{g_a}$ that is furthest from the root of S. The set of locations of locked duplications is D_{locked}.
>
> Step 3: For all locked duplications g_l for which $A_{g_l} \cap D_{absolute} = \emptyset$, if $A_{g_l} \cap D_{locked} = \emptyset$ then g_l is placed at the node $M(g)$, otherwise the duplication is placed such that the total number of locations of gene duplications is minimal.
>
> Step 4: Free duplications g_f for which $A_{g_f} \cap D_{locked} \neq \emptyset$ are placed at the node $s \in A_{g_f} \cap D_{locked}$ that is furthest from the root of S, otherwise they are placed at the root of S.

The result of applying these steps is a clustering of gene duplications into episodes, and a final mapping of duplications onto the species tree. Note that although Guigó *et al.* gave hints about how to minimize the number of gene duplications (Step 3) they did not present a formal algorithm for doing this.

2.4 An alternative formulation

Fellows *et al.* [17] define the MULTIPLE GENE DUPLICATION problem as being the mapping of a set of gene trees $G_1, G_2, \ldots, G_k$ into a species tree S such that the number of multiple gene duplication events is minimal. They go on to show that this problem is *NP*-hard. Their formulation of the problem is somewhat different from Guigó *et al.*'s – those authors aim to minimize the number of locations in S where gene duplications have occurred, but do not postulate any additional duplications over and above those required to reconcile each gene tree G_i with S. Fellows *et al.*, however, will invoke additional duplications if it

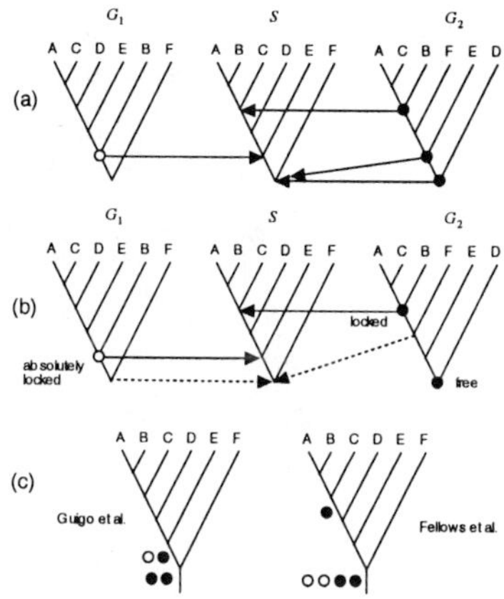

Figure 2: (a)Two gene trees and their species tree with nodes mapped onto S. (b) Node ABCDE in G_1 is absolutely locked, whereas node ABC in G_2 is locked. (c) Comparison of how Guigó *et al.* [16] and Fellows *et al.* [17] would place the duplications on S to minimise the number of multiple gene duplications.

reduces the number of multiple gene duplication events. For example, given the two gene trees in Figure 2a, using the rules of Guigó *et al.* the duplication at node ABCDE in G_1 is absolutely locked and hence cannot be moved. However, Fellows *et al.* move this duplication to the root of the species tree (at the cost of an additional duplication). Similarly, Fellows *et al.* state that "it is not beneficial" to move node ABC in G_2. However, in Guigó *et al.*'s terminology, this duplication is not absolutely locked and could be placed anywhere along the path from ABC to ABCDEF in S. Moving it to node ABCDE in S reduces the number of multiple gene duplications from 4 to 3, the same score as for the Fellows *et al.* reconstruction, but without invoking an extra duplication.

3 Placing duplications using set cover

We can reformulate Guigó *et al.*'s algorithm as a set cover problem. Let D be the set of all nodes $g \in G_i, i = 1, \ldots, k$ that are gene duplications. Each $s \in S$ has associated with it a set of duplications $D_s = \{d : d \in D, s \in A_d\}$. Finding the smallest number of locations at which gene duplication has taken place corresponds to finding the smallest number of sets such that their union is D. The set cover problem is NP-complete, but heuristics are available [19].

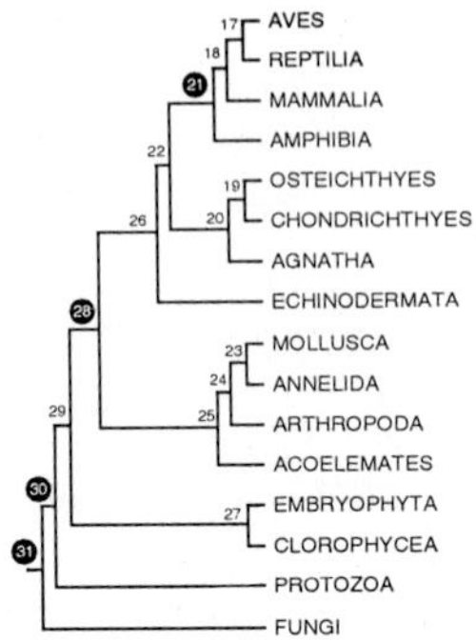

Figure 3: A species tree for 16 eukaryotes from Guigó *et al.* [16]. Internal nodes are labelled 17 – 31 in postorder. The locations of the "genome" duplications inferred by Guigó *et al.* [16] are highlighted.

We illustrate this approach using Guigó *et al.*'s data set. This has played an important role in developing methods of tree reconciliation. Previous work has shown that they miscount the number of gene losses [18] and that their species tree is not optimal for the 53 gene trees [18,20].

The species tree shown in Figure 3 requires 46 gene duplications, which are distributed over 7 nodes in the species tree:

$$D_{21} = \{2, 22, 36, 37, 44, 46\}$$
$$D_{22} = \{8, 9, 13, 32, 33, 35 - 38, 44\}$$
$$D_{26} = \{8 - 9, 13, 32, 33, 35 - 38, 44\}$$
$$D_{28} = \{1, 4, 6, 8, 9, 13 - 17, 19, 20, 25, 26, 29, 32, 33, 35 - 38, 41\}$$
$$D_{29} = \{1, 6, 8, 9, 13 - 17, 19, 20, 24 - 26, 30, 32 - 38, 41\}$$
$$D_{30} = \{1, 7 - 9, 13 - 17, 19, 20, 24 - 26, 30, 32 - 38, 40, 42, 43\}$$
$$D_{31} = \{1, 3, 57 - 18, 21, 23 - 28, 31 - 39, 45\}$$

The duplications are arbitrarily numbered 1–46. The minimal set cover for D is $\{D_{21}, D_{28}, D_{30}, D_{31}\}$. These are the same four locations of the "genome" duplications identified by Guigó *et al.* (Figure 3).

3.1 Final mapping

The minimal set cover might not yield an unambiguous mapping between the gene trees and the species tree; for example, duplication 36 is an element of all four sets in the minimal cover. This node occurs at the root of the gene tree for β-Nerve growth factor precursor (NGF) which has the topology (REPTILIA,(MAMMALIA,(AMPHIBIA,AVES))), and hence in Guigó *et al.*'s terminology is "free." Its set of allowable locations comprises vertex S_{21} and all its ancestors in the species tree (Figure 3). Following Guigó *et al.*, any duplication g which occurs in more than one set in the minimal set cover is mapped onto the node closest to $M(g)$. This can be easily done as follows:

Step 1: Let F be a set of duplications. Initially $F \leftarrow \emptyset$.
Step 2: Process each node in S in postorder. For each node s for which $D_s \neq \emptyset$ go to Step 3.
Step 3: If $F = \emptyset$ then $F \leftarrow D_s$, otherwise $D_s \leftarrow D_s \setminus F$ and $F \leftarrow F \cup D_s$

The result of this procedure is a unique mapping from the gene trees into the species tree, consistent with the minimal set cover. Applying this to Guigó *et al.*'s data we obtain the following mapping, where duplications are labeled by the abbreviated gene family name from Guigó *et al.*'s table 2.

D_{21} = {ACHG, GLUC, NGF, NGF, PAHO, TBB2, TPMA}.
D_{28} = {ACH2, ACT2, ACT3, ACTB, ANFC, COLI, CYLA, CYLA, CYLB, CYLB, G3P, G3P2, H2B, H2B, H4, HBA1, HBA2, PRVA, TBA1}.
D_{30} = {ACT3, H2A3, H4, HMDH, TBA1, TBA1, TBB}.
D_{31} = {ACT, ACT2, ATPB, CATA, CISY, CYLH, G6PI, H2A2, H2B1, H31, H4, RLA2, TOP2}.

This mapping differs from that shown by Guigó *et al.* (their fig. 4), in that those authors assign one duplication in gene NGF to D_{28}, and one duplication of the genes CYLA, CYLB, and TBA1 to D_{30}. However, these placements violate Guigó *et al.*'s own rule that "free duplications are placed at the closest location preceding the node in which the duplication is mapped where a duplication – absolutely locked or locked –, if any, has already been placed" (Step 4 in section ?? above).

3.2 Counting the number of episodes of gene duplication

If more than one duplication in a gene tree G is associated with the same node s in the species tree S (i.e., $|G \cap D_s| > 1$) then we may have to postulate multiple episodes of gene duplication occurring at s. For example, given two nodes g_1 and g_2 where g_1 is ancestral to g_2, if both nodes are in D_s then two duplication episodes are needed. However, if neither g_1 nor g_2 is ancestral to the other then both could be explained by the same event. Let the duplication height, $h(g)$, of a node $g \in G$ be the number of nodes along the path between g and the root of G for which are in D_s. Any duplication $g \in D_s$ with the same height can be explained by the same duplication event. Hence, the minimum number of distinct episodes of duplication at node s in gene family G is then $E_{(G,s)} = \text{MAX}(h(g) : g \in G, g \in D_s) + 1$. The minimum number of episodes of duplication at node s across all k gene families is then $\text{MAX}(E_{(G,s)} : G_1, \ldots, G_k)$.

For the Guigó *et al.* example, we require two episodes of gene duplication at D_{21}, D_{28}, and D_{30}, and one at D_{31}. This differs from their finding single duplications at all locations except D_{30}, where they postulate that a double duplication occurred. This difference stems from their misplacing the duplications for genes NCF, CYLA, CYLB, and TBA1 (see Sec. 3.1).

3.3 Duplication patterns in vertebrates

The locations of the 1380 inferred gene duplications in our 118 gene family data set (Sec. 1.2) were found using the above algorithm (Sec. 3), showing that they can be strongly clustered on the species tree (Fig. 4). Many apparent duplications occur near the tips of the tree in the mouse and human lineages, but the bulk of these "duplications" actually represent multiple alleles at polymorphic loci, rather than gene duplications. Figure 4 shows that substantial numbers of duplication events have occurred throughout vertebrate evolution, often affecting many gene families simultaneously. The largest single such event (duplicating 58 out of 118 families) occurred after the divergence of sharks and rays and prior to the divergence of teleosts and lobe fin fish. Gene duplication is clearly an important feature of vertebrate evolution, but the pattern shown in figure 3 is more complex than that expected from the "2R hypothesis". Some gene families have undergone as many as 11 successive episodes of duplication, and at no point in vertebrate phylogeny can we explain all gene duplications that occurred at that time by a single genome-wide event.

4 Future directions

Further work on this problem is needed. There are two limitations of our algorithm that we are aware of. Our algorithm for the final mapping (Sec. 3.1) minimizes the number of location in the species tree at which gene duplications occur, but it does not guarantee to minimize the total number of episodes of gene duplication. It is possible to construct examples where spreading gene duplications across more locations will reduce the overall number of episodes of duplication.

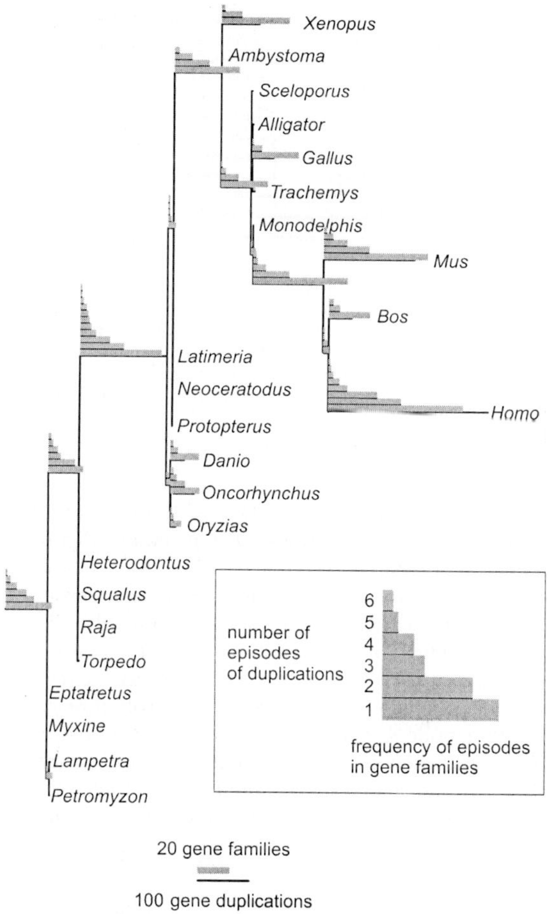

Figure 4: Distribution of gene duplications during vertebrate evolution. The species tree is one of three most parsimonious trees from a GENETREE [15] search. Branch lengths represent the number of separate gene duplications inferred to have occurred along each branch. Stacked bars represent the number of distinct episodes of gene duplication in each of the gene families that have duplicated along the branch. For clarity, bars have been omitted where only a single duplication episode is inferred for each gene family.

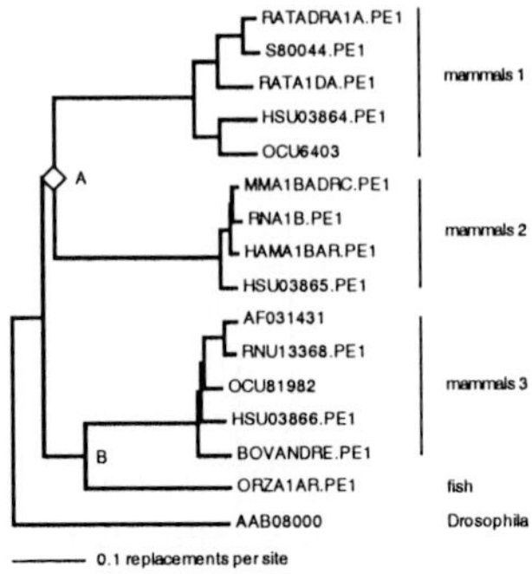

Figure 5: Phylogeny for vertebrate adrenergic receptor $\alpha 1$ sequences. The method for locating gene duplications described in this paper would place node A somewhere after the split of fish and mammals, but prior to the last common ancestor of mammals. Based on relative amount of sequence divergence with respect to node B (the split between fish and mammals), node A in fact pre dates the separation of fish from the ancestors of mammals. Data supplied by Xun Gu [21] . Sequence names are those used in the HOVERGEN database [10], in which ADRA1 is family FAM000048.

Our algorithm uses only the topology of the tree, and hence may make erroneous placements of duplications. For example, Figure 5 shows a gene tree for vertebrate adrenergic receptor $\alpha 1$ (ADRA1). The descendants of the duplication at node A are all mammalian sequences, hence a reconciled tree would place this duplication at the base of mammals. The set of allowed location for this duplication includes the common ancestor of mammals, and every node ancestral to that node that postdates the split between mammals and fish (equivalent to node B in Figure 5)[c]. However, if we consider the branch lengths in the tree, node A is deeper than node B in the tree and hence pre dates the oldest node in its allowed set of locations. One way to address this problem would be to refine the rules for determining sets of allowed location for gene duplications to take into account amounts of molecular sequence divergence (if they are sufficiently clock-like).

[c]This problem will be more prevalent in those gene families that have poorly sampled taxonomically, or have undergone substantial gene loss. Finding a single fish ADRA1 sequence related to either of the group 1 or group 2 mammal sequences would result in the method described here correctly inferring that node A pre dates the split between fish and mammals.

Acknowledgments

This work was supported the NERC, Wolfson Foundation, and the EMBO. Mike Charleston and two anonymous reviewers provided helpful comments.

References

1. R. D. M. Page, J. A. Cotton, in *Comparative Genomics: Empirical and Analytical Approaches to Gene Order Dynamics*, (Kluwer Academic Publishers, 2000)
2. S. Ohno, Cell Devel. Biol. **10**, 517 (1999)
3. A. L. Hughes, J. Mol. Evol. **48**, 565 (1999).
4. L. Skrabanek, K. H. Wolfe, Curr. Opin. Genet. Dev. **8**, 694 (1998).
5. M. Goodman *et al*, Syst. Zool. **28**, 132 (1979).
6. R. D. M. Page, Syst. Biol. **48**, 53 (1994)
7. J. B. Slowinski, R. D. M. Page, Syst. Biol. **48**, 81 (1999).
8. R. D. M. Page, M. A. Charleston, Trends Ecol. Evol. **13**, 356 (1998).
9. L. Zhang, J. Comput. Biol. **4**, 177 (1997).
10. L. Duret, D. Mouchiroud, M. Gouy, Nucleic Acids Res. **22**, 2360 (1994).
11. M. J. Benton, *The Phylogeny and Classification of the Tetrapods* (Clarendon Press, Oxford, 1988).
12. R. Zardoya, A. Meyer, in *Major Events in Early Vertebrate Evolution*, ed. P. Ahlberg (Taylor and Francis, London, 2001).
13. R. D. M. Page, Mol. Phylogent. Evol **14**, 89 (2000)
14. T. Margush, F. R. McMorris, Bull. Math. Biol. **43**, 239 (1981).
15. R. D. M. Page, Bioinformatics **14**, 819 (1998).
16. R. Guigo, I. Muchnik, T. F. Smith, Mol. Phylogenet. Evol. **6**, 270 (1996).
17. M. Fellows, M. Hallett, U. Stege, in *Proceedings of the 9th International Symposium on Algorithms and Computation*, eds. C. Kyung-Yong, O. H. Ibarra (Springer, Heidelberg, 1998).
18. R. D. M. Page, M. A. Charleston, in *Mathematical Hierarchies in Biology*, eds. B. Mirkin, F. R. McMorris, F. S. Roberts, A. Rzhetsky (American Mathematical Society, Providence, RI, 1997).
19. T. H. Cormen, C. E. Leiserson, R. L. Rivest, *Introduction to algorithms* (MIT Press, Cambridge, MA, 1990).
20. M. T. Hallett, J. Lagergren, in *RECOMB '00, Proceedings of the fourth annual international conference on computational molecular biology*, (Association for Computing Machinery, 2000).
21. Y. Wang, X. Gu, J. Mol. Evol. **51**, 88 (2000).

Proteins: Structure, Function and Evolution

Peter Clote
Department of Computer Science, Boston College
Chestnut Hill, MA 02467 clote@bc.edu

Gavin J.P. Naylor
Dept. of Zoology and Genetics
Iowa State University
Ames, Iowa 50011 gnaylor@iastate.edu

Ziheng Yang
Department of Biology, Galton Lab
University College London
London NW1 2HE, United Kingdom
z.yang@ucl.ac.uk

Protein structure prediction is one of the most exciting and difficult problems in computational molecular biology. New computational advances, such as that currently underway with IBM's Blue Gene supercomputer (projected completion in 2005), as well as advances in the understanding of energy potentials and development of statistical methods for recognition of specific folds or protein classes, together promise better methods for protein structure prediction. Indeed, we have recently witnessed enormous progress in the field of protein folding, including successful approaches to computational structure prediction as documented in recent CASP competitions.

On a different front, advances in evolutionary genomics have led to better methods for extracting information from the evolutionary history of proteins, to further our understanding of their structure and function. Just as molecular biologists can study protein function by examining the effects of mutations introduced through site-directed mutagenesis, we can learn to interpret the results of the comprehensive experiment performed by Nature over millions of years of biological evolution. For this purpose novel statistical and computational methods are being developed. Models for detecting amino acid residues under diversifying selective pressure across closely related species are generating interesting hypotheses about the structure and function of the protein, which can be tested in the laboratory. We now have the ability to recreate ancestral proteins, and thus test general trends in protein function and hypothetical correlated functional relationships. In vitro evolution, both random and directed, can attempt to replicate the historical patterns, and further elucidate the details of the adaptive landscape.

There is a growing realization that proper evolutionary analysis is an essential component for optimal extraction of structural and functional prediction from multiple sequences. The patterns of variation and conservation throughout a homologous sequence set provide signals indicating the underlying shared structure. Even neural network methods perform better when multiple and diverse sequences are included in the analysis. Analyzing the presence of co-evolving sites in proteins is beginning to make important contributions to the solution as analytical methods improve, as are more refined estimates of the tendencies of different secondary structures and hydrophobic environments to have different substitution rates. Furthermore, it is equally important to consider structural information when devising evolutionary models of sequence change. In particular, accounting for the heterogeneity of the evolutionary process among sites in the sequence is known to lead to much better fit to the model. We hope that by bringing together scientists working on structural and functional prediction as well as evolutionary analysis, this session will help mutually-beneficial interactions between these fields.

In the current session, *Proteins: Structure, Function and Evolution*, new and cutting-edge approaches are introduced for the determination of protein structure and function; additionally, evolutionary models are introduced, which provide new vantage points for these problems. In "Screened charge electrostatic model in protein-protein docking simulations", J. Fernandez-Recio, M. Totrov and R. Abagyan introduce a new method for treating solvation effects in calculating electrostatics for protein docking, and report the success of their method in screening near-native states from false positives. In "The spectrum kernel: An SVM-string kernel for protein classification", C. Leslie, E. Eskin and W. Stafford Noble introduce a new, easily computable string kernel for support vector machine classification. Their "spectrum kernel" essentially adds up uniform contributions of how many size k subwords are shared (independent of position of subword) between given sequences X, Y, and provides an $O(n \log n)$ algorithm for classifying whether a given protein X belongs to a protein family. In "Detecting positively selected amino acid sites using posterior predictive P-values", R. Nielsen and J.P. Huelsenbeck use a Bayesian approach with the development of "posterior predictive p-values", to identify amino acid residues that are under positive Darwinian selection. Those sites exhibit an excess of replacement (amino acid-altering) substitutions relative to silent (synonymous) substitutions and might be important for functional divergence. In "Improving sequence alignments for intrinsically disordered proteins", P. Radivojac, Z. Obradovic, C.J. Brown and A.K. Dunker measure the performance of different scoring matrices for 55 disordered protein families, and develop an iterative algorithm for realigning sequences and recalculating matrices. Investigating a wide range of gap penalties, the authors obtain an improvement n the ability to detect and discriminate related disordered proteins, when average sequence identity with other members of the same family is below 50%. In "Ab initio folding of multiple-chain proteins", J.A. Saunders, K.D. Gibson, and H.A. Scheraga extend their UNRES force field and Conformational Space Annealing algorithm to handle multiple-chain proteins, illustrating the success of this approach on two homo-oligomeric systems, both of which were targets in the CASP3 experiment (3rd Critical Assessment of Techniques for Protein Structure Prediction). In "Investigating evolutionary lines of least resistance using the inverse protein-folding problem", J. Schonfeld, O. Eulenstein, K Vander Felden and G. Naylor present a polynomial time algorithm for approximating the solution of the inverse protein folding problem using the Sun-Brem-Chan-Dill grand canonical model. The authors give an improvement of J. Kleinberg's application of maximum weighted bipartite matching in this context, and apply their algorithm to the PDB, in order to explore the genotype-phenotype mapping. In "Using evolutionary methods to study *G*-protein coupled receptors", O. Soyer, M.W. Dimmic, R. Neubig, and R. Goldstein develop a model of heterogeneous substitution patterns among partitions of sites in a protein, where the fitness values of amino acids are different in different partition classes. One possible interpretation of the model is that different structural categories (alpha helix or beta sheet, exposed or buried) are under different selective pressures and thus have different substitution rate matrices. The authors find that, in agreement with data, transmembrane regions of G-coupled protein receptors are strongly correlated with hydrophobicity, while non-transmembrane regions are positively correlated with flexibility and negatively correlated with hydrophobicity. In "Progress in predicting protein function from structure: Unique features of O-glycosidases", E.W. Stawiski, Y. Mandel-Gutfreund, A.C. Lowenthal, and L.M.Gregoret describe unique structural features of O-glycosidases, enzymes which hydrolyze O-glycosidasic bonds between carbohydrates. Using these structural characteristics, the authors show that accurate prediction of O-glycosidase function is possible. In "Support vector machine prediction of signal peptide cleavage using a new class of kernels for strings", J.-P. Vert develops a class of SVM kernels which interpolate between the diagonal and product kernel and applies this approach in retrieving up to 47% more true positives in signal peptide cleavage site recognition than that obtained using classical weight matrices.

In "Constraint-based hydrophobic core construction for protein structure prediction in the face-centered-cubic lattice", S. Will develops a new algorithm using constraint programming in order to compute the optimal hydrophobic core for the HP-model on the FCC lattice. The FCC lattice, where each lattice point has 12 nearest neighbors, is much more natural than the 3-dimensional cubic lattice, and though the problem is NP-complete, the author reports in benchmark tests that his algorithm can correctly thread large HP-sequences to a core of size up to 100 within 20 seconds. In "Detecting native protein folds among large decoy sets with hydrophobic moment profiling", R. Zhou and B.D. Silverman use the "hydrophobic ratio" (ratio of radii from the protein centroid where the second order hydrophobic moment and the zero order moment vanishes), as a measure of the extent of a protein's hydrophobic core in successfully distinguishing native protein folds from decoy sets.

We would like to thank the authors of this session for reporting their exciting work on protein evolution and protein structure and function determination, as well as the many other authors, whose submissions could not be reported in the current proceedings. We would like to collectively thank many persons involved in the anonymous reviewing process, and finally to thank Dr. Helen Frame Peters, Dean of the Carroll School of Management at Boston College, for additional financial support for the current session.

SCREENED CHARGE ELECTROSTATIC MODEL IN PROTEIN-PROTEIN DOCKING SIMULATIONS

JUAN FERNANDEZ-RECIO[1], MAXIM TOTROV[2], RUBEN ABAGYAN[1]

[1]*Department of Molecular Biology, The Scripps Research Institute, 10550 North Torrey Pines Road, La Jolla, CA 92037, USA, E-mail: jfrecio@scripps.edu, abagyan@scripps.edu*

[2]*Molsoft, LLC, 3366 Torrey Pines Court, La Jolla, CA 92037, USA, E-mail: totrov@molsoft.com*

A new method for considering solvation when calculating electrostatics for protein docking is proposed. The solvent-exposed charges are attenuated by induced solvent polarization charges. Modified charges are pre-calculated and the correction doesn't affect the speed of the actual simulation. The new Screened Charge Electrostatic Model (SChEM) results in an improved discrimination of near-native solutions from false positives in docking simulations as compared to conventional 'non-solvated' charge assignment. A series of protein-protein complexes were analyzed by running automated rigid-body Monte-Carlo docking simulations using the 3-D coordinates of the unbound components. In all but one case, the use of solvation screened charges for electrostatic calculations helped to improve the rank of the near-native solution after rigid-body simulations. The SChEM also drastically improved the results of the subsequent refinement of the interface side-chains. In all cases the final lowest energy solution was found within 3.0 Å r.m.s.d. of the crystal structure.

1 Introduction

In silico prediction of protein-protein interactions will undoubtedly play an essential role in the structural genomics era. Currently ongoing structural genomics projects will drastically increase the number of available 3-D protein structures[1], and a variety of computational tools will be needed to efficiently use this structural information, with the ultimate goal of understanding the complex network of protein-protein interactions in a living organism. In this context, a number of docking methods have been developed to predict the structure of a protein-protein complex given the 3-D coordinates of its individual components.[2] Although early rigid-body docking methods based on purely geometrical criteria were adequate when the approaching subunits artificially presented the same conformation as in the complex,[3] [4] the prediction results were clearly poorer when using the 3-D coordinates of the uncomplexed subunits.[5] It was soon evident that geometry-based approaches were not accurate enough to model the induced fit of the interacting surfaces upon binding. Treatment of interface flexibility as well as more realistic energy approximations had to be developed.

The inclusion of energy determinants, together with molecular flexibility, can result in more realistic simulations. A global minimization procedure with a complete energy description has been reported and successfully applied in the prediction of a lysozyme-antibody complex[6] and in a blind prediction contest.[7] The method, based on ICM methodology,[8] used a pseudo-brownian Monte-Carlo minimization[9] and a subsequent biased probability Monte Carlo[10] optimization of the interface side-chains. The use of a soft interaction energy function pre-calculated on a grid[11] can drastically increase the speed of the docking simulations, as has been observed for protein-ligand docking.[12]

Electrostatic interactions play an important role in protein-protein docking. The accurate definition of electrostatics, including solvation considerations, is critical for the correct ranking of the near-native solutions. Here we present a method for calculating the solvation-corrected grid electrostatic energy and show that it substantially improves the results of rigid-body docking simulations and side-chain refinement.

2 Methods

2.1 *Energy description*

The interaction energy potentials were pre-calculated on a grid[11] within a 3-D box covering approximately half of the total receptor surface (including the known or hypothetical receptor binding site). The energy estimate used during docking simulations consisted of the following terms:[13]

$$E = E_{\mathrm{Hvw}} + E_{\mathrm{Cvw}} + E_{\mathrm{el}} + E_{\mathrm{hb}} + E_{\mathrm{hp}} \qquad (1)$$

where E_{Hvw} is the van der Waals potential for a hydrogen atom probe, E_{Cvw} the van der Waals potential for a heavy atom probe (a generic carbon of 1.7 Å radius was used), E_{el} an electrostatic potential generated from the receptor with a distance dependant dielectric constant, E_{hb} the hydrogen-bonding potential calculated as spherical Gaussians centered at the ideal putative donor and/or acceptor sites, and E_{hp} a hydrophobicity potential roughly proportional to the buried hydrophobic surface area. Van der Waals potentials were initially truncated to a maximum energy value of 1.0 kcal mol^{-1} to avoid inter-molecule repulsive clashes arising from the rigidity of their side-chains.

2.2 *Solvation correction for electrostatics*

Previously we used a distance-dependent dielectric model $\varepsilon=4{*}r$ [14] to approximate the effects of solvent screening of electrostatic interaction. While this approach is fast and simple, it does not reflect the dependence of the solvation effect on the degree of the solvent exposure.

Charge scaling, i.e. the reduction of solvent-exposed charges, has previously been used to account for solvation effects in certain systems, such as DNA. Polarization of the solvent near the charged solute atom results in the formation of an induced solvent charge of the opposite sign, attenuating the electrostatic interactions of the solute. This effect can be considered as an effective reduction of the solute charges. The optimal degree of charge reduction has to be determined from such factors as solvent accessibility or fit to experimental values. Here we propose a method for charge scaling based on continuum dielectric solvation model.

Continuum dielectric solvation model represents the solvent as a medium of high dielectric constant (ε_{out} =78.5 for water), while the interior of the solute has relatively low dielectric constant (we use ε_{in} =4 in this work). Poisson equation has to be solved to evaluate accurately the electric field in such a system:

$$\nabla(\varepsilon(\mathbf{r})\nabla\phi(\mathbf{r}))=\rho(\mathbf{r}) \tag{2}$$

where ε is the dielectric constant (permittivity), ϕ is the electric potential and ρ is the charge density. The boundary element (BE)[15] method is a popular approach to solving the Poisson equation in continuum dielectric electrostatics calculations. The BE method is based on the representation of electrostatic solvation effects by appropriate induced surface charge density σ_s on the solute/solvent boundary. The full electrostatic field at a point **r** is represented as

$$\mathbf{E}=\sum_i \frac{q_i(\mathbf{r}-\mathbf{r}_i)}{\varepsilon_{in}|\mathbf{r}-\mathbf{r}_i|^3}+\oint\!\!\oint \frac{(\mathbf{r}-\mathbf{r}_s)}{|\mathbf{r}-\mathbf{r}_s|^3}\sigma_s ds \tag{3}$$

where the first term represents the standard Coulomb field of the atomic charges q_i and the second term accounts for the field of the induced surface charge. This representation provides a basis for quantitative evaluation of optimal scaled charges: the induced surface charge density can be projected onto the nearby atoms. The electric field of the resulting corrected atomic charges should approximate closely the exact solution, i.e. the combined field of the original atomic charges and of the induced surface charge distribution.

To obtain numeric solution in the BE method, the boundary is typically split into patches, or boundary elements, and surface charge densities σ_i are evaluated by solving a system of linear equations[15]. Serendipitously, the REBEL (rapid boundary element electrostatics)[16] implementation of the method uses per-atomic BE's, i.e. patches of the molecular surface assigned to the atoms of the solute and generates the surface charge values $\sigma_i S_i$ for these patches (S_i is the area of the *i*-th BE). That allowed us to use a very simple approach to scale the surface charges by adding to each partial atomic charge q_i the corresponding induced solute charge:

$$q_i' = q_i + \sigma_i S_i \tag{4}$$

While the approach obviously simplifies the complex nature of the electrostatic solvation, it largely reproduces the expected effect of solvent on electrostatic interactions in docking: partially buried (but involved in the interaction) charges contribute strongly to the binding, while exposed charges interact relatively weakly.

2.3 Docking simulations

The two-step docking procedure used in this work consisted of a rigid-body docking step followed by side-chain refinement (scheme in Figure 1). The resulting conformations from the first rigid body step were further optimized by an ICM[17] global optimization algorithm, with flexible interface ligand side-chains and a grid map representation of the receptor. In this refinement step, the internal energy for the ligand interface side-chains was also considered, including the van der Waals, hydrogen bonding and torsion energy calculated with ECEPP/3 parameters[18], and the Coulomb electrostatic energy (distance-dependent dielectric constant; ε=4*r).

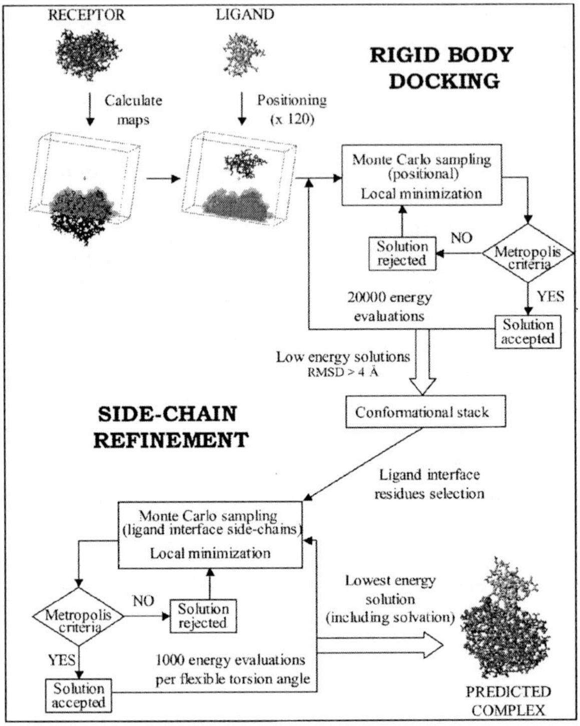

Figure 1. Scheme of the docking protocol used in this work. The ligand molecule was positioned in a random orientation inside the grid potential box and was systematically rotated to generate 120 different starting conformations. The six positional variables of the ligand were sampled by a pseudo-Brownian Monte Carlo optimization, in which each random step was followed by local minimization. New conformations were selected according to the Metropolis criterion[19] with a temperature of 5000K. Each simulation was terminated after 20,000 energy evaluations. Low energy conformations with pairwise r.m.s.d. for the ligand interface C^{α} atoms greater than 4 Å were retained in a conformational stack, and their interface side-chains were further optimized using a Biased Probability Monte Carlo procedure. The simulation temperature for this refinement step was set to 300K, and the total number of energy evaluations was 1,000 times the number of flexible interface torsion angles. The surface-based solvation energy[20] was included in the final energy to select the best refined solutions.

3 Results

3.1 Rigid-body docking simulations

Docking simulations have been applied to the selected protein-protein complexes listed in Table 1. For all of them, the 3-D structures of their unbound subunits are available.

Table 1. PDB codes of the protein-protein complexes and unbound subunits used in this work.

COMPLEX	RECEPTOR		LIGAND	
PDB	Name	PDB	Name	PDB
1ca0	Chymotrypsin	5cha	APPI	1aap
1cbw	Chymotrypsin	5cha	BPTI	1bpi
2sni	Subtilisin	2st1	CI-2	2ci2
1taw	Trypsin (bovine)	5ptp	APPI	1aap
3tgi	Trypsin (rat)	1ane	BPTI	1bpi
1brc	Trypsin mutant (rat)	1bra	APPI	1aap

For all test cases, automated rigid-body docking simulations starting from the unbound subunits were performed, using initially the uncorrected electrostatic energy. Amongst the low energy solutions stored for each complex, we always found at least one conformation within 4 Å r.m.s.d. (calculated for the ligand interface C^{α} atoms when only the receptor C^{α} atoms were superimposed onto the crystallographic structure) from the experimental structure. Rank (according to total energy) and r.m.s.d. values for the near native solutions found in all test cases are shown in Table 2 (column *corrected charges*).

Table 2. Rigid-Body docking results with uncorrected-and corrected charge electrostatic.

Complex	uncorrected charges		corrected charges	
	Rank	r.m.s.d. (Å)	Rank	r.m.s.d. (Å)
1ca0	16 of 233	1.4	6 of 228	1.4
1cbw	10 of 220	1.5	5 of 231	0.9
2sni	43 of 243	2.4	66 of 220	2.5
1taw	3 of 232	3.2	1 of 228	3.5
3tgi	210 of 238	0.3	23 of 243	0.6
1brc	20 of 218	3.5	1 of 237	3.7

In none of the complexes, the near native conformation was found as the lowest energy solution. The best scored near native solution (ranked 3rd) was found for trypsin/APPI complex (PDB 1taw), with an r.m.s.d. of 3.2 Å from the real structure. Interestengly, for the trypsin/BPTI complex (PDB 3tgi), we found a solution very close to the real structure (0.3 Å r.m.s.d.), but unfortunately very poorly scored (ranked 210th).

To evaluate if our SChEM approach could help to remove false positives (e.g. conformations with large interaction surfaces and over-estimated interactions of solvated charges), we performed rigid-body docking simulations for all complexes using the new corrected electrostatics. The results (Table 2; column *corrected charges*) clearly improved by using solvation-corrected electrostatics. The near native solution now ranked first for two of the six test cases (1taw and 1brc), and was found within the 6 lowest energy solutions in more than half of the cases.

3.2 *Interface refinement of docking solutions*

The resulting solutions obtained after rigid-body docking with uncorrected electrostatics, were further refine by optimizing the interface side-chains (using uncorrected electrostatics during the refinement). The results (Table 3; column *uncorrected charges*) show that the refinement of interacting side-chains improved the rank of the near native solution in all cases except one (1brc). Moreover, in two of the complexes, the near native solution is now ranked in first place (2sni and 1taw).

Table 3. Re-evaluation of docking solutions after interface refinement

Complex	*uncorrected charges*		*corrected charges*	
	Rank	r.m.s.d. (Å)	Rank	r.m.s.d. (Å)
1ca0	2 of 233	1.1	1 of 228	1.2
1cbw	3 of 220	1.1	1 of 231	0.7
2sni	1 of 243	2.6	1 of 220	2.7
1taw	1 of 232	2.8	1 of 228	2.9
3tgi	62 of 238	0.5	1 of 243	0.6
1brc	33 of 218	3.1	1 of 237	1.8

When the docking solutions obtained using the corrected electrostatics are refined (also using the solvation-corrected electrostatics), the ranking is further improved (Table 3; column *corrected charges*). In all cases, the near native solution is now the lowest energy conformation.

4 Discussion

Electrostatic forces play an important role in protein-protein interactions. The solvation effects present a major difficulty in modeling electrostatic interactions. We propose and apply a simple screened charge model to account for the attenuation of electrostatic interactions by the solvent in docking simulations. For the six complexes in the test set, the proposed correction results in a dramatic improvement of the ranking of the near-native solution. Already at the rigid-body docking step, the near-native solution is ranked first for two complexes, versus none in the case of uncorrected electrostatics. The influence of the correction on the refinement step is relatively minor. As can be seen in Figure 2 (refinement with uncorrected electrostatics) and Figure 3 (refinement with corrected electrostatics), interface refinement of rigid-body solutions removes clashes from interacting side-chains in the wrong conformation. Refinement is able to mimic the induced fit of the association, and the final conformation of interacting ligand side-chains is very close to the native conformation. However, the effect of the solvation correction is essential for the accurate scoring of the near-native solution. The correction apparently helps to remove false positives, e.g. conformations with extended interacting surfaces and over-estimated interactions of solvated charges. The lowest energy conformation (21.8 Å r.m.s.d.) (Figure 2) found with uncorrected electrostatics presents numerous electrostatic interactions of highly exposed residues for which the attraction is over-estimated. On the contrary, the near-native solution is correctly ranked as the lowest energy conformation when the corrected electrostatic term is used.

As this work shows, failing to consider solvation effects when calculating electrostatics in protein-protein docking simulations often results in numerous misdocked configurations. We show here a way of including such solvation effects in electrostatic energy which greatly improves docking accuracy without significant computational overhead.

A broader docking test using this novel approach for electrostatics is reported in an upcoming publication[21].

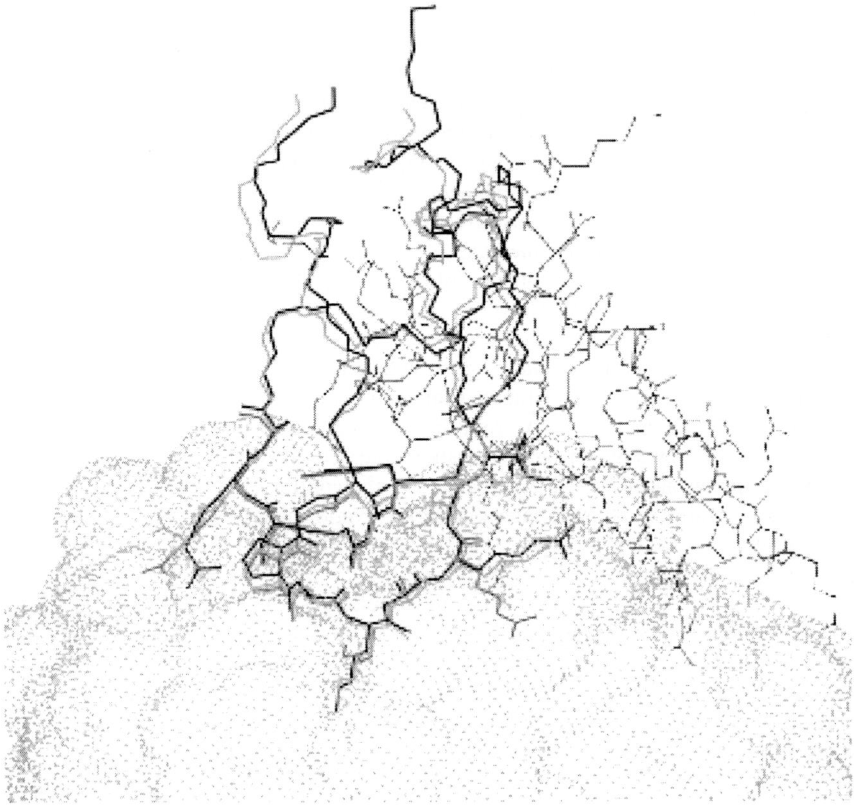

Figure 2. Docking results for trypsin/BPTI (PDB 3tgi) using uncorrected electrostatics. Near native solution obtained after rigid-body docking (dark gray) and after refinement (black) compared to the crystallographic structure (light gray). The lowest energy solution obtained after refinement is shown in thin lines. The trypsin surface is shown as light gray dots.

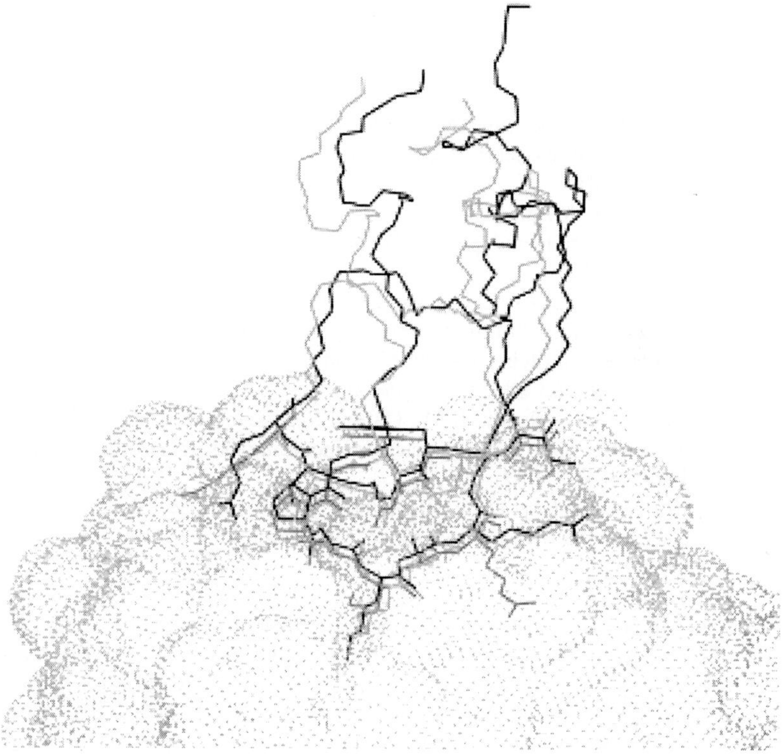

Figure 3. Docking results for trypsin/BPTI (PDB 3tgi) using corrected electrostatics. The near native solution obtained after rigid-body docking (dark gray) and after refinement (black) compared to the crystallographic structure (light gray). The near native solution obtained after refinement is the lowest energy solution. The trypsin surface is shown as light gray dots.

5 Acknowledgements

This work was supported by NIH Grant R01 GM55418. We wish to thank Brian Marsden for his continuous support of the LINUX clusters and Molsoft for making the ICM program available for the project.

References

1. Sanchez, R. & Sali, A. "Large-scale protein structure modeling of the Saccharomyces cerevisiae genome"*Proc. Natl. Acad. Sci. USA* **95,** 13597-602 (1998)
2. Sternberg, M. J., Gabb, H. A. & Jackson, R. M. "Predictive docking of protein-protein and protein-DNA complexes"*Curr. Opin. Struct. Biol.* **8,** 250-6 (1998)
3. Katchalski-Katzir, E., Shariv, I., Eisenstein, M., Friesem, A. A., Aflalo, C. & Vakser, I. A. "Molecular surface recognition: determination of geometric fit between proteins and their ligands by correlation techniques"*Proc. Natl. Acad. Sci. USA* **89,** 2195-9 (1992)
4. Kuntz, I. D., Blaney, J. M., Oatley, S. J., Langridge, R. & Ferrin, T. E. "A geometric approach to macromolecule-ligand interactions"*J. Mol. Biol.* **161,** 269-88 (1982)
5. Norel, R., Petrey, D., Wolfson, H. J. & Nussinov, R. "Examination of shape complementarity in docking of unbound proteins"*Proteins* **36,** 307-17 (1999)
6. Totrov, M. & Abagyan, R. "Detailed ab initio prediction of lysozyme-antibody complex with 1.6 Å accuracy"*Nat. Struct. Biol.* **1,** 259-63 (1994)
7. Strynadka, N. C., Eisenstein, M., Katchalski-Katzir, E., Shoichet, B. K., Kuntz, I. D., Abagyan, R., Totrov, M., Janin, J., Cherfils, J., Zimmerman, F., Olson, A., Duncan, B., Rao, M., Jackson, R., Sternberg, M. & James, M. N. "Molecular docking programs successfully predict the binding of a beta-lactamase inhibitory protein to TEM-1 beta-lactamase"*Nat. Struct. Biol.* **3,** 233-9 (1996)
8. Abagyan, R. & Argos, P. "Optimal protocol and trajectory visualization for conformational searches of peptides and proteins"*J. Mol. Biol.* **225,** 519-32 (1992)
9. Abagyan, R., Totrov, M. & Kuznetsov, D. "ICM: a new method for structure modeling and design: Applications to docking and structure prediction from the distorted native conformation"*J. Comp. Chem.* **15,** 488-506 (1994)

10. Abagyan, R. & Totrov, M. "Biased probability Monte Carlo conformational searches and electrostatic calculations for peptides and proteins" *J. Mol. Biol.* **235,** 983-1002 (1994)
11. Goodford, P. J. "A computational procedure for determining energetically favorable binding sites on biologically important macromolecules" *J. Med. Chem.* **28,** 849-57 (1985)
12. Totrov, M. & Abagyan, R. "Flexible protein-ligand docking by global energy optimization in internal coordinates" *Proteins* **Suppl 1,** 215-20 (1997)
13. Totrov, M. & Abagyan, R. (2001) in *Drug-receptor thermodynamics: Introduction and applications*, ed. Raffa, R. B. (John Wiley & Sons, Ltd., pp. 603-624.
14. McCammon, J. A., Wolynes, P. G. & Karplus, M. "Picosecond dynamics of tyrosine side chains in proteins" *Biochemistry* **18,** 927-42 (1979)
15. Zauhar, R. J. & Morgan, R. S. "A new method for computing the macromolecular electric potential" *J. Mol. Biol.* **186,** 815-20 (1985)
16. Totrov, M., Abagyan, R. "Rapid boundary element solvation electrostatics calculations in folding simulations: successful folding of a 23-residue peptide" *Biopolymers* **60,** 124-133 (2001)
17. MolSoft (2000) (MolSoft LLC, San Diego).
18. Nemethy, G., Gibson, K. D., Palmer, K. A., Yoon, C. N., Paterlini, G., Zagari, A., Rumsey, S. & Scheraga, H. A. "Energy parameters in polypeptides. 10. Improved geometric parameters and nonbonded interactions for use in the ECEPP/3 algorithm, with application to proline-containing peptides" *J. Phys. Chem.* **96,** 6472-84 (1992)
19. Metropolis, N. A., Rosenbluth, A. W., Rosenbluth, N. M., Teller, A. H. & Teller, E. "Equation of state calculations by fast computing machines" *J. Chem. Phys.* **21,** 1087-92 (1953)
20. Wesson, L. & Eisenberg, D. "Atomic solvation parameters applied to molecular dynamics of proteins in solution" *Protein Sci.* **1,** 227-35 (1992)
21. Fernandez-Recio, J., Totrov, M., Abagyan, R. "Soft protein-protein docking in internal coordinates" *Protein Sci.* (In press)

THE SPECTRUM KERNEL: A STRING KERNEL FOR SVM PROTEIN CLASSIFICATION

CHRISTINA LESLIE, ELEAZAR ESKIN, WILLIAM STAFFORD NOBLE[a]
{cleslie,eeskin,noble}@cs.columbia.edu
Department of Computer Science, Columbia University, New York, NY 10027

We introduce a new sequence-similarity kernel, the spectrum kernel, for use with support vector machines (SVMs) in a discriminative approach to the protein classification problem. Our kernel is conceptually simple and efficient to compute and, in experiments on the SCOP database, performs well in comparison with state-of-the-art methods for homology detection. Moreover, our method produces an SVM classifier that allows linear time classification of test sequences. Our experiments provide evidence that string-based kernels, in conjunction with SVMs, could offer a viable and computationally efficient alternative to other methods of protein classification and homology detection.

1 Introduction

Many approaches have been presented for the protein classification problem, including methods based on pairwise similarity of sequences [1,2,3], profiles for protein families [4], consensus patterns using motifs [5,6] and hidden Markov models [7,8,9]. Most of these methods are *generative* approaches: the methodology involves building a model for a single protein family and then evaluating each candidate sequence to see how well it fits the model. If the "fit" is above some threshold, then the protein is classified as belonging to the family. *Discriminative* approaches [10,11,12] take a different point of view: protein sequences are seen as a set of labeled examples – positive if they are in the family and negative otherwise – and a learning algorithm attempts to *learn* the distinction between the different classes. Both positive and negative examples are used in training for a discriminative approach, while generative approaches can only make use of positive training examples.

One of the most successful discriminative approaches to protein classification is the work of Jaakkola et al. [10,11] for detection of remote protein homologies. They begin by training a generative hidden Markov model (HMM) for a protein family. Then, using the model, they derive for each input sequence, positive or negative, a vector of features called Fisher scores that are assigned to the sequence. They then use a discriminative learning algorithm called a support vector machine (SVM) in conjunction with the feature vectors – in the form of a kernel function called the Fisher kernel – for protein family classifica-

[a]Formerly William Noble Grundy, see www.cs.columbia.edu/~noble/name-change.html

tion. A serious drawback of their approach is its computational expense – both for generating the kernel on the training set and for classifying test sequences – since the HMM is required for computing feature vectors both for training and test sequences. Training an HMM, or scoring a sequence with respect to an HMM, requires a dynamic programming algorithm that is roughly quadratic in the length of the sequence.

In this paper, we revisit the idea of using a discriminative approach, and in particular support vector machines, for protein classification. However, in place of the expensive Fisher kernel, we present a new string kernel (sequence-similarity kernel), the spectrum kernel, for use in the SVM. The kernel is designed to be very simple and efficient to compute and does not depend on any generative model, and we produce an SVM classifier that can classify test sequences in linear time. Moreover, the method is completely general in that it can be used for any sequence-based classification problem. In the experiments reported here, we do not incorporate prior biological information specific to protein classification, although we plan to use prior information in future research. We report results for experiments over the SCOP [13] database and show how our method performs surprisingly well given its generality.

When using a kernel in conjuction with an SVM, input sequences are implicitly mapped into a high-dimensional vector space where the coordinates are given by feature values. The SVM produces a linear decision boundary in this high-dimensional feature space, and test sequences are classified based on whether they map to the positive or negative side of the boundary. The features used by our spectrum kernel are the set of all possible subsequences of amino acids of a fixed length k. If two protein sequences contain many of the same k-length subsequences, their "inner product" under the k-spectrum kernel will be large. The notion of the spectrum of a biological sequence – that is, the k-length subsequence content of the sequence – has been used for applications such as sequencing by hybridization [14] and is conceptually related to Fourier-based sequence analysis techniques [15].

We note that recently, Chris Watkins [16] and David Haussler [17] have defined a set of kernel functions over strings, and one of these string kernels has been implemented for a text classification problem [18]. However, the cost of computing each kernel entry is $O(n^2)$ in the length of the input sequences, making them too slow for most biological applications. Our spectrum kernel, with complexity $O(kn)$ to compute each k-spectrum kernel entry, is both conceptually simpler and computationally more efficient.

2 Overview of Support Vector Machines

Support Vector Machines (SVMs) are a class of supervised learning algorithms first introduced by Vapnik [19]. Given a set of labelled training vectors (positive and negative input examples), SVMs learn a linear decision boundary to discriminate between the two classes. The result is a linear classification rule that can be used to classify new test examples. SVMs have exhibited excellent generalization performance (accuracy on test sets) in practice and have strong theoretical motivation in statistical learning theory [19].

Suppose our training set $\mathcal{S}$ consists of labelled input vectors $(\mathbf{x}_i, y_i)$, $i = 1 \ldots m$, where $\mathbf{x}_i \in \mathbb{R}^n$ and $y_i \in \{\pm 1\}$. We can specify a linear classification rule f by a pair $(\mathbf{w}, b)$, where $\mathbf{w} \in \mathbb{R}^n$ and $b \in \mathbb{R}$, via

$$f(\mathbf{x}) = \langle \mathbf{w}, \mathbf{x} \rangle + b$$

where a point $\mathbf{x}$ is classified as positive (negative) if $f(\mathbf{x}) > 0$ ($f(\mathbf{x}) < 0$). Geometrically, the decision boundary is the hyperplane

$$\{\mathbf{x} \in \mathbb{R}^n : \langle \mathbf{w}, \mathbf{x} \rangle + b = 0\}$$

where $\mathbf{w}$ is a normal vector to the hyperplane and b is the bias. If we further require that $|\mathbf{w}| = 1$, then the *geometric margin* of the classifier with respect to $\mathcal{S}$ is

$$m^g_{\mathcal{S}}(f) = \mathrm{Min}_{\{\mathbf{x}_i \in \mathcal{S}\}} y_i f(\mathbf{x}_i).$$

In the case where the training data are linearly separable and a classifier f correctly classifies the training set, then $m^g_{\mathcal{S}}(f)$ is simply the distance from the decision hyperplane to the nearest training point(s).

The simplest kind of SVM is the maximal margin (or hard margin) classifier, which solves an optimization problem to find the linear rule f with maximal geometric margin. Thus, in the linearly separable case, the hard margin SVM finds the hyperplane that correctly separates the data and maximizes the distance to the nearest training points.

In pratice, training sets are usually not linearly separable, and we must modify the SVM optimization problem to incorporate a trade-off between maximizing geometric margin and minimizing some measure of classification error on the training set. See [20] for a precise formulation of various soft margin approaches.

3 Kernels in SVMs

A key feature of any SVM optimization problem is that it is equivalent to solving a dual quadratic programming problem. For example, in the linearly separable case, the maximal margin classifier is found by solving for the optimal "weights" α_i, $i = 1 \ldots m$, in the dual problem:

$$\text{Maximize } \sum_i \alpha_i - \frac{1}{2} \sum_i \sum_j \alpha_i \alpha_j y_i y_j \langle \mathbf{x}_i, \mathbf{x}_j \rangle$$
$$\text{on the region: } \alpha_i \geq 0 \text{ for all } i$$

The parameters $(\mathbf{w}, b)$ of the classifier are then determined by the optimal α_i (and the training data). The dual optimization problems for various soft margin SVMs are similar.

The dual problem not only makes SVMs amenable to various efficient optimization algorithms, but also, since the dual problem depends only on the inner products $\langle \mathbf{x}_i, \mathbf{x}_j \rangle$, allows for the introduction of *kernel techniques.*

To introduce a kernel, we now suppose that our training data are simply labelled examples (x_i, y_i), where the x_i belong to an input space $\mathcal{X}$ which could be a vector space or a space of discrete structures like sequences of characters from an alphabet or trees. Given any feature map Φ from $\mathcal{X}$ into a (possibly high-dimensional) vector space called the *feature space*

$$\Phi : \mathcal{X} \to \mathbb{R}^N,$$

we obtain a kernel K on $\mathcal{X} \times \mathcal{X}$ defined by

$$K(x, y) = \langle \Phi(x), \Phi(y) \rangle.$$

By replacing $\langle \mathbf{x}_i, \mathbf{x}_j \rangle$ by $K(x_i, x_j)$ in the dual problem, we can use SVMs in feature space. Moreover, if we can directly compute the kernel values $K(x, y)$ without explicitly calculating the feature vectors, we gain tremendous computational advantage for high-dimensional feature spaces.

4 The Spectrum Kernel

For our application to protein classification, we introduce a simple string kernel, which we call the spectrum kernel, on the input space $\mathcal{X}$ of all finite length sequences of characters from an alphabet $\mathcal{A}$, $|\mathcal{A}| = l$.

Recall that, given a number $k \geq 1$, the k-spectrum of an input sequence is the set of all the k-length (contiguous) subsequences that it contains. Our feature map is indexed by all possible subsequences a of length k from alphabet $\mathcal{A}$. We define a feature map from $\mathcal{X}$ to $\mathbb{R}^{l^k}$ by

$$\Phi_k(x) = (\phi_a(x))_{a \in \mathcal{A}^k}$$

where

$$\phi_a(x) = \text{number of times } a \text{ occurs in } x$$

Thus the image of a sequence x under the feature map is a weighted representation of its k-spectrum. The k-spectrum kernel is then

$$K_k(x, y) = \langle \Phi_k(x), \Phi_k(y) \rangle.$$

For another variant of the kernel, we can assign to the a-th coordinate a binary value of 0 if a does not occur in x, 1 if it does occur.

Note that while the feature space is large even for fairly small k, the feature vectors are sparse: the number of non-zero coordinates is bounded by length$(x) - k + 1$. This property allows various efficient approaches for computing kernel values.

A very efficient method for computing $K_k(x, y)$ is to build a suffix tree for the collection of k-length subsequences of x and y, obtained by moving a k-length sliding window across each of x and y. At each depth-k leaf node of the suffix tree, store two counts, one representing the number of times a k-length subsequence of x ends at the leaf, the other representing a similar count for y. Note that this suffix tree has $O(kn)$ nodes. Using a linear time construction algorithm for the suffix tree [21], we can build and annotate the suffix tree in $O(kn)$ time. Now we calculate the kernel value by traversing the suffix tree and computing the sum of the products of the counts stored at the depth-k nodes. The overall cost of calculating $K_k(x, y)$ is thus $O(kn)$. One can use a similar idea to build a suffix tree for all the input sequences at once and to compute all the kernel values in one traversal of the tree. This is essentially the method we use to compute our kernel matrices for our experiments, though we use a recursive function rather than explicitly constructing the suffix tree.

There is an alternative method for computing kernel values that is less efficient but very easy to implement. For simplicity of notation, we describe the binary-valued version of the feature map, though the count-valued version is similar. For each sequence x, collect the set of k-length subseqences into an array A_x and sort them. Now the inner product $K_k(x, y)$ can be computed in linear time as a function of length(x) + length(y). Thus the overall complexity of computing the kernel value is $O(n \log(n))$ in the length of the input sequences using this method.

5 Linear Time Prediction

The output of the SVM is a set of weights α_i that solve the dual optimization problem, where $i = 1 \ldots m$ for a set of m training vectors. Training vectors

$\mathbf{x}_i$ for which the corresponding weight α_i is non-zero are called **support vectors**. The parameters $(\mathbf{w}, b)$ of the classifier are determined by the weights and support vectors. Specifically, we have

$$\mathbf{w} = \sum_{\text{support vectors } \mathbf{x}_i} \alpha_i y_i \Phi(\mathbf{x}_i),$$

and in general there is also an expression for b, though in our experiments, we use a version of the SVM algorithm for which $b = 0$. Thus, in the case $b = 0$, test examples are classified by the sign of the expression

$$f(\mathbf{x}) = \Phi(\mathbf{x}) \cdot \mathbf{w} = \sum_{\text{support vectors } \mathbf{x}_i} \alpha_i y_i K(\mathbf{x}, \mathbf{x}_i).$$

For our spectrum kernel K_k, the normal vector $\mathbf{w}$ is given by

$$\mathbf{w} = \left(\sum_{\text{support vectors } \mathbf{x}_i} \alpha_i y_i \phi_a(x_i) \right)_{a \in \mathcal{A}^k}.$$

Note that typically the number of support vectors is much smaller that m, so that the number of non-zero coefficients in the expression for $\mathbf{w}$ is much smaller that mn. We store these non-zero coefficients in a look-up table, associating to each contributing k-length subsequence a its coefficient in $\mathbf{w}$:

$$a \rightarrow \sum_{\text{support vectors } \mathbf{x}_i} \alpha_i y_i \{\#\text{ times } a \text{ occurs in } x_i\}.$$

Now to classify a test sequence x in linear time, move a k-length sliding window across x, look up the current k-length subsequence in the look-up table, and increment the classifier value $f(x)$ by the associated coefficient.

6 Experiments: Protein Classification

We test the spectrum SVM method using an experimental design by Jaakkola et al. [10] for the remote homology detection problem. In this test, remote homology is simulated by holding out all members of a target SCOP [13] family from a given superfamily. Positive training examples are chosen from the remaining families in the same superfamily, and negative test and training examples are chosen from outside the target family's fold. The held-out family members serve as positive test examples. Details of the data sets are available at www.cse.ucsc.edu/research/compbio/discriminative.

Because the test sets are designed for remote homology detection, we use small values of k in our k-spectrum kernel. We tested $k = 3$ and $k = 4$, both for the unnormalized kernel K_k and for the normalized kernel given by

$$K_k^{\mathrm{Norm}}(x,y) = \frac{K_k(x,y)}{\sqrt{K_k(x,x)}\sqrt{K_k(y,y)}}.$$

Our results show that a normalized kernel with $k = 3$ yields the best performance, although the differences among the four kernels are not large (data not shown).

We use a publicly available SVM software implemenation (www.cs.columbia.edu/compbio/svm), which implements the soft margin optimization algorithm described in [10]. Note that for this variant of the SVM optimization problem, the bias term b is fixed to 0. We did not attempt any fine-tuning of the soft margin SVM parameters.

We use ROC_{50} scores to compare the performance of different homology detection between methods. The ROC_{50} score is the area under the receiver operating characteristic curve – the plot of true positives as a function of false positives – up to the first 50 false positives [22]. A score of 1 indicates perfect separation of positives from negatives, whereas a score of 0 indicates that none of the top 50 sequences selected by the algorithm were positives.

For comparison, we include results from three other methods. These include the original experimental results from Jaakkola *et al.* for two methods: the SAM-T98 iterative HMM, and the Fisher-SVM method. We also test PSI-BLAST[3] on the same data. To approximate a family-based homology detection method, PSI-BLAST is run using a randomly selected training set sequence for one iteration, with the positive training set as a database and a very low E-value inclusion threshold. The resulting matrix is then used to search the test set for a maximum of 20 iterations using the default E-value inclusion threshold. The BLOSUM80 substitution matrix is used with PSI-BLAST.

The results for all 33 SCOP families are summarized in Figure 1. Each series corresponds to one homology detection method. Qualitatively, the SAM-T98 and Fisher-SVM methods perform slightly better than PSI-BLAST and the spectrum SVM. However, if we evaluate the statistical significance of these differences using a two-tailed signed rank test [23,24] (including a Bonferroni adjustment for multiple comparisons), only the SVM-Fisher method does better than any other method: SVM-Fisher's performance is better than that of PSI-BLAST with a p-value of 0.000045 and is better than that of the spectrum SVM with a p-value of 0.042. These results suggest that the spectrum SVM performs comparably with some state-of-the-art homology detection methods. In particular, the signed-rank comparison of PSI-BLAST and the spectrum

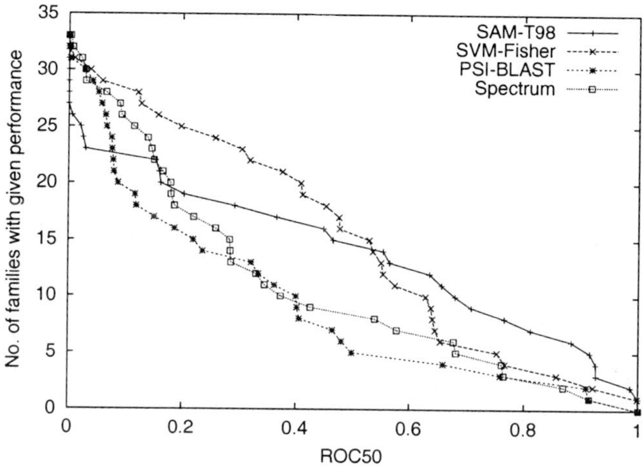

Figure 1: **Comparison of four homology detection methods.** The graph plots the total number of families for which a given method exceeds an ROC_{50} score threshold. Each series corresponds to one of the homology detection methods described in the text.

SVM gives a slight (though not significant) preference to the latter (unadjusted p-value of 0.16).

Figure 2 gives a more detailed view of the performance of the spectrum SVM with the Fisher-SVM method. Here we see clearly the optimization of the Fisher-SVM method for remote homology detection. For relatively easy-to-recognize families (i.e., families with high ROC_{50} scores), the two methods perform comparably; however, as the families become harder to recognize, the difference between the two methods becomes more extreme. Similar results are apparent in a comparison of Fisher-SVM and SAM-T98 (not shown), where SAM-T98 significantly out-performs Fisher-SVM for many of the easier families and vice versa.

7 Conclusions and Future Work

We have presented a conceptually simple, computationally efficient and very general approach to sequence-based classification problems. For the remote homology detection problem, we are encouraged that our discriminative approach – combining support vector machines with the spectrum kernel – performed remarkably well in the SCOP experiments when compared with state-of-the-

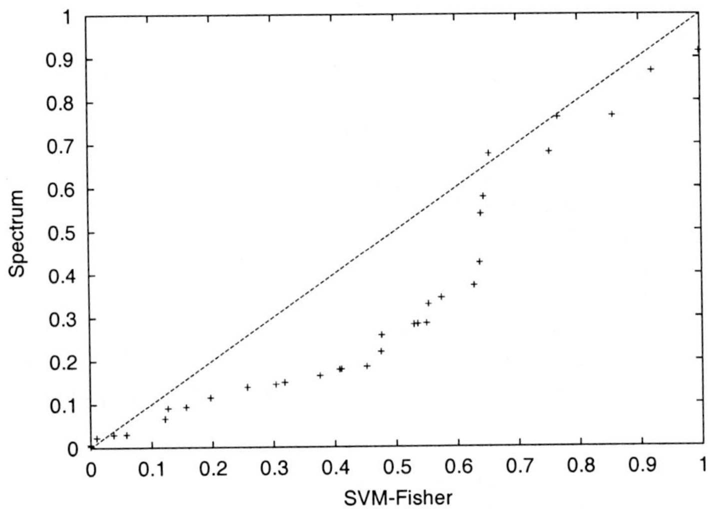

Figure 2: **Family-by-family comparison of the spectrum SVM and Fisher-SVM methods.** The coordinates of each point in the plot are the ROC_{50} scores for one SCOP family, obtained using the spectrum SVM and Fisher-SVM methods.

art methods, even though we used no prior biological knowledge specific to protein classication in our kernel. We believe that our experiments provide evidence that string-based kernels, in conjunction with SVMs, could offer a simple, effective and computationally efficient alternative to other methods of protein classification and homology detection.

There are several directions we plan to take this work. For improved performance in remote homology detection as well as for other discrimination problems – for example, classification problems involving DNA sequences – it should be advantageous to use larger values of k (longer subsequences) and incorporate some notion of mismatching. That is, we might want to change our kernel so that two k-length subsequences that are the same except for a small number of mismatched characters will, when mapped into feature space, have non-zero inner product. For protein classification, we would likely incorporate BLOSUM matrix information [25] into our mismatch kernel. We plan to implement an efficient data structure to enable us to calculate kernel values for a spectrum kernel that incorporates mismatching.

Secondly, in certain biological applications, the k-length subsequence features that are "most significant" for discrimination can themselves be of biological interest. For such problems, it would be interesting to perform feature selection on the set of k-spectrum features, so that we identify a feature subset that both allows for accurate discrimination and gives biologically interesting information about the spectrum differences between positive and negative examples. We are studying feature selection techniques in the context of SVMs, and we hope eventually to apply such techniques to the k-spectrum features.

Acknowledgments: WSN is supported by an Award in Bioinformatics from the PhRMA Foundation, and by National Science Foundation grants DBI-0078523 and ISI-0093302.

1. MS Waterman, J Joyce, and M Eggert. *Computer alignment of sequences*, chapter Phylogenetic Analysis of DNA Sequences. Oxford, 1991.
2. SF Altschul, W Gish, W Miller, EW Myers, and DJ Lipman. A basic local alignment search tool. *JMB*, 215:403–410, 1990.
3. SF Altschul, TL Madden, AA Schaffer, J Zhang, Z Zhang, W Miller, and DJ Lipman. Gapped BLAST and PSI-BLAST: A new generation of protein database search programs. *Nucleic Acids Research*, 25:3389–3402, 1997.
4. M Gribskov, AD McLachlan, and D Eisenberg. Profile analysis: Detection of distantly related proteins. *PNAS*, pages 4355–4358, 1987.
5. A Bairoch. PROSITE: A dictionary of sites and patterns in proteins. *Nucleic Acids Research*, 19:2241–2245, 1991.

6. TK Attwood, ME Beck, DR Flower, P Scordis, and JN Selley. The prints protein fingerprint database in its fifth year. *Nucleic Acids Research*, 26(1):304–308, 1998.
7. A Krogh, M Brown, I Mian, K Sjolander, and D Haussler. Hidden Markov models in computational biology: Applications to protein modeling. *JMB*, 235:1501–1531, 1994.
8. SR Eddy. Multiple alignment using hidden Markov models. In *ISMB*, pages 114–120. AAAI Press, 1995.
9. P Baldi, Y Chauvin, T Hunkapiller, and MA McClure. Hidden Markov models of biological primary sequence information. *PNAS*, 91(3):1059–1063, 1994.
10. T Jaakkola, M Diekhans, and D Haussler. A discriminative framework for detecting remote protein homologies. *Journal of Computational Biology*, 2000.
11. T Jaakkola, M Diekhans, and D Haussler. Using the fisher kernel method to detect remote protein homologies. In *ISMB*, pages 149–158. AAAI Press, 1999.
12. CHQ Ding and I Dubchak. Multi-class protein fold recognition using support vector machines and neural networks. *Bioinformatics*, 17(4):349–358, 2001.
13. AG Murzin, SE Brenner, T Hubbard, and C Chothia. SCOP: A structural classification of proteins database for the investigation of sequences and structures. *JMB*, 247:536–540, 1995.
14. I Pe'er and R Shamir. Spectrum alignment: Efficient resequencing by hybridization. In *ISMB*, pages 260–268. AAAI Press, 2000.
15. D Anastassiou. Frequency-domain analysis of biomolecular sequences. *Bioinformatics*, 16:1073–1081, 2000.
16. C Watkins. Dynamic alignment kernels. Technical report, UL Royal Holloway, 1999.
17. D Haussler. Convolution kernels on discrete structure. Technical report, UC Santa Cruz, 1999.
18. H Lodhi, J Shawe-Taylor, N Cristianini, and C Watkins. Text classification using string kernels. Preprint.
19. VN Vapnik. *Statistical Learning Theory.* Springer, 1998.
20. N Cristianini and J Shawe-Taylor. *An Introduction to Support Vector Machines.* Cambridge, 2000.
21. E Ukkonen. On-line construction of suffix trees. *Algorithmica*, 14:249–260, 1995.
22. M Gribskov and NL Robinson. Use of receiver operating characteristic (ROC) analysis to evaluate sequence matching. *Computers and Chem-*

istry, 20(1):25–33, 1996.

23. S Henikoff and JG Henikoff. Embedding strategies for effective use of information from multiple sequence alignments. *Protein Science*, 6(3):698–705, 1997.
24. SL Salzberg. On comparing classifiers: Pitfalls to avoid and a recommended approach. *Data Mining and Knowledge Discovery*, 1:371–328, 1997.
25. S Henikoff and JG Henikoff. Amino acid substitution matrices from protein blocks. *PNAS*, 89:10915–10919, 1992.

DETECTING POSITIVELY SELECTED AMINO ACID SITES USING POSTERIOR PREDICTIVE P-VALUES

R. NIELSEN
Department of Biometrics, Cornell University, 439 Warren Hall, Ithaca, NY 14853-7801, (rn28@cornell.edu)

J. P HUELSENBECK
Department of Biology, University of Rochester, Rochester, NY 14627-0211, (johnh@brahms.biology.rochester.edu)

Identifying positively selected amino acid sites is an important approach for making inference about the function of proteins; an amino acid site that is undergoing positive selection is likely to play a key role in the function of the protein. We present a new Bayesian method for identifying positively selected amino acid sites and apply the method to a data set of hemagglutinin sequences from the Influenza virus. We show that the results of the new methods are in accordance with results obtained using previous methods. More importantly, we also demonstrate how the method can be used for making further inferences about the evolutionary history of the sequences. For example, we demonstrate that sites that are positively selected tend to have a preponderance of conservative amino acid substitutions.

1 Introduction

1.1 The d_N/d_S ratio

The degree to which an amino acid site is free to vary is strongly dependent on its structural and functional importance. An amino acid that plays a critical role—perhaps as a member in a functionally important structure—is unlikely to change over evolutionary time. In fact, most methods aimed at detecting regions or sites of functional importance in amino acid or DNA sequences are based on detecting regions of low variability. However, very high levels of variability also signify functional importance. For example, many viruses experience positive diversifying selection in their coat proteins to avoid immune recognition[1]. The regions that have been targeted by selection are hypervariable, having an excess of amino acid altering substitutions (nonsynonymous substitutions) compared to what would be expected if all substitutions at the DNA level occur at the same rate. The evidence for selection is in the d_N/d_S ratio. The d_N/d_S ratio is the ratio of the rate of nonsynonymous substitutions per nonsynoymous sites (d_N) to the rate of synonymous substitutions (non-amino acid altering) per synonymous site (d_S). If no Darwinian selection is acting on the DNA sequences, we would expect $d_N = d_S$. If there is negative selection (selection against new amino acid variants) $d_N < d_S$, and if there is positive selection (selection for new variants) $d_N > d_S$. The d_N/d_S

ratio is a proxy for the strength of selection and can, therefore, be used to search for regions of functional importance. For example, in some viruses the amino acid sites that are important for interactions between the virus and the host can be identified by finding the sites undergoing positive selection.

1.2 The maximum likelihood method

Recently, several new methods have been developed for detecting positively selected amino acid sites. The methods of Nielsen and Yang[2] and Yang *et al.*[3] are based on modeling the evolution of a nucleotide sequence as a continuous time Markov chain with state space on the set of possible codons. In these models the d_N/d_S ratio is a parameter and it can be estimated using maximum likelihood (i.e., by finding the value for the d_N/d_S ratio that maximizes the probability of observing the data). The instantaneous rate of change from codon i to codon j in site k is given by

$$q_{ij}^{(h)} = \begin{cases} 0, & \text{if } i \text{ and } j \text{ differ at more than one position,} \\ \pi_j, & \text{if } i \text{ and } j \text{ differ by a synonymous transversion,} \\ \kappa\pi_j, & \text{if } i \text{ and } j \text{ differ by a synonymous transition,} \\ \omega_h\pi_j, & \text{if } i \text{ and } j \text{ differ by a nonsynonymous transversion,} \\ \omega_h\kappa\pi_j, & \text{if } i \text{ and } j \text{ differ by a nonsynonymous transition,} \end{cases} \tag{1}$$

where ω_h is the value of $\omega = d_N/d_S$ in the hth site, κ is the transition/transversion rate ratio, and π_j is the stationary frequency of codon j. The value of ω_h at a site is considered to be unknown, but drawn from some parametric distribution. Parameters of this distribution can be estimated using maximum likelihood. The details of the calculations can be found in Felsenstein[4] and Nielsen and Yang[2]. In brief, the likelihood function is calculated by superimposing the substitution process along the branches of the phylogenetic tree. The pruning algorithm of Felsenstein[4] can then be used to calculate the likelihood function for parameters in the ω distribution and other parameters such as the branch lengths of the tree. Because of the use of a phylogenetic tree, the method is applicable to multiple aligned sequences. However, at this time it can only be applied to one, or a few, trees because of computational limitations.

Formulating the problem of detecting positive selection in an explicitly statistical framework has a number of advantages. An important strength of the likelihood-based methods is the ease with which hypotheses can be tested. Yang *et al.*[3] showed how the method can be used to test if there is evidence for positive selection in a particular data set. For example, in one of the models (M7) in Yang *et al.*[3] it was assumed that ω follows a beta distribution. A beta distributed random variable is defined in the interval between 0 and 1; no positive selection is allowed under this model. A slightly more complicated model can be made by assuming that ω for a site is either drawn from a beta distribution (as before) or is under positive selection ($\omega > 1$). The maximum likelihood values obtained under the more general model and the model assuming a beta distribution can be compared, and the hypothesis of no sites undergoing positive selection can be tested using a likelihood ratio test. If the beta distribution is rejected in favor of a model allowing positively selected sites, it is also possible to predict which sites have values of $\omega >$ 1 using an empirical Bayes method. The positively selected sites can be identified by calculating the posterior probability that a particular site has a value of $\omega > 1$ under the parameter estimates obtained for the model allowing for positive selection.

1.3 Mapping mutations on phylogenies

In this manuscript we will explore an alternative approach. This approach is based on explicitly mapping mutations on a phylogeny. In Nielsen[5] an approach was described in which inferences regarding molecular evolution can be made using the posterior distribution of mappings of mutations. Let D be the data, in our case a set of aligned nucleotide sequences, and let M be a possible mapping of mutations on the phylogeny. Figure 1 shows an example of some observations (our data, D) at the tips of a tree with one possible realization of mutations (a mapping, M) that could have led to the observations at the tips of the tree.

We are typically interested in evaluating some function of the mapping of mutations. For example, we could be interested in the number of nonsynonymous mutations in a particular amino acid (codon) site or the distribution of such mutations along the sequence or along the phylogeny. If we knew the correct mapping of mutations we could easily evaluate this function. The problem is that we do not know which mutational mapping is correct. Performing statistical analyses based on a single mapping, for example a parsimony mapping, might lead to serious biases[5]. Instead a more appropriate approach is to sum a statistic over all possible mappings, weighting by the posterior probability of each.

Let $h(M)$ be the value of the function we are interested in (e.g. the number of nonsynonymous mutations in a site). We assume that this function cannot be calculated directly from the data using some simple method, but that $h(M)$ easily can be evaluated if M is known. We are then interested in evaluating

$$h(D) := E\{h(M) \mid D\} = \sum_{M \in \Psi} h(M) \Pr(M \mid D). \qquad (2)$$

Here Ψ is the set of possible mappings and $\Pr(M \mid D)$ is the posterior probability of a mapping. In a Bayesian framework we can evaluate $\Pr(M \mid D)$ as

$$\Pr(M \mid D) = \int_{\Theta \in \Omega} \Pr(M \mid D, \Theta)\, p(\Theta \mid D) d\Theta . \tag{3}$$

Here Θ is a vector of parameters and Ω is the set of all possible values of this vector. In our case, it will include the topology of the phylogeny, the branch lengths, and parameters of the mutational process which will be detailed later. In other words, the posterior distribution of mappings is evaluated by integrating over the posterior density of Θ, $p(\Theta \mid D)$. This distribution can be specified under a particular model of sequence evolution and using appropriate prior distributions for the parameters. In practice, it is necessary to use Markov chain Monte Carlo (MCMC), as in Larget and Simon[6] or Huelsenbeck *et al*[7]., to evaluate $p(\Theta \mid D)$. In these methods, a Markov chain with state space on the possible values of Θ and stationary distribution $p(\Theta \mid D)$ is simulated using the Metropolis-Hastings algorithm[8,9]. By sampling from this chain at stationarity, (correlated) samples of Θ from $p(\Theta \mid D)$ can be obtained.

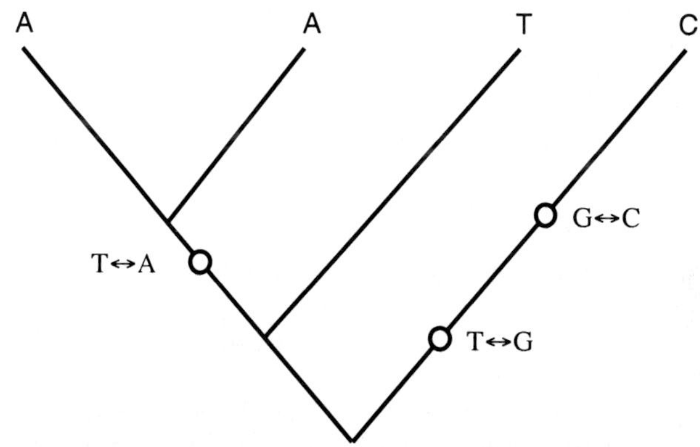

Figure 1. A Mutational mapping on a phylogeny for a single nucleotide site. Circles indicate mutations. For each mutation, the type of mutation (e.g. T↔A), the edge of the tree on which it occurred and, possibly, the time at which it occurred, is recorded in the computer memory. A mutational mapping (M) consists of these recordings for all sites in a data set, and all edges in the tree.

In Nielsen[5] an algorithm for sampling a random mapping of mutations from the distribution $\Pr(M \mid \Theta, D)$ was described. Using this algorithm it is possible to obtain samples of M from $\Pr(M \mid D)$ and thereby stochastically evaluate $h(D)$. First, k values of Θ, $\Theta_1, \Theta_2, \ldots, \Theta_k$, are sampled from $p(\Theta \mid D)$ using the MCMC method. In our particular applications this will be done using the program MrBayes[10]. Then, for each of these values of Θ, a mapping of mutations, M_1, $M_2, \ldots M_k$, is simulated from $\Pr(M \mid \Theta_i, D)$. By a law of large numbers for Markov chains,

$$\frac{1}{k}\sum_{i=1}^{k} h(M_i \mid D) \rightarrow h(D) \tag{4}$$

as $k \rightarrow \infty$. In Nielsen[5] this method was found to be quite computationally efficient. It provides a practical, and statistically well-justified, approach for examining patterns of molecular evolution.

1.4 Posterior predictive distributions

The idea of using posterior predictive distributions for making statistical inferences is well established in Bayesian statistics[11]. The posterior predictive distribution of a statistic is the distribution of a future (predicted) value of the statistic given the observed data. More formally, if D^{rep} denotes a replication of D, the posterior predictive distribution of a statistic, $T(\cdot)$, is given by

$$p(T(D^{rep}) \mid D) = \int_{\Theta \in \Omega} p(T(D^{rep}) \mid \Theta) p(\Theta \mid D) d\Theta \,. \tag{5}$$

Likewise, we can define a posterior predictive p-value[11,12] as

$$p_T = \Pr\{T(D^{rep}) \geq T(D) \mid D\} \tag{6}$$

This probability is evaluated with respect to the probability distribution given by Equation 5. A posterior predictive p-value is a hybrid between Bayesian and frequentist concepts. It involves a p-value, which is a frequentist construction that traditionally is justified by its properties in repeated sampling; however, integration over a posterior distribution of parameters is used to deal with the nuisance parameter problem. Its use can be justified both in a frequentist and a Bayesian setting. Meng[12] showed that the probability of a type I frequentist error of an α-level posterior predictive test is often close to but less than α and will never exceed

2α. Rubin[11] has argued that using posterior predictive p-values is Bayesian justifiable and also Bayesian relevant because of its use in model diagnosis.

A final point worth noticing in the current context, is that to perform hypothesis tests based on posterior predictive p-values, it is only necessary to specify the model under the null-hypothesis. We do not need to explicitly specify an alternative model.

2. Simulations

2.1 Models and data

In the following, we will apply the ideas described above to the identification of positively selected sites in a data set containing 28 sequences of the hemagglutinin protein from the Influenza virus. This data set was previously analyzed in Yang *et al.*[3] where it was shown, using the likelihood methods, that this protein is in fact subject to positive selection. We will use this data set for illustrating the new method to facilitate easy comparison with the likelihood method.

We are interested in testing the hypothesis of $\omega = 1$ against a one sided alternative of $\omega > 1$. Our null model is therefore specified by a codon substitution model in which $\omega = 1$. We notice that such a model is identical to a nucleotide substitution model with state space on the four nucleotides, with the exception of the presence of stop codons and codon usage bias. We will, therefore, for computational reasons, use a nucleotide model to closely approximate the codon substitution model. This will also allow us to use a more complex mutational model than the model assumed in Equation 1. In particular we will use the Generalized Time Reversible model[13] (GTR) to model the mutation process in the DNA sequence. We will use an uninformative prior for the base frequencies: a flat Dirichlet distribution. We will also assume uniform priors for the rest of the parameters: the other remaining parameters of the mutational model, the tree topology, and the branch lengths of the tree. To assure that the resulting posterior distribution is proper, a maximum branchlength of 100 (expected substitutions per nucleotide site) is assumed. We notice that the use of uniform priors ensures that our results are minimally influenced by the choice of priors.

2.2 Estimating the number of nonsynonymous mutations in a site

Using the computer program MrBayes[10] we ran a Markov chain for 1,100,000 cycles. After the first 100,000 cycles, we sampled a value of Θ at every 1000th cycle, eventually obtaining a total of 1000 samples, Θ_1, Θ_2, $\Theta,\ldots,\Theta_{1000}$. These samples are valid, albeit correlated, draws from the posterior probability distribution of Θ. For each of these samples of Θ, we sample a mutational mapping from Pr(M | Θ, D). The resulting set of mutational mappings, M_1, $M_2,\ldots,M_{1000}$, are then distributed as Pr(M | D). The samples will all be correlated because they are sampled from the same Markov chain, but only weakly so, because of the sampling interval of 1000 cycles.

For each of the mappings, we calculate the posterior expectation of the number of nonsynonymous (amino acid altering) mutations in all sites. For site j

$$T(D_j) := E(n_j \mid D) \approx \frac{1}{1000}\sum_{i=1}^{1000} n_j(M_{ij}, D_j), \tag{7}$$

where n_j is the number of nonsynonymous substitutions in site j, and $n_j(M_{ij}, D_j)$ is the number of nonsynonymous mutations that occurred in site j in mutational mapping i. $T(D_j)$ is the statistic we will use to evaluate the hypothesis $\omega = 1$ in site j. Since we are only interested in detecting positive selection, we will make a one sided test, that rejects when

$$p_{T_j} = \Pr\left\{T(D_j^{rep}) \geq T(D_j) \mid D\right\} \leq \alpha. \tag{8}$$

To evaluate this probability, we need to know the distribution of $T(D_j^{rep})$. Notice that $T(D_j^{rep})$ is identically distributed for all j, because of the independence assumption of the substitution process under the null hypothesis. We simply evaluate the predictive distribution for one site-to obtain the distribution for all sites.

2.3 Estimating the posterior predictive distribution

To evaluate the posterior predictive distribution we simulate 10 new sites for each of the previously simulated values of Θ from the distribution Pr(D_j | Θ), for a total of 10,000 new simulated sites, $D_1^{rep}, D_2^{rep},\ldots, D_{10,000}^{rep}$. For each of these 10,000 posterior predictive site patterns, we evaluate the posterior predictive expectation of the number of nonsynonymous substitutions. For site pattern D_j^{rep}, we sample a mutational mapping for each of the 1000 values of Θ, to construct a new simulated set of mappings, sampled from the distribution

$$\Pr_{\Theta|D}(M_j^{rep} \mid D_j^{rep}) = \int_{\Theta\in\Omega} \Pr(M_j^{rep} \mid D_j^{rep}, \Theta)\, p(\Theta \mid D)\, d\Theta, \tag{9}$$

the subscript $\Theta \mid D$ indicates that this probability is evaluated over the posterior distribution of Θ given D. By applying Equation 7 on the set of simulated mappings for a particular predictive site pattern, and repeating this for all of the predictive site patterns, we can construct a sample of 10,000 posterior predictive values of $T(D_j)$. These values approximate the posterior predictive distribution of the test statistic (Equation 7) and the posterior predictive p-value (Equation 8) can be evaluated for all sites.

3. Results

3.1 The number of nonsynonymous mutations along the sequence

The observed distribution of the test statistic (the expected number of nonsynonymous mutations in a site given the site pattern), and its posterior predictive distribution is shown in Figure 2 for the Influenza data set.

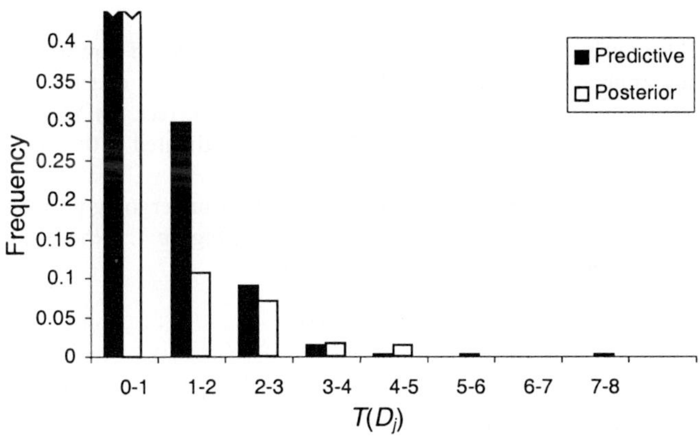

Figure 2. The predictive and the posterior distribution of the number of nonsynonymous substitutions among sites for the Influenza hemagglutinin data set.

The posterior predictive distribution has relatively fewer sites with less than one expected nonsynonymous mutation (60% predicted versus 78% observed). This is not surprising since constraints at the amino acid level will tend to lower the rate of amino acid substitution in many sites. However, we notice from figure 2 that there are also relatively more observed sites with more than 3 amino acid substitutions than expected from the posterior predictive distribution. The posterior predictive p-value in a one-sided test of the hypothesis $\omega = 1$, is shown in Figure 3.

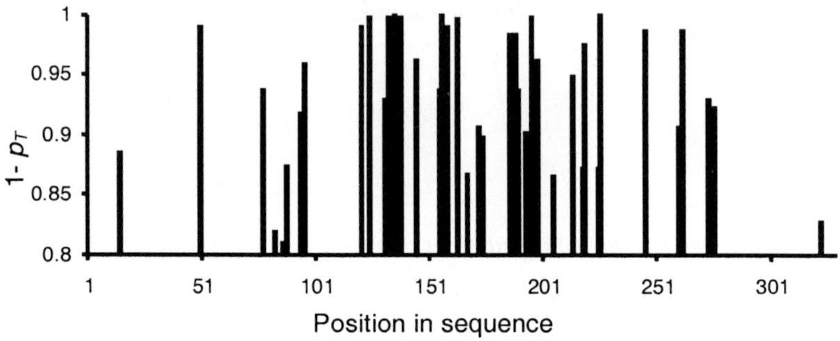

Figure 3. One minus the posterior predictive p-value for the hypothesis $\omega = 1$ in a one-sided test using the expected number of nonsynonymous substitutions given the data, as a test statistic.

We see that there are fairly many sites with p-values close to zero. There are 11 sites with posterior predictive p-values < 0.01. Given the number of sites, we would expect approximately 3 such sites if the null model were correct. In Yang *et al*[3]. the proportion of sites undergoing positive selection was estimated to 0.013, or approximately 4 sites (Model M8[3]).

For comparison, the posterior probability that a site is undergoing positive selection according to model M3 of Yang *et al.*[3] is shown in Figure 4. Notice the very strong correlation between this probability and $1 - p_T$.

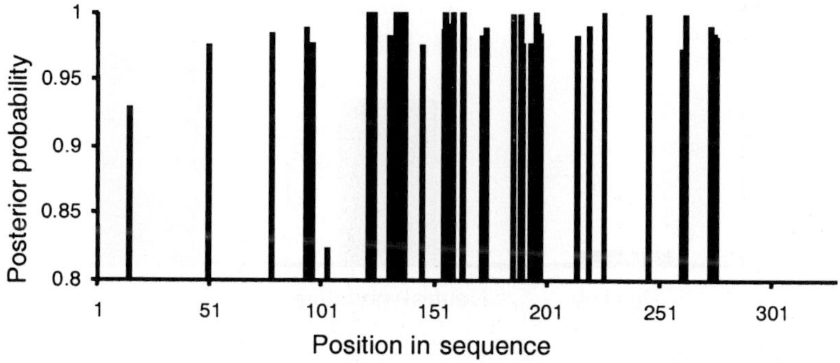

Figure 4. The posterior probability that a site is positively selected according to model M3 of Yang et al (2000).

For example, among the 11 sites with posterior predictive p-values less than 0.01, there are no sites with a posterior probability of being positively selected less than 0.975. The two methods essentially identify the same sites as being positively selected. This is quite remarkable given the differences in model assumptions.

3.2 The number of radical amino acid substitutions along the sequence

One of the advantages of the present approach is that extensions to other problems follow quite easily. For example, it might be of some interest whether positively selected mutations are radical mutations or tend to be conservative amino acid mutations. If positively selected substitutions tend to be radical, we can use this information when trying to identify sites undergoing positive selection. We can estimate the expected number of conservative substitutions given the data for each site in the sequences, using the very same samples of $\Pr(M \mid D)$ obtained for the purpose of identifying positively selected sites. Here, we simply define a substitution to be radical if it has a PAM100 score of less than –2 and conservative if it has a PAM100 score of more than –2. Obviously, many other measures could have been used to divide substitutions into radical and conservative substitutions. We also divide sites into sites with positive selection and sites evolving neutrally or subject to negative selection. The results are shown in Figure 5.

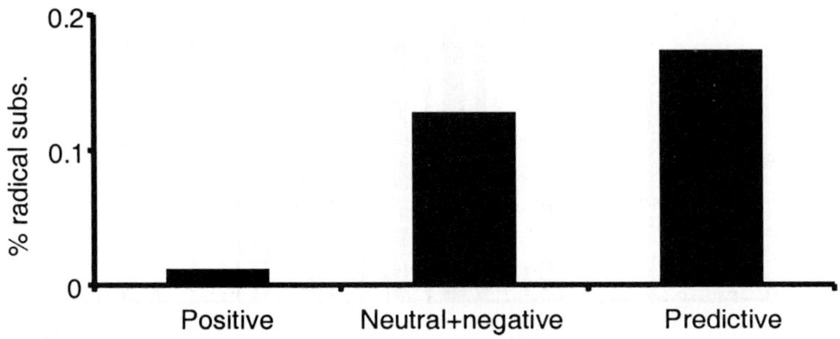

Figure 5. The proportion of radical amino acid changes in positively selected sites (Positive), sites that are evolving neutrally or subject to negative selection (Neutral) and in sites sampled from the posterior predictive distribution (Predictive).

Notice that the proportion of radical substitutions is much lower in the positively selected sites than in other sites. Moreover, the proportion of radical substitutions in positively selected sites is much less than expected from the posterior predictive distribution assuming no selection. It appears that positively selected substitutions tend to be conservative.

Discussion and conclusion

The approach based on posterior predictive *p*-values for identifying positively selected sites differs from the previous approaches[2,3] in several different ways. Most importantly, the model assumptions are quite different. In the present approach we assume prior distributions for all the nuisance parameters, including parameters related to the tree. We then base our inferences directly on the posterior distribution of mutations. In the previous approach parameters are first estimated by maximizing the likelihood function. Estimates of the posterior probabilities are then obtained based on these estimates. The cost in the new approach is an additional set of assumptions, but it does allow a proper treatment of the problem of the unknown tree topology. In the likelihood approach, maximization over tree topologies is not presently computationally feasible.

In the present application, a nucleotide model was used under the null-model to approximate a codon based model with $\omega = 1$. For some data sets it might be worrisome that this model does not take into account codon bias and the existence of stop codons. Also, the null hypothesis of $\omega = 1$ is arguably very simplistic. A more realistic null model that also allows sites in which $\omega < 1$ might be considered in future applications.

Despite the differences between the two approaches, the biological conclusions are remarkably similar in the two studies. There appears to be a small fraction of sites in the data set undergoing positive selection, and the sites identified to be undergoing positive selection are more or less same in the two studies.

The strength of the current approach was best illustrated in the analysis of the proportion of conservative and radical substitutions. It was easily demonstrated that the positively selected amino acid substitutions tend to be conservative substitutions. This is not a trivial result. In fact, it could be hypothesized that in the sites of a coat protein that interacts with a host immune system, any substitution is favorable. In particular, very radical substitutions that would change the binding affinities of antibodies and other components of the host immune system, should be favored. However, as shown here, this does not appear to be the case, at least not in the hemagglutinin protein of the Influenza virus.

Although this question could also have been addressed using explicit modeling in the likelihood framework, this would have required the computer implementation of such a model. In the present case the analysis could be done by simply reading of the results from the simulated mutational mappings. The strength of the present approach is that it allows for this type of exploratory data analysis in a rigorous statistical framework. Earlier studies have also used a mutation-mapping approach, where the mapping is performed using the parsimony method. A parsimony-based approach, however, suffers from a number of problems; the method only considers the mapping with the minimum number of changes (thereby underestimating the total number of changes) and treats the mapping as an observation in further analysis. The Bayesian approach discussed here avoids the many statistical problems associated with using parsimony and focussing on just a single mutational mapping.

Acknowledgments

This research was supported by NSF grants DEB-0089487 awarded to RN and DEB-0075406 awarded to JPH.

References

1. E.C. Holmes, L.Q. Zhang, P. Simmonds, C. A. Ludlam, A.J. Leigh Brown, "Convergent and divergent sequence evolution in the surface envelope glycoprotien of human immunodeficiency virus type 1 within a single infected patient" *Proc. Natl. Acad. Sci.* **89**, 4835 (1992)
2. R. Nielsen. and Z. Yang, "Likelihood models for detecting positively selected amino acid sites and applications to the HIV-1 envelope gene" *Genetics*, **148**, 929 (1998)

3. Z. Yang, R. Nielsen, N. Goldman, and A.-M. K. Pedersen, "Codon-Substitution Models for Variable Selection Pressure at Amino Acid Sites" *Genetics* **155**, 431 (2000)
4. J. Felsenstein, "Evolutionary trees from DNA sequences: a maximum likelihood approach" *J. Mol. Evol.* **17**, 368 (1981)
5. R. Nielsen, "Mapping mutations on phylogenies", *Syst. Biol.* (In review)
6. B. Larget. and D. Simon, 1999 "Markov chain Monte Carlo algorithms for the Bayesian analysis of phylogenetic trees" *Mol. Biol. Evol.* **16**, 750 (1999)
7. J. P. Huelsenbeck,., B. Rannala, and B. Larget, "A Bayesian framework for the analysis of cospeciation" *Evolution* **54**, 353 (2000)
8. N. Metropolis, A. W. Rosenbluth., M. N. Rosenbluth, A. H. Teller, and E. Teller. "Equations of state calculations by fast computing machines" *J. Chem. Phys.* **21**, 1087 (1953)
9. W. K Hastings, "Monte Carlo sampling methods using Markov chains and their applications" *Biometrika* **57**, 97 (1970)
10. J. P. Huelsenbeck F. Ronquist, *MrBayes 2.0.* Available from http://brahms.biology.rochester.edu/software.html (2001)
11. D. B. Rubin, "Bayesianly justifiable and relevant frequency calculations for the applied statisticians" *Ann. Statist.* **12**, 1151 (1984)
12. X.-L. Meng, "Posterior predictive *p*-values" *Ann. Statist.* **22**, 1142 (1994)

IMPROVING SEQUENCE ALIGNMENTS FOR INTRINSICALLY DISORDERED PROTEINS

PREDRAG RADIVOJAC, ZORAN OBRADOVIC
Center for Information Science and Technology, Temple University, U. S. A.

CELESTE J. BROWN, A. KEITH DUNKER
School of Molecular Biosciences, Washington State University, U. S. A.

Here we analyze sequence alignments for intrinsically disordered proteins. For 55 disordered protein families we measure the performance of different scoring matrices and propose one adjusted to disordered regions. An iterative algorithm of realigning sequences and recalculating matrices is designed and tested. For each matrix we also test a wide range of gap penalties. Results show an improvement in the ability to detect and discriminate related disordered proteins whose average sequence identity with the other family members is below 50%.

1 Introduction

Amino acid sequence alignment is the cornerstone of bioinformatics. Alignment algorithms include optimal pairwise comparisons, either global[1] or local[2], as well as heuristic algorithms such as FastA[3] and BLAST[4]. Optimal multiple sequence alignments[5] suffer from exponential complexity with increasing numbers of sequences. Indeed, the multiple alignment problem is NP-complete; furthermore, a scoring system is difficult to define[6]. These facts gave rise to different suboptimal algorithms based on progressive alignments[7, 8]. Finally, there are sequence profiles[9] and hidden Markov models[10], which exploit position-specific dependencies within protein families. All alignment methods require a scoring system, which is typically adjusted to optimize sensitivity and specificity.

In a given twenty-by-twenty scoring matrix, each entry, s_{ij}, is the score when amino acids i and j are aligned opposite one another. Typically, s_{ij} is a function of

$$\log \frac{p(i,j)}{p(i)\,p(j)}, \tag{1}$$

where $p(i,j)$ is the joint probability[f1] of the aligned pair of residues i and j, and $p(i)$ the probability of occurrence of residue i. This expression is called the log-odds ratio (mutual information) and when the logarithm is to base two it is measured in bits. The total score of two aligned sequences is finally calculated as the sum of scores of each aligned amino acid pair, along with empirically determined gap-opening and gap-extension penalties that provide the means to accommodate length variability.

[f1] The scoring function is defined in terms of probabilities that are approximated by observed relative frequencies. We also use terms relative frequency and probability interchangeably.

Two important sets of scoring matrices are the PAM (accepted point mutation) series[11,12] and the BLOSUM (block substitution) series[13]. The initial PAM matrix was based on just 1,572 substitutions. Evolutionary modeling was then used to boost the data and develop a series of matrices, but this modeling was imprecise[14]. The BLOSUM series was based on 2,106 aligned multiple-sequence segments with more than 15 million amino acid pairs and used only segments in highly conserved regions between gaps (e.g. blocks) to calculate substitution probabilities. Grouping the aligned sequences by sequence identity gave the BLOSUM series. After extensive testing, the BLOSUM62 matrix was identified as the best general scoring matrix[13].

Many additional scoring matrices have been developed[15, 16]. These are based on various criteria such as amino acid properties, structural superposition, minimum number of base changes per codon, evolutionary properties, etc. These matrices have idiotypic advantages, but PAM and BLOSUM remain the most widely used.

The development of scoring matrices has focused on ordered proteins that fold into 3-D structures. In contrast, many proteins have functional regions that exist as a structural ensemble at either the secondary or the tertiary level, that is, these regions are intrinsically disordered[17]. The realization that such disorder is not uncommon and is important for the function of essential proteins has led to a call for the reassessment of the view that function always follows from a protein's 3-D structure[18].

Amino acid compositions for ordered and intrinsically disordered protein are clearly different[19]. Also, insertions and deletions are more common in disordered as compared to ordered regions[20]. Thus, the scoring matrices and gap penalties developed from ordered proteins are likely to be inappropriate for disordered protein. Here we report the first attempts to develop disorder-specific scoring matrices with appropriately weighted gap-opening and gap-extension penalties.

2 Materials and Methods

2.1 Databases, hardware and software

A set of proteins with structurally characterized regions of disorder of length ≥ 40 consecutive residues was identified by database and literature searches. Homologous proteins were compiled using the BLAST algorithm[4]. For proteins with both ordered and disordered regions, it was assumed that segments aligning to the disorder were also disordered. The result was the following database[f2]: ldlr (10, 23, 32), 4E binding protein (7, 115, 58), ssDNA binding protein (15, 50, 49), α_tubulin (54, 48, 84), DNA-lyase (7, 40, 80), Bcl-x_L (7, 50, 81), calcineurin (22, 164, 34), cyclin-

[f2] The database is presented in the format: family name (number of sequences, average sequence length, average sequence identity with all family proteins). The latter two numbers were rounded to the nearest integer. For proteins containing both ordered and disordered regions, only disordered regions were used.

dependent kinase inhibitor (4, 162, 80), chloroperoxidase (2, 41, 44), ubiquinol cytochrome C reductase (5, 45, 73), eukaryotic translation initiation factor 4γ (4, 98, 68), carrot embryonic protein 1 (37, 96, 69), epidermal growth factor (8, 38, 78), Phe-tRNA synthetase (14, 88, 34), flagellin (34, 102, 48), negative regulator of flagellin synthesis (8, 98, 45), fibronectin binding protein C (2, 129, 96), oncogene fos (21, 145, 44), Gly-tRNA synthetase (23, 52, 27), glycine methyltransferase (8, 40, 86), gonadotropin (7, 34, 51), transcription factor VP16 (3, 93, 83), histone 5 (9, 114, 67), HMG14 (6, 101, 64), HMG17 (14, 87, 75), HMGI(Y) (10, 153, 39), HMGT (43, 209, 71), I-tRNA synthetase (23, 39, 33), inosine monophosphate dehydrogenase (52, 174, 40), lactose operon repressor (37, 61, 45), metaminopeptidase (3, 123, 86), HIV1 negative factor (3, 119, 25), osteocalcin (20, 47, 61), transcription factor p65 (4, 127, 41), prion (55, 98, 75), prothymosin α (4, 111, 96), *Pvu* II methyltransferase (12, 24, 22), anti-termination protein N (3, 120, 47), regulator of G-protein signalling 4 (17, 80, 38), acidic ribosomal protein P2β (56, 117, 41), replication protein A (7, 61, 21), southern bean mosaic virus capsid (6, 64, 71), translocase sec61 (9, 44, 47), sindbis virus capsid (6, 101, 45), small heat shock protein (6, 40, 42), sulfotransferase (12, 69, 32), α synuclein (22, 134, 62), tomato bushy stunt virus capsid (7, 58, 64), T-cell receptor α (10, 112, 38), telomere bindin gprotein (5, 39, 61), transcription initiation factor IID (3, 59, 53), thyroid transcription factor (10, 187, 47), Topoisomerase II (26, 99, 29), T-tRNA synthetase (24, 95, 28), yeast heat shock protein (2, 195, 23). Overall, this database contains 55 families with 828 segments of disorder containing 81,491 residues in total. Minimum and maximum observed sequence identities between any two aligned sequences were 10% and 99.53%, respectively.

A set of unrelated proteins was taken from reference 21. This set contains 131 proteins and 26,692 residues.

The various experiments were performed on a Windows based 800 MHz Pentium computer using C++ and MATLAB software packages.

2.2 Scoring Matrices

To build scoring matrices we applied a simple iterative algorithm consisting essentially of two steps:

1) for a given scoring matrix, align every protein in every family to all the other proteins belonging to the same family
2) for a given set of alignments calculate a new scoring matrix

Using BLOSUM62 as the initial matrix, these two steps were repeated until the scoring matrices in two successive iterations remained essentially unchanged.

2.3 Aligning sequences

We use both multiple alignment[7] and a series of global pairwise alignments[1] in Step 1. Both methods result in aligning every residue of every sequence in a family opposite only one residue or gap of every other sequence in the same family. Pairs of aligned sequences are used for calculating the final entries in the scoring matrices.

2.4 Assigning weights to sequences

In a scheme that differs from previous approaches, we assign a weight to every sequence as an inverse of its average sequence identity with all proteins of the same family (including itself). We take a soft approach (without thresholds) as compared to reference 13. Note that this method reduces the influence of large families of highly similar sequences, and all families contribute according to their size.

2.5 Counting mutations

No matter which strategy of alignment is used, substitutions are counted as shown in Fig. 1. Note that no counting is done when a residue is aligned to a gap.

This algorithm is applied to all families of disordered proteins and the overall substitution count matrix M is calculated as the sum of all family count matrices.

Input: family f of n aligned proteins $s_1, s_2, \dots s_n$
with corresponding weights $w_1, w_2, \dots w_n$
Output: family count matrix M_f
$M_f \leftarrow 20 \times 20$ zero matrix
for every pair of sequences s_i, s_j
  for every two superimposed amino acids x, y
    $M_f(x, y) \leftarrow M_f(x, y) + w_i \cdot w_j$

Figure 1. *Calculating family count matrix*

2.6 Constructing the scoring matrix

The elements of the scoring matrix S are calculated for each count matrix M, in the following way. The joint probability matrix Q that residue i will be aligned opposite residue j (for every i and j) is computed as

$$Q = \frac{M + M^T}{2\sum_i \sum_j m_{ij}}, \tag{2}$$

where m_{ij} is an element of matrix M. Adding a transpose matrix M^T **and dividing by** two annuls any effects from the counting order. The double sum in the denominator normalizes entries of Q to sum to one. From the elements of Q the conditional probabilities of substitutions $p(i\,/\,j)$ are calculated, yielding the elements of a substitution probability matrix P. In order to adjust the ensuing scoring matrix for longer evolutionary times we can transform matrix P before evaluating expression (1). Modeling the evolution by a discrete time-invariant Markov process with the unknown transition matrix, P can be modified as

$$P = P^{\alpha}, \tag{3}$$

where $\alpha \in (1, \infty)$. This generalizes the idea of Dayhoff *et al.*[11, 12] that models longer evolutionary times by extrapolating from proteins with shorter distances. However, the matrix P is already developed using the available spectrum of divergence from the available data, thus reflecting moderate evolutionary distances. Naturally, since the assumptions about underlying Markov processes do not hold strictly, the model becomes less accurate as α increases

Even with all amino acid exchanges observed in the database, there is no guarantee that P is positive definite. As a result, in order for expression (3) to be well-defined, a test for all positive eigenvalues is performed before raising P to a non-integer power. Failing the test, although such never occurred, would cancel the powering step. Finally, all entries s_{ij} of a scoring matrix are calculated as

$$s_{ij} = \text{round}(C \cdot \log_2 \frac{p(i,j)}{p(i)p(j)}), \tag{4}$$

where we multiply by $C = 2$ before rounding to the nearest integer in order to have s_{ij} entries in half bit units. Multiplying by two increases the resolution of the matrix elements before rounding, which is performed for the convenience of using integer arithmetic during the course of alignment. We use the same gap penalty system as for BLOSUM62. Once chosen at the start, the gap penalties are not changed during the refinement of matrix S, making the construction process far less expensive.

2.7 Evaluating the matrices

Scoring matrices were evaluated by building family-specific hidden Markov models (HMMs) from a set of aligned training proteins as described in Fig. 2. Test family proteins as well as a number of unrelated proteins were aligned to the HMM and the resulting discriminatory capabilities were measured as indicated.

Briefly, this testing procedure consists of two steps. In the first step all proteins from the test families and non-family set are assigned a score for each scoring matrix. Reliable performance assessment is achieved by applying a cross-validation procedure that also emulates real situations where only a small number of known homologues are available. Random division (line two in Fig. 2) was the same for each matrix. In order to build a model we used ClustalW [7] for multiple sequence alignment and the HMMER[10] software for profile HMMs. Aligning a protein to a HMM results in a score and an E-value. While the score reflects the log-likelihood that the query sequence is generated by a HMM the E-value is an estimate of statistical significance of the match. Overall, the best model is the one that provides the smallest overlap between two distributions of scores (family and non-family proteins). However, in a situation when rigid statistical tests are not conclusive, we compared the Z-scores generated by different models. The greater the value of a Z-score the lesser the probability that a query protein is one of the unrelated sequences. The model that best discriminates family from non-family sequences is the one with highest Z-scores. Consequently, in the second step, all length-normalized scores from step one are converted into Z-scores, and the maximum score for each protein is found over all matrices. Then, a cumulative score is calculated for each matrix as indicated in Fig. 2. Note that this score depends on the set of matrices being compared; however it preserves numeric differences and hence the relative order between any two models. Note also that our testing procedure is not optimal for discrimination purposes. As probabilistic models and local optimizers, hidden Markov models can only approximate dependencies in protein sequences. Still, successful application of HMMs to protein representation justifies their use.

```
for each scoring matrix S ∈ 𝑺, corresponding gap penalties, and family f ∈ 𝒇
   randomly divide f into 4 equal sized test groups
   for each test group
      multiply align the other 3 groups using S and gap penalties
      construct a HMM based on the multiple alignment
      align proteins in test group to the HMM and record scores*
      align non-family proteins to the HMM and record scores*
   end
end
for each scoring matrix S and family f
   calculate mean (m) and standard deviation (σ) for non-family protein scores
   for each family protein p (whose score is s), calculate Z_{S,p} = (s − m) / σ
end
for each protein sequence p ∈ f
   max_p = max_{S∈𝑺} {Z_{S,p}}
end
for each scoring matrix S
   cumulative score = Σ_{p∈f} (max_p − Z_{S,p})
end
```

Figure 2. *Testing procedure. *All scores are length normalized.*

3 Results

In the development of our testing procedures, we compared the performance of several previously published matrices and corresponding gap penalties optimized in reference 15 as applied to the 21 largest families of disorder (Table 1). The matrices were ranked by their cumulative scores (see *2.7*) over all of the test proteins up to a threshold sequence identity that was calculated as the average of pairwise sequence identities between a query protein and the set of training proteins used in model construction. A sequence identity threshold of 50% was set to include a reasonable number of proteins while emphasizing the more divergent ones. Furthermore, although both multiple sequence alignments[7] and a series of optimal pairwise alignments[1] were tried in step one of our iterative procedure (section *2.2*), the latter gave slightly better results so that only these results are presented.

Table 1. *Comparative performance of different matrices with given gap penalties for all test proteins whose average sequence identity to the training sequences is less than 50%*

Matrix	Gap-opening penalty	Gap-extension penalty	Cumulative score
GONNET	6	0.8	55.54
BLOSUM30	9	1	65.34
PAM250	12.5	0.1	68.76
BLOSUM62	7.5	0.9	74.55
BLOSUM62	10	0.6	76.03
BENNER74	7	0.8	77.83
BLOSUM30	10	1.5	79.13
BENNER74	9.5	0.8	83.19
IDENT	12	0.5	83.82
BLOSUM80	7	1.5	87.48
PAM300	12.5	0.4	90.78
IDENT	7	1.4	95.40
PAM250	11	0.5	96.45
PAM120	6	1.4	100.29
GONNET	14	0.2	103.47
PAM300	9	2	119.52
BLOSUM80	14.5	0.04	126.44
PAM120	12.5	1	159.28
OPTIMA *	120	20	276.57

*The scale difference in gap penalties for OPTIMA arises from the ten times greater values used to increase alignment sensitivity

In addition to representatives from the BLOSUM and PAM series, we also evaluated two updates of PAM250, Gonnet *et al.*[22] and Benner *et al.*[23]. The matrix IDENT assigns +6 for a match and −1 for a mismatch, and the OPTIMA matrix was taken from reference 24. For each matrix we used gap penalties from reference 15 and both original and modified all-positive scoring matrices. We also tried gap penalties 12/2 exploited in reference 24 but they exhibited poor performance and are excluded from Table 1, except for the OPTIMA matrix for which they are optimal.

Overall, the results of Table 1 suggest that, of the previously published scoring matrices, the matrix of Gonnet *et al.*[22] performed the best on our disordered protein families with less than 50% sequence identity.

The matrix DISORDER was obtained as described in Section 2 with $\alpha = 1.75$ (α values were tested in steps of 0.25 in the interval from 1 to 2). This new, disorder-specific scoring matrix (Fig. 3) differs significantly from all of the other scoring matrices in Table 1. For example, DISORDER differs in 100 out of 210 positions (47.6%) from BLOSUM62, which was used as the initial matrix in the development cycle (section *2.2*). Differences in values were between −3 and 3, and exhibited an almost normal distribution (not shown).

	C	S	T	P	A	G	N	D	E	Q	H	R	K	M	I	L	V	F	Y	W
C	10																			
S	0	3																		
T	1	1	4																	
P	−2	0	−1	6																
A	−1	1	0	−1	3															
G	−3	0	−2	−1	0	5														
N	−1	1	0	−1	−1	0	4													
D	−3	0	−1	−2	−1	−1	1	4												
E	−4	−1	−1	−1	−1	−2	0	2	4											
Q	−3	0	0	−1	−1	−2	1	0	0	5										
H	−1	−1	0	−2	−2	−1	2	−1	−1	1	8									
R	−1	−1	−1	−2	−2	−2	0	−2	−1	1	0	5								
K	−3	−1	0	−1	−1	−2	0	−1	0	0	−1	2	4							
M	0	−2	−1	−2	−1	−4	−2	−4	−3	−1	−2	−1	−2	7						
I	0	−2	−1	−2	−1	−5	−3	−4	−3	−2	−2	−2	−2	1	4					
L	−1	−2	−2	−1	−1	−4	−3	−4	−3	−2	−2	−2	−2	2	2	4				
V	1	−2	0	−1	0	−4	−3	−4	−2	−2	−2	−2	−2	1	3	1	4			
F	−1	−2	−2	−3	−2	−4	−2	−4	−4	−2	0	−3	−3	1	1	1	0	7		
Y	0	−2	−1	−3	−2	−3	−1	−4	−3	0	2	−2	−2	−1	0	0	−1	4	8	
W	−5	−3	−5	−1	−5	−4	−3	−4	−4	−1	−2	0	−3	−1	−2	−2	−4	−1	3	13

Figure 3. *DISORDER matrix*

To compare DISORDER with the other scoring matrices, we measured the performance of each with varying gap penalties and saved the best performing example in each case. The gap-opening penalty was varied over the range from 1 through 14 (in steps of 0.5), and the gap-extension penalty was varied from 0.5 through 2 (in steps of 0.5). The resulting matrices are ranked in Table 2. From the definition given in section *2.7*, it is evident that the cumulative score depends on the set of matrices being evaluated. Thus, the cumulative score values in Tables 1 and 2 cannot be compared directly: only the rankings are important. The DISORDER matrix outperforms the others, but changes in the gap penalties alter the ranking of the other matrices so that BLOSUM62 now becomes better than the others. DISORDER only marginally outperforms BLOSUM62 by the cumulative score measure.

Table 2. *Comparative performance of matrices with optimized gap penalties for all test proteins with average sequence identity with the training sequences less than 50%*

Matrix	Gap-opening penalty	Gap-extension penalty	Cumulative score
DISORDER	3	0.5	56.54
BLOSUM62	3.5	0.5	57.01
BLOSUM30	2	0.5	57.15
PAM250	1.5	0.5	71.32
GONNET	3.5	0.5	71.59
BENNER74	3	0.5	76.21
GONNET	6	0.8	90.61

The distribution of scores for different HMMs were compared (Fig. 4). Shaded bars represent the number of test proteins for which the DISORDER matrix obtained higher scores when aligned to the appropriate HMM, while white bars represent the same number for the BLOSUM62 matrix. These comparisons are plotted as a function of average sequence identity (quantized into 10 bins) as defined above, in the description of Table 1, but without any threshold.

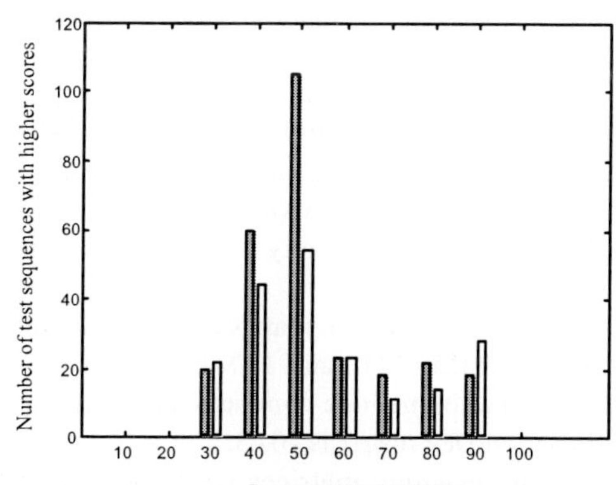

Figure 4. *Comparing DISORDER (shaded) and BLOSUM62 (open)*

Since the DISORDER matrix exhibited the best performance, we further refined its gap penalties. Values of 3.2/0.1 provided the best alignments.

5 Discussion

Although development of a scoring matrix from a set of sequence alignments is straightforward, evaluation of the resulting matrix is not. In reference 13, matrices were tested on an independent dataset of 504 blocks and the matrix that correctly classified a query block to its group the most times for a given level of statistical significance was declared the winner. In reference 24, a scoring matrix was created by maximizing the ability of the system to discriminate between homologous and non-homologous proteins. Performance tests were evaluated on 1,542 pairs of distantly related proteins with less than 40% pairwise sequence identity according to the average confidence value and the probabilities that random scores would be higher than the score for a query homolog. In reference 15 many different matrices were compared using both global and local optimal pairwise alignment algorithms on a database of aligned sequences resulting from superposition of three-dimensional protein structures representing correct alignments. The tests were carried out on 204 structurally aligned proteins from 37 families.

Since our database is rather small as compared to those from references 13 and 24 and since disordered segments are conformational ensembles and so cannot be structurally aligned as was done in reference 15, we developed an alternative method to evaluate the resulting matrices as described herein. The idea behind our evaluation protocol was to mimic how the matrix would likely be used, namely in connection with position-specific modeling.

Reports on new matrices usually contain calculations of the average mutual information (relative entropy, transinformation) and the expected score. The higher values of the average mutual information indicate that the matrix is better adjusted to shorter evolutionary distances. Longer distances, on the other hand, are characterized by smaller differences between diagonal and non-diagonal elements in the transition matrix P, resulting in a smaller relative entropy. The expected score represents an estimate of a per amino acid score of any two aligned proteins with the same distribution of amino acids.

The relative entropy of 0.54 in our matrix is different from that of BLOSUM62 (0.69) and similar to BLOSUM55 (0.56) and PAM180 (0.59). However, these matrices have a different scale so that immediate comparisons are not possible. The expected score obtained for the DISORDER matrix is –0.43.

During the course of designing matrices we have noticed that there is only a small dependence on any individual family in the training set (leaving out any individual family did not change things much), which basically enabled us to test the matrices on the training set. Also, differences in cross-validation steps were small. We have repeated the matrix design procedure with IDENT as the initial matrix and the final results were different at several positions and at most for ±1. The maximum number of iterations was set to 10, but usually a matrix will converge fast to its local optimum in 47 iterations. In the current paper, the matrix was optimized followed by a separate optimization of gap penalties. Future research will explore optimization of a matrix and gap penalties at the same time, a procedure that should lead to improved alignments. Also, we will continue to enlarge the database of intrinsically disordered segments, which at the very least should improve the statistics.

The quality of multiple alignments is improved by using the DISORDER matrix. Even though the new gap penalties are smaller than are typically used for ordered protein sequences, the average number of gaps in aligned disordered sequences actually decreases when the DISORDER matrix is used. When PAM 001 is used to calculate pairwise genetic distances between sequences aligned by either the GONNET or the DISORDER matrices, the average distances of disordered sequences aligned by the DISORDER matrix are smaller than for GONNET (data not shown).

Over the last several years, we have published several predictors of natural disordered regions (PONDRs)[25-28]. We envision an approach in which order/disorder predictions are first carried out using the most appropriate PONDR. During the subsequent HMM (or profile) construction process, BLOSUM62 (or another suitable matrix) would be used as the initial scoring matrix for those regions predicted to be ordered and DISORDER would be used for those regions predicted to be disordered. As the PONDRs and DISORDER are improved over time, this approach should yield improved alignments for proteins containing regions of intrinsic disorder.

Acknowledgement

NIH Grant 1R01 LM06916 awarded to AKD and ZO and NSF Grant CSE-IIS-9711532 awarded to ZO and AKD are gratefully acknowledged. Chris Oldfield, Sachiko Takayama and others did yeoman's work in developing the database proteins with physically characterized regions of intrinsic disorder. Finally, Slobodan Vucetic is thanked for numerous helpful discussions.

References

1. S. B. Needleman, C. D. Wunsh, *J. Mol. Biol.*, **48**, 443 (1970).
2. T. Smith, M. Waterman, *J. Mol. Biol.*, **147**, 195 (1981).
3. W. Pearson, D. Lipman, *Proc. Natl. Acad. Sci. USA*, **85**, 2444 (1988).
4. S. F. Altchul, W. Gish, W. Miller, E. W. Myers, D. J. Lipman, *J. Mol. Biol.*, **215**, 403 (1990).
5. M. Murata, J. Richardson, J. Sussman, *Proc. Natl. Acad. Sci. USA*, **82**, 3073 (1985).
6. D. Gusfield, Algorithms on Strings, Trees, and Sequences, Cambridge University Press, 1997.

7. J. D. Thompson, D. G. Higgins, T. J. Gibson, *Nucleic Acids Research*, **22**, 4673 (1994).
8. R. Durbin, S. R. Eddy, A. Krogh, G. Mitchison, *Biological sequence analysis: probabilistic models of proteins and nucleic acids*, Cambridge University Press, 1998.
9. M. Gribskov, R. Luthy, D. Eisenberg, *Methods in Enzymology*, **183**, 146 (1990).
10. A. Krogh, M. Brown, I. S. Mian, K. Sjolander, D. Haussler, *J. Mol. Biol.*, **235**, 1501 (1994).
11. M. O. Dayhoff, R. M. Schwartz, B. C. Orcutt, *Atlas of protein sequence and structure*, **5**, suppl. 3, 345 (1978).
12. R. M. Schwartz, M. O. Dayhoff, *Atlas of protein sequence and structure*, **5**, suppl. 3, 353 (1978).
13. S. Henikoff, J. Henikoff, *Proc. Natl. Acad. Sci. USA*, **89**, 10915 (1992).
14. W. J. Wilbur, *Molecular Biology and Evolution*, **2**, 434 (1985).
15. G. Vogt, T. Etzold, P. Argos, *J. Mol. Biol.*, **249**, 816 (1995).
16. R. Luthy, A. D. McLachlan, D. Eisenberg, *PROTEINS: Structure, Function, and Genetics*, **10**, 229 (1991).
17. P. E. Wright, H. J. Dyson, *J. Mol. Biol.* **293**, 321 (1999).
18. R. M. Williams, Z. Obradovic, V. Mathura, W. Braun, E. C. Garner, J. Young, S. Takayama, C. J. Brown, and A. K. Dunker. *Pacific Symp. Biocomputing,* **6**, 89 (2001).
19. A. K. Dunker, J. D. Lawson, C. J. Brown, R. M. Williams, P. Romero, J. S. Oh, C. J. Oldfield, A. M. Campen, C. M. Ratliff, K. W. Hipps, J. Ausio, M. S. Nissen, R. Reeves, C. H. Kang, C. R. Kissinger, R. W. Bailey, M. D. Griswold, W. Chiu, E. C. Garner, Z. Obradovic, *J. Mol. Graphics Model.*, **19**, 26 (2001).
20. W. L. Shaiu, T. Hu, T. S. Hsieh, *Pacific Symp. Biocomputing*, **4**, 578 (1999).
21. V. Geetha, V. Di Francesco, J. Garnier, P. J. Munson, *Protein Engineering,* **12**, 527 (1999).
22. G. H. Gonnet, M. A. Cohen, S. A. Benner, *Science*, **256**, 1433 (1992).
23. S. A. Benner, M. A. Cohen, G. H. Gonnet, *Protein Engineering*, **1**, 88 (1994).
24. M. Kann, B. Quiann, R. A. Goldstein, *PROTEINS: Structure, Function, and Genetics*, **41**, 498 (2000).
25. P. Romero, Z. Obradovic, C. R. Kissinger, J. E. Villafranca, A. K. Dunker, *Proc. IEEE. Int. Conf. on Neural Networks,* **1**, 90 (1997).
26. X. Li, P. Romero, M. Rani, A. K. Dunker, Z. Obradovic, *Genome Informatics,* **10**, 30 (1999).
27. P. Romero, Z. Obradovic, X. Li, E.C. Garner, C. J. Brown, A. K. Dunker, *PROTEINS: Structure, Function, and Genetics,* **42**, 38 (2001).
28. S. Vucetic, P. Radivojac, C. J. Brown, A. K. Dunker, Z. Obradovic, *Proc. Int. Joint INNS-IEEE Conf. on Neural Networks*, Washington D.C., **4**, 2718 (2001).

AB INITIO FOLDING OF MULTIPLE-CHAIN PROTEINS

J.A.SAUNDERS, K.D. GIBSON, AND H.A. SCHERAGA
Baker Laboratory of Chemistry and Chemical Biology, Cornell University, Ithaca, NY 14853-1301, U.S.A.

Our previous methodology for *ab initio* prediction of protein structure is extended here to treat multiple-chain proteins. This involved modification of our united-residue (UNRES) force field and our Conformational Space Annealing (CSA) Global Optimization procedure. Good results have been obtained for both a four-and a three-helix protein from the CASP3 exercise.

1 Introduction

Ever since Anfinsen's formulation of the thermodynamic hypothesis for protein folding[1], attempts have been made to compute the structure of a native globular protein as the global minimun of its free energy. These include knowledge-based approaches [such as homology modeling[2-7], threading[3,6,8], and combinations of these together with secondary-structure prediction[9]], and *ab initio* methods[10,11]. The latter are based only on energy, without the aid of knowledge-based information; the purpose of the *ab initio* approach is to provide an understanding of how pairwise and multibody (cooperative) interactions lead to the folded structure.

In the *ab initio* approach, it is computationally impossible at present to predict the 3D structure of an *all-atom* protein by energy minimization, Monte Carlo, or molecular dynamics procedures. However, by breaking the problem into its component parts, it is possible to achieve the realization of this computational goal. Therefore, a hierarchical approach to the computation of protein structure has been developed in our laboratory[10-14]. This hierarchy involves the following six steps, the key stage of which is the global optimization of an off-lattice simplified chain:

1. Using a virtual-bond representation of the polypeptide chain, described by a united-residue potential (UNRES)[13,15-17], and an efficient procedure (Conformational Space Annealing, CSA)[18,19] to search the conformational space of this virtual-bond chain rapidly, we obtain a family of low-energy UNRES structures. The combination of UNRES and CSA narrows the region of conformational space in which the global minimum is likely to lie, which can be achieved at this stage with the simplified virtual-bond model but not with an all-atom model.
2. Next, this very restricted part of conformational space is searched by first converting the virtual-bond chains of the low-energy structures obtained in the united-residue calculations to an all-atom representation, using our dipole-path method[20].

3. The backbone conformation is then optimized, subject to C^{α}- C^{α} distance constraints from the parent united-residue structure, by using the Electrostatically Driven Monte Carlo (EDMC) method[21], which incorporates elements of our Monte Carlo-plus-Minimization (MCM)[22,23] and Self-Consistent Electric Field (SCEF)[24] procedures.
4. All-atom side chains are then attached subject to the condition of non-overlap.
5. Loops and disulfide bonds are treated with the empirical loop-closing potential of ECEPP[25], or with an exact procedure[26,27].
6. Final energy refinement is carried out with the ECEPP/3 all-atom force field[25], plus the SRFOPT surface-hydration model[28], with gradual reduction of the C^{α}- C^{α} distance constraints of the parent united-residue structure (until they vanish at the end of the procedure).

For the success of the hierarchical algorithm outlined above, the united-residue force field must be able to capture the essential features of the interactions in proteins at a coarse-grain level. However, it is equally important to have a reliable all-atom force field that can reproduce the fine details of the all-atom chain in stage 6. Recently, we began to use our algorithms for the global optimization of crystal structures as tools to refine the parameters of all-atom potentials (so that the global minimum of the potential energy function is close to the experimental structure, and its energy is close to the experimental enthalpy of sublimation). These global-optimization algorithms were originally based on the deformation of the potential energy surface, and include the Diffusion Equation Method (DEM)[29], the Distance Scaling Method (DSM)[30] and its improved descendant, the Self-Consistent-Basin-to-Deformed-Basin Mapping (SCBDBM) method[31] which involves back-and-forth deformations and reversals, until self-consistency, with perturbations at each stage of reversal. More recently, we use the Conformation-Family Monte Carlo (CFMC) method[32] for crystal structure calculations.

Our methodology, and other types of approaches such as those of Skolnick and Baker[9,33] (and references cited therein), have had partial success in the CASP3 and CASP4 exercises.

Heretofore, these methods have been applied primarily to single-chain proteins. Recently, we have started to develop methods to treat multiple-chain proteins. Since these methods extend the above single-chain approaches, we begin by summarizing our treatment of single-chain proteins, and then will describe our methodology to treat multiple-chain proteins.

2. Summary of the UNRES Force Field

2.1 UNRES Geometry

UNRES employs a simplified representation of polypeptide chains in which backbone peptide groups and entire side chains are represented by united atoms. The polypeptide backbone is modeled as a sequence of α-carbon atoms linked by virtual bonds 3.8 Å in length (corresponding to *trans* peptide groups), with a united peptide group located at the midpoint of each virtual bond. Attached to each backbone α-carbon is a united side-chain ellipsoid whose size and distance from the backbone are determined by the amino acid identity of the side chain. Each united residue has four degrees of freedom; the angle between successive backbone virtual bonds (θ) and the dihedral angle about each virtual-bond (γ) determine the backbone conformation, and two angles α and β determine the orientation of the side chain relative to the backbone. The united peptide groups and side chains serve as interaction sites for the force field, while the α-carbons are present only to define the geometry of the chain.

2.2 UNRES Force Field

The UNRES force field is derived as a restricted free energy (RFE) function, which corresponds to averaging the energy over the degrees of freedom that are neglected in the united-residue model. It is expressed by eq. (1):

$$U = \sum_{i<j} U_{SC_iSC_j} + w_{SCp} \sum_{i \neq j} U_{SC_ip_j} + w_{el} \sum_{i<j-1} U_{p_ip_j} + w_{tor} \sum_{i} U_{tor}(\gamma_i) + w_b \sum_{i} U_b(\theta_i) + w_{rot} \sum_{i} U_{rot}(\alpha_{SC_i}, \beta_{SC_i}) + \sum_{m=2}^{N} w_{corr}^{(m)} U_{corr}^{(m)} \quad (1)$$

where $U_{SC_iSC_j}$ is the interaction energy between side chains, $U_{SC_ip_j}$ is an excluded-volume potential between side chains and peptide groups, $U_{p_ip_j}$ is the energy of average electrostatic interaction between peptide groups, U_{tor} is the intrinsic energy of rotation about the virtual C^{α}- C^{α} bonds, U_b and U_{rot} are the bending energy of the virtual-bond angles and the energy of different rotameric states of the side chains, respectively, and U_{corr} is the multibody or correlation energy arising from the loss of degrees of freedom when computing the restricted free energy.

The terms $U_{SC_iSC_j}$, U_b, and U_{rot} were parameterized based on distributions within a set of 195 non-homologous structures from the Protein Data Bank. The excluded-volume term $U_{SC_ip_j}$ was parameterized to reproduce the correct backbone geometry in short model helices and sheets. The form of $U_{p_ip_j}$ was obtained by averaging the electrostatic interactions of the peptide groups, and it was parameterized by fitting the average restricted free energy surface of two interacting peptide groups, calculated with the all-atom ECEPP/3[25] force field. The expression for the torsional potential U_{tor} results from the cluster expansion of the RFE, and it has been parameterized by fitting to ECEPP/3[25] RFE surfaces.

Because UNRES is a coarse-grain force field, in which the interactions between the united side chains and peptide groups are mean-field potentials, multibody terms (U_{corr}) that capture the underlying physics of the hidden degrees of freedom are required for successful *ab initio* structure prediction. The RFE is

$$F(\mathbf{X}) = -RT\ln\left(\frac{1}{V_{\mathbf{Y}}}\int_{\mathbf{\Omega}_{\mathrm{Y}}}\exp[-E(\mathbf{X};\mathbf{Y})/RT]dV_{\mathbf{Y}}\right) \tag{2}$$

where $\mathrm{E}(\mathbf{X};\mathbf{Y})$ is the all-atom ECEPP/3 energy function, $\mathbf{X}$ is the set of UNRES degrees of freedom, $\mathbf{Y}$ is the set of degrees of freedom over which the average is computed (e.g. the positions and orientations of solvent molecules, the side-chain dihedral angles, etc.), R is the gas constant, T is the absolute temperature, $\mathbf{\Omega}_{\mathbf{Y}}$ is the region of the $\mathbf{Y}$ subspace over which the integration is carried out, and $V_{\mathbf{Y}}$ is the volume of this region. It can be expressed as a sum of cluster cumulant functions[34], which correspond to increasing order of correlations between component energy terms. The most significant correlation terms are those for electrostatic-local interaction correlations. By expanding these functions into a generalized cumulant series, approximate analytical expressions can be obtained; the lowest order cumulants are sufficient to capture the essential features of the cumulant functions. The multibody term U_{corr} is thus a set of electrostatic-local correlations of up to sixth order, the forms of which were derived from the cumulant expansion, and the parameters of which were found by fitting to appropriate ECEPP/3 RFE surfaces.

The w's are constant weights that balance the contributions of the different kinds of interactions. They are determined by simultaneous Z-score-and-gap optimization[35, 36] on multiple targets. This procedure produces a set of weights that maximize both the ratio of the difference in energy between the native structure and the mean energy of non-native structures to the standard deviation of the energy distribution of the non-native structures, and also the difference in energy between the native structure and the lowest energy non-native structure. The native structure is actually a set of UNRES structures within a certain rmsd cutoff of the experimental structure, and the non-native structures are generated through a global conformational search.

3. Summary of CSA

Conformational Space Annealing is a powerful global optimization method that has been used successfully with UNRES in the CASP blind structure prediction exercises. It is a variation on a genetic algorithm in that it maintains a population of structures and combines portions of the conformations of existing "parent" structures to generate new "offspring" structures, which then replace conformations in the population if they have lower energy.

CSA is distinguished from other optimization methods, however, in its use of a cutoff distance that decreases over the course of a search. Also, conformations are always locally minimized in energy using Gay's Secant-type Unconstrained Minimization Solver (SUMSL)[37]. A distance measure is defined which measures the similarity between two conformations; for UNRES, this distance is defined as the average angular difference between two conformations for all backbone dihedral angles in the chain. If a new structure is produced that is within the cutoff distance of an existing structure, the new one either replaces the old similar structure, if the new one has lower energy, or it is rejected. The effect of this cutoff distance is that, at the beginning of a search, when the cutoff distance is large, CSA will maintain a population that is sparse in the conformation space; i.e. the search will be global. As the cutoff distance decreases, a higher degree of similarity will be allowed between structures in the population, which effectively searches smaller regions for low-energy conformations. CSA thus begins as a global search for potential fold families, and ends as a more local search for the lowest-energy representatives of the best fold families.

4. Performance of Hierarchy in CASP3 and CASP4

The foregoing methodology has been tested in the CASP3[10,13,14] and CASP4[11] exercises. In their evaluation of our performance in CASP3, Orengo et al.[38] cited our predicted submission for HDEA (target 61) as "most impressive…., using more classical *ab initio* methods….(which use) no information from sequence alignments, secondary structure prediction, or threading".

Our predictions in CASP3 were all submitted for largely α-helical targets because we could not predict β structures at that time, i.e. in 1998. Since then, we have improved our UNRES force field with additional terms in the cumulant expansion of the free energy, and have achieved *partial* success in predicting β structures[11] in CASP4, i.e., we were able to predict *part* of the β structure in β and

α/β proteins. Work is now in progress to extend our heirachical procedure (with improvements in our cumulant-based UNRES force field, in our global optimization procedures, and in our all-atom potential function) to try to extend our current predictions for 100-residue α proteins with an rmsd of 4-6Å to 250-residue α, β, and α/β proteins with an rmsd of 2-3Å.

5. Extensions to multiple-chain Proteins

5.1 Extensions to UNRES

To apply the UNRES force field to multiple-chain proteins, new energy terms had to be added to represent the interactions between separate chains. The forms, parameters, and weights of these interchain terms are the same as the single chain terms because the physical bases of the intra- and interchain interactions are the same. However, the purely local interactions (i.e. U_b, U_{rot}, and U_{tor}) have no interchain counterparts. The single-chain multibody term U_{corr} includes correlations for residues that are both adjacent (e.g. electrostatic-local correlations around a β-turn) and separated in sequence (e.g. electrostatic-local correlations between β-sheet strands). The multiple-chain U_{corr} term includes nonadjacent electrostatic-local correlations between chains. The new UNRES energy function is:

$$
\begin{aligned}
U = &\left(\sum_{k}\sum_{i<j} U_{SC_{k,i}SC_{k,j}} + \sum_{k<l}\sum_{i,j} U_{SC_{k,i}SC_{l,j}} \right) \\
&+ w_{SCp}\left(\sum_{k}\sum_{i\neq j} U_{SC_{k,i}p_{k,j}} + \sum_{k<l}\sum_{i,j} U_{SC_{k,i}p_{l,j}} \right) \\
&+ w_{el}\left(\sum_{k}\sum_{i<j-1} U_{p_{k,i}p_{k,j}} + \sum_{k<l}\sum_{i,j} U_{p_{k,i}p_{l,j}} \right) + w_{tor}\sum_{k}\sum_{i} U_{tor}(\gamma_{k,i}) \\
&+ w_b\sum_{k}\sum_{i} U_b(\theta_{k,i}) + w_{rot}\sum_{k}\sum_{i} U_{rot}(\alpha_{SC_{k,i}}, \beta_{SC_{k,i}}) \\
&+ \sum_{m=2}^{N} w_{corr}^{(m)}\left(\sum_{k} U_{corr}^{(m)} + \sum_{k<l} U_{corr,nonadj}^{(m)} \right)
\end{aligned} \tag{3}
$$

where the indices k and l denote chains, the indices i and j denote residues, and the $U_{corr,nonadj}$ term represents the components of U_{corr} corresponding to correlations between nonadjacent residues. In addition to the angles that define the internal conformation of a chain, a set of Euler angles and a translational vector are required for each chain to specify the packing arrangement of the multiple-chain protein.

5.2 Symmetry optimizations

Many multiple-chain proteins contain symmetries in the spatial arrangements of the individual chains. The presence of such symmetries allows optimizations that can greatly increase the speed of energy evaluation, gradient calculation, and local minimization of the energy[39]. For example, in a tetramer of identical chains with four-fold rotational symmetry, the intrachain energy of only a single chain need be computed because each chain has the same internal conformation. Although there are six possible interchain interaction energies between four chains, in this symmetric case only two are unique. Likewise, fewer gradient calculations have to be carried out because some contributions to the gradient can be computed simply by rotating the gradients from other symmetry-related interchain interactions. For identical chains with symmetric packing, the number of independent variables is greatly reduced, and the topology of the energy surface is simplified, which decreases the number of energy evaluations required for local minimization of the energy.

The types of symmetry considered here are rotational, screw, one-chain affine, and two-chain affine symmetries[39]. The packing variables in these symmetries consist of the orientation of the symmetry axis, the displacement of the axis from the origin, the radial distance of the first chain in the group from the axis, the rotation angle to the first chain in the axial frame, the axial shift to the first chain, the Euler angles of the first chain, the axial shift between chains, the axial rotation between chains, and, for two-chain affine symmetry, the Euler angles and position in the axial frame of the second chain. A protein may consist of several independent groups of chains, each with its own symmetry. Also, two or more of these groups can be constrained to share a symmetry axis.

5.3 Extensions to CSA

New ways of generating conformations have been added for the purpose of exploring the space of packing arrangements. The initial, random bank of non-clashing structures is generated with random packing as well as random internal chain conformations. In the stage of CSA when new "offspring" conformations are created by combining the variables of existing conformations, several methods are used to **generate** new chain packings. A new packing is initially copied from a seed conformation and then perturbed by taking packing variables from another conformation. The perturbations can be implemented in several sub-sets of the packing variables.

In this study, the distance between conformations for multiple chains was the same as that defined for single chains, viz., the average angular difference between two conformations for all backbone dihedral angles. The primary goal was to determine the folded structure of a monomer (in its oligomeric state), and this measure reveals the backbone similarity of monomer conformations even in systems with different packings. Also, in the work reported here, the packing variables were highly coupled to the internal conformations being packed; thus, structures with differing internal conformations generally adopted different packing arrangements.

6. Results

Multiple-chain UNRES has been tested successfully on two homo-oligomeric systems that were targets in the CASP3 exercise. In both cases, the search for the native fold was carried out only at the united-residue level, without conversion to an all-atom representation.

6.1 Retro-GCN4 Leucine Zipper

The *retro*-GCN4 leucine zipper[40] (target 84 in CASP3) is an α-helical tetramer with 37 residues per chain. Two independent runs were carried out on both the tetramer and the monomer by itself. For the tetramer, four-fold rotational symmetry was assumed for both runs. The experimental structure shows that the native tetramer is very nearly, but not quite, rotationally symmetric, and consists of two slightly different monomer conformations.

In two runs of the monomer, totaling 76,000 minimizations, the lowest energy structure found (–174.2 kcal/mol) has a C^{α} coordinate rmsd for residues 2–37 (the first residue is not resolved in the experimental structure) of 13.4 Å from the native monomer structure. Of the 100 structures resulting from the two runs, none are closer to the native monomer conformation than 11.6 Å. The native monomer is a single, long helix, and in all cases the single-chain force field breaks the helix into multiple segments.

Two runs of the tetramer, totaling 177,000 minimizations, resulted in a lowest energy structure of –777.3 kcal/mol. This lowest energy structure has an rmsd of only 2.34 Å from the native tetramer (Fig. 1). A tetramer slightly closer to native (1.98 Å) is found as the ninth lowest energy, ~8 kcal/mol higher than the lowest. Comparing the lowest-energy tetramer results with the experimental results at the monomer level yields a monomer rmsd of 2.23 Å or 2.29 Å, depending on the experimental monomer used in the comparison. The monomer of the structure closest to the native tetramer has an rmsd of 1.80 Å or 1.92 Å with the experimental monomers.

While the lowest energy structure from the monomer runs has an energy of –174.2 kcal/mol, the intrachain energy per chain for the lowest-energy tetramer is –149.7 kcal/mol. However, the packing energy due to interchain contacts decreases the total energy per chain in the tetramer to –194.3 kcal/mol, illustrating that the presence of additional chains results in a structure with monomers that are individually less stable than the isolated monomer conformation, but are packed such that favorable interchain contacts more than offset that loss in energy.

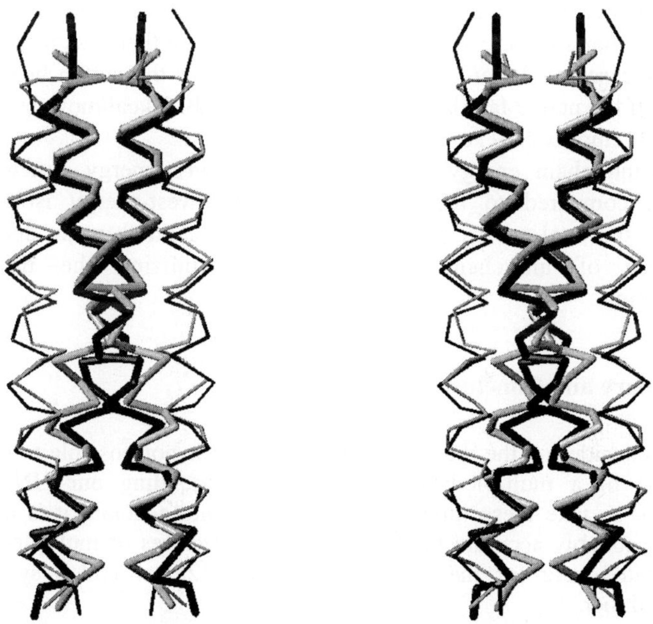

Fig. 1. Stereo view of computed structure of *retro*-GCN4 leucine zipper superposed on the x-ray structure[40].

6.2 *Synthetic Domain-Swapped Dimer*

Target 73 from CASP3 was a synthetic domain-swapped dimer, a designed α-helical dimer of 48 residues per chain that forms a three-helix bundle[41]. Two independent runs were carried out on both the dimer and the monomer by itself. Two-fold rotational symmetry was assumed for both dimer runs.

In two runs of the monomer, totaling 134,000 minimizations, the lowest energy structure found (–245.1 kcal/mol) has an rmsd from the native of 11.1 Å. A minimal-tree clustering of the 150 resulting structures at 3 Å yields 41 conformational families. The native family was found as the eighth lowest of the 41 family clusters. The lowest energy representative of this family, at –237.1 kcal/mol, is the closest to the native, with an rmsd of 2.79 Å. As was the case with the *retro*-GCN4 leucine zipper, the single-chain force field tends to break long helices to achieve better packing of the chain against itself in the absence of interchain contacts from the second monomer.

Two runs of the dimer, totaling 190,000 minimizations, found a lowest energy structure of –526.32 kcal/mol. This lowest-energy structure is a member of the native family, at 5.65 Å from the experimental dimer. Comparing only monomers of the lowest-energy structure and the native gives an rmsd of 3.33 Å. Within this same family, another structure at –517.99 kcal/mol is found that has an rmsd of only 3.16 Å from the native dimer. The closest monomer to the native belongs to another member of the native family, with an energy of –520.2 kcal/mol; the monomer rmsd for this structure is 2.19 Å.

The intrachain energy per chain in the lowest-energy structure is –224.08 kcal/mol, compared to the –245.1 kcal/mol lowest-energy isolated monomer. However, the total energy per chain is –263.16 kcal/mol, again demonstrating the importance of interchain interactions in stabilizing the native monomer conformation.

7. Summary and Conclusions

Without considering the interchain interactions, it is not possible to obtain a correct prediction of a multiple-chain protein. By extending our UNRES and CSA algorithms to take these interchain interactions into account, it is now possible to obtain reasonably accurate predictions of the structures of multiple-chain proteins, as illustrated here for the *retro*-GCN4 leucine zipper and a synthetic domain-swapped dimer.

Acknowledgments

We are indebted to J. Pillardy, C. Czaplewski, A. Liwo and J. Lee for helpful discussions. This work was supported by NIH grant GM-14312 and NSF grant MCB00-03722. Support was also received from the National Foundation for Cancer Research. J. A. Saunders was an NIH Biophysics trainee.

References

1. C.B. Anfinsen, *Science*, **181**, 223 (1973).
2. P.K. Warme, F.A. Momany, S.V. Rumball, R.W. Tuttle and H.A. Scheraga, *Biochemistry*, **13**, 768 (1974).
3. T.A. Jones and S. Thirup, *EMBO J.*, **5**, 819 (1986).
4. D.A. Clark, J. Shirazi and C.J. Rawlings, *Prot. Eng.* **4**, 751 (1991).
5. M.J. Rooman and S.J. Wodak, *Biochemistry*, **31**, 10239 (1992).
6. M.S. Johnson, J.P. Overington and T.L. Blundell, *J. Mol. Biol.*, **231**, 735 (1993).
7. D. Fischer, D. Rice, J.U. Bowie and D. Eisenberg, *FASEB J.*, **10**, 126 (1996).
8. M.J. Sippl, *J. Comput.-Aid. Mol. Des.*, **7**, 473 (1993).
9. J. Skolnick, A. Kolinski and A.R. Ortiz, *J. Mol. Biol.*, **265**, 217 (1997).
10. J. Lee, A. Liwo, D.R. Ripoll, J. Pillardy, J.A. Saunders, K.D. Gibson and H.A. Scheraga, *Intl. J. Quantum Chem.*, **77**, 90 (2000).
11. J. Pillardy, C. Czaplewski, A. Liwo, J. Lee, D.R. Ripoll, R. Kazmierkiewicz, S. Oldziej, W.J. Wedemeyer, K.D. Gibson, Y.A. Arnautova, J. Saunders, Y.-J. Ye and H.A. Scheraga, *Proc. Natl. Acad. Sci., U.S.A.*, **98**, 2329 (2001).
12. A. Liwo, M.R. Pincus, R.J. Wawak, S. Rackovsky, and H.A. Scheraga, *Protein Science*, **2**, 1715 (1993).
13. A. Liwo, J. Lee, D.R. Ripoll, J. Pillardy and H.A. Scheraga, *Proc. Natl. Acad. Sci., U.S.A.*, **96**, 5482 (1999).
14. J. Lee, A. Liwo, D.R. Ripoll, J. Pillardy and H.A. Scheraga, *Proteins: Structure, Function and Genetics*, **Suppl. 3**, 204 (1999).
15. A. Liwo, S. Oldziej, M.R. Pincus, R.J. Wawak, S. Rackovsky and H.A. Scheraga, *J. Comput. Chem.*, **18**, 849 (1997).
16. A. Liwo, M.R. Pincus, R.J. Wawak, S. Rackovsky, S. Oldziej and H.A. Scheraga, *J. Comput. Chem.*, **18**, 874 (1997).
17. A. Liwo, R. Kazmierkiewicz, C. Czapleswki, M. Groth, S. Oldziej, R.J. Wawak, S. Rackovsky, M.R. Pincus and H.A. Scheraga, *J. Comput. Chem.* **19**, 259 (1998).
18. J. Lee, H.A. Scheraga and S. Rackovsky, *J. Comput. Chem.*, **18**, 1222 (1997).
19. J. Lee and H.A. Scheraga, *Intl. J. Quantum Chem.*, **75**, 255 (1999).
20. A. Liwo, M.R. Pincus, R.J. Wawak, S. Rackovsky and H.A. Scheraga, *Protein Science*, **2**, 1697 (1993).
21. D.R. Ripoll, A. Liwo and H.A. Scheraga, *Biopolymers*, **46**, 117 (1998).
22. Z. Li and H.A. Scheraga, *Proc. Natl. Acad. Sci., U.S.A.*, **84**, 6611(1987).
23. Z. Li and H.A. Scheraga, *J. Molec. Str. (Theochem)*, **179**, 333 (1988).

24. L. Piela and H.A. Scheraga, *Biopolymers* **26**, S33 (1987).
25. G. Némethey, K.D. Gibson, K.A. Palmer, C.N. Yoon, G. Paterlini, A. Zagari, S. Rumsey and H.A. Scheraga, *J. Phys. Chem.*, **96**, 6472 (1992).
26. K.D. Gibson and H.A. Scheraga, *J. Comput. Chem.*, **18**, 403 (1997).
27. W.J. Wedemeyer and H.A. Scheraga, *J. Comput. Chem.*, **20**, 819 (1999).
28. J. Vila, R. L. Williams, M. Vasquez and H.A. Scheraga, *Proteins: Structure, Function, and Genetics*, **10**, 199 (1991).
29. L. Piela, J. Kostrowicki and H. A. Scheraga, *J. Phys. Chem.*, **93**, 3339 (1989).
30. J. Pillardy and L. Piela, *J. Phys. Chem.* **99**, 11805 (1995).
31. J. Pillardy, A. Liwo, M. Groth and H.A. Scheraga *J. Phys. Chem.*, **B103**, 7353 (1999).
32. J. Pillardy, C. Czaplewski, W. J. Wedemeyer and H.A. Scheraga, *Helv. Chim. Acta*, **83**, 2214 (2000).
33. P.M. Bowers, C.E.M. Strauss and D. Baker, *J. Biomolec. NMR*, **18**, 311 (2000).
34. A. Liwo, C. Czaplewski, J. Pillardy and H. A. Scheraga *J. Chem. Phys.*, **115**, 2323 (2001).
35. J. Lee, D.R. Ripoll, C. Czaplewski, J. Pillardy, W.J. Wedemeyer and H.A. Scheraga *J. Phys. Chem. B.*, **105**, 7291 (2001).
36. J. Pillardy, C. Czaplewski, A. Liwo, W.J. Wedemeyer, J. Lee, D.R. Ripoll, P. Arlukowicz, S. Oldziej, Y.A. Arnautova and H.A. Scheraga *J. Phys. Chem. B*, **105**, 7299 (2001).
37. D.M. Gay, *ACM Trans. Math. Software*, **9**, 503 (1983).
38. C.A. Orengo, J.E. Bray, T. Hubbard, L.LoConte, and I. Sillitoe, *Proteins: Struct., Funct., Genet.*, **Suppl. 3**, 149 (1999).
39. K.D. Gibson and H.A. Scheraga, *J. Comput. Chem.*, **15**, 1414 (1994).
40. P.R.E. Mittl, C. Deillon, D. Sargent, N. Liu, S. Klauser, R.M. Thomas, B. Gutte and M.G. Grütter, *Proc. Natl. Acad. Sci. USA*, **97**, 2562 (2000).
41. N.L. Ogihara, G. Ghirlanda, J.W. Bryson, M. Gingery, W.F. DeGrado and D. Eisenberg, *Proc. Natl. Acad. Sci. USA*, **98**, 1404 (2001).

INVESTIGATING EVOLUTIONARY LINES OF LEAST RESISTANCE USING THE INVERSE PROTEIN-FOLDING PROBLEM

J. Schonfeld, O. Eulenstein, K. Vander Velden, and G. J. P. Naylor
Bioinformatics and Computational Biology
Iowa State University
Ames, IA 50011

We present a polynomial time algorithm for estimating optimal HP sequences that fold to a specified target protein conformation based on Sun et al's Grand Canonical (GC) model. Application of the algorithm to related proteins taken from the PDB allows us to explore the nature of the protein genotype:phenotype map. Results suggest: (1) that the GC model captures important biological aspects of the mapping between protein sequences and their corresponding structures, and (2) the set of sequences that map to a target structure with optimal energy is affected by minor differences in structure.

1 Introduction

1.1 Background

A fundamental problem in biology is to understand the correspondence between the genotype and phenotype. Understanding the functions that constitute the *genotype: phenotype map (gp-map),* would facilitate the prediction of genetic predisposition to disease, the design of new drugs, and an understanding of the evolutionary origin of the diversity of phenotypes. The complexity of biological systems has made it extremely difficult to elucidate the gp-map at the organismal level. Our current knowledge of this map is generally limited to a few alleles, most of which are associated with human disease conditions. Recent efforts by the biophysics community to understand the gp-map at the level of proteins look particularly promising [1,2], and may provide insight into higher levels of organization.

Proteins represent an excellent system in which to explore the gp-map for two reasons:

1 They exhibit a broad range of functions (phenotypes) including cell signaling, pathogen recognition, structural support, cellular scaffolding, and molecular motors that move components around within the cell.

2 There is a large body of work devoted to solving the *protein folding* problem, which is defined as follows: Given a protein sequence *S*, find the conformation to which *S* folds under physiological conditions. Research in this area has led to an improved understanding of the rules that govern how sequences map to their corresponding protein phenotypes.

1.2 Redundancy and Accessibility

Studies of protein variation in nature indicate that there is extensive redundancy in the mapping from sequence to structure [3,4]. It is likely that the highly redundant mapping between sequence and structure has been important for the evolution and diversification of proteins [5,6,7]. Structures with many sequences mapping to them are predisposed to be more accessible through the evolutionary process than are structures with fewer sequences. For example, consider two sequences A and B, each of which maps to a different structure, and which differ by 20 point mutations. Changing one structure into the other would require the simultaneous mutation at each of the 20 different sites. This is highly unlikely to occur. However, if the same two structures are each represented by several thousand sequences, there is a higher likelihood that some of the sequences mapping to A would be closer to some of the sequences mapping to B. Thus a high degree of redundancy in the mapping between A and B promotes their mutual evolutionary accessibility.

A protein can be thought of as a *cloud* of points in a high dimensional space, where each axis represents a separate amino acid position in the sequence. This space is non-Euclidean and is referred to as *protein space* [6,8]. Each point in the protein space represents a unique sequence while a cloud represents the domain of different sequences that map to a particular protein's function. When clouds representing different proteins come into close proximity, a change from one protein to another is facilitated. Change is unlikely when clouds are distant. Just as spherical clouds will lead to less connectivity than will dendritic clouds in Euclidean space, the multidimensional shape of protein clouds will affect the degree of connectivity and accessibility among proteins. Comparable dendritic clouds have been investigated as sparse random networks [9] and as neutral networks [4,10].

In this paper we set out to use the redundancy in the mapping of sequence to structure to explore the question: "How easy is it to turn one kind of protein into another, and are there paths of least resistance that would allow us to best account for the diversity of proteins observed in nature?" More formally, what is the evolutionary inter-accessibility among proteins? We call this the *minimum evolutionary distance (MED) problem.* To solve the MED problem and examine the structure of these protein clouds we need to have access to a large percentage of all of the sequence variants that occur in nature.

1.3 Previous Results

It is not yet possible to empirically collect all sequence variants that occur in nature to solve MED. Given this we generate sequence variants based on an abstract folding model. There is a natural inverse of the protein folding problem, called the *inverse protein folding* problem *(IPF),* which has been used by many researchers to tackle protein design problems [11]. For our purposes it is sufficient to describe the problem as follows: Given a native protein conformation *C (known as a target conformation)*, find a sequence that folds to *C* with the minimal energy. For more detailed information the reader is referred to references [11,12,13]. A variety of methods have been described that attempt to solve IPF [14,15,16]. Their utility varies with their ability to capture different aspects of the problem and their computational complexity. Only a few of the approaches used are computationally solvable in polynomial time. Almost all are based on Dill's HP-lattice model [17]. In Dill's model a protein conformation is a self-avoiding walk on a regular lattice, and amino acids are classified as either hydrophilic (P), or hydrophobic (H). Two amino acids are assumed to be in contact if they are close in space, but not adjacent on the self-avoiding walk. The energy function rewards only H-H contacts. A considerable body of work suggests that this abstraction can capture important aspects of protein structure [12, 16, 17]. Following earlier models proposed by Shaknovich and Gutin [2], Sun et al. [1] introduced the Grand Canonical (GC) model. This model accommodates important aspects of the 3D conformation of real proteins by relaxing the regular lattice constraint and incorporating solvent accessibility. When applied to real proteins the GC model yields HP-sequences that closely match the HP representation of real sequences [1]. Kleinberg [18] showed that IPF for the GC model can be solved in polynomial time by transforming it into a bipartite network flow problem.

1.4 Contribution of this Paper

In this paper we introduce the MED problem. We present and study aspects of the MED problem as it applies to biological evolution and the "interaccessibility" of proteins. Our approach builds on solutions to IPF based on the GC model and uses network flow techniques initiated by the work of Kleinberg. We apply the mathematical structure and efficient algorithms associated with network flow problems to address issues related to the MED problem. We report the following:

Computational advances:

- An improvement in the running time of Kleinberg's algorithm.
- An efficient representation and algorithm to find all HP-sequences that optimally solve IPF for the GC model.

Evolutionary advances:

- Comparable accuracy to Sun et al.'s model [1] when estimated HP-sequences are contrasted with the corresponding real sequences.
- A demonstration that minor changes in protein structure can have important consequences for the sequences that map to them with minimum energy.

- The surprising finding that the evolutionary information inherent in minor structural differences is reflected in HP-sequences that solve IPF for the GC model.

2. Solving the Inverse Folding Problem in "Computer Space"

The GC model as proposed by Sun et al. [1] is a HP-model that abstracts physico-chemical and geometrical features of real proteins into a contact graph *G*. Amino acids of the protein correspond to the nodes of *G*, and distances between amino acids below a given threshold correspond to edges of *G*. Edges are weighted by distance, and nodes are weighted by the solvent accessibility. The contact graph represents the target conformation. Hs of a HP-sequence map to a subset of the nodes of the contact graph and are referred to as an H-assignment. An H-assignment has an energy value given by an energy function that balances the competing cost of solvent accessibility and H-H contacts. H-H contacts are rewarded proportional to their distance, while H assignments are penalized proportional to their solvent accessibility.

In this section we provide a definition of the GC model that we use as the basis of our work. We then define IPF under the GC model and refer to it as GC-IPF. HP-sequences that solve GC-IPF are referred to as *optimal HP-sequences*. The definitions we give follow Sun et al. [1] and are presented in subsection 2.1. In subsection 2.2 we review the conceptual basis of Kleinberg's algorithm to solve IPF under the GC model, and introduce an algorithm that is asymptotically more efficient. In subsection 2.3 we show that the work of Picard et al. [19] can be applied to reveal the intrinsic structure of all IPF solutions. This intrinsic structure is also used to enumerate all IPF solutions. For a background in standard network flow concepts the reader is referred to [20].

2.1 Grand Canonical (GC) Model

Definition (contact graph): Let *C* be a protein structure. A contact graph $G_c = (V, E, s, d)$ is a simple undirected graph, with vertices *V*, edges *E*, node weights $s: V \rightarrow R$, and edge weights $d: E \rightarrow R$. Vertices in *V* correspond to the amino acids of the conformation. Edges in *E* correspond to non-covalent binding amino acids, whose C_α positions are at most 6.5Å apart. The node weight $s(v)$ represents the "solvent accessibility" of the corresponding amino acid v and the edge weight $d(\{v,w\})$ represents the "distance" between the amino acid residues corresponding to v and w. (In subsection 3.1 we will specify "solvent accessibility" and "distance".) An example for a Contact Graph is shown in Figure 1.

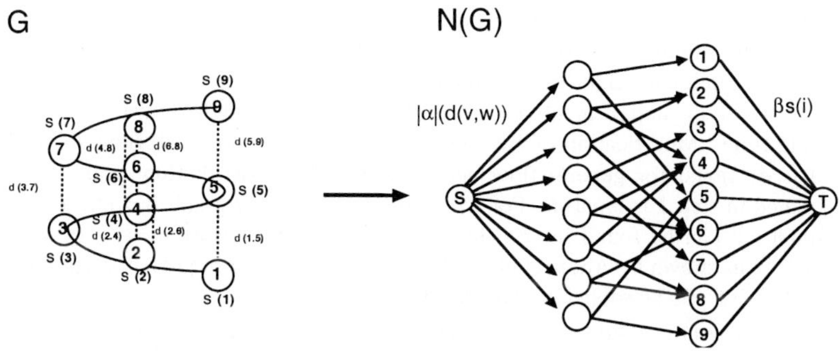

Figure 1: Contact Graph to Network Flow Graph Conversion

Definition (H-assignment): Let $G_C = (V, E, s, d)$ be a contact graph. An element $H \in \wp(V)$ is called an *H-assignment.*

Definition (energy): Let $G = (V,E,s,d)$ be a contact graph. $E_G: \wp(V) \rightarrow R$ is the *energy function* for G, where $E_G(H) = -\alpha \Sigma_{e \in E'} d(e) + \beta \Sigma_{v \in H} s(v)$ and $E' = (H \times H) \cap E$.

Definition (optimal H-assignment): An optimal assignment for a contact graph G is an H-assignment X, such that $E(X)=min\{ E_G(X') \in R \mid X' \text{ is an H-assignment}\}$.

Definition *(GC-IPF):*
Given: A contact graph G_C
Find: An optimal H-assignment for G_C

2.2 Algorithms for GC-IPF

Herein, we review the conceptual basis for Kleinberg's approach and introduce an improved algorithm. Let $G_C=(V, E, s, d)$ be a contact graph.

Overview of Kleinberg's algorithm:

1. The contact graph G is transformed into a bipartite network flow graph $N(G) = (V', E')$, where $V'=\{s,t\} \cup E \cup V$ $s,t \notin E \cup V$ and $E' = (\{s\} \times E) \cup (V \times \{t\}) \cup \{(e,v) \in E \times V \mid \exists w \in e: \{v,w\} \in E\}$. The capacity for an edge $(s,e) \in E'$ is $\alpha d(e)$, the capacity for an edge $(v,t) \in E'$ is $\beta s(v)$, and all remaining capacities are infinite. An example for $N(G)$ is shown in Figure 1.
2. Find a minimum cut $C=(X, Y)$ in $N(G)$, where $s \in X$, $t \in Y$.

It can be shown that the H-assignment $H=X \cap V$ is an optimal assignment [18].

To describe the running time we define $n=|V|$ and $m=|E|$. Step 1 needs $O(n+m)$ time and the running time of step 2 depends on the network flow algorithm used. Kleinberg applies an algorithm designed for general flow networks [21,22]. This approach solves step 2 in $O((n+m)^2 \log(n+m))$ time, which is also the overall running time. Modification of Kleinberg's approach: The network flow algorithm used is designed for general flow networks, but the network flow graph that is calculated in step 1 is bipartite. For a bipartite network flow graph B = (X, Y, E_B) a minimal cut can be found in $O(n_x\ m \log(n_x/m_B))$, where $n_x=|X|$, $n_y=|Y|$, $x\leq y$, and $m_B=|\ E_B\ |$ as shown by Ahuja et al. [23]. Thus by using the Ahuja et al. algorithm in step 2 GC-IPF can be solved in $O(nm \log(n/m+2))$ time.

Kleinberg [18] assumes that in practice volume constraints in three dimensions imply that the number of possible contacts has a constant upper bound. From this assumption follows the running time of $O(n\log n)$ for Ahuja et al., algorithm and $O(n^2 \log n)$ for Kleinberg's algorithm.

2.3 Optimal Assignments: Structure and Algorithm

Clearly, there can be 2^n different optimal H-assignments for a contact graph $G_C = (V,E)$, where $n=|V|$. In practice, it is most unlikely that this would result when the input is based on a contact graph for a PDB structure. As shown by Picard et al. [19], minimal cuts in a network flow graph are structured.

Theorem: If $(S,\overline{S})$ and $(S',\overline{S'})$ are minimal cuts in a flow network, then $(S\cap S',\overline{S\cup S'})$ and $(S\cup S',\overline{S\cap S'})$ are also minimal cuts.

As stated in section 2.2, minimal cuts represent optimal assignments. Thus, for all optimal assignments the assignments with minimal and maximal cardinality, denoted as H_{min} and H_{max}, are unique. All optimal assignments are contained in H_{max} and contain H_{min}.

Biologically this implies that there exists a core pattern of H-assignments that must be present in all of the sequences mapping to a particular structure, and that evolutionary forces will be constrained to maintain this pattern so long as the structure does not change. Conversely it implies that changes in structure can have significant effects on evolutionary opportunity. This is because the constraints associated with maintaining a particular protein structure affect the shape of clouds in protein space thereby affecting MED. Picard et al. [18], proved that all minimal cuts can be computed linear in the output size, when the maximal flow is given.

3. Implementation

In this section we describe details of our implementation for the MED problem. Our implementation takes as input a file from the Protein Data Bank (PDB) [24] containing the 3D coordinates of each C_α atom in a protein, and outputs the set of optimal HP sequences.

3.1 Parameters and Performance

Parameters: For a contact graph $G_C=(V,E,s,d)$ we specify the distance function d, the solvent accessibility s, and the scaling parameters α and β, following Sun et al. [1]. The distance function d is given by the following equation:

$$d(\{i,j\}) = \frac{1}{1+e^{(6.5\text{Å}-d_{ij})}} \qquad (3.1)$$

The value d_{ij} is the distance between the C_α atoms of the amino acids corresponding to the nodes i and j in the contact graph G. The function s is given by DSSP [25]. The range of α is provided by the user, while β=1/3.

Performance: The algorithms were implemented in C++ and compiled with Visual C++ 6.0 using the standard debug mode. We ran our software on a 700 MHz PC with 256 MB of RAM under Windows 2000 Professional Edition. The running time for each of the 15 globin structures was usually less than three minutes.

4. Application to Real Data

In this section we report results for a trial of our approach. We applied the method to a *test set* of 15 globin structures taken from the PDB (Table 1). The test set was chosen to represent a diverse sampling of structures within a protein family and included myoglobins, leghemoglbins, clam hemoglobin, ferric hemoglobin and several hemoglobins from the following animals: human, goose, turtle, trout, shark, skate, lamprey, and the sea cucumber. Selection was restricted to monomeric globins and alpha chains of multimeric globins.

4.1 A Test Run Using Globins

We obtained a total of 19 optimal HP-sequences when we subjected the test set to the GC model at near optimal α values. The number of sequences generated from each structure varied from 1 for several of the taxa to 4 for soybean leghemoglobin. The optimal α values were estimated by comparing α values from 1 to 15 at 0.5 increments. We converted the amino acid sequences associated with each structure in the test set to their HP equivalents following Sun et al. [1]. The resulting HP strings are denoted *transformed sequences*. To explore the effect that changing α had on the output of the model we explored two parameters: The accuracy and the energy difference. The accuracy is measured as the percent sequence similarity between the estimated and transformed sequences and the energy difference is measured as the difference between the energy of the estimated and transformed sequences. When we plotted both parameters against α we found that the energy difference behaved in a concave manner, while the accuracy behaved convexly.

We compared the set of sequences obtained from the GC model with the HP transformed sequences taken from nature for each of the 15 conformations. Our results indicate that many of the estimated assignments are close to those of the transformed sequences. In general the real sequences exhibit a finer grained sequence variation than the HP sequences generated by the GC model. Sequences produced by the GC model tended to have larger uninterrupted "runs" of H's and P's than were observed in the real sequences (sequences available on request).

The accuracies we obtained at optimal α values for each of the 15 globin structures were comparable to those obtained by Sun et al. ranging from 66.4 to 78.4% (Table 1).

Table 1: Globin structures used in test case. The number of optimal HP sequences is indicated in column 4. Percent accuracy reflects the similarity between the optimal HP-sequences and their counterpart transformed sequences.

PDB ID	Structure	Species	# seqs.	% accy
1A3N	human hemoglobin	*Homo sapiens*	1	66.7
1HV4	goose hemoglobin	*Anser indicus*	1	67.4
1OUT	trout hemoglobin	*Onchorhynchus mykiss*	1	69.0
1GCV	shark hemoglobin	*Mustelus griseus*	1	71.1
1CG5	stingray hemoglobin	*Dasyatis akajei*	1	70.2
3LHB	lamprey hemoglobin	*Petromyzon marinus*	1	70.5
1LHT	sea turtle myoglobin	*Caretta caretta*	1	77.8
1VXB	sperm whale myoglobin	*Physeter catodon*	1	78.4
1GDJ	yellow lupine leghemoglobin	*Lupinus luteus*	2	72.2
1BIN	soybean leghemoglobin	*Glycine max*	4	66.4
1MOH	clam ferric hemoglobin	*Lucina pectinata*	1	72.5
1BOB	clam hemoglobin	*Lucina pectinata*	1	73.2
1HLM	sea cucumber hemoglobin	*Caudina arenicola*	1	70.9
1 H97	Trematode hemoglobin	*Paramphistomum*	1	77.6
1DLY	unicellular alga hemoglobin	*Chlamydomonas*	1	70.2

The ClustalW [26] alignment of the 19 optimal HP-sequences was subjected to UPGMA cluster analysis. Optimal HP-sequences clustered according to the structure from which they were derived as shown in Figure 2. These results imply that although the various different globins are structurally similar, the minor

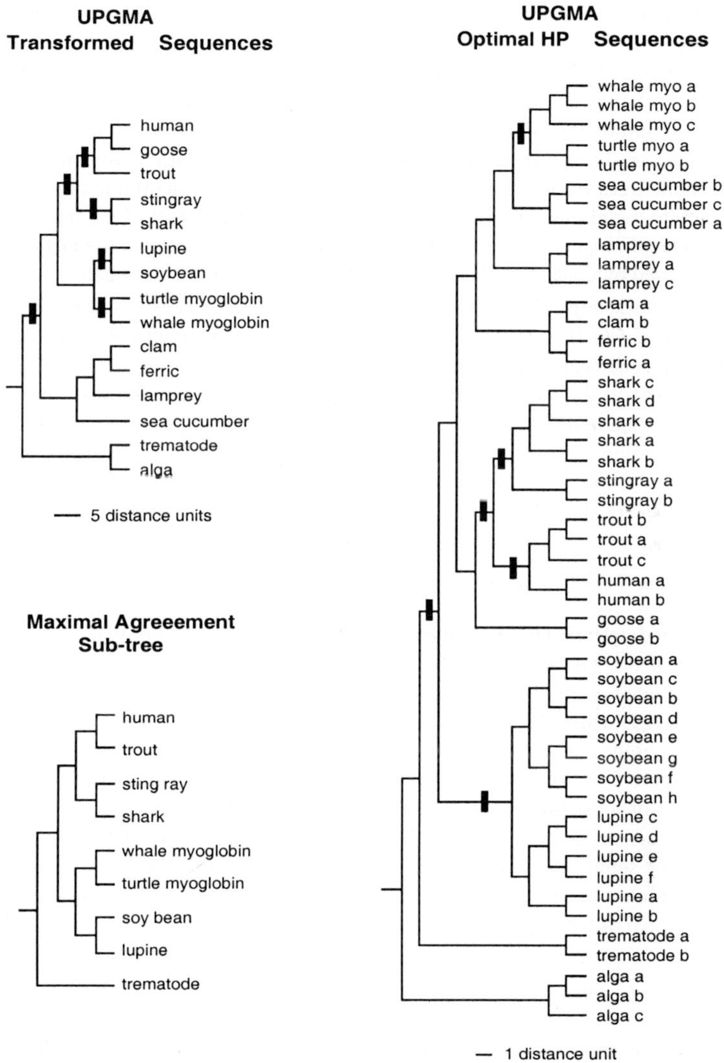

Figure 2: UPGMA of 48 optimal HP assignments resulting from the application of the GC-IPF to 15 globin structures (right). The UPGMA for the transformed sequences is shown at top left. The Maximal Agreement sub-tree shows groupings with a common cluster structure between the two analyses.

structural differences that do exist are sufficient to affect the set of optimal HP-sequences.

This observation is consistent with the theoretical prediction described in section 2.3. This finding has potentially important implications for phylogenetic analysis. If minor differences in protein structure are responsible for significant changes in constraints in nature, then the models of evolution used to estimate phylogenetic trees from naturally occurring sequences must incorporate the changes if they are to avoid yielding misleading results. Put another way, if the models used to estimate phylogenetic trees do not accommodate the non-stationarity in the process caused by constraint changes, the resulting trees will be misleading.

The relationships among the 15 clusters of HP sequences resulting from the application of CG-IPF, mirrored the cluster structure obtained when each of the transformed sequences was examined using a cluster analysis (Figure 2). This adds further weight to the idea that the representation we used is capturing something natural about the mapping between sequence and structure. While we acknowledge that considerable differences exist between the optimal HP-sequences and their transformed counterparts, we find it interesting that the patterns of similarity among the HP-sequences show such correspondence to that of the real sequences.

4.2 Exploring Sub-Optimal α Values

We computed optimal HP-assignments at 90% and 80% of the optimal energy using a α value of 0.1. As expected, the number of optimal HP-sequences increased as the optimal energy criterion was relaxed. 40 optimal HP-sequences resulted at 90% of the energy while 48 resulted at 80%. UPGMA cluster analysis of these sequences yielded a similar cluster structure to that obtained with the optimal energy (Figure 2). Note that the sequences generated at near optimal energy values are a subset of those generated at sub-optimal energy levels. We did not encounter any cases where sequences from different structures clustered together to the exclusion of sequences from the same structure. Presumably, there would be a point at which clusters start to overlap. This point would represent the "bridge point" at which different protein structures become inter-accessible.

4.3 Scope of the Model

This approach and our implementation of it face several potential problems and limitations. Our results show that our model tends to over specify the problem to the point that only a few optimal HP sequences are produced out of an exponential potential number of possible mappings. This problem, however, is tied directly to our implementation and can be solved by reducing the specificity of the distance and solvent accessibility functions.

Our approach is also limited by a dependency on a distance measure for comparing sequences between proteins. Finding an appropriate distance measure for comparing diverse proteins is a challenging problem that has not yet been adequately solved.

5. Outlook

Sequences that map to a particular protein structure form a cloud of points in protein space. The shape of such clouds reflects the structural and functional constraints of the protein [18]. As structures and functions change over the course of evolution, so does the shape of the corresponding cloud. When one cloud comes close to another, the two protein structures represented by the clouds, become evolutionarily inter-accessible. If a mutation occurs that allows a sequence in one cloud to be converted into a sequence in another, a conformational change in the protein will result. Recent empirical evidence suggests that these conversions occur in nature [27,28,29,30].

The protein space representation provides a clear insight into the redundancy of the gp-map. It explains how resilience in the phenotype can be reconciled with a genotype that is free to explore different configurations [31]. Redundant mappings promote evolutionary access to new phenotypes and are predisposed to yield discrete changes at the phenotype level. We conjecture that some of the abrupt morphological changes seen in the fossil record represent discrete transitions caused by redundant mappings at higher levels of biological organization. An understanding of the gp-map at the organismal level will likely remain beyond our reach for some time to come. However, we believe that advances in algorithmic approaches combined with representations that meaningfully capture biology at the molecular level, promise to bring an understanding of the gp-map at the protein level within grasp in the near future.

Acknowledgements

We are grateful to Cecilia Clementi for pointing out empirical work consistent with the predictions of quantum evolution in protein space. During the review process we learned that a similar approach had been explored independently by Aspnes et al. [32] at Yale University. Their contribution is to appear in ISAAC. NSF grant DEB 9707145 to GJPN and DEB 0075319 to OE supported the current work.

References

1. S. Sun, R. Brem, H. Chan and K. Dill, *Protein Eng.*, **8,** 12, 1205-1213 (1995)
2. E.I. Shakhnovich and A.M. Gutin, *Protein Eng.*, **6**, 8, 793-800 (1993)
3. L. Holm, C. Sander, *Science,* **273**, 595 (1996)
4. P. Schuster, W. Fontana, P.F. Stadler and I.L. Hofacker, *Proc. R. Soc.,* **255**, 279-284 (1994)
5. H. Li , R. Helling, C. Tang and N. Wingreen, *Science,* **273**, 666-669 (1996)
6. G.J.P. Naylor and M. Gerstein, *J. Mol. Evol.*, **51**, 223-233 (2000)
7. D.J. Lipman and W.J. Wilbur, *Proc. R. Soc.*, **245**, 7-11 (1991)
8. J. Maynard Smith, *Nature*, **225**, 563-564 (1970)
9. C. Reidys, P.F. Stadler, and P. Schuster, *Bull. Math. Bio.,* **59**, 339 (1997)
10. M. A. Huynen, P.F. Stadler and W. Fontana, *PNAS*, **93**, 397-401 (1996)
11. K.A. Dill, S. Bromberg, K. Uayue, K. Fiebig, D. Yee, P. Thomas and H. Chan, *Protein Science*, **4**, 561 (1995)
12. K. Yue and K.A. Dill, *PNAS*, **89**, 4163-4167 (1992)
13. J.M. Deutsch and T. Kurosky, *Phy. Rev. Lett.*, **76**, 2, 323-326 (1996)
14. W. Hart, *Proc. RECOMB*, 128-136 (1997).
15. S. Kamtekar, J. M. Schiffer, H. Xiong, J.M. Babik and M.H. Hecht, *Science,* **262**, 1680-1685 (1993)
16. E.I. Shakhnovich and A.M. Gutin, *PNAS*, **90**, 7195-7199 (1993)
17. Lau, K.F. and K.A. Dill, *Macromolecules*, **22**, 3986-3997 (1989)
18. J.M. Kleinberg, *Proc. RECOMB*, 226-2367 (1999)
19. J. Picard and M. Queyranne, *Math. Prog. Study*, **13**, 8-16 (1980)
20. R. Ahuja, T. Magnanti and J. Orlin in *Network Flows*, (Prentice Hall 1993)
21. D. Sleator and R.E. Tarjan, *JCSS*, **26**, 362–391 (1983)
22. A.V. Goldberg and R.E. Tarjan, *Journal of ACM,* **35**, 921–940 (1988)
23. R.K. Ahuja, J.B. Orlin, C. Stein and R.E. Tarjan, *SIAM*, (1994)
24. H.M. Berman, J. Westbrook, Z. Feng, G. Gilliland, T.N. Bhat, H.Weissig, I.N. Shindyalov and P.E. Bourne, *Nuc. Acids Research*, **28**, 235-242 (2000)
25. W. Kabsch and C. Sander, *Biopolymers*, **22**, 2577-2637, (1983)
26. J. D. Thompson, D. G. Higgins and T. J. Gibson, *Nuc. Acids Research*, **22**, 4673-4680 (1994)
27. R.D. George, *Nature Struct. Biol.*, **4**, 7, 512-514, (1997)
28. S. Dalal, S. Balasubramanian and L. Regan, *Nature Struct. Biol.,* **4**, 7, 548-552 (1997)
29. S. Dalal and L. Regan, *Protein Science*, **9**, 9, 1651-1659 (2000)
30. M.H. Cordes, R.E. Burton, N.P. Walsh, C.J. McKnight and R.T. Sauer. *Nature Struct. Biol.*, **7**, 12, 1129-1132 (2000)
31. B. Küppers, *Molecular Theory of Evolution*, (Springer-Verlag, NY 1983)
32. J. Aspnes, J. Hartling, M.Y. Kao, J. Kim, and G. Shah in *Lecture Notes in Computer Science: 12 Ann. Inter. Symp. on Algorithms and Computation*, Eds. P. Eades and T. Takaoka, (Springer-Verlag, NY 1983)

USING EVOLUTIONARY METHODS TO STUDY G-PROTEIN COUPLED RECEPTORS

ORKUN SOYER[*], MATTHEW W. DIMMIC[#], RICHARD R. NEUBIG[&]
RICHARD A. GOLDSTEIN[*#]

[*]Department of Chemistry, [#]Biophysics Research Division, [&]Department of Pharmacology
University of Michigan, Ann Arbor, MI 48109-1055

A novel method to analyze evolutionary change is presented and its application to the analysis of sequence data is discussed. The investigated method uses phylogenetic trees of related proteins with an evolutionary model in order to gain insight about protein structure and function. The evolutionary model, based on amino acid substitutions, contains adjustable parameters related to amino acid and sequence properties. A maximum likelihood approach is used with a phylogenetic tree to optimize these parameters. The model is applied to a set of Muscarinic receptors, members of the G-protein coupled receptor family. Here we show that the optimized parameters of the model are able to highlight the general structural features of these receptors.

1 Introduction

One of the current main challenges in life sciences is understanding the machinery of biological systems. In the core of this machinery lies proteins; our understanding of biological systems is bound to our knowledge of protein structure and function. There has been increasing interest in obtaining information on structure and function from the rapidly-increasing databases of protein sequences, often through the comparison of related sequences. Despite these efforts, there are currently no generally-applicable methods to derive detailed structural and functional information from such investigations.

Like all biological entities, proteins are a result of evolution. They have developed their current structure and function under the influence of billions of years of selective pressure. Analyzing a family of proteins from different species can unravel information regarding this selective pressure. Such a study might allow one to detect and interpret the selective pressures that have acted on these proteins, providing insight about their structure and function.

In order to be fruitful, such examination requires careful choice of the protein family to be examined and the model of evolution to be used. Here we present the application of an evolutionary model based on amino acid substitution to a family of G-Protein coupled receptors (GPCRs). Located in the cell membrane, these receptors activate the associated G-protein bound to their intracellular part upon binding a ligand on their extracellular side. GPCRs constitute one of the largest protein families, making up 3% to 5% of the coding regions in the human genome[1]. They are associated with many signaling pathways in different cells ranging from neurons to muscle cells, and are targeted by more than 50% of all drugs[1]. In addition to their biological significance, the high structural and functional resemblance among family members is another reason to use GPCRs in evolutionary studies. Throughout the family the general topology of seven transmembrane helices connected by extra- and intracellular loops is highly conserved (see Fig.1). Despite this conserved general topology, GPCRs are able to achieve different functions by coupling to different ligands and/or G-proteins. The sequence similarity among family members is low due to varying composition and length of loops, while highly conserved transmembrane regions allow for reliable sequence alignments. Currently there are more than 3500 known GPCR sequences with only one crystal structure solved, that of bovine rhodopsin (PSB code 1F88)[2].

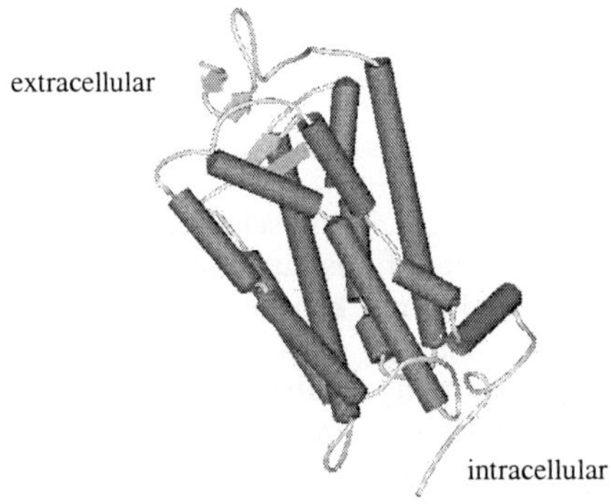

Figure 1: Representation of the crystal structure of bovine rhodopsin. The helical parts, including the seven transmembrane helices, are shown as red cylinders

All these properties make GPCRs a good candidate for sequence based studies and there are many examples of such studies in the literature. Most important of these are techniques based on pattern recognition[3] and correlation analysis[4]. These analyses are mainly focused on defining key residues responsible for ligand and/or G protein binding. While these methods have provided important information about specific residues, they are unable to generate more general information about how the observed protein properties are determined by the sequence. In addition, the correlation analysis explicitly neglects evolutionary relationships between the proteins, making them susceptible to misinterpreting correlations induced by the phylogenetic relationships.

2 Model

Evolution proceeds from the fixation of errors occurring during DNA replication. This is generally represented by a substitution matrix, encoding the relative rate at which every possible amino acid or nucleotide substitution occurs on the evolutionary timescale[5,6]. These matrices generally assume that all locations in all proteins can be represented by the same model. Despite the success of these matrices, there are shortcomings both in their creation and use. Derived from a particular set of proteins these matrices might not be able to mimic the substitutions in a different protein family. Their use for different locations of a given protein is also questionable. Given the structural and functional constraints on a protein both the rate and nature of substitutions among different locations should vary.

There have been attempts to incorporate absolute rate heterogeneity among locations by having the substitution rates multiplied by a site-specific scaling factor[7]. While these models are better able to represent biological data, they cannot account for qualitative variations in the type of selection pressure at various locations. Other models have been developed that allow for different locations to be under different types of selective pressure, either due to differences in local structure[8-10] or by allowing every location to be described by a different model[11]. The former method ignores differences in selective pressure due to other factors than local structure, while the latter is limited by the amount of available data. We (and others) have developed methods that allow for variation at different locations by postulating that there are a number of different types of locations, each describable with a specific substitution model, where the assignment of locations to different types is not known a priori[12-15]. In our model this is achieved by using the notions of amino acid fitness and site classes.

The basis of our evolutionary model based on amino acid substitution has been described previously[14,15]. In brief, we encompass the distribution of selective pressures at different locations in the protein by assuming that each location under consideration can be described by one of a number of possible site classes; each with its own set of parameters defining the substitution rates. The model does not assign locations to site classes, instead we define an unknown prior probability $P(k)$, that any given location belongs to site class k. As all locations must belong to a site class, $\Sigma_k P(k)=1$. We also imagine that there is a relative fitness $F_k(A_i)$ of amino acid A_i for any location described by a particular site class k. For example, at the core of a protein we expect a hydrophobic amino acid to have a high fitness value, however we do not impose such expectations on the model *a priori*. We further argue that the probability of substitution between two amino acids should directly depend on the change in fitness values resulting from such substitution. Thus for each site class we define a matrix for all possible substitutions based on the fitness values. Our particular model uses a function, composed of Gaussian and sigmoidal distributions, to calculate the substitution matrix for a small interval of time:

$$Q(k)_{ij} = v_k e^{-\lambda_k (\Delta F_k)^2} \frac{\beta_k e^{\Delta F_k / 2}}{\beta_k e^{|\Delta F_k / 2|} + e^{-|\Delta F_k / 2|}} \tag{1}$$

where v_k is the substitution rate for site class k, λ and β are parameters of the function, and

$$\Delta F_k = F_k(A_j) - F_k(A_i) \tag{2}$$

The use of this so-called "gaumoidal" function allows us simulate two different ideas about the process of evolution simultaneously. For small values of λ, the above function will approach a sigmoidal function where substitutions are accepted with v_k if favorable and tolerated with a decaying probability if unfavorable. For large values of λ, the function will approach a Gaussian distribution where conservative substitutions are favored.

To determine the substitution matrix M, representing the possible substitutions from amino acid A_i to A_j for any particular amount of evolutionary time t, the Q matrix is exponentiated:

$$M(k)(t) = e^{tQ(k)} \tag{3}$$

At each location l, the likelihood L_l can be calculated as the probability of data given the model's parameters θ and the evolutionary tree topology and branch lengths. Since each location can be represented by any of the site classes and each site class has distinct parameters θ_k we have to sum over all possible site classes to calculate this likelihood:

$$L_l = \sum_k P_l(Data \mid \theta_k, T) P(k) \tag{4}$$

with the likelihood for the tree calculated as the product of the likelihood at each location.

The parameters of the model can be optimized using a maximum likelihood approach on a given tree. To summarize there are 23 parameters per site class: 20 amino acid fitness values F_k (with one held constant since the fitness values are relative), substitution rate v_k, gaumoidal function parameters λ and β, and the prior probability for that site class $P(k)$. One of the site class prior probabilities will depend on the others since all priors must add to 1. The initial values for substitution rates are derived from a gamma distribution as described by Yang[16], while the other parameters are set to arbitrary initial values.

While we do not know to which site class a location belongs *a priori*, we can calculate *a posteriori* probabilities. The conditional probability that a location l belong to site class k is given by:

$$P(k \mid Data_l) = \frac{P(Data_l \mid \theta_k)P(k)}{\sum_k P(Data_l \mid \theta_k)P(k)} \qquad (5)$$

This equation allows us to group locations in the protein that are under similar selective pressure; the parameter values give us insight into the nature of the selective pressure at these locations.

3 Data and Methods

The model explained above is used to predict the structural and functional properties of a subfamily of the GPCR family. The selected subfamily was that of Muscarinic Receptors. These receptors are activated upon binding of acetylcholine and initiate a set of diverse events in the cell through the associated G protein. There are five known types of Muscarinic Receptors that couple to two different G proteins. The data set contained twenty-two receptor sequences from eight different species, representing all five types of Muscarinic Receptors. The sequences and the corresponding multiple alignment of length 530 are obtained from GPCRdb[17].

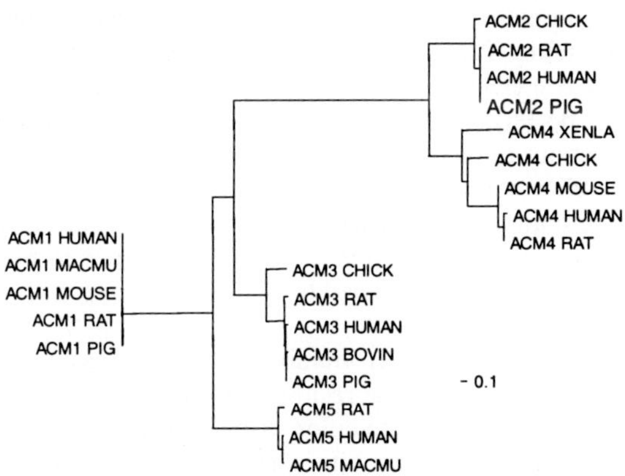

Figure 2: Unrooted phylogenetic tree of ACM receptors. Branch lengths are scaled to number of substitutions along each branch, with the given scale representing 1 substitution per 10 sites.

In order to optimize the parameters of the model a phylogenetic tree of the selected proteins, shown in Fig. 2, was created using PROTML[18]. This software uses a maximum likelihood approach to search for the most likely tree topology. We used this program with the default settings, which use automatic search and the JTT matrix of Jones et. al.[6]. The branch lengths of the resulting tree were further optimized using PAML[19], along with the alpha parameter of gamma distribution used in determining rate variation among site classes.

We optimized our model using the tree with optimized branch lengths for increasing number of site classes. For each run the initial rates for each site class are derived from the gamma distribution using the alpha parameter optimized with PAML. The software optimizes all the parameters for each site class and calculates the posterior probability of each location being represented by any site class.

4 Results

The results of this study consist of optimized fitness values for each site class and the posterior probabilities for each alignment position. We ran the program with two to eight site classes. The resulting optimal parameters are listed in Table 1. In order to interpret the resulting fitness functions, we determined the correlation coefficient between the values of $F_k(A_i)$ and 145 selected amino acid indices from the AAindex database[20]; the two most highly-correlated indices are also listed in Table 1.

The posterior assignment of site classes was mapped onto the structure of the Muscarinic type 3 receptor from human (ACM3_Human). Fig 3 shows the results for the optimization of two site classes. The same plot also contains the hydropathy plot for this receptor. Hydropathy plots are generally used to detect transmembrane regions of membrane proteins. These plots show the seven transmembrane helices of GPCRs clearly and are generally used to predict their sequence location. The correlation between the posterior assignments into the two site classes and the hydropathy plot show that our model assigns the residues into site classes according to their location. Almost all non-transmembrane residues are assigned to site class 2, while almost all transmembrane residues are assigned to site class 1. The results also show that certain conserved non-transmembrane residues such as residues from 2nd and 3rd intracellular loops are also assigned to site class 1 along with most of the transmembrane residues.

The posterior information for larger numbers of site classes is harder to interpret with such simple plots. To see the results from runs with more site classes we converted the posterior information into color strips, where each color represents an assignment of the location to a given site class. Fig.4 shows these strips for different numbers of site classes.

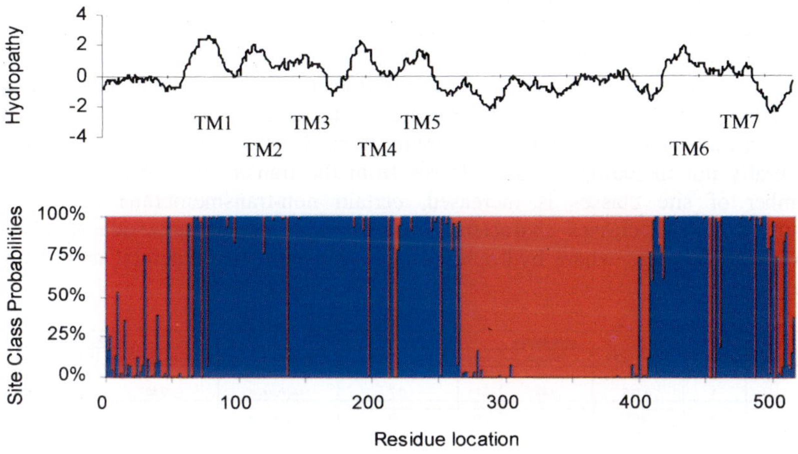

Figure 3: Correlation between hydropathy plot and posterior probabilities for ACM3_Human. Top plot: Kyte-Dolittle hydropathy index. Bottom plot: Relative probability that a location is assigned to site class 1 (blue) or 2 (red). Putative transmembrane helices, identified as the peaks on the hydropathy plot, are marked.

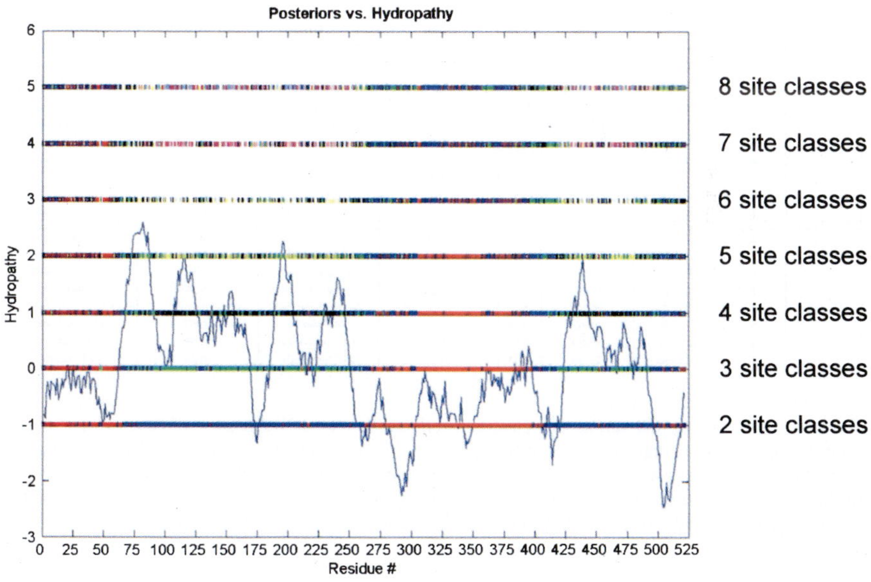

Figure 4: Color strips from posterior probabilities. The strips are matched to the hydropathy plot. Different colors represent different site classes.

As with two site classes, these color strips also indicate a distinct classification of certain residues to different site classes. This classification seems to follow the general topology of GPCRs, which is seven transmembrane helices connected with intra- and extracellular loops. Most of the transmembrane residues fall into same site class regardless of the number of site classes in the model. The residues from the non-transmembrane regions distribute themselves among a set of site classes, generally not including the site classes from the transmembrane helices. As the number of site classes is increased, certain non-transmembrane residues are assigned to site classes characteristic of the transmembrane residues, possibly involving locations where hydrophobicity is important for structural or functional reasons.

# of site	site	parameters			1st correlation		2nd correlation	
classes	class #	subst. rate	lambda	beta	Corr. Coeff.	Property Code	Corr. Coeff.	Property Code
2	2	**1.76**	0.45	0.84	0.81	**26-Flexibility**	-0.77	**9-Hydro(β)**
	1	0.116	3.83	2.31	0.7	10-Volume	0.68	16-Inner beta sheet
3	3	**2.07**	0.25	0.98	0.81	**26-Flexibility**	-0.77	**9-Hydro(β)**
	2	0.351	2.18	2.31	0.77	10-Volume	-0.63	5-Beta turn freq.
	1	0.0255	20.85	4.18	-0.71	**24-Polarity**	0.60	1-Extended Str.
4	4	**2.37**	0.14	7.31	0.83	**26-Flexibility**	-0.78	**9-Hydro(β)**
	3	0.649	0.85	0.01	-0.63	22-Principal z3	-0.63	3-Free E of soln.
	2	0.209	4.26	2.67	-0.54	4-Chg. Transfer	-0.52	14-Heat capacity
	1	0.00992	12.36	3.25	-0.60	**24-Polarity**	-0.48	6-C term helix
5	5	**2.74**	0.19	10.15	0.81	**26-Flexibility**	-0.77	**9-Hydro(β)**
	4	0.914	1	1.04	0.75	27-Flexibility	0.71	17-Hydrophobicity
	3	0.324	3.51	2.45	-0.73	15-S bend freq.	0.61	10-Volume
	2	0.0789	9.36	2.86	-0.53	**24-Polarity**	0.53	11-Beta sheet freq.
	1	0.00611	20.28	4.22	-0.64	**24-Polarity**	-0.48	6-C term helix
6	6	**3.67**	1.25	2.45	-0.73	2-Polarizability	0.71	**26-Flexibility**
	5	1.24	0.54	0.04	0.72	**26-Flexibility**	-0.70	**8-Hydro(α)**
	4	0.523	3.46	2.25	-0.69	3-Free E of soln.	-0.67	22-Principal z3
	3	0.192	5.13	2.93	0.56	16-Inner beta sheet	0.50	12-Width of side chain
	2	0.0474	14.68	3.37	-0.65	**24-Polarity**	-0.62	21-Hydropathy loss
	1	0.00368	21.68	4.71	-0.67	13-Positive charge	-0.51	6-C term helix
7	7	**3.65**	1.25	2.35	-0.73	2-Polarizability	-0.72	4-Chg. Transfer
	6	1.44	0.41	0.05	0.78	**26-Flexibility**	-0.77	**9-Hydro(β)**
	5	0.714	2.97	2.56	-0.77	3-Free E of soln.	-0.69	22-Principal z3
	4	0.333	3.61	2.37	-0.72	15-S bend freq.	0.65	12-Width of side chain
	3	0.13	12.36	3.75	-0.54	4-Chg. Transfer	0.47	18-Beta sheet freq.
	2	0.0336	18.76	3.61	-0.63	**24-Polarity**	0.59	11-Beta sheet freq.
	1	0.00277	11.75	3.05	-0.61	13-Positive charge	-0.45	6-C term helix
8	8	**4.51**	1.81	2.51	-0.76	4-Chg. Transfer	-0.71	20-Accessible area
	7	1.75	0.46	0.09	-0.81	**9-Hydro(β)**	0.78	**26-Flexibility**
	6	0.915	0.66	0.08	-0.63	3-Free E of soln.	-0.54	22-Principal z3
	5	0.478	8.57	3.24	-0.60	3-Free E of soln.	0.56	25-Isoelectric point
	4	0.228	6.12	3.36	-0.66	7-C term non beta	0.62	23-Hydrophobicity
	3	0.0898	13.02	3.48	-0.72	**24-Polarity**	-0.67	19-Hydrophobicity
	2	0.0233	15.89	3.57	0.58	1-Extended Str.	-0.58	24-Polarity
	1	0.00193	22.39	4.58	-0.57	6-C term helix	-0.49	24-Polarity

Table 1: Parameters for the various site classes. λ reflects the relative importance of conserving a given property (large λ) vs. improving that quantity (small λ). Also shown are the two physico-chemical properties with the largest absolute value of correlation coefficients (cc) with $F(A_i)$. The most important of these correspond to flexibility, polarity, hydrophobicity for beta proteins (Hydro (β)), hydrophobocity for alpha proteins (Hydro (α)). The full citation of these and all other properties are given in Table 2.

The most important of the site class parameters are the fitness values. These values show which amino acids are favored in a given site class. In order to interpret this information we searched for correlation between fitness values and amino acid properties. Table 1 shows these correlation coefficients for runs with different numbers of site classes. Looking at these values, we see two main site classes with high correlation to certain properties regardless of the number of site classes. These properties are flexibility and hydrophobicity in one case and polarity in the other. Interpreting these results together with the color strips we see that the fitness values for the site class that holds the non-transmembrane residues show a strong positive correlation to flexibility and negative correlation to hydrophobicity. The fitness values of the site class that is mainly occupied by residues from transmembrane regions show a negative correlation to polarity. These correlations are in agreement with the general expectation of non-transmembrane residues being hydrophilic and transmembrane residues being hydrophobic. The positive correlation with high flexibility also makes sense since the non-membrane regions have to be highly flexible in order to accommodate the movements of the helical regions during activation of the receptor.

Code	AAindex Code	Property
1	BURA740102	Normalized frequency of extended structure
2	CHAM820101	Polarizability parameter
3	CHAM820102	Free energy of solution in water, kcal/mole
4	CHAM830108	A parameter of charge transfer donor capability
5	CHOP780101	Normalized frequency of beta-turn
6	CHOP780205	Normalized frequency of C-terminal helix
7	CHOP780211	Normalized frequency of C-terminal non beta region
8	CIDH920101	Normalized hydrophobicity scales for alpha-proteins
9	CIDH920102	Normalized hydrophobicity scales for beta-proteins
10	COHE430101	Partial specific volume
11	CRAJ730102	Normalized frequency of beta-sheet
12	FAUJ880105	STERIMOL minimum width of the side chain
13	FAUJ880111	Positive charge
14	HUTJ700101	Heat capacity
15	ISOY800105	Normalized relative frequency of bend S
16	KANM800104	Average relative probability of inner beta-sheet
17	LEVM760101	Hydrophobic parameter
18	PALJ810112	Normalized frequency of beta-sheet in alpha/beta class
19	PRAM900101	Hydrophobicity
20	RADA880106	Accessible surface area
21	ROSM880103	Loss of Side chain hydropathy by helix formation
22	WOLS870103	Principal property value z3
23	ZIMJ680101	Hydrophobicity
24	ZIMJ680103	Polarity
25	ZIMJ680104	Isoelectric point
26	VINM940102	Normalized flexibility parameters (B-values) for each residue surrounded by none rigid neighbours
27	VINM940103	Normalized flexibility parameters (B-values) for each residue surrounded by one rigid neighbour

Table 2. Amino acid indices from AAindex database[20].

5 Discussion

The preliminary results presented in this paper show that our model is capable of detecting general structural and functional differences among different locations of a protein from sequence data. The key points in this approach are the degree of relation among the proteins of interest and the phylogenetic tree that relates them. There is much evidence for a strong probability of kinship among all GPCRs. They are believed to share a common ancestor, which is believed to have given rise to general topology of seven transmembrane helices upon gene duplication[21]. Given this evidence of evolutionary relation among GPCRs and the strength of maximum likelihood methods in phylogenetics, we believe that the tree used in this study can be considered "reasonable", if not the most likely tree. We believe that this is a good enough tree to optimize the parameters of the model, given the other studies showing low variation among estimated parameters of a model using a set of possible phylogenetic trees[22].

Besides the structural information, the model used also gives insight about the process of evolution. For all of the site classes the optimized parameters of the gaumoidal fitness function weight it towards a Gaussian distribution. This might indicate that evolution favors mutations resulting in small fitness changes, as would be expected if multiple substitutions in nearby residues favor the current amino acid type.

The most important point is that we impose no structural or functional information into the model *a priori*. All results regarding fitness and posterior values are a result of the optimization process and their correlation to structural features is a validation of the model. There is still much to do using larger data sets and site class numbers and as the resolution increases we should be able to pick more detailed structural and functional information from such studies. Currently we are only able to detect correlation and classification of general structural features such as transmembrane/non-transmembrane, but as we develop the method further we hope to be able to detect distinct site classes, holding functionally important residues. It will be interesting to compare the results of runs from other subfamilies of GPCRs and see whether these can account for their different functional properties such as coupling to different G-proteins.

Acknowledgments

Thanks to Sarah E. Ingalls for her contributions in writing the software and Todd Raeker for computer support. The financial support was provided by NIH Grant LM0577, NSF Grant 9726427, and NSF equipment grant BIR9512955.

References

1. Iismaa PT, B.J., Shine J, *G-Protein Coupled Receptors.* 1995: Springer Verlag.
2. Palcewski K, K.T., Hori T, Behnke CA, Motoshima H, Fox BA, Trong IL, Teller DC, Okada T, Stenkamp RE, Yamamoto M, Miyano M, *Crystal Structure of Rhodopsin: A G-protein Coupled Receptor.* Science, 2000. **289**: p. 739-745.
3. Attwood, T., *Acompendium of specific motifs for GPCR subtypes.* Trends in Pharmacological Sciences, 2001. **22**(4): p. 162-165.
4. Horn F, W.M., Oliviera L, Ijzerman AP, Vriend G, *Receptors Coupling to G proteins: Is There a Signal Behind Sequence?* Proteins: Structure, Function and Genetics, 2000. **41**: p. 448-459.
5. Dayhoff M. O., E.R.V., *Atlas of Protein Sequence and Structure,* . 1966, National Biomedical Research Foundation.
6. Jones D.T., T.W.R., Thornton J.M., *The rapid generation of mutation data matrices from protein sequences.* Computational Applications in Bio Sciences, 1992. **8**(3): p. 275-282.
7. Yang Z., *Maximum-Likelihood Estimation of Phylogeny from DNA Sequences When Substitution Rates Differ over Sites.* Molecular Biology and Evolution, 1993. **10**(6): p. 1396-1401.
8. Koshi JM, G.R., *Context Dependent Optimal Substitution Matrices.* Protein Engineering, 1995. **8**: p. 641-645.
9. Goldman N., T.J., Jones DT., *Assessing the Impact of Secondary Structure and Solvent Accessibility on Protein Evolution.* Genetics, 1998. **149**: p. 445-458.
10. Lio P, G.N., *Using Protein Structural Information in Evolutionary Inference: Transmembrane Proteins.* Mol. Biology and Evolution, 1999. **16**(12): p. 1696-1710.
11. Halpern AL, B.W., *Evolutionary Distances for Protein-Coding Sequences: Modeling Site-Specific Frequencies.* Mol. Biology and Evolution, 1998. **15**(7): p. 910-917.
12. Yang Z, *Relating Physicochemical Properties of Amino Acids to Variable Nucleotide Substitution Patterns Among Sites.* Pacific Symposium on Biocomputing, 2000.
13. Yang Z., N.R., Goldman N., Pedersen AM., *Codon-substitution Models for Heterogeneous Selection Pressure at Amino Acid Sites.* Genetics, 2000. **155**: p. 431-449.
14. Koshi JM, M.D., Goldstein RA, *Using Physical-Chemistry-Based Substitution Models in Phylogenetic Analyses of HIV-1 Subtypes.* Mol. Biology and Evolution, 1999. **16**(2): p. 173-179.

15. Dimmic MW, M.D., Goldstein RA, *Modeling Evolution at the Protein Level Using an Adjustable Amino Acid Fitness Model.* Pacific Symposium on Biocomputing, 2000: p. 18-29.
16. Yang Z., *Maximum Likelihood Phylogenetic Estimation from DNA Sequences with Variable Rates over Sites: Approximate Methods.* Journal of Molecular Evolution, 1994. **39**(306-314).
17. Horn F, W.J., Beukers MW, Horsch S, Bairoch A, Chen W, Edvardsen O, Campagne F, Vriend G, *GPCRDB: an information system for G protein coupled receptors.* Nucleic Acid Research, 1998. **26**(1): p. 275-279.
18. Adachi J., H.M., *Maximum Likelihood Inference of Protein Phylogeny,* . 1992: Tokyo.
19. Yang Z., *Phylogenetic Analysis by Maximum Likelihood,* . 2000, University College, London: London.
20. Kawashima S., O.H., Kanehisa M., *AAindex: amino acid index database.* Nucleic Acid Res., 1999. **27**: p. 368-369.
21. Taylor EW, A.A., *Sequence Homology Between Bacteriarhodopsin and G-protein Coupled Receptors: Exon Shuffling or Evolution by Duplication.* FEBS, 1993. **325**(3): p. 161-166.
22. Yang Z., *A Space-Time Process Model for the Evolution of DNA Sequences.* Genetics, 1995. **139**: p. 993-1005.

PROGRESS IN PREDICTING PROTEIN FUNCTION FROM STRUCTURE: UNIQUE FEATURES OF O-GLYCOSIDASES

E.W. STAWISKI*, Y. MANDEL-GUTFREUND*, A.C. LOWENTHAL, L. M. GREGORET

Departments of Chemistry and Biochemistry and Molecular, Cell and Developmental Biology
University of California
Santa Cruz, CA 95060

The Structural Genomics Initiative promises to deliver between 10,000 and 20,000 new protein structures within the next ten years. One challenge will be to predict the functions of these proteins from their structures. Since the newly solved structures will be enriched in proteins with little sequence identity to those whose structures are known, new methods for predicting function will be required. Here we describe the unique structural characteristics of O-glycosidases, enzymes that hydrolyze O-glycosidic bonds between carbohydrates. O-glycosidase function has evolved independently many times and enzymes that carry out this function are represented by a large number of different folds. We show that O-glycosidases none-the-less have characteristic structural features that cross sequence and fold families. The electrostatic surfaces of this class of enzymes are particularly distinctive. We also demonstrate that accurate prediction of O-glycosidase function from structure is possible.

1 Introduction

1.1 Structural genomics

The completion of the sequencing of entire genomes has tremendously increased our knowledge of biology. It has also revolutionized our thinking about the scale at which is possible to make inquiries and has created new challenges. One of these challenges is the determination of the three dimensional structures of a large, representative set of proteins. This endeavor, called *structural genomics*, aims to 1) understand the basis of disease at the atomic resolution level of detail, 2) provide homology modeling scaffolds for all proteins, 3) obtain structures for the targets of structure-based drug discovery, and 4) map out protein fold space to better understand sequence-structure relationships[1]. Within the next ten years, between 10,000 and 20,000 new structures are expected to be solved as the result of the structural genomics initiative[2].

It is well established that the proteins that are similar in sequence are likely to have evolved from a common ancestor and thus retain similar functions. However, in order to sample protein structure space broadly, the structures selected as targets for structural genomics will primarily be those with little or no sequence identity to

*These authors contributed equally to this work

proteins whose structures have already been solved. Many of these proteins are expected to have unknown functions.

1.2 Predicting function from structure

In the absence of sequence identity, how is it possible to predict function from structure? It has been noted that proteins that adopt similar folds often have similar functions[3]. However, there are many well-known examples, such as the SH3 fold and the TIM barrel[4], which are used for many different functions. It is also often possible to predict the functions of enzymes based on the spatial arrangement of catalytic residues[5]. However, some functions (e.g. proteolysis) have evolved multiple times independently and cannot be described by any single set of residues.

To account for the likelihood that entirely new folds and catalytic mechanisms will be discovered through structural genomics, it will be necessary to develop new methods of function prediction that do not rely on existing examples of known sequences, folds or catalytic residue arrangements. We have demonstrated recently that proteases as a group (including those with very different folds and catalytic mechanisms) share common structural features[6]. These features include smaller surface areas, higher packing densities, and less helical structure. We speculate that these features arose independently as mechanisms to avoid autolysis. More importantly, we are able to use these identifying features to train a neural network to predict protease function with high accuracy even in the absence of related structures being present in the training set[6]. We have recently tackled the problem of nucleic acid binding protein function prediction[7]. In this case, discrimination of the nucleic acid binding proteins (again at high accuracy) relied on a novel method using positive electrostatic patch analysis.

1.3 The O-glycosidases

We now turn our attention to identifying the characteristic features of O-glycosidases. These enzymes hydrolyze linkages between carbohydrate molecules. We chose this class of enzymes for several reasons. First, oligosaccharides play critical roles in a variety of biological processes including viral invasion and cell signaling events[8]. Thus O-glycosidases represent an important class of enzymes for drug discovery, especially with regard to antiviral and anticancer agents. Second, the O-glycosidases are structurally diverse and include members of at least six different folds ranging from all alpha to all beta[9, 10]. This enzyme family is therefore an excellent test case for fold independent function prediction methods. O-glycosidases are also well represented in the existing Protein Data Bank[11], allowing us to build a substantial representative data set. Finally, O-glycosidases were frequent false positives in our protease prediction effort[6]. We speculated that this as hydrolases, they may have been subjected to similar anti-autolysis evolutionary pressures. Identification of the unique features of O-glycosidases should improve prediction on both classes of enzymes.

We have gleaned a set of identifying features of O-glycosidases. Like for the nucleic acid binding proteins, we again used electrostatic patch analysis, this time concentrating on the negatively-charged surface patches to help characterize O-glycosidases. Using an ensemble of features, we were able to train a neural network to predict O-glycosidase function with over 87% accuracy. We are also now able to better discriminate between O-glycosidases and proteases.

2 Methods

2.1 Data Set Construction

Representative data sets of both O-glycosidase and non-O-glycosidase protein structures from the Protein Data Bank (PDB)[11] were constructed. These data sets consisted of proteins that were solved by X-ray crystallography and had an atomic resolution of better than 2.5 Å. Within the sets, sequence identity cutoffs were used such that no two members in any one data set contained more than 25% sequence identity within the non-O-glycosidase data set and 35% sequence identity for the O-glycosidases. The O-glycosidase set was constructed by mining the PDB for the Enzyme Commission (EC) number 3.2.1.x (x stands for substrate specificity) and consisted of 39 proteins. The non-O-glycosidase data set consisted of 258 monomeric proteins and was constructed from Hobohm's and Sander's "pdb select" list of proteins[12]. A description of the glycosidase data set along with the PDB codes for the non-glycosidase data set can be found at:
http://www.chemistry.ucsc.edu/gregoret/PSB_supp.html.

2.2 Electrostatic Patch Analysis

The UHBD[13] program was used to derive a continuum electrostatic description for each protein, using the Poisson-Boltzmann equation. For all UHBD calculations, the grid dimensions were set to 65x65x65 with 2.0 Å distance from each grid point to another. Dielectric constants of 2.0 and 80.0 were used for the protein and the solvent, respectively. Other parameters for UHBD were set to their default values. Patches were constructed based on UHBD output with an in-house program, PATCHFINDER. Continuum negative electrostatic patches, mapped to the protein surface, were constructed by assembling adjacent surface points, which possessed a negative potential less than or equal to -5 *kT*. The patch size was defined as the number of surface points within a continuous cluster of points. The largest negative patch was used to extract sequence and structural information.

2.3 *Sequence Conservation*

For each protein in the data set, a multiple sequence alignment (MSA) was constructed using PSI-BLAST version 2.1.1[14] to search the non-redundant NCBI data base for similar sequences that were significant (E-value <0.001). Since we wanted to include only sequences that are likely to be structurally related to the representative sequence we eliminated sequences with <35% identity. In addition, to reduce redundancy from very close homologous sequences, only sequences with <90% identity were included in the MSAs.

The conservation of specific residues within the negative electrostatic patch was analyzed, including residues which were occupied by Glu, Asp and Asn. In addition to the simple amino acid conservation we also calculated the conservation of aromatic residues as a group. A residue position was considered to be conserved when ≥75% of the sequences in the MSA contained the same amino acid (for amino acid conservation) or property (for property conservation) as in the representative sequence. For each of the amino acids above, the normalized frequency of conserved residues in the electrostatic patch was calculated.

2.4 *Cleft Detection*

The program SURFNET[15] was used to analyze protein clefts. The residues identified within the largest two clefts of each protein were examined and the number of residues that overlapped with the largest negative patch were calculated. The cleft identified as having the largest overlap with the patch was further examined to confirm whether it had potential residues that could participate in a glycolysis reaction. Only Asp and Glu were considered for donation of the acid or base atoms for the reaction. All distances between carboxylate groups of ASP and GLU residues were calculated. The two carboxylate atoms with a distance closest to either 5 or 9.5 Å of each other, were identified as possible catalytic residues.

2.5 *Calculation of Other Structural Features*

A protein's solvent accessible surface area was calculated by using Lee and Richards method[16] as implemented in the program CALC-SURFACE[17] with a default probe radius of 1.4 Å. The program DMS under the UCSF MidasPlus software package[18] was used to calculate molecular surface. The roughness, or fractal dimension, D, of the surface[19] was calculated using equation 1:

$$1.\quad D = 2 - \frac{d\log A_S}{d\log R}$$

where R is the probe radius and A_S is the molecular surface area. In this case, radii of 1.25, 1.5, 1.75 and 2.0 Å were used. The area of a perfectly smooth surface will not depend on the probe size, and will thus have a fractal dimension of two.

2.6 *Machine learning*

For function prediction, we applied the Nevprop4 neural network package[20]. The neural network consisted of a single hidden layer with 3 nodes and a single output node. All training was performed with a standard feedforward, error backpropagation algorithm. The cross validation scheme used was to train on all but one member of the data set, which was withheld from training and subsequently tested. This was done for each of the members of the O-glycosidase data set. For the non-glycosidase set, in each training session a random sample of 10% of the data set was withheld and subsequently tested. For the false set the average performance over all runs was reported.

3 Results and Discussion

3.1 *Identifying Characteristics of O-glycosidases*

Based on general properties of the proteins structures, we wanted to identify unique features that could be used to distinguish O-glycosidase proteins from other classes of proteins. To do this, two representative data sets of crystallographically-determined three-dimensional protein structures were constructed as described in Methods. The O-glycosidase data set consisted of 39 proteins and the non-O-glycosidase proteins (including a full spectrum of different proteins except those with glycosidase function) had 258 members. A list of the proteins in the glycosidase data set and their structural classification (based on SCOP classification) is given at: http://www.chemistry.ucsc.edu/gregoret/PSB_supp.html. For the two data sets, we calculated both global and local structural features that could potentially distinguish between them. Although most of the features analyzed did not show statistically significant differences between the O-glycosidases and other proteins, when combined, these inputs were successfully used in predicting glycosidase function.

3.1.1 *Surface features*

We have previously shown that electrostatic patch analysis, a combination of structural and sequence features extracted from a distinct region on the protein surface defined by electrostatic potential, can help to distinguish nucleic acid binding proteins from other proteins[7]. Surface electrostatics have been used previously to help indicate potential protein functions based on their structure[21]. O-glycosidases use mostly Glu and Asp residues for catalysis, and as a result usually have a negatively-charged surface associated with their active sites[9]. To characterize the O-glycosidase protein family, we performed our analysis on negatively-charged electrostatic patches on the protein surface (similar to the analysis performed for nucleic acid binding proteins[7]). As a first step, the continuum electrostatic potential was calculated for the whole protein using UHBD[13] software package, negative surface electrostatic patches were then assembled using an in house program PATCHFINDER (as described in methods) and the largest negative patch of every protein structure was analyzed.

O-glycosidases were found to have, on average, larger negative surface patches than non-O-glycosidases. The average size of the negative patch is 229 ± 163 surface points (compared to 86 ± 110 surface points for the non-O-glycosidases). In hydrolysis by O-glycosidases, one negatively-charged residue typically acts as a base and another as an acid[22]. We found that in 33 out of 39 proteins (85%), the largest negative patch contained at least one of the two residues (base or acid), known to be involved in catalysis. In some cases, the patch did not contain either the acid or the base These patches, however, were found to be in close proximity to other functionally important residues. The high overlap between the surface patch and the O-glycosidases active site allows us to conclude that the region we are analyzing is directly involved in the function of these proteins.

As shown in Figure 1, the amino acid distribution in the negatively-charged patches of O-glycosidase proteins differs from that of other proteins. The differences observed are mostly found for the aromatic amino acids Trp and Tyr and for Asn. All three amino acids occur more frequently in the O-glycosidases than in all other proteins. Aromatic amino acids are known to commonly act as docking sites for the non-polar inner portion of cyclic carbohydrate molecules[23]. (Since our negative patches are derived from Asp and Glu residues on the protein surface, the normalized percentage of these two amino acids was roughly the same in all proteins.)

Because of the high correlation between the active site and the negative patch, we also expected the residues within the patch to be functionally important and thus more conserved[7, 24]. To examine this, we analyzed the frequency with which residues were conserved within the negative patch. Specifically we were interested in the conservation of Asn, Glu, Asp, Tyr and Trp, which are frequent in the negatively-charged patches of O-glycosidases. In addition to analyzing the conservation of the individual residues, we also looked at the conservation of the aromatic amino acids when grouped together. As summarized in Table 1, we found that Asp, Glu, Asn, Tyr and Trp are on average more conserved (though not significant) within the negative patches of the O-glycosidase family than in the non-O-glycosidases. These differences were more obvious when we grouped the aromatic residues together. O-glycosidases have on average more conserved aromatics (6.9±4.6 residues) than do the non-O-glycosidases (1.3±2.0 residues).

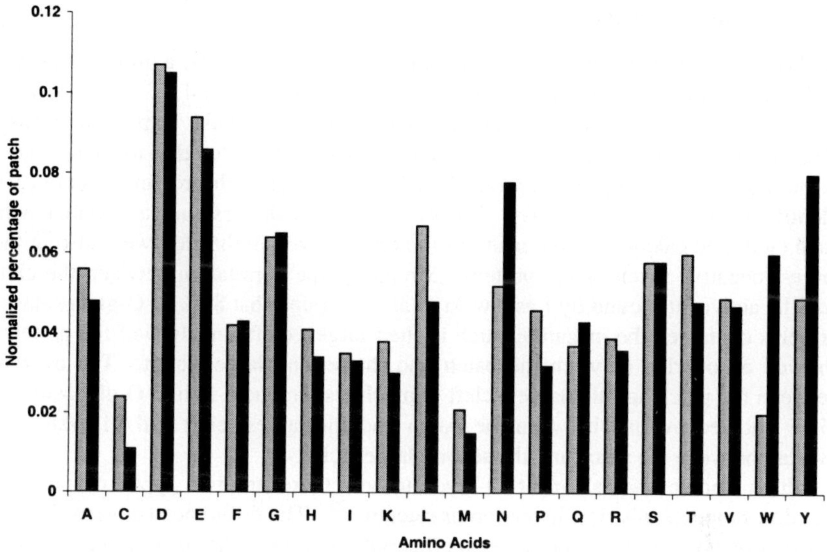

Figure 1. The normalized frequency of the 20 amino acids within the largest negative patch for O-glycosidases (black) and non-O-glycosidases (gray).

Table 1. Frequency* and conservation of specific amino acids in the negative surface patches of O-glycosidase and non-O-glycosidase proteins

	O-glycosidase		Non-O-glycosidase	
	Total number of residues in patch	Number of conserved residues in patch	Total number of residues in patch	Number of conserved residues in patch
GLU	4.5 (2.9)	2.3 (1.6)	2.1 (2.5)	0.7 (1.2)
ASP	4.8 (3.9)	2.5 (2.3)	2.2 (3.1)	1.0 (1.5)
ASN	3.4 (2.0)	1.7 (1.3)	1.1 (1.9)	0.4 (0.9)
TRP	3.4 (2.0)	1.7 (1.1)	1.1 (1.8)	0.4 (0.9)
TYR	3.7 (2.8)	2.0 (1.5)	1.1 (1.8)	0.5 (0.9)

*Frequency is averaged over all proteins, the average number and standard deviation are shown.

3.1.2 Structural features

Laskowski et al., have previously shown that amongst enzymes, a protein's largest and second largest clefts are bound to the ligand 84% and 9% of the time, respectively[25]. To see if information on protein clefts could help us to further discriminate O-glycosidases, we first identified the clefts belonging to each protein in both data sets, using the program SURFNET[15]. For each protein structure, we identified the two largest clefts. We then extracted the residues associated with each cleft, and calculated how many of the residues within the cleft were also in the largest negative patch of the protein. Similar to the correlation between the cleft and the active site found by Laskowski et al., we found that 82% of O-glycosidases proteins contained the negative patch in their largest cleft, and 10% of the proteins showed an overlap between the patch and the second largest cleft. The overlap between the patch and the largest cleft is much less frequent in non-O-glycosidases (58% showed overlap between the patch and the largest cleft and 21% showed overlap between the patch and the second largest cleft).

It is known that amongst O-glycosidases, there are two types of possible reaction stereochemistries: inversion or retention[22]. The difference between them is whether they invert or retain the anomeric configuration of the sugar at the cleavage site. The acid and base residues involved in the enzymatic reaction have been found to be predominantly Glu and Asp residues, although exceptions involving His and Ser have been found[26]. For a retention reaction, the average spacing between the two participating carboxyl groups is 5.5 Å, and for an inversion reaction, the spacing is 9.5 Å. Although there are exceptions to these reaction geometry distances[27], they serve as a general guideline for glycosidic hydrolysis.

We examined the clefst that had the highest patch overlap and identified the candidate residues that could be involved in a glycosidic reaction (see Methods). In general, the O-glycosidases have only slightly "better" putative catalytic residues (determined by the smallest difference in distance between two negative residues in either 5 or 9.5 Å) within their clefts (90% less than 0.24 Å) than other classes of proteins (59% less than 0.24 Å). The relatively large number of potential binding residues in non-O-glycosidase proteins are most likely due to the high frequency of Asp and Glu in the negative patches.

Lewis and Rees originally proposed that roughness may be associated with ligand binding, since a greater surface area allows more possibilities for van der Waals contacts[19]. Indeed, it has been found that functional sites in proteins (e.g. enzyme active sites) are rougher than other parts of the protein's surface[28]. To see if roughness could further discriminate the negative patches within our data set, we calculated the fractal dimension (D) for each protein patch. Although not significant, the O-glycosidase patches had slightly rougher surfaces (2.67 ± 0.25) than the non-O-glycosidase patches (2.52 ± 0.41). Interestingly, the 6 O-glycosidases whose catalytic residues were not in the patch showed a lower degree of a roughness – 2.46 ± 0.05 – which is comparable to the roughness of the non-O-glycosidases. This again strengthens the idea that the patches are associated with an O-glycosidase active site.

We have previously shown that other hydrolases (the proteases), have less accessible surface area per molecular weight as compared to other types of proteins[6]. Since a large number of the false positives in this experiment were O-glycosidases, we examined the accessible surface area per molecular weight amongst our glycosidase data set. Interestingly, O-glycosidases were also found to have strikingly less accessible surface area per molecular weight than non-O-glycosidases. As shown in Fig. 2, nearly 84% of the O-glycosidase data set falls below the line that represents the best fit for the non-O-glycosidase proteins. Moreover, the majority of the O-glycosidase proteins fall at the lower limit for accessible surface area per molecular weight. For proteases, we speculated that the lower solvent accessible surface area per molecular weight may have evolved to prevent self-cleavage. It has previously been proposed that different types hydrolases evolved from a common ancestor[29]. Moreover, O-glycans are also known to inhibit proteolysis[30]. O-glycosidases may therefore have to work in coordination with proteases to remove O-glycans from proteins in order to make them accessible to protein degradation in the same milieu with proteases. We again speculate that O-glycosidases may be under similar types of evolutionary pressure and hence show similar structural properties to proteases.

3.2 Prediction vs. Non-O-Glycosidases

Although most of the features characterized above were not individually sufficient to discriminate O-glycosidases from other non-O-glycosidase proteins, we wanted to see if we could use them collectively to infer function. To do this we created a feature vector consisting of the following 10 inputs: surface area per molecular weight, negative patch size, patch-cleft overlap booleans (1 for true) for the largest and second largest cleft, sequence conservation within the patch (aromatics, Asn, Glu, and Asp), roughness of patch, and the best distance for a putative active site. We used this feature vector to train a neural network (see Methods), to distinguish O-glycosidase proteins from non-O-glycosidase proteins.

Using the cross validation scheme described in Methods, we were able to predict O-glycosidase proteins with 87% accuracy and non-O-glycosidase proteins with 93% accuracy. When we examined the relevance of the inputs, defined as the sum of the square weights for a given input group, divided by the sum of all input groups, the most relevant inputs were the frequency of conserved aromatics, the SA/MW, patch size, conservation of Glu, and geometry of the putative active site.

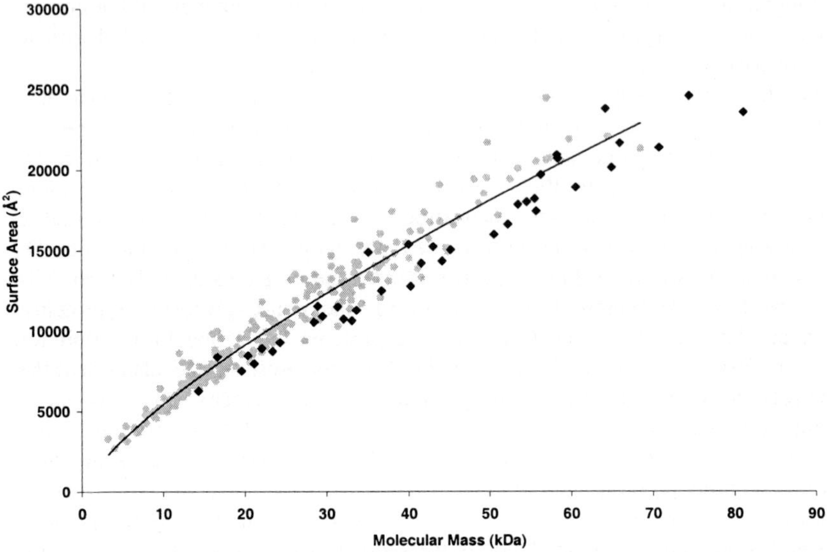

Figure 2. Molecular mass vs. surface area for O-glycosidases (♦) and non-O-glycosidases (*). The solid line is the best fit for the non-O-glycosidases. The average surface area to molecular mass ratio for O-glycosidases is 0.36 ± 0.04 and 0.46 ± 0.08 for the non-O-glycosidases.

3.3 Discriminating Glycosidases and Proteases

Since we found some similar structural properties to exist between proteases and O-glycosidases, we were interested to see if we are now able to distinguish between these two classes of proteins. From the output of the neural net discussed above, we were able to correctly identify over 80% of the proteases as being non-O-glycosidases. The biggest difference between the two data sets arises from the electrostatic patch analysis (data not shown). One feature that differs between the two groups, that was not addressed earlier, is the presence of the TIM fold. The TIM barrel is a common fold amongst O-glycosidases, but is absent amongst proteases. It is interesting to speculate why this fold, which is obviously compatible with hydrolase function, was never recruited for protease function. We speculate the high helical content may be incompatible with protease stability. Regardless of the reason, this feature could further help to differentiate between the two. We predict that similar approaches could be used for distinguishing between other protein classes which possess different functions, yet retain some similar structural characteristics.

4 Conclusions

We have shown that O-glycosidases possess some unique global and local properties that can be distinguished from other proteins with different functions. Like proteases, O-glycosidases have less accessible surface area per unit molecular weight. We speculate that as for the proteases, this structural property can be involved in preventing proteolysis. Beyond this we have shown that O-glycosidases have characteristic electrostatic features. Most O-glycosidases have large negative patches on their surface that are highly correlated with active sites. This could help to identify putative active sites, which could be targeted for molecular based drug design. We have also shown for the second time that electrostatic patch analysis, using a combination of structural and sequence analysis, can be used to characterize a class of proteins. We believe that electrostatic patch analysis should therefore be a common tool in the characterization of the relationship between a protein's structure and its function.

Using this combination of global and region-specific features, we were able to successfully train a neural network to predict O-glycosidase function with high accuracy. Since the O-glycosidase data set used in this study was comprised of many different folds and had little sequence homology amongst them, we propose that the neural net would potentially be able to identify a protein having an O-glycosidase function even if it possessed a novel fold. This could be particularly useful for products of the structural genomic initiative. Here, for the first time, we show that prediction of protein function from structure can be improved by the ability to identify a class of proteins that was a common false positive prediction of another protein family. We suggest that protein function prediction, may be improved by analyzing common themes amongst classes of proteins that are incorrectly identified. Multi-class automated protein function prediction based on structure, may therefore be close at hand.

5 Acknowledgements

We thank Janet Newman for helpful disussions. This work was supported by NIH grant GM52885 to LMG, an award from the University of California Cancer Research Coordinating Committee to LMG, and an American Cancer Society California Division postdoctoral fellowship to YMG.

References

1. A. Sali, *Nat Struct Biol* **5**, 1029 (1998); C. A. Orengo, A. E. Todd, J. M. Thornton, *Curr Opin Struct Biol* **9**, 374 (1999).
2. D. Vitkup, E. Melamud, J. Moult, C. Sander, *Nat Struct Biol* **8**, 559 (2001).

3. W. A. Koppensteiner, P. Lackner, M. Wiederstein, M. J. Sippl, *J Mol Biol* **296**, 1139 (2000).
4. J. A. Gerlt, P. C. Babbitt, *Annu Rev Biochem* **70**, 209 (2001).
5. A. C. Wallace, N. Borkakoti, J. M. Thornton, *Protein Sci* **6**, 2308 (1997); J. S. Fetrow, J. Skolnick, *J Mol Biol* **281**, 949 (1998).
6. E. W. Stawiski, A. E. Baucom, S. C. Lohr, L. M. Gregoret, *PNAS*, **97**, 3954 (2000).
7. E. W. Stawiski, Y. Mandel-Gutfreund, L. M. Gregoret, *Submitted*, (2001).
8. K. von Figura, *Curr Opin Cell Biol* **3**, 642 (1991); R. C. Wade, *Structure* **5**, 1139 (1997).
9. G. Davies, B. Henrissat, *Structure* **3**, 853 (1995).
10. H. Hegyi, M. Gerstein, *J Mol Biol* **288**, 147 (1999).
11. H. M. Berman *et al.*, *Nucleic Acids Res* **28**, 235 (2000).
12. U. Hobohm, C. Sander, *Protein Sci* **3**, 522 (1994).
13. M. E. Davis, Mandura, J.D., Luty, B.A., McCammon, J.A.,, *Comp. Phys. Comm.*,, 187 (1991).
14. S. F. Altschul *et al.*, *Nucleic Acids Res* **25**, 3389 (1997).
15. R. A. Laskowski, *J Mol Graph* **13**, 323 (1995).
16. B. Lee, F. M. Richards, *J Mol Biol* **55**, 379 (1971).
17. M. Gerstein, *Acta Crystallogr A* **48**, 271 (1992).
18. T. E. Ferrin, C. C. Huang, L. E. Jarvis, R. Langridge, *J Mol Graph* **6**, 13 (1988).
19. M. Lewis, D. C. Rees, *Science* **230**, 1163 (1985).
20. P. Goodman. (University of Nevada, Renvo, NV, 1996).
21. T. J. Boggon, W. S. Shan, S. Santagata, S. C. Myers, L. Shapiro, *Science* **286**, 2119 (1999); B. Honig, A. Nicholls, *Science* **268**, 1144 (1995).
22. M. L. Sinnott, *Chem Rev* **90**, 1171 (1990).
23. T. N. Petersen, S. Kauppinen, S. Larsen, *Structure* **5**, 533 (1997).
24. R. Landgraf, I. Xenarios, D. Eisenberg, *J Mol Biol* **307**, 1487 (2001).
25. R. A. Laskowski, N. M. Luscombe, M. B. Swindells, J. M. Thornton, *Protein Sci* **5**, 2438 (1996).
26. B. Henrissat, G. Davies, A. Bairoch. *http://afmb.cnrs-mrs.fr/~pedro/CAZY/ghf.html* (2001).
27. R. Pickersgill, D. Smith, K. Worboys, J. Jenkins, *J Biol Chem* **273**, 24660 (1998).
28. F. K. Pettit, J. U. Bowie, *J Mol Biol* **285**, 1377 (1999).
29. D. L. Ollis *et al.*, *Protein Eng* **5**, 197 (1992).
30. B. Garner *et al.*, *J Biol Chem* **276**, 22200 (2001).

SUPPORT VECTOR MACHINE PREDICTION OF SIGNAL PEPTIDE CLEAVAGE SITE USING A NEW CLASS OF KERNELS FOR STRINGS

J.-P. VERT

Bioinformatics Center, Institute for Chemical Research, Kyoto University
Uji, Kyoto 611-0011, JAPAN
Jean-Philippe.Vert@mines.org

A new class of kernels for strings is introduced. These kernels can be used by any kernel-based data analysis method, including support vector machines (SVM). They are derived from probabilistic models to integrate biologically relevant information. We show how to compute the kernels corresponding to several classical probabilistic models, and illustrate their use by building a SVM for the problem of predicting the cleavage site of signal peptides from the amino-acid sequence of a protein. At a given rate of false positive this method retrieves up to 47% more true positives than the classical weight matrix method.

1 Introduction

Probabilistic models for strings play a central role in computational biology nowadays[1]: they include weight matrix for short signal sequences, Markov or hidden Markov models for DNA and protein sequences, stochastic context-free grammars for RNA or Bayesian graphical models for gene expression data[2]. Their strength comes from their ability to catch subtle regularities in the sequences and to quantify various biological features in a sound theoretical framework. They can integrate valuable biological knowledge (such as the possibility of mutations, insertions or deletions in hidden Markov models) difficult to handle otherwise.

A classical use of probabilistic models is to combine them with Bayes rule to classify sequences into one out of several competing categories. However experimental evidences in other areas as well as in computational biology[3] suggest that the resulting classifier can perform poorly compared to other discriminative methods, such as support vector machines (SVM).

There is therefore an incentive currently to adapt efficient discriminative methods to handle objects such as strings or graphs. For SVM this can be done by defining a kernel function $K(x, y)$ between any two objects x and y, which can be thought of as a dot product between objects. An question of interest is therefore: how to derive a kernel $K(x, y)$ from a probability density $p(x)$, in order to integrate in the kernel biological features caught by the probabilistic model?

Several methods have been proposed recently[4,5,6] with interesting experimental results[3]. This paper introduces a new way to derive a kernel from a probability distribution, motivated by the intuition that the "closeness" of two strings should increase when they share common substrings which appear with small probability under the probabilistic model.

For several widely-used probabilistic models on strings the kernel can be factorized and therefore efficiently computed. In this paper we explicitly show how to compute the kernel derived from either (i) independent probabilities or from (ii) Markov models. The resulting kernel can then be used by any kernel-based method, including SVM or non-linear principal component analysis[7].

As an application we derive an SVM algorithm to predict the cleavage site of signal peptide from the amino acid sequence of proteins. This algorithm is shown to significantly outperform the classical method based on a weight matrix.

2 SVM and kernels

SVM[8,9,10] are a family of algorithms remarkably efficient in many real-world applications. A growing interest for SVM in bioinformatics has emerged recently and resulted in powerful methods for various tasks[11,3,12].

A SVM basically learns how to classify objects $x \in \mathcal{X}$ into two classes $\{-1, +1\}$ from a set of labelled training examples $\{x_1, \ldots, x_m\}$. The resulting classifier is based on the decision function:

$$f(x) = \sum_{i=1}^{m} \lambda_i K(x_i, x), \tag{1}$$

where x is any new object to be classified, $K(.,.)$ is a so-called *kernel function* and the coefficients $\{\lambda_1, \ldots, \lambda_m\}$ are learned during training by solving a constrained optimization problem.

The kernel K can be considered as a dot product between objects, or more precisely between the images of the objects after a mapping to a high-dimensional Hilbert space. As a result it defines the metric properties of the space of objects, namely the "size" of each object and the "angle" between any two objects.

3 Kernels and probability distributions

Typical objects in computational biology are strings, for which many clever probabilistic models have been developed over the years to integrate biologically relevant information. In order to combine these clever models with efficient kernel-based methods it is important to develop general principles to derive kernels $K(x, y)$ from probability densities $p(x)$.

Obviously there is not a single way to do that[4,5,6]. In this paper we investigate particular kinds of kernels, which are probability densities on the product space $\mathcal{X} \times \mathcal{X}$, i.e., which satisfy:

$$\begin{cases} \forall (x, y) \in \mathcal{X} \times \mathcal{X}, \quad 0 \leq K(x, y) \leq 1, \\ \sum_{(x,y) \in \mathcal{X} \times \mathcal{X}} K(x, y) = 1. \end{cases} \tag{2}$$

Such kernels are called *P-kernels*[5]. An example of a P-kernel is the *product kernel*

$$K_{\text{prod}}(x, y) = p(x)p(y). \tag{3}$$

The decision function corresponding to the product kernel is simply $f(x) = a.p(x) + b$ with $a = \sum_{i=1}^{m} \lambda_i p(x_i)$, which shows that the resulting classifier simply classifies a new object x depending on whether $p(x)$ is above or below the threshold $-b/a$. The corresponding feature space is the 1-dimensional line, and each point is mapped to the number $p(x)$; two objects are "close" when their probabilities are close.

A second P-kernel example is the *diagonal kernel*:

$$K_{\text{diag}}(x, y) = p(x)\delta(x, y), \tag{4}$$

where $\delta(x, y)$ is 1 if $x = y$, 0 otherwise. The corresponding decision function tests whether the object has been seen in the training set, in which case it assigns it to the most probable class. In the corresponding feature space the set of objects $\mathcal{X}$ forms an orthogonal basis.

These two kernels are extreme and show how the choice of the kernel influences the metric properties of the space of objects. To derive an interesting kernel from a probability model, the product kernel is a good starting point because the resulting SVM classifier maps any object to a class based on the probability of that object, just like the classical Bayes classifier. In order to enhance the ability of the SVM to discriminate between two objects, it is natural to allow the "angle" between two objects to vary in order to reflect some notion of "closeness", which can not be handled by the diagonal nor the product kernel.

4 A P-kernel based on rare common subsets

Let us now introduce the main contribution of this paper, namely the definition of a new general kernel for discrete objects. We propose to consider two strings as "close" when they share rare common substrings. Here "rare" refers to the probability of the substring under the model p. As an example, if a particular sequence of amino acids is very rare in a training database, then it is natural to think that two proteins which share it could be particularly related to each other.

Let us introduce some notations to formalize this intuition. Let S be a finite set (usually $\{0, 1, \ldots, N\}$ for sequences), $\mathcal{A}$ a finite set called alphabet, and $X = (X_s)_{s \in S}$ a family of random variables defined on a probability space $(\Omega, \mathcal{F}, P)$ and indexed by the elements of S with values in $\mathcal{A}^S$. For any subset $T \subset S$ we note $X_T = (X_s)_{s \in T}$. For any subset $T \subset S$ and realization $x_T \in A^T$ we note $p_T(x_T) = P(X_T = x_T)$. If there is no ambiguity we simply write $p(x_T)$ for $p_T(x_T)$. We define similarly $p(x_T, y_U) = P(X_T = x_t, X_U = y_U)$ and $p(x_T \,|\, y_U) = P(X_T = x_T \,|\, X_U = y_U)$ for any two subsets $T \subset S$ and $U \subset S$ and realizations $x_T \in A^T$ and $y_U \in A^U$. Finally let $\mathcal{P}(S)$ be the power set of S (i.e., the set of subsets of S) and $\mathcal{V} \subset \mathcal{P}(S)$ be a particular set of subsets.

Using these notations we can define a kernel as follows:

Definition 1 *For any probability density p on $\mathcal{X}$ and any set of subsets $\mathcal{V} \subset \mathcal{P}(S)$ we define the $(p,\mathcal{V})$-common subset kernel $K_{p,\mathcal{V}}$ by the formula:*

$$K_{p,\mathcal{V}}(x, y) = \frac{p(x)p(y)}{|\mathcal{V}|} \sum_{T \in \mathcal{V}} \frac{\delta(x_T, y_T)}{p(x_T)}, \tag{5}$$

for any two realizations $(x, y) \in A^{2S}$, where $\delta(x_T, y_T)$ is 1 if $x_T = y_T$, 0 otherwise.

The main properties of this function are summarized here:

Proposition 1 *1. For any probability density p on $\mathcal{X}$ and set of subsets $\mathcal{V} \subset \mathcal{P}(S)$ the function $K_{p,\mathcal{V}}$ is a valid P-kernel on $\mathcal{X} \times \mathcal{X}$.*

2. When $\mathcal{V}$ only contains the empty set $\emptyset$, we have $K_{p,\{\emptyset\}} = K_{prod}$.

3. When $\mathcal{V}$ only contains the full set S, we have $K_{p,\{S\}} = K_{diag}$.

4. For a general set of subsets $\mathcal{V} \subset \mathcal{P}(S)$ we have for any $(x, y) \in A^{2S}$:

$$K_{p,\mathcal{V}}(x, y) = \frac{1}{|\mathcal{V}|} \sum_{T \in \mathcal{V}} \sum_{z_T \in A^T} p(z_T) p(x|z_T) p(y|z_T). \tag{6}$$

This proposition, whose proof can be found in the Appendix, shows that the kernel $K_{p,\mathcal{V}}$ interpolates between the diagonal kernel and the product kernel. Equation (5) shows that correlations are introduced between strings through their common substrings indexed by $\mathcal{V}$. The contribution of a common substring is inversely proportional to its probability, so the rarer a common substring the more it increases the similarity between the strings.

For a general density p and a general set $\mathcal{V}$ there is usually no way of computing $K(p.\mathcal{V})$ without computing the $|\mathcal{V}|$ terms in the sum defining the kernel. This might be prohibitive as soon as the set $\mathcal{V}$ becomes large, which happens when one considers for instance the set $\mathcal{V} = \mathcal{P}(S)$ with size $2^{|S|}$. In the next two sections we provide two examples where the kernel can be factorized and computed in linear time with respect to $|S|$. All proofs can be found in the Appendix.

5 Independent variables

In this section we compute the kernel derived from a product probability density, corresponding to modeling the variables as independent. Examples of such models include many probabilistic profiles for signal sequences or transcription factors binding sites in DNA. The corresponding kernels can be computed as follows:

Proposition 2 *Let $\{p_i, i \in S\}$ be a family of probability densities on A, and let p be the product distribution on A^S, i.e.,*

$$\forall x \in A^S, \quad p(x) = \prod_{i \in S} p_i(x_i). \tag{7}$$

Then the kernel $K_{p,\mathcal{V}}$ derived from p when $\mathcal{V} = \mathcal{P}(S)$ is the set of all subsets of S can be computed in linear time with respect to $|S|$ by:

$$K_{p,\mathcal{V}}(x, y) = \frac{1}{2^{|S|}} \prod_{i \in S} \phi_i(x_i, s_i), \tag{8}$$

with:

$$\phi_i(x_i, y_i) = \begin{cases} p_i(x_i) + p_i(x_i)^2 & \text{if } x_i = y_i, \\ p_i(x_i)p_i(y_i) & \text{if } x_i \neq y_i. \end{cases} \tag{9}$$

6 Markov chain and common blocks

In this section we suppose that $S = \{0, \ldots, N\}$ and that the density p is first-order Markovian, i.e.:

$$\forall x \in A^S, \quad p(x) = p_0(x_0) \prod_{i=1}^{N} p_i(x_i \,|\, x_{i-1}), \tag{10}$$

for a density p_0 on A and a set of conditional densities $p_i(x_i \,|\, x_{i-1})$ for $i = 1, \ldots, N$. The kernel resulting from such a Markov distribution is not easily computed for a general set of subsets $\mathcal{V}$ because the computation of $p(x_T)$ can be tricky for a general subset $T \subset S$. However it is possible to get a factorized expression of the kernel if we restrain the set $\mathcal{V}$ to be the set of integer intervals:

$$\mathcal{V} = \{[k, l] : 0 \leq k \leq l \leq N\} \cup \{\emptyset\}. \tag{11}$$

Observe that two realizations x and y have the same value on an interval $T = [k, l]$ (i.e., $x_T = y_T$) if and only if they share the common block $x_k \ldots x_l = y_k \ldots y_l$.

The corresponding kernel can be computed as follows:

Proposition 3 *For a Markov probability density as defined by Eq. (10) and for the set of integer intervals $\mathcal{V} = \{[k, l] : 0 \leq k \leq l \leq N\} \cup \{\emptyset\}$, the kernel between two strings $(x, y) \in A^{2S}$ can be computed in linear time with respect to N as:*

$$K_{p,\mathcal{V}}(x, y) = \phi_0(N) + \phi_1(N) + \phi_2(N), \tag{12}$$

where ϕ_0, ϕ_1 and ϕ_2 are defined recursively by:

$$\begin{cases} \phi_0(0) &= p_0(x_0)p_0(y_0), \\ \phi_1(0) &= p_0(x_0)\delta(x_0, y_0), \\ \phi_2(0) &= 0 \end{cases} \tag{13}$$

and for $i = 1, \ldots, N$:

$$\begin{cases} \phi_0(i) = p_i(x_i \,|\, x_{i-1})p_i(y_i \,|\, y_{i-1}) \times \phi_0(i-1), \\ \phi_1(i) = p_i(x_i \,|\, x_{i-1})\delta(x_i, y_i) \times \left[\phi_1(i-1) + \frac{p_i(y_i \,|\, y_{i-1})}{p_i(x_i)}\phi_0(i-1)\right], \\ \phi_2(i) = p_i(x_i \,|\, x_{i-1})p_i(y_i \,|\, y_{i-1}) \times [\phi_1(i-1) + \phi_2(i-1)]. \end{cases} \tag{14}$$

Remark 1 *The term $p_i(x_i)$ which appears in the recursive definition of $\phi_1(i)$ can itself be computed using the classical recursive algorithm:*

$$p_i(x_i) = \sum_{x_{i-1} \in A} p_{i-1}(x_{i-1})p_i(x_i \,|\, x_{i-1}). \tag{15}$$

7 Experiment : cleavage site prediction of signal peptides

As an application to test the performance of the kernels introduced in this paper we consider the problem of predicting the cleavage site of protein signal sequences. These sequences, also called signal peptides, play a central role in the process of directing each newly created polypeptide to a particular destination in the organism[13]. They comprise the amino terminus of the amino-acid chain and are cleaved off while the protein is translocated through the membrane.

The identification of signal peptides and their cleavage site is of interest to the development of new effective drugs. The rapid increase in the number of available protein sequences in databases requires the use of effective prediction tools to reduce the time and cost of experimental verifications.

A simple weight matrix method[14] is known to be quite efficient to recognize cleavage sites. Indeed many cleavage sites are strongly characterized by a set of simple rules which are quantified by the weight matrix methods, e.g., the residues at positions -3 and -1 relative to the cleavage site are usually small and neutral. Computing scores from a weight matrix method is equivalent to computing the probability of a sequence under an independent model[1], so it is possible to use the kernel presented in Sec. 5 instead of the classical scoring function in order to recognize cleavage site.

In order to evaluate the performance of the kernels defined in this paper we focus on the following problem: given a window of amino acids, predict whether cleavage will occur at a given position of the window. In our experiments we chose to predict a cleavage site from the observation of 8 amino acids before the site (i.e., which should belong to the signal part) and 2 amino acids after the site (i.e., which should belong to the mature part of the protein). Hence a basic window is a sequence $x = x_{-8}x_{-7}\dots x_{-1}x_1x_2$ of length 10.

We experimented on the database of proteins used by Nielsen et al.[15a]. We used a total number of 1418 non-redundant sequences (1011 from eukaryotes, 266 from Gram-negative prokaryotes and 141 from Gram-positive prokaryotes) made of the signal peptide and the first 30 amino acids of the mature protein. We extracted all possible amino acid windows of size 10, resulting in 66,634 windows, divided into 1418 "positive" windows (i.e., with a cleavage site between the amino acids x_{-1} and x_{+1}) and 65,216 "negative" windows.

We randomly split this database into a training set (80 % of the windows) and a test set (20%). From the training set we built:

- a weight matrix as $w_i(x_i) = \log p_i^+(x_i) - \log p_i^{\text{total}}(x_i)$, where $p_i^+(x_i)$ is the probability that amino acid x_i occurs at position i estimated from the positive training set (using pseudocounts[1]), and p_i^{total} is the probability that amino acid x_i occurs at position i estimated from the total training set (usually referred to as the background model);
- a SVM classifier based on the product probability $p^+ = \prod_i p_i^+$ and

[a] Available at ftp://virus.cbs.dtu.dk/pub/signalp

trained on the training set. We used the public domain implementation of mySVM[1b] where we implemented a user-defined kernel as presented in Sec. 5. All parameters and files necessary to reproduce this experiment can be downloaded from the author's web page[c].

This results in two competing classification functions for amino acid windows. The first one is the score function:

$$s(x) = \sum_{i=-8}^{2} w_i(x_i), \tag{16}$$

and the second one is the classification function used by the SVM:

$$f(x) = \sum_{x^{(j)} \text{ in the training set}} \lambda^{(j)} K_{p^+,\nu}(x^{(j)}, x). \tag{17}$$

At a given threshold δ, each of these functions classifies a new example as positive or negative depending on whether the function is above or below the threshold. By varying the threshold and classifying the examples in the test set, we can build a curve of true positive versus false positive for each function, and compare them.

The curves (averaged over a number of random training/test set splits) are shown on Fig. 1 and Fig. 2. The curve of the weight matrix method shows that about 44% of the sequences can be very "easily" recognized by that method, because they exhibit strong characteristics of typical cleavages sites; for the remaining sequences, the curve increases smoothly. The curve of the SVM method, on the other hand, is above the first curve, and increases smoothly from "easy" examples to "hard" examples. The difference between the two curves is particularly important for small false positive ratios, which is the most important part of the curve for concrete application. As an example, if one is ready to have 3% of false positive, then the weight matrix method would retrieve on average 46% of true positives, while the SVM method would retrieve 68% of true positive. This corresponds to an increase of 47% in terms of true positive retrieval, and illustrates the discriminative power of SVM compared to simple score functions

8 Conclusion and future work

We introduced a new class of kernels for strings which we think can be of interest for many applications in computational biology, and showed on the

[b]Available from `http://www-ai.cs.uni-dortmund.de/SOFTWARE/MYSVM/`
[c]`http://web.kuicr.kyoto-u.ac.jp/~vert/`

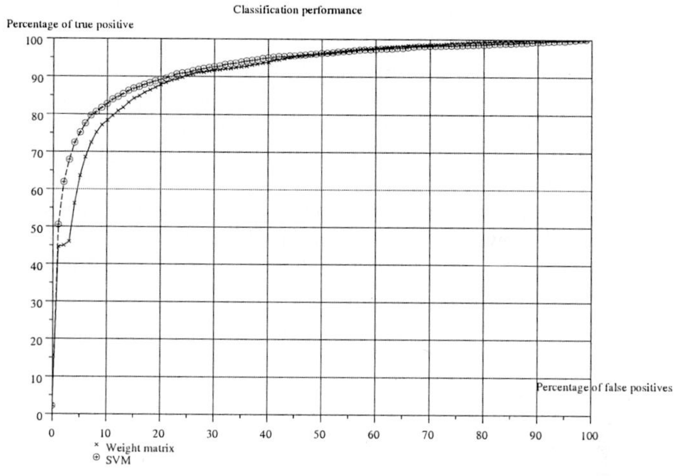

Figure 1: Classification performance of the weight matrix method and the SVM method

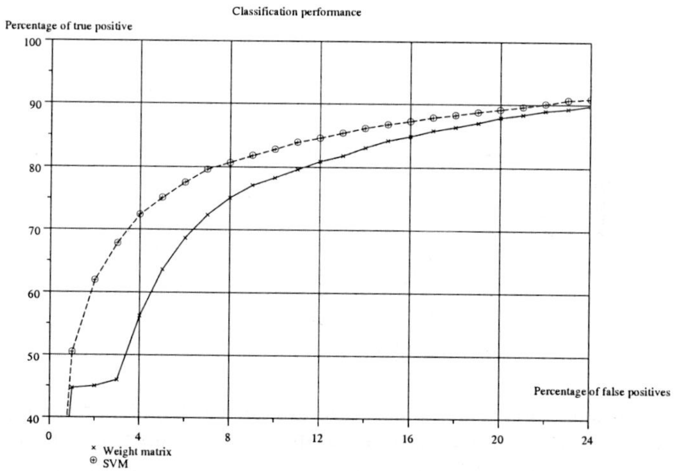

Figure 2: Classification performance for small false positive rates

example of signal sequence cleavage site prediction how a SVM using this kernel significantly outperforms a classical weight matrix method, providing a new evidence that SVM will play a more and more important role in the coming years in computational biology.

Much research remains to be done to fully exploit the capacities of kernel-based methods in bioinformatics. On the theoretical point of view, the development of new kernels for specific objects (strings, graphs...) and specific applications is an important research topic today. The particular class of P-kernels should be more deeply investigated and might give birth to interesting theoretical links between probability theory and machine learning. On the practical point of view, kernel-based methods could be tested on new tasks; moreover the class of kernel-based methods is not limited to SVM, but also includes algorithms such as non-linear principal component analysis[7], which might be of great use for data mining of biological databases.

Appendix : proofs

Proof of Proposition 1

The points 2 and 3 follow directly from the definition of the kernel in Eq. (5). To prove point 4 observe that for any subset $T \subset S$ the following holds for any $x \in A^S$ and $z_T \in A^T$:

$$p(x|z_T) = \frac{p(x)}{p(z_T)}\delta(x_T, z_T). \tag{18}$$

As a result we can use the definition of the kernel in Eq. (5) to compute, for any $(x, y) \in A^{2S}$ and $\mathcal{V} \subset \mathcal{P}(S)$:

$$\begin{aligned} K_{p,\mathcal{V}}(x,y) &= \frac{p(x)p(y)}{|\mathcal{V}|}\sum_{T\in\mathcal{V}}\frac{\delta(x_T,y_T)}{p(x_T)} \\ &= \frac{p(x)p(y)}{|\mathcal{V}|}\sum_{T\in\mathcal{V}}\sum_{z_T\in A^T}\frac{\delta(x_T,z_T)\delta(y_T,z_T)}{p(z_T)} \\ &= \frac{1}{|\mathcal{V}|}\sum_{T\in\mathcal{V}}\sum_{z_T\in A^T} p(z_T)p(x|z_T)p(y|z_T), \end{aligned} \tag{19}$$

which concludes the proof of point 4. Summing this expression for all possible x and y easily shows that $K_{p,\mathcal{V}}$ is a probability density, i.e., $\sum_{(x,y)} K_{p,\mathcal{V}}(x,y) = 1$. Hence it is a conditionally independent probability density (as defined in[6]), and it is therefore a valid P-kernel (from the main result of[6]).

Proof of Proposition 2

For a product density $p(x) = \prod_{i \in S} p_i(x_i)$, the following holds for any subset $T \subset S$:

$$\forall x_T \in A^T, \quad p(x_T) = \prod_{i \in T} p_i(x_i). \tag{20}$$

Therefore, we can compute for any $(x, y) \in A^{2S}$:

$$\frac{p(x)p(y)\delta(x_T, y_T)}{p(x_T)} = \prod_{i \in T} p(x_i)\delta(x_i, y_i) \times \prod_{i \notin T} p(x_i)p(y_i). \tag{21}$$

Using Eq. (5) and the fact that $|\mathcal{V}| = 2^{|S|}$ we can therefore compute:

$$\begin{aligned} K(x, y) &= \frac{1}{2^{|S|}} \sum_{T \subset S} \frac{p(x)p(y)\delta(x_T, y_Y)}{p(x_T)} \\ &= \frac{1}{2^{|S|}} \sum_{T \subset S} \left\{ \prod_{i \in T} p(x_i)\delta(x_i, y_i) \times \prod_{i \notin T} p(x_i)p(y_i) \right\} \\ &= \frac{1}{2^{|S|}} \prod_{i \in S} \Big\{ p(x_i)\delta(x_i, y_i) + p(x_i)p(y_i) \Big\}. \end{aligned} \tag{22}$$

Proof of Proposition 3

First observe that by mapping any interval $[k, l]$ (with $0 \le k \le l \le N$) into the sequence $(s_1, \ldots, s_N)$ defined by $s_i = 0$ if $i < k$, $s_i = 1$ if $k \le i \le l$ and $s_i = 2$ if $i > l$, and by mapping the empty set $\emptyset$ to the constant sequence $(s_1, \ldots, s_N) = (0, \ldots, 0)$, we define a bijection between $\mathcal{V}$ and the set of sequences $\mathcal{S}$ with values in $0, 1, 2$, starting value $s_0 \in \{0, 1\}$ and increments $s_{i+1} \in \{s_i, s_i + 1\}$ for $i = 1, \ldots, N$.

Now, for any interval $T = [k, l]$ and corresponding sequence $(s_0 \ldots s_N)$, the Markov property of p yields the following equality for any realization $x_T \in A^T$:

$$p(x_T) = p_k(x_k) \prod_{i=k+1}^{l} p_i(x_i \,|\, x_{i-1}). \tag{23}$$

As a result it is easy to check the following relation:

$$\forall (x, y, T) \in A \times A \times \mathcal{V}, \quad \frac{p(x)p(y)\delta(x_T, y_T)}{p(x_T)} = g_0(s_0) \prod_{i=1}^{N} g_i(s_{i-1}, s_i), \tag{24}$$

where g_i is defined by $g_0(0) = p_0(x_0)p_0(y_0)$, $g_0(1) = p_0(x_0)\delta(x_0, y_0)$ and for $i = 0, \dots, N$:

$$\begin{cases} g_i(0,0) = g_i(1,2) = g_i(2,2) = p_i(x_i|x_{i-1})p_i(y_i|y_{i-1}), \\ g_i(0,1) = \frac{p_i(x_i|x_{i-1})p_i(y_i|y_{i-1})\delta(x_i,y_i)}{p_i(x_i)}, \\ g_i(1,1) = p_i(x_i \mid x_{i-1})\delta(x_i, y_i), \end{cases} \tag{25}$$

By definition of the kernel in Eq.(5) and using Eq. (24) we therefore get:

$$K_{p,\mathcal{V}}(x_S, y_S) = \sum_{(s_1 \dots s_N) \in \mathcal{S}} \left\{ g_0(s_0) \prod_{i=1}^{N} g_i(s_{i-1}, s_i) \right\}. \tag{26}$$

The set of equations in Proposition 3 is now the classical forward algorithm corresponding to the dynamic programming computation of this sum.

References

1. R. Durbin *et al, Biological sequence analysis : Probabilistic models of proteins and nucleic acids* (Cambridge University Press, 1998).
2. N. Friedman *et al., Journal of Computational Biology,* **7**, 601 (2000).
3. T. Jaakkola *et al., Journal of Computational Biology,* **7**, 95 (2000).
4. T. Jaakkola and D. Haussler in *Advances in Neural Information Processing Systems* **11**, 1998.
5. D. Haussler, *Technical report UCSC-CRL-99-10* (1999).
6. C. Watkins, *Technical report CSD-TR-98-11* (1999).
7. B. Schölkopf *et al.*, in *Advances in kernel methods: support vector learning*, 327 (The MIT Press, 1999).
8. V. Vapnik, *Statistical learning theory* (Wiley, 1998)
9. N. Christianini and J. Shawe-Taylor, *An introduction to Support Vector Machines and other kernel-based learning methods* (Cambridge University Press, 2000).
10. C. Burges,*Data Mining and Knowledge Discovery,* **2**, 121 (1998).
11. M.P.S. Brown *et al., Proc. Natl. Acad. Sci. USA*, **97**, 262 (2000).
12. P. Pavlidis *et al, Proceedings of the Pacific Symposium on Biocomputing 2001*, 151 (2001).
13. L.M. Gierarch,*Biochemistry,* **28**, 923 (1989).
14. G. von Heijne, *Nucleic Acids Res.*, **14**, 4683 (1986).
15. H. Nielsen *et al., Protein Eng.* **10**, 1 (1997).
16. S. Rüping, *mySVM - Manual* (2000).

CONSTRAINT-BASED HYDROPHOBIC CORE CONSTRUCTION FOR PROTEIN STRUCTURE PREDICTION IN THE FACE-CENTERED-CUBIC LATTICE

SEBASTIAN WILL[a]
Institut für Informatik, Ludwig-Maximilians-Universität München, Oettingenstraße 67, D-80538 München, Germany
wills@informatik.uni-muenchen.de

We present an algorithm for exact protein structure prediction in the FCC-HP-model. This model is a lattice protein model on the face-centered-cubic lattice that models the main force of protein folding, namely the hydrophobic force. The structure prediction for this model can be based on the construction of hydrophobic cores. The main focus of the paper is on an algorithm for constructing maximally and submaximally compact hydrophobic cores of a given size. This algorithm treats core construction as a constraint satisfaction problem (CSP), and the paper describes its constraint model. The algorithm employs symmetry excluding constraint-based search [6] and relies heavily on good upper bounds on the number of contacts. Here, we use and strengthen upper bounds presented earlier.[8] The resulting structure prediction algorithm (including previous work [8,7]) handles sequences of sizes in the range of real proteins fast, i.e. we predict a first structure often within a few minutes. The algorithm is the first exact one for the FCC, besides full enumeration which is impracticable for chain lengths greater than about 15. We tested the algorithm succesfully up to sequence length of 160, which is far beyond the capabilities even of previous heuristic approaches.

1 Introduction

Protein structure prediction is one of the most important unsolved problems of computational biology. It can be specified as follows: Given a protein by its sequence of amino acids (more generally monomers), what is its native structure? NP-completeness of the problem has been proven for many different models, among them lattice and off-lattice models.[10,12]

To tackle structure prediction and related problems, simplified models have been introduced. For this aim, they are used in hierarchical approaches for protein folding.[25] Here, see also the meeting review of CASP3,[17] where some groups have used lattice models. Furthermore, simplified models are a major tool for investigating general properties of protein folding.

Most important are the so-called lattice models, where protein structure is modeled as a self-avoiding walk on a lattice. In the literature, many different lattice models (each specified by a lattice and an energy function) have been

[a]Supported by the PhD programme GKLI of the "Deutsche Forschungsgemeinschaft".

used. It was shown how such models can be used for predicting the native structure or for investigating principles of protein folding.[24,1,15,23,16,2,19,25]

Of course, the question arises which lattice and energy functions should be preferred. There are two aspects that have to be evaluated when choosing a model: 1) the accuracy of the lattice in approximating real protein conformations (aka structures), and the ability of the energy function to discriminate native from non-native conformations, and 2) the availability and quality of search algorithms for finding minimal (or nearly minimal) energy conformations. Obviously, the two aspects are somewhat conflicting. While the first aspect is well-investigated in the literature[20,13] the second aspect is neglected.

In this paper, we follow the proposal of Agarwala et.al.[3] to use a lattice model with a simple energy function, namely the HP (hydrophobic-polar) model (which has been introduced by Lau and Dill[18] using cubic lattice), but on a better suited lattice (namely the face-centered cubic one). The resulting model is called the *FCC-HP-model.* In the HP-model, the 20 letter alphabet of amino acids is reduced to a two letter alphabet {H, P}. H represents *hydrophobic* amino acids, whereas P represent *polar* or hydrophilic amino acids. The energy function for the HP-model simply states that the energy contribution of a contact between two monomers is -1 if both are H-monomers, and 0 otherwise. Two monomers form a *contact* in some specific conformation if they occupy positions of minimal distance. A conformation with *minimal energy* (called *native conformation*) is just a conformation with the maximal number of contacts between H-monomers. Even for the HP-model, the structure prediction problem was shown as NP-complete.[10,12]

There are two reasons for using the FCC-HP-Model: First, the FCC can model real protein conformations with good quality (up to coordinate root mean square deviation below 2 Å).[20] Second, the HP-model models the aspect of hydrophobicity. Its energy function enforces compactification due to the hydrophobic force, while polar residues and solvent molecules are not explicitly regarded. Hydrophobicity is very important, since one assumes that the hydrophobic effect determines the overall configuration of a protein.[18,13]

Once a search algorithm for minimal energy conformations is established for the FCC-HP-model, one can employ it as a filter step in an hierarchical approach. This way, one can improve the energy function to achieve better biological relevance and go on to resemble amino acid positions more accurately.

Related Work In this paper, we describe a successful application of constraint-programming for finding native conformations in the FCC-HP-model. There, the situation as given in the literature was not very promising. Although the importance of the FCC-HP-model is widely known, exact algorithms for finding native conformations were known only for cubic lattice models. Even for the cubic lattice, there are only three exact algorithms known [26,1,9] that are able to enumerate minimal (or nearly minimal) energy conformations. However, the ability of this lattice to approximate real protein conformations is poor. Especially the parity problem was pointed out as drawback of the cubic lattice.[3] This problem is that every two monomers with chain positions of equal parity cannot form a contact.

So far, besides heuristic approaches (e.g., the hydrophobic zipper,[14] the genetic algorithm by Unger and Moult,[22] and the chain growth algorithm by Bornberg-Bauer [11]), there is only one approximation algorithm [3] for the FCC. It finds conformations whose number of contacts is guaranteed to be 60% of the number of contacts of the native conformation. The situation was even worse, since the main ingredient needed for an exact method was missing, namely bounds on the number of contacts between hydrophobic monomers given some partial information about the conformation. This changed with recent work,[5,8] where such a bound was introduced and applied for finding maximally compact hydrophobic cores. Given a conformation of an HP-sequence, the *hydrophobic core* of this sequence is the set of all points occupied by H-monomers. A hydrophobic core of n points is *maximally compact* if no packing of n points in the FCC has more contacts. Hydrophobic cores were used for structure prediction in the HP-model and HPNX-model on the cubic lattice before.[26,9]

Contribution and Use for Structure Prediction The goal of structure prediction in the FCC-HP-model can be achieved via the construction of hydrophobic cores (i.e. point sets) in the FCC. For predicting optimal structures of a sequence s, we will proceed as follows. First, search for the optimal number of contacts in any core of size $|s|$. Then, construct the set of all cores of size $|s|$ with optimal number of contacts. Try to thread the sequence s to all cores in this set. Now, possibly we cannot thread s to any of the cores. In this case, we iterate the process going on to suboptimal numbers of contacts.

The problem of threading a sequence s to a given core C (a set of lattice points) is finding a tuple $(\vec{p}_i)_{1\leq i\leq |s|}$ of lattice points, called a *structure* for s, subject to the constraints $\forall 1 \leq i < |s| : \vec{p}_i$ and $\vec{p}_{i+1}$ have minimal distance, $\forall 1 \leq i < j \leq |s| : \vec{p}_i \neq \vec{p}_j$, and $\forall 1 \leq i \leq |s| : s_i = H \rightarrow \vec{p}_i \in C$. As the problem is strongly constrained, we can solve it by constrained search.

The main contribution of this paper is an algorithm for constructing the maximally (and specified submaximally) compact hydrophobic cores of a given size in the FCC. A key idea of our method is to slice a core into layers orthogonal to the coordinate axis in every dimension. In previous work, upper bounds were given on the number of contacts for sequences of certain layer parameters.

As a result of this, only a very restricted number of layer parameter sequences has to be considered in a search for compact cores. Thus, the missing step is to search those candidate layer parameter sequences, here done by constraint-based search. We give a symmetric constraint model for the problem, which on the one hand permits to use the precomputed candidate layer parameter sequences for the layers in each of the three dimensions and on the other hand enables us to apply general symmetry exclusion.[6] A constraint-based algorithm is presented, suited for implementation in the constraint language Oz.[21]

2 Preliminaries and Basic Definitions

A *lattice* L is the minimal set of points that contains so called *generating vectors* $\vec{v}_1, \ldots, \vec{v}_n$ and where $\forall \vec{u}, \vec{v} \in L$, both $\vec{u} + \vec{v} \in L$ and $\vec{u} - \vec{v} \in L$ holds. The *face-centered cubic lattice* (FCC) is defined as the point set $D_3 = \{ \left(\begin{smallmatrix} x \\ y \\ z \end{smallmatrix}\right) \mid \left(\begin{smallmatrix} x \\ y \\ z \end{smallmatrix}\right) \in \mathbb{Z}^3 \text{ and } x + y + z \text{ is even}\}$.

An (FCC-)core f is just a set of points in D_3. Define MV as the set of vectors $\vec{p} \in D_3$ with length $\sqrt{2}$ (which is the minimal euclidian distance of two lattice points). That is, $MV = \left\{ \left(\begin{smallmatrix} 0 \\ \pm 1 \\ \pm 1 \end{smallmatrix}\right), \left(\begin{smallmatrix} \pm 1 \\ 0 \\ \pm 1 \end{smallmatrix}\right), \left(\begin{smallmatrix} \pm 1 \\ \pm 1 \\ 0 \end{smallmatrix}\right) \right\}$. The *number of contacts* contacts(f) of a core f is defined as $\frac{1}{2} \left| \{ (p, q) \in f^2 \mid \vec{p} - \vec{q} \in \text{MV} \} \right|$.

A *caveat in a point set* f is a k-tuple of points $(\vec{p}_1, \ldots, \vec{p}_k)$ such that $\exists \vec{v} \in \text{MV}\, \forall 1 \leq j < k : ((\vec{p}_{j+1} - \vec{p}_j) = \vec{v})$, $\{\vec{p}_1, \vec{p}_k\} \in f$ and $\forall 1 < j < k : \vec{p}_j \notin f$. Implicitly, we only handle cores without caveats. This restriction was also used in analogous cases by others[26] and is acceptable, since the presented ideas can be extended to handle caveats as well.

We define notations for certain point sets of $\mathbb{R}^3$, namely *lines* and *planes*. For vectors $\vec{a}, \vec{u} \in \mathbb{Z}^3$, let $\text{LN}(\vec{a}, \vec{u})$ denote the set $\{\vec{p} \in \mathbb{R}^3 \mid \exists \lambda \in \mathbb{R} : \vec{p} = \vec{a} + \lambda \vec{u}\}$ and for $\xi \in \{x, y, z\}$ define $\text{PL}(\xi, c)$ as the set $\{\vec{p} \in \mathbb{R}^3 \mid \xi = c\}$. For $a \in D_3$, we are interested in so-called *lattice lines* $\text{LN}(\vec{a}, \vec{u})$, where $\vec{u} \in \text{MV}$, and further so-called *non-lattice lines* $\text{LN}(\vec{a}, \vec{u})$, where $\vec{u} \in \mathbb{Z}^3$ has length 2.

For a core f and $c \in \mathbb{Z}$, the set $f_{\xi=c} = f \cap \text{PL}(\xi, c)$ is called the *ξ-layer of f in plane $\xi = c$*. For an x-layer f in plane $x = c$, define olines(f) as the tuple (a, b), where $a = \left| \left\{ l \;\middle|\; \exists \vec{a} \in D_3 : l = \text{LN}(\vec{a}, \left(\begin{smallmatrix} 0 \\ 1 \\ 1 \end{smallmatrix}\right)) \wedge l \cap f \neq \emptyset \right\} \right|$ and analogously $b = \left| \left\{ l \;\middle|\; \exists \vec{a} \in D_3 : l = \text{LN}(\vec{a}, \left(\begin{smallmatrix} 0 \\ -1 \\ 1 \end{smallmatrix}\right)) \wedge l \cap f \neq \emptyset \right\} \right|$. Define olines$(f)$ analogously for y-layers and z-layers f. Let min be the minimal number such that $\text{PL}(\xi, \min) \cap f \neq \emptyset$ and m the maximal number such that $\text{PL}(\xi, m + \min - 1) \cap f \neq \emptyset$. Define the *layer parameter sequence of a core f in dimension $\xi \in \{x, y, z\}$* as the finite sequence $\mathcal{L}_\xi(f) = (n_i, a_i, b_i)_{1 \leq i \leq m}$, such that $\forall 1 \leq i \leq m : |f_{\xi=\min+i-1}| = n_i$ and $\text{olines}(f_{\xi=\min+i-1}) = (a_i, b_i)$.

Due to previous work,[8] we are able to compute for a layer parameter sequence $\mathcal{L}$ the upper bound on the number of contacts in the class of cores f with $\mathcal{L}_\xi(f) = \mathcal{L}$ for $\xi \in \{x, y, z\}$. This bound is denoted by $\mathrm{BMC}(\mathcal{L})$. Any core f with $\mathcal{L}_\xi(f) = \mathcal{L}$ has the same size determined by $\mathcal{L}$. Denote this size by $\mathrm{size}(\mathcal{L})$. Moreover, we are able to compute the sets of all layer parameter sequences, which have the same upper bound for a fixed core size. Define these sets by $S(n, \mathrm{bnd}) := \left\{ \mathcal{L} \mid \mathrm{size}(\mathcal{L}) = n \wedge \mathrm{BMC}(\mathcal{L}) = \mathrm{bnd} \right\}$.

3 Problem Specification

For core construction, it remains to solve the following *core construction problem*. Given a core size n and the sets $S(n, \mathrm{bnd})$, construct a set $\mathrm{Cores}(n, \mathrm{con})$ of the cores of size n with at least con contacts modulo geometrical symmetry. First, note that it does not suffice to construct only maximally compact cores, if we want to use the cores for protein structure prediction, since there may be sequences which do not fit to all of the optimal cores. Second, by *modulo geometrical symmetry* we express that $\mathrm{Cores}(n, \mathrm{con})$ contains only one representative of every equivalence class due to translations, rotations and reflections. To abstract from at least translation symmetries is essential, since otherwise the set $\mathrm{Cores}(n, \mathrm{con})$ is trivially infinite.

Actually, we are going to solve the following (more general) problem. Given a set of layer parameter sequences S for cores of size n and a number of contacts con, compute the set of all cores f with at least con contacts which have layer parameter sequences from the set S in every dimension, i.e. compute

$$\mathrm{Cores}(n, S, \mathrm{con}) = \left\{ f \subset D_3 \;\middle|\; \begin{array}{c} \mathrm{contacts}(f) \geq \mathrm{con} \\ \wedge\ \forall \xi \in \{x, y, z\} : \mathcal{L}_\xi(f) \in S \end{array} \right\}$$

modulo geometrical symmetry.

As an abbreviation define $S_{con}(n, \mathrm{con}) = \bigcup_{\mathrm{bnd} \geq \mathrm{con}} S(n, \mathrm{bnd})$. Due to the equality $\mathrm{Cores}(n, \mathrm{con}) = \mathrm{Cores}(n, S_{con}(n, \mathrm{con}), \mathrm{con})$, the general problem solves the former problem of core construction. A difficulty remains with $\mathrm{Cores}(n, \mathrm{con})$, since $S_{con}(n, \mathrm{con})$ is not necessarily finite. However, in general there are finitely many cores $\mathrm{Cores}(n, S, \mathrm{con})$ only for finite input. Unfortunately, for sufficiently low numbers bnd, there may be layer parameter sequences $\mathcal{L} \in S(n, \mathrm{bnd})$, such that there is an $\min_j(n_j \neq 0) < i < max_j(n_j \neq 0)$ where $n_i = 0$. We say the sequence $\mathcal{L}$ has a *gap*. In this case, there are infinitely many layer parameter sequences expanding the gap in the sequence, which have the same bound bnd. Cores with gaps consist of separated sub-cores, instead of one connected set of points. Note that for structure prediction, this case occurs very rarely.

Nevertheless, we can cope with this problem, by a certain kind of symmetry exclusion. To generate cores in $\text{Cores}(n, S_{con}(n, \text{con}), \text{con})$, we will first consider only the set S of layer parameter sequences in $S_{con}(n, \text{con})$ without gaps. This guarantees that the set $\text{Cores}(n, S, \text{con})$ is finite. It is now possible to find the cores $f \in \text{Cores}(n, S, \text{con})$, which can be split into non-empty sub-cores $f' = f \cap \{\vec{p} \mid p_\xi \geq c\}$ and $f'' = f \cap \{\vec{p} \mid p_\xi < c\}$ along a plane $\xi = c$, such that $\text{contacts}(f') + \text{contacts}(f'') \geq \text{con}$. Those cores can be used to generate an infinite number of elements of $\text{Cores}(n, S_{con}(n, \text{con}), \text{con})$ by translation of one of the sub-cores. To be complete this has to be done recursively. Finally, note that for structure prediction we can even in this case restrict ourselves to finite sets of cores due to the restriction introduced by the chain length. However, in most cases, where we are interested in optimal or only slightly suboptimal cores, we can easily conclude that there are no such cores.

4 Description of the Algorithm

We solve the problem of constructing the cores in $\text{Cores}(n, S, \text{con})$ given a set of layer parameter sequences S without gaps for cores of size n and a number of contacts con. Our algorithm follows the constrain-and-generate principle. By large, this approach is to state constraints on solution variables and then enumerate values of the variables by branching to generate the solutions. At each branching, insert in the left branch a constraint c and in the right branch $\neg c$ to split the search space. In constraint programming, the branching is done concurrently to propagation of the stated constraint to prune the search tree.

We introduce variables together with data structures to organize them and constraints to express dependencies on the variables. It is useful to introduce auxiliary variables (instead of only the solution variables) and necessary to introduce redundant constraints for efficiency. Finally, we apply symmetry excluding constrained search [6] for enumerating the cores in $\text{Cores}(n, S, \text{con})$.

The main idea of our approach is getting as much knowledge as possible on the distribution of points from the layer parameter sequences. Therefore, we model layers and so called lattice lines of the layers to express the constraints by the a and b parameters. Further, it is crucial to employ the dependencies between layers of different dimensions. To express those dependencies we have to model non-lattice lines of the layers. The number of contacts con, yields further constraints, which are non-redundant to the fórmer ones, since not every core satisfying the layer sequences has necessarily at least con contacts.

Variables All the variables are *finite domain variables (FD-variables)*, which means that their assigned values are restricted to values of finite integer do-

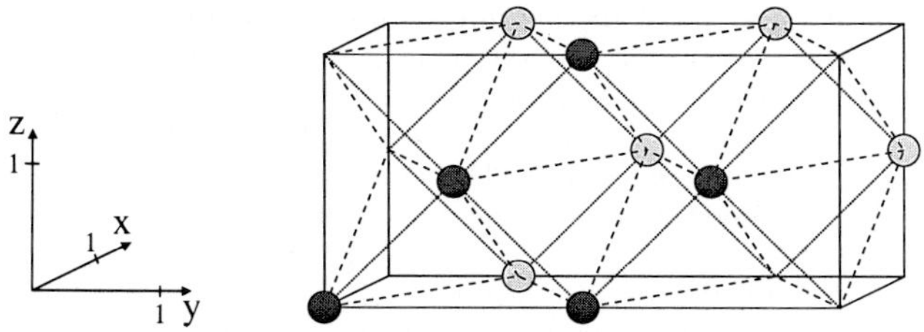

Figure 1: The cube for $m_x = 2$, $m_y = 5$, $m_z = 3$ and $\min_x + \min_y + \min_z$ even. The contacts within each x-layer are shown by dotted lines and the interlayer contacts between the two x-layers by dashed lines. The circles give an example core within the cube.

mains. Denote the number of non-empty layers in the dimension $\xi \in \{x, y, z\}$ by m_ξ. All points of the core will be placed in a $m_x \times m_y \times m_z$ *surrounding cube*. We can nearly fix the absolute coordinates of this cube to exclude translation symmetries. However, since D_3 contains only points of $\mathbb{Z}^3$ with even coordinate sum, the cube can only be fixed up to the minimal x, y, and z coordinate being one of $\{0, 1\}$. Store these coordinates in FD-variables $\min_x$, $\min_y$, and $\min_z$ respectively. Fix the surrounding cube to consist of the points $\mathrm{CB} = \{\min_x, \ldots, \min_x + m_x - 1\} \times \{\min_y, \ldots, \min_y + m_y - 1\} \times \{\min_z, \ldots, \min_z + m_z - 1\} \cap D_3$. Please see Figure 1 for an illustration.

For every point $\vec{p} \in \mathrm{CB}$, maintain a boolean FD-variable $\mathbf{pnt}(\vec{p}) \in \{0, 1\}$ that has value 1 iff the point $\vec{p}$ is element of the core. Let $\xi \in \{x, y, z\}$. For every layer $\xi = c$, where $\min_\xi \leq c \leq \min_x i + m_\xi - 1$, we have FD-variables $\mathbf{lay}(\xi, c).n$, $\mathbf{lay}(\xi, c).a$, and $\mathbf{lay}(\xi, c).b$ for the layer parameters.

Further, we have variables for all lattice and non-lattice lines within layers that intersect with the cube. For $\vec{v}$ in $\mathrm{LV} = \mathrm{MV} \cup \left\{ \begin{pmatrix} \pm 2 \\ 0 \\ 0 \end{pmatrix}, \begin{pmatrix} 0 \\ \pm 2 \\ 0 \end{pmatrix}, \begin{pmatrix} 0 \\ 0 \\ \pm 2 \end{pmatrix} \right\}$, there are FD-variables $\mathbf{ln}(\vec{a}, \vec{v})$, for every set $\mathrm{LN}(\vec{a}, \vec{v})$ which has a non-empty intersection with CB. We identify variables $\mathbf{ln}(\vec{a}, \vec{v})$ and $\mathbf{ln}(\vec{a}', \vec{v})$ if $\mathrm{LN}(\vec{a}, \vec{v}) = \mathrm{LN}(\vec{a}', \vec{v})$. $\mathbf{ln}(\vec{a}, \vec{v})$ is the number of occupied points in $\mathrm{LN}(\vec{a}, \vec{v}) \cap D_3$. Finally, we introduce variables $\mathbf{con}(\vec{p}, \vec{q}) \in \{0, 1\}$ for $\vec{p}, \vec{q} \in \mathrm{CB}$, such that $\vec{p} - \vec{q} \in \mathrm{MV}$.

Basic Constraints Before giving the constraints, we introduce a notation to express *reified constraints*. Let c be a constraint, fix a mapping δ, $\delta(c) \in \{0, 1\}$, such that $\delta(c) = 1$ iff c holds. The FD-variables are subject to the following constraints. First of all, we get

$$\sum_{\vec{p} \in \mathrm{CB}} \mathbf{pnt}(\vec{p}) = n \quad \text{and} \quad \sum_{\vec{p}, \vec{q} \in \mathrm{CB}\, \vec{p} - \vec{q} \in \mathrm{MV}} \mathbf{con}(\vec{p}, \vec{q}) \leq \mathrm{con}\,.$$

Any core must have one of the parameter layer sequences in each dimension $\xi \in \{x, y, z\}$. This is expressed by the (constructive) disjunction over all layer parameter sequences $\mathcal{L} = (n_i, a_i, b_i)_{1 \leq i \leq |\mathcal{L}|}$ in S of

$$\forall 1 \leq i \leq |\mathcal{L}| : \begin{array}{l} \mathbf{lay}(\xi, min_\xi + i - 1).n = n_i \\ \wedge\ \mathbf{lay}(\xi, min_\xi + i - 1).a = a_i \wedge \mathbf{lay}(\xi, min_\xi + i - 1).b = b_i. \end{array}$$

Whereas, in general constructive disjunction is inefficient, here one can easily propagate information, e.g. the domains of the layer variables.

It remains to constrain relations between the variables to get the basic constraint formulation for our problem. First, we relate lines to points by $\mathbf{ln}(\vec{a}, \vec{v}) = \sum_{\vec{p} \in \mathrm{LN}(\vec{a}, \vec{v}) \cap \mathrm{CB}} \mathbf{pnt}(\vec{p})$ for all line variables. Then, we relate layer parameters to their layers by $\sum_{\vec{p} \in \mathrm{CB} \cap \mathrm{PL}(\xi, c)} \mathbf{pnt}(\vec{p}) = \mathbf{lay}(\xi, c).n$ for all $\xi \in \{x, y, z\}$ and $\min_\xi \leq c < \min_\xi + m_\xi$ and further, for x-layers and lattice lines in direction $\begin{pmatrix} 0 \\ 1 \\ 1 \end{pmatrix}$ introduce the constraints

$$\sum_{r \in \mathbb{Z},\ \mathrm{LN}\left(\begin{pmatrix} c \\ r \\ 0 \end{pmatrix}, \begin{pmatrix} 0 \\ 1 \\ 1 \end{pmatrix}\right) \cap \mathrm{CB} \neq \emptyset} \delta\left(\mathbf{ln}\left(\begin{pmatrix} c \\ r \\ 0 \end{pmatrix}, \begin{pmatrix} 0 \\ 1 \\ 1 \end{pmatrix}\right) > 0\right) = \mathbf{lay}(x, c).a$$

and the analogous constraints for $\mathbf{lay}(x, c).b$, the y-layers, and z-layers.

Now, relate contacts to points and to the total number of contacts. For any contact variable $\mathbf{con}(\vec{p}, \vec{q})$, introduce $\mathbf{con}(\vec{p}, \vec{q}) = \delta(\mathbf{pnt}(\vec{p}) = 1 \wedge \mathbf{pnt}(\vec{q}) = 1)$. Finally, state $\sum \mathbf{con}(\vec{p}, \vec{q}) = \mathrm{con}$. The previous constraints define the problem non-redundantly. For sufficient constraint propagation, we need to introduce redundant constraints like the following ones.

For example, the surrounding cube has to be large enough to include the core. Therefore, we introduce the constraints $\lceil \frac{m_x m_y m_z}{2} \rceil \geq n$ if the sum $\min_x + \min_y + \min_z$ is even and $\lfloor \frac{m_x m_y m_z}{2} \rfloor \geq n$ otherwise. Further, the line variables are connected to the layer parameter n by constraints

$$\sum_{r \in \mathbb{Z},\ \mathrm{LN}\left(\begin{pmatrix} c \\ r \\ 0 \end{pmatrix}, \begin{pmatrix} 0 \\ 1 \\ 1 \end{pmatrix}\right) \cap \mathrm{CB} \neq \emptyset} \mathbf{ln}\left(\begin{pmatrix} c \\ r \\ 0 \end{pmatrix}, \begin{pmatrix} 0 \\ 1 \\ 1 \end{pmatrix}\right) = \mathbf{lay}(x, c).n \quad \text{and analogous ones.}$$

Using Local Upper Bounds on the Number of Contacts The number of contacts within each layer is determined by the layer parameters, since we exclude caveats.[8] Thus, we can constrain the number of these (intra)layer contacts. We use also a constraint to forbid caveats directly. It constrains the core points along each lattice line to be connected.

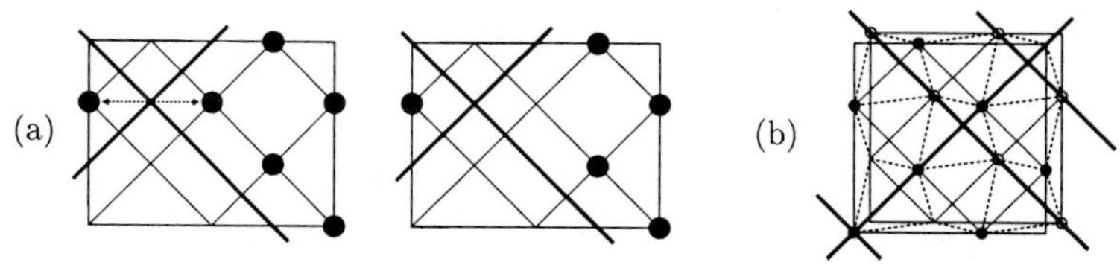

Figure 2: (a) The thick lines are drawn between non-overlapping pairs of lines. In both layers, we count one non-overlap and in the right layer one non-connect, since there is no connection (as shown in left layer by the arrow). (b) represents an example situation in the search. The thick lines are already known to intersect the core. Assume in each layer there are 5 core points, the beads mark remaining potential positions. The line constraints restrict the number of contacts, hence this additional knowledge is exploitable for the contacts bound.

Furthermore, we introduce redundant constraints that employ the upper bounds on the number of contacts between successive layers, called interlayer contacts. From earlier work,[8] we know non trivial upper bounds on the number of layer and interlayer contacts given parameters of the layers, namely the layer size, the previously defined olines(f) and the number of *non-connects* and *non-overlaps*. For an illustration of the latter terms, please see Figure 2(a).

Now, for $\min_\xi \leq c_1, c_2 < \min_\xi + m_\xi$ and $c_2 = c_1 \pm 1$, introduce FD-variables $\mathbf{ilay}(\xi, c_1, c_2).con$ to hold the number of interlayer contacts between layers $\xi = c_1$ and $\xi = c_2$. This variable is constrained to the sum of the corresponding contact variables and the total number of contacts is constrained to the sum of the layer contacts and the variables for interlayer contacts. The bound is strengthened and recomputed during the enumeration as more and more information, e.g. which lines intersect the core (see Figure 2(b)), becomes known. Therefore, variables to hold the additional parameters, the number of non-overlaps and non-connects, are introduced for each layer and corresponding constraints are stated. Furthermore, we introduce FD-variables $\mathbf{ilay}(\xi, c_1, c_2).i$ to hold the number of core points in $\xi = c_2$ with at least $i = 1, 2, 3$, or 4 contacts to core points in $\xi = c_1$. Such points were called *i-points*. Finally we can bind $\mathbf{ilay}(\xi, c_1, c_2).con$ to the sum $\sum_{1 \leq i \leq 4} \mathbf{ilay}(\xi, c_1, c_2).i$.

Search strategy We start a search by enumerating the variables m_x, m_y, m_z, $\min_x$, $\min_y$, and $\min_z$. This fixes the surrounding cube and allows in an implementation to construct all data structures. Afterwards, we distribute over the point variables to fix the core. To exclude rotation and reflection symmetries, we employ symmetry excluding search.[6] This search is a special form of constrained search, which only finds solutions modulo given symmetries

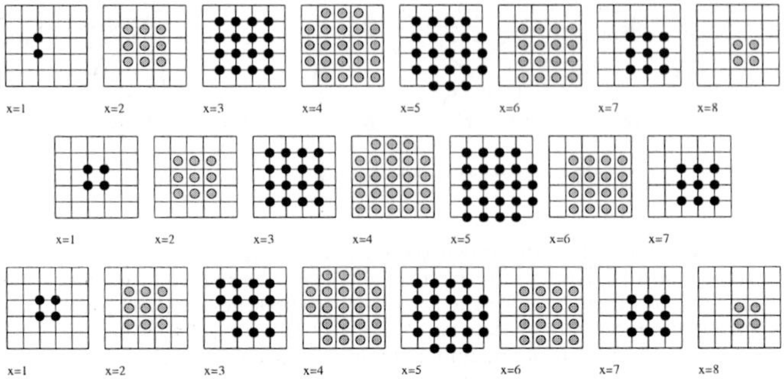

Figure 3: Plane sequence representations of 3 optimally compact cores of size $n = 100$.

and employs this to prune the search tree.

5 Results

All sets of layer parameter sequences $S_{\text{con}}(n, \text{con})$ without gaps for $n \leq 100$ were computed in about ten days on a standard PC. After this precomputation, which has to be done only once, the set of all optimally compact cores usually is found within a few seconds to minutes by our search program. Some results are shown in Table 1. Currently, the search program implements most of the presented ideas as well as additional redundant constraints.

Further, some optimal cores for $n = 100$ elements are shown in Figure 3. The cores are shown in plane sequence representation. This representation shows a core by the sequence of its occupied x-layers rotated by 45°. For each x-layer $x = x_0$ the lower left corner of the grid has coordinates $(x_0, 0, 0)$. The grid-lines are parallel to the lattice lines in x-layers and have distance $\sqrt{2}$. The core points in each x-layer are shown as filled circles.

Finally, we are able to thread sequences to hydrophobic cores for structure prediction, which is described in detail elsewhere.[7] There, we experimentally evaluate the ability of our algorithm to predict the structure for random sequences with 100 H-monomers and chain lengths of up to 160. We are able to find structures for 60% of the sequences with length 160 within 15 minutes. This percentage increases to 82%, when we allow one hour search time.

1. V. I. Abkevich, A. M. Gutin, and E. I. Shakhnovich. Impact of local and non-local interactions on thermodynamics and kinetics of protein folding. *Journal of Molecular Biology*, 252:460–471, 1995.

Table 1: Search for all optimally compact cores with n elements, given the layer sequences. We list the number of contacts, the number of nodes and depth of the search tree, and time of the constraint search for every core size n.

n	# contacts	# search-nodes	depth	time
40	152	167	17	17.2 s
60	243	182	72	4.6 s
82	349	220	37	14.2 s
102	447	54	20	8.2 s

2. V.I. Abkevich, A.M. Gutin, and E.I. Shakhnovich. Computer simulations of prebiotic evolution. In *PSB'97*, pages 27–38, 1997.
3. Richa Agarwala, Serafim Batzoglou, Vlado Dancik, Scott E. Decatur, Martin Farach, Sridhar Hannenhalli, S. Muthukrishnan, and Steven Skiena. Local rules for protein folding on a triangular lattice and generalized hydrophobicity in the HP-model. *Journal of Computational Biology*, 4(2):275–296, 1997.
4. Rolf Backofen. Constraint techniques for solving the protein structure prediction problem. In *Proceedings of* 4th *International Conference on Principle and Practice of Constraint Programming (CP'98)*, volume 1520 of *Lecture Notes in Computer Science*, pages 72–86. Springer Verlag, 1998.
5. Rolf Backofen. An upper bound for number of contacts in the HP-model on the face-centered-cubic lattice (FCC). In *Proc. of the 11th Annual Symposium on Combinatorial Pattern Matching (CPM2000)*, volume 1848 of *Lecture Notes in Computer Science*, pages 277–292, Berlin, 2000. Springer–Verlag.
6. Rolf Backofen and Sebastian Will. Excluding symmetries in constraint-based search. In *Proceedings of* 5th *International Conference on Principle and Practice of Constraint Programming (CP'99)*, volume 1713 of *Lecture Notes in Computer Science*, pages 73–87, Berlin, 1999. Springer–Verlag.
7. Rolf Backofen and Sebastian Will. Fast, constraint-based threading of HP-sequences to hydrophobic cores. In *Proceedings of* 7th *International Conference on Principle and Practice of Constraint Programming (CP'2001)*, Lecture Notes in Computer Science, Berlin, 2001. Springer–Verlag. To appear.
8. Rolf Backofen and Sebastian Will. Optimally compact finite sphere packings — hydrophobic cores in the FCC. In *Proc. of the 12th Annual Symposium on Combinatorial Pattern Matching (CPM2001)*, Lecture Notes in Computer Science, Berlin, 2001. Springer–Verlag.
9. Rolf Backofen, Sebastian Will, and Erich Bornberg-Bauer. Application of constraint programming techniques for structure prediction of lattice proteins with extended alphabets. *J. Bioinformatics*, 15(3):234–242, 1999.
10. B. Berger and T. Leighton. Protein folding in the hydrophobic-hydrophilic (HP) modell is NP-complete. In *Proc. of the Second Annual International Conferences on Compututational Molecular Biology (RECOMB98)*, pages 30–39, New York, 1998.

11. Erich Bornberg-Bauer. Chain growth algorithms for HP-type lattice proteins. In *Proc. of the* 1st *Annual International Conference on Computational Molecular Biology (RECOMB)*, pages 47 – 55. ACM Press, 1997.
12. P. Crescenzi, D. Goldman, C. Papadimitriou, A. Piccolboni, and M. Yannakakis. On the complexity of protein folding. In *Proc. of STOC*, pages 597–603, 1998. Short version in *Proc. of RECOMB'98*, pages 61–62.
13. K.A. Dill, S. Bromberg, K. Yue, K.M. Fiebig, D.P. Yee, P.D. Thomas, and H.S. Chan. Principles of protein folding – a perspective of simple exact models. *Protein Science*, 4:561–602, 1995.
14. Ken A. Dill, Klaus M. Fiebig, and Hue Sun Chan. Cooperativity in protein-folding kinetics. *Proc. Natl. Acad. Sci. USA*, 90:1942 – 1946, 1993.
15. Aaron R. Dinner, Andreaj Šali, and Martin Karplus. The folding mechanism of larger model proteins: Role of native structure. *Proc. Natl. Acad. Sci. USA*, 93:8356–8361, 1996.
16. S. Govindarajan and R. A. Goldstein. The foldability landscape of model proteins. *Biopolymers*, 42(4):427–438, 1997.
17. Patrice Koehl and Michael Levitt. A brighter future for protein structure prediction. *Nature Structural Biology*, 6:108–111, 1999.
18. Kit Fun Lau and Ken A. Dill. A lattice statistical mechanics model of the conformational and sequence spaces of proteins. *Macromolecules*, 22:3986 – 3997, 1989.
19. Hao Li, Robert Helling, Chao Tnag, and Ned Wingreen. Emergence of preferred structures in a simple model of protein folding. *Science*, 273:666–669, 1996.
20. Britt H. Park and Michael Levitt. The complexity and accuracy of discrete state models of protein structure. *Journal of Molecular Biology*, 249:493–507, 1995.
21. Gert Smolka. The Oz programming model. In *Computer Science Today*, Lecture Notes in Computer Science, vol. 1000, pages 324–343. Springer-Verlag, Berlin, 1995.
22. R. Unger and J. Moult. Genetic algorithms for protein folding simulations. *Journal of Molecular Biology*, 231:75–81, 1993.
23. Ron Unger and John Moult. Local interactions dominate folding in a simple protein model. *Journal of Molecular Biology*, 259:988–994, 1996.
24. A. Šali, E. Shakhnovich, and M. Karplus. Kinetics of protein folding. *Journal of Molecular Biology*, 235:1614–1636, 1994.
25. Yu Xia, Enoch S. Huang, Michael Levitt, and Ram Samudrala. Ab initio construction of protein tertiary structures using a hierarchical approach. *Journal of Molecular Biology*, 300:171 – 185, 2000.
26. Kaizhi Yue and Ken A. Dill. Forces of tertiary structural organization in globular proteins. *Proc. Natl. Acad. Sci. USA*, 92:146 – 150, 1995.

DETECTING NATIVE PROTEIN FOLDS AMONG LARGE DECOY SETS WITH HYDROPHOBIC MOMENT PROFILING

RUHONG ZHOU and B. DAVID SILVERMAN
IBM Thomas J. Watson Research Center
Route 134 & PO Box 218, Yorktown Heights, NY 10598

Abstract

A new hydrophobic score will be presented in this paper for detecting native-like folds from a large set of decoy structures. A recent paper (B. D. Silverman, PNAS 98, 4996, 2001) had revealed that for globular proteins there exists a relatively universal constant of 0.74 for a hydrophobic ratio, which is defined as the ratio of radii from the protein centroid at which the second order hydrophobic moment and the zero order moment vanishes. This paper further defines a new hydrophobic score which will be used to examine protein decoys, in particular, the Holm & Sander, Park & Levitt and Baker decoy sets. It will be shown that the hydrophobic score and profile shapes can provide useful information that should be complementary to the information provided by other procedures, such as free energy calculations.

1 Introduction

The ability to recognize native protein conformations from misfolded ones is a problem of fundamental importance in the development of methods for protein structure prediction. Decoy structures of proteins have provided test sets for the evaluation of scoring functions in threading and homology modeling, as well as energy functions used in ab-initio predictions. While an ideal objective would be the determination of a free energy function that selects structures that are native structures or minimally displaced spatially from the native structures, success has not been forthcoming. One suspects that the difficulty in the determination of an appropriate free energy function is related to the approximate manner in which the calculations treat the entropic character of solvation. One global structural feature arising from solvation is the ubiquitous hydrophobic core and hydrophilic exterior of soluble globular proteins. This feature has been used to identify protein structures that might be candidates that approximate the native structure or used to eliminate candidate structures that might not.[1–3] Considerations of hydrophobicity together with free energy approaches[4,5,8] can provide a more selective procedure than the use of either alone.

The universal spatial transition from the hydrophobic core to the hydrophilic exterior of globular proteins motivated detailed spatial profiling calculations of

Table 1: Eisenburg hydrophobicity consensus values for each amino acid.

res	hydro.	res	hydro.	res	hydro.	res	hydro.
ARG	-1.76	GLU	-0.62	TYR	0.02	TRP	0.37
LYS	-1.10	HIS	-0.40	CYS	0.04	LEU	0.53
ASP	-0.72	SER	-0.26	GLY	0.16	VAL	0.54
GLN	-0.69	THR	-0.18	ALA	0.25	PHE	0.61
ASN	-0.64	PRO	-0.07	MET	0.26	ILE	0.73

this transition. With an ellipsoidal characterization of protein shape, an appropriate scaling of residue hydrophobicity and a second-order ellipsoidal moment, it was shown that thirty or more diverse globular proteins shared detailed spatial features of this transition, with a quasi-invariant hydrophobic ratio (defined in next section) of 0.74 ± 0.05 for the protein structures examined. Since the small protein decoys of Park and Levitt, and those of the Baker group, have been central to ab-initio procedures in discriminating decoys from native structures, it is of interest to see if moment profiling could yield useful supplemental information, even in the regime of profile irregularities due to the small size of proteins. The Holm & Sander decoy sets, which include larger-sized proteins, are also examined. The results show that useful discrimination between native and decoy structures can be obtained.

2 Methodology

For proteins, each residue exhibits a different degree of hydrophobicity or hydrophilicity, based upon its solubility in water. A value of hydrophobicity, h_i, can then be assigned to each residue of type, i. Table 1 lists the Eisenberg hydrophobicity consensus values for each amino acid.[6]

Since the distribution of hydrophobicity is profiled from the protein interior to the exterior of globular proteins, an ellipsoidal profiling shape had been chosen with axes determined by the moments-of-geometry of the residue distribution,

$$I_{jk} = \int_V \rho(\vec{r})(r^2\delta_{jk} - x_j x_k)dV, \tag{1}$$

where I_{jk} is the moment-of-geometry terms and $\rho(\vec{r})$ is the density of the residue centroids of unit mass. Diagonizing the moment-of-geometry matrix, one obtains the three principal axes as well as the moments of geometry. The x, y, and z axes are then aligned with the principal axes. The moments of geometry are designated as g_1, g_2 and g_3, with $g_1 < g_2 < g_3$. The ellipsoidal representation generated by these moments will be,

$$x^2 + g_2' y^2 + g_3' z^2 = d^2, \tag{2}$$

where $g_2' = g_2/g_1$, $g_3' = g_3/g_1$. The value d is the major axis of the ellipsoid and can be considered as a generalized ellipsoidal radius.

The Eisenburg hydrophobicity distribution is shifted such that the net hydrophobicity of each protein vanishes. The distribution is then normalized to yield a standard deviation of one. Shifting the residue hydrophobicity distribution for each protein selects a common structural reference and thus enables the quantitative comparison of protein profile shapes and profile features such as the hydrophobic ratio. The zero-order hydrophobic moment of the residue distribution within the ellipsoidal surface specified by d is then written,

$$H_0(d) = \sum_{r<d} h_i^{'} = \sum_{r<d} (h_i - \bar{h}) / < (h_j - \bar{h})^2 >^{1/2}, \tag{3}$$

The prime designates the value of hydrophobicity of each residue after shifting and normalizing the distribution, and $\bar{h}$ is the mean of the h_i for all residues in the protein. Therefore, when the value of d is just sufficiently large enough to collect all of the residues, the net hydrophobicity of the protein vanishes. This value of d_0, for which $H_0(d)$ vanishes assigns a surface as common structural reference for each protein.

Second-order moments amplify the differences between hydrophobic and hydrophilic residues that contribute to the spatial profile of the hydrophobicity distribution. The second-order hydrophobic moment is defined as,

$$H_2(d) = \sum_{r<d} h_i'(x_i^2 + g_2' y_i^2 + g_3' z_i^2), \tag{4}$$

where the (x_i, y_i, z_i) denote the position of a residue centroids. For native globular protein structures, the zero and second-order moments are positive when d is small. Both increase with distance, d, within the region of the hydrophobic core. The increase of both the zero- and second-order moments with distance then slows and turns around decreasing with increasing d. Since the second order moment amplifies differences in the distribution, this moment will cross zero, becoming negative at a distance below the value of, d_0, at which the zero-order moment vanishes. The location at which the second-order moment vanishes is defined as d_2. The hydrophobic-ratio is then defined as,

$$R_H = d_2/d_0. \tag{5}$$

The paper by Silverman[7] had shown the hydrophobic-ratio to be a quasi-invariant, 0.74, for all of the native protein structures that had been examined. This ratio will also be shown to characterize native and near native structures in the following section. Such ratio, however, cannot always be defined for arbitrary protein structures. This is particularly true if the second-order moment profile does not exhibit the smooth generic behavior expected. The hydrophobic ratio would then be unable to provide a continuous score with respect to how deviant a decoy profile would be with respect to its native profile. To provide such continuous ranking of each decoy profile with respect to its native profile, a new scoring function will have to be defined.

3 Results and Discussion

The Holm & Sander, Park & Levitt[2] and Baker decoy sets[8] examined in this study have been downloaded from the web (http://dd.standford.edu for the Holm & Sander and Park & Levitt set, and http://depts.washington.edu/bakerpg for the David Baker set). Since the hydrophobic moments and ratios involve the spatial profiling of the residue distribution, and this distribution is discretely distributed in space, a typical window of 1Å in generalized ellipsoidal radius, d, had been used to generate the nested ellipsoidal surfaces. This provided reasonable resolution in obtaining the generally smooth moment profiles over the range of protein sizes previously investigated. Protein size imposes a constraint upon the ability to generate relatively smooth profiles. It is found that a relatively smooth profile can be obtained for proteins with residue number greater than 100. This number of residues, namely 100, was the lower limit chosen in the investigation of the Holm Sander decoys, which resulted in a total of 14 decoy sets out of total 26, with protein size ranging from 107 residues to 317 residues. The Park & Levitt and Baker decoys range in size well below this limit so proteins chosen for the present study are limited to have a residue number of no less than 60.

For the Baker decoy sets, we have also applied two other criteria to eliminate decoy sets from the total of 92: (1) those decoy sets where 10% or less of the decoys have RMSD's from the native structure that are less than 8 Å will be eliminated; and (2) those decoy sets having the smallest RMSD larger than 4 Å will be eliminated. The objective is to examine decoys with a broad range of RMSD's and hence a broad range of "similarity" to their native structures. Decoys significantly displaced in RMSD from their native structures are not included. This selects the decoys that should be more difficult to distinguish from their native structure. This decoy set elimination together with the residue number limitation reduces the number of Baker sets studied to 11 from the total of 92. The residue number restriction imposed on the Park & Levitt decoy sets reduces the number of sets examined to 4 from a total of 7 (one decoy set has outdated native PDB structures, which has also been eliminated). The numbers of residues of these proteins range from 60 to 75. These protein sizes are insufficient, in most cases, to yield smooth hydrophobic moment profiles. It will, however, be shown that even subject to this limitation, the moment profiling can provide useful complementary information to that obtained from energy minimization procedures. The RMSD values for the Park & Levitt decoy sets are supplied by the authors on their web site. These are RMSDs for C_α atoms. The RMSD values for the Baker decoy sets are not available from the web site and are, therefore, recomputed with the IMPACT program[9] for all backbone atoms. The RMSD values based on the C_α atoms, backbone atoms, or all of the atoms will be slightly different, but for the case at hand, they should be equally instructive.

All native second-order moment profiles of the proteins in Holm & Sander, Park & Levitt, and Baker decoy sets selected here (total 29) show a hydrophobic core and a sharp plunge to negative values in the transition from hydrophobic core to hydrophilic exterior. Similar to previous results,[7] the native decoy structures have R_H values that range from 0.640 to 0.791, with a mean of 0.73.

Holm Sander decoys had been generated to test their solvation preference method[1] designated to distinguish native(correct) from decoy(incorrect) structures. Figure 1 shows the second-order hydrophobic moment profiles for 6 such decoys (all 14 decoys show the same behavior basically). All native structures exhibit a second-order profile shape that had been previously found for thirty PDB structures of diverse size and fold. All of the decoy structures, on the other hand, do not show the significant separation between the hydrophobic residues forming the native core and hydrophilic exterior. The second-order moments fluctuate around zero on the radial axis. The hydrophobic ratio cannot be defined for these decoy structures.

The second order moment profiles of the thousands of Park & Levitt and Baker decoy structures do not, however, always exhibit easy patterns to be discriminated against as in the Holm & Sander single decoy sets. It is also not feasible to plot thousands of profiles in one or a few figures. Therefore, a new scoring function is needed to quantitatively rank each decoy profile with respect to an expected native profile. Examination of a few of the decoy profiles will reveal several of the issues involved in defining the scoring function. In general, the structures of smaller RMSD ($< 2.0Å$, native-like) approximate the native profile more closely. The decoy structures with larger RMSD have hydrophobic peaks that are either not well-defined as shown in Holm & Sander's decoy profiles (Figure 1) or less pronounced than their native structures. The hydrophilic ranges are also generally extended out to greater distances and are not as negative as seen in the native structure profiles. This suggested that the area under the hydrophobic peak and that under hydrophilic exterior could play a role in discriminating the native from the decoy structures. On the other hand, a significant increase in protein extent of the decoy could yield a spurious contribution from the area under the negative moment profile. This contribu-

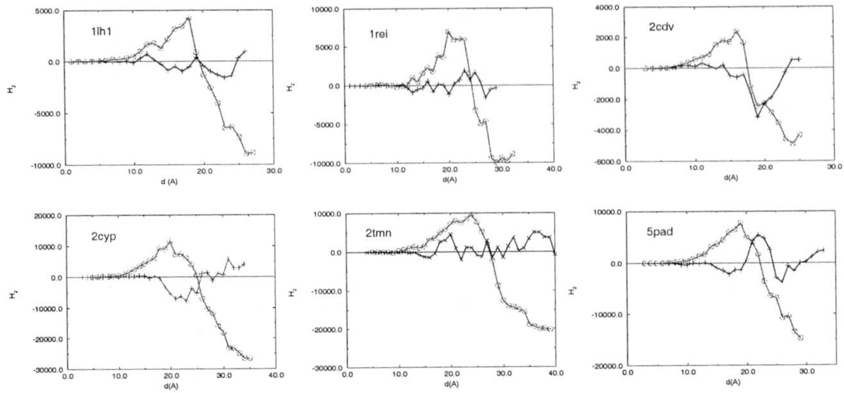

Figure 1: Second-order moments for the native and decoy structures of the Holm & Sander single decoy sets (circles: native; plus: decoy).

tion could, however, be reduced by scaling the native and decoy structures by the value of protein extent, namely, by d_0. The abscissa on the moment plot was, therefore, divided by d_0 and the second-order moment divided by d_0^2. Such comparison does not take differences in residue number into account. For the present case, however, the decoys and their corresponding native structures have the same number of residues.

The proposed hydrophobic score, S_H, which ranks the quality of the decoys with respect to an expected native profile, is then chosen as the integral of the area under the normalized second-order hydrophobic moment profiles,

$$\begin{aligned} \tilde{H}_2 &= H_2/d_0^2 \\ \tilde{r} &= r/d_0. \end{aligned} \tag{6}$$

The absolute value of $\tilde{H}_2$ is integrated over the normalized distance,

$$S_H = \int_0^1 |\tilde{H}_2| d\tilde{r}. \tag{7}$$

This score not only measures the prominence of the hydrophobic core, but also the prominence of the hydrophilic exterior by taking into account the rapidity of decrease of the profile in the hydrophilic exterior.

Figure 2 shows the hydrophobic scores versus the RMSDs for the four Park & Levitt decoy sets. Almost all decoys have lower hydrophobic scores than their corresponding native structures. Table 2 lists the number and percentage

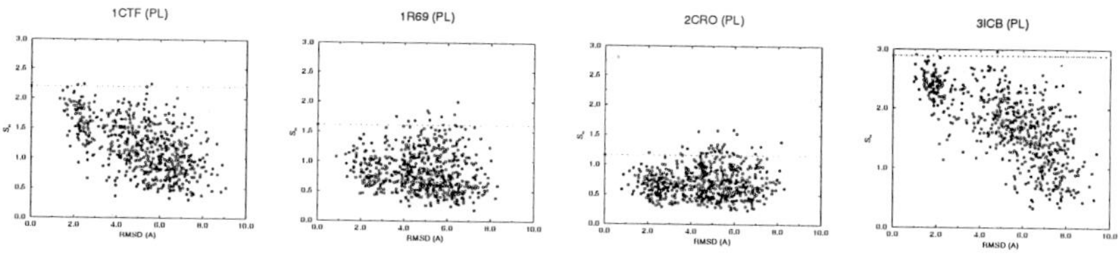

Figure 2: Hydrophobic score versus RMSD for Park & and Levitt decoys.

Table 2: Performance of the hydrophobic score: the percentage of decoy structures which have lower hydrophobic score than their native ones (denoted as "low scores" in the table).

Decoy Set	PDB entry	low scores	total decoys	%
Park/Levitt	3icb	651	654	99.5
	1ctf	627	631	99.4
	1r69	664	676	98.2
	2cro	637	675	94.4
Baker	2ezh	957	1000	95.7
	1mzm	864	1000	86.4
	1nkl	848	1000	84.8
	1ctf	816	1000	81.6
	1r69	656	1000	65.6
	2fow	627	1000	62.7
	2ptl	619	1000	61.9
	1sro	559	1000	55.9
	1c5a	493	991	49.8
	1hsn	245	970	25.4
	1leb	253	1000	25.3

of decoys out of the total which have lower hydrophobic scores than their native proteins. 99.5%, 99.4%, 98.2%, and 94.4% of the decoys have hydrophobic scores below their native benchmarks for 3icb, 1ctf, 1r69, and 2cro, respectively. Proteins 3icb and 1ctf which show native profiles accentuating the hydrophobic and hydrophilic regions have fewer than 0.5–0.6% of decoys with a score that is greater than that of the native structures. One also notes significant correlation in their decoy distributions, namely, decoys with greater RMSD generally have smaller hydrophobic area or score. Proteins 2cro and 1r69 with native profiles that do not accentuate the hydrophobic and hydrophilic regions as observed for proteins 1ctf and 3icb show slightly greater numbers of decoys with greater scores than their native structures, and their distribution of decoy scores does not exhibit the correlation found for 1ctf and 3icb. The decoy scores of 1r69 and 2cro appear to be essentially uniformly distributed about the RMSD values.

Little or no correlation of hydrophobic score with RMSD might arise from

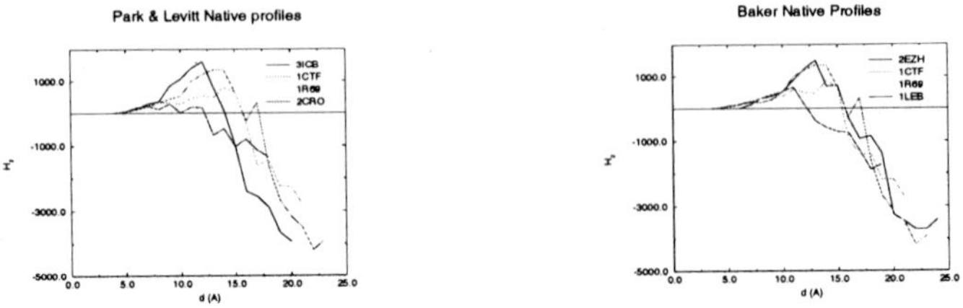

Figure 3: (a) Left: the four native structure profiles in the Park & Levitt decoy set, 3icb, 1ctf, 1r69, and 2cro. (b) Right: the four native structure profiles in the David Baker decoy set, 2ezh, 1ctf, 1r69, and 1leb.

native structures with profiles that do not accentuate the core and hydrophilic regions. It is then less restrictive for a decoy to score well with respect to the native structure. Figure 3(a) shows the native profiles of the four decoy sets of Park & Levitt, namely, 3icb, 1ctf, 1r69 and 2cro. It is clear that 1r69 and 2cro have native profiles with hydrophobic and hydrophilic regions of lesser prominence than found for 1ctf and 3icb.

Figure 4 shows the hydrophobic scores for the four representative Baker decoy sets: two decoys 1ctf and 1r69 which are also in the Park & Levitt set, and the other two 2ezh and 1leb which have the highest and lowest percentage of decoys with scores below their native structure scores respectively (see Table 2). In contrast to the Park & Levitt decoy sets, the Baker decoy sets show a much broader distribution of hydrophobic scores. The percentage of decoys which have scores below their native benchmark scores ranges from 25.3% (1leb) to 95.7% (2ezh), with the majority in the range of 60 – 80%. Also, most of these decoy sets do not exhibit the correlation with RMSD that the 1ctf and 3icb Park & Levitt decoys show. The four shown decoy sets 2ezh, 1ctf, 1r69, and 1leb have a percentage of decoys with scores below the native to be 95.7%, 81.6%, 65.6% and 25.3%, respectively. Interestingly, 2ezh and 1ctf (higher percentages 95.7% and 81.6%), show a more prominent native structure profile than 1r69 and 1leb (lower percentages 65.6% and 25.3%), as can be seen from the Figure 3(b). Other decoys in the Baker set show similar behavior. The numbers of decoys with a higher percentage below the native score (2ezh, 1mzm, 1nkl, 1ctf, etc) show more pronounced native structure profiles than decoys with a lower percentage (1hsn, 1leb, etc). As discussed earlier, for the Park & Levitt decoys 1r69 and 2cro, this correspondence between a higher percentage scoring poorly with the less prominent native profiles appears reasonable. It is easier for decoys to score well against native structures that exhibit reduced separation of hydrophobic and hydrophilic residues with consequent low score.

The relatively large number of Baker decoys with high hydrophobic score

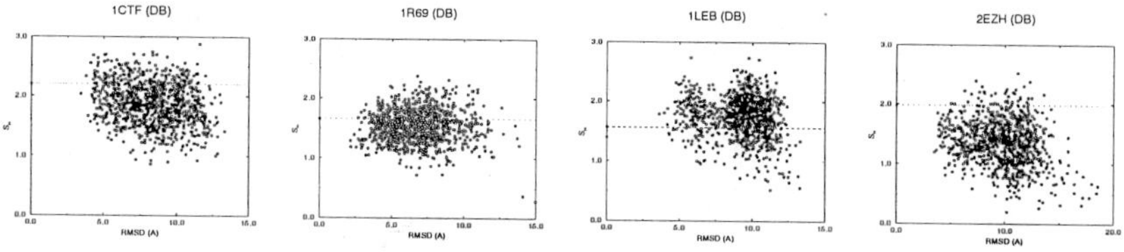

Figure 4: Hydrophobic score versus RMSD for Baker decoys.

compared with the Park Levitt decoys might be related to the manner in which the decoys were generated and selected. In particular, a significant fraction of the decoys of 1leb Baker set clearly show greater spatial segregation of the hydrophobic and hydrophilic residues than observed for the native structure. The scores of the 1r69 and 1ctf Park & Levitt and Baker decoy sets can be compared from Figure 2 and Figure 4. The Baker decoys clearly show a greater number of structures with scores that are higher than their native scores when compared with the Park & Levitt decoy scores. Calculation of the radii of gyration (Rg) found the Baker sets to have slightly larger Rg's compared with the Park and Levitt sets. Decoys of 1r69 have Rg of 12.00 ± 0.81 Å in Baker set and 10.99 ± 0.53 Å in the Park & Levitt set. Similarly, decoys of 1ctf have Rg of 11.65 ± 0.66 Å in the Baker set and 11.19 ± 0.59 Å in Park & Levitt set. Perhaps larger radius of gyration provides greater spatial freedom to segregate the hydrophobic from hydrophilic residues. A point of greater relevance is related to the way Baker and coworkers have selected these ab-initio decoys. One of the fundamental assumptions underlying their program Rosetta[3,8] is that the distribution of conformations sampled for a given nine residue segment of the chain is reasonably well approximated by the distributions in known protein structures in the PDB Databank. Fragment libraries for each 3- and 9-residue segment of the chain are extracted from the protein structure database using a sequence profile-profile comparison method. The conformational space defined by these fragments is then searched using a Monte Carlo procedure with an energy function that favors compact structures with paired β strands and buried hydrophobic residue. The favoring of buried hydrophobic residues in the energy function should provide the Baker sets with greater segregation of hydrophobic and hydrophilic residues from the protein core to exterior and consequently provide higher hydrophobic scores than achieved by the Park & Levitt decoy sets.

It is also interesting to note that there are low RMSD structures which have low hydrophobic scores even among the decoys of the well correlated sets, such as 3icb. Figure 5(a) shows several hydrophobic moment profiles for 3icb decoy

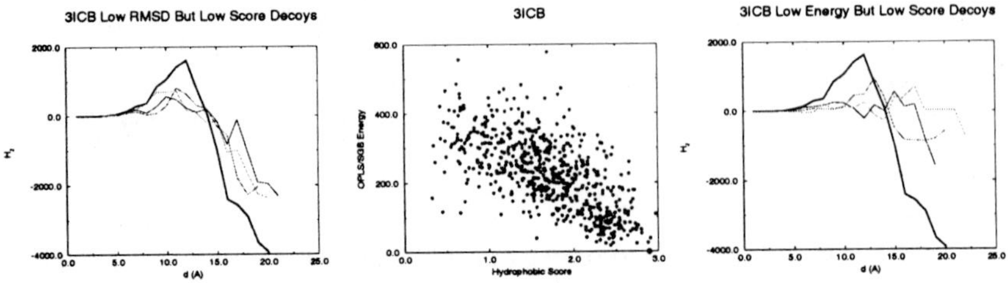

Figure 5: For Park & Levitt decoy set 3icb: (a) left: Hydrophobic moment profiles for some low RMSD structures but with low hydrophobic scores (the thick dark line is from the native structure for comparison). (b) Middle: hydrophobic score versus OPLSAA/SGB energy[5] (the native one is marked with a bigger circle). (c) Right: Hydrophobic moment profiles for some of the low OPLSAA/SGB energy structures but with low hydrophobic scores.

structures with less than a 3.0Å RMSD and less than a 1.5 hydrophobic score (decoy index a587, a591, and a8110, to name a few). The native score is 2.89 for this case. These decoy structures have fewer hydrophobic residues in the protein interior and consequently fewer hydrophilic residues in the protein exterior than expected for native structures. The hydrophobic residues and hydrophilic residues are more spatially mixed. Might these structures be less favorable candidates as near native structures? From the reported OPLSAA/SGB free energies,[5] they are indeed energetically unfavorable structures. The three decoys plotted, a587, a591, and a8110, are 206.98, 116.94, 110.14 kcal/mol higher than the native structure. The OPLSAA/SGB energies are obtained from Levy and coworkers (see below for more details). This indicates that the simple hydrophobic score should provide useful information in discriminating decoy structures from native structures.

Levy and coworkers[5] have calculated the energies of the Park & Levitt decoys using the OPLSAA force field[10] and a Surface Generalized Born (SGB) model[11] for a continuum solvent. They found that without the continuum solvation free energy, the OPLSAA gas phase energies are not sufficient to distinguish native-like from non native-like structures. Figure 5(b) plots the OPLSAA/SGB energy (the energy of the native structure is set zero) versus the hydrophobic score for the protein 3icb of the Park & Levitt set. The OPLSAA/SGB energies have been kindly supplied by the Levy group. It should be noted that in the Levy energy calculations, the decoy structures are minimized first to remove bad contacts (otherwise the energies could be huge and meaningless). Thus, the structures used in the Levy energy calculations are slightly different from ours; however, we don't believe that this should affect the hydrophobic scores. This is an advantage of the method of hydrophobic scoring. Differences in structure that would affect the free energy values significantly will not affect the hydrophobic scores much. One need not even add hydrogen atoms to the PDB structures for most of the calculations. Free energy calculations, on the other hand, are not only sensitive to the presence or absence of hydrogen atoms, but extremely sensitive to small differences in structure. Figure 5(b) shows the correlation between the OPLSAA/SGB energy and the hydrophobic score, i.e., decoys with smaller or poorer scores have higher energies compared with the native energy, and those with higher or better scores are closer in energy to the native structures. Similar to 3icb, protein 1ctf also shows a significant correlation between the OPLSAA/SGB energy and the hydrophobic score, whereas 1r69 and 2cro show a weaker correlation. This weak correlation for the 1r69 and 2cro decoys reflects their weak correlation between hydrophobic score and RMSD as described previously.

Interestingly, there are decoy structures with low OPLSAA/SGB free energies that do not have high hydrophobic scores. This is found even for the decoys of 3icb, which show a strong correlation between hydrophobic score and RMSD. The decoy sets showing poorer correlation have a greater number of decoys exhibiting this behavior. Figure 5(c) also shows several representative profiles of 3icb decoy structures with low free energies and also low hydrophobic scores. These decoys are not the same as those with low RMSD and low score as discussed previously and shown in Figure 5(a). The bad or low hydrophobic scores indicate that the structures have a poorly formed hydrophobic core and hydrophilic exterior even though the free energy is low. By comparison with the native profile (the thick dark line in the figure), it is evident that the hydrophobic core of these decoys has been "damaged". The region of positive moment that might be identified as a core region is shifted out to greater distances than found for the native structure. Furthermore, none of the decoys exhibit the sharp plunge to negative values in the protein exterior expected for a native structure. This yields a low score or unfavorable protein structure. This example demonstrates the value of the hydrophobic score in providing complementary information to that obtained from the free energy calculations. Previously we had shown that a low RMSD does not necessarily guarantee a good hydrophobic score, and here we have shown that a low free energy does not guarantee a good hydrophobic score either.

It should be pointed out that the resulting second-order profiling pattern, hydrophobic ratio and hydrophobic score are for globular proteins only. Other type of proteins, such as DNA-binding proteins and membrane proteins might have hydrophobic residues in the exterior region, thus the profiling and ratios will be different.

This approach should be very useful for pre-screens in various structure prediction or refinement algorithms, since it is extremely fast. It takes less than a minute on a typical PC to calculate the second-order moment profiling and the hydrophobic score for each structure.

4 Conclusions

The present paper has examined the utility of molecular moment hydrophobicity profiling in discriminating between near native protein structures and incorrect decoy structures for the widely used Holm & Sander, Park & Levitt and Baker decoy sets. Subject to the conditions that limit the type of small structures examined, the moment profiling and a subsequent hydrophobic score, which is the integral of the area under the normalized second-order hydrophobic moment profile, enable one to distinguish the decoys from near native structures of globular proteins. It is also found that the hydrophobic score can suggest that certain structures with small RMSD from the native structure should be eliminated as candidates due to profiles displaced significantly from their native hydrophobicity profiles. Interestingly, some decoys with low free energies, such as OPLSAA/SGB energy, can also be eliminated by the hydrophobic moment profiling and consequent hydrophobic score, since they show little or no hydrophobic core and hydrophilic exterior compared with their native profiles. This shows that the simple hydrophobic score can provide information that complements that obtained by the more rigorous free energy approach.

References

1. L. Holm and C. Sander. *J. Mol. Biol.*, 225:93–105, 1992.
2. E. S. Huang, S. Subbiah, and M. Levitt. *J. Mol. Biol.*, 252:709–720, 1995.
3. R. Bonneau, C. E. M. Strauss, and D. Baker. *Proteins: Structure, Function and Genetics*, 43:1–11, 2001.
4. B. Park and M. Levitt. *J. Mol. Biol.*, 258:367–392, 1996.
5. A. K. Felts, A. Wallqvist, E. Gallicchio, R. Levy, D. Bassolino, and S. R. Krystek. submitted, 2001.
6. D. Eisenberg, R. M. Weiss, and T. C. Terwilliger. *Nature (London)*, 299:371–374, 1982.
7. B. David Silverman. *Proc. Natl, Acad. Sci. USA*, 98:4996–5001, 2001.
8. K. T. Simons, R. Bonneau, I. Ruczinski, and D. Baker. *Proteins: Structure, Function and Genetics*, 37 S3:171–176, 1999.
9. F. Figueirido, R. Zhou, R. Levy, and B. J. Berne. *J. Chem. Phys.*, 106:9835, 1997.
10. D. Maxwell W. L. Jorgensen and J. Tirado-Rives. *J. Am. Chem. Soc.*, 118:11225–11236, 1996.
11. A. Ghosh, C. S. Rapp, and R. A. Friesner. *J. Phys. Chem.*, 102:10983–10990, 1998.